nteractions between sediments and fresh water

The symposium was organized with cooperation and support from the Societas Internationalis Limnologiae and from UNESCO's programs Man and the Biosphere and the International Hydrological Decade

pudoc

# Interactions between sediments and fresh water

Proceedings of an international symposium held at Amsterdam, the Netherlands, September 6-10, 1976

Editor: H.L. Golterman

Dr W. Junk B.V. Publishers
The Hague

Centre for agricultural publishing and documentation
Wageningen

1977

ISBN 90 06193 563 6
ISBN 90 220 0632 8

Cover design: Pudoc, Wageningen
Printed in the Netherlands

ONTENTS

THEME 3 - ROLE OF SEDIMENTS IN ELEMENT CYCLES

## International Steering Committee

H.L. Golterman (convenor)
C. Serruya
G.F. Lee

## Local Organizing Committee

H.L. Golterman
H. Verdouw
H. de Haan
Symposium secretariate:
M.F.L. Golterman
R.B. Golterman

## Editorial advisors

A.J. de Groot
L. Lyklema
R.D. Schuiling

## PREFACE

H.L. Golterman, Limnological Institute, Nieuwersluis, the Netherlands
A.B. Viner, DSIR Freshwater Station, Taupo, New Zealand
G. Fred Lee, University of Texas at Dallas, Richardson, Texas

### History and acknowledgements

During the 19th conference of the International Association of Limnology (Winnipeg, 1974), a proposal was accepted to arrange a small meeting on the interactions between sediments and water, with emphasis on those interactions which have a consequence for the aquatic ecosystems and on water quality. It was felt that limnology, which arose originally from the study of the deeper lakes, had to a large extent overlooked the significance of sediments, whose function was described by some, until recently, merely as a "sink". At present, knowledge is rapidly increasing on the many fundamental processes in lakes and impoundments which are controlled to a smaller or larger extent by the sediments.

In addition to advancing the field of limnology, justification for this symposium is derived from the increased importance being given to sediment-associated contaminants. Water pollution control agencies in several parts of the world are developing water quality standards which are based on the total concentration of contaminants present in water. As discussed in a subsequent section of this report, it is generally found that contaminants associated with sediments are much less available to affect water quality than contaminants in "true" solution. There are examples, however, where sediment-associated contaminants have had significant detrimental effects on water quality. The objective of this knowledge and to identify the areas of greatest concern and gaps in knowledge.

It was thought that cooperation with other scientists working in the freshwater field would be usefull and therefore contact was made with the UNESCO program's "Man and Biosphere" and the "International Hydrological Decade". From both these programs support was received and the meeting was announced through their channels as well.

The result was that 169 registered participants convened in Amsterdam from September 5-12, 1976, for which meeting the following five themes were outlined:

1. Sources and composition of sediments;
2. Sediment formation;
3. Role of sediments in element cycles;
4. Transport phenomena in and from sediments;
5. Water quality, health and management aspects of sediments.

Participants came from 32 countries from all continents; 16 European, 5 African (among which were Egypt, Sudan, Nigeria), 7 Asiatic (among which were India, Thailand, Shri Lanka and Turkey), 3 American (Canada, USA, Brazil), Australia and New Zealand were represented. During the plenary morning sessions, 20 invited lectures were given, while during parallel sessions in the afternoon 64 communications were presented. Travel and accomodation was paid for six invited speakers and for four scientists from developing countries. The proceedings are published in this volume.

The editor wishes to thank several persons and institutions for their help with the symposium or with the publication of this volume.
The editor acknowledges the help from his colleagues Dr A. de Groot, Dr L. Lyklema and Dr R.D. Schuiling for their advice. The Director of the Freshwater Biological Association helped to correct the English of several papers. Mr Th. Dekkers drafted several of the figures and Mr J. Landstra helped with the general organization. Mrs L. Willeboordse-Ruiter retyped several of the papers with great skill.
The editor thanks the publishers for their advice and patience and for their willingless to allow the camera-ready copy to be edited, hardly usual practice.
Grants were provided by the International Association of Limnology, Unesco (Man and Biosphere and International Hydrological Decade) and the Netherlands Ministeries of Education and Science, of Health and Environmental Protection and of Transport and Public Works.

The editor feels that publication of this kind of proceedings with camera-ready copy keeps the price of scientific publications within reach and speeds publication. As long as authors submitting their papers do not completely appreciate the consequences, editorial work remains necessary. The editor of this volume enjoyed working on many papers and could accept without any changes the greater part. He remains fully responsible for remaining flaws.

After the meeting, a workshop was organized with the objective to prepare an outline for a publication on the role of sediments in the transport and availability of contaminants in natural water systems. The convening of the workshop was requested by UNESCO. The publication that will result from the completion of the outline by the attendees will give basic information on the role of sediments in the pollution of lakes and will give information on procedures for measurement. The workshop in which 20 scientists participated, lasted three days and completely fulfilled its tasks. The prepared outline was submitted to UNESCO for approval and further action.

## Evaluation of the presentations*

Phenomena associated with sediments in lakes and rivers are intimately related not only to the internal living and nonliving processes in the water body but also to the almost infinite array of external processes occurring in the catchment areas, including those processes directly attributable to man. The actual sediment formation is an outcome of combined physical, chemical and biological effects. It is not surprising therefore that such a meeting as this should see the presence of a diversity of specialists, such as engineers, applied chemists, physicists, geologists as well as biologists. Inevitably with such a gathering, the tone of the proceedings was appropriately towards eutrophication and pollution issues, although not entirely so.

For sediments to be in such a pivotal position, intellectually entertaining as it may be in some respects, has a disadvantage in that they are rather difficult to define, both physically and functionally, and consequently i is difficult also to provide a logic framework upon which to base a discussion about them at a symposium. The orientation of topics which crystallized from the sessions was an in dicator of the things which are curr ently exercising the minds of invest igators into freshwater sediment-wat interactions. In other words, the sy posium must represent a description of "the state of the art". The follo ing is a brief summary of this description. The topics into which this account has been subdivided attempts to be more nearly an approximation t events; but of course, it is an arti icial classification of convenience, and there was a great deal of overla of subject material actually discuss ed.

### Physical Transportation and Distribution (12 papers)

Sedimentologists have understood for a long time that to interpret proper] the processes of suspension and trans portation of sediment, sophisticated investigation is required, even in re latively confined lake basins; the same is not always appreciated by limnologists. One still sees reporte estimations of sediment formation bas ed upon one, or very few, sediment traps. This issue was debated during discussions, but various papers dealt directly or indirectly with the problem, particularly the influence of horizontal movement of suspended matt er by turbulence and by the peculiari ties of the basin configuration ("sed iment focusing" appears to be fashion able phrase). Various examples of local horizontal differentiation were given, some of the best worked studie being from Canada (Lakes Erie, St. Claire, Ontario). These tended to be elegant manipulations of empirical re sults to enable differentiation of or sediment type from another. The importance was illustrated of distinguishing sediments within a water body according to their particle sizes so as to relate transportability to ener gy properties; a fundamental discrimination which was brought up again in other contexts.

Fundamental theory was also illustrat ed by an excellent review of factors controlling the precipitation and resuspension along the coasts of S.W. Netherlands, as for instance, the shearing and collision responses of clay particles under various kinetic and mechanical energy regimes.

*Based on the oral presentations, not on the submitted manuscripts.

Mineralogical Reactions, Diagenesis (6 papers)

Apart from demonstrating well the difficulty of knowing where a sediment ends and the soluble material of the water begins, contributors to this topic provided useful descriptions of mineralogical processes to enable a better interpretation of, for example, elemental associations to be found in sediment cores. A review at a global level showed the sort of mineralogical information which is of value, e.g., distinguishing marine from freshwater sedimentary environments, and the chemical typing of source material.

The precipitation of minerals in Lake Chad was shown to prevent the increase of salinity which should otherwise occur in this closed basin. Other papers discussed the mobility of elements within the sediments, glauconite formation, ferromanganese nodule formation and shifts in heavy metal associations through time in response to cultural changes.

Phosphate Exchange Dynamics (21 papers)

Not surprisingly, the largest proportion (about 25 percent) of contributions dealt with various aspects of phosphorus transfers from and to the sediments, sometimes in association with other chemical components. Unfortunately this bulk of information contributed disproportionately little to the further understanding of the dynamics of this nutrient in natural waters. The papers which concerned themselves with experimentally deciphering physico chemical factors influencing release appeared essentially to substantiate what was originally proposed by the classical work of Einsele and Mortimer and subsequently elaborated by many other workers, and which indeed has become the basis of many lake restoration programs. However, as discussed in a subsequent section, the role of iron in controlling the overall phosphorus dynamics in the lakes was demonstrated to be much less general than often thought.

The more important aspect is that the different extents to which those factors, such as forms of iron, oxygen, pH, calcium carbonate, etc., operate in different limnological environments cannot readily be predicted quantitatively without prior investigation of the particular water body. Thus when one is faced with a management or restoration problem, one can presume little on the basis of other work before one may usefully develop a model of phosphorus transfers.

A reflexion of this limitation for extrapolation was the reported large range of recovery lag periods for lakes subjected to restoration programs. The amount of phosphorus which continues to be recycled from the sediment (= part of the "internal loading") after restriction of "external loading" is of overriding importance in this. Estimation of the lag period, in formal presentation and informal discussion, ranged from one year to 50 years. An obviously important factor which was discussed is the flushing rate, which will be dependent upon the lake's morphometry as well as a host of external features. But microbial conditions, frequently a function of fertility levels were, not unexpectedly, shown to be of influence.

Two contributors in this group illustrated how interstitial water (= pore water) chemical profiles can be usefully related to the chemical budget in the overlying water. More innovative was an attempt to use the radioisotope $^{222}$radon as an exceedingly sensitive tool to measure eddy diffusion near the sediment surface. The results in this presentation were applied to phosphorus transferred by such a mechanism.

Nutrient Effects of Artificial Sediment Additions or Removals (4 papers)

An all too small group of papers was applied to this topic. Two extreme examples were mentioned of the chemically stabilizing effect of riverine sediments; a small experimental chalk stream system and the Nile. Very much pertinent to modern practical problems, the effects of dredging in two contrasting situations in the Canadian Great Lakes were described for which this operation in a small harbor showed no long term phosphate resuspension but did so for a much larger water mass. Calculations regarding an industrial pollutant input of silt, taconite, again for a locality in the Great Lakes, showed apparent small nutrient pollutant effects.

Microbial Transformations (17 papers)

Aspects of the nitrogen cycle were accentuated, and in particular provided useful information regarding denitrification. Unfortunately only one study attempted the use of $^{15}N$, even though (apart from the difficult to arrange $^{13}N$) it is the only adequate tool at the moment for demonstrating a definite loss of nitrogen from an experimental system. The measurement of losses of nitrate, interesting in itself, does not imply denitrification since a great deal may be immobilized in the sediment either as ammoniacal nitrogen or in some other form, and is particularly misleading if artificially high nitrate concentrations, which can distort the normal metabolism, have to be used. Nevertheless, it is well to have some record of efforts to measure the process accumulated in one place, as in these proceedings, since the ecological requirements for denitrification could be of some practical importance for predicting the importance of nutrient loadings, and even for nutrient stripping.

The present shortcomings of trying to relate experiments such as these to the field was clearly manifest, although good attempts at <u>in situ</u> measurements were made. So much depends upon variations in microclimate, extremely difficult to predict in most lakes. Two situations were mentioned where the relative simplicity of these environmental factors allowed for some prediction. In one case, a stream with its obviously dominant one way flow system and maximal physico-chemical constancy allowed for the rigorous application of mathematical principles to predict the denitrification of stream nitrate. In another case the constancy due to the equatorial climate of a tropical lake allowed for an indirect estimation of nitrogen and phosphate liberation to be made.

Efforts to relate microbial transformations to sediment ATP were of interest but showed much variability and were not especially convincing. Muramic acid has been used with greater success as a biochemical indicator of bacterial biomass.

In a new twist given to the old concept of formation of particles of organic material in water, it was shown, by scanning electron microscope pictures, how bacteria can use an important part of their organic production to agglomerate sedimentary particles by investing them with secretions.

Influence of Meiofauna (2 papers)

A shortcoming of a symposium which has such a high proportion of nonbiologists is that scant attention is paid to sediment as an ecosystem in itself, and to the inseparable interaction between the living and nonliving components that the world ecosystem implies. Microbial transformations tend to be formalized incorrectly in the same manner as are chemical reactions, and disregard that what one looks at in such transformations is a final manifestation of far more random and less predictable events than those for chemical reactions.

Part of the ecosystem which conditions the activities of the microorganisms in this way is the meiofauna (mainly insect larvae but also other invertebrates). Largely because of the intimate association between the ecosystem components very few studies have been able to devise controlled experiments to quantify meiofaunal influence, but in a thorough review of the literature the meeting was given a number of examples of when and how these organisms could well have a substantial effect on the sediment environment, not necessarily by direct transfer of primary nutrients but by, for instance, adjusting the redox potential to favorable microbial requirements.

In another paper a specific study was reported which showed considerable insect manipulation of material, but the relative importance of this aspect of sediments remains to be elucidated for a diversity of aquatic systems.

Organic Material: Significance and Discrimination
(5 papers)

In the past, a limited insight into the gross workings of aquatic systems has been gained by analysis of total dissolved organic matter (DOM) both in the water column and the sediment, and a certain amount of such information was presented at the symposium. It would now seem more appropriate to be able to discriminate many more details. The typing of allochthonous and autochthonous material is obviously relevant here. Two papers approached the subject of molecular "finger printing" the plant origins of the deposits. Such refined techniques are perhaps more appropriate to palaeolimnology (for which they have largely been developed) rather than the environmental control general orientation of the current symposium. Nevertheless, these approaches, and also the correlation, which was pointed out in another paper, between purines and pyrimidine concentrations and cultural organic eutrophication, are potentially additional weapons in the armoury of techniques which can be brought to bear upon interpreting the relationships between land and water.

Heavy Metals and Other Pollutant Associations (11 papers)

A good review of the importance of sediment investigations in this respect was given appropriately during the first session of the symposium. This was supported by a description of what must be a classic pollution study, that of the Rhine Delta, which showed the manner in which heavy metals are accumulated to very high levels in the bed sediments with increasing industrial utilization of the water along the length of the Delta. The point was made that the much lower concentrations of these contaminants in the water give no indication of the real levels of contamination involved.

These two contributions provided a very suitable platform upon which to base the various other specialized topics affecting, especially, heavy metal distribution which were aired by other speakers. Particularly important were those which dealt with particle size and uptake ability, and related closely in this to other papers considering mechanical transport phenomena. Close attention to this, and the use of ordination statistical techniques, for Lake Erie sediments could not only distinguish the differential transportation of mercury and PCB's but could also be related to the known hydrodynamic regime.

A well worked investigation with $^{65}Zn$ nicely showed the manner in which this metal may be transported by clays and to what extent it could be available to algal metabolism. The recycling effects of bacteria, and chelating effects of humic material were also mentioned.

Methodology (5 papers)

The few contributions in this field were nevertheless of value. A summary was given of the use of a statistical treatment of particle size data with respect to energy properties so as to distinguish one sediment spectrum from another, each being the outcome of definable hydrodynamic and basin configuration features which impart the environment under which sorting of particles will occur. The idea was to describe the sedimentary basis for the ultimate distribution of sediment physicochemical interaction, and had considerable relevance to the report on transportation processes in S.W. Netherlands (see above). In spite of the presentation being somewhat excessively theoretical there may well be substantial application for the approach but more in large, complex lakes than in other waters.

An illustration of the relationship between theory and practical application was given in a discussion of the manner in which copper, zinc and iron existed in the sediment, and a scheme for fractionating these forms, an account of value to a laboratory newly embarking upon heavy metal investigations.

Three descriptions were presented of technological innovations of very great interest to those who have to attempt studies on large water bodies with a small number of personnel. A polarographic system for simultaneously measuring oxygen, sulfide and divalent iron and manganese was shown to be extremely successful in laboratory arrangements and perhaps adaptable to <u>in situ</u> measurements, thereby considerably simplifying the following of significant chemical changes during seasonal fluxes of nutrient liberation from sediments.

Seasonal metabolic changes in sediments, at least in the situation presented, were shown to be traceable by the degree of sediment transparency to echo sounding wave frequencies. This was thought to be related to the amount of metabolic gas bubble formation.

It is apparently possible to assess suspended load with surprising ease to within 20 percent of other calculations of it by remote sensing of the amount of reflective solar radiation from the water surface.

Other papers described useful techniques, or modifications of existing methods, or ones not commonly used in lake studies, in conjunction with a more dominant topic. For those interested in long term changes recorded in sediment cores, reference could be made in the published proceedings to the chlorophyll separation used in cores from Lake Lucerne, showing correlations to eutrophication, and the use of $^{137}$caesium for dating. Also discussed was the discrimination of allochthonous from autochthonous material in Lake Kinneret by fractionation of the stable isotopes $^{12}$carbon and $^{18}$oxygen.

New Bearings Arising?

With some exceptions, extremely little was said that was really new or surprising. However, taken as a whole, some reported results brought out well the significance of certain topics which perhaps are not always fully appreciated.
These can be listed:

a. The necessity of detailing the sediment particle sizes and other physical properties so as to relate these to the chemical components, using types of cluster and factor analyses;

b. The considerable importance of horizontal sediment movement, the realization that this must be properly distinguished for sedimentation rate assessments, and the present lack of direct means of doing so;

c. That input nutrient forms of nitrogen can be immobilized, or lost as gaseous nitrogen by denitrification, to a very substantial proportion, and is a factor to be considered on both lake and river enrichment studies. But for whole water bodies assessments are presently indirect and rather crude approximations unless the microbial environment can be safely predicted;

d. Mud interstitial water analysis can give revealing insight into the depth of interaction between mud and water and can discriminate potential rates of efflux of phosphate from that of nitrogen; need was shown to exist for a chemical fractionation of phosphate in sediments by less aggresive extractants than normally used.

e. Lakes even of similar dimensions can differ very greatly in their ability to recover from phosphate induced excess fertility, and so in their sediment influence, and consequently must be investigated individually;

f. Sediment cores can now be inspected from a number of different viewpoints to make them more valuable than ever for interpreting shifts in catchment area use. These data are suitable for waters with slower environmental response rates, usually a large lake. Information can come from organic molecule discriminations, heavy metal vs organic association, conservative vs labile elements, $^{137}$caesium, $^{13}$carbon $^{18}$oxygen.

Subsidiary, but new to most:

g. Sediment mineralogical reactions can have considerable moderating influence on water chemistry;

h. Bacteria can assist particle formaion to a greater degree than is normally considered.

i. New techniques can be used to markedly lighten field work load, for large lakes investigations in particular. These include adaptations of polarographic, echo sounding and remote sensing methods.

In general, the meeting can be said to have provided an impression of the direction of thrust of effort into sediment-water interactions, and a good cross comparison of work in Europe and North America. There were also limitations. It was demonstrated that some government and other organizations realize the revealing information that sediments can provide for environment management policies, and are prepared to support appropriate investigations and monitoring. From the point of view of someone having to carry out such work it should be possible to be able to recommend a systematic approach that takes into consideration the best methods currently available. Certainly this is a difficult task; the methods are almost as diverse as water bodies and sediments themselves, but it should be possible to recommend what the optimal approach should be for different situations, such as large or small rivers and lakes, water dominated by a particular type of surrounding land use or effluent, etc., and under what circumstances one may usefully extrapolate from restricted observations as opposed to whole water body studies.

If the symposium was a "state of the art" description, then it must be said that the state does not yet provide for such a function as just defined. But it does enable one to perceive where the deficiencies lie. The succeeding remarks suggest some of these.

In this context it would seem highly cogent that the symposium produced a workshop session for the devising of a methodological handbook (for UNESCO). Even though analogous schemes have been produced elsewhere, these are not generally known nor necessarily broadly applicable.

ʒpecial problem areas

Vhile in general the quality of the )apers presented at this symposium was ;ood, there were some problems related ›o the fact that the investigator used ‸nalytical procedures which, while ;ommonly used by some individuals in he limnology field, are generally re- arded as being inadequate today. The rime example of this type of situation s the use of the loss of weight on ombustion (ignition) as a measure of he organic matter present in sedim- nts. Many years ago this procedure as commonly used for measurement of rganic content of sediments. However, t was found that for some sediments, specially those containing high car- onates, the method produced inaccur- te results. Unfortunately, the method as continued to be used even though t is clear that other methods are eadily available which can produce a ıch more reliable estimate of the or- anic matter present in natural water ediments than loss on ignition. This rror can be especially significant ı those investigations such as some f those presented at the symposium, ıere the researcher attempts to ex- ain the behaviour of contaminants ı natural water sediments based on loss of ignition estimate of the or- nic matter present in the sediments.

oblems of this type are not restrict- to sediment work, but unfortunately e frequently encountered in many en- ronmental fields, where individuals th limited backgrounds in chemical chniques will utilize these tech- ques in their investigations without aluating whether the analytical thod properly measures the parameter concern. This approach frequently ads to a tremendous waste of effort d often produces data which are in- rrect, resulting in inappropriate nclusions being drawn about the sig- ficance of a particular process or diment characteristic involvement in diment-water exchange reactions.

is imperative that any investigator operly evaluate the applicability of particular analytical procedure to system he/she is investigating to certain that the analytical method asured the parameter with the desired gree of reliability. This is especi- y important for individuals working h natural water sediments. Many iments contain contaminants which l cause significant interference in analysis of other constituents.

Another problem encountered with some of the papers presented at the sympos- ium which is common with many studies conducted on the significance of con- taminants in sediments is that some investigators attempted to utilize ex- perimental techniques which would iso- late some part of the natural water system such as in a core, and study or attempt to study sediment-water ex- change processes. Approaches such as this can usually, if conducted pro- perly, indicate the direction in which the exchange reaction will proceed, i.e., will the chemical of concern move to or from the sediments. However, these studies generally are grossly inadequate to properly characterize the magnitude or the rate of the ex- change.

One of the primary, yet unresolved problems, in understanding the signif- icance of contaminants associated with sediments based on laboratory studies, is that it is virtually impossible to simulate in the laboratory the de- grees of mixing that occur in natural water systems within the sediments, overlying waters, and between the sed- iments and water. This problem causes most laboratory studies on sediment- water exchange processes to be of lim- ited validity for estimating effects of contaminants present in sediments on overlying water quality. Water quality is primarily determined by the concentrations of contaminants avail- able to aquatic organisms or man. In order for a contaminant present in the sediment to be of significance to af- fect water quality, it must be released in sufficient amounts to cause the con- centration in the overlying waters to exceed a critical level for some bene- ficial use of the water and its organ- isms or in sediment-associated organ- isms. Therefore, not only is the rate of release of importance, but also the rate of dilution of the released contam- inants in the overlying waters must be considered as well as the background levels of the contaminant. It is rare that information is available on rates of release from sediments as well as rates of dilution that occur in natural water systems.

One of the best examples of the potent- ial problems associated with extrapol- ating from laboratory systems, which attempt to isolate part of the lake sediment system for study purposes to the whole system, was shown in the work of several investigators concerned with phosphorus exchange in lake sediments. One of the most important conclusions from this symposium was that the classical story of Einsele and Mortimer of iron-phosphate interactions was convincingly shown to be of limited importance in governing the overall phosphate exchange processes in several, especially calcium rich, lakes.

Several investigators presented data which showed that there was significant phosphorus release from lake sediments under aerobic conditions. In the classical limnological literature, ferric iron is supposed to tie up any phosphorus released from lake sediments under aerobic conditions and prevent it from becoming available in the water column. However, when the whole lake study approach is used where the total mass of phosphorus present in the lake is examined in the light of phosphorus inflow and outflow from the water body, it is concluded that while on an annual cycle lake sediments generally represent significant sinks for phosphorus, there are periods of the year when there is significant aerobic release of phosphorus from the sediments to the overlying waters. This release has been shown to be related to mineralization processes where bacterial mineralization of organic phosphorus produces soluble orthophosphate which is not tied up with iron and therefore is available to stimulate further algal growth in the water body.

One of the real advantages of a symposium of this type was that a number of independent investigators were able to learn that other investigators had come to similar conclusions on the lack of significance of the ferric iron-phosphate system in controlling phosphorus release from lake sediments. As a result, it can now be stated that aerobic phosphorus release may play a dominant role in the aqueous environmental chemistry of phosphorus in lakes and impoundments.

It was further brought out at this symposium that when examining the potential significance of sediment-water interactions in lakes, consideration must be given to the role of the thermocline in trapping sediment released contaminants and thereby preventing their availability to the surface waters during the summer stratification period when lakes and impoundments tend to have their greatest recreational use. For example, it appears that the phosphorus release from sediments that occurs under anoxic conditions in eutrophic lakes may be of min significance in many water bodies in affecting water quality since this release occurs in the presence of the thermocline that tends to trap any released phosphorus in the hypolimnetic waters. Aerobic release, however, occurs above the thermocline or when it is not present and thereby directly contributes to internal fertilization of the water body during the periods when most water bodies receive their greatest use by the public.

It became clear from several of the papers presented at this symposium that much greater attention has to be devoted to overall lake dynamics in evaluating the significance of a particular release of contaminant from the sediments on water quality.

<u>Future importance of sediment-associa ed contaminants</u>

Increasing emphasis will have to be placed on determining the significanc of sediment-associated contaminants i affecting the limnological characteristics of water bodies and water qual ity. While sediments in many cases ar sinks for chemicals, there are situat ions where there is sufficient releas of chemicals to play a major role in determining the limnological characte istics of the water body. This is especially true in shallow lakes and in poundments. Water pollution control agencies throughout the world are ad ting water quality criteria and stan ards with ever increasing lower crit ical concentrations. Often, the tota contaminant present in the water is specified as the critical concentrat ion. This will require that greater emphasis will have to be placed on t significance of particulate contamin ants because at microgram or submicr gram per liter levels, most contamin ants which are of significance in na ural waters tend to become associate with solid materials present in natu water systems.

Failure to base water quality standa on available forms of contaminants w mean that large amounts of funds wil be spent in the name of water pollut control with little or no benefit in improved water quality. Not only wil attention have to be given to determining the amount of immediately ava able forms, but also to transformati of these forms within the sediments changes in the characteristics of th water and sediments. This type of in formation will require a substantial increase in the fundamental and applied research on the environmental chemistry of sediment-associated contaminants in order to make wise use of the pollution control funds available.

Based on the success of this meeting, it is felt that every three to five years similar meetings of this type should be held in which scientists and engineers working on the significance of sediment-associated contaminants in limnology and water quality should meet to discuss the results of their investigations.

November 1976

# Theme 1 - Sources and composition of sediment

R.D.Schuiling

Vening Meinesz Laboratory, Department of Geochemistry, State University of Utrecht, Netherlands

## Geographical distribution of lakes

In order to gain insight in the distribution of lakes, the total area occupied by lake waters will be calculated for zones of 5 degrees latitude; the ratio between land and lakes then can be obtained.

Halbfass (1922) listed about 1500 lakes, geographically unevenly distributed. For poorly covered areas, an additional 800 lakes have been measured planimetrically.

Figure 1, based on all these data, is considered a fairly representative picture of the geographical distribution. Two distinct maxima at high latitudes north and south can be distinguished, one maximum near the equator, and two minima at about 25$^o$ north and south. Most lakes without outlet

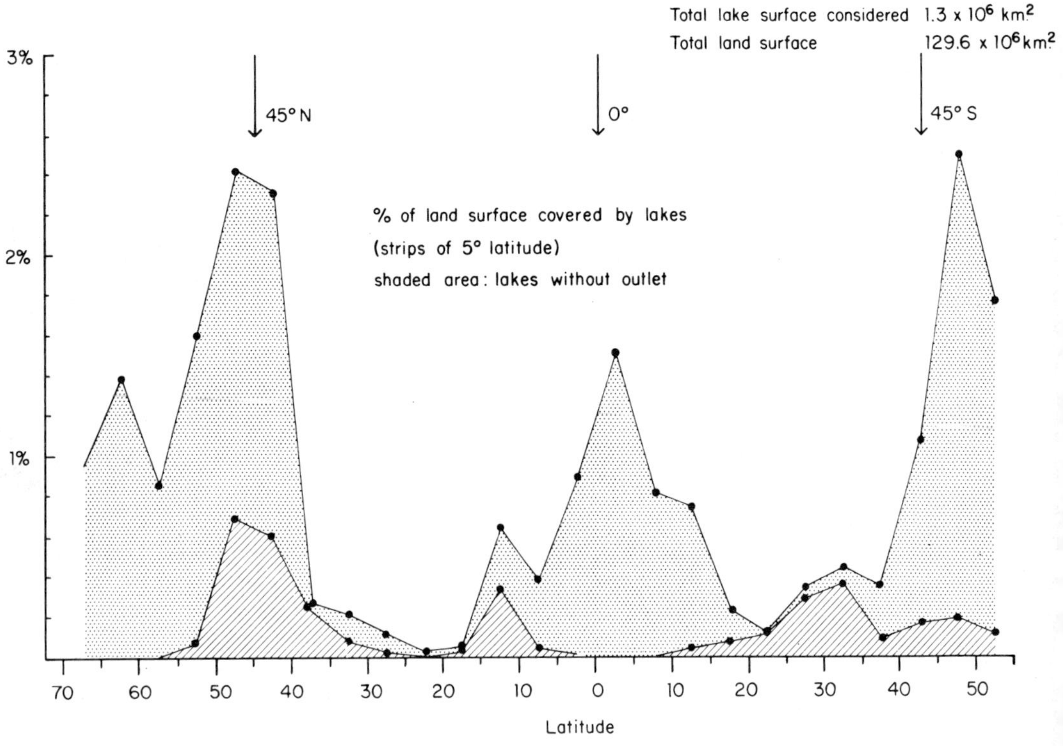


Fig.1 Distribution of lakes as a function of geographic latitude

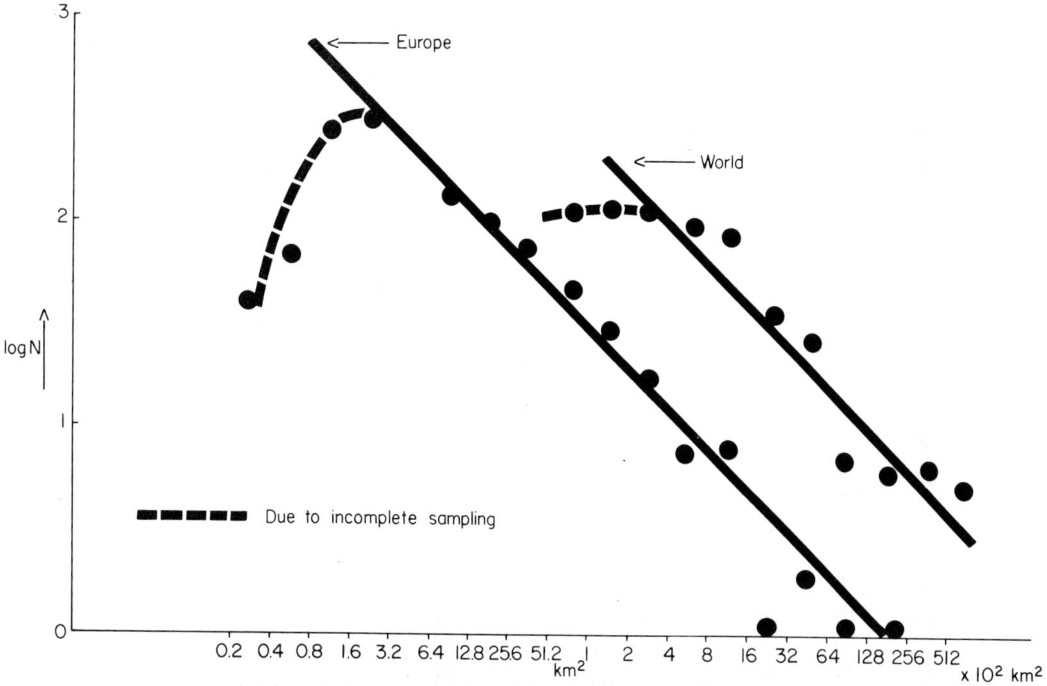


Fig.2 Frequency distribution of lake size

·e found in a wide zone, located about 10-20 ·grees north and south of the mentioned nima.

The analysis shows that the total area of ·e considered lakes is slightly more than ·e percent of the total land area. Except r Europe, only lakes larger than 100 $km^2$ ve been included. For Europe, more formation about lakes is available than r the rest of the world (fig. 2). The milarity in slope of the two distribution rves, however, indicates that the stribution in the two regions is not sentially different.

From fig.3 it can be concluded that lakes rger than 100 $km^2$ together occupy a rger area than all smaller lakes combined.

## dimentation in closed-lake systems

The endorheic lakes are restricted to ·eas where evaporation exceeds precipitation.

In the subsystem, formed by an endorheic lake and its drainage basin, the weathering products are transported from land towards the lake, just as in the general system continent-ocean. The question arises why endorheic lakes do not follow a chemical evolution similar to that of the sea. Instead, most of these lakes become high pH soda lakes, from which, in an advanced stage, typical minerals may be precipitated, such as sodium carbonates and bicarbonates (trona, natron, thermonatrite, nahcolite), mixed sodium calcium carbonates (gaylussite and pirssonite), sodium sulfates (thenardite, mirabilite) and sodium silicates (analcite, magadiite, kenyaite). Under exceptional geologic and climatic conditions, nitrate and borate minerals may be formed.

Isothermal evaporation of almost any natural surface water, which owes its chemical composition mainly to weathering reactions and which remains in equilibrium

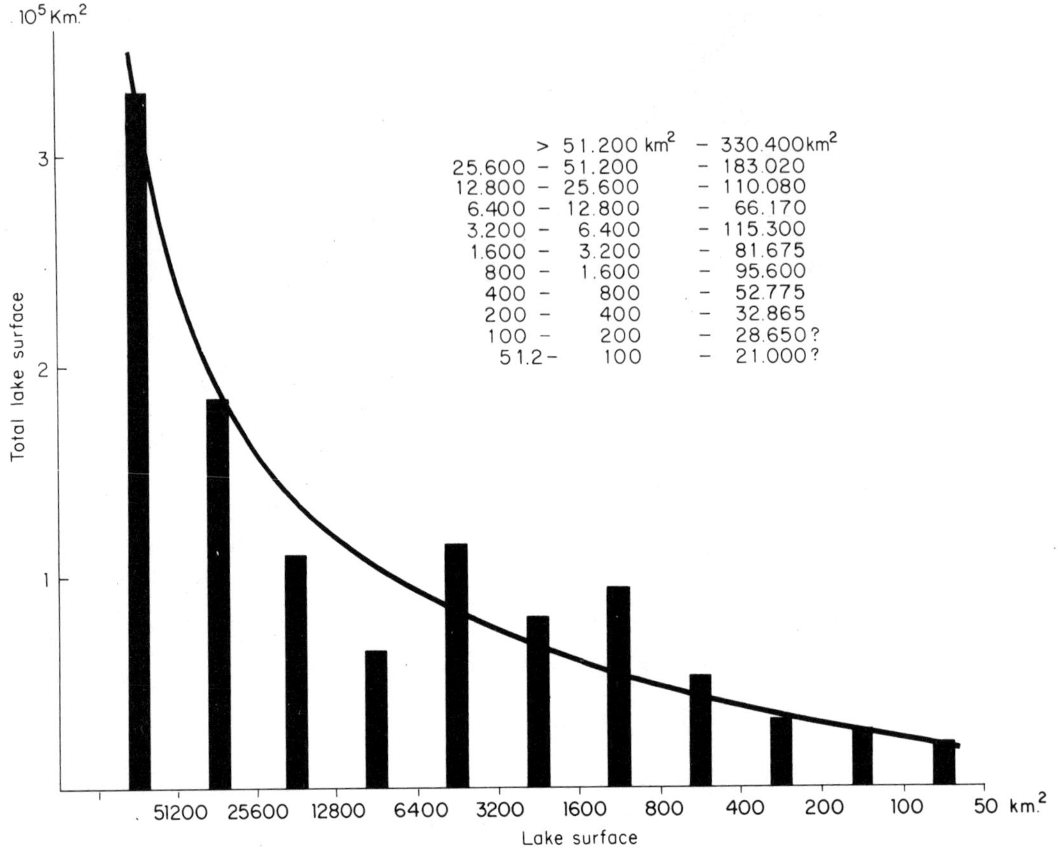


Fig.3 Sum of lake surfaces as a function of lake size.

with the atmosphere, will lead to the development of dominantly sodium bicarbonate lakes with high pH values. The simplest model is plagioclase (generally the determining mineral in many rock-systems) weathering to kaolinite:

$$2NaSi_3AlO_8.CaSi_2Al_2O_8 + 6CO_2 + 17H_2O \longrightarrow$$
$$3Al_2Si_2O_5(OH)_4 + 4H_4SiO_4 + 2Na^+ + 2Ca^{2+} + 6HCO_3^-$$

Upon evaporation calcite will be precipitated first (or, depending on the amount of Mg in the water, high Mg-calcite, aragonite or dolomite). After most of the Ca is precipitated as calciumcarbonate, or in sulfate-rich waters as gypsum, the remaining cations (in our example only sodium) are essentiall balanced by bicarbonate ions. At a given pH the sum of bicarbonate and carbonate is limited by the solubility of $CO_2$. As the waters become more concentrated during evaporation, part of the $CO_2$ escapes. The rising Na-concentration must increasingly b balanced by $OH^-$ and by a shift from bicarbo nate to carbonate (Garrels & Mackenzie, 196 The pH increases until the water becomes saturated with sodium carbonate or sodium bicarbonate, at pH values of about 11. Such high pH values are accompanied with a rise in silica solubility, which may reach several grams per liter and causes the formation of sodium silicates as magadiite and kenyaite, discovered in recent years.

:Eugster, 1967).

The major difference between endorheic .akes and the ocean, and the answer to the juestion why the sea does not behave like a :losed-type lake system, is that the veathering products, more or less amorphous aluminum silicates, do not back-react under acustrine conditions, as in a marine nvironment. It is this slow reconstitution f cation aluminum silicates - aptly called everse weathering - that determines the hemistry of ocean water. Either because the eathering products never reach the saline akes, or because they are too rapidly isoated by overlying new sediments, this everse weathering is only of minor mportance in determining the chemical omposition of lakes. Several studies on ake sediments, e.g. by Förstner (1973), ndicate a certain degree of reconstitution f chlorite and illite in saline endorheic eas.

## edimentation in open-lake systems

The study of closed-type lakes is extremely nteresting from a geochemical point of view, it the total area of these lakes is only out 15% of the total lake area in the orld; it is understandable that limnologists e more interested in the study of open-type kes. Unfortunately, generalizations about diments and sedimentation in open lakes can rdly be made, because they cover such a de range of conditions: climatic and relting weathering conditions, geology and emical composition of the surrounding cks, and physical and chemical conditions the lakes. The combination of these ctors governs the quality and quantity of e biomass in the lakes, and the eventual eservation of the organic material.

Although a large part of the world's lakes e found at moderately high latitudes, dicating that glaciation has played a major role in the formation of many of them, open-type lakes are also present near the equator (see fig.1). The lakes at high latitude are subject to large variation in physical conditions, due to contrasts between summer and winter conditions; lakes in lower latitudes in areas with pronounced wet and dry seasons are also subject to rather marked seasonal variation. During the wet season a high flux of particulate matter is carried to the lake and settles on the bottom. During the dry season, organic matter is predominant; these contrasting sedimentary regimes may give rise to finely laminated sediments, called varves. In many places the lamination may completely disappear as a result of bioturbation or wind action.

The sedimentation rate is a function of the ratio between the land area from which material is eroded and the size of the water body in which this material is deposited, and of the erosion rate proper. The ratio between the area of a lake and the size of its drainage basin is invariably much smaller than the ratio between the total area of the oceans and the total size of the continents, which is roughly 7:3. Consequently, the sedimentation rate in lakes is much higher than the average sedimentation rate in the oceans, which is usually 1-20 mm/1000 yr. Taking 10 mm/1000 yr. as an average value for sedimentation in the deep oceans, and allowing for the much higher sedimentation rates in submarine deltas and shelf areas, we arrive at an erosion rate for the continents of 3-4 cm/1000 yr., which is compatible with figures based on quantitative studies on world wide river discharges. In order to calculate in the same manner the average sedimentation rate in lakes, the ratio between lake area and size of drainage basin feeding the lake must be taken into consideration. For open-type lakes this ratio is approximately 1:5.5, which would give sedimentation rates of about 0.2 mm/yr.; in many places, the values are 5 to 10 times higher.

For endorheic lakes this ratio is much smaller, in the order of 1:32, but most of these lakes are found in areas with low erosion rates. This must be attributed to the dry climate and the scarce vegetation. This can be translated into a virtual absence of mature soils and a very low partial pressure of $CO_2$ in the soil atmosphere. The aggressiveness of soil water is mainly determined by the $CO_2$ pressure of the soil atmosphere. Low $CO_2$ pressures cause low rates of weathering; this is one of the reasons why most closed-type lakes are found in areas with low erosion rates.

The rapid sedimentation and the exposed situation render the lakes, geologically speaking, shortlived phenomena. From the distribution of lakes, as presented in fig.1, it may be concluded that recent glaciation is probably the major cause of the presence of lakes at this moment. Periods of glaciation were rare in geologic history, and it may be expected that lakes in general were less common in the geologic past than at present. Nevertheless, some of the largest lakes existed in the Triassic and in the early Tertiary in the United States; the Eocene lakes covered large areas of Utah and Wyoming.

## Source of lake sediments

In the concept of the geochemical cycle (fig.4), it is assumed that the different geological reservoirs changed neither their volume nor their composition with time; this should hold for the earth's crust, for the soil cover (a small transitory reservoir), for the rivers, and for the oceans. If this steady state situation is approximately realized, the following relations can be

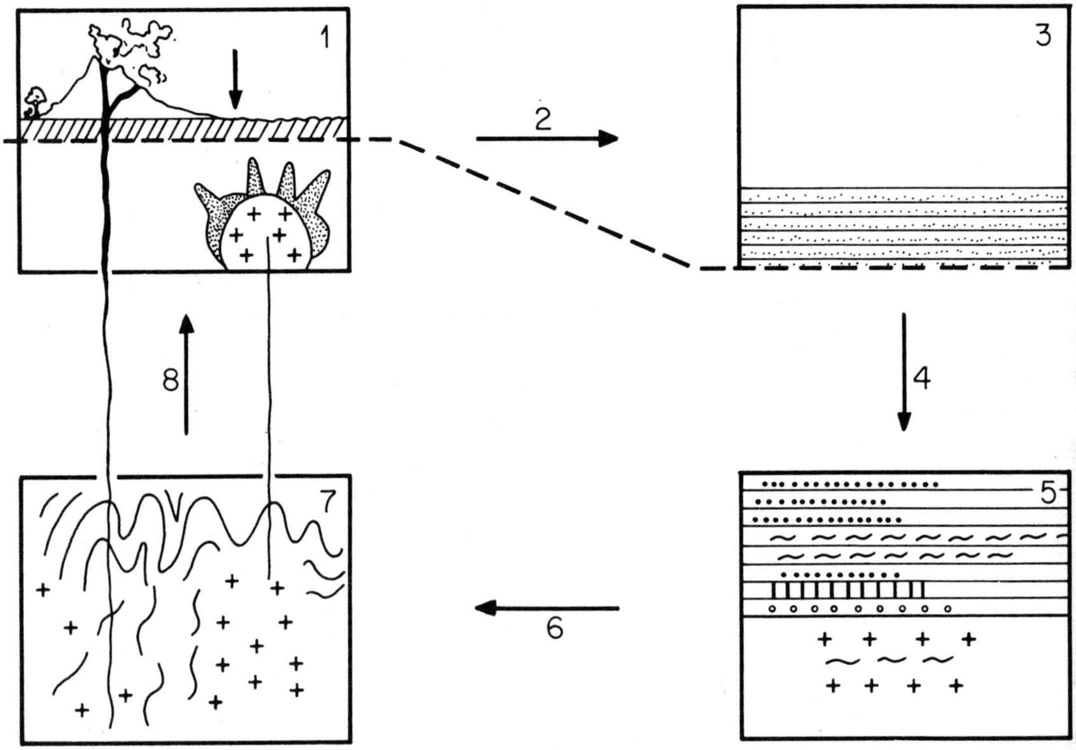


Fig.4 The geochemical cycle

defined:

erosion rate x composition of eroded material = concentration of dissolved and particulate matter in rivers x flow rate of rivers = composition of ocean sediments x rate of sedimentation.

To the question whether these relations can also be applied to lakes, the answer is no. Whereas in the oceans the composition of the average sediment should equal the composition of the material being eroded (the ocean is in a steady state, and its composition remains constant with time) this is not true for closed lakes or for open lakes. The composition of closed lakes changes with time, not only owing to annual variations in precipitation, but also as a result of building up a salinity in the course of their short history. In addition, they are usually present in regions where at least part of the solids transported towards the lakes never reach their destination. Sediments in closed-lake systems are therefore quite often biased towards the material transported in dissolved form. The reverse holds true for open lakes. Here the particulate matter is deposited almost quantitatively, whereas a large part of the dissolved matter passes through the lake and out again, unless precipitated by biologic agents; this is the mechanism by which a considerable part of the dissolved silica and calcium carbonate may be retained in lake sediments.

## Composition of lake sediments

The composition of lacustrine sediments is largely determined by the chemistry of the drainage area. If drainage areas are sufficiently large, individual geochemical characteristics of the different rock types become smoothed out. In such cases it is possible to calculate the total flux of a certain element by measuring the fluxes of a few major elements into the lake basin. This procedure can be illustrated with the following example. In the river Rhine, the total fluxes were measured of those major elements which are not strongly affected by human activity. The geochemical character of the Rhine catchment area can be approximated by a two-component model, based on "average carbonate" and "average shale" (including slates, micaschists, gneisses). The contribution of each of these classes can be found by solving the following mass balance equation:

$$X_i^A \cdot p \cdot T + X_i^B (1-p) T = t_i$$

in which $X_i$ is the weight percentage of element i (Mg, Al, Si, Ca, Ti, Fe) in A or B; A denotes shale and B carbonate rock. T is the total amount of material eroded per year, p its fraction of shale, and $t_i$ is the total flux of element i.

The major rock categories and their compositions were listed by Horn & Adams (1966). In the ideal case, the lines representing the 6 equations should intersect in one point in the T vs. pT plane, but in practice the intersections delineate a small area (for further details, see Schuillng et al., in preparation). From the contributions of average shale and carbonate, the relative contributions of trace elements can be calculated as well, and these quantities can be compared with those actually found. Quantities largely in excess of the calculated base line contributions as a result of natural erosion must be attributed to human activities.

The application of this method to the chemistry of lake sediments is not without difficulty, as the conditions imposed by the method are not fullfilled in closed-lakes or in open-lakes. By measuring fluxes entering the lake and leaving the lake, and applying the method with due regard to the chemical composition of the drainage basin, it may be decided whether the fluxes of trace elements can be attributed to natural processes only or to human activities.

The method can be applied to other cases, in which the small size of the drainage basin precludes the use of "average" rock types. Then it becomes possible, either at the inlet of a lake, or by studying lake sediment composition, to answer the question whether the amounts of trace elements are "normal" when we have related them, as described above, to the major element load, or that we are dealing with marked geochemical anomalies in the drainage basin of a lake. The successful use of lake sediment composition in exploration geochemistry shows that this is possible.

Key words: Lake
Sediment
Endorheic
Latitude dependance
Size distribution

References

Eugster, H.P., (1967). Hydrous sodium silicates from Lake Magadi, Kenya: precursors of bedded chert. Science, 157, p.p. 1177-1180.

Förstner, U. (1973). Petrographische und geochemische Untersuchungen an afghanischen Endseen. N.Jhb.Min.Abh., 118, p.268-312.

Garrels, R.M. & F.T.Mackenzie, (1967). Origin of the chemical composition of some springs and lakes. In: Equilibrium Concepts in natural water systems, Am. Chem. Soc., Washington, p.222-242.

Halbfass, W. (1922). Die Seen der Erde. Peterm. Mitt., Ergänzungsheft 185, p.1-169.

Horn, M.K. & J.A.S.Adams (1966). Computer derived geochemical balances and elemental abundances. Geochim. Cosmochim. Acta 30, p. 279-297.

Schuiling, R.D., C.H.van der Weijden, H. de Hek & A.K.van Helden . Natural erosion as reference for the transport of elements in polluted river systems (in preparation).

# DEPOSITION, TRANSPORTATION AND EROSION OF MUD

J.H.J. Terwindt

Rijkswaterstaat-Deltadienst, Middelburg, The Netherlands[1]

## Summary

Flocculation affects the fall velocity of the mud and the rate of deposition, especially in flowing water having bed shear stresses below a certain critical value. In the latter case finally all suspended sediment will settle down. In flowing water which exhibits bed shear stresses above the critical value, the rate of deposition is determined by the initial concentration in the fluid. In a steady state there is an equilibrium mud concentration which remains in suspension for a given value of the bed shear stress.

Deposited mud undergoes a relatively rapid initial consolidation which results in an increasing shear strength of the mudlayer. So in time there is an increase in the critical bed shear stress necessary to initiate erosion of the mudlayer. The value of this critical bed shear stress and the rate of erosion largely depends on the chemical state of the particles and the surrounding fluid.

## Introduction

It is the intention of this paper to introduce some physical aspects of the deposition, transport and erosion of mud to those who are not familiar with this field of study. To this purpose a small literature survey is presented, illustrating some recent trends in the investigations.

Mud is defined here as all material, suspended in water, which has a physical size, smaller than 50 microns.

## Flocculation

Flocculation is a process, operative in river- and seawater, which favours settling of the mud. Two conditions are necessary for flocculations:

- the particles must collide; depends on their concentration
- the particles must cohere; depends on the ion-content of the water and the physico-chemical forces around the clay particles.

Collisions may be due to the Brownian motion, to internal shearing and to differential settling. Internal shearing occurs if there is a velocity gradient in the fluid. By turbulent exchange particles with higher velocity may reach areas of lower velocity and vise versa. Particles with different velocities can collide more easily. According to Krone (1972) as aggregates become larger, or as the shearing rate increases, internal shearing rapidly becomes the dominant mechanism for collision.

Differential settling is important if the aggregation is already far advanced. In that case larger and faster settling particles may collide with smaller and slower ones.

Krone (1972) distinguished several orders of aggregation of particles. Zero order aggregates form when individual clay particles stick together. These aggregates have the strongest bonds, the highest shear strength and density. First order aggregates result from collision of zero order ones. The bonds are less strong and the shear strength and density lower. Second and third order aggregates develop from collisions of first and second order aggregates and have increasing weakness and decreasing shear strength and density.

Higher order aggregates can be broken up into lower ones by local high shearing rates and later reaggregate repeatedly. The maximum floc size depends on the velocity gradient or shear stress in the

1. Present address: Dept. of Physical Geography, State University, Heidelberglaan 2, Utrecht, The Netherlands

surrounding fluid. It should be noted, that the weakness of the bonds, the shear strength and the density also depends on the nature of the sediment; the clay-mineralogical composition and the chemical characteristics.

Fall velocity

Obviously the degree of aggregation has a definite effect on the fall velocity of the mud. The median fall velocity of flocculated suspensions have a range of 0,1 - 0,5 mm/sec or 0,4 - 1,8 m/hour, according to measurements made by many authors (fig. 1).

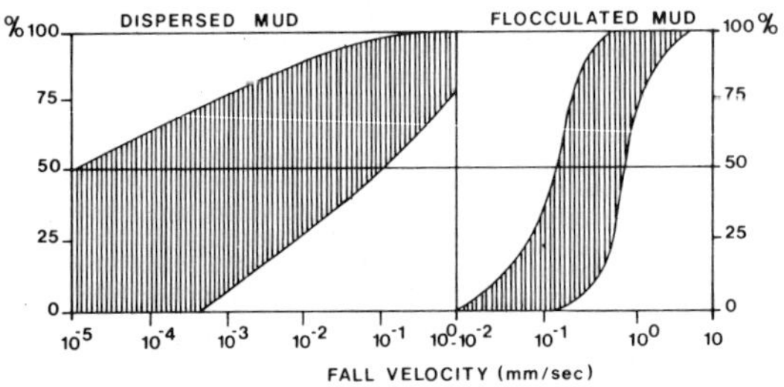


Fig. 1. Range of fall velocity distributions for dispersed and floculated mud suspensions.

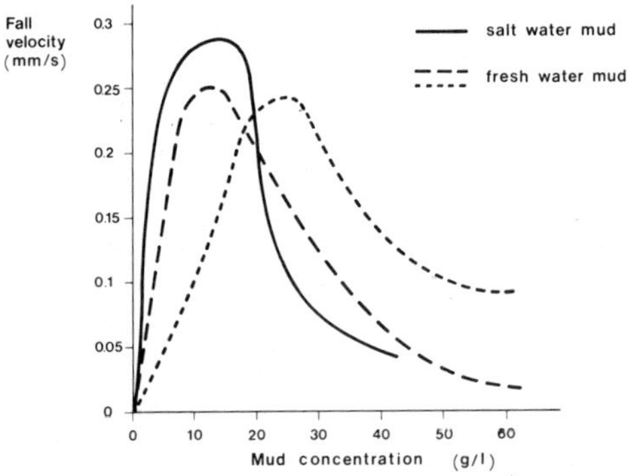


Fig. 2. The relationsship between fall velocity and mud concentration (Redrawn after Migniot, 1968).

Migniot (1968) points tot the fact, that also the concentration influences the fall velocity (fig. 2). There is an increase in fall velocity up to a concentration of 10-30 gr/l. At higher concentrations some kind of fluid mud comes into being resulting in a hindered settling and lower fall velocities. A similar observation was reported by the Delft Hydraulics Laboratory (1962).

Deposition

The rate of deposition depends not only on trivial factors as depth of flow, initial suspended concentratio and type of sediment but also on the shear stress exerted by the fluid flow to the bed. Below a certain minimum value of the bed shear stress particles in the flowing fluid canno remain in suspension but will all gradually sink to the bed (fig. 3).

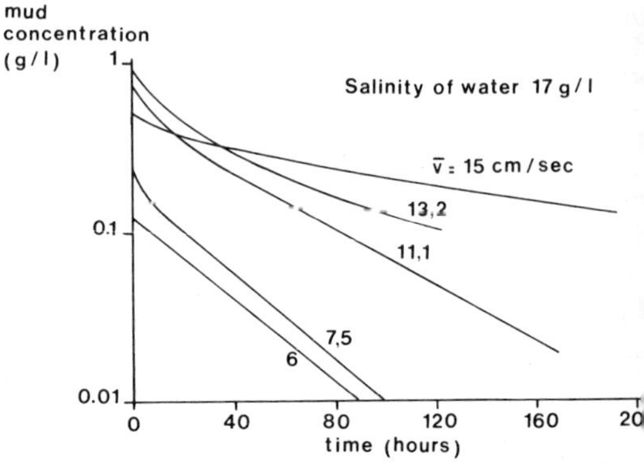


Fig. 3. The reduction of the mud con centration with time for several mea velocities ($\bar{v}$) in a saline fluid (flume experiments). (Redrawn after Krone, 1976).

Experiments made by Krone (1976) and by Metha (1973) indicate that this minimum value $\tau_{b\,min}$ = 0.6-0.8 dynes/cm$^2$ for some natural muds. Metha & Partheniades (1975) found somewhat higher values in flume experiments ($\tau_{min} \backsimeq 1.8$ dyne/cm$^2$). The final mud concentration tends to ward zero after some time for bed shear stresses below the minimum value.

Metha & Parteniades (1975) investigated the rate of deposition at higher values of the bed shear stres (fig. 4). They observed that after a period of relatively rapid depositio the suspended sediment concentration attains a constant steady state valu an equilibrium concentration. The value of this equilibrium concentratio was zero for $\tau_b < \tau_{b\,min}$ but had increa sing real values for higher bed shea stresses.

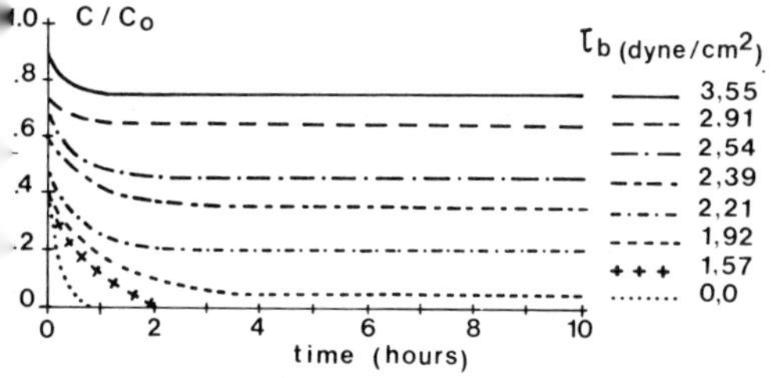


ig. 4. The reduction of the initial oncentration ($C_o$) towards an equilirium concentration ($C_{eq}$) as a funcion of time for different values of he bed shear ($\tau_b$) in a flume. Redrawn after Metha & Partheniades, 975).

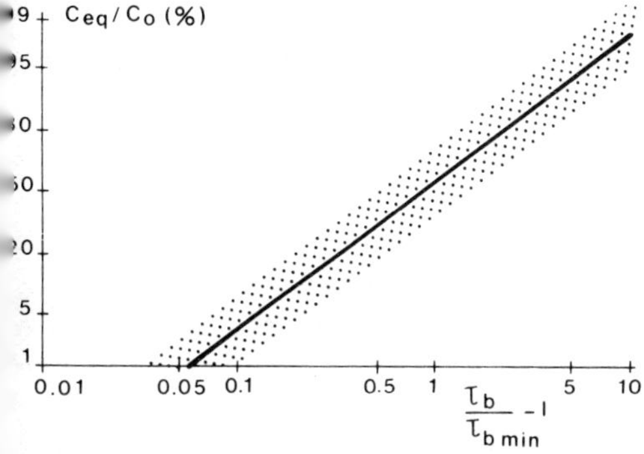


ig. 5. The relationship between the quilibrium concentration $C_{eq}/C_o$ and ne bed shear stress ($\tau_b$) above the inimum value ($\tau_{b\ min}$). Redrawn after Metha & Partheniades, 75).

Figure 5 implies that for a given ed shear stress and sediment type ne ratio $C_{eq}/C_o$ remains constant and oes not depend on the depth of flow the initial concentration. This ery interesting observation means or instance that for $\tau_b = 2\ \tau_{b\ min}$ e equilibrium concentration ($C_{eq}$) 50% of the initial concentration $_o$). So if $C_o$ = 2000 mg/l then $C_{eq}$ ll be 1000 mg/l, but if $C_o$ = 20.000 /l, $C_{eq}$ will be 10.000 mg/l. parently the amount of sediment, at can be maintained in suspension a steady state for a given $\tau_b$ dends solely on the initial concenation.

Metha & Partheniades (1975) stated at $C_{eq}$ is not a true equilibrium ncentration but rather represents constant amount that indefinitely mains in suspension at steady state at a given flow condition. This implies further that any exchange of particles between the bed and the suspension, as occurs in the case of cohesionless materials, must be precluded for cohesive sediments. There is no exchange between bed and suspension of mud as is the case for cohesionless sands.

The fraction 1- $C_{eq}/C_o$ represents the degree of deposition at steady state.

Metha & Partheniades came to the following hypothesis as to explain the observed behaviour of the mud deposition. Near the bed the strongest shear and lift forces occur and these control the shear strength of the flocs. A floc, which is strong enough to withstand the maximum shearing force acting on it near the bed will settle on the bed and will attach itself to the bed by cohesive bonds. On the other hand a floc with a relatively low shear strength will be broken into two or more smaller units and these will be reentrained in the suspension by the strong hydrodynamic lift forces. The steady state suspension represented by $C_{eq}$ contains flocs that are too weak and small to be able to withstand the shear stresses near the bed. In estuaries where varying flow conditions occur, the described mechanism may be operative as well, especially during decellerating flow.

## Consolidation

Once attached to the bed the mud flocs undergo a process of initial consolidation: a squeezing out of the water between the particles and a gradual increase in concentration, density and shear strength (fig. 6). This initial consolidation depends on the sand content of the mud, and the initial concentration. A higher sand content favours the velocity of the initial consolidation, like a lower initial concentration does. (Migniot 1968, Terwindt and Breusers, 1972).

Deposited sediments create an overburden to the already present sediments. The pressure of this overburden gives a further consolidation resulting in higher concentration and shear strength (fig. 7).

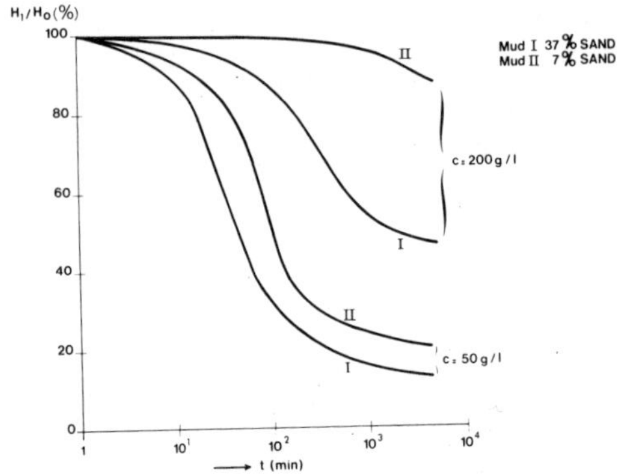


Fig. 6. The lowering of the water-mud interface ($H_1$) with time, concentration and sand content of the mud in suspension.

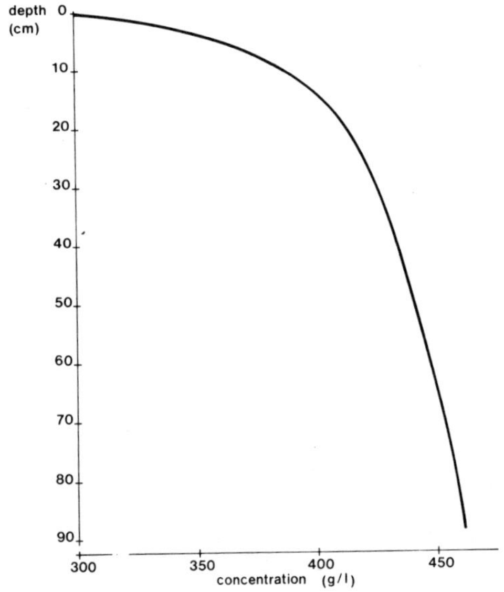


Fig. 7. Example of the increase of the concentration of mud with depth in the mud layer.
(Redrawn after Migniot, 1968).

## Erosion

As the applied shear by the fluid flow exceeds the bulk shear strength of the deposit erosion may start. Deposits having increasing shear strength with depth would fail to the depth where the strength equals the applied stress. Erosion may occur by failure of a part of the bed layers or particle by particle or aggregate by aggregate removal. Initially consolidating mudlayers exhibit an increase in the critical fluid shear stress to start erosion (fig. 8).

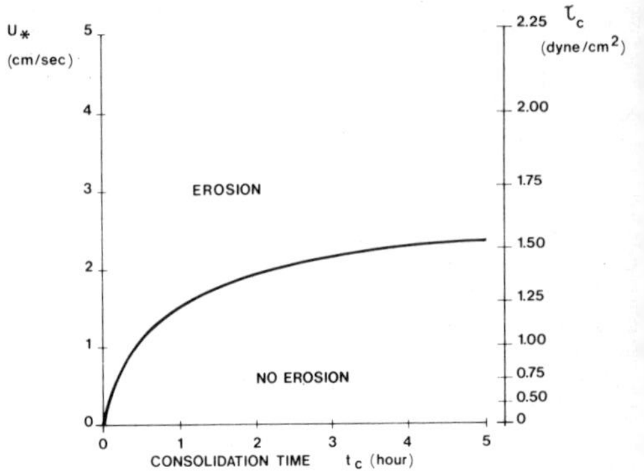


Fig. 8. The increase of the critical bed shear stress ($\tau_c$) or shear velocity ($U_* = \sqrt{\tau_{b/\rho}}$) with consolidation time of the mudlayer.
(After Terwindt and Breusers, 1972)

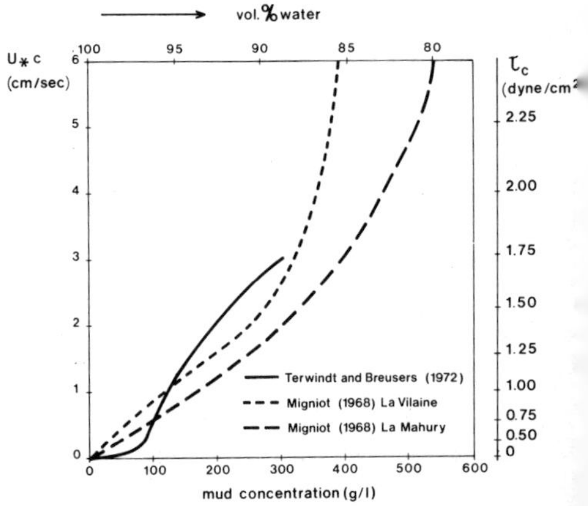


Fig. 9. The relationship between mu concentration in the bed and the cr tical shear stress ($\tau_c$) or shear ve locity ($U_{*c}$).

Several factors influence the begin ning and the rate of the erosion. Figure 9 shows that the critical shear velocity or the bed shear stress depends on the concentration of the mud deposit. In this figure the effect of the formation of gas the deposit by means of decompositi of organic material is not consider Gas formation has a decreasing effe on the critical shear velocity valu

It should be noted, that most proba-
oly there is a strong gradient in shear strength in the topmost layer of the bed. This gradient is very difficult to measure and this hampers the establishment of the exact relationship between the applied fluid stress and the shear strength of the water-bed interface.

Kandiah et al. (1975) pointed out that the mechanical structure of the deposited mud is related to the salt concentration, the sodium adsorption ratio (SAR) = $Na^+/\sqrt{\frac{Ca^{2+} + Mg^{2+}}{2}}$ and the Exchangeable Sodium Percentage ESP. The ESP is defined as (Exchangeable $Na^+$/cation exchange capacity) x 100 of the fluid. These chemical parameters influence the swelling, dispersion and permeability of the deposit.

Kandia et al. (1975) stated that in general Ca and Mg clays are bonded more firmly than sodium clays. Divalent clays can form a strong, relative rigid structure with an extremely thin water interlayer. At high SAR repulsive forces between the particles will predominate and thus producing significant swelling and a decrease in the interparticle bonding force. This results in a low strength. This may be illustrated by their results on a specific loam (fig. 10). The effect of SAR on the critical shear stress to initiate erosion for different salt concentrations in the pore fluid is demonstrated in figure 11. Higher salt concentrations in the pore fluid result in stronger bonds and increasing values of the critical shear stress.

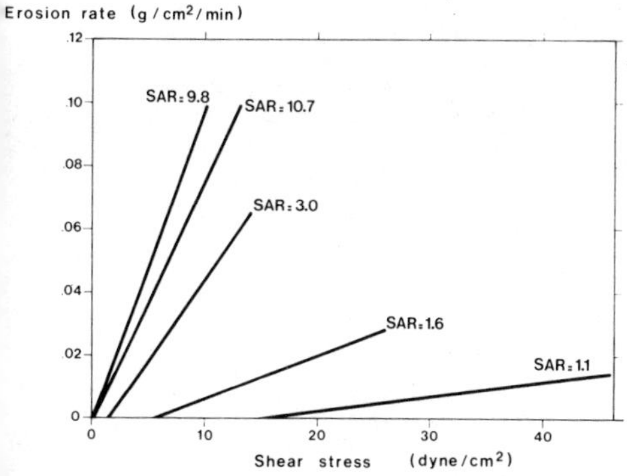


Fig. 10. Relationship between erosion rate and shear stress for different SAR for a specific loam.
(Redrawn after Kandiah et al., 1975).

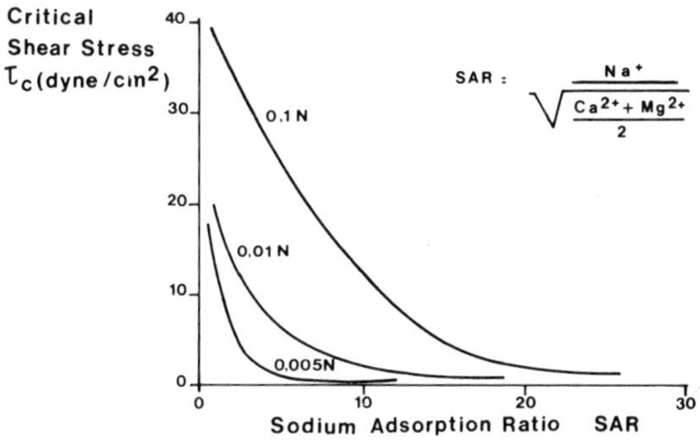


Fig. 11. The relationship between the critical shear stress ($\tau_c$) and the SAR for different salt concentrations in the pore fluid.
(Redrawn after Kandiah et al., 1975).

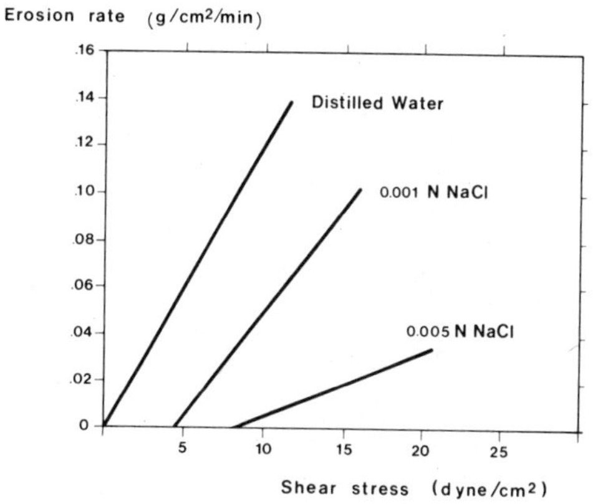


Fig. 12. The relationship between erosion rate and shear stress for different salt concentrations of the eroding fluid for a specific loam.
(Redrawn after Kandiah et al., 1975).

The erodability of a mud layer also depends on the salt concentration of the eroding fluid, decreasing with increasing NaCl concentration (fig. 12).

From Kandiah et al's explanatory experiments it appears that the structure of the mudlayer and the shear strength is determined by the sodium adsorption ratio (SAR) and the salt concentration of the pore fluid. As a result these parameters are reliable to the values of the critical shear stress and the erosion rate.

Acknowledgement.

The author greatfully acknowledges the permission of several authors to reproduce some of their original graphs.

References

Delft Hydraulics Laboratory 1962. Report on the siltation of Demarara Bar channel and coastal erosion in British Guiana. Delft, 237 pp.

Kandiah A., Loganathan P. and Krone, R.B., 1975. Pore and eroding fluid influences on surface erosion of soil. Journ. Geotechn. Engineering Division ASCE, 101, GTI:51-65.

Krone, R.B., 1962. Flume studies of the transport of sediment in esturial shoaling processes. Hydraulics Engineering Laboratory, Un. of California, Techn. Report.

Krone, R.B., 1976. Engineering interest in the benthic boundary layer. In: The Benthic Boundary Layer. I.N. Mc Cave, editor. Plenum Press, New York: 143-156.

Metha, A.J., 1973. Depositional behaviour of cohesive sediments. Ph.D. Diss. Un. of Florida.

Metha, A.J. & Partheniades, E., 1975. An investigation of the depositional properties of flucculated fine sediments. Journ. of Hydraulic Research, 13, 4: 361-382.

Migniot, C., 1968. Etude des propriétés physique de differents sediments très fins et de leur comportement sous des actions hydrodynamiques. La Houille Blache, 23, 7: 591-620.

Terwindt, J.H.J. & Breusers, H.N.C., 1972. Experiments on the origin of flaser, lenticular and sand-clay alternating bedding. Sedimentology, 19: 85-98.

Keywords:

Mud
Deposition
Transport
Erosion

# ISSOLVED AND SUSPENDED MATTER CARRIED BY RIVERS : COMPOSITION, TIME AND SPACE VARIATIONS, AND ORLD BALANCE

. Meybeck

aboratoire de Géologie, Ecole normale supérieure, Paris, France

Rivers are by far the most important supply f matter from land to ocean : the total load f rivers (suspension + solution) is around 0 000.$10^{12}$g.$year^{-1}$. The other forms of ransport are mainly the glacial scour 3 000.$10^{12}$g.year including Antarctica) and he continental dust fallout over the oceans a few 100 $10^{12}$g, Goldberg, 1975). Now rivers re also the major pathway of pollutants rom man to the ocean. Studies of river disolved and suspended material have begun a undred years ago (Forel, 1892) and have volved since in two main directions : i) study of fundamental geochemical processs on small representative basins such as the ubbard Brook, New Hampshire (Johnson et al., 969) (ii) estimate of the river elemental nputs to the ocean, derived mainly from tudies on major rivers (Alekin & razhnikova, 1962 & 1968; Durum et al., 1960; ivingstone, 1963). Since these works a lot f new data on river transport has been ublished concerning mostly dissolved aterial (Gibbs, 1967; Roche, 1973; Carbonnel Meybeck, 1975) but also suspended matter oad (Fournier, 1960; Holeman, 1968; Jansen Painter, 1974) and quality (Brunskill et al. 975; Fortsner & Muller, 1975). However, ittle information is available on the espective contribution of the dissolved and olid loads to the total or elemental transorts.

The concern of this paper is (i) to sum up ome of the major characteristics of river issolved and solid transports, (ii) to xamine their variations with environmental actors, (iii) to estimate the world balance f major elements carried either in solution r in suspension,(iv) to examine man influnce on the dissolved matter balance.

There is no sharp limit between the dissoled and the solid states and some elements, mong which Fe and Al, can be carried as olloïdal hydroxides (Stumm & Bilinski, 1972). he dissolved content of these elements is hen related to the filter pore-size which s used. However, in rivers, the great ajority of Fe and Al is transported as olid material - or as material bound to articles. The other major elements i.e. C, , Cl, Ca, Mg, Na, K, Ti and Si, representng more than 95% of the total load, are arried either as dissolved or as solid aterial.

As land denudation material is not entirely carried by rivers, and as rivers can transport material from other sources (wet and dry fallouts, atmospheric $CO_2$, various wastes resulting from human activities), river transport rate is always very different from the true erosion rate used by geomorphologists and geologists. Thus it is difficult to estimate land erosion from river transport. However, an attempt of such computation is presented here for the chemical denudation. There is evidence of changes of river solid load either by daming or by accelerated erosion (Meade, 1969) but also of dissolved load in highly industrialized countries (Weiler & Chawla, 1969). Therefore in order to study the natural processes, the less damaged rivers were taken into account or corrections were applied.

## 1. Relative variations of river dissolved and solid transport

### 1.1. Variation of concentrations with water discharge

The suspended matter content $C_s$(g.$m^{-3}$) is usually related to the water discharge Q ($m^3.s^{-1}$) as :

$$C_s = a'Q^{b'} \quad \text{with } b'>1$$

where a' varies according to the general turbidity of the river (Müller & Förstner, 1968). The total dissolved content (i.e. the sum of major ions and silica) Cd (g.$m^{-3}$) is usually linked to Q as :

$$Cd = a\ Q^{b} \quad \text{with generally } -1<b<0$$

where a express the usual salinity of the river. This trend is well known since Forel (1882) and has been extensively studied since Durum (1953). It can be explained by dilution of underground waters by much less mineralized surface runoff waters.

A typical example is given for the alpine Rhone entering lake Léman (figure 1) (Meybeck, 1972). It also shows that Cs to Q relationship is often more complex : here two hydrological periods have been separated and for the same Q the highest turbidities are measured in winter during snow melt and rain floods, while glacier melt water is less turbid. It is important to note that as a rule for all rivers Cs to Q relationship is also less defined than Cd to Q and that the turbidity range - here two orders of magnitude - is much greater than the salinity range.

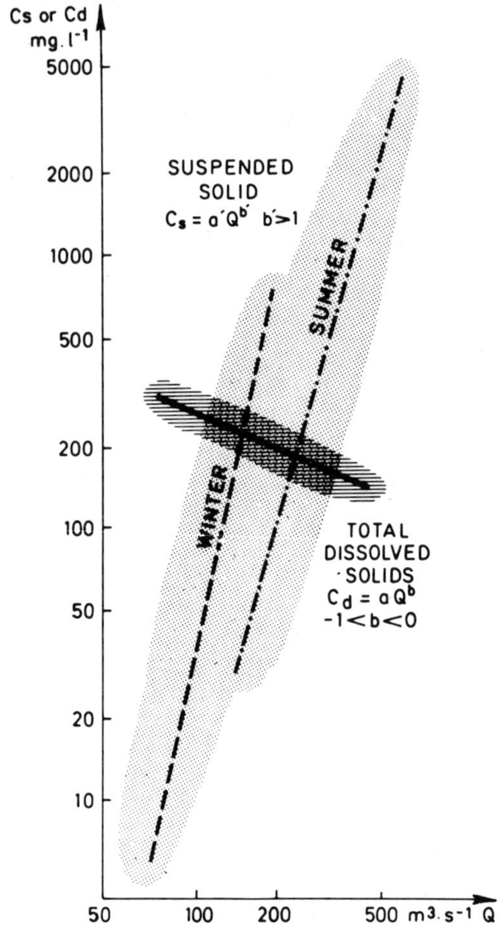


Figure 1. Range of variations of suspended matter concentration Cs (mg/l) and total dissolved content Cd (mg/l) with water discharge $Q(m^3.s^{-1})$. Rhone river (Switzerland).

1.2. Variation of transport with water discharge

Total dissolved and solid transports Qd and Qs $(kg.s^{-1})$ are simply derived from Cd and Cs : Qd = Cd Q , Qs = Cs Q . Most of the time the transport is normalized by the watershed area A. This defines the specific transports or transport rates $Td = Qd/A(t.km^{-2}.year^{-1})$ and $Ts = Q_s/A$, which allow comparison between rivers of different sizes. Qd and Qs are linked to Q as: $Qd = a\,Q^{\beta}$ with generally $0<\beta=b+1<1$ and $Q_s = a'\,Q^{\beta'}$ with $\beta'>>1$
If both forms of transport increase with the discharge, the solid load increases much more rapidly. This is well observed for the Dranse river, another mountain tributary of Lake Léman (figure 2). In this example $Q_s$ is dominant only during the high-water period $(Q>50m^3.s^{-1})$, caused by a high total dissolved content in this river (250 to 500 $g.m^3$).

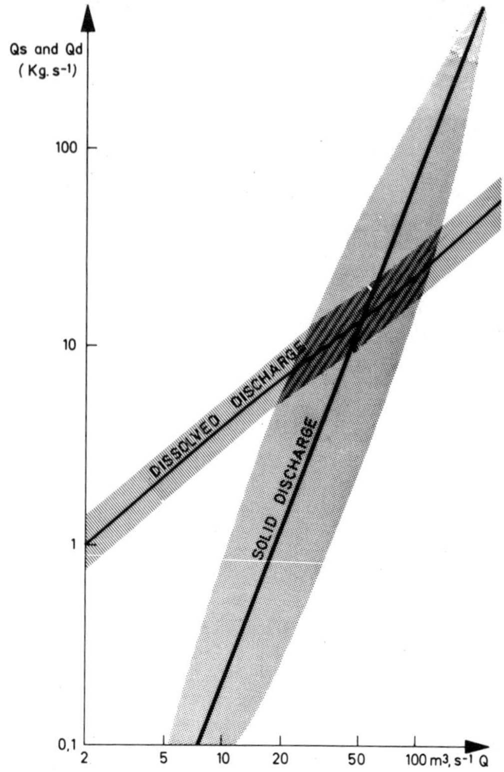


Figure 2. Range of variations of suspended matter transport $Q_s$ $(kg.s^{-1})$ and total dissolved transport Qd $(kg.s^{-1})$ with water discharge $Q(m^3.s^{-1})$. Dranse river (France).

In all rivers the ratio $Q_s$/Qd increases whe the water discharge increases, here from 0.01 to 10.

These relationship are the simplest ones and other variations have been observed. Well-known are the hysteresis cycles of $C_s$ and Q during floods or even during the whole hydrc logical cycle. Less known are the Cd and Qd cycles, which can be either clock wise (Gunneroon, 1967) or counter-clock wise (Roche, 1973). Some important Cd scattering from th general trend caused by secondary floods in the Annamitic Range, have also been describe for the Mekong river (Carbonnel & Meybeck, 1975). Similar relationships with water discharge have been observed for most major dissolved elements such as $Ca^{++}$, $Mg^{++}$ $HCO_3^-$, $SO_4^{--}$... Exceptions are noted for nutrients and sometimes for silica, for which the dissolved content can increase during floods or varies according to the bi logical cycle (Johnson et al., 1969; Edward 1973).

1.3. Seasonal variations of river transport

As both forms of transport are directly r

ated to the water discharge it is possible to define a transport regime which will be closely related to the hydrological regime itself. In the temperate regions Qd, expressed as Td, is dominant during the low-water period. Examples for the Green river, Kentucky (Douglas, 1964) and for the alpine Rhone river are given in figure 3. In the plains of Siberia and of North-East Canada Qd is

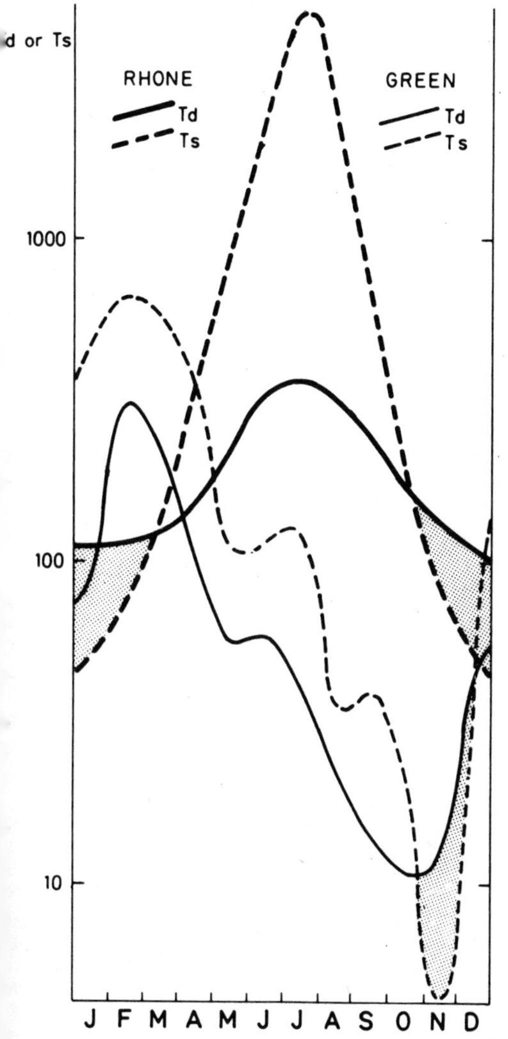


gure 3. Seasonal variations of specific lid transport Ts($t.km^{-2}.year^{-1}$) and ecific total dissolved transport Td ($t.km^{-2}.ar^{-1}$) for the Rhone river (Switzerland) and e Green river (Kentucky).
aded area : periods of dominant Td.

minant throughout the year while the verse is observed in semi-arid regions. pical examples i.e. the case of Harricana ver (data from Quebec Ministère des Ri-esses Naturelles) and of Rio Puerco, a non-perennial river in New Mexico (data from US geological Survey) are presented in figure 4

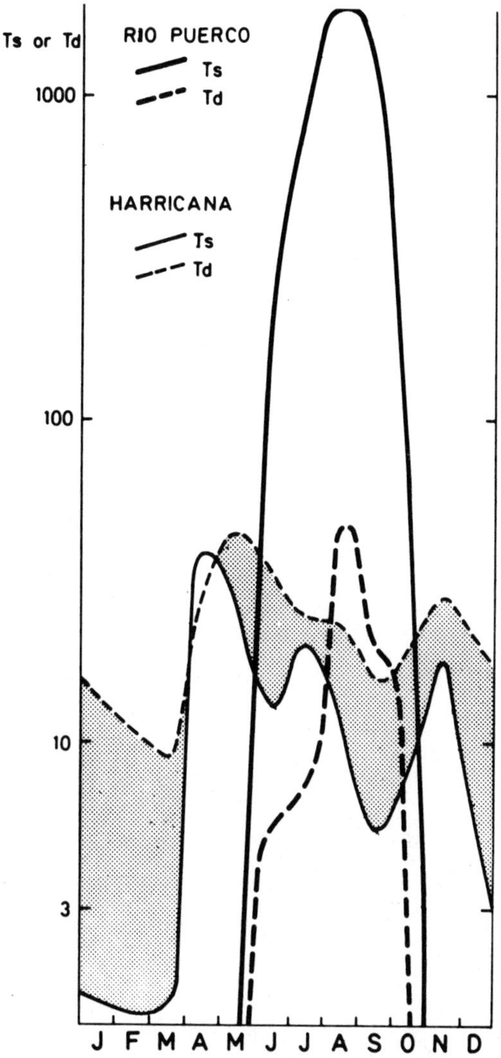


Figure 4. As for figure 3. Harricana river (Québec) and Rio Puerco (New Mexico).

2. Environmental factors governing river transport.

Td and Ts vary widely across the world according to the different environmental factors : climate, topography, geology, etc. These variations are studied here for the world major rivers ($Q>5000 m^3.s^{-1}$ or $A > 400\ 000\ km^2$). Some homogeneous regions such as Québec's Canadian Shield or Finland were also taken into account. Most of the data and detailed discussion have been previously presented (Meybeck, 1976). Some other data have been added in this paper (table 1).

Table 1. Rivers transport rates (8)

| | A $10^3 km^2$ | Q $m^3.s^{-1}$ | Td $t.km^{-2}.year^{-1}$ | Ts | Ref. |
|---|---|---|---|---|---|
| Burma | 710 | 14900 | 142 | 700 | (1) (2) |
| Philippines | 300 | 10500 | 252 | - | (1) |
| New Guinea | 400 | 25900 | 263 | - | (3) |
| Texas | 830 | 880 | 20 | 200 | (4) (5) |
| Alps | 325 | 5380 | 163 | 300 | (6) (5) |
| Baltic | 600 | 5000 | 32 | 5 | (7) (5) |

(1) recomputed from Kobayashi's data (1966). (2) Ts value as Irrawady value (U.N. Ecafe, 1968). (3) Q and Td computed from data of the P.N.G. Water Dept.
(4) U.S. geol. Survey.
(5) Approximate average.
(6) Po+Rhone+Rhine in 1900.
(7) Soviet part of Baltic Sea watershed, from Soviet Atlas Myra.
(8) Additional data to Meybeck, 1976.

2.1. Variation with runoff

Td is directly related to runoff $q(l.s.^{-1}.km^{-2})$ which is mostly related to the climatic features of watersheds (figure 5).

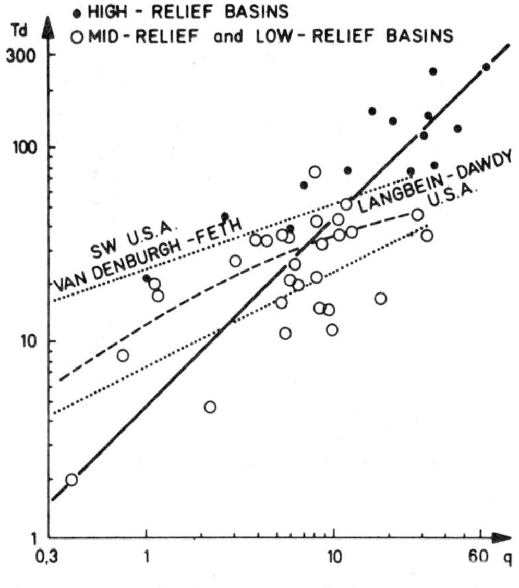


Figure 5. Variation of specific total dissolved transport Td $(t.km^{-2}.year^{-1})$ with runoff $q(l.s^{-1}.km^{-2})$ for the world major rivers (from Meybeck, 1976 and table 1 additional data).

If climatic conditions are the major environmental factors affecting Td, the topographic conditions are also very important and the maximum Td values (greater than 100 $t.km^{-2}.year^{-1}$) are always observed in mountain regions (Colorado, Alps, Magdalena river).The lowest values (less than 10 $t.km^{-2}.year^{-1}$) are characteristic either of very low runoff (Rio Grande), of high runoff in tropical plains (Lower Amazon, Congo) or of both low runoff and low salinity (East Siberia). This relationship between Td and q has been previously observed in certain regions (Van Denburgh & Feth, 1965; Langbein & Dawdy, 1964). As Tabuteau (1960) noted for the mediterranean watersheds, it is not possible to correlate Ts with q as this variation is masked by the relief influence.

2.2. Relationship between Td and Ts

This is a much debated question, as for american authors (Judson & Ritter, 1964) Td is inversely related to Ts while for soviet authors (Alekin & Brazhnikova, 1962) Td is directly related to Ts. This latter trend is clearly observed on figure 6 and the ratio Ts/Td increases from 0.1 in plains to more than 10 in semi-arid mountains.

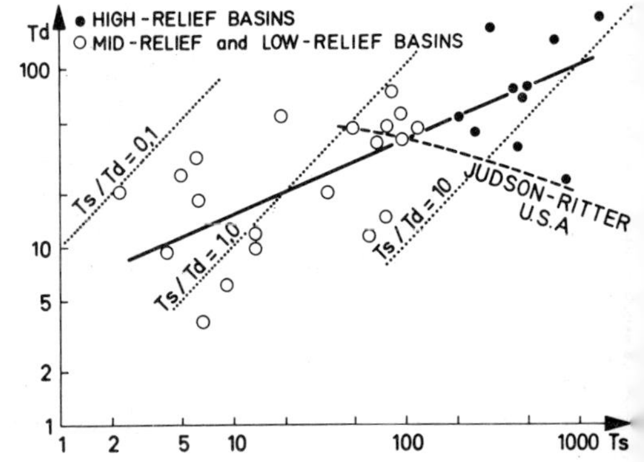


Figure 6. Variation of specific total dissolved transport Td$(t.km^{-2}.year^{-1})$ with specific solid transport Ts $(t.km^{-2}.year^{-1})$ for the world major rivers (from Meybeck 1976 and table 1 additional data).

he relief factor is therefore the major one ut the climate influence may cause important eviations from this general trend and even reverse figure such as it was observed in he USA by Judson and Ritter. The importance f relief signifies that most of material arried by the rivers in both forms of transort is coming from mountains denudation. For nstance the upper Amazon basin (13% of total rea) is responsible for 34% of Qd and for 6% of Qs (from Gibbs data, 1967). Gibbs also ound that relief was by far the dominant nvironmental factor acting on river transort.

Lithological factors were not considered ere. Jansen and Painter (1974) observed on any rivers that lithology has no major ffect on Ts. As Td is concerned, this nfluence is well-known for small basins, ut is secondary when extended areas are tudied. However there is an important excepion for the dissolved silica which highly epends upon climatic and lithologic factors.

.3. River transport typology.

A rough typology of Ts and Td, according to elief, average temperature and runoff, is resented in table 2. There remain some ypes of morphoclimatic environments for hich no data are available. It must be kept n mind that these figures are valid for exnded areas ( 400 000 $km^2$) and that great eviation from this picture can be found for ch smaller watershed. Although this typogy is given for the total dissolved solids, silica included, this trend can be extrapolated to most major ions. As a matter of fact the order of abundance of major ions is the same for more than 90% of the rivers considered here which correspond to the rock-dominated type described by Gibbs (1970). The silica variation does not always follow this figure.

3. World average quality and balance of material carried by rivers

3.1. Dissolved matter

Collected data on major rivers, corresponding to more than 50% of the world water discharge, and transport typology for the remaining areas have been used to compute the world balance and the average dissolved content of rivers (table 3). Corrections were applied in polluted regions in order to approach the natural values as close as possible. The relative abundance of elements is like in previous studies (Alekin and Brazhnikova, 1962 & 1968; Livingstone, 1963) but world Td is slightly higher even as compared to our previous estimate (Meybeck, 1976). This may be caused by under estimation of South East Asia contribution to the balance : the region between Ganges and Yang Tse Kiang (10 $10^6$ $km^2$) could contribute to 35% of world Qd and to 50% of world Qs.

3.2. Suspended matter

Since suspended matter composition has been much less studied, an approximation

| | q > 15 HIGH RUNOFF | | | 5 < q < 15 MEDIUM RUNOFF | | | q < 5 LOW RUNOFF | | |
|---|---|---|---|---|---|---|---|---|---|
| | t<5 | 5<t<15 | t>15 | t<5 | 5<t<15 | t>15 | t < 5 | 5 < t<15 | t>15 |
| IGH ELIEF | | ALPS Td ≃135 | MAGDAL. Td≃140 | | GANGES Td ≃ 85 | BLUE NILE Td≃50 | | COLORADO Td ≃ 28 | |
| | Ts > 100 | | | Ts > 150 | | | Ts > 250 | | |
| EDIUM ELIEF | | | AMAZON Td≃40 | YUKON Td≃40 | DANUBE Td≃50 | PARANA Td≃17 | | ARAL B. Td≃18 | NILE Td ≃ 5 |
| | 10 < Ts < 100 | | | 20 < Ts < 150 | | | 50 < Ts < 250 | | |
| OW ELIEF | QUEBEC Td ≃ 30 | | | YENISSEI Td ≃ 25 | BALTIC B. Td ≃ 35 | CONGO Td≃13 | NELSON Td ≃ 23 | DNIEPR Td≃25 | MURRAY Td ≃ 5 |
| | Ts < 10 | | | Ts < 20 | | | Ts < 50 | | |

ble 2. Typology of river total dissolved transport (Td) and solid transport (Ts) in $km^{-2}.year^{-1}$ according to water runoff q in $l. s^{-1}.km^{-2}$ and to average watershed temperature in °C.
rld major rivers $Q > 5000\ m^3.s^{-1}$ or $A > 400\ 000\ km^2$.

| | $SiO_2$ | $Ca^{++}$ | $Mg^{++}$ | $Na^+$ | $K^+$ | $Cl^-$ | $SO_4^{--}$ | $HCO_3^-$ | $\Sigma^{+-}$ | S |
|---|---|---|---|---|---|---|---|---|---|---|
| Ci | 11.6 | 14.6 | 3.8 | 5.1 | 1.35 | 5.3 | 8.5 | 57.7 | 96.4 | 108 |
| Tdi | 4.35 | 5.5 | 1.4 | 1.9 | 0.5 | 2.0 | 3.15 | 21.6 | 36.0 | 40.4 |

Table. 3. River natural dissolved content Ci ($mg.l^{-1}$) and dissolved transport Tdi ($t.km^{-2}.year^{-1}$). $\Sigma^{+-}$ is total ionic content and S is total dissolved content (in $mg.l^{-1}$)

| | Si | Al | Fe | Mg | Ca | Na | K | Ti | P | Mn |
|---|---|---|---|---|---|---|---|---|---|---|
| Suspended matter | 32.6 | 6.5 | 4.1 | 1.1 | 3.4 | 0.9 | 1.95 | 0.38 | 0.07 | 0.04 |
| Surficial rocks | 27.5 | 6.9 | 3.6 | 1.6 | 4.5 | 1.4 | 2.45 | 0.38 | 0.06 | 0.07 |

Table 4. Average composition of river solid material in percent as compared to average surficial rock. (Based on Nile, Mississipi, Mac Kenzie, Colorado and Danube rivers).

based on Mississipi and Nile (Clarke, 1924), Mac Kenzie (Brunskill et al., 1975), Colorado (Gould, 1960), Danube (Georgescu et al., 1973) is presented in table 4. This sample, only 5% of world Qs, may not be quite representative of the world average since there is evidence that tropical river material is much more depleted incalcium. However there is here a significant depletion in Na, Ca, K, Mg which are the major cations found in solution.

### 3.3. World elemental river balance

For this computation the total river solid load to the ocean - 15 500 $10^{12}g.year^{-1}$ - has been estimated mainly with data collected by the IASH and published in the Hydrol. Sci. Bull. This value is slightly less than previous ones (Fournier, 1960; Holeman, 1968; Jansen & Painter, 1974). From table 5 it is clear that S, Cl, Na and Ca are mostly carried as soluble matter while Al, Si and Fe are in particulate form. Dissolved contents for Al and Fe have been taken as 400 ppb (average from Durum et al., 1960) and 40 ppb (Gibbs, 1975). The ratio of elemental dissolved load/elemental total load, which may be called the soluble transport index, is very useful for the comparison of the geochemical behaviour of each elements in rivers. Environmental factors ma cause a considerable variation of this index from the average values presented here.

When the geographic distribution of Td and Ts is examined it appears that Td is dominan for 35% of the continental area drained to t ocean and that Ts/Td is between 1 et 2 for another 25%. This figure is based on weighte Ts and Td averages for the whole watershed o major rivers, were it based on smaller basi than the importance of Td would have been even greater.

## 4. Origins of material carried by rivers

Evidence of suspended matter pollution has been found in some industrial countries (Förstner & Müller, 1975) but it is very difficult to separate the various sources of solid material (dust fallout, soil and rock erosion, solid wastes, adsorbed elements on solids, etc...) For the total dissolved load one can attempt to separate the different components. The total fallout contribution including marine salts, dust and atmospheric pollution was estimated to be around 11% on the basis of fallout quality in the USA, Europe and USSR. This figure is similar to Zverev estimate (14%) for the Soviet Union

| | S | Cl | Na | Ca | Mg | K | $SiO_2$ | Al | Fe |
|---|---|---|---|---|---|---|---|---|---|
| Dissolved | 105 | 197 | 190 | 547 | 143 | 51 | 435 | 15 | 1.5 |
| Solid | < 25* | < 70* | 122 | 480 | 151 | 274 | 9860 | 900 | 580 |
| Dissolved % Total | 85 | 75 | 61 | 53 | 49 | 16 | 4 | 1.5 | 0.3 |

Table 5. Natural elemental budgets of river input to the ocean, in $10^{12}g.year$. * Estimated

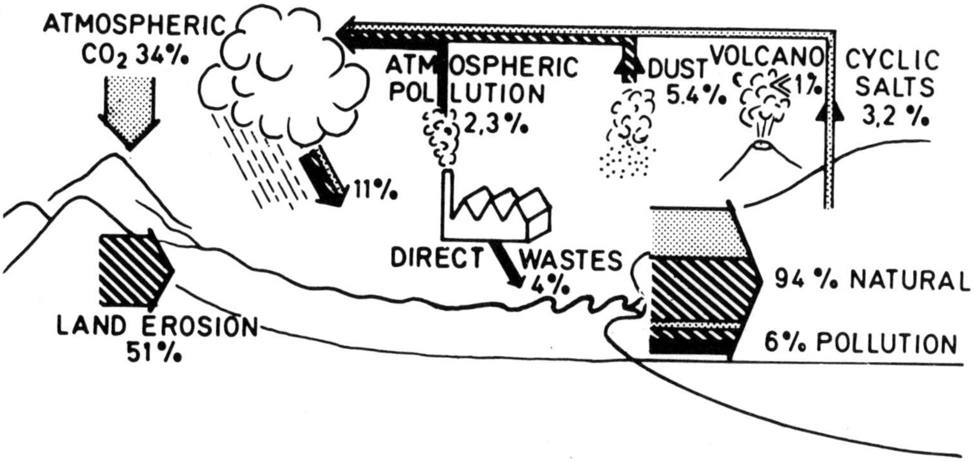


Figure 7. Estimates of various sources and pathways of dissolved matter carried by rivers to the ocean.

erritory. The atmospheric $CO_2$ probably onstitutes 70% of the natural dissolved icarbonates, that is about 34% of the total issolved load. This value is based on the mount of dissolved $Ca^{++}$ in rivers resulting rom the carbonate rocks dissolution (75% ccording to Turekian, in Berner, 1971). The iver pollution either through direct waste elease, through surface runoff or through tmospheric pollution has been estimates by er capita loadings computed from river evo- ution in industrial countries (Mississipi, aint Lawrence, Rhine). The contribution of an is now estimated to be around 6% and his figure is much higher for $Cl^-$ (34%) and or $SO_4^{--}$(25%). The remaining 50% of the otal load were attributed to land erosion nd correspond to an average chemical denuda- ion rate of $20t.km^{-2}.year^{-1}$ over the conti- ental area drained to the ocean $100.10^6$ km$^2$).

eferences


lekin, O.A. & L.V. Brazhnikova, 1962. The correlation between ionic transport and suspended sediment. Dokl. Akad. Nauk. SSSR. 146 : 203-206.

lekin, O.A. & L.V. Brazhnikova, 1968. Dissolved matter discharge and mechanical and chemical erosion. Int. Ass. Sci. Hydrol. publ. n° 78 : 35-41.

rner, R.A., 1971. Worldwide sulfur pollution of rivers. J. Geophys. Res., 76, 6597-6600.

unskill, G.J. et al., 1975. The chemistry, mineralogy and rates of transport of sediments in the Mackenzie and Porcupine watersheds, 1971-1973. Techn. Rpt. n° 566, Fisheries and Marine Service, Environment Canada.

Carbonnel, J.P. & M. Meybeck, 1975. Quality variations of the Mekong river at Pnom Penh, Cambodia, and chemical transport in the Mekong basin. J. Hydrol. 27 : 249-265.

Clarke, F.W., 1924. Data of geochemistry, Vth edition U.S. Geol. Surv. Bull. 770.

Douglas,I., 1964. Intensity and periodicity in denudation processes with special references to the removal of material in solution by rivers. Zeitschrift Geomorphologie 8 : 453-472.

Durum, W.H., 1953. Relationship of the mineral constituants in solution to streamflow, Saline river near Russel, Kansas. Am. Geoph. Union Trans., 34 : 435-442.

Durum, W.H., G. Heidel & L.J. Tison, 1960. Worldwide runoff of dissolved solids. Int. Ass. Sci. Hydrol. publ. 51 : 618-628.

Edwards, A.M.C., 1973. The variation of dissolved constituants with discharge in some Norfolk rivers. J. Hydrol. 18 : 219-242.

Forel, F.A., 1892-1904. Le Léman, monographie limnologique. Tome I, II & III. F. Rouge, Lausanne & Slatkine reprints, Genève.

Förtsner, U. & G. Müller, 1975. Schwermetalle in Flüssen und Seen als Ausdruck der Umweltverschmutzung. Springer-Verlag, Berlin

Fournier, F., 1960. Climat et érosion. Presses Universitaires de France, Paris.

Georgescu,I.I., S. Lupan, M. Sâlâgean & M. Oancea, 1973. Chemical composition of Danube water, sea water, algae and sediments of the Black Sea. Thalassia Yugoslavica 9 : 87-99.

Gibbs, R.J., 1967. The geochemistry of the Amazon river. Geol. Soc. Am. Bull. 78 : 1203-1232.

Gibbs, R.J., 1970. Mechanism controlling world water chemistry. Science 170 : 1088-1090.

Gibbs, R.J., 1975. In Assessing potential ocean pollutants. National Academy of Sciences, Washington D.C.

Goldberg, E.D., 1971. Man's role in the major sedimentary cycle. In Nobel Symposium, Dyrssen and Jagner editors, 267-283.
Gould, H.R., 1960. Character of the accumulated sediment. In : Comprehensive study of sedimentation in lake Mead 1948-1949. U.S. Geol. Surv. Prof. Pap. 295.
Gunnerson, C.G., 1967. Streamflow and quality in the Columbia River basin. Proc. Am. Soc. Civ. Eng., J. Sanit. Eng. Div. 39 : 5626-5636.
Holeman, J.N., 1968. The sediment yield of major rivers of the world. Wat. Res. Res. 4 : 737-747.
Jansen, J.M.L. & R.B. Painter, 1974. Predicting sediment yield from climate and topography. J. Hydrol. 21 : 371-380.
Johnson, N.M., G.E. Likens, F.H. Bormann, D.W. Fisher, & R.S. Pierce, 1969. A working model for the variation in streamwater chemistry at the Hubbard Brook experimental forest, N.H. Wat. Res. Res. 5 : 1353-1363.
Judson, S. & D.F. Ritter, 1964. Rates of regional denudation in the United States. J. Geophys. Res. 69 : 3395-3401.
Kobayashi, J., 1966. Silica in freshwaters estuaries. Proc. IBP Symp. Amsterdam - Nieuwerluis, 1966, 41-55.
Langbein, W.B. & D.R. Dawdy, 1964. Occurence of dissolved solids in surface waters in the United States. U.S. Geol. Surv. Prof. Pap. 501 D.
Livingstone, D.A., 1963. Chemical composition of rivers and lakes. U.S. Geol. Surv. Prof. Pap. 440 G.
Meade, R.H., 1969. Errors in using modern stream load data to estimate natural rates of denudation. Geol. Soc. Am. Bull. 80 : 1265-1274.
Meybeck, M., 1972. Bilan hydrochimique et géochimique du lac Léman. Verh. Int. Ver. Limnol. 18 : 442-453.
Meybeck, M., 1976. Total dissolved transport by world major rivers. Hydrol. Sci. Bull. 21 : 265-284 and erratum in 21, 4.
Müller, G. & U. Förstner, 1968. General relationship between suspended sediment concentration and water discharge in the Alpenrhein and some other rivers. Nature 217 : 244-245.
Roche, M.A., 1973. Geochemistry and natural ionic tracing : two complementary ways to study the natural salinity regime of the hydrological system of lake Chad. J. Hydrol. 26 : 153-171.
Stumm, W. & H. Bilinski, 1972. Trace metals in natural waters; difficulties of interpretation arising from our ignorance of their speciation. Advances in Water Pollution Research, 1972: 39-49.
Tabuteau, M., 1960. Etude graphique pour les conséquences hydro-érosives du climat méditerranéen. Ass. Géographes Français Bull. 294-295 : 130-142.
U.N.-ECAFE, 1968. A compendium of major international rivers in the ECAFE region. Water Resources Series n°29.
van Denburgh, A.S. & J.H. Feth, 1965. Solute erosion and chloride balance in selected river basins of the western conterminous United States. Wat. Res. Res. 1 : 537-541.
Weiler, R.R. & V.K. Chawla, 1969. Dissolved mineral quality of the Great Lakes waters. Proc. of the 12th Conf. on Great Lakes research, Int. Ass. Great Lakes Res. : 801-818.
Zverev, V.P., 1971. Hydrochemical balance of the USSR territory. Dokl. Akad. Nauk. SSSR 198 : 161-163.

Key words: rivers
dissolved and suspended matter
world balance
river water quality
world mass balance

# THE NEOFORMATIONS OF SEDIMENTS IN LAKE CHAD; THEIR INFLUENCE ON THE SALINITY CONTROL

J.P. Carmouze, O.R.S.T.O.M., Section d'hydrobiologie, 24, rue Bayard, 75008-Paris

H.L. Golterman, Limnologisch Instituut, Nieuwersluis, The Netherlands

G. Pedro, I.N.R.A., Département de Science du Sol, C.N.R.A., 7800-Versailles

Summary

From geochemical and mineralogical analyses of the suspended matter of the Chari and of the surface sediments of the lake Chad we could present evidence of a neoformation of different types of smectite in the lake. The following smectites were identified:

a) a montmorillonite in the area with diluted waters in the south near the delta of the Chari;
b) a ferriferous montmorillonite in the central zone;
c) and a magnesium montmorillonite in the north region, where the concentration of dissolved compounds is high.

We could establish that, in the main water body, a part of the newly formed material is derived from dissolved silica, $Ca^{2+}$ and $Mg^{2+}$. Similar phenomena have been invoked to explain the stability of the oceans and the control of their pH (Sillen, 1961; Holland, 1965; Kramer, 1965; Helgeson and Mackenzie, 1970), and are also responsible for the reactions of silica in river mouths (Wollast and de Broere, 1971). These processes have not yet been reported for the lacustrine environment. In this respect hydrochemistry of Lake Chad is not so unusual and thus it is not unlikely that these processes also occur in other freshwater lakes.

The smectites are formed from amorphous constituents (hydroxides of iron and aluminium) and from cations of the alkaline earth metals. The explanation of Wollast and de Broere (1971) about the origin of dissolved silica in the estuaries of the Scheldt by neoformation of smectite from kaolinite - does not seem to be true for Lake Chad.

The neoformation of magnesium smectite and of calcite takes place in the same region of Lake Chad. From our hydrochemical data one could easily misinterpret the sedimentation in Lake Chad by assuming a precipitation of a dolomite (magnesium-calcium carbonate), which process in other situations must not be overlooked as possibility.

## 1. Introduction

Lake Chad is a large lake of 21.000 $km^2$, has a mean depth of 3 to 4 m. It is situated in the centre of the Central African plain, which extends from 5° to 25° latitude north (fig. 1). The main inflow is the river Chari, and there is no surface outflow. Due to it being an enclosed basin situated in a subarid climate the lake should act as an evaporation pan for concentrating the imported dissolved salts of the river. However, the total salinity remains at a mean value of 350 mg/l; the main constituents being $Na^+$, $K^+$, $Ca^{2+}$, $Mg^{2+}$, $H CO_3^-$, $CO_3^{--}$ and $Si (OH)_4$.

This relatively low salinity can be explained by several factors:

- The water from the Chari has a low salinity (65mg/l).
- The volumes of rain falling on the lake and of water lost by infiltration are relatively important when compared with the evaporation, i.e., 15% and 7.5% respectively.
- Except for sodium (Carmouze, 1976) the dissolved compounds participate in chemical precipitation as appeared from the geochemical budget of sodium, by which means considerable quantities of the annual input are eliminated from the water amounting to: $Ca^{2+}$, 45%; $Mg^{2+}$, 41%; $K^+$, 19.5%; $H CO_3^-$, 30.5% and $Si (OH)_4$, 81.5%.

## 2. Geochemical modifications in the lake basin

The Chari has a mean discharge into the lake of about 40 x $10^9 m^3$. Typically, there is a high flow in the period mid-August until mid-January and a low flow from March until June. The amount of transported suspended matter is 2.8 x $10^6$ tonnes per year of which the lake receives 60% during June until September. The mean mineralogical composition of the suspended matter - of which kaolinite is the

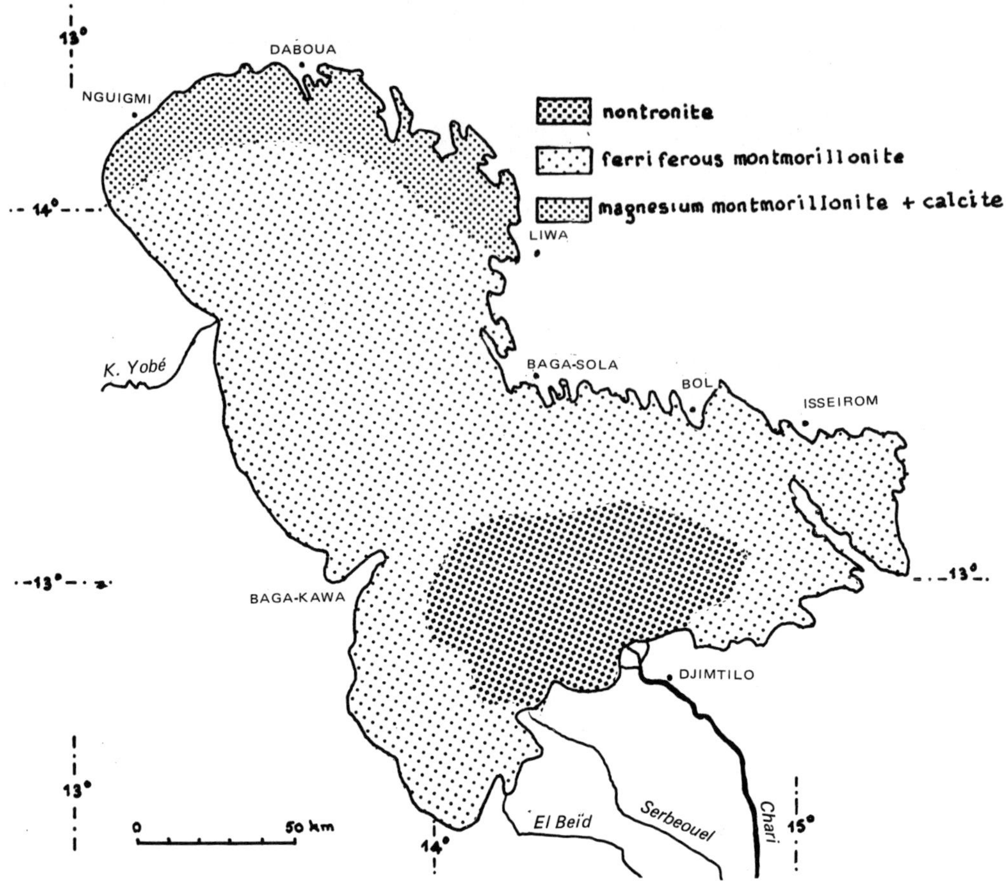


Fig. 1 Lake Chad - Domains of neoformations of sediments.

Table I - Mean annual mineralogical composition of suspended matter in the Chari and mean mineralogical composition of the lake sediments in weighted percentage.

| | Quartz | Kaolinite | Illite | Feldspars | Smectites | Amorphous Compounds $SiO_2$ | $Fe_2O_3$ | $Al_2O_3$ | Loss o[n] igniti[on] |
|---|---|---|---|---|---|---|---|---|---|
| Chari | 9.45 | 38.60 | 11.35 | 5.83 | 10.7 | 2.30 | 2.75 | 1.30 | 18.1 |
| Lake* | 11.2 | 26.55 | 7.85 | 4.25 | 21.0 | 4.85 | 1.39 | 0.18 | 22.8[illegible] |

*In addition, sediments located in the northern part of the lake contain 5 to 10% of calcite

major component - is given in table I. The river smectite is a ferriferous beidellite of which the composition approximates to:

$$\left[Si_{3.47}Al_{0.53}\right]O_{10}\left[Al_{0.33}Fe^{III}_{1.17}Ti_{0.29}Mg_{0.21}\right](OH)_2Mg_{0.06}Ca_{0.14}Na_{0.02}$$

The lake bottom has mainly a clay - silt appearance, while the region near the delta of the Chari (2000 km$^2$) has a more granular appearance, and at some places sand may be found. The mineralogical study of the bottom re- vealed the same minerals as those im- ported by the Chari, but their percen tages are different from those of the suspended matter in the Chari (table I), where the smectite content is con siderably smaller.

These results can be explained by ge

chemical modifications. In order to distinguish these modifications it is necessary to have a baseline against which to assess the degree of modification. For this purpose we have used the ratio of the percentages of illite, kaolinite and feldspar, which are homogeneously distributed over the lake bottom and which do not undergo any mineralogical modification during sedimentation. Indeed the ratios of the contents of these three components in the suspended matter in the river are near to those in the sediments.

With this starting point, and realising the limitation that the river suspended matter may have varied during the period corresponding to the period of the formation of the suspended sediments, both quantitatively as well as qualitatively, we have calculated the gains and losses of the other mineral constituents after the sedimentation of the suspended load. The results show (table II) a relative decrease of the contents of the amorphous hydrates of iron- and aluminium oxides, but an increase in the content of amorphous silica and of smectite.

On a yearly basis, this represents a loss of 21.3 x $10^6$ kg of $Fe_2O_3$ and of 29.3 x $10^6$ kg of $Al_2O_3$, or about 30% and 80% of the mean annual input of these minerals which must be used in the formation of other compounds On the other hand the lake sediments are enriched with 130 x $10^6$ kg of $(SiO_2)_{am}$ and of 545 x $10^6$ kg of smectite, representing 200% and 175% of the mean annual input. Carmouze (1976) explained that in the past the content of smectite could not be higher than at present, so we may conclude that in situ neoformation of smectite has taken place. Assuming that the mineralogical composition of the river sediments has remained constant in the period during which the surface sediments in the lake have been formed, we can calculate that 35% of the smectite in the lake is allochthonous and 65% autochthonous.

### 3. Geochemical characterisation of the newly formed smectite.

In order to know whether the newly formed smectite has the same composition whether it be found in the granular of silty sediments, near to or distant from the river delta, we distinguished for comparison three major types:
a) smectite from the granular sediments
b) smectite in the central lake zone
c) smectite in the north region, far away from the delta.

#### a) The smectite of the granular sediments

These sediments, the so-called pseudo-sands, contain granules with a mean diameter of 0.200 till 0.285 mm and are generally deposited on the banks of the lake. The granules have different forms, colours and hardness, changing from black, manually friable, spherical ones to oval - sometimes truncated with marked irregularities at the surface, and with a greenish - yellow colour, easily broken by hand. Classical mineralogical analysis

Table II - Contents of amorphous constituents and of smectite relative to the sum of kaolinite-illite-feldspar in the river (KIF) and lake sediments $(KIF)_L$. Changes of these components (i) in relative, $(\Delta i_r)$, and absolute value, $\Delta i$.

| i | $i / (KIF)_C$ (in %) | $i / (KIF)_L$ (in %) | $\Delta i_r$ (in %) | $\Delta i$ (in $tx10^3$) |
|---|---|---|---|---|
| $SiO_2$ am | 4.1 | 12.55 | + 8.45 | + 130 |
| $Fe_2O_3$am. | 4.95 | 3.59 | - 1.36 | - 21.20 |
| $Al_2O_3$am. | 2.34 | 0.46 | - 1.88 | - 29.3 |
| Smectite | 19.27 | 54.3 | +35 | + 545 |

showed these granules to be composed of two components: goethite and nontronite. Carmouze (1976) showed that the black granules could develop into the greenish-yellow ones, with an associated enrichment of the granules with silica, which corresponds to a transformation of goethite into nontronite with the following formula:

$$\left[Si_{3.83}Al_{0.06}Fe^{III}_{0.11}\right] O_{10} \left[Fe^{III}_{1.76}Mg_{0.20}\right] (OH)_2Ca_{0.22}Na_{0.02}K_{0.02}$$

b) The smectite in the central lake zone

Mineralogical analysis of a representative mixture of several sediments from this zone indicated that they contained smectites approximately 35% derived from the rivers and 65% authigenic, the latter having the formula of a ferriferous montmorillonite:

$$\left[Si_{4.00}\right] O_{10} \left[Al_{1.32}Fe^{III}_{0.43}Ti_{0.02}Mg_{0.18}\right] (OH)_2Ca_{0.14}Na_{0.01}$$

c) The smectites newly formed in the sediments in the north

These consist of a mixture of locally formed smectite enriched with magnesium, smectites from other parts of the lake and riverine smectite. The ratios of these different representations could not be evaluated, and so a formula for the authigenic portion could not properly be derived, although it can be suggested that the formula approximates to that of montmorillonite:

$$\left[Si_{3.92}Al_{0.08}\right] O_{10} \left[Al_{0.57}Fe^{III}_{0.37}Ti_{0.10}Mg_{1.21}\right] (OH)_2Ca_{0.175}Na_{0.05}$$

because Cheverry (1974) showed that the soils formed in those isolated pieces of water which were once part of the main lake consist of magnesium montmorillonite. Assuming that the two other fractions in the northerly sediments occur in the same ratio as they do in the central lake, we obtained the following formula for the locally formed smectite:

$$\left[Si_{4}\right] O_{10} \left[Al_{0.24}Fe^{III}_{0.12}Ti_{0.11}Mg_{1.99}\right] (OH)_2Ca_{0.195}Na_{0.07}$$

which is indeed a magnesium montmorillonite.

Thus, the newly formed smectites are a ferriferous montmorillonite (central zone), a magnesium rich one (north zone) and a strongly ferriferous beidellite of a nontronite type (delta zone). The riverine smectite is a not very pronounced beidellite, rich in iron.

4. Reactions of smectite formation. Hydrochemical modifications.

a) The nontronite

The granules go through an initial stage of formation and deposition of oval particles of goethite followed by a stage of a progressive silicification. In this later stage the grains consolidate and transform into nontro ite. It seems that the ferric hydroxide, coming into the lake as non-crys tals, form small granules from the point of their input in the lake onwards. The silica and the magnesium which are built into the nontronite are derived from the dissolved compounds. The source of the aluminium might be the riverine amorphous aluminium or from the small portion of al uminium from the goethite. The exchangeable cations, $Ca^{2+}$, $Na^{+}$, and $K^{+}$, we may assume come from the dissolved st te.
The reaction scheme may be the follow ing: (see next page nr. I)

By this process a considerable amount of dissolved silica disappears from the water together with smaller amoun of alkaline - earth elements. $CO_2$ is produced with a concomitant possible decrease of the pH which in Lake Chad is balanced by the progressive decrea in salinity. Theoretical consideratio of this process are discussed by Sill (1961).

b) The ferriferous montmorillonite

We assume that the iron and the alumir ium constituents of the montmorillonit are derived, respectively, from the amorphous iron hydrate or the goethite and from the amorphous alumina coming from the River Chari. This would cause the impoverishment of these elements in the lake basin (Table 2). We can nc completely exclude the possibility tha a small fraction of the newly formed nontronite is destroyed in the reducing environment of the sediment, in which case the iron may participate in the formation of the montmorillonite.

Nr. I:

$$1.87\ FeO(OH)_{s} + 0.06\ Al(OH)_{3\ s} + 3.83\ Si(OH)_4 + 0.20Mg^{2+} + 0.22Ca^{2+} + 0.02Na^{+} + 0.02K^{+} \longrightarrow \left[Si_{3.83}Al_{0.06}Fe^{III}_{0.11}\right]O_{10}\left[Fe^{III}_{1.76}Mg_{0.20}\right](OH)_2Ca_{0.22}Na_{0.02}K_{0.02} + 0.88H^{+} + 7.24H_2O$$

followed by: $0.88H^{+} + 0.88HCO_3^{-} \longrightarrow 0.88CO_2 + 0.88H_2O$

---

Nr. II:

$$4.00Si(OH)_4 + 1.32\ Al(OH)_{3\ s} + 0.43\ Fe(OH)_{s} + 0.02TiO(OH)_2 + 0.18Mg^{2+} + 0.13Ca^{2+} + 0.01Na^{+} \longrightarrow \left[Si_{4.000}\right]O_{10}\left[Al_{1.32}Fe^{III}_{0.43}Ti_{0.02}Mg_{0.18}\right](OH)_2Ca_{0.13}Na_{0.01} + 0.63H^{+} + 7.9H_2O$$

---

Nr. III:

$$4Si(OH)_4 + 0.24\ Al(OH)_{3\ s} + 0.12\ FeO(OH)_{s} + 0.11TiO(OH)_2 + 1.99Mg^{2+} + 0.195Ca^{2+} + 0.07Na^{+} \longrightarrow \left[Si_{4.0}\right]O_{10}\left[Al_{0.24}Fe^{III}_{0.12}Ti_{0.11}MG_{1.99}\right](OH)_2Ca_{1.95}Na_{0.02} + 4.44H^{+} + 4.40H_2O$$

followed again by the liberation of $CO_2$.

---

The small amounts of Ti are certainly derived from the (dissolved(?)) compounds transported by the river. Silica and the magnesium of the crystal lattice and the exchangeable cations are derived from their dissolved forms. The following reaction scheme may be proposed: (see above nr. II)

As in the case of the neoformation of nontronite considerable quantities of silica are withdrawn from the dissolved state as are, also as before, lesser quantities of alkaline earth metals. Again there is a liberation of $CO_2$, tending to decrease the pH.

c) The magnesium montmorillonite

With the same reasoning as that for the ferriferous montmorillonite we can assume that the aluminium of this mineral comes from the amorphous aluminium hydroxide. We may again assume that the small amounts of iron and titanium are taken up from the free oxides. The silica, magnesium and the exchangeable cations are used from the dissolved forms. The reaction scheme is: (see above nr. III). This reaction withdraws considerable quantities of dissolved silica and magnesium from the open lake waters, while large amounts of escaping $CO_2$ counteract the potential increase of salinity caused by the evaporation.

5) Location of smectite neoformation

The neoformation of clays takes place at the water - sediment interface and probably in the top most layer and it seems likely that these reactions occur more in the interstitial waters than in the open water. In figure 2 we have depicted the domains of the three types of neoformation and the chemical characteristics of the associated waters.

In general, clays appear to form in dilute solutions, but rich in iron, in agr-

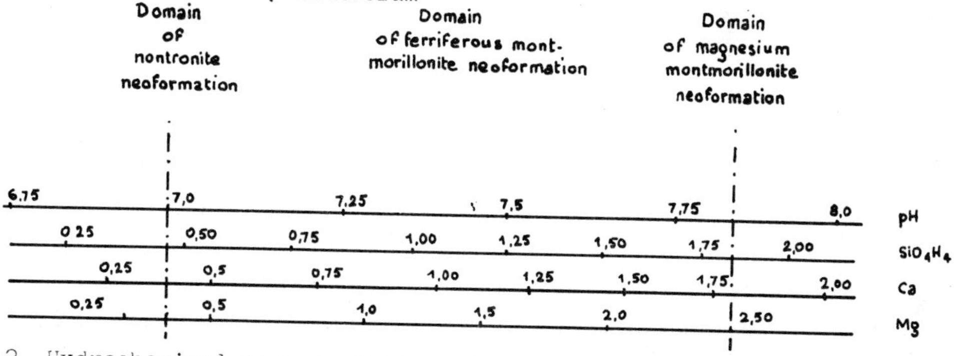


Fig. 2 Hydrochemical characteristics of the interstitial waters in which neoformation of clays take place (conc. in mole $l^{-1}$).

eement with the results of Tardy, Trescases and Fritz (1974). The neoformation of nontronite takes place in interstitial waters with a relatively low concentration (100-150 mg/l) of dissolved components.

## 6. Formation of calcite. Conditions of precipitation.

In addition to the neoformation of clays, precipitation of calcite takes place in the north (fig. 1). As noted in table I, sediments of this region contain 5 to 15% of calcite. A detailed study about this phenomenon has been made (Carmouze, 1976) which shows that precipitation occurs when the product of the activities of calcium and carbonate dissolved as free $Ca^{++}$ and $CO_3^{--}$, $Ca^{2+} \times CO_3^{2-}$, exceeds 18-19 times the product of solubility of the calcite. Characteristics of waters are mostly as follows: pH = 8.9-9.1; $CO_{3\,T}$ = 1.5-2.5 mm/l; $Ca_T$ = 0.40-0.90 mm/l. $CO_{3\,T}$ and $Ca_T$ correspond to the values analyzed including $CaH^+CO_3$, $CaCO_3$ $MgCO_3$ and $CO_3^{--}$ for the first one, and $CaH^+CO_3$ and $CaCO_3$ for the second term.

This important degree of supersaturation of water with calcite could be due to the high rate of concentration of the waters in this shallow lake submitted to a strong evaporation.

## 7. Contribution of the neoformation of clays to chemical sedimentation

The fractions of the dissolved element met in the lake due to the neoformation of silt cannot be determined by the present hydrochemical data, but can be derived from the geochemical and mineralogical analyses of suspended matter brought in by the Chari when compared to the sediments of the lake. The data thus produced are based on the assumption that clay formation during the period represented by the sampled silt has been constant. The period concerned has been estimated to be about two centuries.
The results are compiled in table III.

From table III one can see that the sedimenting silica contributes about 75% to the total quantity of newly formed silts, while the calcium contributes about 20%. The remaining part is due to precipitation of calcite (50%) and due to the action of organisms (especially molluscs, 20%). The amount of magnesium that precipitates is two thirds incorporated by way of the neoformation processes and one third is fixed by organisms, especially the macrophytes. Potassium is nearly completely incorporated in the sediments through the macrophytes, while the carbonates are deposited in the form of calcite.

## 8. Conclusions

From geochemical and mineralogical analyses of the suspended matter of the Chari and of the surface sediments of the Lake Chad we could present evidence of a neoformation of different types of smectite in the lake. The following smectites were identified:
a) a montmorillonite in the area with diluted waters in the south near

Table III. Mean, specific chemical sedimentation in Lake Chad of dissolved silica, calcium, magnesium and potassium and of carbonates in the last two centuries.

| Dissolved salts | Total chemical sedimentation(1) (in $10^9$ Mole per year) | Geochemical sedimentation (in %) smectite | calcite | Biochemical sedimentation (in %) |
|---|---|---|---|---|
| $Si(OH)_4$ | 7.65 | 73.2 | - | 26.8 |
| $Ca^{2+}$ | 1.46 | 20.5 | 59 | 20.5 |
| $Mg^{2+}$ | 0.95 | 71.5 | - | 28.5 |
| $K^+$ | 0.18 | - | - | 100 |
| $CO_3^{2-}$ | 0.86 | - | 100 | - |

(1) The present regime of riverine transport is used as the basis for such calculations. The total quantities have been assessed using a proportionality coefficient, because the regime might have changed in the period under discussion.

the delta of the Chari;
b) a ferriferous montmorillonite in the central zone;
c) and a magnesium montmorillonite in the north region, where the concentration of dissolved compounds is high.

We could establish that, in the main water body, a part of the newly formed material is derived from dissolved silica plus $Ca^{2+}$ and $Mg^{2+}$. Similar phenomena have been invoked to explain the stability of the oceans and the control of their pH (Sillen, 1961; Holland, 1965; Kramer, 1965; Helgeson and Mackenzie, 1970), and are also responsible for the reactions of silica in river mouths (Wollast and de Broere, 1971). These processes have not yet been demonstrated for the lacustrine environment. In this respect hydrochemistry of Lake Chad is not unusual and thus it is not unlikely that these processes also occur in other freshwater lakes. Wiersma (1970) reviewed the genesis of smectites in different environments (Soil, marine and lacustrine) and mentioned a few publications suggesting these processes to occur also in lakes.

The smectites are formed directly with ions coming from amorphous constituents (hydroxides of iron and aluminium) and solution (alkaline earth metals). The explanation of Wollast and de Broere (1971) about the origin of dissolved silica in the estuaries of the Scheldt by neoformation of smectite from kaolinite - does not seem to be true for Lake Chad.

The neoformation of magnesium smectite and of calcite takes place in the same region in Lake Chad. From our hydrochemical data one could easily misinterpret the sedimentation in Lake Chad by assuming a precipitation of a dolomite (magnesium-calcium carbonate), which process in other situations must not be overlooked as possibility.

## References

J.P. Carmouze, 1976. La régulation hydrogéochimique du lac Tchad. Contribution à l'analyse biogéodynamique d'un système lacustre endoreïque en milieu continental. Thèse Univ. Paris VI - Paris; ORSTOM-Paris. 419 p. multigr.

Cl. Cheverry, 1974. Contribution à l'étude des polders du lac Tchad. La dynamique des sels lors de la transformation de sédiments argileux et organiques en milieu continental sub-aride. Thèse Univ. L. Pasteur, Strasbourg; ORSTOM-Paris, 275 p. multigr.

H.L. Golterman, 1975. Physiological limnology, Elsevier-Amsterdam, 489 p.

H.C. Helgeson & F.T. Mackenzie, 1970. Silicate - sea water equilibrium in the ocean system. Deep-sea Research, 17, 877-892.

H.D. Holland, 1965. The history of ocean water and its effect on the chemistry of the atmosphere. Proc. Nat. Acad. Sci., 53, 825-854.

J.R. Kramer, 1965. History of sea water Constant temperature pressure equilibrium models compared to liquid inclusion analyses. Geochim. Cosmochim. Acta 29, 921-946.

F.T. Mackenzie & Garrels, R.M., 1966. Chemical mass balance between rivers and oceans, Am. J. Sci. 264, 507-525.

L.G. Sillen, 1961. The physical chemistry of sea water. In: Oceanography. Pub. Am. Assoc. Adv. Sci., 67, 549-581

Y. Tardy, Trescases, J.J. & Fritz, B., 1974. Evaluation de l'enthalpie libre de formation de montmorillonite ferrifère. C.R. Acad. Sc. Paris, sér. D, No. 13, pp 1665-1668.

J. Wiersma, 1970. Provenance, genesis and paleogeographical implications of microminerals occurring in sedimentary rocks of the Jordan Valley area. Publ. Fys. Geogr. Lab. Univ. Amsterdam, 15.

R. Wollast & F. de Broere, 1971. Study of the behaviour of dissolved silica in the estuary of the Scheldt. Geochim. Cosmochim. Acta 35, 613-620.

Key words: neoformation
clay
sediments
Lake Chad
salinity control

# BACTERIAL SEDIMENT FORMATION IN LAKES: TROPHIC IMPLICATIONS

Hans W. Paerl

Freshwater Section, Department of Scientific and Industrial Research, Taupo, New Zealand

## Summary

Numerous studies have revealed interactions between bacteria and suspended sediments in a variety of lakes. The action of bacteria on sediments is usually examined through measurements of mineralisation. This approach seems logical, since heterotrophic oxidisers of organic matter are widely distributed on suspended particles. Furthermore it is not surprising that heterotrophic bacteria are typified as mineralisers, given experimental evidence that as much as 80% of isotopically-labelled dissolved organic matter can be oxidised. However, it is shown here that suspended sediments are sites of dissolved organic carbon (DOC) scavenging by attached microbes, leading to particulate organic carbon (POC) production during sedimentation processes. This is achieved by bacterial conversion of non-respired DOC to structural components added to particles while the particle core remains unaltered. In this manner DOC "lost" by phytoplankton through extracellular production, lysis and partial decomposition may be reintroduced into grazer food chains. Suspended sediments appear to represent significant sites for bacterial POC production. I therefore urge that a more flexible definition of aquatic heterotrophic bacteria be considered, including their ability to quantitatively produce particles in addition to being strict decomposers.

## Introduction

Suspended sediments are complex matrices of non-living and living organic matter, often containing inorganic particles. The mechanisms by which sediment aggregates are formed remain unclear. Physical dissection or disruption is often impossible, thus leaving a heterogeneous aggregate for investigation. The inability to clearly identify sediment components obscures our understanding of processes involved in suspended sediment formation.

Evidence exists citing the importance of both physical or biological aggregation processes. In lakes and streams having strong pH and dissolved chemical gradients physical aggregation and flocculation appear to be chief mechanisms involved in particle formation (Lush and Hynes 1973). Baylor and Sutcliffe (1963) showed that particles may be formed by bubbling air through seawater. It has also been shown that bubble surfaces act as concentrating sites for colloidal and small particulate matter, including bacteria (Carlucci and Bezdek 1972).

In aquatic ecosystems not possessing strong chemical gradients or sizeable physical perturbations, it is difficult to relate sediment formation to distinct physical or chemical processes. Light microscopic (Rodina 1963; Riley 1963) and electron microscopic (Paerl 1973a, 1973b, 1975; Ferrante and Parker 1976) examinations of both natural and cultured samples reveal bacterial-particle interactions, and bacteria commonly inhabit particle surfaces (Waksman and Carey 1935; Stark <u>et al</u> 1938; Stotzky 1967). However, recent studies have linked bacteria with particle aggregation (Rodina 1963; Seki 1972; Paerl 1973) and some investigations suggest that heterotrophic products appear to make up the bulk of fine particulate matter (Tilzer 1972; Paerl 1974; Monheimer 1975).

Bacterial particle formation is of ecological importance for several reasons;

I. In oligotrophic systems particle surfaces provide micro-environments with relatively concentrated nutrients. Particles being formed or degraded by bacteria are important sites for nutrient uptake and/or release in aquatic ecosystems. It has been suggested that biological nitrification, denitrification, ammonification and phosphate uptake and release are accelerated by particle surfaces (Alexander, 1961).

II. Particle sedimentation rates are usually increased by bacterial aggregation.

III. Attached bacteria may be a more readily available food source to aquatic grazers, because individual free-floating bacteria may not be commonly retained by most copepod and cladoceran zooplankton filtration mechanisms (Boyd 1976).

IV. The chemical composition of particles may undergo radical changes as a result of combined effects of attached bacterial uptake of ambient soluble nutrients while releasing particulate nutrients by mineralisation.

In this presentation experimental evidence

coupled with microscopic observations, is used to show interconnections among the above categories. It is maintained that bacterial growth on and in sediments cannot be solely discussed in terms of mineralisation or nett nutrient loss from these sediments. Although aquatic bacteria mineralise and excrete significant fractions (at times exceeding 50%) of incorporated organics (Hobbie and Crawford 1969; Allen 1973), much of the residual "uptake" is transferred into extracellular particulate material. The role of non-respired uptake in sediment formation is an important process which has thus far been largely ignored in studies of sediment production. Scanning electron microscope (SEM) observations in conjunction with *in situ* and laboratory sediment formation studies, indicate that bacterial biomass and extracellular deposition are substantial components of aquatic particulate matter.

## Materials and methods

Bacterial sediment production was assessed by:

(a) dialysis bag experiments,
(b) micro-autoradiographic experiments, and
(c) SEM observations of sediments.

### Dialysis bag experiments

Cellulose dialysis tubing, $5\frac{3}{4}$ inch wide (approximately 10 inch diameter when completely filled) was used to construct 1 liter containers allowing for diffusion of soluble materials while retaining bacteria-size particles. Care was taken to thoroughly clean the tubing prior to experimentation, since it contains glycerol and trace concentrations of sulphur as lubricants and conditioners. Detailed methodology is given elsewhere (Paerl 1975). The incubation of dialysis tubing in Lake Tahoe, followed by microscopic observations, revealed only slight amounts of bacterial growth (less growth than on glass slides) on its surface. All dialysis tubing experiments were carried out in darkened double-walled containers having 2 cm non overlapping holes to allow for ample diffusion of soluble materials.

Sterile control bags were made by immersing closed bags, having the ends previously dipped in sealing wax, into solutions of both .001 M $HgCl_2$ and .001 M $NaN_2$ for 24 hours prior to lake incubation. These bags remained sterile for periods up to 2 weeks as confirmed by plate and direct counts and adenose triphosphate (ATP) analysis of contents.

All dialysis experiments were conducted *in situ* in Lake Tahoe using natural microflora sampled from depths corresponding to incubation depths. Stream-borne sediments taken from the mouth of Ward Creek, a tributary of Lake Tahoe, were used as starting sediment content. SEM observations on these sediments showed a mixture of inorganic and organic particles. Inorganic particles were largely feldspar and mica while organic particles constituted eroded and decaying periphyton and leaf litter. Ward Creek sediments were dispensed into dialysis bags at a dilution of 1 to 10 parts of lake water, which approximates the sediment content of Lake Tahoe in sediment plume studies performed in the lake proper (Paerl and Goldman 1972; Goldman *et al* 1974). During the 7 day experimental period in May, 1973, particle aggregation and relative sinking rates were tested in triplicate sterile vs non-sterile bags suspended at 10 meters. In a following 7 day period in June 1973, triplicate sets of bags were incubated at 50, 100 and 200 meters in a pelagic section of the lake approximately 1.8 km offshore. A third experiment in July 1973 was designed to assess the breakdown of $^{14}C$ and $^{32}P$ labelled detritus, composed of Lake Tahoe seston photosynthetically grown on either $^{14}C$ bicarbonate or $^{32}P$-orthophosphate for 4 days. (This growth period allowed for thorough incorporation and distribution of the isotopes into particulate material). The labelled detritus was then heat-sterilised and dispensed into dialysis bags along with natural lake microflora.

An additional experiment was conducted in a darkened constant-temperature incubator. Ward Creek sediment was added to 1 liter Erlenmeyer flasks. Duplicate flasks containing: (a) sediment and lake water - $NaN_2$ sterilised, (b) sediment and lake water - untreated, (c) sediment and lake water and 200 µg/l glucose-carbon, and (d) sediment and lake water and 200 µg/l glucose-carbon, 20 µg/l $NO_3$-N and 10 µg/l $PO_4$-P were incubated for 5 days. The flasks were agitated by hand twice a day to resuspend sediments.

Relative aggregation and sinking rates of sediment particles were examined by withdrawing samples from both dialysis bags and flasks and placing them in 4 cm quartz cuvettes. Changes in U.V. absorbance at 300 nm (the maximum absorbance of Ward Creek and Lake Tahoe sediments) were monitored over a 20 minute period by taking a reading at 1 minute intervals on a Beckman DBG spectrophotometer equipped with a hydrogen U.V. lamp (Tenney & Stumm 1965; Teot & Daniels 1969). Relative slopes of decreases in U.V. absorbance imparted by particles remaining in solution were then plotted. Free-floating bacteria at concentrations encountered in bags had little effect on U.V. absorbance. When non-aggregated bacteria grown in culture were added to lake waters at non-turbid concentrations no significant effect on U.V. absorbance was encountered. Thus this method proved to be sensitive in specifically discriminating relative degrees of sediment aggregation.

### Autoradiography

The utilisation and fate of dissolved organic compounds in bacterial sediment formation processes were observed by applying micro-autoradiography to a variety of experiments conducted on Lake Tahoe, California, and Lakes Taupo and Rotongaio, New Zealand waters. Trace concentrations of high specific activity 6-$^3$H glucose, 2-$^3$H glycine and sodium 2-$^3$H acetate were added to lake water samples, followed by 2 to 24 hour incubations at sampling temperatures in a darkened incubator. All samples were slowly stirred by a magnetic stirrer placed in the incubator. At 2 hour intervals 10 ml subsamples were withdrawn from 1 liter Erlenmeyer flasks containing $^3$H-incubated lake water and prepared for autoradiography as described by Paerl (1974).

Samples were also taken from the abovementioned dialysis experiments. These samples were incubated with the same $^3$H labelled compounds previously described. It was shown that glucose was effectively utilized as a precursor for aggregate formation in Lake Tahoe waters (Paerl 1974). It thus seemed worthwhile to follow the uptake and conversion of this dissolved substrate in a study of bacterial sediment formation, where extracellular excretions suspected of containing carbohydrates (Floodgate 1972) may be important structural components.

### SEM observations

Suspended sediments from dialysis and flask experiments as well as lake waters sampled from a range of North American and New Zealand lakes were examined by SEM. Methods for sample preparation and observations were basically those given by Paerl and Shimp (1973).

### Discussion

In all in situ experiments the sterile contents of dialysis bags showed the least amount of particle aggregation as measured by U.V. absorbance. Untreated controls consistently revealed more rapid buildup of large particles. SEM observations on both sets of bags indicated that the most profound structural difference between the contents of sterile and non-sterile bags was in the magnitude of bacterial growth and deposition of extracellular and non-living cellular material left behind from successive periods of bacterial growth.

Figure 1 shows the aggregation of labelled particulate matter under sterile vs non-sterile conditions.

Figure 1. Results of a week-long incubation of isotopically-labelled particulate matter under sterile vs non-sterile (Tahoe microflora added) conditions. The % C-14 and P-32 refers to the % of the original activity of each isotope left in dialysis bags at the end of the experiment.

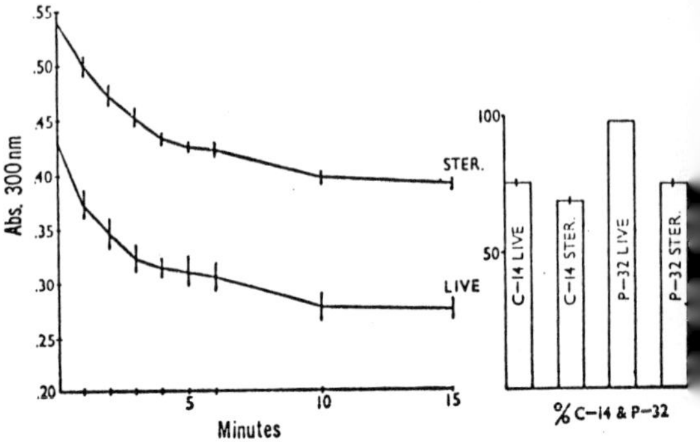


Typically, the initial slope of decreasing absorbance was steeper in non-sterile bags, indicating that large particles were settlin out of solution rapidly and that these particles were derived from smaller particles which had aggregated. These findings were confirmed by both light microscopy and SEM observations on filtered sediments. With the absence of smaller size particles, the absorbance measured in non-sterile bags was much less than measurements from sterile bag Surprisingly, leaching of both $^{14}$C and $^{32}$P from particles in non-sterile bags was less than losses from corresponding sterile bags during this 7 day experiment. This indicated that while they were actively building aggregates and depositing particulate matter on sediments, bacteria were maintaining $^{14}$C and $^{32}$P in a particulate state. It thus appeared that during favourable growth periods bacteria may retard if not complete obscure nutrient release processes from suspended sediments. Dialysis bags containing a dilute suspension of Ward Creek sediments in Tahoe water were suspended in darkened containers at 50, 100 and 200 meters in Lak Tahoe. An additional set of $NaN_3$- sterilis bags was included to check for physical aggregation of sediments at each depth. Afte a one-week incubation period the bags' contents were examined for relative sinking rates and observed by SEM. As greater dept were used for incubation a lesser degree of aggregation was observed (Fig.2). The uppe two depths were located at the middle and lowest depth of the euphotic zone respectiv ly. Samples suspended at 200 m were well i to aphotic waters.

Figure 2. Relative sinking rates of Ward Creek sediments incubated at indicated depths with water taken from those depths. Sterile samples (STER) incubated alongside each set of non-sterile bags proved to be similar for each depth. Therefore the error bars given for (STER) include results from replicates at all depths.

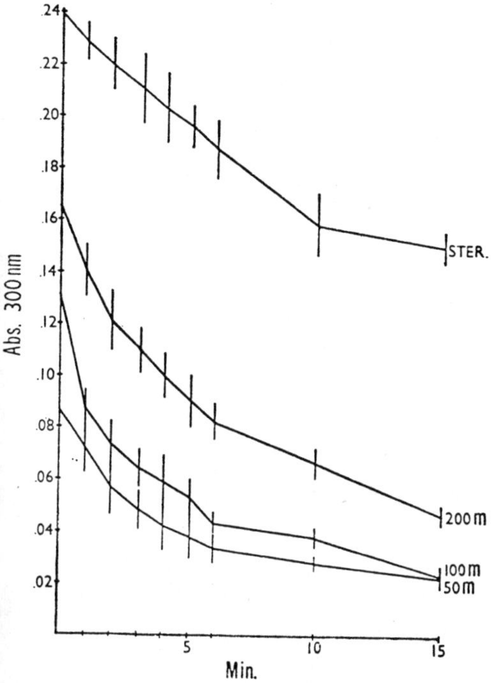


During this thermally stratified period dissolved organic carbon content of the euphotic zone of Lake Tahoe was measurably higher than in the aphotic zone (Paerl 1973b), although the composition of the DOC remains unknown. Furthermore, in situ bacterial biomass at this time was about 3 times higher in euphotic than in aphotic waters (Paerl 1973b). It is likely that a larger standing stock of more active heterotrophic bacteria in euphotic waters, perhaps utilising slightly higher levels of ambient DOC, were responsible for more extensive aggregates in these waters. SEM observations on sediments from 50 vs 200 meter dialysis experiments clearly revealed much more extensive bacterial growth on the particles from 50 m. Other investigators have found similar occurrences in marine systems, where near surface waters contained a more extensive and active microflora than deeper waters (Jannasch et al 1971). The environmental parameters responsible for these events remain unclear.

The process of bacterial scavenging of DOC dissolved phosphorus and nitrogen can be further illustrated by use of dialysis bag incubation of suspended sediment. Table 1 shows results of chemical analyses performed on Ward Creek sediments entering Lake Tahoe, following erosion by heavy thunderstorm activity in the catchment.

Table 1. Changes in carbon and phosphorus composition of particulate material originating from Ward Creek. Initial samples (1200 on 10/22) were taken before a severe thunderstorm, followed by later (1700, 2100 on 10/22, 0900 on 10/23) samples taken after thunderstorms had led to freshly eroded silts in the creek. The results of a 6 day dialysis bag incubation of the eroded silts with Lake Tahoe microflora are listed at the bottom.

| Time | Sediment Load ($mg.l^{-1}$) | % C | % P |
|---|---|---|---|
| 1200 10/22 | 0.30 | 51.0 | 5.34 |
| Rainfall | | | |
| 1700 | 2.10 | 21.9 | 2.10 |
| 2100 | 2.40 | 20.9 | 3.08 |
| 1900 10/23 | 1.70 | 26.0 | 2.24 |
| Dial.ster. 10/29 | 1.45 | 29.0 | 2.39 |
| Dial.live 10/29 | 2.00 | 48.0 | 4.90 |

Freshly eroded sediments which accompanied increased runoff in this tributary were characteristically low in organic carbon and phosphorus content and contained virtually no attached microflora. However, after entering the lake such sediments rapidly became populated by bacteria as can be shown by samples taken in a sediment plume in Lake Tahoe (see photograph 2), or by dialysis bag incubations of suspended sediments. With the introduction of the lake's microflora came a rapid shift in organic carbon and phosphorus content of sediments. Sterile bags failed to show such shifts.

The addition of possible biochemical precursors of bacterial extracellular production can have profound effects on the extent of sediment aggregation. When relatively small (relative to mM concentrations normally used in laboratory experiments) additions of D-glucose were made to suspended sediments sedimentation rates and average sizes of sediment particles were increased (Fig. 3).

Figure 3. Relative sinking rates of Ward Creek sediments in response to additions of of D-glucose (GLU), nitrate (N) and phosphate (P) in sterile vs. non-sterile Lake Tahoe water. Samples were taken 24 and 48 hours after inoculation with lake water and nutrients.

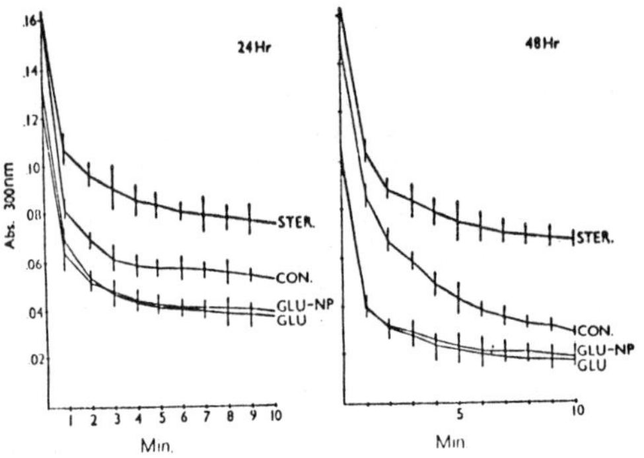


The most obvious morphological change which occurred in response to glucose additions was the increased appearance of slime material produced by attached microbes. This microbial "glue" can be shown by autoradiography to be derived from soluble precursors, and it appears to be an important element regulating sediment aggregation in aquatic ecosystems (Paerl 1973a, 1974). Phosphate and nitrate enhancement complementing glucose additions revealed no increase in aggregation above glucose alone (Fig. 3). These experiments, show that adequate supplies of soluble organics are crucial in particle formation by heterotrophic bacteria.

Autoradiographs of Lake Taupo (oligotrophic) and Lake Rotongaio (eutrophic) samples readily show the fate of DOC converted to bacterial particulate matter associated with suspended sediments. Photo 1 shows the progressive labelling of detrital aggregates in Lake Rotongaio water to which trace concentrations of uniformly labelled $^3$H-glucose were added. At 2 hours of incubation the label (clumps of dark silver grains covering bacteria) was associated with individual bacteria located in mucoid matrices (Photo 1A). Following 2 hours of incubation the aggregates were intensively labelled, with the label being dispersed throughout most of the aggregates (Photo 1B, C). Most of the label was still concentrated near bacterial cells, but a spreading of the label due to $^3$H incorporation into surrounding extracellular material was evident. In this preparation free-floating $^3$H labelled bacteria filtered onto the surface of the filters could be seen as well, in contrast to clumped bacteria in aggregates. Virtually all of the free floating bacteria failed to show dispersion of the label into the surrounding medium.

When trace concentrations of uniformly labelled $^3$H-glycine were added to the same water sample and incubated identically to $^3$H-glucose experiments, uptake after 2 hours was again associated with individual free-floating and aggregated bacteria. However, when 24 hour incubations were examined the labelling failed to reveal dispersal in aggregates. In this case particulate extracellular production by detrital bacteria did not contain significant amounts of $^3$H derived from glycine.

Initial work at Lakes Tahoe, Taupo and Rotongaio indicate that sugars are rapidly transformed to detrital materials which appear outside bacterial cell walls. These findings correspond to earlier electron microscopic examination of bacterial adhesion products. In these studies Floodgate (1972) concluded that acidic polysaccharides were major constituents of extracellular pads used to maintain bacterial adhesion to surfaces. Extensive chemical studies also point to polysaccharides as being important aggregating substances (Harris & Mitchell 1973). Sterile incubations failed to show either uptake or dispersal of labelled substances, thereby indicating that the eventual incorporation of soluble compounds into detrital aggregates is bacterially-mediated.

Bacterial extracellular material laid down in aggregates escapes detection by light microscopy. However, SEM observations reveal delicate matrices in which bacteria are often embedded. These were found in freshly sampled aggreg[illegible]es from a variety of lakes (Photos 3, 4 ) SEM examinations of particulate matter during favourable bacterial growth periods reveal copious bacterial secretions on surfaces. Often excretion products can be seen to exceed the intact bacterial content of particles, indicating that the turnover of bacterial populations and deposition of cellular debris is an important factor to consider in the overall process of particle production. During late summer months at Lake Tahoe the buildup of POC in a 450 meter water column is far in excess of the POC production calculated from primary productivity measurements (Paerl <u>et al</u> 1975).

The importance of bacterial sediment production in terms of food value for grazing organisms remains obscure. Much work has described the inability of Copepod and Cladoceran zooplankton to feed on particles less than 3µm in size. This has been summarized by Boyd (1976). Such findings would mean that free-living bacteria escape consumption by predominant herbivores. However, attached bacteria or bacterial particles ranging between 5 and 100 µm (Photo 5) may be ingested by grazers. Approximately 75% of the sediment particles discussed and shown in

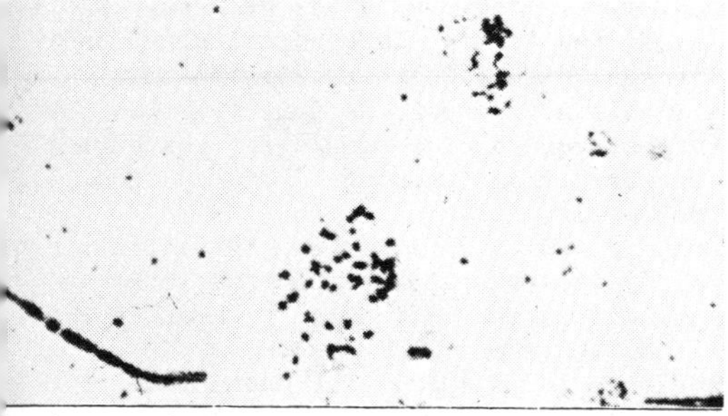

. Autoradiographs demonstrating labelling
f individual bacteria in aggregates at 2
ours of incubation with $^{3}$H glucose.

3. Bacterial $^{3}$H labelling at 12 hours showing
ispersal of radioactivity in an aggregate.

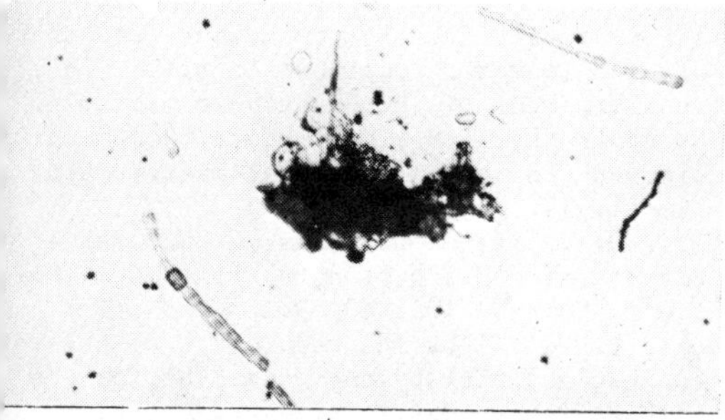

. Bacterial labelling at 24 hours showing
rther dispersal of radioactivity in an
gregate.

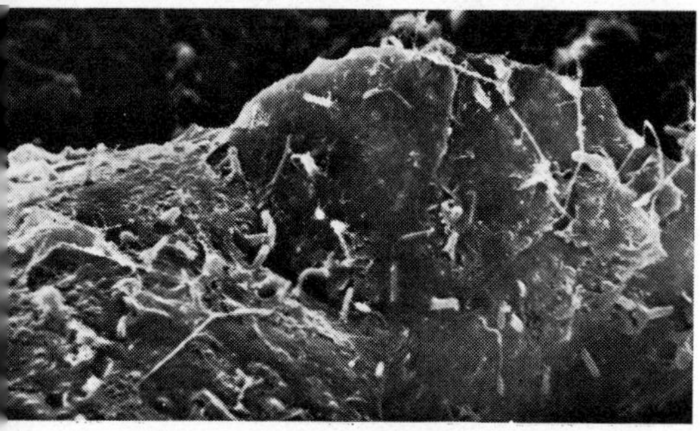

Initial stages of bacterial colonisation
a mineral sediment particle washed into
ke Tahoe from Ward Creek (a tributary).
proximate size of bacteria is 1µ.

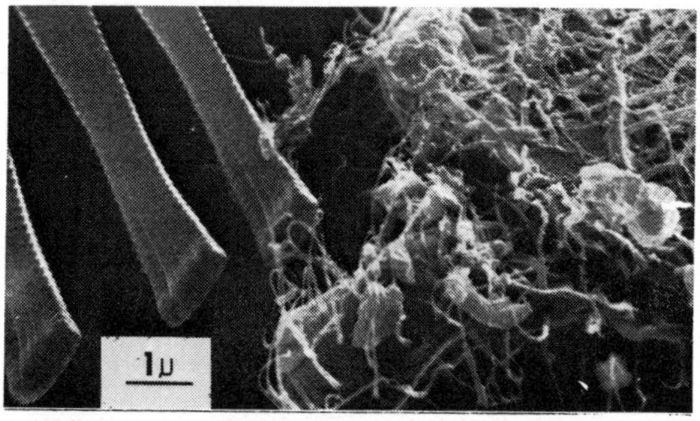


3. Filamentous bacteria forming aggregate partly composed of the diatom Fragilaria. Sample taken at Lake Tahoe, U.S.A.

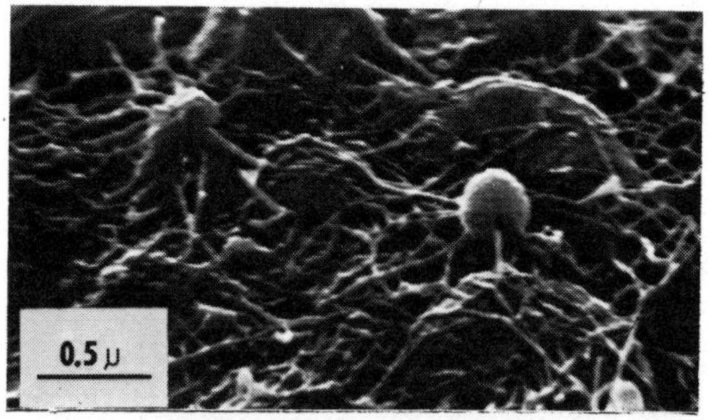


4. High magnification view of bacterial "webbing" laid down on a detrital particle. Sample taken at Heart Lake, Canada.

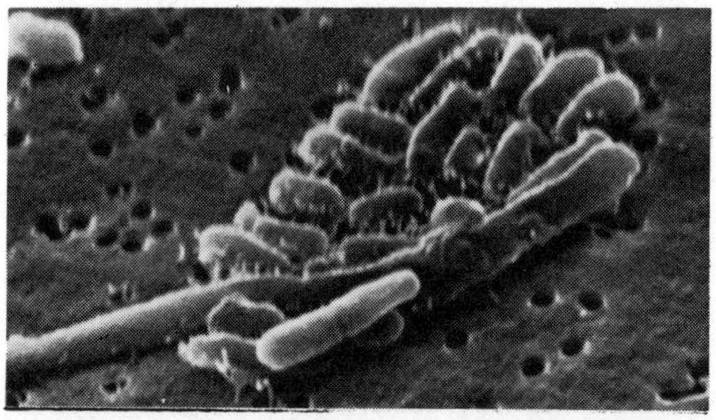

5. Small bacterial particle (approximately 10 µm length) commonly observed in Lake Tahoe, U.S.A.

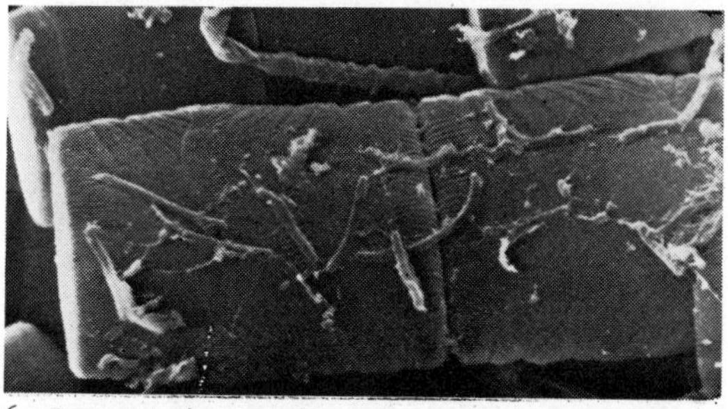

6. Decomposing periphyton cells colonised by bacteria. These cells were washed into Lake Tahoe during a period of heavy rainfall and stream bottom scouring.

this paper fall into this size range.

Bacterial aggregation is not limited to a narrow range of substrates. In Lake Tahoe bacterial growth on mineral particles as well as on decaying algal cells (Photo 6) and wood particles is commonly observed. In Lake Taupo bacterial attachment and particle building has been observed in eroded microscopic pumice particles as well as on diatoms, decaying bits of aquatic weeds and zooplankton fecal pellets.

In this presentation I have tried to present a case for closer scrutiny of heterotrophic bacterial POC production. Heterotrophic POC production occurs simultaneously with mineralisation of organic matter. Much bacterial POC production is associated with suspended sediments, whether they be organic or inorganic. Consideration of bacterial production and modification of particles is of two-fold importance. First, it may be helpful to relate bacterial excretion and deposition products to the chemical content of sediments in future analyses, as bacterial products can constitute an important fraction of the sediments. Secondly, sediments harboring bacteria are likely to be important food reservoirs in the transfer of particulate organic carbon in aquatic food chains.


## Acknowledgments

I would like to thank all the individuals who assisted in data collection and helpful discussions at the Division of Environmental Studies, University of California, Davis, the Lake Tahoe Research Group, and the Freshwater Section, Department of Scientific and Industrial Research, Taupo, New Zealand. I am also grateful to the D.S.I.R. for funding my travels and taking an active interest in both my North American and New Zealand research findings. Many thanks to Drs J.Gibb and E.White of D.S.I.R. Ecology Division for their encouragement.

This research was funded by the D.S.I.R., New Zealand and National Science Foundation - R.A.N.N. Grant GI-22, U.S.A.


## References


Alexander, M., 1961. Introduction to soil microbiology. Wiley.

Allen, H.L., 1973. Dissolved organic carbon: patterns of utilization and turnover in two small lakes. Int.Rev.ges.Hydrobiol.58: 617-24.

Baylor, E.R., & W.M. Sutcliffe, 1963. Dissolved organic matter in seawater as a source of particulate food. Limnol.Oceanogr. 8: 369-71.

Boyd, C.M., 1976. Selection of particle sizes by filter feeding Copepods: A plea for reason. Limnol.Oceanogr.21: 175-80.

Carlucci, A.F. & H.F. Bezdek, 1972. On the effectiveness of a bubble for scavenging bacteria from seawater. J.Geophys.Res. 77; 6608-10.

Ferrante J.G. & J.I. Parker, 1976. Transport of diatom frustules by Copepod fecal pellets to the sediments of Lake Michigan. Limnol.Oceanogr. In press.

Floodgate, G.D., 1972. The mechanisms of bacterial attachment to detritus in aquatic systems. Mem.Ist.Ital.Idrobiol. 29 (suppl. 309-23.

Goldman, C.R., R.C. Richards, H.W. Paerl, R.C. Wrigley, V.R. Overbeck and W.L. Quaide 1974. Limnological studies and remote sensing of the Upper Truckee sediment plume in Lake Tahoe, California-Nevada. Remote Sensing Environ. 3: 49-67.

Harris, R.H. & R. Mitchell, 1973. The role of polymers in microbial aggregation. Ann.Rev. Microbiol. 27: 27-50.

Hobbie, J.E. & C.C. Crawford, 1969. Respiration corrections for bacterial uptake of dissolved organic compounds in natural waters. Limnol.Oceanogr. 14, 528-32.

Jannasch, H, K. Eimjellen, C.O. Wirsen & A. Farnanfarmaian, 1971. Microbial degradation of organic matter in the deep sea. Science 171: 672-75.

Lush, D.L. & H.B.N. Hynes, 1973. The formation of particles in freshwater leachates of dead leaves. Limnol.Oceanogr.18: 968-77.

Monheimer, R.H., 1975. Planktonic microbial heterotrophy: Its significance to community biomass production. Verh.Internat.Verein. Limnol. 19: 2658-63.

Paerl, H.W., 1973a. Detritus in Lake Tahoe: Structural modification by attached microflora. Science 180: 496-98.

Paerl, H.W., 1973b. The regulation of heterotrophic activity by environmental factors at Lake Tahoe, California-Nevada. Ph.D. thesis, Univ. Calif. Davis. 139 p.

Paerl, H.W., 1974. Bacterial uptake of dissolved organic matter in relation to detrital aggregation in marine and freshwater systems. Limnol.Oceanogr.19: 966-72.

Paerl, H.W., 1975. Microbial attachment to particles in marine and freshwater ecosystems. Microb.Ecol.2: 73-83.

Paerl, H.W. & C.R. Goldman, 1972. Heterotrophic assays in the detection of water masses at Lake Tahoe, California. Limnol. Oceanogr. 17: 145-48.

Paerl, H.W. & S.L. Shimp, 1973. Preparation of filtered plankton and detritus for study with scanning electron microscopy. Limnol. Oceanogr.18: 802-5.

Paerl, H.W., R.D. Thomson & C.R. Goldman, 19
The ecological significance of detritus formation during a diatom bloom in Lake Tahoe, California-Nevada. Verh.Internat. Verein.Limnol. 19: 826-34.

Riley, G.A., 1963. Organic aggregates in seawater and dynamics of their formation and utilization. Limnol.Oceanogr. 8: 372-80.
Rodina, A.G., 1963. Microbiology of detritus of lakes. Limnol.Oceanogr. 8: 388-93.
Seki, M, 1972. The role of microorganisms in the marine food chain with reference to organic aggregates. Mem.Ist.Ital.Idrobiol. 29: (Suppl.) 245-59.
Stozky, G. 1967. Clay minerals and microbial ecology. Trans.N.Y.Acad.Sci. 30: 11-21.
Tenney, M.W. & W. Stumm, 1965. Chemical flocculation of microorganisms in biological waste treatment. Wat.Poll.Contr.Fed.J. 37: 1370-88.
Teot, A.S. & S.L. Daniels, 1969. Flocculation of negatively charged colloids by inorganic cations and anionic polyelectrolytes. Env.Sci.Tech. 3: 825-29.
Tilzer, M.M., 1972. Bacterial productivity of a high mountain lake. Verh.Internat. Verein.Limnol. 18: 188-96.

Key words: heterotrophic bacteria
dissolved organic carbon (DOC)
particulate organic carbon (POC)
dialysis
autoradiography
scanning electron microscopy
aggregation

# RATES OF SEDIMENTATION AND RESUSPENSION IN LAKE KINNERET

C. Serruya

Kinneret Limnological Laboratory, P.O.B. 345, Tiberias, Israel.

## Introduction

In principle, a lake of medium size is an ideal ecosystem where inputs and outputs of all kinds can be measured and the flow pattern of nutrients determined. This, however, assumes that all the parameters required for such an overall balance can be measured with a reasonable accuracy. Although much effort has been invested in the measurement of the "standing crop" type of parameters (chemical concentrations, biomass of algae....) precise techniques for estimating the "integrative" type of parameters are lacking in many fields and particularly in studies of the rate of sedimentation. The reliability of the measurement of this parameter by sediment traps has been questioned by Kleerekoper (1952) and Golterman (1973) on the grounds that the decrease of turbulence caused by the very presence of the trap may increase the rate of sedimentation. Conversely, Kirchner (1975) evaluated the influence of turbulence by comparing the catch per unit area for sediment traps of different collecting surface. This author found that the variances from different size traps were not significantly different.

In spite of the possible disadvantages of the technique, sediment traps were placed at eight stations in Lake Kinneret during the hydrological cycle 1972-1973.

The amounts of trypton collected in the traps showed that the actual rate of sedimentation in Lake Kinneret is much higher than that found by Stiller (1974) from her study of Cesium 137 distribution. We shall see in this paper that this is only an apparent contradiction due to an intensive process of resuspension of bottom material taking place mainly during the turnover period. It then appears that, in Lake Kinneret, the trap technique is not only able to give the correct rate of "net sedimentation" but also supplies valuable information concerning the quantitative importance of resuspension which affects many other limnological processes.

## General Background

Lake Kinneret is located at 209 m below MSL; it has a surface of 170 $km^2$ and its maximum and mean depths are respectively 42 m and 25 m. The lake drains a watershed of 2760 $km^2$. The considerable differences in altitude (+2800 m to -209 m) offer natural favourable conditions for erosion. The partial destruction of the vegetal cover of the slopes of Mount Hermon and the Golan Heights during the last millenium accentuated this tendency. However, according to Schumacher (1888), dense forests still existed, a century ago, on the Golan Heights; the systematic cutting of trees which took place at the beginning of this century, created wide barren areas an accelerated the erosion process.

The River Jordan is the main source o detrital material, the concentration of suspended solids in the river wate ranges from 10 to a few hundred mg $l^{-}$ It is maximum during the peaks of winter floods and minimum in winter between the floods (Avnimelech, persona communication).

The flow pattern of the Jordan water in the lake has been studied by S. Serruya (1974). In early winter, the water of the Jordan is cooler tha the lake water and sinks down immediately after its entrance into the lake. Its solid load is deposited in a NS tongue-shaped area which is clea ly seen in the maps of detrital elements of the lake bottom (Serruya, 1971).

In late winter and spring, the water of the Jordan is warmer than the lake it therefore floats on the lake surfa and is deviated to the western coast the lake with the general circulation of the top layers of the lake. In summer, during the stratification period which extends from June to December, the Jordan water flows at the upper limit of the thermocline.

The hydrological cycle 1972-1973 was unusually dry; the total yield of the Jordan River was $270.10^6$ $m^3$ in comparison with the long term average value of $550.10^6$ $m^3$. An interesting feature of this cycle is that, in contrast with other years, the only flood of the winter took place very late, in early March. It follows that the concentration of suspended solids* in the river water remained very low during the period September 1972 to February 1973 (Fig. 1).

## Methods

The sediment traps utilized in this study were built at the Kinneret Limnological Laboratory according to the type described by Brunskill (1969). Each trap is composed of four plastic funnels (surface = 314 $cm^2$) attached to vertical tubes (30 mm diameter 32 cm length). The traps were always installed and retrieved by a diver in order to maintain the apparatus vertical and bring it back without disturbing the sediments in the tube or on the funnel wall. The traps were placed at about 1 meter to 1.5 m above the lake bottom. In general, the traps were retrieved once a month, the material was dried, weighed and analysed.

The carbonates were determined by gasometric measurements of $CO_2$ in a Collins calcimeter and expressed as $CaCO_3$. The organic matter was determined by the loss-on-ignition technique at 500°C, and the alumino silicate fraction was estimated to be the difference between the total material and the cumulated amounts of $CaCO_3$ and organic matter. The organic carbon was determined by reduction of potassium dichromate and the results reported as the equivalent amount of carbon. Organic nitrogen was measured by the micro-Kjeldahl technique. For total phosphorus analysis, the sample was digested with a mixture of three acids ($H_2SO_4$, $HNO_3$ and $HClO_4$); the reagent used was ammonium molybdate reduced with ascorbic acid.

## Results

1. Spatial variations of the annual sedimentation rate (S.R.):

The annual amount of trypton collected varied from 640 to 2158 g $m^{-2}$ $y^{-1}$. (Fig. 2 and Table 1). The S.R. was found to be maximum at station G the northernmost station and minimum in the central and deep parts of the lake. We note the high values measured at station L, the southern most station of the lake.

2. Seasonal distribution of S.R. (Fig. 3):

In all the deep stations, there is a considerable difference between winter and summer S.R.. The winter-spring rates ranged from 8 to 12 g $m^{-2}$ $d^{-1}$ whereas the summerfall values seldom exceeded 2 g $m^{-2}$ $d^{-1}$. It follows that approximately 85% of the material sedimented during the whole hydrological cycles reaches the bottom from November to April. At station G, the S.R. is already high in November-December. At station B, it is maximum in January. Conversely at station C and M, the maximum occurs later in February-March. The symetrical shape resulting from this type of distribution is not observed at shallow stations such as station L (11 m depth) where the S.R. is nearly constant throughout the year.

3. Interannual variations:

In Fig. 4, we can compare the amount of trypton collected at station F during two consecutive years. In 1971-1972, an "average year" from meteorological point of view, the annual S.R. was 1950 g $m^{-2}$ $y^{-1}$; in 1972-1973 only 927 g $m^{-2}$ $y^{-1}$ were deposited at the same station. Although we have no data for the unusually rainy year 1968-1969, it is reasonable to think that it was higher than in 1971-1972. It follows that, if we consider the S.R. of 1972-1973 as a minimal value, the interannual variations at station F range from 900 to over 2000 g $m^{-2}$ $y^{-1}$.

4. Chemical composition:

The chemical composition of trypton is shown in Table 1. On a yearly basis, the trypton is composed of 48-60% $CaCO_3$, 8-14% organic matter and 30-43% alumino silicates. The organic carbon varies from 4.1 to 5.3% except at station A, organic N from 0.4 to 0.6% and total P from 0.13 to 0.34%.

Although the ratio of organic matter to organic C is variable (Alexander and Byers, 1932; Broadbent, 1953), a value of 2:1 is generally accepted for this ratio. In our determinations, we found values of 3:1. This high value of the organic matter: organic C ratio

* These data have been kindly supplied by Professor Avnimelech, Mekorot Water Co. Ltd.

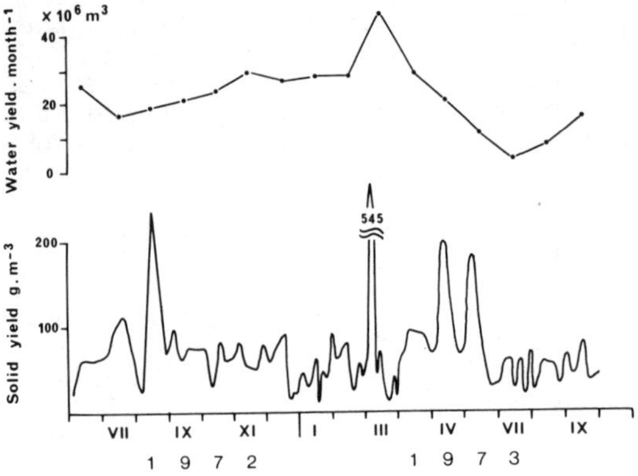


Fig. 1. Water yield and solid yield of River Jordan for hydrological cycle 1972-1973.

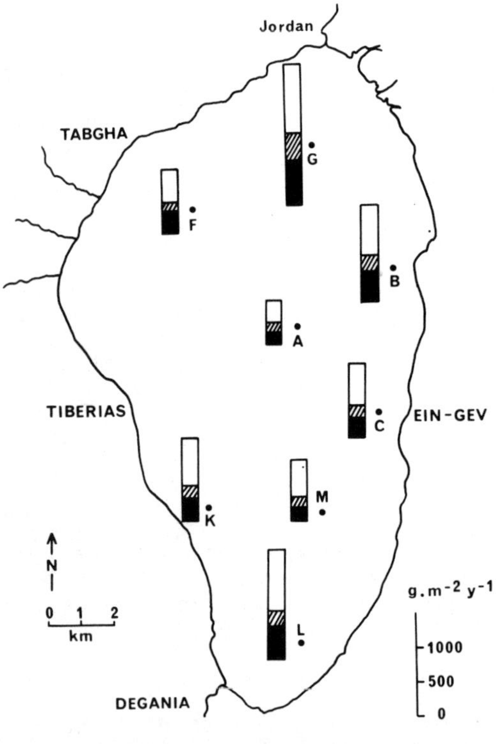


Fig. 2. Rates of sedimentation at various stations in Lake Kinneret for the period October 1972, September 1973. Results in g $m^{-2}$ $y^{-1}$. Black area = alumino silicate striped area = organic matter white area = calcium carbonate

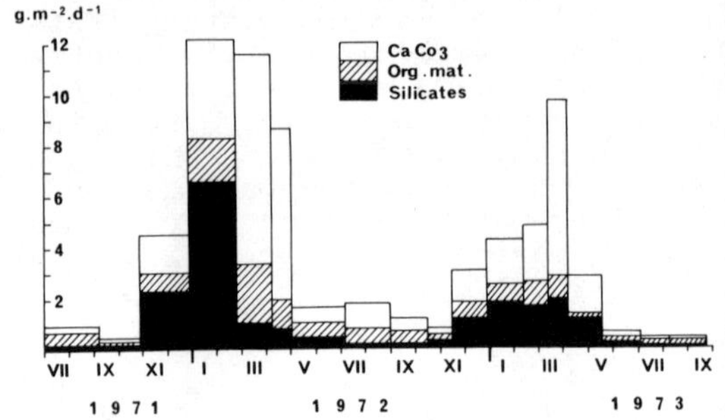


Fig. 4. Rates of sedimentation at station F during two consecutive hydrological cycles.

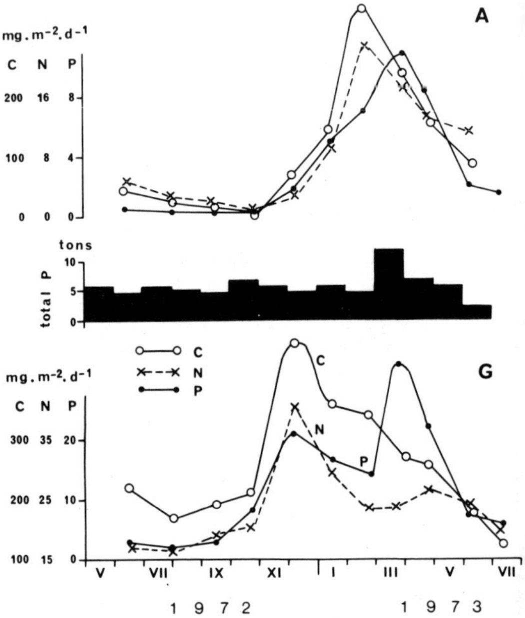


Fig. 5. Rates of sedimentation of organic carbon (C), organic nitrogen (N), total phosphorous (P) at station A (upper graph) and station G (lower graph). Monthly amounts of tot phosphorus carried into the lake by t River Jordan (center graph).

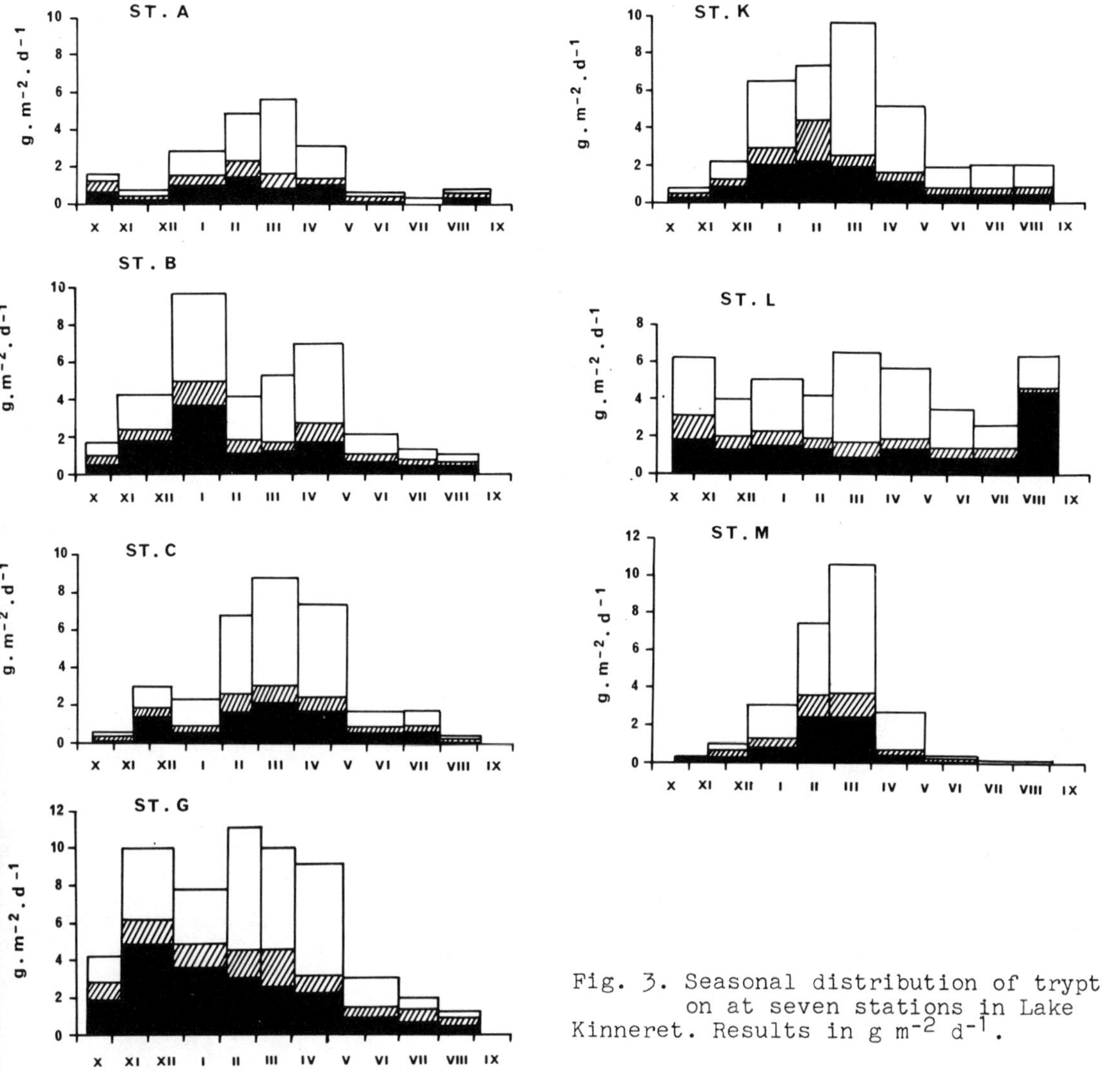


Fig. 3. Seasonal distribution of trypton at seven stations in Lake Kinneret. Results in g $m^{-2}$ $d^{-1}$.

Table 1. Sedimentation rates.
Period: X.1972-IX.1973. Results in g $m^{-2}y^{-1}$ percentages are indicated in brackets.

| S | D | Total | alumino silicates | Organic matter | $CaCO_3$ | C | N | P |
|---|---|---|---|---|---|---|---|---|
| G | 23 | 2204 | 954 (43) | 196 ( 9) | 1054 (48) | 97.8 (4.4) | 9.0 (0.41) | 5.0 (0.2[illegible] |
| F | 23 | 927 | 362 (39) | 98 (11) | 467 (50) | 49.0 (5.3) | 3.9 (0.42) | 1.9 (0.2[illegible] |
| B | 24 | 1467 | 588 (40) | 134 ( 9) | 745 (51) | 67.1 (4.6) | 6.0 (0.41) | 3.1 (0.2[illegible] |
| C | 29 | 1208 | 396 (33) | 100 ( 8) | 712 (59) | 49.7 (4.1) | 4.9 (0.40) | 2.1 (0.1[illegible] |
| A | 41 | 660 | 222 (34) | 94 (14) | 344 (52) | 47.0 (7.1) | 4.0 (0.60) | 1.5 (0.2[illegible] |
| K | 23 | 1294 | 423 (33) | 124 ( 9) | 747 (58) | 61.6 (4.8) | 5.5 (0.42) | 3.5 (0.2[illegible] |
| M | 31 | 916 | 276 (30) | 92 (10) | 548 (60) | 46.0 (5.0) | 3.9 (0.43) | 1.2 (0.1[illegible] |
| L | 11 | 1862 | 743 (40) | 184 (10) | 935 (50) | 92.0 (4.9) | 8.5 (0.45) | 6.4 (0.3[illegible] |
| Average: | | 1317 | 495 (37) | 128 (10) | 694 (53) | 63.8 (4.8) | 5.7 (0.43) | 3.09 (0.[illegible] |

S = Stations; D = Depth; in meters

might be due to a partial decomposition of the carbonates of the mud by heat or to a partial decomposition of hydrated alumino silicates (Broadbent, 1965). Edelstein (unpublished data) ran a series of measurements concerning the variation of weight with temperature of pure calcium carbonate. These experiments showed clearly that the decomposition of carbonates begins at 600°C. Therefore, since the loss-on-ignition measurements were carried out at 500°C it is improbable that carbonates were decomposed. In contrast, 65% of the detrital clay fraction is composed of montmorillomite (Banin et al, 1972). The differential-thermal-analysis curves obtained by these authors on the detrital fraction indicated two clear endothermic breaks at 150 and 220°C corresponding to the disappearance of water adsorbed on hydrated clays. This strongly suggests that the values concerning the organic matter of the trypton are overestimated and consequently the alumino silicates are underestimated. We therefore did not utilize the results obtained by the loss-on-ignition technique and calculated the amount of organic matter from the organic carbon values assuming a ratio of 2:1 between both parameters.

## Discussion

### 1. Sedimentation rates and resuspension:

The unusually dry winter 1972-1973 presented unique conditions to study the amount of trypton reaching the lake floor when the allochthonous contribution was minimum.

The total amount of suspended matter brought by the River Jordan from October 1972 to September 1973 was, acc ding to the data presented in Fig. 1, of an order of magnitude of 20.000 tons. The total amount of trypton co lected in the trap during this perio was approximately 200.000 tons. This considerable difference cannot be accounted for by any other additiona sources of detrital matter and it is suggested that the resuspension of in situ sediments was the main source o trypton during this year.

The study of the seasonal distributi of the trypton components strengthen this hypothesis. The increase of S.R. of alumino silicates at station G (Fig. 3), occurred in November-December at a period when there was no increase whatsoever of the Jordan soli yield. Since there are no authogenic clays in Lake Kinneret (Banin et al, 1972), the previously mentioned peak of alumino silicates can only originate from resuspension of in situ sediments.

On Fig. 5, we have represented the S.R. of the trypton carbon, nitrogen and phosphorus at stations A and G together with the monthly supply of phosphorus of the River Jordan.

At Station G, simultaneous peaks of C, N, P are observed in November-December; the peaks do not correspond t any increase of the total P supply o the river. In March, a second peak o trypton P occurs but no concomittant increase of trypton C and N is noted.

In order to clarify the origin of th

trypton of the first peak, we compared its N:P ratio with the N:P of the Jordan suspended material collected during this period and the N:P ratio of the in situ sediments. The results are shown in Table 2.

Table 2. N:P ratio of trypton material (1), Jordan suspended matter (2), and in situ sediments (3)

| | (1) | (2) | (3) |
|---|---|---|---|
| N:P atomic ratio | 4.2:1 | 8.8:1<br>10.0:1 | 3.8:1 |

These results indicate that the material of the first peak has the chemical characteristics of the in situ sediments and cannot be identified with the Jordan suspended matter. Conversely, the second peak is related to the March flood of the Jordan. It has already been shown (Serruya, 1974) that approximately 75% of the total phosphorus carried by the Jordan is composed of detrital particulate phosphorus associated with the fraction in suspension. During the flood of March 1973, this latter fraction rose to 90%. Moreover, nearly 80% of the total nitrogen brought by the river water during this period belonged to dissolved forms of nitrogen, mainly nitrate.

These special features of the chemistry of the Jordan nutrient load during the flood period explain that the second peak of trypton has a high concentration of phosphorus and a relatively low concentration of nitrogen. There are no available data concerning the organic carbon of the Jordan suspended matter during this period. It followed that the N:P ratio of the flood material of the Jordan was very low and ranged from 0.6:1 to 1.4:1 and compares with the value of 1.6:1 of the N:P ratio of the second peak of the trypton. It is then established that the November-December peak of trypton at Station G should be related to resuspension and the March peak to the river flood.

The timing of the first peak at station G corresponds to the turnover at this station. It seems that the vertical currents associated with the destratification process are able to mix thoroughly the in situ sediments. If this hypothesis is correct, the resuspension peak at station A should take place later since the thermocline reaches the lake floor about one month later at station G. The results shown on Fig. 5, indicate that the amount of trypton C, N, P started to increase in January and was maximum in February which corresponds to the period of full turnover at this station.

2. Allochtonous, contribution, autochtonous sedimentation and resuspension:

The relative composition of the suspended matter of the Jordan, of the trypton and of the in situ sediments is given in Table 3. The total contribution of the Jordan during the investigated period was about 20.000 tons and represented a S.R. of 118 g per square meter of lake per year. It follows that an approximate amount of alumino silicates of 14.000 tons (82 g $m^{-2}y^{-1}$) were brought from the watershed. The average values of the eight traps show that an average of 495 g $m^{-2}y^{-1}$ of alumino silicates were actually collected in traps. Since no clays are formed in the lake (Banin et al, 1972), the difference 495-82 = 413 g $m^{-2}y^{-1}$ should be related to resuspension.

Table 3. Relative composition of suspended matter of the River Jordan, of the trypton and of the in situ sediments. Results in %.

| | Jordan | Trypton | In situ Sediments |
|---|---|---|---|
| Alumino silicates | 70 | 37-35 | 46 |
| Organic matter | 5 | 10-12 | 7 |
| $CaCO_3$ | 25 | 53 | 47 |

The alumino silicates represent approximately 46% of the in situ sediments (Table 3). It follows that an amount of 897 g $m^{-2}$ is yearly resuspended. This allows us to give in Table 4 the order of magnitude of contribution of

Table 4. The different sources of the trypton.
Results mg $m^{-2}y^{-1}$.

| Total trypton | Jordan contribution | Authigenic sediment | Resuspension |
|---|---|---|---|
| 1317 | 118 | 302 | 897 |

the various sources of the trypton during the hydrological cycle 1972-1973.

It becomes clear that we should refer to two types of sedimentation rates; the "trap sedimentation rate $S.R_t$." corresponding to the total trypton collected in the traps and the "net sedimentation $S.R_n$." corresponding to the actual addition of matter to the lake bottom during the considered period.

3. Net sedimentation rate in Lake Kinneret:

During the hydrological cycle 1972-1973, $S.R_t$. was found to be equal to 1317 g $m^{-2}y^{-1}$ (Table 1) and $S.R_n$. is the difference between this later value and the rate of resuspension (897 g $m^{-2}y^{-1}$). It follows that $S.R_n$. was equal to 420 g $m^{-2}y^{-1}$ in 1972-1973. This is obviously a minimum value since in "normal" years, the Jordan brings 80 to 120.000 tons of suspended matter (Serruya, 1971; Inbar, 1976). These values correspond to a detrital sedimentation rate of 470 to 705 g per square meter of lake and per year.

The algal bloom in Lake Kinneret does not show large variations from year to year (Pollingher and Berman, 1975) and the precipitations of calcium carbonate and organic matter associated to the bloom can be taken as constant.

It follows that, in first approximation, the value of 302 g $m^{-2}y^{-1}$ (Table 4) can be taken as a constant authogenic S.R. which is very little influenced by meteorological fluctuations. The $S.R_n$. is then equal to the authogenic sedimentation plus the detrital sedimentation: 302 + 470 = 772 g $m^{-2}y^{-1}$ for a detrital annual input of 80.000 tons and 302 + 705 = 1007 g $m^{-2}y^{-1}$ for a detrital input of 120.000 tons per year.

These values are in good agreement with Stiller's determination of the sedimentation rates. Stiller (1974) considers that, in general, the sedimentation rate at station F is near to the lake average value.

The values of the S.R. obtained by Stiller at station F from her study of Cesium distribution in the upper layers of cores taken at this station, varied from 1060 ± 170 to 1250 ± 260 g $m^{-2}y^{-1}$ (M. Stiller, personal communication). We can then conclude that the present rate of "net sedimentation" in Lake Kinneret is of an order of magnitude of 1000 g $m^{-2}$ $y^{-1}$.

4. Authogenic precipitation of $CaCO_3$:

If, during the period investigated 897 g $m^{-2}$ $y^{-1}$ were resuspended, an amount of 420 g $m^{-2}$ $y^{-1}$ of $CaCO_3$ collected in the traps came from in situ sediments. Since the average S.R. of $CaCO_3$ was 694 g $m^{-2}y^{-1}$ (Table 1) and the Jordan contribution was equivalent to 29 g $m^{-2}y^{-1}$, the rate of authogenic $CaCO_3$ sedimentation was of an order of magnitude of 245 g $m^{-2}$ $y^{-1}$.

During the bloom period, corresponding to the main period of authogenic precipitation of $CaCO_3$, a decrease of 9 g $m^{-3}$ is observed in the trophogenic zone of the lake which is then 15 m deep. This corresponds to the formation of 337 g $m^{-2}$ $y^{-1}$ of $CaCO_3$. Considering that part of the $CaCO_3$ precipitated in the upper layers is dissolved in the lower ones where an increase of calcium is then observed, we can conclude that the value calculated from the trap studies represents a correct order of magnitude of the authogenic precipitation of $CaCO_3$ or 42.000 tons for the whole lake.

Assuming that the average interannual Jordan supply of suspended matter is of an order of magnitude of 80.000 tons per year, an amount of 20.000 tons of allochtonous $CaCO_3$ is introduced yearly into the lake (the Jordan material contains 25% of $CaCO_3$, Table 3). It follows that the authogenic precipitation represents 68% of the total annual sedimentation of calcium carbonate.

5. The importance of resuspension:

The results obtained from the traps represent the first direct evidence of the occurrence of resuspension of upper sediments during the turnover period and of its quantitative importance in Lake Kinneret. Taking into account a water content of 300% in the surface sediment and a density of 1.2 of the wet weight material, a yearly resuspension of 897 g $m^{-2}$ is equivalent to the resuspension of the upper 2.9 millimeters, as a lake average.

The winter resuspension of sediments explains various observations which were not, until now, satisfactorily understood. In an ecosystem where the detrital material reaches the bottom only in winter and where the bulk of

$CaCO_3$ precipitation occurs in April-May, distinct varvae should have been observed. Such varvae were observed in lakes of similar depths: Vivier and Serruya (1966) and Brunskill (1969) were able to count the sedimentation cycles in Lake Nantua, France (40 m maximum depth) and in Green Lake, U.S.A. (maximum depth 52 m). We note that Lake Nantua is protected from N and S winds by high cliffs and in winter the total wind run is four times lower than in Lake Kinneret. Green Lake sediments remain undisturbed because of the presence of a permanent monimolimnion. In Lake Kinneret, where the input of mechanical energy is considerable (S. Serruya, 1975), varvae were never observed in the thousands of sediment samples taken all over the lake.

The nutrient balances (Serruya, 1975) indicated that the lake sediments behave as a phosphorus trap except in winter when they supply phosphorus to the lake water. A turbulent mixing of the sediments is the only mechanism which can explain the increase of the total P storage of the lake water in December-January.

The fluctuations of the nitrification potential (NP) in sediments and lake water in winter (Cavari, 1977) can only be accounted for by turbulent mixing. The NP of the sediments which remains high during the whole stratification period drops suddenly at the turnover period whilst the NP of the lake water increases. It is clear that the large inocolum of nitrifying bacteria of the sediments is redistributed in the water allowing the rapid nitrification of ammonia.

Many benthic lake species have been found in the plankton during the turnover period, particularly the Copepoda *Eucyclops serrulatus* (M. Gophen, personal communication), and various species of benthic diatoms (U. Pollingher, personal communication).

In her description of the seasonal variations of organic $^{13}C$ content, in plankton, Jordan material and trypton, Stiller (1977) noted that the isotope composition of the trypton collected at station G "during October-November and February.... are somewhat confusing since they are more depleted in $^{13}C$ than Jordan material and plankton by about 2‰". In fact the $^{13}C$ value of this winter trypton is close to -30‰ which is the value found by Stiller for the $^{13}C$ content of the *in situ* sediments (M. Stiller, personal communication). It is doubtless that this "turnover value" of the $^{13}C$ content of the trypton is due to the fact that at this period, the trypton is composed mainly of resuspended material.

The fluctuations of the lake level largely affects the rate of resuspension. Ben Harim (1976), in his study of the inorganic suspended matter in Lake Kinneret found unusually high values during the winter 1974 which he relates to the very low levels of this period.

The existence of an active process of resuspension in Lake Kinneret has considerable bearing on management problems. In particular, it is clear that, even if the external nutrient load is reduced to a minimum, the upper layers of the sediments can supply enough nutrients to support algal blooms in this lake for several years, thus decreasing the short term rentability of watershed management. However, the nutrients and especially the phosphorus stored in and released from the sediments at the turnover period originate also in the watershed. It is then obvious that the long term management program of the lake will have to deal with both watershed and lake sources of nutrients.

## Acknowledgements

I wish to thank M. Hatab, skipper of the RV/Hermona and Y. Edelstein, diver for their skillful handling of the traps. I am grateful to M. Edelstein for the chemical analysis and useful discussions. I am indebted to Dr. M. Stiller and Professor Y. Avnimelech for making available unpublished data and for fruitful discussions and criticisms. I am grateful to Miss R. Hershcovitch for her careful typing.

## References

Alexander, L.T. & H.G. Byers, 1932. A critical laboratory review of methods of determining organic matter and carbonates in soil. U.S. Dept. Agr. Tech. Bull. 317.

Banin, A., A. Singer & M. Gal, 1972. Amounts, composition and physicochemical properties of clays in Lake Kinneret sediments. Tech. Rep. Fac. Agr. Hebrew Univ. Jerusalem. 87 p. (in Hebrew).

Ben Harim, I., 1976. Study of suspended matter in the National Water System. Mekorot Internal Report. (in

Hebrew).
Broadbent, F.E., 1953. The soil organic fraction. Advan. Agrom. 5: 153-183.
Broadbent, F.E., 1965. Organic matter. In: Methods of soil analysis, part 2. C. A. Black (ed.).
Brunskill, G.F. & S.D. Ludlam, 1969. Fayetteville Green Lake, New York. I. Physical and chemical limnology. Limnol. Oceanogr. XIV(6): 817-829.
Cavari, B.Z., 1977: Nitrification potential and factors governing the rate of nitrification in Lake Kinneret. Oikos (in press).
Golterman, H.L., 1973. Vertical movement of phosphate in freshwater, 509-538. In: E.J. Griffith, et al (eds.), Environmental phosphorus handbook. Whiley.
Inbar, M., 1976. Movements of suspended matter in the northern Jordan River. Haifa University. Internal Report No. 3.
Kirchner, W.B., 1975. An evaluation of sediment trap methodology. Limnol. Oceanogr. (20)4: 657-660.
Kleerekoper, H., 1952. A new apparatus for the study of sedimentation in lakes. Can. J. Zool. 30: 185-190.
Pollingher, U. & T. Berman, 1975. Temporal and spatial patterns of Dinoflagellate blooms in Lake Kinneret (1969-1974). Verh. Intern. Verein. Limnol. 19: 1370-1382.
Schumacher, G.,1888. The Jaulan. London.
Serruya, C.,1971. Lake Kinneret: The nutrient chemistry of the sediments. Limnol. Oceanogr. 16(3): 510-521.
Serruya, C., 1974. Nitrogen and phosphorus balances in Lake Kinneret. Verh. Internat. Ver. Limnol. 19: 1357-1369.
Serruya, C., M. Edelstein, U. Pollingher & S. Serruya, 1974. Lake Kinneret sediments: Nutrients composition of the pore water and mud water exchanges. Limnol. Oceanogr. 19(3): 489-508.
Serruya, S., 1974. The mixing pattern of the Jordan River in Lake Kinneret. Limnol. Oceanogr. 19: 175-181.
Serruya, S., 1975. Wind, water temperature and motions in Lake Kinneret: general pattern. Verh. Internat. Verein. Limnol. 19: 73-87.
Stiller, M., 1974. Rates of transport and sedimentation in Lake Kinneret. Ph.D. thesis. The Weizman Inst. Science. Tel Aviv. 240 p. (in Hebrew)
Stiller, M., 1977. Origin of sedimentation components in Lake Kinneret, traced by their isotopic composition. Int. Symp. Interaction between sediments and freshwater. Amsterdam. (This Volume, page 57).
Vivier, P. & S. Serruya, 1966. La pollution organique dans le lac de Nantua. L'Eau. 53(4): 167-171.



Key words: Sedimentation rate
Resuspension
Sediment traps
Lake Kinneret
Internal nutrient load

Abstract

Eight stations were equipped with sediment traps during the exceptional-ly dry hyrological cycle 1972-1973. Although the amount of suspended mate-rial brought by the rivers was minimur a steep increase of the sedimentation rate was observed at the turnover period and during the mixed period. Then, the chemical compositions and $^{13}C$ content of the trap material compared well with the *in situ* sediment and was different from the suspended matter of the rivers. Since the alumi silicate fraction cannot be of authigenic origin, it was possible to estimate the rate of "net sedimentation" and the rate of "resuspension". This calculation was done for the dry cycl 1972-1973 and for average years when the contribution of the rivers in sus-pended material is a significant frac-tion of total sedimentation. It appea that the "net sedimentation" rate ranges from 700 to 1000 g.m$^{-2}$.y$^{-1}$ whi is in excellent agreement with Stille results obtained from the study of di tribution of Cesium. The rate of resu pension is of the same order of magni-tude. The importance of resuspension on the supply of nutrients to algal blooms is discussed.

# ORIGIN OF SEDIMENTATION COMPONENTS IN LAKE KINNERET TRACED BY THEIR ISOTOPIC COMPOSITION

M. Stiller

Isotope Department, Weizmann Institute of Science, Rehovot, Israel

## Summary

The stable isotopic composition of allochthonous and autochthonous $CaCO_3$ and organic matter was defined in Lake Kinneret in an attempt to use this information for the estimation of the relative contributions of these components to recently deposited sediments.

Due to wide seasonal variations in the isotopic composition of plankton and in the expected isotopic composition of authigenic carbonate it became evident that in Lake Kinneret the relative contribution of these components cannot be accurately estimated from the isotopic composition found in the sediments. The isotopic balances were performed on a short-term, monthly or bimonthly basis, by relying on the isotopic composition of material collected by sediment traps.

A reasonable estimate for the percent of carbonate of autochthonous origin, which settles annually to Lake Kinneret sediments, was obtained from $^{18}O$ balances in carbonates. Based on $^{13}C$ data, the contribution of authigenic organic carbon to organic matter deposition in Lake Kinneret, can be estimated only for the period March-September.

## Introduction

The presence of $CaCO_3$ and organic matter of authigenic origin in the sediments of a lake are a valuable indicator of its trophic state.

Gorham & Sanger (1972) used the ratios of chlorophyll derivatives to carotenoids, in sediments, to reveal changes in the balance between autochthonous and allochthonous organic contributions. Nissenbaum et al. (1972) measured the abundance of lower molecular weight hydrocarbons ($C_{15}$ and $C_{17}$), originating from photosynthetic bacterial and algal populations and of longer chain hydrocarbons, which are indicative of contributions from terrigenous plant material.

Serruya (1973a) attempted an evaluation of the contribution of $CaCO_3$ originating from the watershed to the trypton collected in Lake Kinneret, by making a carbonate balance based on several assumptions, one of them being that the watershed is the sole source of silicates.

In this investigation it was attempted to quantify the amounts of carbonate and of organic matter which are formed in Lake Kinneret and of those which are brought into the lake from its watershed by defining the stable isotopic compositions of the above mentioned sedimentation components and their relative contribution to the isotopic composition of material collected in sediment traps.

## Methods

Samples of suspended matter from the Jordan River were collected at Arik Bridge, which is situated about 1 km north of the Jordan inflow into the lake (Fig. 1). Samples of plankton, from Lake Kinneret surface waters,

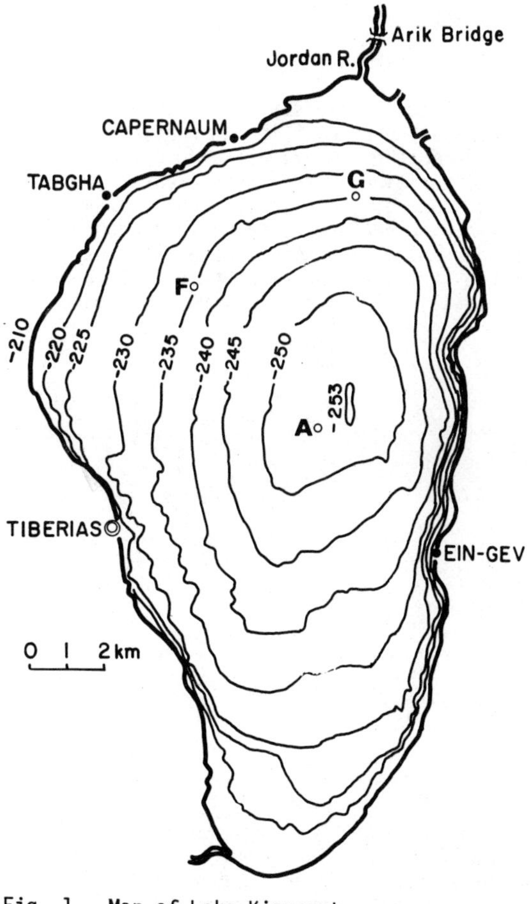


Fig. 1. Map of Lake Kinneret

and samples of trypton, collected by sediment traps which were submerged at 1 m above the sediment water interface, at stations F and G (Fig. 1), were obtained from the Kinneret Limnological Laboratory. Samples of lake water were collected for $^{18}O$ and $^{13}C$ analysis. Total dissolved carbonate ($\Sigma C$) was fixed on board as $BaCO_3$. $CO_2$ was extracted from carbonates by phosphoric acid (specific gravity 1.82) and from organic carbon by ignition at 900°C under a stream of oxygen. Previous to the organic carbon analysis, suspended matter and trypton samples were treated with dilute HCl, to remove carbonate, washed and dried. The isotopic composition of $CO_2$ was measured on a M86 Atlas Mass Spectrometer. The isotopic abundances for $^{18}O$ in carbonate samples and $^{13}C$ in all samples are given in $\delta$ units which are per mill deviations relative to the PDB Chicago standard. In water samples, $^{18}O$ abundances are given in per mill deviations relative to the SMOW standard. The analytical reproducibility was ±0.15% for carbonate samples and for most organic carbon samples. For some of the plankton samples the error is higher.

## Carbonates

### Authigenic carbonate

The isotopic composition of carbonate, precipitating at equilibrium conditions depends on the temperature and on the isotopic composition of the environment in which it was formed. In a lake, then, it may be possible to differentiate between autochthonous and allochthonous carbonate due to the fact that the allochthonous carbonate is usually formed in an environment quite different from that of the lake into which it is transported and would be characterized by a different set of isotopic values.

The isotopic composition of allochthonous carbonate was measured in samples of suspended matter collected from the Jordan, while that of authigenic $CaCO_3$, which cannot be collected separately, was predicted by calculations.

The oxygen isotopic composition of authigenic carbonate was evaluated according to the temperature scale given by Craig (1965):

$$t = 16.9 - 4.2(\delta^{13}_{PDB} - \delta^{18}_{SMOW}) + 0.13(\delta^{13}_{PDB} - \delta^{18}_{SMOW})^2$$

where t is temperature of the lake water and $\delta^{13}_{PDB}$ and $\delta^{18}_{SMOW}$ (in °/₀₀ of the denoted scales) refer to the $^{18}O$ contents of the carbonate and water, respectively, by substituting the temperature measured in Lake Kinneret surface waters (Serruya, 1973) and their $^{18}O$ content (Fig. 2).

The expected isotopic composition of oxygen in autochthonous carbonate displays a seasonal variation of about 3°/₀₀ (Fig. 2) which results mainly from temperature fluctuations rather than from variations of the $^{18}O$ content of lake water. It follows that, in the sediments, carbonate of authigenic origin will have an isotopic composition which is a weighted average of these

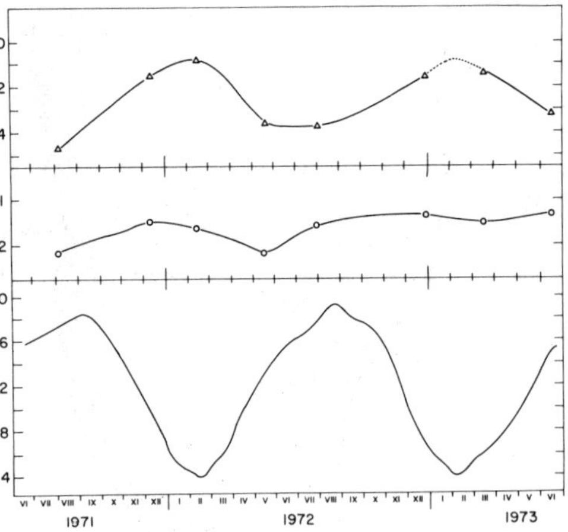


Fig. 2. Measured seasonal variations of temperature and $^{18}O$ content o——o in Lake Kinneret surface waters and the estimated $^{18}O$ content of authigenic carbonate Δ——Δ.

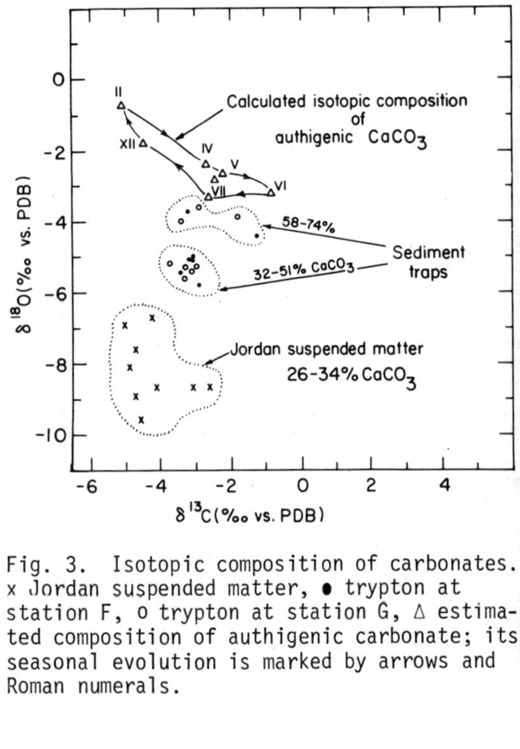


Fig. 3. Isotopic composition of carbonates. x Jordan suspended matter, • trypton at station F, o trypton at station G, Δ estimated composition of authigenic carbonate; its seasonal evolution is marked by arrows and Roman numerals.

seasonal variations. Thus, it becomes obvious that the isotopic composition of oxygen in the sediments cannot be used for accurately deriving the relative contribution of authigenic carbonate. Such estimates should be done on trypton samples collected monthly, thus allowing for the predicted seasonal variations in the isotopic composition of oxygen in authigenic carbonate.

For the estimates of the isotopic composition of carbon in authigenic carbonate, the measurements (Stiller & Magaritz, 1974; Stiller, 1974) of the $^{13}C$ content in the dissolved species of Lake Kinneret surface waters were used. Assuming that the dissolved carbon species are at isotopic equilibrium, the temperature dependent fractionations for carbon isotopic exchange equilibria (Mook, 1968) were applied. However, the carbon-13 compositions of authigenic carbonate which were thus calculated (and are presented in Fig. 3) are quite imprecise estimates. The daily variations observed in the $^{13}C$ content of surface waters (Fig. 4) may induce an error of about 1°/₀₀ to any calculated carbon-13 composition of autochthonous carbonate. Therefore, carbon-13 balances in trypton were not attempted. It can be said only qualitatively that in spring, when the highest deposition rates of autochthonous carbonate occur, its $^{13}C$ composition is expected to be enriched by about 2°/₀₀ as compared to carbonate precipitating in winter.

## Allochthonous carbonate and trypton carbonate

The carbon and oxygen isotopic composition of carbonates of allochthonous origin, i.e. in suspended matter collected from the Jordan River, are summarized in Table 1. The samples JSM 2 and JSM 3 were obtained from fortnightly and weekly samplings, respectively. The isotopic compositions of carbonates in suspended matter which settles to Lake Kinneret sediments (trypton) and which contains a mixture of carbonates of allochthonous and autochthonous origin are also listed in Table 1.

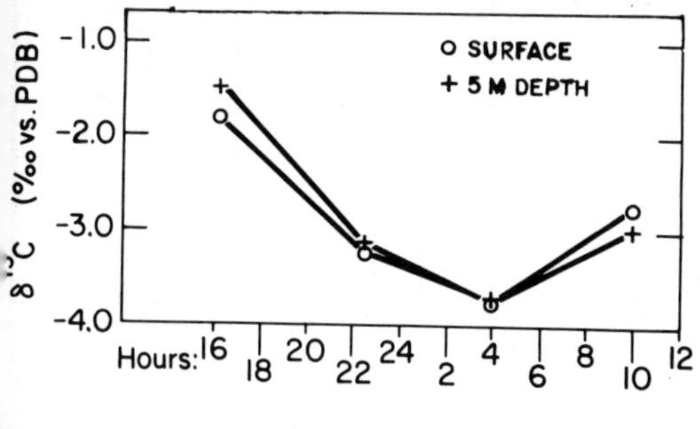


Fig. 4. Daily variation in $^{13}C$ content of dissolved carbonate (ΣC) in Lake Kinneret. Sampling: May 8, 1975. o surface, + 5 m depth.

Table 1. Isotopic composition of carbonates in Jordan suspended matter and in Lake Kinneret trypton.

| Sample | Date of sampling | Isotopic composition of carbonate $\partial^{13}C$(°/₀₀ PDB) | $\partial^{18}O$(°/₀₀ PDB) | $CaCO_3$ % |
|---|---|---|---|---|
| Jordan suspended matter | | | | |
| JSM 1 | Dec.7, 1971 | -4.7 | -7.6 | 27.7 |
| JSM 2 | Feb.10-Apr.13, 1972 | -5.0 | -6.9 | 33.9 |
| JSM 7 | May 8, 1972 | -2.6 | -8.7 | |
| JSM 3 | May 8-June 22, 1972 | -3.1 | -8.7 | 26.8 |
| JSM 8 | June 22, 1972 | -4.1 | -8.7 | |
| JSM 4 | July 22, 1972 | -4.7 | -8.9 | 26.2 |
| JSM 5 | Aug. 23, 1972 | -4.9 | -8.1 | |
| JSM 6 | Sept. 25, 1972 | -4.6 | -9.6 | |
| JSM 9 | Oct. 6, 1972 | -4.2 | -6.7 | |
| average | | -4.2 ± 0.8 | -8.2 ± 1.0 | |
| L. Kinneret trypton, station F, 1 m above sediment | | | | |
| F23 I | Jul.19-Sept.2, 1971 | -3.1 | -5.1 | 50.7 |
| F23 II | Sept.2-Oct.25, 1971 | -3.4 | -5.4 | 32.6 |
| F23 III | Oct.25-Dec.24, 1971 | -3.1 | -5.0 | 34.6 |
| F23 IV | Dec.24,1971-Feb.20,1972 | -3.1 | -5.1 | 32.4 |
| F23 V | Feb.20-Apr.3, 1972 | -3.2 | -3.7 | 71.3 |
| F23 VI | Apr.3-Apr.27, 1972 | -1.2 | -4.4 | 78.3 |
| F23 VII | May 15-Jul.3, 1972 | -2.9 | -5.8 | 34.0 |
| weighted average | | -2.7 | -4.4 | 48.9 |
| L. Kinneret trypton, station G, 1 m above sediment | | | | |
| G1 | May 15-Jul.6, 1972 | -3.1 | -5.4 | 35.3 |
| G2 | Jul.6-Aug.23, 1972 | -3.0 | -5.3 | 43.3 |
| G3 | Aug.23-Oct.9, 1972 | -3.3 | -5.3 | 37.9 |
| G4 | Oct.9-Nov.11, 1972 | -3.7 | -5.2 | 31.7 |
| G5 | Nov.11-Dec.21, 1972 | -3.3 | -5.6 | 37.0 |
| | Dec.21,1972-Feb.7,1973 | sample missing | | 38.0 |
| G6 | Feb.7-Mar.3, 1973 | -3.4 | -4.0 | 58.0 |
| G7 | Mar.3-Apr.2, 1973 | -2.9 | -3.6 | 61.2 |
| G8 | Apr.2-May 13, 1973 | -1.8 | -3.9 | 73.7 |
| weighted average | | -2.8 | -4.4 | 49.5 |

For most samples, the $CaCO_3$ content was measured at the Kinneret Limnological Laboratory (Serruya, 1973a; 1973b).

The isotopic data from Table 1 are presented in a $^{13}C$-$^{18}O$ plane in Fig. 3 and the seasonal variations in $^{18}O$ are plotted in Fig. 5.

As shown in Fig. 3, the isotopic composition of allochthonous carbonate is different from that which was collected in Lake Kinneret. The isotopic composition of trypton is more enriched both in $^{13}C$ and in $^{18}O$ than that of Jordan suspended matter. The trypton samples showing the heaviest isotopic composition are those which were collected during February-April (Fig. 5), and which also have the highest $CaCO_3$ content. This fact suggests that in Lake Kinneret authigenic $CaCO_3$, which precipitates at the peak of the Peridinium bloom in February-April, has an isotopic composition which is enriched in both $^{13}C$ and $^{18}O$ compared to that of allochthonous carbonate introduced by the Jordan River and this is in reasonable agreement with the isotopic composition of authigenic carbonate predicted by calculations. It is interesting to note that during this period the measured $^{18}O$ values in trypton seem to reflect the trend (Fig. 5) of the expected $^{18}O$ composition of autochthonous carbonate: most enriched $^{18}O$ carbonate should form in the coldest water, in February,

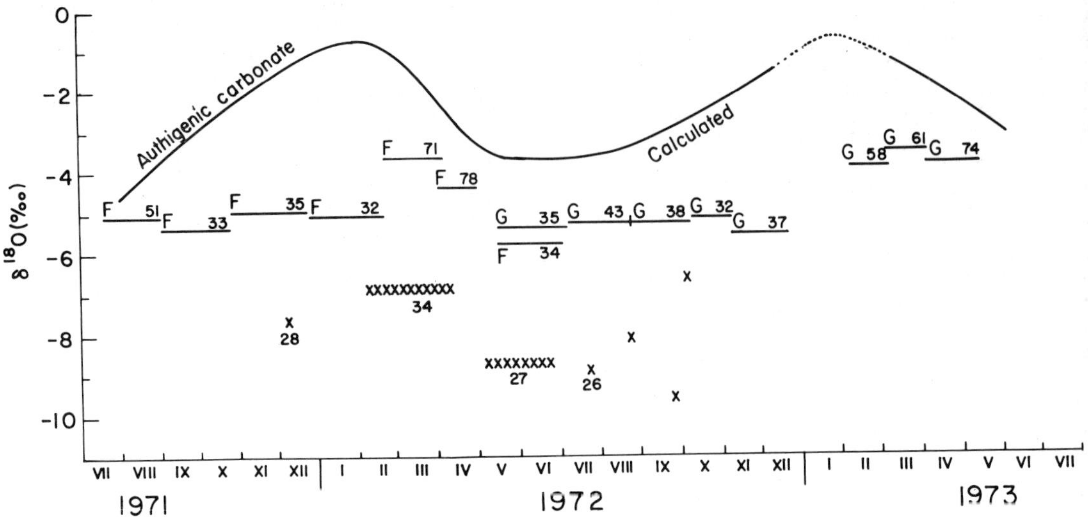


Fig. 5. Seasonal variation of the isotopic composition of oxygen in carbonates. x Jordan suspended matter, collecting site and % $CaCO_3$ are given for each trypton —— sample. The expected variation in $^{18}O$ content of authigenic carbonate is also indicated.

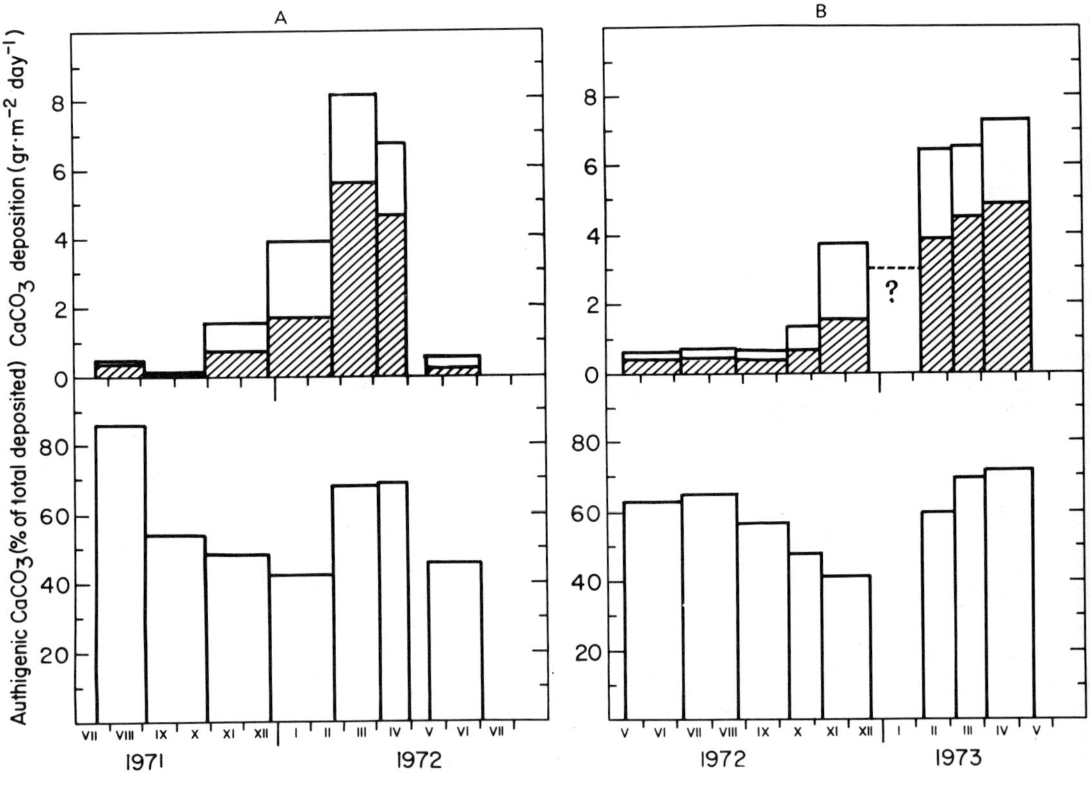


Fig. 6. Percent of authigenic carbonate in trypton $CaCO_3$ (bottom) and rates of $CaCO_3$ deposition (top). A. at station F, 71/72; B. at station G, 72/73. ▨ authigenic $CaCO_3$, ▭ allochthonous $CaCO_3$.

and trypton deposited in February-March has the heaviest $^{18}O$ composition

Tentative carbonate balances

An attempt to estimate the relative contribution of autochthonous $CaCO_3$ to the total $CaCO_3$ content in each trypton sample, by relying on $^{18}O$ balances is now described. For this purpose, a constant value of $\delta^{18}O = -8.2°/_{oo}$ (see averages in Table 1) was assumed for $CaCO_3$ of allochthonous origin. The curve representing the expected seasonal variations in the isotopic composition of oxygen in authigenic carbonate (Fig. 5) was used for selecting $\delta^{18}O$ values, which correspond to (about the middle of) the collecting period, of each trypton sample.

The estimates of the relative contributions of authigenic carbonate to total carbonate deposition are given in the lower part of Fig. 6. In all trypton samples more than 40% of the carbonate is of authigenic origin and in February-April it is about 70%. Further, the rate of deposition of authigenic carbonate was evaluated by multiplying these relative contributions by the daily rate of sedimentation of $CaCO_3$, as measured by Serruya (1973a; 1973b). These evaluations are shown in the upper part of Fig. 6. It is obvious that the highest daily rates of deposition of authigenic carbonate accompany the Peridinium bloom and occur in March-April 72 and February-April 73. By summing up the deposition rates on a yearly basis, it follows that 59% and 62% of the $CaCO_3$ collected by sediment traps at stations F and G, respectively, is of authigenic origin. (For station G the estimate may be somewhat incorrect, because one sample is missing.) At station F, an autochthonous carbonate contribution of 59% is in reasonable agreement with 69% as evaluated by Serruya (1973a).

The total annual deposition rate measured by collecting trypton at station F was 1823 gr sediment/$cm^2$ with a content of 48.9% $CaCO_3$ (Serruya, 1973a) and at station G 2180 gr/$m^2$ with 49.5% $CaCO_3$ (Serruya, 1973b). This means that 528 grs of $CaCO_3/m^2$ of authigenic origin were deposited at station F during 1971/1972 and 665 grs of $CaCO_3/m^2$ at station G during 1972/73.

Organic carbon

The isotopic composition of authigenic organic carbon in lakes - the phytoplankton - depends mainly on the temperature and on the isotopic composition of dissolved carbonate. It was found that the fractionation between cellular carbon and bicarbonate also depends on pH, concentration of carbonic species and the growth rate of organisms (Degens et al., 1968). Therefore, the physical, chemical and biological seasonal changes in a lake are reflected in the isotopic composition of the plankton.

Particulate organic carbon introduced by rivers into lakes represents the plants which prevail in the watershed. Depending on their metabolic pathway of carbon fixation, the plants divide into two main groups, each of them having a characteristic carbon-13 content of the organic carbon.

The $^{13}C$ content of organic carbon of allochthonous origin was measured in Jordan suspended matter collected close to the river's inflow into the lake; it is quite uniform: $\delta^{13}C = -27.4 \pm 0.9°/_{oo}$ (Table 2 and Fig. 7). Similar values, in the range $-25°/_{oo}$ to $-28°/_{oo}$, with an average of about $-27°/_{oo}$, were obtained in sediments of rivers on the Eastern coast of the United States (Hunt, 1967) and in tributaries of the Mississippi (Sackett and Thompson, 1963).

Table 2. Isotope composition of organic carbon in Jordan suspended matter and Lake Kinneret trypton.

| Sample | Organic carbon C % | Organic carbon $\partial^{13}C$ °/oo PDB | Autochthonous organic C estimated % |
|---|---|---|---|
| Jordan suspended matter* | | | |
| JSM 1 | 3.1 | -26.6 | |
| JSM 2 | | -27.8 | |
| JSM 3 | 2.7 | -27.6 | |
| JSM 4 | | -28.8 | |
| JSM 5 | | -26.5 | |
| JSM 9 | 2.2 | -26.9 | |
| average | | -27.4±0.9 | |
| L. Kinneret trypton, station G, at 1 m above sediment* | | | |
| G 1 | 8.5 | -19.7 | 87 |
| G 2 | 6.9 | -25.3 | 48 |
| G 3 | 7.1 | -26.2 | 46 |
| G 4 | 4.2 | -29.2 | |
| G 5 | 2.8 | -30.6 | |
| G 6 | 2.3 | -30.0 | |
| G 7 | 2.3 | -26.5 | 22 |
| G 8 | 2.8 | -24.0 | 63 |
| weighted average | 3.3 | -26.9 | |
| L. Kinneret trypton, station F, at 1 m above sediment** | | | |
| F 23 | | -25.7 | |

*For dates of sampling, see Table 1.

**Samples taken Jul. 19, 1971-Jul. 3, 1972.

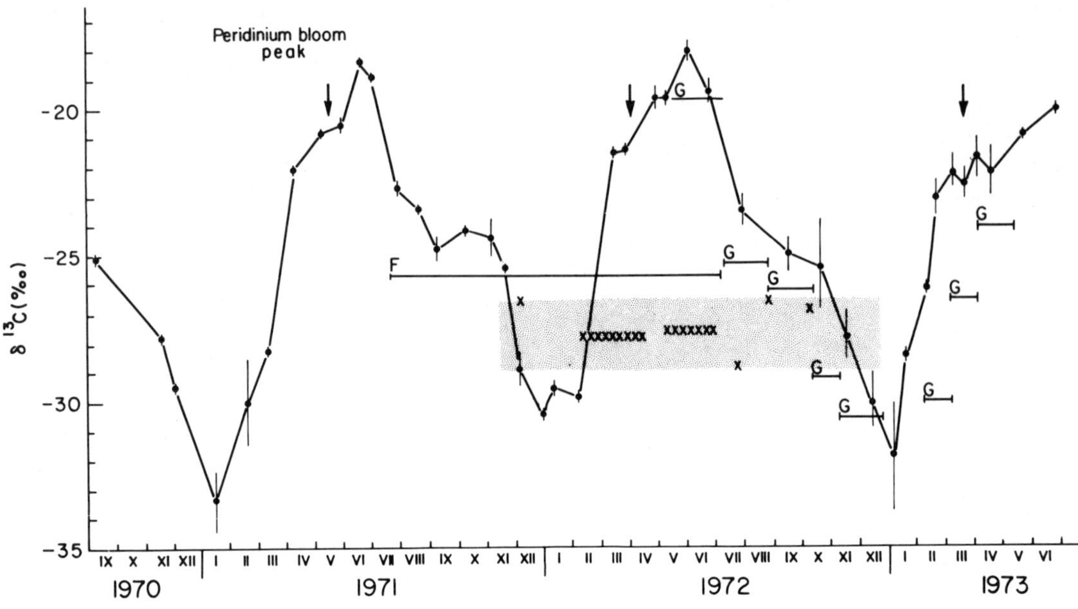


Fig. 7. Seasonal variations in organic $^{13}C$ content. o Lake Kinneret plankton, x Jordan suspended matter, —— Lake Kinneret trypton. The peak of the Peridinium bloom is marked by arrows.

Plankton samples contain mixed populations of phytoplankton and zooplankton and for our purpose they represent the isotopic composition of autochthonous organic matter.

The carbon-13 content in plankton from Lake Kinneret, collected once a month from September 1970 to June 1973 is shown in Fig. 7. Wide seasonal variations with a fairly reproducible annual pattern were observed: It seems that plankton becomes most enriched in $^{13}C$ in June ($\delta^{13}C$ values of up to about -18°/∘∘) and most depleted in December-January (down to about -33°/∘∘).

During March-May Peridinium makes up 94%-97% of the algal biomass and 70%-80% of the total biomass (Serruya, 1975), so that $^{13}C$ variations measured during this period more or less characterize the Peridinium bloom. During the growth period of Peridinium, the isotopic composition of plankton becomes gradually enriched in $^{13}C$ and at the biomass peak (Fig. 7) it is in the range -20°/∘∘ to -22°/∘∘. In May-June there is a further enrichment, up to -18°/∘∘, which characterizes the waning of the bloom.

The wide seasonal variations in the isotopic composition of plankton indicate that the source of organic carbon cannot be identified by measuring organic carbon-13 in the sediments, and again, similar to the carbonate system, such estimates should be done on short-term collections of trypton.

The comparison of organic $^{13}C$ data in Jordan suspended matter and in lake plankton shows that from March till September, allochthonous organic matter has an isotopic composition which is significantly more depleted in $^{13}C$ than that of the plankton. In October-November and in February the isotopic composition of organic matter of allochthonous origin resembles that of Lake Kinneret plankton. For December-January, because of inadequate sampling of Jordan suspended matter, no comparison can be made. Consequently, the relative contributions of allochthonous and autochthonous organic carbon can be identified in Lake Kinneret trypton only from March to September.

The organic $^{13}C$ content in trypton collected at station F (Table 2) could be analysed only on a combined sample (due to insufficient available material) representing the deposition of a one year period. The value obtained, -25.7°/∘∘, suggests the contribution of plankton debris from the Peridinium bloom, which are enriched in $^{13}C$, to organic matter introduced by the Jordan.

The isotopic composition of organic carbon in trypton collected at station G displays a seasonal variation which, at first sight, resembles the general pattern observed in the plankton samples (Fig. 7), indicating the presence of freshly deposited authigenic organic matter. But, during the period when the carbon-13 content of allochthonous organic carbon is similar to that of plankton, during October-November and February, the $^{13}C$ values in trypton (samples G4-G6) are somewhat confusing since they are more depleted in $^{13}C$ than both sources by about 2°/₀₀. This observation may suggest that the isotopic composition of sinking dead plankton is not equivalent to that of living plankton in the euphotic layer. Also, it might be possible that, during the collecting period, alteration of the organic matter takes place in the trap. Sampling of plankton debris, below the euphotic zone (which is a difficult task), and subsequent $^{13}C$ analysis are therefore desirable.

Keeping in mind the above difficulties, the organic $^{13}C$ balances which have been undertaken for May-September 1972 and March-May 1973 (Table 2) must be examined with great care. In these balances, the values chosen for the $^{13}C$ content in plankton were those corresponding to about one month before the end of the collecting period of each trypton sample, attempting thus to allow for the sinking rate of planktonic debris; for allochthonous organic matter, the average value of -27.4°/₀₀ was used. The highest autochthonous organic carbon contributions, 87% for May-June 1972 and 63% for April-May 1973, correspond to the end of the bloom.

## Trypton compared with sediments

The average isotopic composition of carbonate in trypton, representing a one-year collection, was calculated (Table 1) by weighing for the amounts of trypton deposited during each collecting period (Serruya, 1973a; 1973b). The weighted averages should coincide with the isotopic composition of carbonates of freshly deposited sediments.

In the uppermost 2 cm layer of sediment at station F, the isotopic composition of carbonate is $\delta^{13}C$ = -2.4°/₀₀ and $\delta^{18}O$ = -4.6°/₀₀ which is within 0.3°/₀₀ of the average value found for trypton at station F during 1971/1972. This comparison seems satisfactory. At station G, the situation is different: in the uppermost 5 cm of sediment, the isotopic composition of carbonate is $\delta^{13}C$ = -4.2°/₀₀ and $\delta^{18}O$ = -6.2°/₀₀, differing by 1.4°/₀₀ and 1.9°/₀₀, respectively, from the average value of trypton during 1972/1973. But in layers below 5 cm, at station G, the isotopic composition of carbonates is within 0.5°/₀₀ of the average value of trypton. In the sediments of station G, the stable isotopic data taken together with $^{210}Pb$ and $^{137}Cs$ data (Stiller, 1976) suggest that the upper 5 cm do not represent the usual sedimentation pattern, but a recent major intrusion of material from the watershed; the floods of winter 1968/69 might be responsible.

The isotopic composition of organic carbon in the sediments is not expected to match with that measured in trypton. In the anaerobic environment of the sediments which is also devoid of sulphate ions, it seems that the decomposition of organic matter preferentially acts on substances more enriched in $^{13}C$. As a result, in the sediments of both stations, the isotopic composition of organic carbon gradually becomes more depleted in $^{13}C$ with depth (Stiller & Magaritz, 1974; Stiller, 1976). This fact also leads to the conclusion, already stated in a previous section (Organic carbon), that identification of the origin of sedimenting organic carbon by isotopic measurements can be done only on freshly deposited material.

## Acknowledgements

I wish to thank the management of the Kinneret Limnological Laboratory at Tabgha for generously providing the plankton and trypton samples and to Dr. C. Serruya for making available data as yet unpublished. I am indebted to Drs. J.R. Gat and A. Nissenbaum for helpful criticism and to Dr. M. Magaritz for suggestions which initiated this study. Many thanks to Mrs. C. Aini who carefully typed the manuscript.

## References

Bender, M.M., 1968. Mass spectrometric studies of $^{13}C$ variations in corn and other grasses. Radiocarbon 10:468-472.

Craig, H., 1965. The measurement of oxygen isotope paleotemperatures. In: E. Tongiorgi (Ed.): Proc. Spoleto Conf. Stable Isotopes in Oceanographic Studies and Paleotemperatures. CNR Lab. Geol. Pisa. p. 165-182.

Degens, E.T., R.R.L. Guillard, W.M. Sackett & J.A. Hellebust, 1968. Metabolic fractionation of carbon isotopes in marine plankton. I. Temperature and respiration experiments. Deep Sea Res. 15:1-9.

Hunt, J.M., 1967. The significance of carbon isotope variations in marine sediments. In: G.D. Hobson (Ed.): Advances in Organic Geochemistry, Proceeding of Third Intern. Conf. p. 27-35.

Mook, W.G., 1968. Geochemistry of the stable carbon and oxygen isotopes of natural waters in the Netherlands. Ph.D. thesis, Rijksuniversiteit. Groningen. 157 p.

Nissenbaum, A., M.J. Baedecker & I.R. Kaplan, 1972. Organic geochemistry of Dead Sea sediments. Geochim. Cosmochim. Acta 36: 709-727.

Sackett, W.M. & R.R. Thompson, 1963. Isotopic organic carbon composition of recent continental derived clastic sediments of eastern Gulf Coast, Gulf of Mexico. Bull. Am. Assoc. Petrol. Geologists 47:525-528.

Sanger, J.E. & E. Gorham, 1972. Stratigraphy of fossil pigments as guide to the post-glacial history of Kirchner Marsh, Minnesota. Limnol. Oceanogr. 17:840-854.

Serruya, C., 1973a. Sediments. In: Lake

Kinneret Data Record, Kinneret Limnological Laboratory. NCRD 13-73. p. 39-46.
Serruya, C., 1973b. Personal communication.
Serruya, C., 1975. Lake Kinneret Report: 1974.
Serruya, S.,1973. Physical limnology of the Kinneret. In: Lake Kinneret Data Record, Kinneret Limnological Laboratory. NCRD 13-73. p. 13-30.
Stiller, M. & M. Magaritz, 1974. Carbon-13 enriched carbonate in interstitial waters of Lake Kinneret sediments. Limnol. Oceanogr. 19:849-853.
Stiller, M., 1974. Rates of transport and sedimentation in Lake Kinneret. Ph.D. thesis (in Hebrew), Weizmann Institute of Science, Rehovot. 240 p.
Stiller, M., 1976. Rates of sedimentation in Lake Kinneret (in preparation).

Key words: carbonate sources
organic C sources
carbon-13
oxygen-18
Lake Kinneret

# SEDIMENTATION RATES AND SEDIMENT CORES IN TWO SWISS LAKES OF DIFFERENT TROPHIC STATE

J. Bloesch

Swiss Federal Institute of Water Resources and Water Pollution Control (EAWAG), Swiss Federal Institute of Technology, 8600 Dübendorf, Switzerland

## Summary

In 1969-1970 sedimentation rates in the mesotrophic Lake of Lucerne (Horw Bay) and the highly eutrophic Rotsee have been measured to be 3.7 and 2.5 mm/yr, respectively. The collected sediments have been analyzed for planktonic and chemical composition (total organic substance [loss on ignition], organic C, clay minerals [HCl-insoluble fraction], Ca, Mg, P, N, Fe and Mn). According to the calculated yearly increase of bottom deposits, the sediment of 1969-1970 is now at a level lying 2.6 cm and 1.8 cm, respectively, below the sediment surface. In spring 1976 two sediment cores at the same stations have been taken in both lakes. Chemical analyses of the cores and the comparison with the collected sediments indicated that most components have been released during the last 7 years. Thereby Fe and P show a quite similar behaviour, but differe from Mn; moreover differences between the lakes are obvious. The Mn accumulation in the sediments of the eutrophic lake, decreasing the Fe:Mn ratio from 32:1 to 9:1, is discussed.

## Introduction

Occurrence of release and accumulation at the mud-water-interface of lakes are of great importance for the nutrient cycles and for the progress of eutrophication. Exchange processes may be directly measured in experiments (e.g. Heyes et al., 1958; Olsen, 1964; Banoub, 1975), with the disadvantage that conclusions drawn from in vitro tests are difficult to transfer to in situ conditions. On the other hand, release of dissolved compounds may be quantified indirectly in the lake itself by measurements of accumulation in the hypolimnion and redox potentials (e.g. Tessenow, 1975). In the investigations presented here another possibility of using sediments is shown, namely the comparison of recent sediments collected earlier with actual sediment cores.

In 1969-1970 the settling material in two Swiss lakes of different trophic state has been collected with sediment traps at different depths (Bloesch, 1974). The lakes in question are the mesotrophic Horw Bay of Lake of Lucerne and the eutrophic Rotsee, the main characteristic data for which are compiled in Table 1.

Table 1. Characteristic data for the lakes (compiled from Bloesch, 1974).

| | Lake of Lucerne (Horw Bay) | Rotsee |
|---|---|---|
| Surface area ($km^2$) | 1.69 | 0.47 |
| Mean depth (m) | 42.6 | 9 |
| Sampling depth (m) | 60 | 14 |
| Trophic state | meso-trophic | highly eutrophic |
| Primary production ($gCm^{-2}yr^{-1}$) | 415 | 511 |
| Sedimentation above lake bottom | | |
| organic C ($gCm^{-2}yr^{-1}$) | 67.0 | 141.8 |
| total N ($gNm^{-2}yr^{-1}$) | 13.4 | 16.3 |
| total P ($gPm^{-2}yr^{-1}$) | 1.4 | 2.1 |
| Oxygen in the hypolimnion (mg $O_2$ liter$^{-1}$) | 6.5 | anoxic |
| pH above lake bottom | 7.5-8.2 | 6.9-7.2 |
| Maximum concentrations above lake bottom | | |
| orthophosphate (μg liter$^{-1}$) | 30-40 | 1,620 |
| $NH_4$-N (μg liter$^{-1}$) | 70-90 | 9,560 |
| dissolved Fe (μg liter$^{-1}$) | 10-19 | 153 |
| dissolved $SiO_2$ (mg liter$^{-1}$) | 3 | 8 |

The collected sediments have been analyzed for their planktonic and chemical composition (total organic substance [loss on ignition], organic C, clay minerals [HCl-insoluble fraction], Ca, Mg, P, N, Fe and Mn). For chemical analysis titrimetric (Ca, Mg) and colorimetric (P) methods, a CHN analyzer (org. C, N) and an atomic absorption spectrophotometer (Fe, Mn) have been used. The details of analytical methods have been reviewed by Bloesch (1974).

The measured sedimentation rates correspond to a yearly increment of bottom sediments of 3.7 mm in Lake of Lucerne and 2.5 mm in the Rotsee, whereby the compacting effect was considered approximately. Therefore the sediments collected in 1969-1970 are situated now, 7 years later, at a level of about 2.6 cm and 1.8 cm in Horw Bay and in the Rotsee, respectively.

In March 1976 two parallel sediment cores have been taken at the same sampling places, using the sediment corer of Ambühl & Bührer (1975) and the preparation technique reported by Staub (1974). The upper 5 cm of each core were divided into 0.5 cm sections and analyzed chemically in the same way as the collected sediments. Consequently it is possible to find again the sediments of 1969-1970 approximately and to test them for changes; moreover we may look at the cores themselves and examine their vertical varves, which are visible mainly in the sediments of Horw Bay. Both ways of investigation should give information about chemical transformation, i.e. whether the different components have been released from or accumulated in the sediments in the last 7 years. We may assume that release processes are effective, if concentrations in the mud are lower than those in the collected sediments. If higher values are measured in the bottom sediments, an accumulation of the compounds in question has to be considered.

## Results

The chemical analysis of the sediments collected in 1969-1970 at different levels showed that relative small differences exist in the vertical distribution, but seasonal changes were rather considerable (Bloesch, 1974). Concerning our exchange problems at the mud-water-interface we restrict us to the deepest trap level, which is situated directly above lake bottom, i.e. at 60 m and 14 m in Horw Bay and in the Rotsee, respectively. Now, the different components of the sediment cores taken in 1976 will be compared each with the mean percentage concentration in these collected sediments.

The water content (Fig. 1), which was - by methodical reasons - determined only in the sediment cores, shows that both parallel cores can be regarded as more or less identical. In the vertical distribution the water content decreased downwards from 79 % to 50 % in Horw Bay and from 88 % to 83 % in the Rotsee.

Loss on ignition (Fig. 2), which mostly represents the sedimented biomass, and organic carbon (Fig. 3) show a similar curve in both lakes. The loss on ignition is about twice as great as the content of organic C. Due to the higher trophic state of the Rotsee organic compounds have a higher concentration in this lake than in Lake of Lucerne. Only a small decrease can be found in the vertical distribution of the cores; Horw Bay however shows two greater changes in the first and fifth centimeter. In comparison with the collected sediments the content in the mud is only few lower in Horw Bay but remarkably more lower in the Rotsee.

The HCl-insoluble part of the residue on ignition (Fig. 4) contains allochthonous clay minerals (mainly in Horw Bay) and autochthonous frustules of diatoms (mainly in the Rotsee). Whereas this fraction in the sediment core of the Rotsee is constantly 30 %, we can find in the fifth centimeter of the Horw Bay core an increase of about 10 - 20 % from 50 - 55 % to 62 - 69 % of d.w.. In both lakes the concentrations are distinctly higher than in the collected sediments.

Calcium (Fig. 5) and magnesium (Fig. 6) are similar in both lakes. Only small differences occur between the collected sediments and the cores; the latter however have rather lower concentrations. The contents of Ca are 10 - 20 % of d.w., those of Mg, which vary remarkably more, are 0.3 - 0.8 % of d.w..

Total phosphorus content (Fig. 7) is different in both lakes. In Lake of Lucerne, in the hypolimnion of which more than 6 mg $O_2$/ liter were measured throughout the whole year (Bloesch, 1974), we can find a decrease of 50 % in the top half centimeter of sediment. This decrease retards downwards but increases again in the fifth centimeter. The values at the top of the core are higher than those found in the sedimenting material, the values in the corresponding layer at 2.6 cm depth are smaller. This distribution may be explained by redox processes: dissolution of P compounds occurs in the deeper sediment layers (below 1 cm) according to the low redox potential that we may expect there. The dissolved phosphates diffuse upwards into the sediment-water-interface, where they are precipitated again due to the high redox potential originating in the great hypolimnetic oxygen concentrations. Moreover amounts of mostly oxidized P compounds are added sinking down the lake. These two opposite processes lead to an accumulation of P in the bottom sediment, which however is limited to the top millimeters of the mud. In the Rotsee, where at all times anoxic conditions exist above lake bottom, P decrease in the core is gradual and small. The values are about half of those in the collected sediment. In this sediment a continuous P release is most probably due to a low redox potential.

The total nitrogen content (Fig. 8) continually decreases from the top of the core

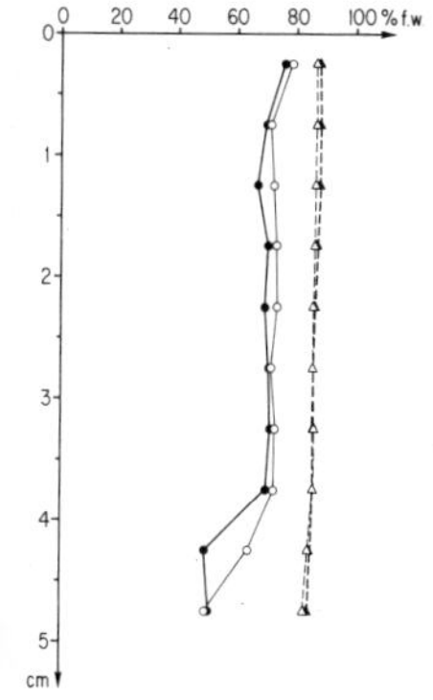

Fig. 1. $H_2O$-content

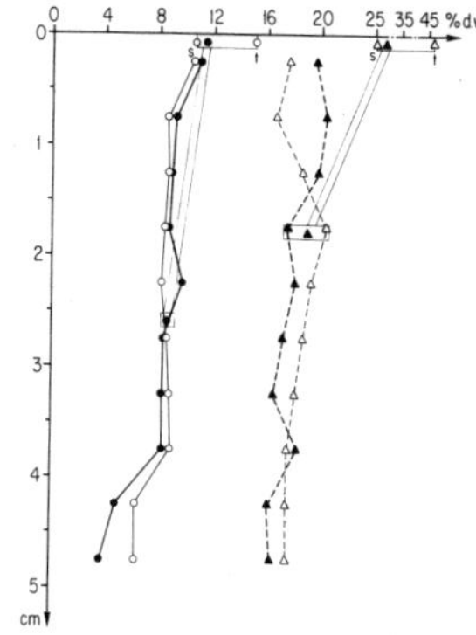

Fig. 2. Organic substance (loss on ign.)

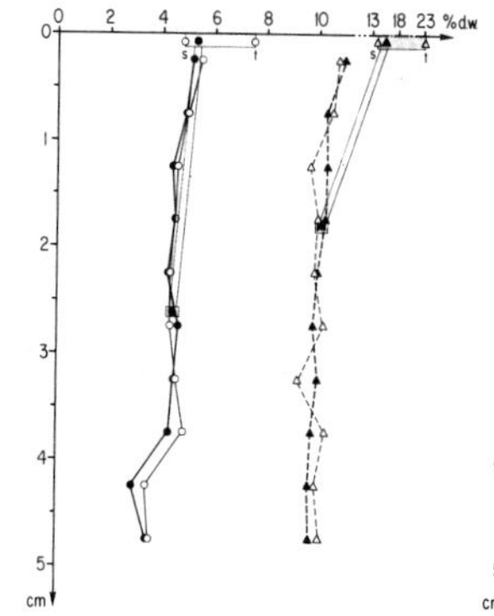

Fig. 3. Organic C

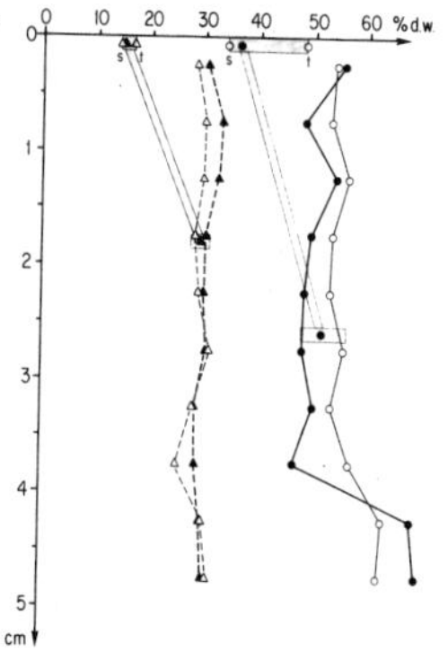

Fig. 4. HCl-insoluble fraction

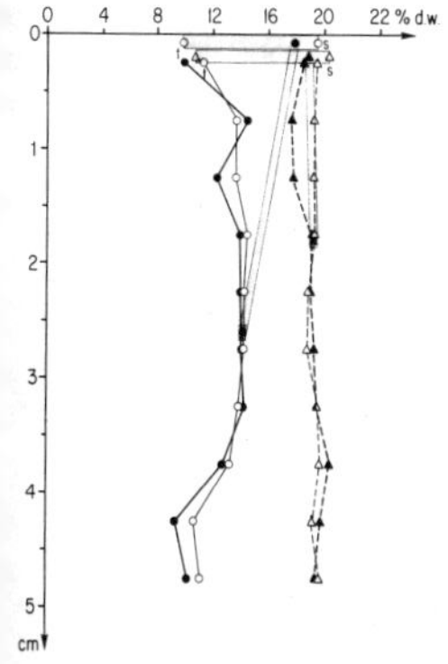

Fig. 5. $Ca^{2+}$

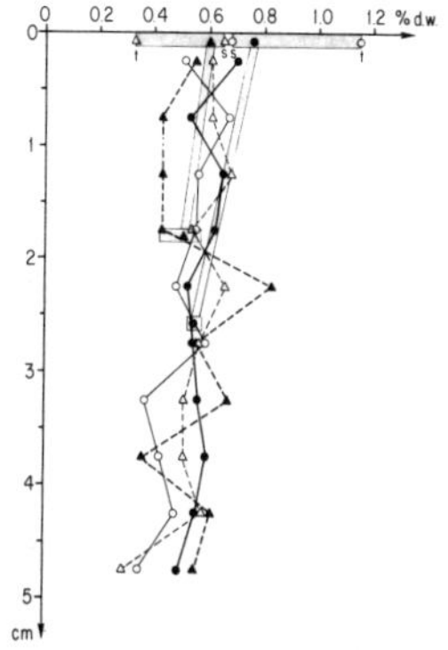

Fig. 6. $Mg^{2+}$

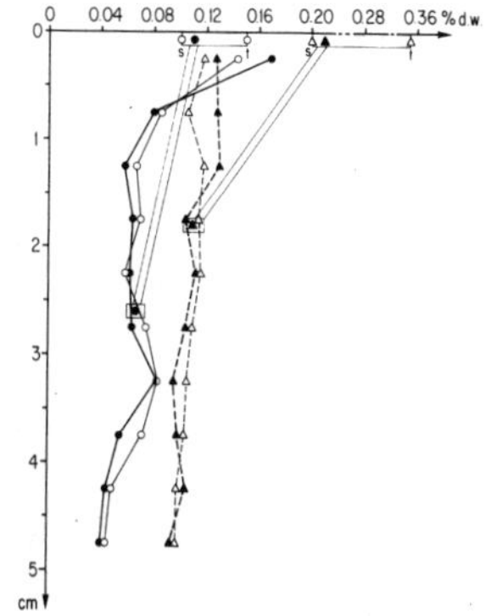

Fig. 7. Total P

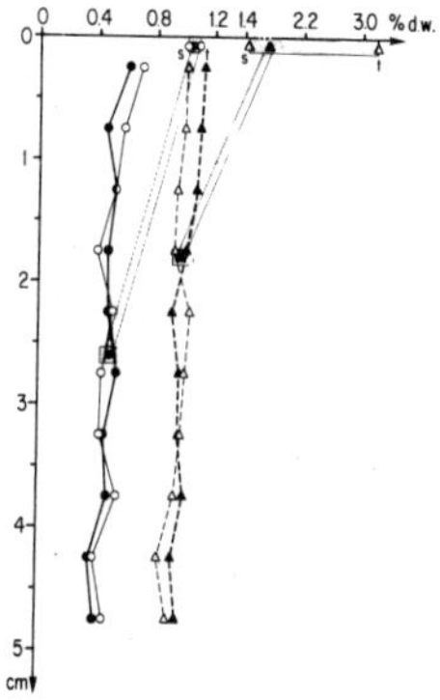

Fig. 8. Total N

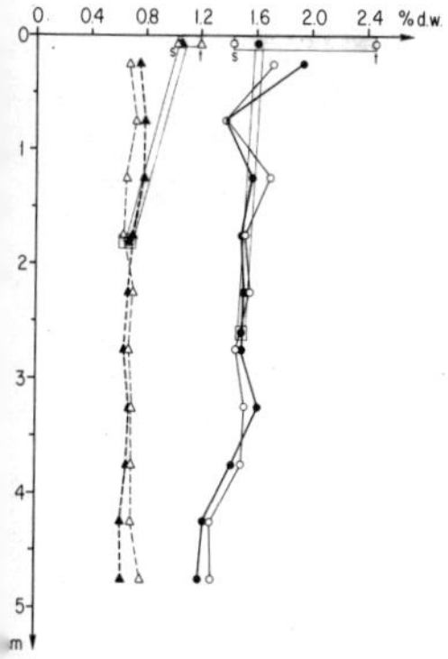

Fig. 9. Total Fe

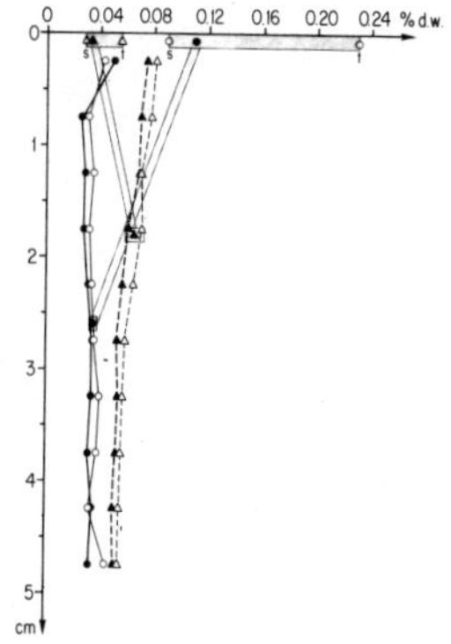

Fig. 10. Total Mn

Figs. 1-10. Amounts of water and different compounds in the sediment cores and in the collected sediments expressed as percentage of fresh weight and dry weight, respectively.

Sediment cores (1976)

●—● Lake of Lucerne, Horw Bay: Core No 1
○—○ Lake of Lucerne, Horw Bay: Core No 2
▲---▲ Rotsee: Core No 1
△---△ Rotsee: Core No 2

Sediments collected 1969-1970 above lake bottom (Horw Bay: 60 m, Rotsee: 14 m), shown at the top of the cores.

- ● ,▲ total average for the year (Horw Bay, Rotsee)
- $○_t$, $△_t$ average for turnover period (Horw Bay), Rotsee)
- $○_s$, $△_s$ average for stagnation period (Horw Bay, Rotsee)

downwards in both lakes and is distinctly lower as in the collected sediment.

Iron (Fig. 9) shows similar behaviour as P, but Fe concentrations are lower and P concentrations are higher in the cores of the Rotsee as in those of Lake of Lucerne. Moreover the difference between collected sediments and mud is less.

The differences of manganese (Fig. 10) only are small in the cores of both lakes and vary from 0.026 to 0.050 % of d.w. and 0.050 to 0.092 % of d.w. in Horw Bay and in the Rotsee, respectively. The comparison with the collected sediments shows a contrary behaviour of Mn due to the different trophic state of both lakes. In Lake of Lucerne the concentrations in the bottom sediment are remarkably lower, which points towards release processes. In the Rotsee higher concentrations are found in the sedimenting material, which may be a sign of accumulation. This unexpected finding will be discussed later.

## Discussion

The values given as percentage of dry weight are not fully suitable, if we want to decide, whether release or accumulation of a certain component takes place. We therefore have to relate the measured data to a conservative element or compound that does not undergo chemical and biological reactions and hence does not change markedly. Among the analyzed fractions introduced before we may choose the HCl-insoluble part, which fulfills our demand. Now, if any degradable substances are liberated from a given mass of collected sediment, the conservative HCl-insoluble fraction will be found in a greater percentage of d.w. in the equivalent amount of bottom sediment. We can find a similar relative augmentation of the salt content of waters in aride zones where high evaporation occurs. Further, let us take the measured percentages of d.w. of the HCl-insoluble part, which are

- 36.3 % in the collected sediments of Horw Bay (mean value of the year at 60m)
- 52.1 % in the bottom sediments of Horw Bay (mean value of two cores 0-4cm)
- 14.6 % in the collected sediments of the Rotsee (mean value of the year at 14m)
- 29.7 % in the bottom sediments of the Rotsee (mean value of two cores 0-5cm),

then we may calculate the total loss of degradable and soluble substances in bottom deposits, using equation (1) and (1a), respectively,

$$(1) \quad \left(\frac{52.1 - 36.3}{52.1}\right) \cdot 100\ \%$$

$$(1a) \quad \left(\frac{29.7 - 14.6}{29.7}\right) \cdot 100\ \%$$

to be 30±7 % in Horw Bay and 51±5 % in the Rotsee. To calculate the release or accumulation of some substance x we may use equation (2) and (2a), respectively,

$$(2) \quad T_x = \left(\frac{-S_{cx} + \frac{36.3}{52.1} \cdot S_{bx}}{S_{cx}}\right) \cdot 100\ \%$$

$$(2a) \quad T_x = \left(\frac{-S_{cx} + \frac{14.6}{29.7} \cdot S_{bx}}{S_{cx}}\right) \cdot 100\ \%,$$

whereby $T_x$ = transformation of the substance x, $S_{cx}$ = substance x in the collected sediment and $S_{bx}$ = substance x in the bottom sediment. The results of these calculations are compiled in Table 2.

Table 2. Release (-) and accumulation (+) in bottom sediments of Lake of Lucerne (Horw Bay) and the Rotsee, expressed as percentage of the amount collected in 1969-1970.

| Chemical component | Lake of Lucerne (Horw Bay) | | Rotsee | |
|---|---|---|---|---|
| | * | † | * | † |
| HCl-insoluble fraction (no transformation) | + 5 | - 2 | 0 | - 1 |
| Loss on ignition | -34 | -47 | -68 | -68 |
| Organic C | -30 | -42 | -65 | -65 |
| $Ca^{2+}$ | -59 | -45 | -51 | -50 |
| $Mg^{2+}$ | -45 | -51 | -53 | -60 |
| Total N | -56 | -67 | -69 | -72 |
| Total P | - 1 | -59 | -73 | -76 |
| Total Fe | -21 | -35 | -67 | -69 |
| Total Mn | -70 | -79 | +18 | - 2 |

* collected sediment compared with top of the core

† collected sediment compared with core at 2.6 cm depth (Horw Bay) and at 1.8 cm depth (Rotsee), respectively.

According to the explanations given before, we can see that the HCl-insoluble fraction shows, by definition, no change upon burial (percentages less than approximately 5 % are not significant). Nearly all components have been released from the sediments during the last 7 years, whereby in the eutrophic Rotsee release was greater than in the mesotrophic Horw Bay, except release of Mn. In Horw Bay Mn is more released than Fe, and P is only released in the deeper layers of the sediment, indicating the redox interface at the top of the sediment, which prevents P release to the hypolimnion (Mortimer, 1949). In the Rotsee P and Fe are mostly released, but Mn is even accumulated at the sediment surface. These differences indicate that no correlation exist between P, Fe and Mn (r = 0.43 - 0.78) in the upper 5 cm of the sediment cores in opposition to the patterns of one-meter cores (Bortleson & Lee, 1974).

Whereas P and Fe show more or less the expected results, the behaviour of Mn in both lakes is rather astonishing and not easy to explain. According to Stumm & Morgan (1970) Mn participates more in redox reactions than Fe does, and therefore is reduced at higher redox potentials. Moreover Fe reacts with P in a complicated way by forming complex iron-phosphorus compounds and by adsorption phenomena. As eutrophication in Lake of Lucerne yearly increases (Ambühl, 1975), the measured release of Mn may be the first sign of the beginning of at least temporarily anoxic conditions at the mud-water-interface. This assumption is supported by Mortimer (1971), who found that at oxygen concentrations below 2 mg/liter Mn is mobilized from the sediment before Fe. In the Rotsee, where we have this anoxic hypolimnion, we would rather expect a great release of Mn. Hence Tessenow (1975) recently has shown that dissolved Mn can be transported from the near-shore sediments to pelagic areas and precipitated again in the sediments of the deepest zones of a lake, possibly as Mn(II)phosphate. This secondary accumulation may be also efficient in the Rotsee. Moreover, most of the decomposable part of the particulate Mn, settling down the lake, is dissolved before reaching the lake bottom (Bloesch, 1974). Thus only resistent Mn compounds are found at the bottom sediment.

The strikingly contrasting behaviour of Fe and Mn in both lakes is also demonstrated by the Fe:Mn ratio shown in Fig. 11. Changes are rather due to Mn than to Fe. In the particulate material collected in the Rotsee, we find low concentrations of Fe (approximately 1 % of d.w.) and high concentrations of Mn, which however diminish in the deep hypolimnion from about 0.35 % to 0.03 % of d.w., according to the anoxic conditions. At the top of the bottom sediment Fe contents are lower and Mn values increase to approximately 0.08 % of d.w., decreasing again continually to 0.05 % of d.w. down to the fifth centimeter of the core. In Horw Bay Fe concentrations in the particulate material are 2.3 - 2.6 % of d.w., whereas Mn reaches only 0.14 - 0.23 % of d.w., increasing slowly from the upper to the lower layers of the lake. In the bottom sediment Mn is more released than Fe in the upper 1.5 cm, augmenting the ratio up to 50:1; then, being more or less constant, Fe decreases diminishing the ratio to 32:1.

Fig. 11. Fe:Mn ratio in the particulate material collected in traps and in the bottom sediments of Lake of Lucerne (Horw Bay, solid lines) and the Rotsee (dashed lines). Heavy line = mean of 1969-1970, thin lines = mean of turnover (t) and stagnation period (s).

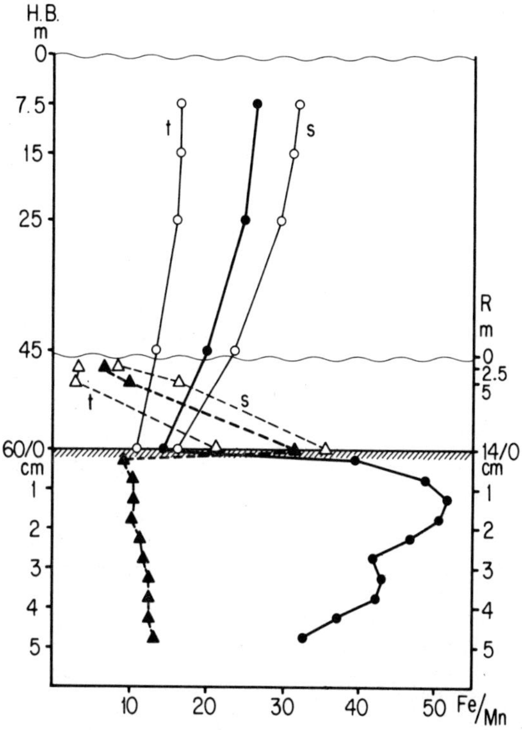


If we compare these findings with results obtained by Ungemach (1960), some discrepances appear, which may give us some information about long time changes in the deeper layers of the sediments. According to the measured and extrapolated yearly sedimentation rates, the top of Ungemach's cores are in the year 1976 approximately at a depth of 7 cm and 4.5 cm in Horw Bay and in the Rotsee, respectively. As can be seen from Table 3, water content, nutrients (N and P) and earth alkaline ions (Ca and Mg) in the Rotsee sediments did not change markedly, whereas greater parts of organic material (loss on ignition, organic C) indicate the progressive eutrophication of this lake. The comparison makes obvious the Fe release and the Mn accumulation, diminishing the Fe:Mn ratio from 50:1 down to 13:1.

Table 3. Comparison of the cores with those made by Ungemach (1960). $H_2O$-content = % of fresh weight, other components = % of dry weight.

| Chemical component | Rotsee | | | Lake of Lucerne | | |
|---|---|---|---|---|---|---|
| | Ungemach 1960 | Bloesch 1976 | | Ungemach 1960 § | Bloesch 1976 † | |
| | 0-4 cm | 4-5 cm* | | 0-4 cm | 4-5 cm* | |
| $H_2O$-content | 84 | 84 | ± 0.9 | 81 | 53 | ± 7 |
| HCl-insoluble fraction | 36 | 30 | ± 0.5 | 68 | 66 | ± 3.7 |
| Loss on ignition | 11 | 17 | ± 0.7 | 11 | 5.4 | ± 1.3 |
| Organic C | 5.7 | 9.8 | ± 0.2 | 2.6 | 3.4 | ± 0.3 |
| Total N | 0.80 | 0.90 | ± 0.06 | 0.30 | 0.39 | ± 0.04 |
| Total P | 0.12 | 0.10 | ± 0.005 | 0.10 | 0.045 | ± 0.003 |
| $Ca^{2+}$ | 22 | 20 | ± 0.3 | 4.6 | 10 | ± 0.8 |
| $Mg^{2+}$ | 0.64 | 0.51 | ± 0.15 | 1.0 | 0.47 | ± 0.08 |
| Total Fe | 1.7 | 0.68 | ± 0.07 | 4.2 | 1.2 | ± 0.04 |
| Total Mn | 0.034 | 0.053 | ± 0.003 | 0.15 | 0.037 | ± 0.006 |
| Fe:Mn ratio | 50 | 13 | | 28 | 35 | |

* mean from two cores, § Kreuztrichter, † Horw Bay

In Horw Bay, where more changes occurred, the comparison is subjected to some uncertainities, because sediments do not come from the same sampling site, and because the 7 cm level has not been analyzed in our core. However the increase of organic C, N and Ca and the decrease of Fe and Mn may be as well as signs of an intensified production and autochthonous sedimentation (biogenic precipitation), and of beginning Fe and Mn release. P on its part is released only in the lower areas of the cores as mentioned before. The contrarious behaviour of loss on ignition and organic C as well as of Ca and Mg cannot be explained.

Further investigations in this field of sedimentology, such as heavy metals, should be done to demonstrate the holding capacity and source capacity, respectively, of the lakes' sediments. I think that comparison of recent sediments and mud can help us to confirm and verify the release and accumulation mechanisms at the mud-water-interface we always know.

Acknowledgment

I would like to thank Dr. H.R. Hegi, Dr. H. Leidner and E. Szabo for chemical analyses, H. Bolliger for drawings, T. Oberholzer for typing, Prof. A. Lerman for reviewing this paper, and E. Staub for the improving discussions.

References

Ambühl, H., 1975. Versuch der Quantifizierung der Beeinflussung des Oekosystems durch chemische Faktoren: Stehende Gewässer. Schweiz. Z. Hydrol. 37(1):35-52.

Ambühl, H. & H. Bührer, 1975. Zur Technik der Entnahme ungestörter Grossproben von Seesediment: Ein verbessertes Bohrlot. Schweiz Z. Hydrol. 37(1):175-186.

Banoub, M.W., 1975. Experimental studies on material transactions between mud and water of the Gnadensee (Bodensee). Verh. Internat. Verein. Limnol. 19(2):1263-1271.

Bloesch, J., 1974. Sedimentation und Phosphorhaushalt im Vierwaldstättersee (Horwer Bucht) und im Rotsee. Schweiz. Z. Hydrol. 36(1):71-186.

Bortleson, G.C. & G.F. Lee, 1974. Phosphorus, iron, and manganese distribution in sediment cores of six Wisconsin lakes. Limnol. Oceanogr. 19(5):794-801.

Heyes, F.R., B.L. Reid & M.L. Cameron, 1958. Lake water sediments II. Oxidation-reduction relations at mud-water interface. Limnol. Oceanogr. 3:308-317.

Mortimer, C.H., 1949. Seasonal changes in chemical conditions near the mud surface in two lakes of the English Lake District. Verh. Internat. Verein. Limnol. 10:353-356.

Mortimer, C.H., 1971. Chemical exchanges between sediments and water in the Great Lakes - Speculations on probable regulatory mechanisms. Limnol. Oceanogr. 16(2):387-404

Olsen, S., 1964. Phosphate equilibrium between reduced sediments and water. Laboratory experiments with radioactive phosphorus. Verh. Internat. Verein. Limnol. 15: 333-341.

Staub, E., 1974. Biologisch-stratigraphische Untersuchungen an jungen Sedimenten des Vierwaldstättersees. Diplomarbeit EAWAG/ ETH, Zürich (unpublished).

Stumm, W. & J.J. Morgan, 1970. Aquatic Chemistry. An introduction emphasizing chemical equilibria in natural waters. Wiley-Interscience, New York.

Tessenow, U., 1975. Solution, diffusion and sorption in the upper layer of lake sediments. V. The differentiation of the profundal sediments of an oligotrophic mountain lake (Feldsee, Hochschwarzwald) by sediment-water-interaction. Arch. Hydrobiol./Suppl. 47(3):325-412.

Ungemach, H., 1960. Sedimentchemismus und seine Beziehung zum Stoffhaushalt in 40 europäischen Seen. Doctoral Dissertation, Univ. Kiel, 420 pp.

Key words: sedimentation rates
sediment cores
release processes
nutrients
heavy metals

# COMPARISON OF TRIPTON SEDIMENTATION IN FOUR SMALL LAKES

Z. Kajak & W. Ławacz

Institute of Ecology, 05-150 Łomianki, Dziekanów Leśny near Warsaw, Poland

Summary

Comparing four lakes with various trophic and mictic status it was found that the amount of sedimenting tripton was highest in a polymictic eutrophic and lowest in a humic rich lake. The pattern of sedimentation differed significantly: there was a summer maximum in 2 eutrophic (dimictic and polymictic) lakes, and autumn one in the third eutrophic lake; there was no distinct maximum in the humic rich lake. The percentage of C and N of the dry weight was higher in the sedimenting tripton and bottom deposits of the humic rich lake, than in all other lakes. In all lakes % of C and N of the dry weight was higher in the sedimenting tripton than in bottom deposits. Sedimentation was higher than primary production only in the polymictic lake; this could result both from resedimentation or from an underestimation of primary production or an overestimation of sedimentation.

## Introduction

The purpose of this paper is to characterise the quantity of sedimenting tripton and its C and N contents in 4 small lakes of various trophic and mictic status, at Mazurian Lakeland, north-east of Poland.

Every 2 weeks the materials were collected in the lakes: Smolak, Piecek, Dgał Mały and Czarna Kuta, between February 25 and November 10, 1971. All these lakes of which the characteristics are given in table I were artificially fertilized starting from 1971 (Zdanowski et al, 1975). Some comparison has been made with the data from the second year of fertilization (1972), as well as with data from a greater and deeper eutrophic dimictic lake Mikołajskie (460 ha, 27.8 m max. depth).

The collected sediments are practically total yearly sediments, since winter sedimentation in this region is negligible (Ławacz, 1969).

Tripton traps were located in the centre of each lake, 1 m above the bottom.

Table I. Characteristics of the lakes under study and of sedimenting tripton for the period March-November 1971.

| | Smolak | Piecek | Dgał Mały | Czarn Kuta |
|---|---|---|---|---|
| surface (ha) | 6.4 | 8.3 | 15.8 | 24.7 |
| max.depth (m) | 5.8 | 23.1 | 14.3 | 3.3 |
| status | humic rich | eutrophic with somc humus comp. | eutrophic | polytrophi polymiotic |
| Dry weight $g/m^2$ x day | 1.0 | 5.7 | 7.7 | 30.0 |
| % in dry weight | | | | |
| C | 32.9 | 12.3 | 7.8 | 12.0 |
| N | 3.8 | 2.2 | 1.6 | 2.0 |
| R $\frac{max.\ \%}{min.\ \%}$ | | | | |
| C | 1.8 | 2.4 | 1.6 | 1.4 |
| N | 2.7 | 2.9 | 2.8 | 1.6 |
| R $\frac{max.abs.}{min.abs.}$ | | | | |
| dry weight | 3.7 | 16.2 | 6.5 | 5.8 |
| C | 3.6 | 12.9 | 7.5 | 3.3 |
| N | 8.0 | 12.5 | 8.3 | 5.7 |

R $\frac{max.\%}{min.\%}$ = ratio of maximal to minima daily % share

R $\frac{max.abs.}{min.abs.}$ = ratio maximal to minimal daily absolute value

In the deepest lake Dgał Mały an additional set of traps was located at a depth of 5 m, just below epilimnion. Each set consisted of 3 jars 30 cm high, 102 $cm^2$ catchment area (11.4 cm of diameter at the top and broader - 20 cm of diameter in the lower part). Usually 2 sets of jars were located at the same time at each station. The collected material was dried at 105°C to constant weight; C and N contents were determined with a CHN analyzer.

## Results

The rate of sedimentation in time varies significantly in particular lakes (Figure 1).

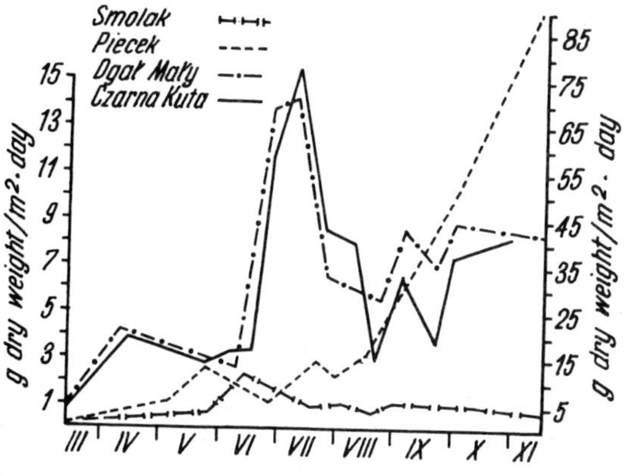


Figure 1. Sedimentation in 4 lakes, 1971. (Right scale for the lake Czarna Kuta, left one for other 3 lakes.)

There is always a minimum in winter and spring values are low. Lake Dgał Mały and Czarna Kuta exhibited a summer maximum; absolute values of sedimenting matter were several times higher in Czarna Kuta. Lake Piecek showed a very high autumn maximum whereas the values were comparatively low during the rest of the year. In the lake Smolak there always were small values with a very weakly pronounced June maximum.

Average sedimentation values (dry weight) differed 30 times being maximal for lake Czarna Kuta and minimal for lake Smolak. The difference in sedimented C and N is smaller - about 12-15 times (Table I).

The content of C was in lake Smolak 3 to 4 times higher than in other lakes and the N content 1.5 to 2 times what obviously shows lesser decomposition in this lake.

The fluctuations of the amount of the sedimenting tripton (expressed as the ratio between the maximal daily values and the minimal daily values, winter period excluded) were highest in lake Piecek, due to a very high autumn peak of sedimentation. In the other 3 lakes the mentioned ratio was much smaller (Table I).

The ratio of the maximal to minimal % contents of N was for individual lakes slightly higher than that of C. The lowest fluctuations of these elements were observed in lake Czarna Kuta (Table I).

There was no clear correlation between sedimentation and primary production or chlorophyll, although in the 2 lakes where the course of sedimentation was similar, the summer peak of sedimentation correlated with the one of primary production (Figure 2).

Sedimentation was 3.5 times higher than primary production in lake Czarna Kuta; in all other lakes it was smaller (Table II).

Table II. Comparison of gross primary production in pelagial and sedimentation for the period without ice in 1971.

| Lake | Production | Sedimentation | S/P |
|---|---|---|---|
| | g C per m$^2$ | | % |
| Smolak | 203 | 72 | 35 |
| Piecek | 182 | 146 | 80 |
| Dgał Mały | 256 | 126 | 49 |
| Czarna Kuta | 256 | 893 | 349 |
| Mikołajskie | 278 | 203 | 73 |

The percentage of N of the dry weight was the highest in lake Piecek in spring, but in lake Smolak in summer. Values for spring (III-V), summer (VI-VIII) and autumn (IX-XI) for the lake Piecek were 3.7; 1.9; 2.2, and for lake Smolak 3.5; 5.3; 2.6. For the other two lakes fluctuations were smaller and not clearly dependent on the season.

In the second year of fertilization, both the amounts of sedimenting tripton on the % contents of C and N were higher than in the first year of fertilization (Table III).

The sedimenting tripton was in all lakes richer in C and N than the bottom deposits, although sometimes only slightly (Table IV).
The amount of sedimenting tripton and the percentage of N at different depths was often similar but some times differences were high (Table V).

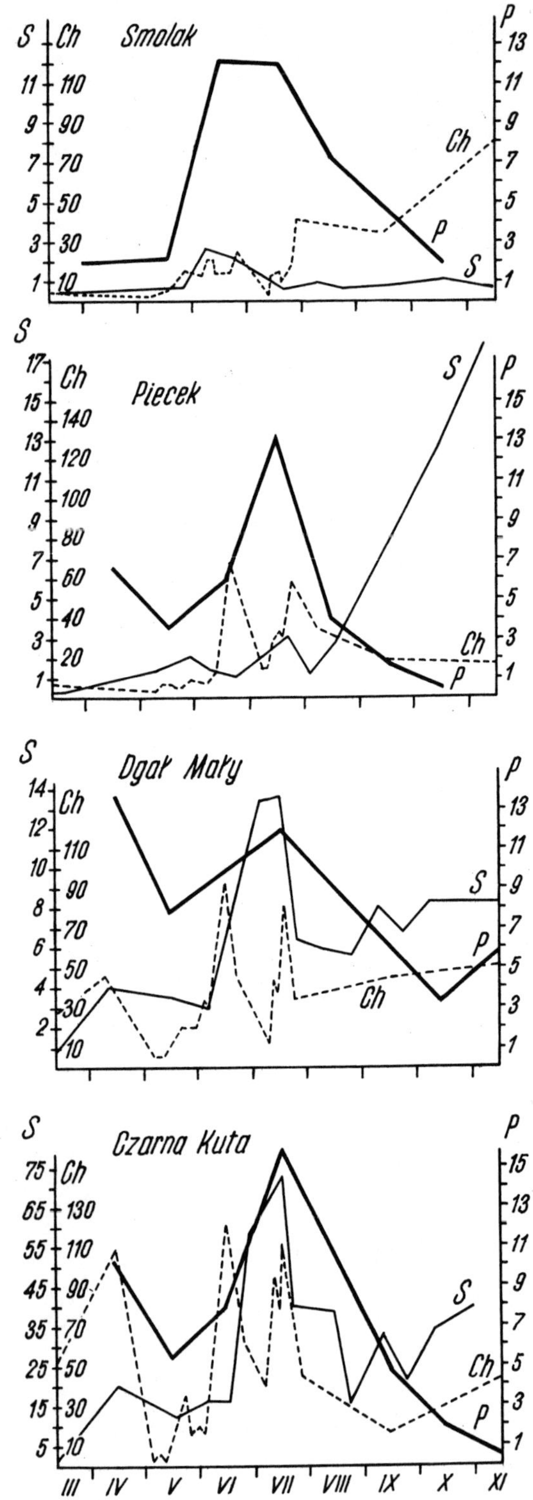


Figure 2. Primary production (P, $O_2$ g $m^{-2}d^{-1}$), chlorophyll (Ch, µg $dm^{-3}$) (Zdanowski et al, 1975) and sedimentation (S, g dry weight $m^{-2}d^{-1}$); 1971. Note change of scale for sedimentation in lake Czarna Kuta.

Table III. Comparison of tripton sedimentation in chosen periods of first (1971) and second (1972) year of fertilization.

| Characters | Lake & period | | |
|---|---|---|---|
| | Piecek | | Dgał Mały |
| | VII-VIII | VIII-X | VIII-X |
| g dry w./$m^2$xday | | | |
| 1971 | 2.5 | 12.7 | 6.6 |
| 1972 | 5.1 | 14.3 | 10.3 |
| % C | | | |
| 1971 | 10.2 | 11.4 | 9.0 |
| 1972 | 13.8 | 12.6 | 11.4 |
| % N | | | |
| 1971 | 1.7 | 1.9 | 1.5 |
| 1972 | 2.3 | 2.1 | 1.9 |

Table IV. Comparison of C and N in % of dry weight of sedimenting tripton (Sed.) and bottom deposits (Dep.)

| Lake | C | | N | |
|---|---|---|---|---|
| | Sed. | Dep. | Sed. | Dep. |
| Smolak | 32.9 | 32.2 | 3.8 | 3.0 |
| Piecek | 13.2 | 11.5 | 2.2 | 1.2 |
| Dgał Mały | 10.4 | 8.9 | 1.6 | 1.1 |
| Czarna Kuta | 12.0 | 10.8 | 2.0 | 1.4 |

(Tripton - own data, bottom deposits - after Bnińska & Zdanowski in press.)

## Conclusions and discussion

The amounts of sedimenting tripton in lake Smolak were much lower than in a of the other . It had the highest con tents of C and N both in sedimenting tripton and in bottom deposits. The difference in % contents of C and N between sedimenting tripton and botto deposits was in this lake much smalle than in all other lakes; the % of N was only about 20% higher in the sedimenting tripton than in bottom deposits, while in the other lakes it wa 50 to 80% higher. In lake Czarna Kuta the amount of sedimenting tripton was at least several times higher than in the other lakes and 3.5 times higher than the gross primary production. On the other hand the percentages of C and N in sedimenting tripton in this lake were similar to that in the othe 2 eutrophic lakes (but not to the hum

Table V. Comparison of sedimentation at 5, 10 or 15 m depth.

| | Lake Mikołajskie | | | | Lake Dgał Mały (centre) | | | | | | | |
|---|---|---|---|---|---|---|---|---|---|---|---|---|
| | August | | September | | 7-20 July | | 21 July-19 Aug. | | 31 Aug.-15 Sept. | | 20 March-5 May | |
| Depth | Dr.w. | % N | Dr.w. | % N | Dr.w. | % N | Dr.w. | % N | Dr.w. | % N | Dr.w. | % N |
| near shore | | | | | | | | | | | | |
| 5 | 5.0 | 1.3 | 7.4 | 2.5 | | | | | | | | |
| 10 | 7.1 | 3.8 | 10.0 | 1.8 | | | | | | | | |
| centre | | | | | | | | | | | | |
| 5 | 2.9 | 1.5 | 8.3 | 2.7 | 15.8 | 1.2 | 7.6 | 1.2 | 15.6 | 1.6 | 4.3 | 1.2 |
| 15 | 3.3 | 1.4 | 9.3 | 1.9 | 14.0 | 1.3 | 6.5 | 1.5 | 8.1 | 1.5 | 4.3 | 1.2 |

(Dr.w. = dry weight in $g\ m^{-2}d^{-1}$)

ic lake); also the percentage of C and N in the bottom sediments in this lake was similar to that in the other 2 lakes. The amount of sedimenting tripton during autumn or spring was not high, as one could expect if there was significant resuspension and resedimentation.

Although the interest to the process of sedimentation, and the recognition of its role in the ecosystem functioning is increasing, some essential questions are still to be solved. To what extent is the tripton collected in the sedimentation traps reflecting the real sedimentation process in the water body, or what is the influence of resuspension and resedimentation. These processes seem to occur everywhere (at least in shallow zones) but in different sometimes insignificant degree (Kajak, in press). The second question is the fate of the sedimented material, i.e. the amount returning to the water, and its annual pattern.

Sedimentation values higher than gross primary production, are published by Steele & Baird (1972) and Stephens & Sheldon (1967). This may be the result either from resedimentation or from underestimation of the primary production. Underestimation of primary production seems to be a possibility in lake Czarna Kuta.

High values of sedimentation (such as found in lake Czarna Kuta) can be found also in the literature - e.g. data from meromictic Green lake (Ludlam, 1969) and from a marine bay (up to 3000 g m$^2$ - Stephens & Sheldon, 1967). In some cases, e.g. that of Stephens & Sheldon (1967) they undoubtedly result from resedimentation, but in most cases, the data from traps probably reflect real tendencies of sedimentation.

## References


Bnińska, M. & B. Zdanowski, in press. Chemical composition of bottom sediments in lakes with induced eutrophication. Pol.Arch.Hydrobiol.

Kajak, Z., in press. Sedymentacja tryptonu i miarodajność jej oceny. Wiadomości Ekologiczne.

Ludlam, S.D., 1969. Fayetteville Green Lake, New York. III. The laminated sediments. Limnol.Oceanogr. 14: 848-857.

Ławacz, W., 1969. The characteristics of sinking materials and the formation of bottom deposits in a eutrophic lake. Mitt. int. Ver. Limnol. 17: 319-331.

Steele, J.H. & I.E. Baird, 1972. Sedimentation of organic matter in a Scottish, sea Loch. Mem. Ist. Ital. Idrobiol. 29 suppl.: 73-88.

Stephens, K., R.W. Sheldon & T.R. Parsons, 1967. Seasonal variations in the availability of food for benthos, in a coastal environment. Ecology 40: 852-855.

Zdanowski, B., 1976. The influence of mineral fertilization on phytoplankton production in lakes of various trophic types. Ekol. pol. 24: 167-195.

Zdanowski, B., M. Bnińska, A. Korycka & J. Sosnowska, 1975. The effect of mineral fertilization on ecosystem structure and functioning in lakes of different trophic type. I. Pol. Arch. Hydrobiol. 22: 217-232.



Key words: sedimentation
bottom deposits
lakes
carbon
nitrogen

# SEDIMENTARY ENVIRONMENTS IN THE GREAT LAKES

P.G. Sly

Canada Centre for Inland Waters, Burlington, Ontario, Canada

## Summary

Studies on bottom sediments near Kingston and the Niagara River mouth in Lake Ontario, and in Georgian Bay, have provided a basis for characterising sedimentary and geochemical relationships in both shallow and deep Great Lakes environments.

The distribution of particle size reflects a rapid decrease in dynamic energy with water depth and the dominance of fetch is clearly evident. The distribution and accumulation of fine particulates is strongly influenced by circulatory patterns and by minor variations in bathymetry. Sand-size materials are limited mostly to water depths of 20m or less, coincident with the depth of maximum wave influence.

The geochemistry of 'conservative elements' (Kemp et al., 1976) such as Al, K, Na, and Si, shows strong correlations with depth and reflects particularly the distributions of clay minerals, feldspars, micas, and quartz.

'Enriched elements', for example Hg, show somewhat different relationships; the influence of cultural loading is significant in Kingston and Niagara, whereas background situations are observed in Georgian Bay.

## Introduction

Throughout the Great Lakes there have been severe fluctuations in water level, largely as a result of ice-ponding and isostatic adjustment (Sly and Thomas, 1974). In both Lake Ontario and Georgian Bay water levels have remained at about their present elevation for little more than the past 2,000 years. Because of this and because accumulative rates differ within the lakes, many surface sediments are related to present-day sedimentary conditions or incorporate materials related to previous conditions when lake levels or local morphology may have been different (Sly, 1973a and b).

For comparative purposes, therefore, it is necessary to develop a common approach in different lakes, or in different parts of a lake, for the purpose of defining the existing sedimentological regimes and to point to the occurrence or reworking of previous materials.

Three concepts underlie the following discussions:

1. Sediment distributions are usually dominated by physical processes.
2. Sediment geochemistry is strongly influenced by the character of source materials but may be modified by authigenic or diagenetic processes.
3. In many environments the effects of human development may superimpose changes upon the composition of recent sediments.

Bottom sediment sampling was carried out in three localities in the Great Lakes, each representing different sedimentological environments (Sly, 1969a). The sampling was designed in the form of nested grids with sample spacing ranging between 30m and 3,000m. A Shipek bottom sampler (Sly, 1969b) was used to recover surficial sediments which were sub-sampled for both textural and geochemical analysis (Sly, 1975).

## Sediment-energy relationships

In referring to Lakes Huron and Ontario, Thomas et al., (1973) stated, "Both of these large lake systems demonstrate the control exerted by water depth on energy dissipation and lacustrine sedimentation processes.

Since the energy is derived predominantly from wind stress then initial energy will be a function of mean wind strength and direction related to lake surface area and fetch. It is likely, therefore, that all lake sediment systems will respond as a direct function of total initial energy and water depth".

The significance of these comments may be seen in Figure 1 which shows the relationship between mean particle size in phi units ($\bar{x}\phi$) and percent clay content with depth, for Kingston and Niagara in Lake Ontario and Tobermory in Georgian Bay. ($\phi$ is defined as $^{-}\log_2$ x diameter in mms.) A mean size of 1$\phi$ occurs at a depth of 5m at Niagara, 17m at Kingston, and 20m at Tobermory; conversely a mean size of 8$\phi$ occurs at a depth of 120m at Niagara, 45m at Kingston and 120m at Tobermory. A comparable trend, also, may be seen in percent clay content. These data imply that surface wave action is strongest at Tobermory and least at Niagara (despite the length of fetch at Niagara this site is dominated by offshore winds). The greatest rate of loss of dynamic energy with depth occurs at Kingston. The least rate of loss, at Niagara, suggests strongly that sustained transport of silt and clay particulates is associated with relatively strong current activity.

Data in Figure 2 shows that samples having a mean size less than 4 phi occur in less than 20m of water, these are mostly sands whose distribution is closely related to wave energy (Sly, 1973b). Figure 2 also shows the use of size/depth relationships to resolve anomalies which may be produced by local bathymetric controls or which may be related to the presence of sediments deposited under different conditions at an earlier time.

Plots of standard deviation/mean size (Figure 3) provide an indication of sediment sorting in the central part of their size distribution curve. Similar curves have been developed for many lake sediment analyses (J.-P. Vernet, et al., 1972, Thomas et al., 1973, and Damiani and Thomas 1974), and confirm that best sorting is achieved in the sand fraction and that lesser improvements may be seen in the coarser and finer fractions.

This distribution conforms to that anticipated from the erosion-transportation-deposition regimes as defined by Hjulström (1939), Sundborg (1956) and others. Surface sediments having equivalent size and sorting characteristics may be used, therefore, to show that the integrated dynamic energy levels to which they respond are sensibly the same; and use of this approach has been made by Lewis and Sly (1971), in a study of the Toronto Waterfront, and by Coakley (1972) in studies on Point Pelee in Lake Erie.

The distribution of skewness in Great Lakes sediments has been interpreted by Thomas et al., (1973) as the mixing of silt with the end member size populations of sand and clay, where silt in sand produces a positive skew and silt in clay a negative skew. An approach to zero skewness is suggestive of mixed populations. In Figure 4 skewness/mean size are shown for both Kingston and Tobermory. The similarity of these relationships for mean particle sizes finer than 3 phi is strong, whilst at Kingston the influence of additional gravel materials is made evident by reduced skewness in the coarser fractions. This latter trend is also shown by sediments from the Bay of Quinte (Damiani and Thomas, 1974), where the eventual dominance of gravel leads to increasing positive skewness.

Whilst the significance of mean particle size, standard deviation and skewness may be readily understood, the use of the 4th moment measure, kurtosis, is often less appreciated. Measurements of standard deviation define the dispersion in the central 2/3 of the distribution curve; kurtosis, on the other hand, provides a comparison of the sorting in the tails of the distribution curve, relative to the central portion, and is very sensitive to even minor departures from normal distribution. As stated by Folk and Ward (1957) the introduction of silt into the end member populations of sand or clay (Spencer 1963) results in deterioration in the sorting of the tails of distribution curves, whilst sorting in the central part remains good; this is referred to as a Leptokurtic condition. Mixed population samples, with a higher standard deviation, are typically Platykurtic.

Figure 5 shows kurtosis/mean size data for Niagara sampling; Platykurtic conditions result from mixtures of sand and gravel, and from silts with sand or clay. The sand fraction, however, is strongly Leptokurtic (following previous practice (Thomas et al., 1973) kurtosis measures have been normalized using the SEDAN size analysis program of Coakley and Beal, 1972). This relationship is typical of Great Lakes sediments where gravel and clay (end member) populations also show a trend towards Leptokurtosis.

On the basis of the moment measures, as described, mean size/standard deviation (Figure 3) may be used for a general assessment of sediment-energy relationships. However, two important deficiencies must be accepted: (i) the standard deviation is rather insensitive to variations in the tails of size distribution curves (ii) mean size may retain the same value despite changes in the mixture. As a complement, kurtosis/skewness relationships may be used to refine sediment-energy response in the form of a non-dimensional expression.

In publications by Thomas, et al., (1972, 1973) and by Damiani and Thomas (1974), kurtosis/skewness relationships were separated into four components (ABCD) in order of decreasing energy level. However, as a result of the author's more detailed sampling it has been possible to develop a further refinement, where additional components (EFGH) have been established by separating those samples, from the previous sectors, which contain gravel, Figure 6. These additional components (E-H) therefore likely include both lag deposits (produced by erosion rather than deposition) and the higher energy gravel fractions. The use of the sectors (A-H) as a basis for facies discrimination follows a discussion of the geochemical characteristics of the Niagara, Kingston and Tobermory locations.

## Sediment Geochemistry

Geochemical analysis of major elements provides a means to establish sediment mineral phases (Sly and Thomas, 1974), which may provide an indication of sediment dispersal and may suggest the manner in which trace metals and other components may be bound. Typically, bulk sediment composition of Lake Ontario and Lake Huron (and by analogy, Western Georgian Bay) includes: quartz, feldspar, clay, carbonate, organic matter, hydrated oxides of iron and manganese and sulphide (Sly and Thomas Op. Cit.).

Bearing in mind that the more 'active' geochemical parameters are related to the presence of clay particles, hydroxides (colloids), and particulate organics, examples of selected element concentrations have been related to water depth, in order

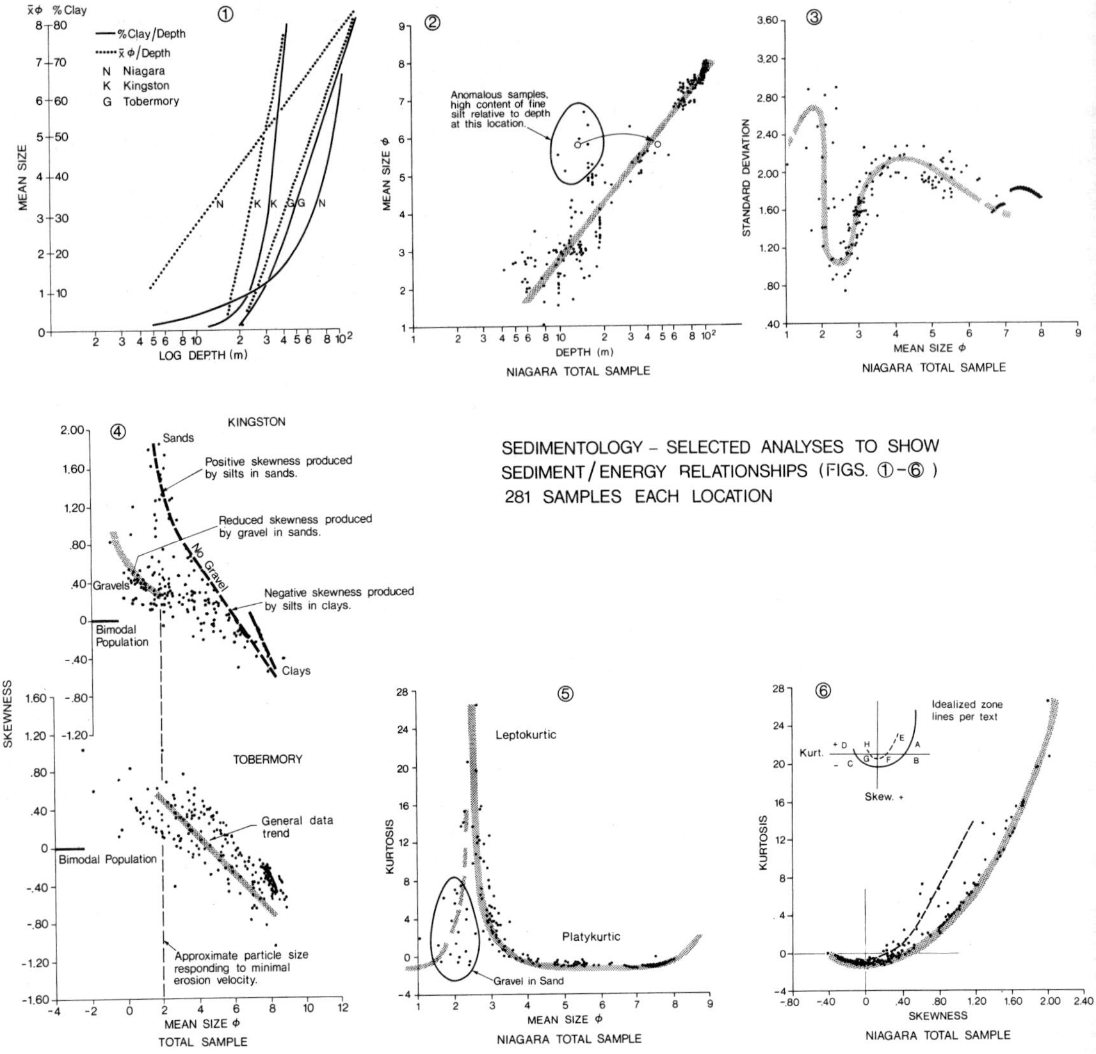
①
x̄φ %Clay
—%Clay/Depth
••••• x̄ φ/Depth
N Niagara
K Kingston
G Tobermory
MEAN SIZE
LOG DEPTH (m)
②
Anomalous samples, high content of fine silt relative to depth at this location.
MEAN SIZE φ
DEPTH (m)
NIAGARA TOTAL SAMPLE
③
STANDARD DEVIATION
MEAN SIZE φ
NIAGARA TOTAL SAMPLE
④
KINGSTON
Sands
Positive skewness produced by silts in sands.
Reduced skewness produced by gravel in sands.
No Gravel
Gravels
Negative skewness produced by silts in clays.
Bimodal Population
Clays
SKEWNESS
TOBERMORY
General data trend
Bimodal Population
Approximate particle size responding to minimal erosion velocity.
MEAN SIZE φ
TOTAL SAMPLE
SEDIMENTOLOGY – SELECTED ANALYSES TO SHOW SEDIMENT/ENERGY RELATIONSHIPS (FIGS. ①-⑥)
281 SAMPLES EACH LOCATION
⑤
Leptokurtic
Platykurtic
Gravel in Sand
KURTOSIS
MEAN SIZE φ
NIAGARA TOTAL SAMPLE
⑥
Idealized zone lines per text
Kurt.
Skew. +
KURTOSIS
SKEWNESS
NIAGARA TOTAL SAMPLE

Fig. 1 - 6

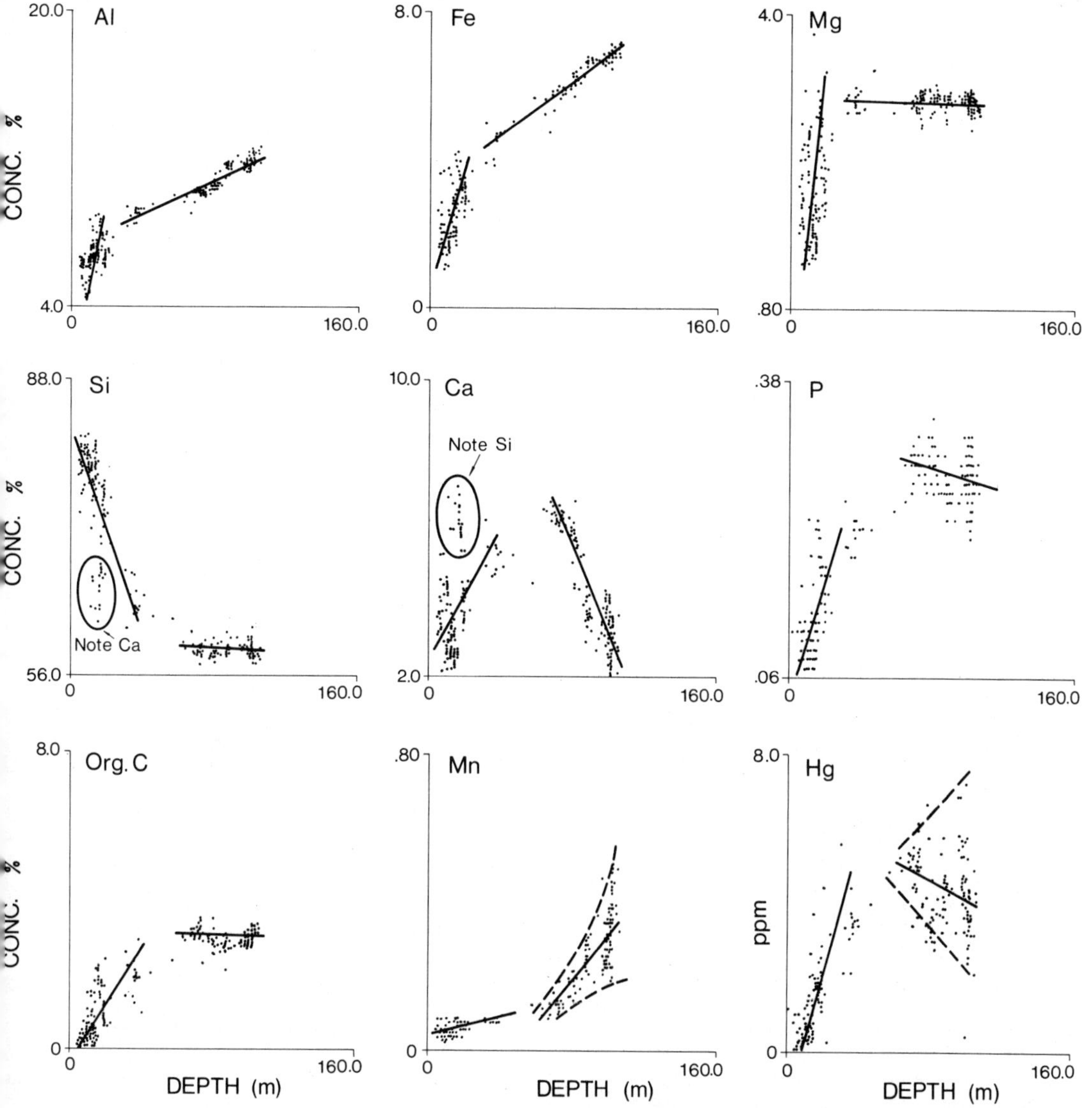


SEDIMENT GEOCHEMISTRY– SELECTED ELEMENTS/DEPTHS
NIAGARA LOCATION

Fig. 7

to provide a simple common reference at each of the sampling locations (Figure 7).

At Niagara, Al, clay and K, typically show an increase in concentration with depth whilst Na and Si, associated with quartz and feldspar in the sand fraction, show a corresponding decrease. An anomaly in the Si content corresponding to 15/20m depth is related to an equivalent increase in Ca content; this effect results from a local 'dilution' of quartz by calcite. In these data plots, while clay and K content show a continuous increase with depth, both Al and Si show two-component trends, for Al the break appears at 15/20m (corresponding to a mean size of about 3 phi) whereas the break for Si occurs at about 40/50m (equivalent to a mean size of nearly 6 phi). Quartz in finer silt and clay sizes appears to decrease only slightly.

The trend of Fe concentration versus depth is similar to that of Al and it has been suggested by Vernet and Thomas (1972 and by Thomas (1973), that the iron is mostly in the form of hydrated oxide which, with complexed inorganic P, occurs as coatings on individual clay particles.

The trend for Organic-C is opposite to that of Si with, again, the two-component break between 40/50m; it shows a strong affinity with the silt and clay sized fractions. Although the clay minerals of Lakes Ontario and Huron (Sly and Thomas 1974) are mainly composed of illite and chlorite (K and Mg) the Niagara sediment concentrations of K and Mg do not correspond. The concentration trend of Mg shows a distinct break at 15/20m (as does Fe and Al) and concentrations at greater depth, in the silt and clay fractions appear constant. Mg in the coarser fractions like Ca, is probably related to carbonate mineralogy.

At Niagara, Ca is strongly associated with the carbonate phase and similarity between concentration trends of Ca and S strongly suggests that S is associated with the sulphate phase. The presence of local gypsum sources also supports this. The concentrations of both Ca and S reach maximum values at 40/50m and correspond to sediments with a mean particle size of about 6 phi, typical of coarser silts.

The concentration trend for P, at Niagara, departs from the more widely observed conditions as noted by Sly and Thomas (1974). In shallow water P/Fe relationships remain similar to depths of 15/20m. Between this depth and 40/50m, P content continues to increase but with considerable scatter. In deeper waters P is unrelated to clay mineral content but appears closely related to Organic-C. Presumably the scatter between 20 and 50m is related to mixed mineralogical and organic forms.

Mn and Hg show interesting concentration trends with depth, and typically show a two-component break at 40/50m. Both show a remarkable increase in the scatter of data points in deep water. Mn concentrations show little increase in coarser sand and silt fractions but the mean of data scatter in deep water appears to follow increasing clay content. The mobility of Mn under variable Eh/pH conditions may account for such data scatter. The mean of Hg data points appears to follow, more closely, the trend for organic carbon; the reasons for data scatter in deep water are not clear. Much of the Hg is related to various industrial and urban inputs and some variation may relate to differences in loadings. However, the scatter appears to increase with depth and, perhaps on this basis, other post-depositional changes may be in effect.

Sample data from Kingston is similar to that at Niagara though with important differences. At Kingston no marked variation is seen in the trends of clay and Organic-C, and strong correlations between silt and Ca suggest the presence of precipitated calcite in this fraction. Fe, Mn and P concentrations all show strong similarities which are correlated to the presence of surface nodules. Ca and S show no correlation, and S is probably associated with the sulphide phase.

Tobermory samples show more geochemical similarity with Kingston than Niagara though again, notable differences mark this location as distinct. At Tobermory the presence of glacial materials induces anomalous relationships in respect to clays, K, and Mg and Ca carbonates. High Mn concentrations are associated with nodules and coatings but show little affinity with Fe or P. Hg levels are generally low suggesting a lack of 'cultural impact', but a scatter of high concentration values may be related to the effects of local mineralization (sphalerite deposits).

In a study to establish the effects on the geochemistry of sediments by 'cultural impact', Kemp and Thomas (1976) compared compositional data above and below pre-industrial time horizons from Lake Erie cores. By using a sediment enrichment facto (SEF) based upon an Al standard the authors were able to show compositional groupings having similar characteristics.

$$SEF = \frac{\frac{E_s}{Al_s} - \frac{E_a}{Al_a}}{\frac{E_a}{Al_a}}$$

Where E is the observed elemental concentration, and Al is the observed aluminium concentration.
Subscript s is surface value and subscript a is value below *Ambrosia* time horizon.

Such groupings, applied to the data reported herein, would mostly include: Al, K, Mg, Na and Si as conservative elements, Hg as an enriched element, Organic-C and P as nutrient elements, Ca as a carbonate element, and Fe, Mn and S as mobile elements (at Niagara, S linked with sulphates would be comparable to the carbonates; also, in some areas, Mg would be included in the carbonates).

Such groups provide a useful means of comparison for geochemical characteristics of recent sediments, between different localities and environments.

## Sediment facies

The term facies is used by geologists to express a form of common resemblence or aspect relating to lithological, geochemical and, or, biological content of sediments. For example, a simple use of the term could be to distinguish the distribution of sands silts and clays, as various litho-facies.

On the basis of data described from Niagara, Kingston and Tobermory localities, however, it is proposed that a more refined approach may provide a mechanism by which lithological and geochemical characteristics are combined.

In the localities discussed, a primary discrimination based upon physical characteristics would seem most appropriate.

Table I. Data/zone relationships (ABCD).

| Location | | A | B | C | D |
|---|---|---|---|---|---|
| Niagara | mean of | 11.3 | 33.7 | 88 | - |
| Kingston | mean | 22.4 | 29.9 | 36 | 33.5 |
| Tobermory | depths(m) | 35.8 | 42.9 | 95.2 | 102.4 |
| Niagara | mean of stan- | 1.6 | 2.0 | 1.8 | - |
| Kingston | dard devia- | 2.1 | 2.7 | 2.2 | 2.4 |
| Tobermory | tion(size $\phi$) | 2.4 | 3.0 | 2.3 | 2.0 |
| Niagara | mean of | 2.9 | 5.3 | 7.4 | - |
| Kingston | mean | 3.0 | 5.2 | 6.5 | 6.0 |
| Tobermory | size $\phi$ | 3.0 | 4.4 | 7.1 | 7.4 |

Table I shows the values for the mean of mean depths (m), the mean of standard deviations and the mean of mean size ($\phi$) for each of the localities, based upon the four zone separations (ABCD) of skewness/kurtosis relationships used by Thomas et al., (1973). Although trends are established by this data treatment, continuity within such data sets is not consistent. Instead, it is proposed to use the zonal separation of skewness/kurtosis referred to earlier in this work, and defined by the identifiers A-H. For simplicity, Table II shows only a mean size ($\phi$) data treated in this manner, in which the zones have been generally grouped according to particle size.

Table II. Data/zone relationships (A-H)

| Location | E | A | F | H | B | G | C | D |
|---|---|---|---|---|---|---|---|---|
| (in approx. increasing order of $\phi$ value) | | | | | | | | |
| Niagara | 2.5 | 2.9 | 3.5 | - | 5.5 | 5.8 | 7.5 | - |
| Kingston | 1.9 | 3.4 | 3.8 | 5.0 | 5.6 | 5.6 | 7.3 | 8.5 |
| Tobermory | 2.1 | - | 3.2 | 6.5 | 5.3 | 5.9 | 7.8 | 8.3 |

All values quoted as mean $\phi$ size

Two important observations relate to this data: (i) although mean size usually shows a significant change from one zone to the next, this is not always true; a facies distinction based on mean size could miss subtle changes; (ii) trends in mean size for zones B,G,C, and D are similar at all three locations and suggest that sedimentary regimes are equivalent.

While these zonal groupings may reflect useful information about the physical regimes of sedimentary processes do they reflect geochemical composition in a useful way? To test this, a secondary discrimination of sediments using geochemical characteristics grouped as per Kemp et al., (1976) and Kemp and Thomas (1976) is shown in Table III.

Table III. Zonal relationships with geochemistry - Niagara.

| Zone | Conservative | | | | | Mobile | | Carb | Nut | | Enr |
|---|---|---|---|---|---|---|---|---|---|---|---|
| | Al | K | Mg | Na | Si | Fe | Mn | Ca | P | Org C | Hg |
| E | 5.9 | 1.7 | 2.4 | 1.2 | 77 | 2.4 | 0.1 | 4.0 | 0.1 | 0.7 | 1.0 |
| A | 6.6 | 1.9 | 1.7 | 1.3 | 78 | 2.4 | 0.1 | 3.0 | 0.1 | 0.4 | 0.7 |
| F | 6.8 | 1.8 | 2.5 | 1.2 | 74 | 2.8 | 0.1 | 4.4 | 0.1 | 1.1 | 1.6 |
| H | no samples | | | | | | | | | | |
| B | 8.8 | 2.3 | 2.9 | 1.2 | 65 | 4.4 | 0.1 | 5.6 | 0.2 | 2.1 | 3.1 |
| G | 9.2 | 2.4 | 3.1 | 1.1 | 63 | 4.6 | 0.1 | 5.8 | 0.2 | 2.4 | 3.1 |
| C | 11 | 3.1 | 3.1 | 1.0 | 59 | 6.5 | 0.3 | 4.2 | 0.3 | 3.0 | 4.3 |
| D | no samples | | | | | | | | | | |

Carb = carbonate; Nut = nutrient; Enr = enriched

Note: Major elements by XRF analysis quoted in % oxide. Hg by flameless AAS quoted in ppm. Values > 10 rounded to whole number.

Unquestionably the element concentration/depth (size) relationships, as discussed previously, are well reflected by this tabulated data. In fact, additional resolution in the coarser particle size is demonstrated in the data for Mg and Ca, and Si, where both Ca and Mg (carbonates) appear to dilute the Si (quartz) content in the (coarsest mean size) E zone.

Detailed discussions of sediment-energy relationships and associated geochemistry will appear, elsewhere, in print. However, as a guide to the implications of these (A-H) zonal separations it is suggested that zones E and A represent high energy regimes in which relatively coarse materials are subject to both erosion and deposition. Zones B, C and D are mostly depositional; zones H, G and F include lag deposits, reworked or otherwise modified materials.


Acknowledgements

I would like to acknowledge the interesting and constructive discussions with my colleagues; in particular Drs. A.L.W. Kemp, N.A. Rukavina and R.L. Thomas.

I am also greatly indebted to Mr. R.G. Sandilands without whose assistance it would have been impossible to reduce the vast quantities of original data into manageable form.

The Director, Canada Centre for Inland Waters, is acknowledged for permission to publish.


References


Coakley, J.P., 1972. Nearshore sediment studies in western Lake Erie. Proc. 15th Conf. Great Lakes Res. Internat. Assoc. Great Lakes Res. 330-343.

Coakley, J.P. and G.S. Beal, 1972. SEDAN - a computer program for sediment particle size analysis. Inland Waters Branch (DOE) Report Series #20. Ottawa. 33 p.

Damiani, V. and R.L. Thomas, 1974. The surficial sediments of the Big Bay Section of the Bay of Quinte, Lake Ontario. Can. Jour. Earth Sci. 11: 1562-1576.

Folk, R.L. and W.C. Ward, 1957. Brazos River Bar: A study in the significance of grain size parameters. Jour. Sediment. Petrol. 27: 3-26.

Hjulström,F.,1939.Transportation of detritus by moving water. In Recent Marine Sediments, a Symposium. Ed. P.D. Trask, Tulsa.

Kemp, A.L.W. and R.L. Thomas, 1976. Cultural impact on the geochemistry of the sediments of Lakes Ontario, Erie and Huron. Geoscience Canada 3: 191-207.

Kemp, A.L.W., R.L. Thomas, C.I. Dell and J.-M. Jaquet, 1976. Cultural impact on the geochemistry of sediments in Lake Erie. Jour. Fish. Res. Board Can. 33: 440-462.

Lewis, C.F.M. and P.G. Sly, 1971. Seismic profiling and geology of the Toronto Waterfront area of Lake Ontario. Proc. 14th Conf. Great Lakes Res. Internat.Assoc. Great Lakes Res. 303-354.

Sly, P.G., 1969a. Sedimentological studies in the Niagara area of Lake Ontario, and in the area immediately north of the Bruce Peninsula in Georgian Bay. Proc. 12th Conf. Great Lakes Res., Internat. Assoc. Great Lakes Res. 341-346.

Sly, P.G., 1969b. Bottom sediment sampling. Proc. 12th Conf. Great Lakes Res., Internat. Assoc. Great Lakes Res. 883-898.

Sly, P.G., 1973a. Sediment processes in Great Lakes. In Fluvial Processes and Sedimentation. Proc. of Hydrology Symp. Univ. Alberta, Edmonton. Pub. Nat. Res. Council Can., Ottawa. 464-492.

Sly, P.G., 1973b. The significance of sediment deposits in large lakes and their energy relationships. Proc. Symp. Hydrology of Lakes, Helsinki. IAHS-AISH pub #109. 383-396.

Sly, P.G., 1975. Statistical evaluation of recent sediment geochemical sampling. Proc. IX Internat. Congress of Sedimentology, Nice.

Sly, P.G. and R.L. Thomas, 1974. Review of geological research as it relates to an understanding of Great Lakes limnology. Jour. Fish. Res. Board Can. 31:795-825.

Spencer, D.W., 1963. The interpretation of grain-size distribution curves of clastic sediments. Jour. Sediment. Petrol. 24: 151-158.

Sundborg, A., 1956. The River Klarälven, a study of fluvial processes. Geog. Annaler XXXVIII-2: 127-316.

Thomas, R.L., 1973. The distribution of mercury in the surficial sediments of Lake Huron. Can. Jour. Earth Sci. 10: 194-204.

Thomas, R.L., A.L.W. Kemp and C.F.M. Lewis, 1973. The surficial sediments of Lake Huron. Can. Jour. Earth Sci. 10:226-271.

Vernet, J.P. and R.L. Thomas, 1972. The occurrence and distribution of mercury in the sediments of the Petit Lac (Western Lake Geneva). Eclogae Geol. Helv. 65: 307-316.

Vernet, J.-P., R.L. Thomas, J.-M. Jaquet and R. Friedli, 1972. Texture of the sediments of the Petit Lac (Western Lake Geneva). Eclogae Geol. Helv. 65: 591-610.



Key words: Lakes
Sediments
Sedimentology
Geochemistry
Facies

# TRACE METAL COMPOSITION AND FRACTIONATION OF Mn, Fe, S, P, Ba and Si IN THE BAY OF QUINTE FRESH WATER FERRO-MANGANESE CONCRETIONS, LAKE ONTARIO

V. Damiani[1], A. Ferrario[2], G. Gavelli[1] and R. L. Thomas[3]

Istituto Richerche, "G. Donegani" (Montedison), Novara, Italia[1]

Istituto di Mineralogia, Universita di Milano, Milano, Italia[2]

Canada Centre for Inland Waters, Burlington, Ontario Canada[3]

## Abstract

Nodules recovered from seven locations in the Bay of Quinte were analysed for mineralogical and chemical properties. The distribution of elements within nodules was further analysed by microprobe. The nodules occurred on topographic highs downstream of depositional basins from which the soluble Fe, Mn and associated elements were derived. Nodules were structured in alternating iron and manganese-rich concentric bands around a central nucleating clast and were composed of goethite, lepidocrocite, psilomelane and rhodochrosite. Small nodules were found to be iron-rich compared to larger concretions due to changing conditions in the Bay of Quinte relative to the period of nodule growth. The Bay of Quinte nodules proved to be low in trace metals when compared to concretions from the open waters on the north shore of Lake Ontario.

## Introduction

A number of ferro-manganese concretion locations were found in the Bay of Quinte, Lake Ontario during a sediment survey of the region carried out in 1972 and 1973. A description and the chemical composition of some of these nodules recovered from the Big Bay Section of the Bay of Quinte was previously reported by Damiani et al. (1973).

The occurrence of ferro-manganese concretions in fresh water lakes in Canada has been well documented by Kindle (1932, 1936). The first finding of such deposits in the Great Lakes was made in Lake Michigan by Rossmann and Callender (1968). Since that time, a number of publications have appeared on the occurrence, mineralogy and chemistry of the Lake Michigan nodules (e.g., Moore and Meyer, 1969; Edgington and Callender, 1970; Rossmann et al., 1972).

Ferro-manganese concretions were subsequently described in Lake Ontario by Cronan and Thomas (1970 and 1972). Sly and Thomas (1975) described the general distribution and occurrence of known ferro-manganese coated sands and nodules for the entire Great Lakes system. The objective of the present study is to investigate in depth the mineralogy and chemical composition of the nodules from the Bay of Quinte.

## Methods

Sediment samples and associated ferro-manganese nodules were collected by Shipek sampler as described by Thomas et al. (1972). Chemical analyses of the nodules were carried out by A.A.S. (Techtron AA-5) for Fe, Mn, Ca, Mg, K, Sr, Be, Cd, Cr, Cu, Pb, Ni, V, and Zn after digestion in concentrated HCl. Selected nodules were mounted in "epoxy-resin", cut and polished and used firstly for reflected light ore microscopy and secondly for microprobe analysis. For microprobe analysis, the specimens were coated in high purity aluminium to avoid "charge-up" and beam damage. A Jeol-50A C.C.S. instrument was used with operating conditions of 25 KeV and 30-80 μA absorbed current. Two techniques were used for the microprobe analysis:

### Wavelength Scanning

At selected points on the electron micrographs, semi-quantitative X-ray analysis was carried out. Computer-controlled wavelength scanning enabled rapid selection of peaks above background level for subsequent automatic counting. Count comparison was used to evaluate compositional characteristics both within and between nodules.

### Electron Beam Scanning

Element distribution maps of selected areas of nodule sections were prepared at 300X magnification to investigate compositional variations within nodules and to assess elemental co-variance. Exposure times ranged from 250 to 1000 seconds with element detection limits on the X-ray maps ranging between 300 and 1000 ppm. Elements occurring above background levels and analysed by the two methods above included Fe, Mn, Ba, P, S, K and Mg.

## Location and Occurrence

The location of the ferro-manganese nodule

samples reported here, other than the Brothers' Island nodules, are shown in Fig. 1. The mode of occurrence of the nodules recovered in Big Bay [P-28 (Alpha-numeric designation of sediment sampling location based on a grid sampling pattern defined by Damiani and Thomas (1974)) and Q-29], Telegraph Narrows (S-41) and Brothers' Island was reported by Damiani et al. (1973). The locations for the additional nodule sites reported here may be summarized briefly as follows:

Green Point:

T-54, Q-53, located in 7.0 and 14.5 m of water respectively. T-54 nodules occurred in association with a lag sand with shells over glacio-lacustrine clay whereas Q-53 nodules occurred directly on a glacio-lacustrine clay surface. The nodules at these locations are small ranging from a few mm to 1.0 cm. They are generally spherical in nature but are not well formed.

Mallory Bay:

Station BB-51 in 3.0 m of water. The nodules occur embedded in a glacio-lacustrine surface. The nodules are flattened, poorly formed, discoidal ranging up to 3.0 cm in size.

Glenora Ferry:

Stations GG-52 and HH-53 in 23.0 and 15.0 m of water, respectively. This location is, without question, the most abundant though restricted nodule location discovered in the Great Lakes to date. At GG-52, Shipek grab samples indicated that the thickness of the deposit is in excess of 10 cm. The interstices between nodules at this deposit contained recent silty clays and the substrate was not observed. At HH-53, the substrate consisted of a lag sand over glacio-lacustrine clay. The physical geological characteristics of the Glenora Ferry location is well shown in the echogram given in Fig. 2. The central portion of the echogram shows a topographic high composed of eroded compact material, probably glacial till veneered with glacio-lacustrine clay, flanked on either side by accumulations of recent muds. The nodules occur on the topographic high devoid of modern deposits.

The Glenora Ferry nodules are generally of the spherical type ranging up to 5 cm in size.

Adolphus Reach:

Station II-60 in 22 m of water. The nodules are similar in character to the Glenora Ferry nodules and occur on an eroded surface of glacio-lacustrine clay.

Indian Point:

Station I-89 in 11 m of water; nodules occur in association with a sandy gravel overlying glacial till. The nodules were sparse, small, poorly formed, ranging up to 1.0 cm in size.

Fig. 1, in addition to showing the nodule location, illustrates the distribution of sediment types. In all cases, it can be seen that the ferro-manganese nodules occur in regions characterized by glacio-lacustrine clays and glacial tills. This supports the general hypothesis for the formation of fresh water ferro-manganese deposits in regions of non-deposition of modern sediment material (e.g., Sly and Thomas, 1975). It is further felt that the occurrence of the nodules on non-depositional "highs" downstream of depositional basins is highly significant in terms of the source of iron and manganese. As postulated by Cronan and Thomas (1973) and discussed in more depth by Sly and Thomas (1975), it is felt that the iron and manganese are released to the bottom water from fine-grained reducing mud deposits. These enriched bottom waters are subsequently moved downstream across the bathymetric highs where, under the correct conditions (Eh and pH) and the availability of nucleation sites, the iron and manganese precipitates to form discrete concentrations.

## Mineralogy

Damiani et al. (1973) have reported some preliminary findings on the mineralogy of the Big Bay and Telegraph Narrows nodules. Goethite was recorded in the X-ray diffractograms for all the nodules examined. Rhodochrosite was identified in a single nodule from station P-28. The authors have subsequently noted the occurrence of rhodochrosite in Lake Michigan nodules reported by Rossmann (1973). In addition to rhodochrosite, Rossmann (1973) reported the presence of goethite, todorokite, birnessite and psilomelane in the Lake Michigan nodules.

## Petrography

Reflected light microscopic examination of the Glenora Ferry nodules enabled the identification of internal structure and the mineralogy of inclusions or clasts occurring predominantly in the central part of the nodules. The most common structure is in the form of concentric, alternating layers varying in thickness from a few microns to many millimeters. The concentric layering generally appears to form around a core nucleus of one or more clasts. The following minerals were identified by

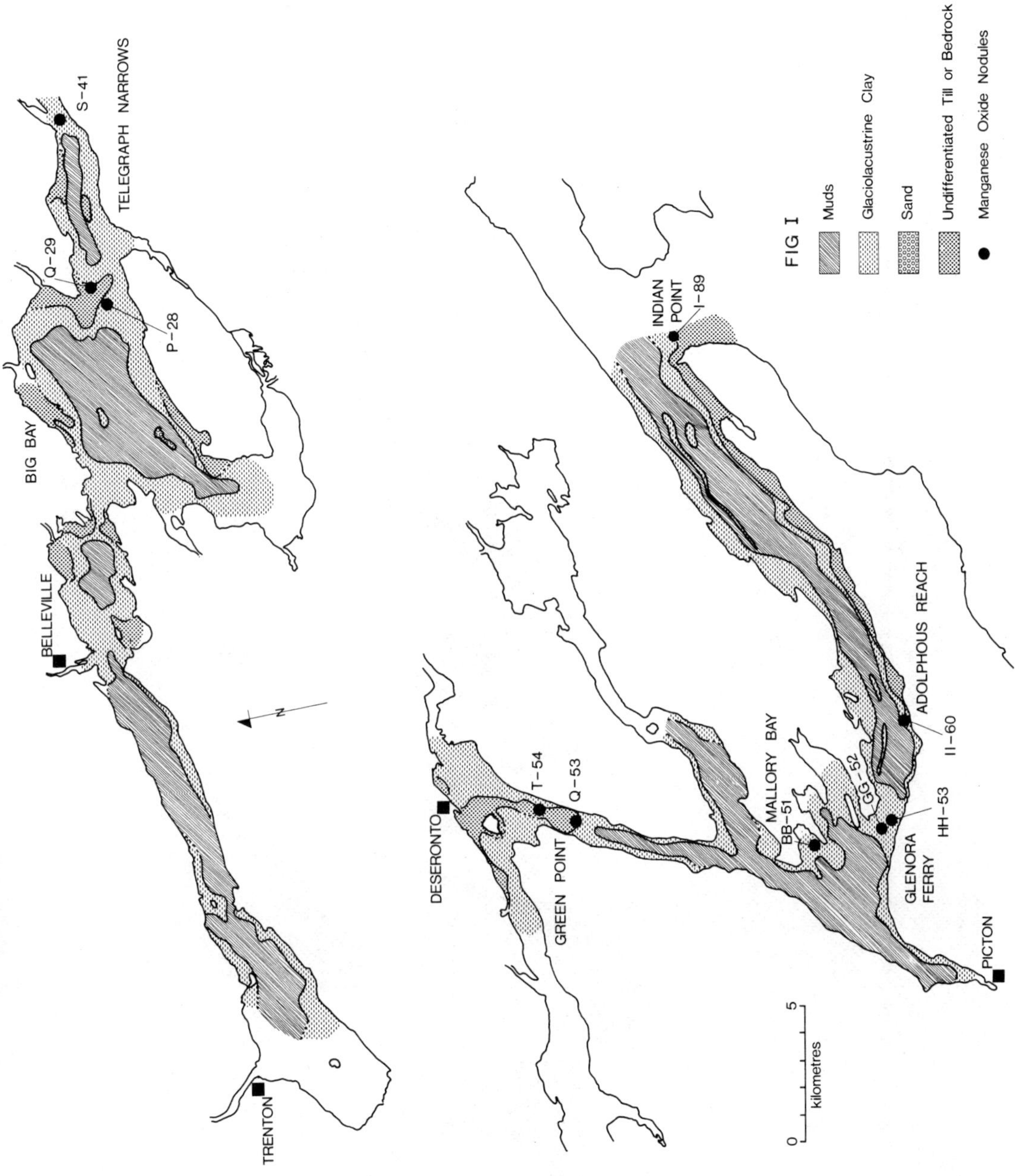


Figure 1. Locations of Fe/Mn nodules and sediment distribution.

Figure 2. Physical geological characteristics of the Glenora Ferry location.

microscopic examination of the major composition of the layers:

Goethite: reflection pleochroism with colour ranging from brown to brownish-grey, variable internal reflections usually yellowish-red in colour which may on occasion mask the anisotropic effects.

Lepidocrocite: grey metallic colour with stronger pleochroism than goethite. Internal reflection weaker with readily distinguishable anistotropic effects.

Psilomelane: clearly recognized by microscope and subsequently confirmed by microprobe. In comparison to goethite, pleochroism is less pronounced and of grey colouration. Strongly anisotropic with no internal reflections.

The clasts contained within the nodules (Fig. 3B, C, D) show considerable variability in size and shape and are angular or sub-angular in character. Two types of clasts were recognized, terrigenous silicates and carbonates and metallic minerals. Of the former, quartz, carbonate (mainly calcite with subsidiary dolomite) and feldspars were recognized. The metallic mineral clasts included ilmenite, rutile, hematite, pyrrhotyte with pentlandite exsolutions and complex minerals of nickel. A mineral of the sperrilyte type ($PtAs_2$) was also detected. The general disposition of the clasts within the nodule matrix indicated that the quartz, carbonates and feldspars formed the nodule nucleating clasts whereas the metallic clasts appeared to be incorporated during nodule growth.

X-ray diffraction

Powdered nodules showed poor crystallinity which, however, was considerably improved by pre-heating at 800°C for four hours. Diffraction peaks for goethite, psilomelane and possibly todorokite were recognized.

Electron Microprobe

Element distributions within the Glenora nodules are illustrated in Fig. 4. 4A shows a scanning electron micrograph of the nodule section at the standard 300X magnification used here. This section shows a number of dehydration cracks and a variation in colour related to composition and structure. Fig. 4B and C show the distribution of Mn and Ba. The distribution of these elements parallels the distribution of the lighter-coloured areas of the electron micrograph (Fig. 4A). When compared to the iron distributions in Fig. 4D, the alternating Fe and Mn concentric nature of the nodules is confirmed. The obvious co-variance of Mn and Ba further substantiates the occurrence of psilomelane in these nodules. This mineral has previously been recorded in Lake Michigan by Rossmann (1973) and in marine nodules by Price and Calvert (1971). Psilomelane has the composition $(Ba, H_2O)_2Mn_5O_{10}$ and may frequently contain significant amounts of adsorbed metals, Cu, Co, Ni, Mg, Ag, Ca, V. At the levels of detection, no element other than Ba was observed to co-vary with Mn. Fig. 4D, E and F show the distribution of Fe, Si and S, respectively. An obvious co-variance of Fe and Si can be observed. S values are low in that they are close to the detection limits but tend to show a relationship with Fe. Rossmann (1973) quotes Si values for goethite from the Lake Michigan nodules. He showed that Si ranged from 5 to 7% by weight and had a correlation coefficient with Fe of 0.93.

Rossmann (1973), again using microprobe analysis, observed a correlation of (r=0.68) between Fe in goethite and P. This relationship is confirmed in our analyses and the distribution of P relative to Fe can be observed in Fig. 5A and B in a different nodule to that given in Fig. 4. Fig. 5C and D shows the distribution of K and Si in the same section as given in Fig. 5A and B. The relationship of Si to Fe is again obvious but additional small patches of high silica can be seen. These represent the clasts, the distribution of which can be compared to the distribution of K in Fig. 5C. Many of the clasts are composed of K and Si and represent inclusions of potassic feldspars. Many of the clasts, however, are composed only of silica and represent quartz inclusions.

Three locations for semi-quantitative microprobe analysis are shown on Fig. 4A and identified numerically. Location 1 represents a high manganese location, 2 a high iron location and 3 a location in which neither iron nor manganese appear to be significantly high based upon the elemental distribution patterns. The semi-quantitative analysis expressed as total counts are summarized in Table 1. Location 1 shows high Mn and Ba with low Si, Ca and K. This confirms the Mn, Ba relationship associated with psilomelane. Location 2 shows high Fe and Si, moderate Mn and Ca with low but significant counts for S and P. The Fe and Si appear to represent the significant relationship with respect to goethite with adsorbed Mn and P. The form of the S is unknown and the higher Ca probably represents the presence of some calcite. Location 3 is high in Si, Ca and Mg with moderate Fe and low Mn. The mineralogy of this location is unknown but the high Si, Mg and Ca may well indicate a complex mixture of quartz, calcite and dolomite.

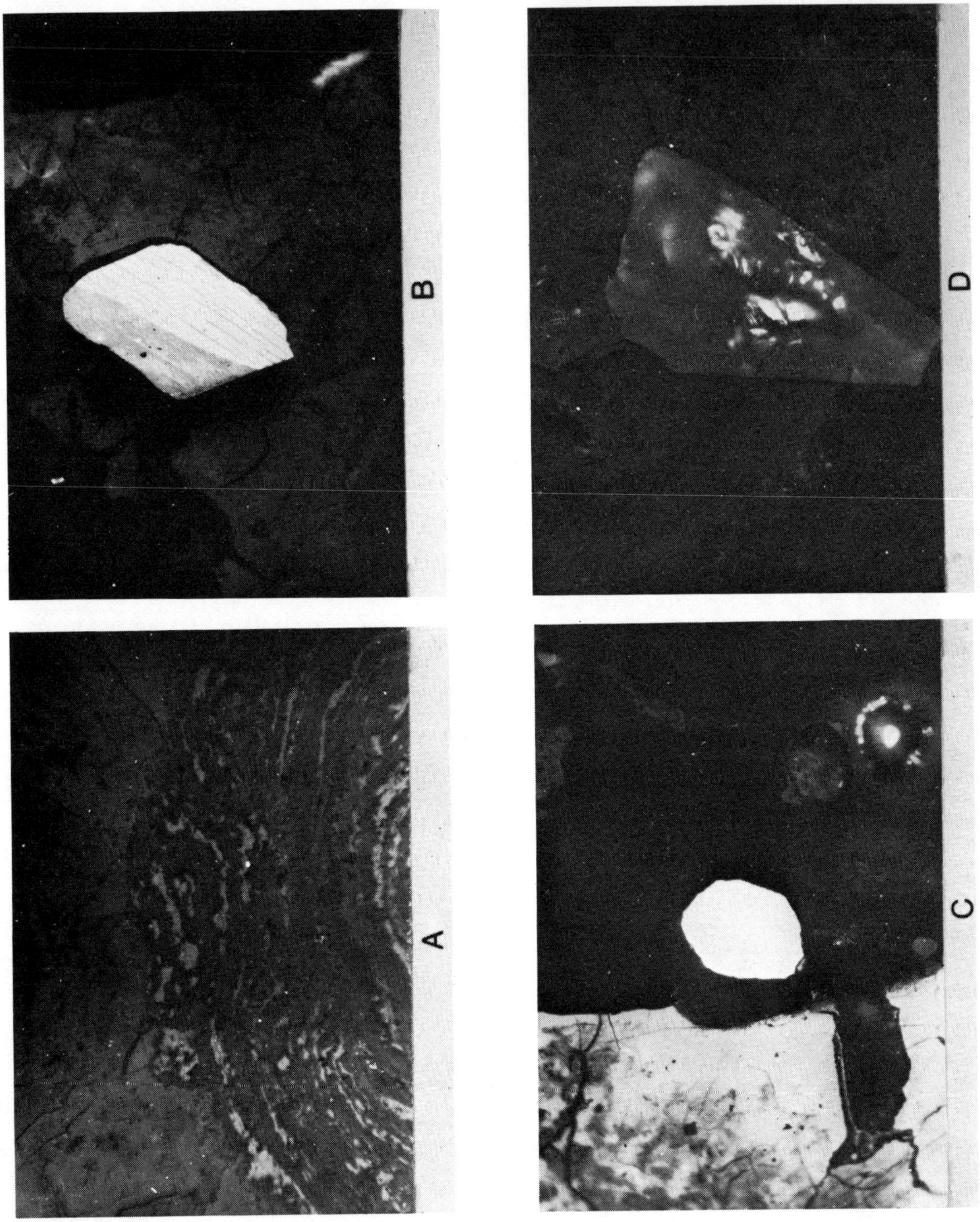


Figure 3. Reflected light micrographs. A, Internal structure. B,C,D, individual clasts.

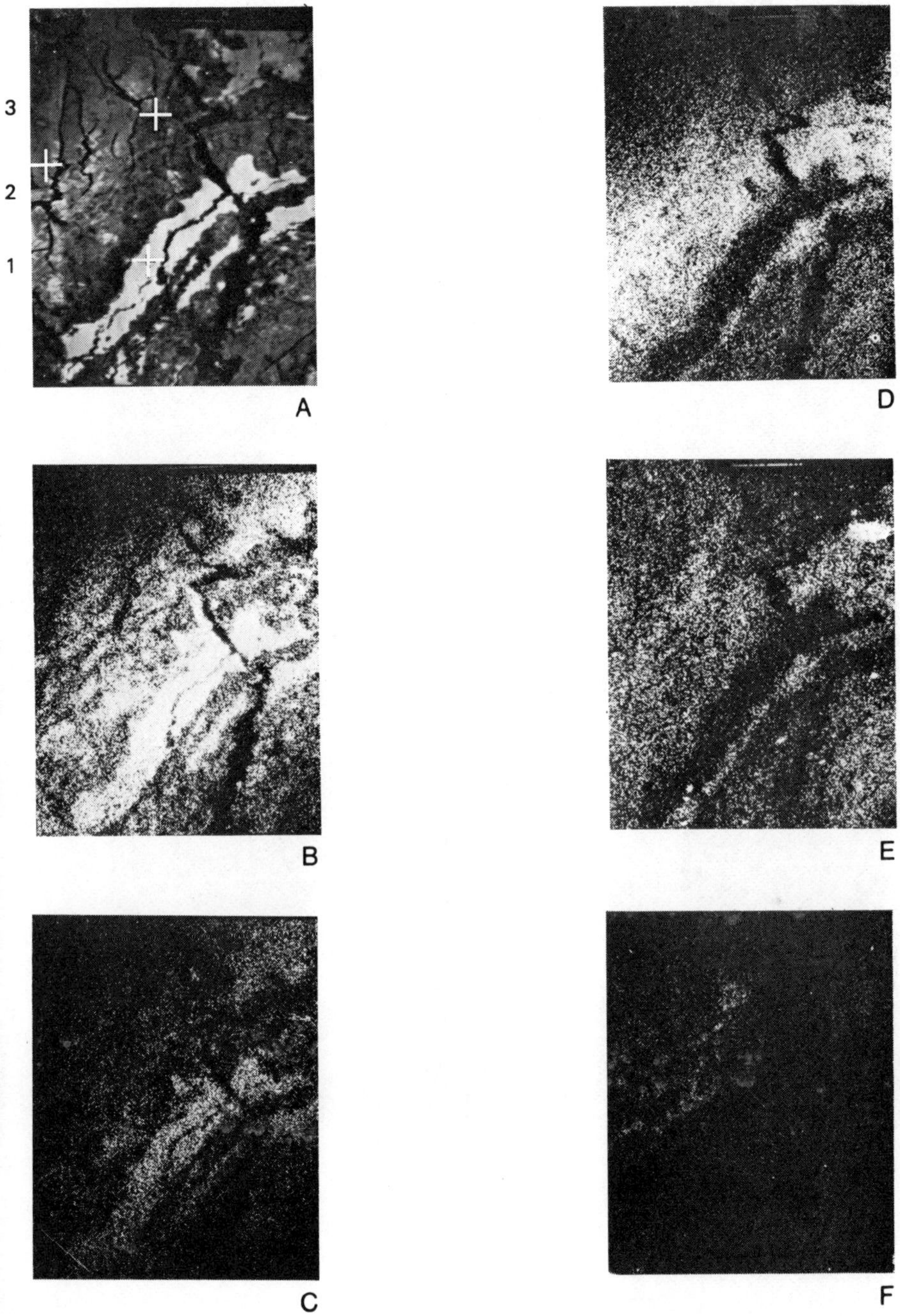


Figure 4. Element distribution by microprobe. A, electron micrograph; B, Mn distribution; C, Ba distribution; D, Fe distribution; E, Si distribution; F, Sulphur distribution.

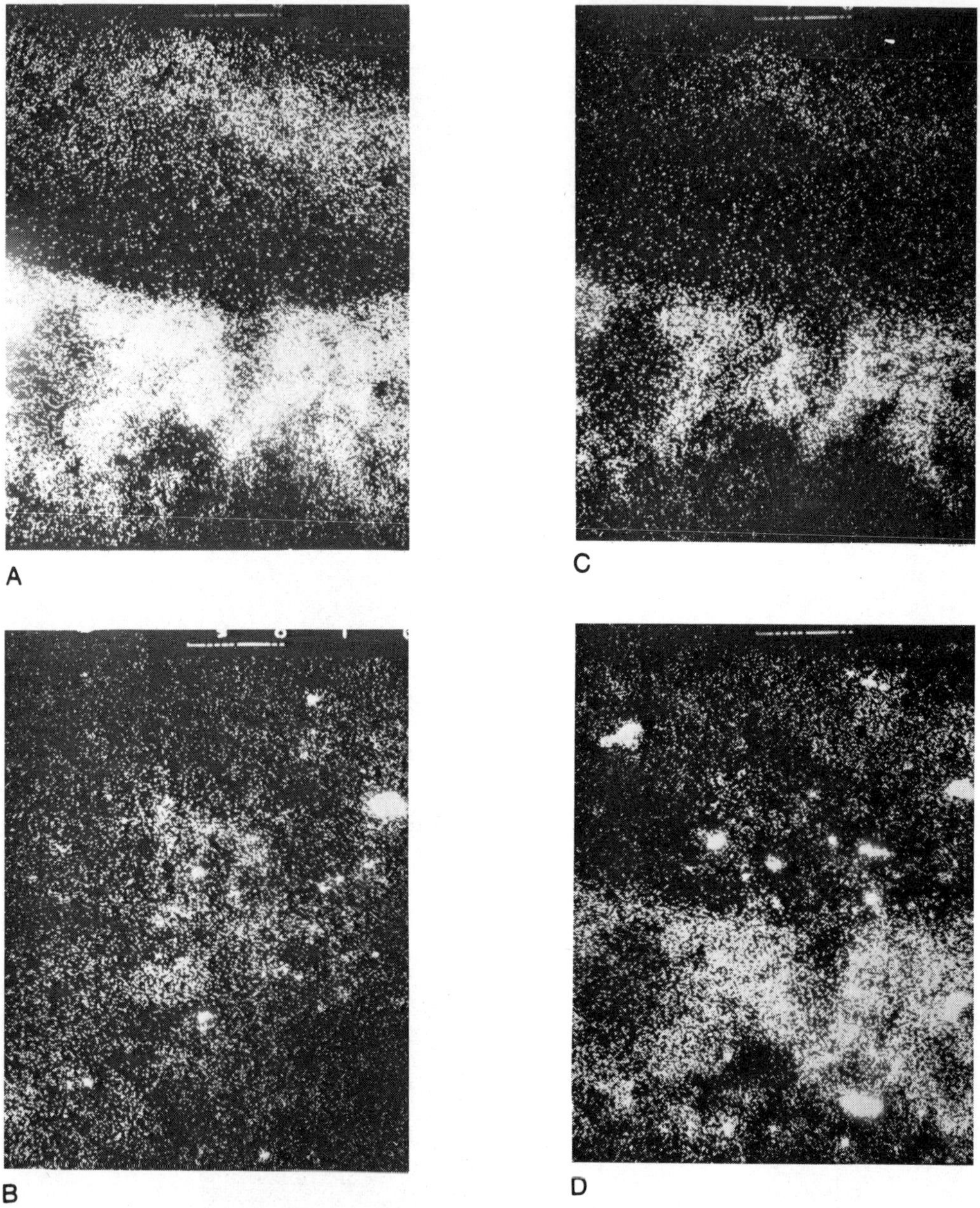


Figure 5. Element distribution by electron microprobe.
A, Fe distribution; B, P distribution; C, K distribution, D, Si distribution.

Table 1. Semi-Quantitative microprobe analysis for 3 locations in nodule S-41 from Telegraph Narrows, Bay of Quinte. Analysis values given as total counts.

| Element | Location | | |
|---|---|---|---|
| | 1 | 2 | 3 |
| Fe | 234 | 9,488 | 1,225 |
| Mn | 10,291 | 1,404 | 90 |
| Ba | 5,873 | 267 | Background |
| Si | 144 | 2,156 | 8,418 |
| Ca | 684 | 1,335 | 12,496 |
| K | 70 | 47 | Background |
| Mg | Background | Background | 2,925 |
| S | 11 | 19 | Background |
| P | 11 | 42 | Background |

## Chemical Composition

The average compositions of nodules from all known locations in the Bay of Quinte are summarized in Table 2. This table, in addition, presents analyses of concretions and ferro-manganese coated sands from the open waters of Lake Ontario (Cronan and Thomas, 1970, 1972).

Fe and Mn values show considerable variation in concentration resulting in variable Fe/Mn ratios. Average Bay of Quinte levels for Fe are comparable to Lake Ontario whereas the average Mn concentration is considerably lower. From the Fe/Mn ratios given in Table 2, it can be stated that, in general, the Bay of Quinte nodules are iron-rich. However, large nodules in the range of 1-5 cm size at both Glenora Ferry and Adolphus Reach are of a manganese-rich type, whereas nodules <1.0 cm are iron-rich. Assuming that size of nodule is related to time of formation, the composition of the younger nodules infer a change in depositional conditions which favour the precipitation of iron relative to manganese. This could well be related to the historical trend for increasing eutrophic conditions in the Bay of Quinte relative to increased nutrient loadings due to man's activities in the watershed (Johnson and Owen, 1971). This could result in decreasing redox potentials at the sediment-water interface to a level which effectively maintains manganese in solution resulting in the formation of nodules of an iron-rich nature.

In terms of trace metal levels, the Bay of Quinte nodules are low when compared to concretions from Lake Ontario. Levels in the nodules within the Bay of Quinte are generally comparable though higher values in Ni, Cr and V can be observed in the Mallory Bay and Brother's Island locations. Additionally, Co and Cd are significantly higher in the Brother's island nodules and the Adolphus Reach nodules are substantially lower in Be, Co, Cr, V, Pb and Zn (Table 2).

Inter-element relationships are summarized by the correlation matrix given in Table 3. This matrix is subdivided into two, showing the correlation coefficients for nodules having Fe/Mn ratios below and above 1.0.

Nodules with Fe/Mn greater than 1.0

From a correlation analysis, four groups of element associations can be defined.

1. Mn, Pb, Cd, V.
2. Fe, Be.
3. Cr, Ni, Cu, V. (Fe, Mn)
4. Zn, Co, V. (Fe, Mn)

Groups 1 and 2 represent the well-defined relationships with Mn and Fe, respectively. The remaining two groups, even though highly correlated, cannot be directly related to either Mn or Fe, suggesting that uptake by the hydrated oxides of both elements is taking place.

Nodules with Fe/Mn less than 1.0

Two groups of elements can be defined.

1. Mn, Cu, Cr, Cd, Be, V.
2. Zn, Ni, Co, Cd, Pb. (Fe, Mn)

Group 1 clearly defines those elements directly associated with manganese. Group 2 elements are highly correlated with each other and, in general, show some relationships to both Fe and Mn.

## Conclusions

Based on the results of the present study on ferro-manganese nodules in the Bay of Quinte, a number of descriptive and interpretive conclusions may be made, as follows:

1. The nodules occur in non-depositional areas of the Bay of Quinte on topographic highs downstream of depositional basins from which the soluble Fe, Mn and associated elements forming the nodules are derived.

2. Nodule-forming minerals identified included goethite, lepidocrocite, psilomelane, and rhodochrosite with clasts of calcite, dolomite, feldspar, quartz, ilmenite, rutile, hematite, pyrrhotite, nickeliferrous minerals and sperrilyte.

3. Nodules were formed as concentric layers of alternating iron and manganese bands around a central clast. Microprobe element distribution patterns showed the covariance of Mn and Ba in psilomelane and Fe, Si, P and S in goethite.

Table 2. Bay of Quinte: Nodule composition

| Location | | % | | | | ppm | | | | | | | | | | | |
|---|---|---|---|---|---|---|---|---|---|---|---|---|---|---|---|---|---|
| (No. of analysis) | | Fe | Mn | Fe/Mn | Ca | K | Mg | Sr | Be | Cd | Co | Cr | Cu | Pb | Ni | V | Zn |
| Average Bay of Quinte | | 17.7 | 10.5 | 4.02 | 2.0 | 1,160 | 2,820 | 1,090 | 1.0 | 3.0 | 71 | 40 | 25 | 32 | 71 | 82 | 164 |
| Average Big Bay | (4) | 38.0 | 5.3 | 7.17 | 2.0 | 319 | 616 | 282 | 1.9 | 3.7 | 58 | 29 | 23 | 42 | 40 | 100 | 146 |
| Average Telegraph Narrow | (5) | 19.2 | 6.5 | 2.95 | 6.6 | 1,461 | 3,252 | 882 | 1.3 | 2.6 | 77 | 28 | 19 | 57 | 35 | 56 | 157 |
| Average Green Point | (3) | 28.5 | 2.3 | 12.39 | 3.5 | 582 | 1,862 | 1,900 | 1.5 | 2.0 | 58 | 28 | 16 | 34 | 38 | 56 | 149 |
| Average Mallory Bay | (2) | 10.7 | 2.5 | 4.28 | 0.2 | 5,690 | 9,879 | 396 | 1.5 | 0.8 | 78 | 332 | 36 | 26 | 269 | 109 | 131 |
| Average Glenora Ferry 1 to 5 cm large | (6) | 10.7 | 18.8 | 0.57 | 1.3 | 884 | 2,104 | 1,729 | 0.7 | 3.4 | 81 | 22 | 25 | 24 | 91 | 78 | 190 |
| Average Glenora Ferry <1 cm | (2) | 24.0 | 4.5 | 5.33 | 0.8 | 282 | 1,315 | 525 | 1.8 | 2.1 | 88 | 24 | 24 | 39 | 49 | 148 | 221 |
| Average Adolphus Reach 1 to 5 cm large | (2) | 6.7 | 12.3 | 0.54 | 0.9 | 243 | 1,617 | 625 | 0.5 | 2.1 | 38 | 11 | 16 | 12 | 40 | 21 | 64 |
| Average Adolphus Reach <1 cm | (1) | 14.8 | 3.0 | 4.93 | 0.8 | 75 | 1,022 | 433 | 0.4 | 1.5 | 20 | 11 | 30 | 7 | 48 | 78 | 108 |
| Average Indian Point | (1) | 21.1 | 1.4 | 15.07 | 0.3 | 5,985 | 10,679 | 390 | 1.2 | 2.4 | 81 | 53 | 36 | 35 | 93 | 94 | 252 |
| Average Brothers' Islands | (2) | 14.7 | 8.9 | 1.65 | 0.9 | 2,403 | 5,082 | 766 | 1.1 | 6.4 | 121 | 32 | 47 | 53 | 175 | 171 | 210 |
| Lake Ontario K-2 | | 20.00 | 20.50 | 0.98 | 1.3 | 2,400 | 9,000 | 430 | - | - | 305 | - | 90 | 70 | 725 | - | 460 |
| Average Lake Ontario Coated Sands | | 20.53 | 17.10 | 1.20 | 3.1 | 2,380 | 21,300 | 160 | - | - | 650 | - | 360 | 1,890 | 2,410 | - | 2,020 |

Table 3. Correlation matrix for the Bay of Quinte nodules

| | Fe | Mn | Cu | Zn | Ni | Co | Cr | Cd | Be | V | Pb | |
|---|---|---|---|---|---|---|---|---|---|---|---|---|
| Fe | | .192 | -.217 | -.012 | -.556 | -.297 | -.372 | .200 | .652 | .176 | .244 | Fe/Mn ratio greater than 1.0 (21 nodules) |
| Mn | .110 | | .250 | .224 | -.060 | -.252 | .241 | .683 | .269 | .491 | .712 | |
| Cu | .522 | .645 | | .178 | .624 | .399 | .261 | .333 | -.123 | .434 | .006 | |
| Zn | .334 | .403 | .644 | | -.066 | .426 | -.196 | .303 | .025 | .402 | -.050 | |
| Ni | .250 | .288 | .560 | .847 | | .398 | .789 | -.044 | -.157 | .352 | -.132 | |
| Co | .256 | .312 | .403 | .746 | .646 | | .113 | .187 | .045 | .465 | .110 | |
| Cr | .656 | .661 | .853 | .600 | .562 | .538 | | -.372 | .084 | .131 | -.148 | |
| Cd | .433 | .680 | .810 | .759 | .753 | .481 | .727 | | -.019 | .456 | .415 | |
| Be | .268 | .522 | .622 | .605 | .631 | .539 | .607 | .752 | | .342 | .338 | |
| V | .488 | .694 | .768 | .586 | .681 | .357 | .744 | .897 | .700 | | .292 | |
| Pb | .555 | .179 | .424 | .750 | .702 | .873 | .635 | .546 | .534 | .414 | | |

Fe/Mn ratio less than 1.0 (17 nodules)

4. The Bay of Quinte nodules are iron-rich other than large nodules occurring at Glenora Ferry and Adolphus Reach. The nodules are low in trace metals when compared to other occurrences reported from Lake Ontario.

5. Small nodules were found to be iron-rich compared to large concretions. This is felt to be indicative of historical changes in the Bay of Quinte, representing increasing productivity within the Bay.

Acknowledgements

The authors would like to thank the Coxswain and seamen of the C.S.L. LEMOYNE and T.W. Morton of C.C.I.W. for their capable assistance in sampling, Drs. P.G. Sly and C.I. Dell for critical review of the manuscript and Dr. P.G. Sly, Chief, Process Research Division and Prof. G. Sironi, Director, Donegani Research Institut (Montedison), for permission to publish this material. We would further like to thank Prof. Bassi for evaluating X-ray diffractograms.

References

Cronan, D.S., & R.L. Thomas, 1970. Ferro-manganese concretions in Lake Ontario. Can. J. Earth Sci. 7: 1347-1349.

___________ 1972. Geochemistry of ferro-manganese oxide concretions and associated deposits in Lake Ontario. Geol. Soc. Amer. Bull., 83: 1493-1502.

Damiani, V., T.W. Morton, & R.L. Thomas, 1973. Freshwater ferro-manganese nodules from the Big Bay section of the Bay of Quinte, Northern Lake Ontario. Proc. 16th Conf. Great Lakes Res.

Damiani, V., & R.L. Thomas, 1974. The surficial sediments of the Big Bay Section of the Bay of Quinte, Lake Ontario. Can. J. Earth Sci. 11: 1562-1576.

Edgington, D.N. & E. Callender, 1970. Minor element geochemistry of Lake Michigan ferro-manganese nodules. Earth & Planet. Sci. Letters 8: 97-100.

Johnson, M.G. and G.E. Owen, 1971. Nutrients and nutrient budgets in the Bay of Quinte, Lake Ontario. J. Water Poll'n Control Fed. 43(5): 836-853.

Kindle, E.M., 1932. Lacustrine concretions of manganese. Amer. J. Sci. 24: 496-504.

___________ 1936. The occurrence of lake bottom manganiferous deposits in Canadian lakes. Econ. Geol. 31: 755-760.

Moore, J.R. & R.P. Meyer, 1969. Progress report on the geological-geophysical survey of Green Bay 1968. Univ. of Wisconsin Sea Grant Program Tech. Report No. 1(1): 16 p.

Price, N.B. & S.E. Calvert, 1970. Compositional variations in Pacific Ocean's ferro-manganese nodules in its relationship to sediment accumulation rates. Marine Geol. 9: 145-171.

Rossmann, R. & E. Callender, 1968. Manganese nodules in Lake Michigan. Science 162: 1123-1124.

Rossmann, R., E. Callender & C.J. Bowser, 1972. Interelement geochemistry of Lake Michigan ferro-manganese nodules - 24th Internat'l Geol. Conf., section 10: 336-341.

Rossmann, R., 1973. Lake Michigan ferro-manganese nodules. The Univ. of Michigan, Ph.D., 1973. University Microfilms, A Xerox Company, Ann Arbor, Mich.

Sly, P.G., & R.L. Thomas, 1974. Review of geological research as it relates to an understanding of Great Lakes Limnology. J. Fish. Res. Board Can. 31: 795-825.

Thomas, R.L., A.L.W. Kemp, & C.F.M. Lewis, 1972. Distribution, composition and characteristics of the surficial sediments of Lake Ontario. J. Sedim. Petrol. 42(1): 66-84.

Key words: Trace metals Lake Ontario
Ferro Manganese Nodules
Mineralogy Chemistry of Sediments

# METAL CONCENTRATIONS IN FRESHWATER SEDIMENTS - NATURAL BACKGROUND AND CULTURAL EFFECTS

Ulrich Förstner

Laboratorium für Sedimentforschung der Universität Heidelberg
P.O. Box 104040, d-69 Heidelberg, Federal Republic of Germany

## Summary

A review of lake samples of widely diversified geological, climatic and hydrological origin, reveals the general factors and processes governing the metal enrichment of Recent freshwater sediments. Man-made effects on aquatic ecosystems may be distinguished from other influences by analysis of surface and core sediments, and are revealed in particular, by the levels of metals such as zinc, lead, cadmium, mercury, copper and arsenic. The grain-size fraction of the sample meterial is an important parameter for determining the actual metal concentrations. In consequence, fraction separation and selective extraction procedures are being increasingly applied to determine metal concentrations and to differentiate between various types of metal bonding.

## Introduction

Hutchinson and co-workers (1943) and Züllig (1956) drew attention to a major application of geochemical research in the study of Recent freshwater sediments. Since "sediments may be regarded as a response of the conditions in an aquatic system" (Züllig, 1956), the study of freshwater sediments could play a key role in ascertaining the complex hydrochemical interactions pertaining to lakes and rivers. Also, approximately 20 years ago, the study of freshwater sediments was initiated in Canada as an aid to geochemical reconnaissance i.e. for tracing mineral deposits of economic value (Schmidt, 1956; Hawkes, et al. 1957). However, during the last two decades, sediment analysis has acquired a new dimension, by being employed as a tool to trace man-made pollution influences in inland waters and the coastal marine environment.

## Natural ("background") composition

Account must be taken of a number of biological, geological, chemical and hydrological processes controlled both by internal and external factors when interpreting sedimentological data as a function of the aquatic ecosystems (Table 1):

The natural influences - both introduced from outside and occurring within the freshwater system - are illustrated by the study of more than 70 lakes during the past 5 years, the lakes being situated in South and Western Australia, South America, East and South Africa, Central and Eastern Europe and from Western Asia (Förstner, 1977a).

The lakes were chosen from widely diversified areas with regard to climates ranging from extremely humid (e.g. the Amazon Basin) to arid (for example Western Australia). Correspondingly the salinity of the lake waters ranges from less that 20mg/l to more than 400 g/l. Lake water of extremely low salinity has a composition similar to rain water with predominance of sodium, chloride and bicarbonate ions. With increasing salinity, the calcium content becomes more important at very high salt concentrations, either sodium or magnesium predominates over calcium.

Table 1: Factors influencing the concentration of metals in lacustrine sediments.

I. Allochthonous influences
   1. Natural effects
   2. Cultural effects

II. Autochthonous influences
   1. Synsedimentary mechanisms (e.g. precipitation, sorption)
   2. Postdepositional transformations (diagenetic effects)

The mineralogical composition of the lake sediments reflects the lithological and hydrochemical influences: Clastic sedimentation prevails both in extremely arid and humid zones. The formation and diagenesis of carbonate minerals must be regarded as real criteria for distinguishing between the different types of lacustrine evolution in moderately humid and semi-arid climatic zones (Müller et al., 1972).

## Average metal values

A comparison of the median values (statistical mean) of metal concentrations from the pelitic fractions (particles with a diameter of less than two microns ) of the investigated lake sediments with the average shale composition for fine-grained sedimentary rocks by Turekian & Wedepohl (1961), indicates a wide agreement in the metal concentrations of recent and ancient clayey sediments (Table 2, columns 1 & 2). There exist, however, large

Table 2. Mean metal concentrations (parts per million) and major influencing factors in lacustrine sediments.

| | Metal contents of pelitic sediments: | | | Major influencing factors: | | | |
|---|---|---|---|---|---|---|---|
| Metal | Average shale (Turekian & Wedepohl '61) | Lacustrine deposits median n= 74 | Lacustrine deposits range for 90% of the data | Bedrock lithology granitoid n = 10 | Bedrock lithology gabbroid n = 2 | Ca-Mg-Carbonates absent n = 50 | Ca-Mg-Carbonates present n = 6 |
| Iron | 46,700 p.p.m. | 43,000 | (12,000-69,000) | 49,000 | 58,000 | 47,100 | 16,900 |
| Manganese | 850 p.p.m. | 750 | ( 100- 1,800) | 660 | 1,100 | 902 | 475 |
| Strontium | 300 p.p.m. | 142 | ( 55- 600) | 113 | 70 | 118 | 1,311 |
| Zinc | 95 p.p.m. | 110 | ( 45- 220) | 102 | 85 | 124 | 63 |
| Chromium | 90 p.p.m. | 59 | ( 25- 180) | 42 | 490 | 91 | 42 |
| Nickel | 68 p.p.m. | 66 | ( 30- 250) | 69 | 1,000 | 119 | 46 |
| Lithium | 66 p.p.m. | 45 | ( 15- 200) | 40 | 19 | 60 | 81 |
| Copper | 45 p.p.m. | 43 | ( 20- 80) | 31 | 44 | 46 | 34 |
| Lead | 20 p.p.m. | 28 | ( 8- 75) | 32 | 27 | 29 | 21 |
| Cobalt | 19 p.p.m. | 15 | ( 3- 40) | 7 | 68 | 19 | 16 |
| Mercury | 0.4 p.p.m. | 0.31 | ( 0.15- 1.20) | 0.44 | 0.23 | 0.41 | 0.46 |
| Cadmium | 0.3 p.p.m. | 0.35 | ( 0.10- 1.20) | 0.49 | 0.20 | 0.42 | 0.39 |

differences in the extent of deviation from the mean values. Variations are relatively low for iron (90% of the data fall between 12,000 and 69,000 parts per million in dry mass samples), for zinc (90% between 45 and 220 p.p.m.) and copper (commonly between 20 and 80 p.p.m. in the pelitic sediment fraction). Manganese and strontium, on the other hand, cover particular wide ranges of values due to the effects of diagenetic remobilization (for Mn) and co-precipitation with carbonate minerals (for Sr).

Allochthonous influences

In the search for simplified conditions to evaluate representative background concentrations of minor and trace elements in lacustrine sediments, arid regions appeared to be most promising. According to Grim (1968), "the clay mineral composition of desert saline sediments is controlled almost entirely by the composition of the source area". Metal concentrations in lacustrine deposits from such regions, therefore, should rather reflect the primary chemical anomalies pertaining to the particular watershed, than the secondary processes occurring within the lake area itself.

In Table 2, a comparison of metal data from two lake areas in Australia, situated in different geological surroundings is shown (Förstner, 1977b). The sediments of 10 arid lakes in the South Australian Eyre Peninsula, which are largely derived from granitoid rocks, show metal contents closely corresponding to the composition of the shale standard. On the other hand, high enrichment of cobalt, nickel and chromium, by factors to ten or more, occurs in the lake deposits of the ultrabasic "greenstone belt" around Kalgoorlie, Western Australia.

Allochthonous dilution of metal contents in lake sediments resulting from weathering effects is clearly demonstrated by examples from the Amazon "River" Lakes near the confluence of the Rio Negro and Rio Solimoes, near Manaus (Irion & Förstner, 1975): The deposits contributed by rivers from regions with deeply leached soil consist mainly of kaolinite minerals and are significantly depleted in iron, manganese, zinc, copper and cobalt. Natural allochthonous influences on lake sediment composition include the discharge of metals by hydrothermal solutions, should they have occurred. Recent sulphide ore formation is reported from late stage volcanic lakes in Hokkaido, Japan (Wauschkuhn, 1974). Geothermal mercury "pollution" by fumarolic emanations was determined in lakes of New Zealand (Weissberg & Zobel, 1973).

Autochthonous influences

Intralacustrine accumulation and dilution of metals is chiefly controlled by sorption processes on minerals and organic substances, and by co-precipitation with carbonate minerals and hydrous Fe/Mn-oxides. The effects of autochthonous dilution of metals by Ca-Mg-carbonates are shown in table 2 by comparing carbonate-free lake sediments with deposits of the sub-brackish and playa lakes which on a average generally contain approximately 50% Ca-Mg-carbonates in the pelitic fractions. Data in table 2 (6th and 7th column) indicate that dilution by calcite, dolomite (and huntite in the Tuz Gölü, Turkey) is the highest for iron ( a reduction of 75%), less strong for nickel, chromium, zinc and manganese (50%). On the other hand, strontium contents are enriched by a factor of 40. Similar effects are observed from aragonite precipitations in Anatolian Lakes (Irion, 1973) and by sulfates, e.g. by alunite, in Australian desert lakes. In Lake Malawi, metal contents of sediments are diluted by silica from diatoms, which form almost pure "diatomite" in the southern part of the lake, overlain by a few centimeters of diatom-free sediments (Müller & Förstner, 1973). The dilution is very effective for the iron and manganese concentrations; reduction for zinc, copper, chromium, nickel lead and cobalt is between 30% and 60%.

A high degree of positive correlation between the content of organic materials and metal concentrations has often been observed in lake sediments. This, however, may not always be interpreted as being due to direct bonding of heavy metals to organic substances. In highly polluted areas, for instance, contamination by characteristic heavy metals simply coincide with the accumulation of organic substances, both derived from urban, industrial or agricultural sewage effluents. Extraction of metals from fulvic and humic acids in sediments from Lake Malawi indicated that only zinc, copper and vanadium were accumulated in organic material, whereas iron, manganese, chromium, lead, cobalt and nickel are more or less diluted by the organic substances.

Diagenetic effects

The distribution of manganese, iron and cobalt in both lake sediments and water is controlled to a much greater extent by redox conditions than is the case with other minor or trace elements, such as copper, nickel and zinc (Stumm & Morgan, 1970). Manganese, in particular, reveals a great diagenetic mobility and "can be supplied to the surface sediments from above and from below"(Delfino & Lee, 1968). Enrichment of manganese in the near-surface sediments commonly results from syndiagenetic dissolution of Mn-compounds within the lower reducing part of the sedimentary column, upward migration due to sediment compaction and subsequent precipitation at the oxidizing sediment/ water interface. This process is also considered as an important mechanism in the formation of manganese concretions in the lacustrine environment (Kindle, 1932; Rossmann & Callender, 1968).

The behaviour of other metals (besides manganese) during redox changes is a subject which is still open to a great deal of controversy. Some metals, such as uranium, chromium and molybdenum, can be transferred under oxidative conditions from the sediment into the water. There are indications, that other metals, e.g. copper, zinc and cadmium, are released from reducing sediments into surface water at low oxygen contents (Reinhard & Förstner, 1976). These processes, however, seem to be controlled rather more by bacterial activity and subsequent complex formations than by unidirectional ion migration, since the calculated stability constants for the sulfide compounds are drastically exceeded by the concentrations of Cu, Cd, Zn, Pb and other metals in the pore water (Brooks et al 1968; Cline & Upchurch, 1973;Nissenbaum & Swaine, 1976). Studies by Elderfield & Hepworth (1975) on the fluxes of metals in anoxic sediments in the Conway Estuary revealed that ~2% (Zn) to 22%(Ni) of the metal enrichment of the surface of sediments compared with deeper sections result from diffusion processes. Diagenetic contributions, therefore, should be taken into account when calculating the pollution input rates.

Man's impact on the lacustrine environment

Freshwater lakes have formed the centres of important cultural developments ever since the earliest days of civilization. As a con sequence of increased population and industrial densities, the threat of pollution has become most acute in such areas. A comparison of uncontaminated subsurface deposits (earlier than 1800 AD) from Lake Constance + Tegernsee with surface sediments from lakes of Upper Bavaria in Southern Germany clearly indicates man-derived accumulation of lead, mercury, cadmium, zinc and copper which have increased four-fold some distance from the industrial, municipal and agricultural sources of pollution. Civilizational influences can be traced both by means of regional distribution patterns and also chronologi variations of characteristic metals in lake sediments.

Metal distribution in surficial sediments

Lateral variations of metal concentrations in narrow grid sediment samples were originally studied for the assessment of both mineralization zones and pollution sources in Canadian lakes (Allan, 1971; Thomas, 1972). Similar investigations were undertaken in lakes in Sweden, Finland, Poland, Germany, Switzerland, Italy, the United States, Australia and other countries (for compilation see Förstner, 1976).

An example of the interface of lithologic influences and man-made effects is given in Figure 1 (after data of Wittmann & Förstner, 1975), indicating the concentrations of two selected metals, chromium and zinc, in the pelitic fraction of sediment grab samples from the Hartbeespoort Dam: Situated 40 km to the west of Pretoria between the Daspoort Hills and the Magaliesberg Mountains, the Hartbeespoort Dam represents one of the most eutrophied impoundments of South Africa From the western catchment area the dam is fed by the Magalies and Skeerpoort Rivers, which drain a predominantly agricultural basin in gabbroid rocks. In contrast, the eastern Crodocile inflow stems from the highly industrialized Kempton Park/ Isando area, north of Johannesburg. The increases of

Figure 1. Contents of Cr and Ni in sediments of Hartbeespoort Dam, Republic of South Africa.

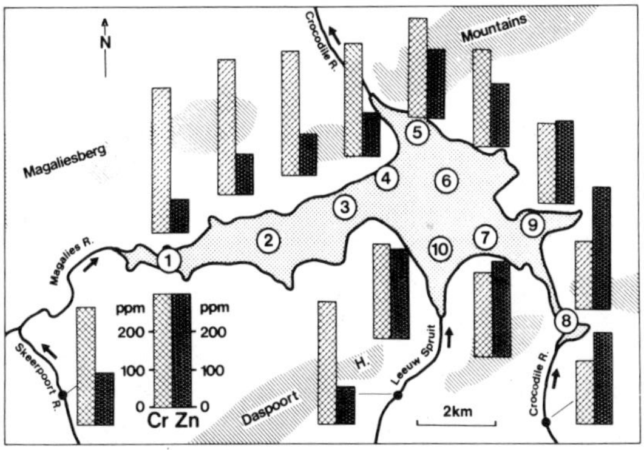


of chromium ( and similarly of nickel, cobalt and iron)levels in the western zone are largely due to the lithology of the catchment area. On the other hand, the enriched mercury, lead, cadmium, and zinc values encountered in the southeastern zone of the Hartbeespoort Dam are due to civilisational effects. The high zinc and lead values can possibly be ascribed to domestic waste effluents whilst the cadmium, and in particular, the mercury enrichment could have resulted from industrial discharges (Wittmann & Förstner, 1976).

Chronological variations in sediment cores

The evaluation of pollution effects in comparison to a global standard, e.g. the shale composition, was often not satisfactory due to the presence of regional anomalies in lithology. In such a situation, the investigation of dated sedimentary cores proved to be very useful, since it provides a historical record of the various influences on the aquatic system by indicating both the natural background levels and the man-induced accumulation of elements over a extended period of time. Figure 2 summarizes some data from a sediment core study from Lake Constance which is not highly polluted by heavy metals (Förstner et al., 1974). Similar trends in both nitrogen / phosphorous and zinc / lead concentrations indicate that these substances mostly originate within the public sewage system. The increase of cadmium and chromium within the upper part of the sediment core is caused to a considerable extent by industrial emissions.

The example of Lake Constance represents the very common influence of contamination by mixed sewage effluents. Similar and other effects on lake sediments are briefly summarized from North American examples in Table 3: Beside the diffuse ("non-point") source pollution exemplified by the cores from Lake Michigan, there is an added effect of metal-bearing pesticides,which cause an increase of copper and arsenic in certain Wisconsin lakes by factors ranging from 12 to 25 as

Figure 2. Contents of N, P, Cd, Pb, Cr, Zn in a sedimentary core from Lake Constance.

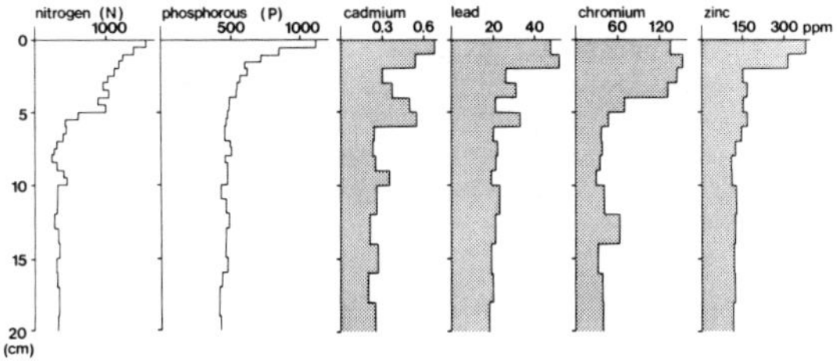


compared with pre-cultural metal concentrations. Air-borne particulate fallout from smelters result in a 20-fold enrichment compared with the natural background levels of lead and arsenic in sediments from Lake Washington. High rates of enrichment of a characteristic metal are commonly observed where industrial point sources release their concentrated effluents. The sediment core example of Lake Erie near Cleveland in the Table 3 shows distinct increases of cadmium and mercury, due to emissions from electroplating and chlorine-alkali industries.

Textural and compositional variations

Even from the study of sedimentary cores, there may, however, arise difficulties in the interpretation of analytical data, particularly owing to the influence of grain size effects and of changes in the chemical association of the metals.

Grain size effects

Figure 3. Concentrations of nickel and zinc in various grain size fractions of sediment samples from Lower Rhine and Lake Constance.

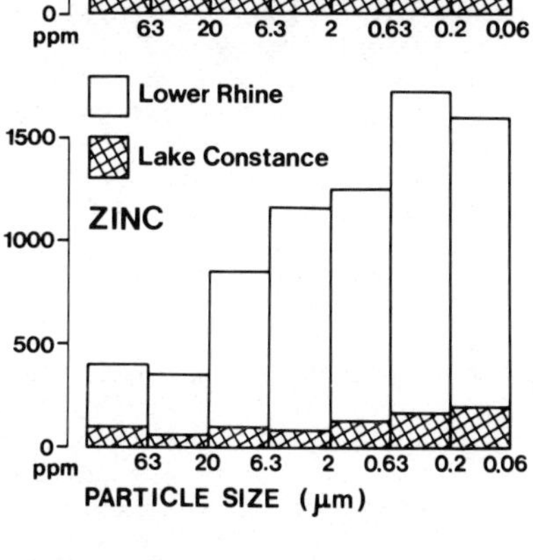

Table 3. Metal pollution sources (examples) derived from sediment core studies on lakes in North America. Data of background and maximum value in p.p.m.; F = factor of increase.

| Metal | Lake Michigan[1] | | | Wisconsin Lakes[2] | | | Lake Washington[3] | | | Lake Erie[4] | | |
|---|---|---|---|---|---|---|---|---|---|---|---|---|
| | back-ground | max. value | F | back-ground | max. value | F | back-ground | max. value | F | back-ground | max. value | F |
| Zinc | 129 | 317 | 2.5 | 15 | 92 | 6 | 60 | 230 | 4 | 7 | 42 | 6 |
| Chromium | 77 | 85 | 1 | 7 | 49 | 7 | -- | n.d. | -- | 13 | 16 | 4.5 |
| Nickel | 54 | 44 | 1 | 34 | 50 | 1.5 | (iron: | | 1) | 40 | 95 | 2.5 |
| Copper | 44 | 75 | 1.5 | 22 | 268 | 12 | 15 | 50 | 3 | 18 | 58 | 4 |
| Lead | 40 | 145 | 3.5 | 12 | 124 | 9 | 20 | 400 | 20 | -- | n.d. | -- |
| Arsenic | 11 | 22 | 2 | 2 | 51 | 25 | 10 | 200 | 20 | 0.6 | 3.2 | 5.5 |
| Mercury | 0.04 | 0.2 | 5 | 0.24 | 1.12 | 5 | 0.1 | 1.0 | 10 | 0.04 | 0.48 | 12 |
| Cadmium | -- | n.d. | -- | 2.5 | 4.6 | 2 | -- | n.d. | -- | 0.14 | 2.4 | 17 |
| Source: | Diffuse sources | | | Algicides and herbicides | | | Atmospheric fall-out from smelters | | | Industrial effluents | | |

[1] Ruch et al. (1970); Shimp et al.,(1971); Kennedy et al., (1971); Fryc & Shimp, (1972); [2] Shukla et al., (1972); Syers et al. (1973); [3] Barnes & Schell, Crecelius & Piper (1973); Schell (1974); Crecelius (1975); [4] Walters et al., (1974)

Figure 3 shows the heavy metal distribution within two sediment samples from different sites of the Rhine River system. The first sample stems from the relatively unpolluted deposits of Lake Constance (shaded signature), the second is from the Lower Rhine area, where the sediments are heavily polluted by the large industrial and urban centres of the Rhine/ Main and Rhine/ Ruhr areas. In both samples, the highest metal concentrations occur in the finer-grained material. Maximum concentrations of the highly polluted Rhine sediments are found in the grain size fraction between 0.2 and 0.6 μm diameter, decreasing both with larger and finer-grained particles. These effects can be attributed to the low metal contents of the quartz and felspar components, mainly occurring in the coarser grain fractions. The examples of nickel and zinc each represent typical behaviour of heavy metals: whereas the polluted sediment of the Lower Rhine is only 30% - 50% higher in nickel, it still exceeds the zinc deposits of Lake Constance by a factor of 10.

It is clearly demonstrated, that without a correction for the grain size, metal analyses from bulk sediment samples may be of little significance for monitoring purposes, particularly from river deposits undergoing strong granulometric changes. In practice, there exist several methods to reduce grain size effects; examples are given in Table 4.

Distribution of chemical forms of metals

A study of the chemical associations of metal species present in sediments may give indications and reveal the availability of these substances for metabolic processes with toxic effects under certain environmental conditions. Furthermore, a knowledge of the particular bonding type of the different metals in sediment samples may be the key for the detection of specific pollution sources in our aquatic systems.

In spite of their different origin, environmentally related types of metal bonding are comparable with the natural forms. Table 5 gives a compilation of the most common associations of heavy metals in particulate matter. It should be mentioned, however, that there are numerous forms of humic substances, clay minerals and hydroxides, particularly via sorption and coating processes.

Table 4. Reduction of grain size effects

| | |
|---|---|
| **Grain size distribution** | |
| extrapolation (<16 μm ) | De Groot et al., '71 |
| separation (< 2 μm ) | Banat et al., 1972 |
| (<63 μm ) | Allan et al., 1972 |
| specific surface area | Oliver, 1973 |
| **Acid soluble fraction** | |
| 0.1 M hydrochloric acid | Piper, 1971 |
| 0.1 M nitric acid | Jones, 1973 |
| aqua regia/ignit. loss | Literáthy & László, 1975 |
| **Mineral separation** | |
| quartz correction | Thomas, 1972, 1973 |
| **Conservative elements** | |
| ratio X/ aluminum | Bruland et al., 1974 |
| sediment enrich. factor | Kemp et al., 1976 |

Several attempts have been undertaken for the selective extraction of different metal compounds from sediments. A first detailed study by Hirst & Nicholls (1958) investigated

Table 5. Metal associations in sediments

| | | |
|---|---|---|
| Mineral detritus (mainly silicates) | | Metal bonding mostly in inert positions |
| Heavy metal -hydroxides -carbonates -sulphides | | Precipitation as a result of exceeding solubility product in the water course |
| Clay minerals (sorption) | pH | Physico-sorption (electr. attraction) |
| | | Chemical sorption (exchange $H^+$ in SiOH, AlOH and $Al(OH)_2$) |
| Bitumen, lipids Humic substances Residual organics | pH | Physico-sorption Chemical sorption (COOH-, OH-groups) |
| | | Complexes |
| Hydrous Fe/Mn-oxides | pH | Physico-sorption Chemical sorption |
| | | Co-precipiation by exceeding the solubility product |
| Calcium carbonate | pH | Physico-sorption Pseudomorphosis (supply and time) |
| | | Co-precipitation |

the differentiation of the metal contents in the detrital and non-detrital fraction of carbonate rocks by acid treatment (25% acetic acid), using organic precipitate for spectrochemical analysis (Scott & Mitchell, 1943 and 1947; Heggen & Strock, 1953). The separation of the carbonate fraction by acid treatment (see also Table 4), however, is rather problematic: Ray et al. (1957) have shown that dilute hydrochloric acid will attack the lattice structure of certain clay minerals, e.g. Fe-chlorite. Although this is not the case with dilute acetic acid, it was demonstrated by Hogson (1960) that 2.5% acetic acid will strip the exchangeable fraction of adsorbed ions from the surface of montmorillonite minerals. Even by differentiation in exchanger columns (Lloyd, 1954) or by means of complexation (Hill & Runnels, 1960; Glover 1961), the exchangeable cation contents and (especially with strongly complexing EDTA and NTA) other phases are more or less strongly influenced here. In our laboratory, Patchineelam (1975) tried to bring about a selective release of the metals in carbonates by introducing $CO_2$ gas into water suspensions. Although this is probably the most careful procedure, it involves a large amount of time and the possibility of new precipitations must be considered.

Methods for differentiation of metals in sorbed, hydroxidic and organic phases are compiled in Table 6. A combination of extraction procedures was first used by Gibbs (1973)

Table 6. Extraction of metals associated with different chemical phases in sediments

| Chemical phase | Extraction procedure | Authors |
|---|---|---|
| Adsorption & cation exchange | (a) $BaCl_2$, (b) $MgCl_2$, (c) $NH_4OAc$ | Jackson, 1958; Gibbs, 1973 |
| Detrital/authigenic phases in pelagic sediments | EDTA treatment | Goldberg & Arrhenius, 1958 |
| Reducible phases | 1 M $NH_2OH \cdot HCl$; 25%v/v acetic acid | Chester & Hughes, 1967 |
| Moderately reducible phases (hydrous Fe-oxides) | Reduction with sodium dithionite complexing with sodium citrate | Holmgren, 1967 |
| Easily reducible phase (Mn-oxide & amorph. Fe-oxide) | 0.1 M $NH_2OH \cdot HCl$; 0.01 M nitric acid | Chao, 1972 |
| Organics, sulphides | 30% $H_2O_2$ at 95 $^0$ C, extract with (a) 1 N $NH_4OAc$ or (b) 0.01 M $HNO_3$ | Jackson, 1958; Engler et al., 1974; Gupta & Chen, 1975 |
| Lipids, bitumen | fat solvents: e.g. chloroform, ether, gasoline, benzene, carbon disulfide | Bergmann, 1963; Welte, 1969 |
| Humic & fulvic acids | 0.5 N NaOH; 0.1 N $NaOH/H_2SO_4$ | Rashid, 1971; Volkov & Fomina, 1974 |
| Solid organic material | Na-hypochlorite, dithionate/citrate | Gibbs, 1973 |
| Detrital silicates | Digestion with $HF/HClO_4$; $HNO_3$ (a) Lithium metaborate (1000 $^0$C), $HNO_3$ (b) | e.g. Bruland et al., 1973 Gibbs, 1973 |

for analyses of trace metals associated with different carrier substances in suspended materials of the Amazon and Yukon Rivers (procedures 1b-4-9-10b in table 6). Engler et al. (1974) developed a separation technique, particularly for dredged materials, precluding atmospheric oxidation at sensitive steps (1c-5-6a-4-10a); further modification of that scheme was performed by Gupta & Chen (1975; 1c-5-6b-3-10a). Patchineelam (1975) differentiated the metal contents associated with cation exchange positions (1a), humic acids (8), carbonates and hydrous Fe/Mn-oxides (3), respectively.

Figure 4. Distribution of bonding types (a) and sources (b) of metals in Rhine sediment.

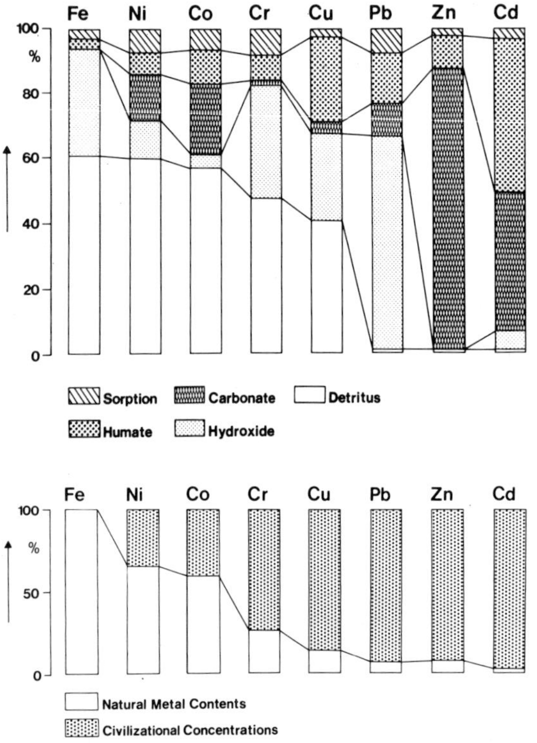


As an example, Figure 4a shows the application of our procedure to a representative sediment sample from the Lower Rhine (see Figure 3). Due to the large catchment area, all essential pollution may be considered to be represented in the sample under consideration. Starting with the metal contents in the detrital-mineral association, two groups of metals can be distinguished: Firstly iron, nickel, lead and cadmium with very low content of detrital mineral bonding. The same sequence has been obtained, when we established an order of man-made effects on the metal concentration of the Lower Rhine sediments, by comparing the data from recent material with analyses of ancient Rhine sediments obtained from drilling cores in the Cologne area. According to our calculations, cultural influences are relatively low for iron, nickel and cobal moderately high for chromium and copper, and very high for lead and zinc (Figure 4b). Onl two percent of the cadmium content in the recent sediments of the Lower Rhine originate fr natural sources, 98% are derived from industri and urban effluents as well as from agricultu and atmospheric pollution.

Figure 4a shows, that lead, copper and chrom ium are preferentially bonded to the hydroxid phases. This finding can be explained by the specific sorption of lead to iron hydroxide (Jenne, 1968) and the effective co-precipitati of copper on hydrous Fe/Mn oxides (Groth, 197 Furthermore, the absence of independent carbonate phases of Cr in natural aquatic systems On the other hand, the preferential carbonate bonding of zinc and cadmium can be attributed to a relative high stability of zinc carbonat and cadmium carbonate under the existing Eh-pH conditions of common inland waters as well as to the characteristic co-precipitation of these components with calcium carbonate. The metal contents thus sorbed consitiute less than 10 percent of the total metal contents of all examples studied. This fact is of special interest in connection with remobilization effects under the influence of elevated salt concentrations, e.g. for the interaction of seawater with suspended river sediments in the estuarine mixing zone (Müller & Förstner, 1975).

More detrimental effects, however, may be expected to arise from changes of the redox conditions and by increased inputs of natural and synthetic chelating agents, whereby mechanisms are effected to remobilize the metals bonded to hydroxides and carbonates into the aqueous phase. In addition, long-term changes of the pH conditions which have been observed from atmospheric $SO_2$ emissions and acid mine drainage, can lead to the dissolution of carbonate and hydroxide minerals and also - as a result of hydrogen ion competition - to an increased desorption of metal cations. The release of metals from pore solutions is enhanced by mechanical effects on the aquatic deposits, e.g. by erosion, dredging and bioturbation. Additional remobilization of metals can be caused by biological activities; organ mercurial transformations exemplify the fact that these mechanisms may increase metal toxic ities.

## Acknowledgements


The financial support of the investigations of freshwater sediments by the Deutsche Forsch ungsgemeinschaft (German Research Society) is gratefully appreciated. I wish to express my thanks to Prof. Dr. G. Müller for discussions, and to Dr. S.R. Patchineelam, R. Deurer (Laboratorium für Sedimentforschung, Universität Heidelberg) and Dr. G.T.W. Wittmann (Department of Chemistry, University of Pretoria) for provision of data and valuable suggestions

References


Allan, R.J., 1971. Lake sediments: a medium for regional geochemical exploration of the Canadian Shield. Can. Inst. Mining Metall. Bull. 64: 43-59.

Allan, R.J., E.M. Cameron & C.C. Durham, 1972. Reconnaissance geochemistry using lake sediments of a 36,000 square-mile area of the northwestern Canadian Shield. Geol. Surv. Can. Pap. 72-50: 70 p.

Arrhenius, G.O.S. & J. Korkish, 1959. Uranium and thorium in marine minerals. Intern. Ocean. Congr., Am. Assoc. Adv. Sci., preprints, 497 pp.

Banat, K., U. Förstner & G. Müller, 1972. Schwermetalle in Sedimenten von Donau, Rhein, Ems, Weser und Elbe im Bereich der Bundesrepublik Deutschland. Naturwissenschaften 69: 525-528.

Barnes, R.S. & W.R. Schell, 1973. Physical transport of trace metals in the Lake Washington watershed. In: Cycling and Control of Metals, Proc. Environm. Resources Conr., Columbus/Ohio (Oct. 31- Nov. 2, 1972). p. 45-53.

Bergmann, W., 1963. Geochemistry of lipids. In: I.A. Breger, Organic Geochemistry, Pergamon Press, New York.

Brooks, R.R., B.J. Presley & I.R. Kaplan, 1968. Trace elements in the interstitial waters of marine sediments. Geochim. Cosmochim. Acta 32: 397-444.

Bruland, K.W., K. Bertine, M. Koide & E.D. Goldberg, 1974. History of metal pollution in Southern California Coastal Zone. Environ. Sci. Techn. 8: 425-432.

Chao, L.L., 1967. Selective dissolution of manganese oxides from soils and sediments with acidified hydroxylamine hydrochloride. Soil. Sci. Soc. Am. Proc. 36: 764-768.

Chester, R., & M.J. Hughes, 1967. A chemical technique for the separation of ferro-manganese minerals, carbonate minerals and adsorbed trace elements from the pelagic sediments. Chem. Geol. 2: 249-262.

Cline, J.T. & G.B Upchurch, 1973. Mode of heavy metal migration in the upper strata of lake sediments. Proc. 16th Conf. Great Lakes Res. p. 349-356.

Crecelius, E.A., 1975. The geochemical cycle of arsenic in Lake Washington and its relation to other elements. Limnol. Oceanogr. 20: 441-451.

Crecelius, E.A. & D.Z. Piper, 1973. Particulate lead contamination recorded in sedimentary cores from Lake Washington, Seattle. Environ. Sci. Techn. 7: 1053-1055.

Delfino, J.J. & G.F. Lee, 1968. Chemistry of manganese in Lake Mendota, Wisconsin. Environ. Sci. Techn. 2: 1094-1100.

Engler, R.M., J.M. Brannon, J. Rose & G. Bigham, 1974. A practical selective extraction procedure for sediment characterization. 168th ACS Natl. Meeting, Atlantic City, N.J. (Sept. 1974).

Elderfield, H. & A. Hepworth, 1975. Diagenesis, metals and pollution in estuaries. Mar. Poll. Bull. 6: 85-87.

Förstner, U., 1976. Lake sediments as indicators of heavy-metal pollution. Naturwissenschaften 63: 465-430.

Förstner, U., 1977a. Metal concentrations in Recent lacustrine sediments. Arch. Hydrobiol. (in press).

Förstner, U. 1977b. Mineralogy and geochemistry of sediments in arid lakes of Australia. Geol. Rundschau 66: 146-156.

Förstner, U. & S.R. Patchineelam, 1976. Bindung und Mobilisation von Schwermetallen in fluviatilen Sedimenten. Chemiker-Zeitung 100: 49-57.

Förstner, U., G. Müller & G. Wagner, 1974. Schwermetalle in Sedimenten des Bodensees. Naturwissenschaften 61: 270.

Frye, J.C. & N.F. Shimp, 1973. Major, minor and trace elements in sediments of late Pleistocene Lake Saline compared with those in Lake Michigan sediments. Illinois Geol. Surv. Environ. Geol. Note 60: 13 pp.

Gibbs, R.J., 1973. Mechanisms of trace metal transport in rivers. Science 180: 71-73.

Glover, E.D., 1961. Method of solution of calcareous materials using the complexing agent EDTA. J. Sediment. Petrol. 31: 622-626.

Goldberg, E.D. & G.O.S. Arrhenius, 1958. Chemistry of Pacific pelagic sediments. Geochim. Cosmochim. Acta 13: 153-212.

Grim, R.E., 1968. Clay Mineralogy. 2nd ed., McGraw-Hill, New York, 596 pp.

De Groot, A.J., J.J.M. de Goeij & C. Zengers, 1971. Contents and behaviour of mercury as compared with other heavy metals in sediments from the rivers Rhine and Ems. Geologie en Mijnbouw 50: 393-398.

Gupta, S.K. & K.Y. Chen, 1975. Partitioning of trace metals in selective chemical fractions of nearshore sediments. Environ. Lett. 10:129-158.

Hawkes, H.E., H. Bloom & J.E. Riddell, 1957. Stream sediment analysis discovers two mineral deposits. 6th Commonwealth Mining Metallurgy Congr., p. 259-268.

Heggen, G.E. & L.W. Strock, 1953. Determinations of trace elements combining chemical enrichment and sprectrochemical methods. Anal. Chem. 25: 859-863.

Hill, W.R.jr. & R.T. Runnels, 1960. Versene, new tool for study of carbonate rocks. Bull. Amer. Assoc. Petrol. Geol. 44: 631-632.

Hirst, D.M. & G.D. Nicholls, 1958. Techniques in sedimentary geochemistry. 1. Separation of the detrital and non-detrital fractions of limestones. Jour. Sediment. Petrol. 28: 468-481.

Hogson, J.F., 1960. Cobalt reactions with montmorillonite. Soil Sci. Soc. Amer. Proc. 24: 165-168.

Holmgren, G.S., 1967. A rapid citrate-dithionite extractable iron procedure. Soil Sci. Soc. Amer. Proc. 31: 210-211.

Hutchinson, G.E., A Wollack & J.K. Setlow, 1943. The chemistry of lake sediments from

Indian Tibet. Amer.Jour.Sci.241: 533-542.
Irion, G., 1973. Die anatolischen Salzseen, ihr Chemismus und die Entstehung ihrer chemischen Sedimente.Arch. Hydrobiol. 71: 517-557.
Irion, G. & U. Förstner, 1975. Mineralbestand und Chemismus amazonischer Seetone. Naturwissenschaften 6: 179.
Jackson, M.L., 1958. Soil Chemical Analysis. Prentice Hall, Englewood Cliffs, 498 pp.
Jenne, E.A., 1968. Controls on Mn, Fe, Co, Ni, Cu, and Zn concentrations in soils and water: the significant role of hydrous Mn and Fe oxides. In: Trace Inorganics in Water. Am. Chem. Soc. Publ., Adv. Chem. Ser. 73: 337-387.
Jones, A.S.G., 1973. The concentration of copper, lead, zinc and cadmium in shallow marine sediment, Cardigan Bay (Wales). Mar. Geol. 14: M1-M9.
Kemp, A.L.W., R.L. Thomas, C.I. Dell & J.-M. Jaquet, 1976. Cultural impact on the geochemistry of sediments in Lake Erie. J.Fish. Res. Bd. Can. 33: 440-462.
Kennedy, E.J., R.R. Ruch & N.F. Shimp, 1971. Distribution of mercury in unconsolidated sediments from Southern Lake Michigan. Illinois Geol. Surv. Environ. Geol. Note 44: 18 pp.
Kindle, E.M., 1932. Lacustrine concretions of manganese. Amer. Jour. Sci. 24: 496-504.
Literâthy, P. & F. Lâszlô, 1975. New method for characterizing the chemistry of bottom sediments. Int. Sym. Geochem. Natural Water, Burlington, Ontario. August 18-20. 1975.
Lloyd, R.M. A technique for separating clay minerals from limestones. Jour. Sediment. Petrol. 24: 218-220. (1954).
Mitchell, R.L. & R.O. Scott, 1947 . Recovery of trace constituents in plant materials and soil extracts by mixed organic reagents. Soc. Chem. Ind. Journ.(Trans.) 66:330-336.
Müller, G. & U. Förstner, 1973. Recent iron ore formation in Lake Malawi, Africa. Mineral Deposits 8: 278-290.
Müller, G. & U. Förstner, 1975. Heavy metals in sediments of the Rhine and Elbe estuaries: mobilization or mixing effect? Environm. Geol. 1: 33-39.
Müller, G. & G. Irion & U. Förstner, 1972. Formation and diagenesis of inorganic Ca-Mg-carbonates in the lacustrine environment. Naturwissenschaften 59: 158-164.
Nissenbaum, A. & D.J. Swaine, 1976. Organic matter-metal interactions in Recent sediments: the role of humic substances. Geochim. Cosmochim. Acta 40: 809-816.
Oliver, B.G., 1973. Heavy metal levels of Ottawa and Rideau river sediments. Environ. Sci. Technol. 7: 135-137.
Patchineelam, S.R., 1975. Untersuchungen über die Hauptbindungsarten und die Mobilisierbarkeit von Schwermetallen in fluviatilen Sedimenten. Unpubl. Dr. thesis, Heidelberg, 137 pp.
Piper, D.Z., 1971. The distribution of Co, Cr, Cu,Fe, Mn, Ni, and Zn in Framvaren, a a Norwegian anoxic fjord. Geochim. Cosmochim. Acta 35: 531-550.
Rashid, M.A., 1971. Role of humic acids of marine origin and their different molecular weight fractions in complexing di- and trivalent metals. Soil Sci. 111: 298-306.
Ray. S., H.R. Gault & C.G. Dodd, 1957. The separation of clay minerals from carbonate rock. Am. Mineralogist 42: 681-686.
Reinhard, D.& U. Förstner, 1976. Metallanreicherungen in Sedimentkernen aus Stauhaltungen des mittleren Neckars. N.Jb.Geol.Paläont.Mh. 1976: 301-320.
Rossmann, R. & E. Callender, 1968. Manganese nodules in Lake Michigan. Science 162: 112[illegible] 1124.
Ruch. R.R., E.J. Kennedy & N.F. Shimp, 1970. Distribution of arsenic in unconsolidated sediments from Southern Lake Michigan. Illinois Geol. Surv. Environ. Geol. Note 37: 16 pp.
Schell, W.R., 1974. Sedimentation rates and mean residence times of Pb and 210-Pb in Lake Wahington, Puget Sound estuaries and a coastal region. Abstr. Meet. Amer. Soc. Limnol. Oceanogr., June 23-27, 1974.
Schmidt, R.C., 1956. Adsorption of Cu, Pb, and Zn on some common rock forming minerals and its effect on lake sediments. Unpubl. Ph.D. thesis, Montreal, 181 pp.
Scott, R.O. & R.L. Mitchell, 1943. Recovery of cobalt, nickel, molybdenum, copper, and zinc from plant materials and soil extracts by 8-hydoxyquinoline. Soc. Chem. Ind. Jour (Trans.) 62: 4-8.
Shimp, N.F., J.A. Schleicher, R.R. Ruch, D.B. Heck & H.V. Leland, 1971. Trace element an organic carbon accumulation in the most recent sediments of Southern Lake Michigan. Illinois Geol. Surv. Environ. Geol. Note 41 25 pp.
Shukla, S.S., K.J.K. Syers & D.E. Armstrong, 1972. Arsenic interference in the determination of inorganic phosphate in lake sed iments. J. Environ. Qual. 1:292-295.
Stumm, W. & J.J. Morgan, 1970. Aquatic Chemistry. Wiley, New York, 583 p.
Syers, J.K., I.K. Iskandar & D.R. Keeney, 1973. Distribution and background levels in sediment cores from selected Wisconsin Lakes. Water Air Soil Poll. 2:105-118.
Thomas, R.L., 1972. The distribution of mercury in the sediments of Lake Ontario. Car J.Earth Sci. 9: 636-651.
Thomas, R.L., 1973. The distribution of mercury in the surficial sediments of Lake Huron, Can.J.Earth Sci. 10: 194-204.
Turekian, K.K. & K.H. Wedepohl, 1961. Distribution of the elements in some major units of the earth's crust. Bull. Geol. Soc. Am 72: 175-192.
Walters, L.J.jr., T.J. Wolery & D.R. Myser, 1974. Occurrence of As, Cd, Co, Cr, Cu, F[illegible] Hg, Ni, Sb, and Zn in Lake Erie sediments. Proc. 17th Conf. Great Lakes Res. p. 219-2[illegible]
Wauschkuhn, A., 1973. Rezente Sulfidbildung

in vulkanischen Seen auf Hokkaido (Japan). Geol. Rundschau 62: 774-785.
Weissberg, B.G. & M.G.R. Zobel, 1973. Geothermal mercury pollution in New Zealand. Bull. Environ. Contam. Toxicol. 9: 148-155.
Welte, D., 1969. Organic geochemistry of carbon. In: K.H. Wedepohl, Handbook of Geochemistry, II-6-L, Springer Verlag, New York.
Wittmann, G.T.W. & U. Förstner, 1975. Metal enrichment of sediments in inland waters - the Hartbeespoort Dam Water SA 1: 76-82.
Wittmann, G.T.W. & U. Förstner, 1976. Metal enrichment of sediments in inland waters - the Jukskei and Hennops River drainage systems. Water SA 2: 67-72.
Züllig, H., 1956. Sedimente als Ausdruck des Zustandes eines Gewässers. Schweiz. Z. Hydrol. 18: 7-143.

Key words: Lakes, Sediments, Metals, Pollution, Phase partitioning

A.J. de Groot

Institute for Soil Fertility, Haren (Groningen), The Netherlands

## Summary

The river Rhine and to a less extent the rivers Meuse and Scheldt transport large quantities of heavy metals to the Dutch coastal area. The river Ems, on the contrary, can be regarded as relatively clean. Upstream of the rivers large proportions of the heavy metals are bound to the suspended matter. This especially holds for Cr, Pb and Cu. Elements like Zn and Ni, on the other hand, show a preferred occurrence in the dissolved state. The bond of the main part of the heavy metals to the suspended matter means that the distribution of these metals in the aquatic environment is highly determined by the hydrodynamic behaviour of the suspended matter. Furthermore, for a full understanding of the consequence of the presence of the heavy metals in the environment, the study of the geochemical behaviour and of the uptake by organisms is necessary. In this connection an integrated study is carried out in the Netherlands by the Delft Hydraulics Laboratory (hydrodynamics), the Institute for Soil Fertility, Haren, Groningen (geochemistry) and the Central Laboratory TNO, Delft (biology).

The distribution of heavy metals as components of deposited mud over the Dutch Delta has been given schematically in Fig.1. The contents of the heavy metals in the deposited material have been expressed as percentages of the contents of the relevant metals in the suspended matter upstream of the Rhine river. From this figure it is obvious that the main influence on the heavy metal composition is originating from the Rhine river.

The Rhine has still a natural pathway to the sea via the Rotterdam harbour area (Waalhaven and Europoort). Within these areas mixing processes occur of heavily polluted Rhinesediments with less polluted material coming from the North Sea. The mixing processes could be described quantatively by means of the application of the isotope geochemistry (C12/C13- and O16/O18-ratios) on several components of the sediments (carbonates, clay minerals and organic matter).

The other mouthing areas of the Rhine have been enclosed. This happened first by building the Enclosure Dike in 1932. As a consequence the IJssel (distributary of the Rhine river) was mouthing without the interaction of tidal effects into the Lake IJssel. The Ketelmeer (as a part of Lake IJssel) is now the main deposition area of the IJssel sediments. It is obvious from Fig. 1 that these deposits are from a view-point of heavy metal pollution completely equal to those of the upper Rhine river.

Another important mouthing area of the Rhin was originally the estuary via Hollands Diep and Haringvliet. The Haringvliet has been enclosed in 1970 as a part of the well-known Delta-plan. It can be seen from Fig. 1 that the deposits in the Haringvliet are more contaminated than those of the Europoort. We expect that the removal of the tidal effects in this area will give rise to the deposition of more and more contaminated sediments. In the far future a situation can be expected, which equals that of the Ketelmeer.

A similar effect can be seen in the Grevelingen area, which has been enclosed in 1972. This area, however, has no more an open connection with the Rhine river.

The Oosterschelde will be only partially enclosed in the near future. This will lead to an enhanced deposition of sediments. The degree of pollution of these sediments will mainly depend on the situation on the open sea.

A part of the suspended matter of the river Rhine will reach the Dutch Wadden Sea. Although the heavy metal contents of these sediments represent the lowest values in Fig. 1, they may not be regarded as unpolluted. For elements like Cd, Zn and Pb the values are at least twice as heigh as their base lir levels.

From the above mentioned data it is obvious that large civil-engineering projects (enclosure dams) largely influence the retention of polluted sediments within the Dutch aquatic territory. The same holds for the deepening and extension of large sea-harbours. By these actions the retention of sediments is favoured. Especially from those harbours in the Rotterdam area which are located more upstream of the river (e.g. Waalhaven), large quantities of polluted sediments have to be dredged. The dredge spoils are either dumped at sea or used for land-filling. Under both conditions the material can give rise to severe environmental problems. The dredge spoils brought on land are often used for agricultural and horticultural purposes. Especially high amounts of cadmium, which are easily taken up by crops, restrict the use o these soils for plant growth.

In the preceding paragraphs much attention has been given to the deposited material. Heavy metals, however, can be present in

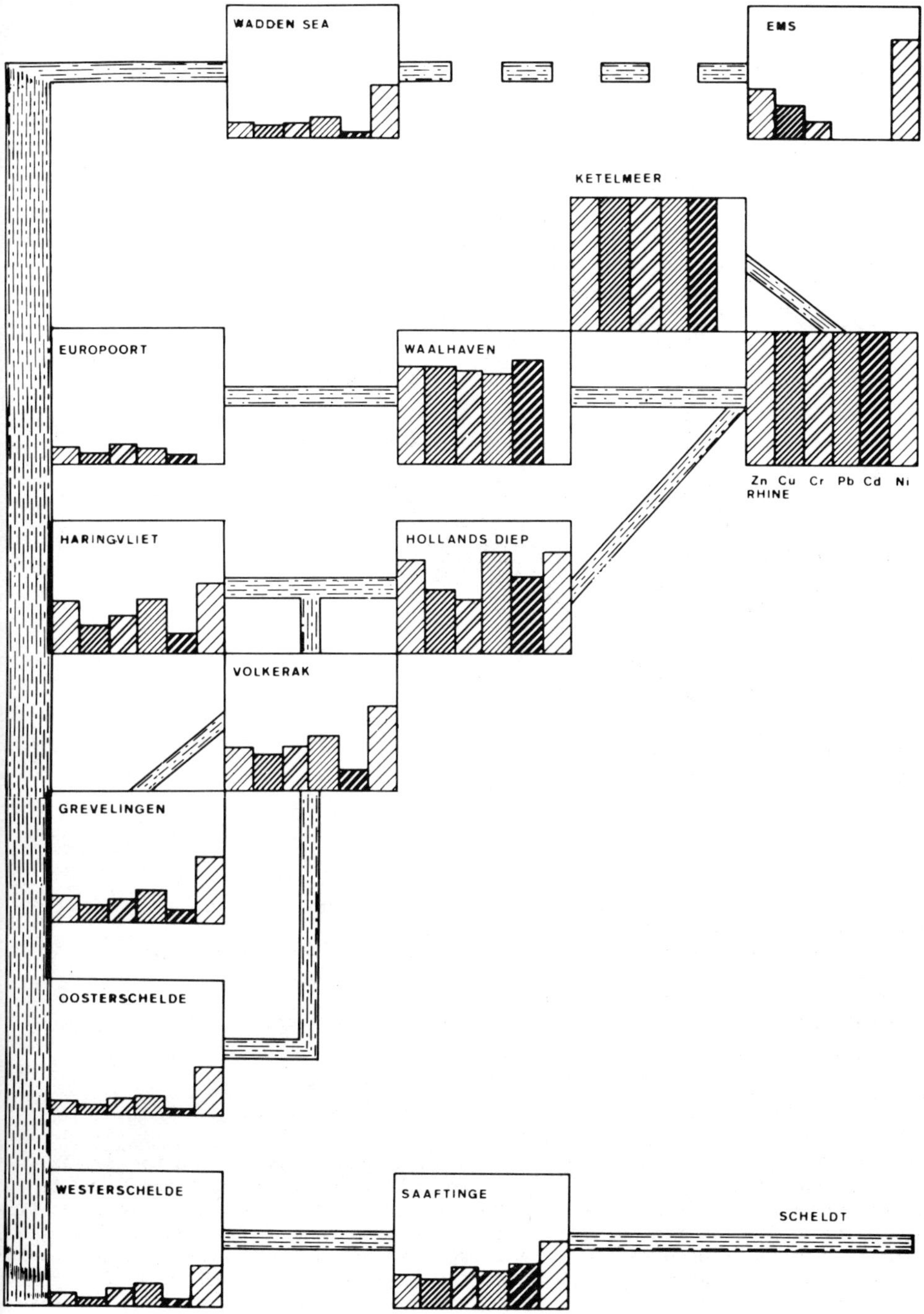


Fig. 1. Heavy metals in deposits of the Dutch Delta. Rhine = 100

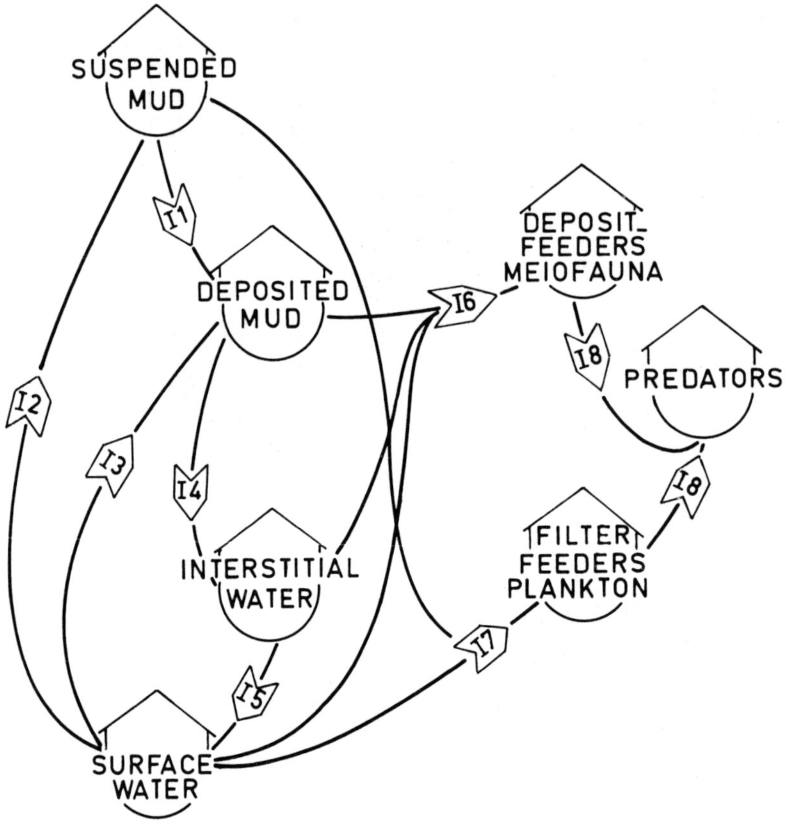


Fig. 2. Review of the role of heavy metals in an aquatic ecosystem.
I1: Sedimentation, transport, erosion.
I2: Adsorption of metals to suspended matter. Precipitation of dissolved metals.
I3: Reactions between surface water and deposited mud.
I4: Provision of the interstitial water with heavy metals.
I5: Escape of metals from the interstitial water to the surface water.
I6, I7: Uptake of metals by organisms.
I8: Accumulation in the food chain.

several reservoirs of on ecosystem. A survey of these reservoirs and their interactions (representing the integrated study as mentioned before) has been given in Fig. 2. Much of the present work is focused on the study of this system. We will restrict us in this summary to a few remarks in this respect.

As abiotic reservoirs the deposited mud, the interstitial water, the suspended material and the surface water can be distinguished. Under reduced conditions the deposited material can provide the interstitial water with rather high concentrations of heavy metals. The latter can escape to the overlying waters by means of processes of diffusion, consolidation, erosion and bioturbation. In polluted rivers there is not always an equilibrium between the metals dissolved in the water and those bound to the suspended matter. A longer residence time of the river water in a reservoir like Lake IJssel may favour the fixation of the metals to the suspended matter and subsequently give rise to more polluted deposits, as a consequence of sedimentation.

The contribution of the 4 abiotic reservoirs to the heavy metal burden of organisms depends on their way of life. In Fig. 2 some biotic reservoirs have been indicated, as they are investigated by the Central laboratory TNO.

As a representative organism for the water phase (surface water and suspended matter) e.g. the mussel (*Mytilus edulis*) is used in both laboratory and field experiments. As deposit feeders certain worms (e.g. Arenicola maritima) can be used. Under fresh water conditions similar organisms are available. These studies have proved that organisms which accumulate heavy metals are very useful for the characterization of the significance of heavy metals in the aquatic environment.

## Literature


Groot, A.J. de, 1966. Mobility of trace elements in deltas. Int. Soil Sci. Soc., Trans. Comm. II & IV, Aberdeen, 1966, pp. 267-279.

Groot, A.J. de, 1973. Occurrence and behaviour of heavy metals in river deltas, with special reference to the Rhine and Ems rivers. In: E.D. Goldberg (Editor), North Sea Science. MIT Press, Cambridge (Mass), pp. 308-325.

Groot, A.J. de, and Allersma, E., 1975. Field observations on the transport of heavy metals in sediments. In: P.A. Krenkel (Editor), Heavy Metals in the Aquatic Environment. Proceedings International Conference Nashville, Tenn., 1973 - Prog. Water Technol. 7: 85-95.

Groot, A.J. de, Zschuppe, K.H., Bruin, M. de, Houtman, J.P.W. and Amin Singgih, P., 1968. Activation analysis applied to sediments from various river deltas. Reactor Inst., Delft, Rep. 133-68-08: 16 pp. Also: Trans Int. Conf. Modern Trends activation Anal., Gaithersburg, 1968 (1971): 62-71.

Groot, A.J. de, Allersma, E., Bruin, M. de, and Houtman, J.P., 1970. Cobalt and tantalum tracers measured by activation analysis in sediment transport studies. In: Isotope Hydrology 1970. Int. Atom. Energy Agency, Vienna, pp. 885-898.

Groot, A.J. de, Goey, J.J.M. de, and Zegers, C., 1971. Contents and behaviour of mercury as compared with other heavy metals in sediments from the rivers Rhine and Ems. Geol. Mijnbouw 50: 393-398.

Groot, A.J. de, Allersma, E., Bruin, M. de, and Houtman, J.P.W., 1973. Application of activation analysis in sediment transport studies, particularly cobalt and tantalum. In: Tracer Techniques in Sediment Transport - IAEA Tech. Rep. Ser. 145: 151-166.

Hueck, H.J., 1975. Active surveillance and use of bioindicators. Sci. Eur. Colloq. "Principles and Methods for Determining Ecological Criteria on Hydrobiocenoses". Comm. Eur. Communities, Luxembourg.

Hueck, H.J. and Hueck-van der Plas, E.H., 1975. A system of tests for the assessment of potential effects of chemicals and quality data in the aquatic environment. Cent. Lab. TNO, Rep. CL 75/119: 58 pp.

Kock, W.Chr. de, 1975. Milieutoxicologische waarnemingen over de verontreiniging van het Oosterschelde-areaal met zware metalen. Cent. Lab. TNO, Rapp. CL 75/83: 53 pp.

Salomons, W., 1975a. Chemical and isotopic composition of carbonates in recent sediments from Western Europe. J. Sediment. Petrol. 45: 440-449.

Salomons, W., Hofman, P., Boelens, R. and Mook, W.G., 1975. The oxygen isotopic composition of the fraction less than 2 microns (clay fraction) in recent sediments from Western Europe. Mar. Geol. 18: M23-M28.

Wolf, P. de, 1975. Mercury contents of mussels from West European Coasts. Mar. Pollut. Bull. 6: 61-63.



Key words: Heavy metals
Rhine
Delta
Sediments
River pollution

RELATIVE ATOMIC VARIATION (RAV) OF ELEMENTS IN LAKE SEDIMENTS: LAKE WINNIPEG AND OTHER CANADIAN LAKES

R. J. Allan and G. J. Brunskill

Freshwater Institute, Winnipeg, Canada

Summary

From 1968 to 1971 and again in 1975, lake sediment samples were obtained for geochemical study from sites throughout Lake Winnipeg Manitoba. The objectives were twofold. We wished to determine the influence of agricultural, industrial and urban-domestic development of the 414,000 square mile drainage basin of the lake on the element concentrations in bottom sediments. A second objective was to investigate the validity of making qualitative assessments of cultural contamination on element concentration differences alone. Although concentrations alone have proved a valid means of identifying contamination at sites of extreme additions, various parameters including sedimentation rate, sediment type, particle size, and analytical methods make comparisons between sites difficult. It is proposed here that a dimensionless index based on ratios for various element combinations is a useful tool for detection of geochemical imbalances in lake sediments. This dimensionless index is called the Relative Atomic Variation (RAV) and is the slope of the significant linear regression line for a particular element pair. Similarity in RAV implies similarity or eventual homogeneity in a large number of processes in the geochemical cycle including weathering, transport, deposition and diagenesis. RAV for various element pairs was calculated for sediments from the two basins of Lake Winnipeg and for sediments from lakes in geographically and geologically dissimilar areas of Canada. Surprisingly there was a great deal of correspondence in many of the values. Most exceptions could be related to cultural activities. Although the approach advocated has been applied here exclusively to lake sediments, it is thought to have greater significance to the study of river sediment geochemistry or contamination, because of the relative difficulty in obtaining meaningful cores of river sediments.

Introduction

Western Canada has three of the 13 largest lakes in the world. Lake Winnipeg is 11th in surface area (23,745 $km^2$). Of significance is the size of the lakes's drainage basin (953,250 $km^2$). Extending from the Rocky mountains of British Columbia to the Canadian Shield of Ontario, it includes all of the Prairie Region of Canada. The main rivers draining the basin are the Saskatchewan, flowing east from the Rockies and entering the north basin on its western shore, the Winnipeg flowing west from Ontario into the south basin at the eastern shore, and the Red flowing north from the U.S.A. into the south basin at the southern shore. Most of the water entering the lake originates in the Winnipeg river drainage basin. Alternatively, most of the suspended sediment load originates from the Red River watershed.

Lake Winnipeg operates physically and biologically as two separate basins connected by a 3 to 5 mile wide channel referred to as the Narrows. Water flows out of Lake Winnipeg through the Nelson River to Hudson Bay. The lake is located between 96°20' and 99°20' W and 50°20' and 53°51' N in south central Canada. It has a surface area of 23,750 $km^2$, a maximum depth of 32 m, mean depth of 10.6 m, a volume of 252-284 $km^3$, and a theoretical residence time between 3 and 4 years. The lake is a remnant of glacial Lake Agassiz (Elson, 1967). It lies at the boundary of Palaeozoic and Mesozoic sedimentary strata to the south and west, and granitic-gneissic and volcanic bedrock to the east and north.

Previous study of Lake Winnipeg sediments has been confined to particle size distribution analysis and mineralogy (Kushnir, 1971). He examined some 60 dredge samples from the south basin and showed that off-shore samples were 70% clay ($<4\mu$) and that the clay fraction was mineralogically uniform throughout the basin. The clay composition was 80-85% montmorillonite-illite with 10-20% kaolinite and chlorite. This mineralogy was similar to that of surficial deposits of western Canada, and implies a minimum of diagenetic changes in the lake environment.

Data and calculations relevant to nutrient loadings to Lake Winnipeg have been published by Brunskill (1973). A more recent assessment of the limnology of Lake Winnipeg is in preparation (Brunskill et al., in prep.), which will contain all of the raw data on metals and nutrients presently determined on bottom sediments. Analytical results from other study sites in Canada are used in the text. These sites have been described in detail elsewhere (Allan et al., 1973; Allan and Timperley, 1975). Brief summaries are given in Table 1.

Table 1. Brief descriptions of lake sediment sampling sites and sub-sites[1].

| Site | Sub-Site | Brief Description | Cultural Influence | Geological Influence |
|---|---|---|---|---|
| Lake Winnipeg, Manitoba | South Basin | Smaller basin of large lake in south central Manitoba | Possible industrial agricultural and urban-domestic | None |
| | North Basin | Larger basin, as above | As above, possible airborne metals from smelters | None |
| Red Lake Area, Ontario | Red Lake | Large lake in forested Canadian Shield | Industrial gold mining and refining | Possibly Au ores |
| | Snib Lake | Small lake, as above | As above | As above |
| | Sully Lake | As above | As above | As above |
| | Gullrock Lake | As above | As above | As above |
| Sudbury Area, Ontario | Sudbury Lakes | Numerous small lakes, in forested Canadian Shield | Industrial Ni-Cu mining and refining | Possibly Ni-Cu ores |
| Bear-Slave Area, N.W.T. | Agricola | Numerous small lakes, barren tundra of northern Canadian Shield | None | Multielement anomalies from sulphide sources |
| | Bear | Numerous small lakes, forested northern Canadian Shield | None | Multielement anomalies of unknown source |
| | Hepburn | Numerous small lakes, forest/tundra transition of northern Canadian Shield | None | Multielement anomalies of unknown source |
| | Granite | As above | None | None |
| | Gneiss | As above | None | None |
| | Volcanic | As above | None | None |

1. For detailed descriptions of sites other than Lake Winnipeg see publications referred to in text.

## Sampling and Analysis

A set of Lake Winnipeg sediment samples was collected by Skipek, Eckman dredge, and Benthos corers between 1968 and 1971 (Figure 1).

Figure 1. Location and sample numbers of dredged and cored sediments, Lake Winnipeg

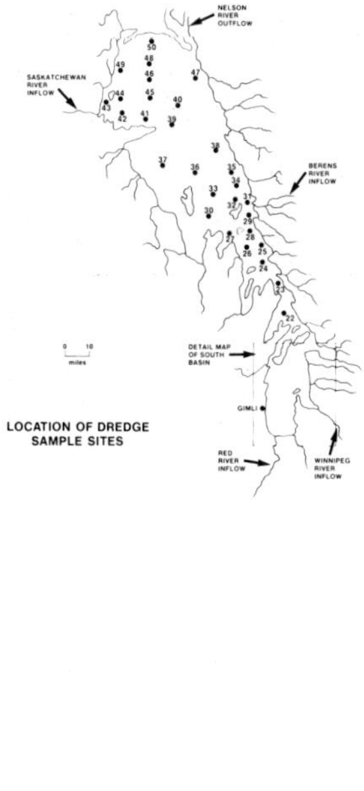


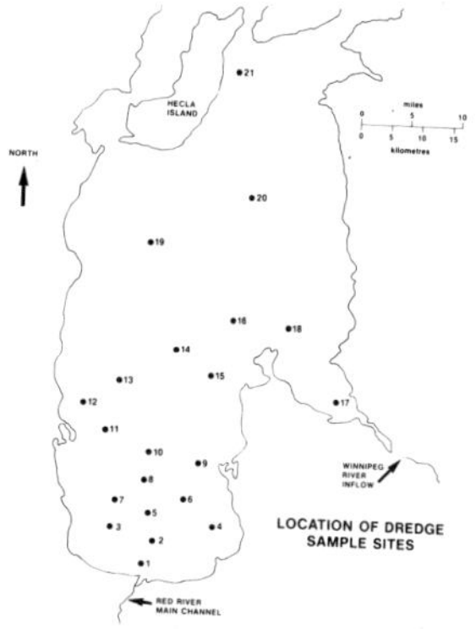


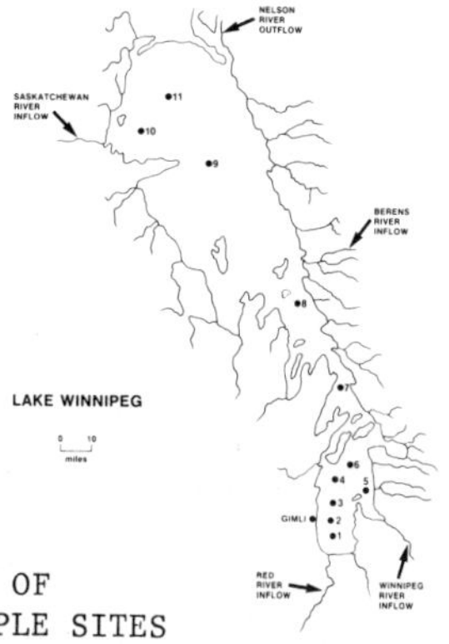


LOCATION OF CORE SAMPLE SITES

A second set of samples was collected using similar devices during 1975. All of the samples exhibited a surface oxidised zone (2-4 cm) underlain by greyish-blue sediment.

The samples were analysed by Radio Frequency Plasma Emission Spectroscopy (RFPES). The sediment was refluxed with HF and $HNO_3$-$HClO_4$. After evaporation to dryness, the residue was taken up in 4% $HNO_3$. This technique determines total concentrations of elements. Mercury was determined by the stannous sulphate reduction technique followed by cold vapour atomic absorption spectroscopy. Arsenic was determined by hydride generation, silver DDC complexation, and colorimetric measurement at 520 nm using a Spectronic 20.

The results are given in Figure 2 and in Tables 2, 3 and 4.

Figure 2. Location and sample numbers of dredged and cored sediments, Lake Winnipeg. Mercury and arsenic distributions.

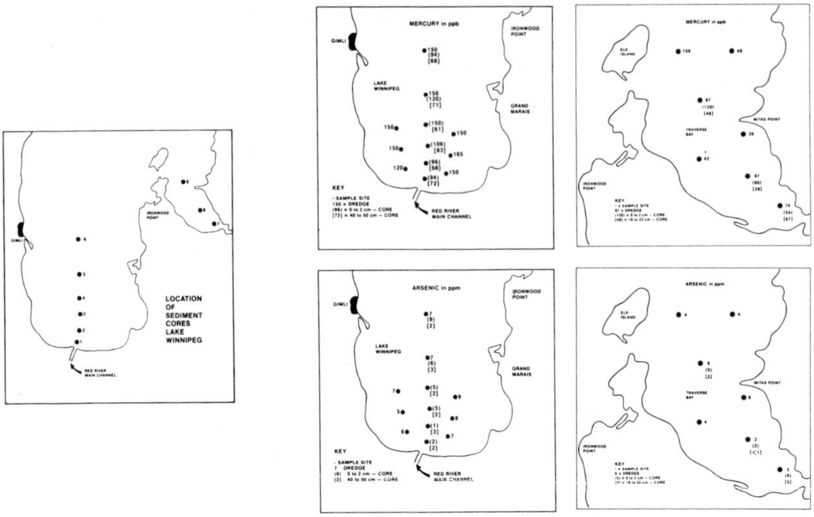


Table 2. Concentrations of elements in dredged sediments from Lake Winnipeg. Station localities will be found in Brunskill et al. (in prep.)

| Sample Location | Sample Number | Ni | Cu | Co | Cr | Zn | Cd | Pb | V | Ti | Fe | Mn | Be | Na | Mg | Ca | Sr | P | Al |
|---|---|---|---|---|---|---|---|---|---|---|---|---|---|---|---|---|---|---|---|
| | | (ppm) | (ppm) | (ppm) | (ppm) | (ppm) | (ppm) | (ppm) | (ppm) | % | % | (ppm) | (ppm) | % | % | % | (ppm) | (ppm) | % |
| South Basin | 1 | 29 | 16 | 20 | 75 | 123 | 1 | 16 | 88 | 0.21 | 1.4 | 510 | 1.1 | 0.6 | 1.7 | 4.5 | 184 | 440 | 4.1 |
| | 2 | 47 | 41 | 29 | 133 | 206 | 2 | 22 | 203 | 0.36 | 2.8 | 860 | 1.9 | 0.4 | 2.6 | 4.6 | 154 | 600 | 6.9 |
| | 3 | 47 | 38 | 30 | 120 | 209 | 2 | 32 | 195 | 0.39 | 2.8 | 1050 | 2.0 | 0.4 | 2.6 | 4.5 | 151 | 640 | 6.6 |
| | 4 | 44 | 34 | 27 | 115 | 199 | 2 | 18 | 178 | 0.33 | 2.4 | 790 | 1.7 | 0.4 | 2.5 | 4.1 | 146 | 570 | 6.0 |
| | 5 | 54 | 37 | 33 | 159 | 226 | 2 | 20 | 232 | 0.40 | 3.4 | 1150 | 2.3 | 0.4 | 2.3 | 3.3 | 138 | 650 | 7.6 |
| | 6 | 47 | 34 | 29 | 135 | 204 | 2 | 24 | 199 | 0.35 | 2.7 | 1030 | 1.9 | 0.4 | 2.3 | 4.1 | 136 | 590 | 6.5 |
| | 7 | 53 | 35 | 32 | 147 | 216 | 2 | 18 | 218 | 0.40 | 3.2 | 1050 | 2.0 | 0.4 | 2.4 | 2.5 | 137 | 590 | 7.3 |
| | 8 | 54 | 38 | 32 | 156 | 209 | 2 | 18 | 229 | 0.42 | 2.9 | 1140 | 2.3 | 0.3 | 1.8 | 1.7 | 139 | 580 | 8.0 |
| | 9 | 48 | 38 | 31 | 146 | 211 | 2 | 28 | 214 | 0.39 | 2.5 | 950 | 2.0 | 0.4 | 2.2 | 2.8 | 144 | 590 | 7.4 |
| | 10 | 53 | 39 | 31 | 157 | 205 | 2 | 30 | 229 | 0.42 | 3.0 | 1240 | 2.3 | 0.3 | 1.6 | 1.4 | 132 | 610 | 8.0 |
| | 11 | 56 | 40 | 31 | 152 | 199 | 2 | 38 | 218 | 0.43 | 2.9 | 1220 | 2.3 | 0.3 | 1.8 | 1.7 | 131 | 550 | 7.8 |
| | 12 | 49 | 33 | 30 | 130 | 190 | 2 | 36 | 173 | 0.35 | 2.3 | 1060 | 1.7 | 0.4 | 3.4 | 4.5 | 139 | 510 | 6.9 |
| | 13 | 55 | 41 | 33 | 155 | 204 | 2 | 26 | 231 | 0.44 | 3.1 | 1650 | 2.5 | 0.4 | 1.6 | 1.2 | 137 | 640 | 8.3 |
| | 14 | 50 | 38 | 29 | 142 | 193 | 2 | 108 | 210 | 0.39 | 2.8 | 970 | 2.1 | 0.4 | 1.4 | 1.1 | 133 | 540 | 7.8 |
| | 15 | 56 | 37 | 33 | 148 | 255 | 2 | 56 | 229 | 0.42 | 3.6 | 1080 | 2.4 | 0.4 | 1.9 | 1.7 | 142 | 630 | 7.9 |
| | 16 | 54 | 34 | 35 | 146 | 199 | 2 | 12 | 202 | 0.42 | 3.2 | 900 | 2.0 | 0.4 | 2.1 | 2.2 | 174 | 590 | 7.6 |
| | 17 | 37 | 21 | 24 | 86 | 155 | 1 | 8 | 97 | 0.24 | 1.2 | 380 | 1.2 | 0.9 | 3.6 | 5.8 | 194 | 470 | 5.3 |
| | 18 | 51 | 32 | 31 | 118 | 188 | 2 | 32 | 147 | 0.32 | 1.8 | 670 | 1.7 | 0.7 | 2.8 | 4.0 | 181 | 540 | 6.3 |
| | 19 | 57 | 42 | 33 | 159 | 210 | 2 | 30 | 228 | 0.45 | 3.2 | 1320 | 2.4 | 0.4 | 1.6 | 1.2 | 143 | 610 | 8.7 |
| | 20 | 61 | 39 | 35 | 151 | 216 | 2 | 26 | 223 | 0.42 | 3.0 | 1480 | 2.3 | 0.5 | 1.5 | 1.1 | 152 | 590 | 8.4 |
| | 21 | 27 | 15 | 17 | 62 | 86 | 1 | 28 | 63 | 0.13 | 0.9 | 600 | 1.1 | 0.8 | 0.6 | 1.1 | 192 | 330 | 4.1 |
| Narrows | 22 | 54 | 37 | 31 | 125 | 161 | 2 | 16 | 146 | 0.34 | 2.0 | 880 | 1.7 | 1.4 | 1.2 | 2.0 | 211 | 600 | 7.1 |
| | 23 | 49 | 37 | 28 | 113 | 155 | 2 | 22 | 125 | 0.32 | 2.1 | 800 | 1.6 | 1.4 | 1.2 | 1.8 | 211 | 580 | 7.4 |
| | 24 | 56 | 42 | 33 | 134 | 163 | 2 | 32 | 155 | 0.38 | 2.5 | 1090 | 1.9 | 1.0 | 1.4 | 1.6 | 202 | 560 | 7.4 |
| | 25 | 55 | 50 | 32 | 138 | 167 | 2 | 30 | 161 | 0.40 | 2.6 | 780 | 1.9 | 1.0 | 1.4 | 1.6 | 201 | 560 | 7.8 |
| | 26 | 57 | 44 | 31 | 153 | 176 | 2 | 18 | 181 | 0.40 | 2.5 | 560 | 2.0 | 0.7 | 1.5 | 1.3 | 191 | 510 | 7.6 |
| | 27 | 40 | 31 | 20 | 103 | 122 | 1 | 22 | 102 | 0.23 | 1.7 | 590 | 1.3 | 0.9 | 1.0 | 1.5 | 206 | 450 | 5.6 |
| | 28 | 57 | 44 | 33 | 136 | 175 | 2 | 74 | 165 | 0.38 | 2.3 | 660 | 2.0 | 1.1 | 1.3 | 1.6 | 207 | 570 | 7.4 |
| | 29 | 41 | 31 | 24 | 107 | 144 | 2 | 52 | 110 | 0.30 | 1.8 | 530 | 1.7 | 1.9 | 1.1 | 2.0 | 213 | 550 | 7.6 |
| North Basin | 30 | 62 | 42 | 33 | 162 | 197 | 2 | 28 | 168 | 0.40 | 2.9 | 590 | 1.9 | 0.7 | 2.2 | 2.9 | 191 | 590 | 7.9 |
| | 31 | 45 | 30 | 31 | 130 | 173 | 2 | 42 | 124 | 0.33 | 2.5 | 1130 | 1.5 | 1.0 | 1.2 | 1.6 | 200 | 590 | 7.2 |
| | 32 | 55 | 42 | 32 | 146 | 178 | 2 | 118 | 162 | 0.42 | 2.9 | 670 | 1.9 | 1.1 | 1.6 | 1.8 | 203 | 580 | 8.5 |
| | 33 | 63 | 47 | 37 | 164 | 204 | 2 | 22 | 181 | 0.37 | 2.6 | 650 | 2.0 | 0.9 | 1.7 | 1.5 | 191 | 590 | 7.3 |
| | 34 | 66 | 47 | 39 | 161 | 194 | 2 | 48 | 157 | 0.43 | 3.2 | 830 | 1.8 | 1.5 | 2.1 | 2.0 | 208 | 560 | 9.1 |
| | 35 | 69 | 46 | 39 | 175 | 216 | 2 | 24 | 186 | 0.42 | 2.9 | 730 | 2.3 | 1.0 | 1.7 | 1.5 | 202 | 630 | 8.1 |
| | 36 | 60 | 50 | 36 | 155 | 199 | 2 | 26 | 164 | 0.39 | 2.9 | 700 | 1.9 | 0.9 | 2.2 | 2.2 | 187 | 570 | 7.8 |
| | 37 | 61 | 49 | 33 | 156 | 199 | 2 | 36 | 166 | 0.40 | 3.0 | 650 | 1.9 | 0.8 | 2.1 | 1.9 | 187 | 600 | 8.1 |
| | 38 | 65 | 30 | 32 | 116 | 172 | 2 | 24 | 126 | 0.32 | 2.2 | 750 | 1.9 | 2.6 | 1.2 | 2.3 | 214 | 620 | 10.1 |
| | 39 | 59 | 47 | 33 | 152 | 184 | 2 | 26 | 164 | 0.40 | 2.9 | 740 | 1.9 | 0.8 | 1.9 | 1.7 | 188 | 560 | 7.9 |
| | 40 | 72 | 48 | 40 | 178 | 237 | 2 | 12 | 184 | 0.40 | 3.0 | 1110 | 2.4 | 0.8 | 2.6 | 2.8 | 191 | 670 | 7.7 |
| | 41 | 78 | 53 | 42 | 195 | 241 | 2 | 14 | 214 | 0.47 | 3.3 | 930 | 2.6 | 0.8 | 2.0 | 1.5 | 195 | 670 | 8.7 |
| | 42 | 70 | 45 | 39 | 174 | 224 | 2 | 278 | 177 | 0.41 | 3.1 | 1140 | 2.3 | 0.8 | 2.4 | 2.5 | 191 | 650 | 8.2 |
| | 43 | 52 | 34 | 31 | 123 | 176 | 2 | 54 | 112 | 0.34 | 2.0 | 880 | 1.6 | 0.8 | 4.5 | 6.8 | 189 | 520 | 6.4 |
| | 44 | 68 | 42 | 36 | 165 | 217 | 2 | 30 | 162 | 0.39 | 2.8 | 930 | 2.1 | 0.8 | 3.0 | 3.7 | 187 | 670 | 7.6 |
| | 45 | 81 | 52 | 42 | 197 | 238 | 3 | 18 | 213 | 0.46 | 3.5 | 1170 | 2.6 | 0.8 | 1.9 | 1.4 | 193 | 700 | 8.7 |
| | 46 | 80 | 52 | 43 | 188 | 242 | 3 | 50 | 203 | 0.45 | 3.4 | 1350 | 2.5 | 0.9 | 1.9 | 1.5 | 198 | 720 | 8.5 |
| | 47 | 55 | 28 | 32 | 126 | 161 | 2 | 12 | 130 | 0.33 | 2.2 | 660 | 1.9 | 1.8 | 1.6 | 2.7 | 211 | 550 | 7.5 |
| | 48 | 66 | 42 | 37 | 159 | 208 | 2 | 26 | 157 | 0.38 | 2.5 | 820 | 2.1 | 0.9 | 2.7 | 3.5 | 195 | 610 | 7.0 |
| | 49 | 34 | 11 | 23 | 82 | 135 | 1 | 72 | 53 | 0.15 | 1.0 | 580 | 1.0 | 1.1 | 5.2 | 9.3 | 194 | 380 | 4.3 |
| | 50 | 54 | 32 | 37 | 134 | 178 | 2 | 2 | 114 | 0.31 | 1.0 | 600 | 1.7 | 0.8 | 3.1 | 10.6 | 200 | 490 | 5.7 |

Table 3. Concentrations of elements in cored sediments from Lake Winnipeg.

| Core No. | Core Location | Depth in Core | Ni | Cu | Co | Cr | Zn | Cd | Pb | V | Ti | Fe | Mn | Be | Na | Mg | Ca | Sr | P | Al |
|---|---|---|---|---|---|---|---|---|---|---|---|---|---|---|---|---|---|---|---|---|
| | | Cm | (ppm) | (ppm) | (ppm) | (ppm) | (ppm) | % | (ppm) | (ppm) | % | % | (ppm) | (ppm) | % | % | % | (ppm) | (ppm) | % |
| 1. | South end of | 0-5 | 62 | 35 | 29 | 171 | 239 | 2 | 36 | 239 | 0.38 | 2.5 | 810 | 2.4 | 0.2 | 1.9 | 2.2 | 131 | 580 | 7.3 |
| | south basin - | 10-15 | 49 | 998 | 29 | 148 | 216 | 2 | 24 | 226 | 0.39 | 2.6 | 860 | 2.3 | 0.3 | 2.0 | 2.3 | 133 | 670 | 7.5 |
| | near Red River | 20-25 | 50 | 32 | 29 | 149 | 203 | 2 | 10 | 221 | 0.39 | 2.5 | 970 | 2.1 | 0.3 | 2.0 | 2.3 | 136 | 540 | 7.1 |
| | | 30-35 | 46 | 29 | 29 | 129 | 214 | 2 | 16 | 202 | 0.40 | 2.3 | 1060 | 2.4 | 0.3 | 2.1 | 3.0 | 142 | 520 | 6.8 |
| | | 50-55 | 48 | 29 | 29 | 139 | 197 | 2 | 24 | 210 | 0.40 | 2.2 | 1190 | 2.3 | 0.4 | 2.2 | 2.8 | 149 | 550 | 6.8 |
| | | 70-75 | 49 | 29 | 29 | 142 | 189 | 2 | 16 | 223 | 0.40 | 2.3 | 1410 | 2.2 | 0.3 | 2.0 | 2.8 | 140 | 540 | 6.9 |
| 2. | South of centre | 0-5 | 60 | 49 | 29 | 166 | 210 | 3 | 24 | 239 | 0.43 | 3.1 | 960 | 2.0 | 0.3 | 2.0 | 1.8 | 131 | 690 | 7.9 |
| | of south | 10-15 | 61 | 37 | 31 | 145 | 207 | 2 | 24 | 238 | 0.42 | 2.9 | 1170 | 2.2 | 0.4 | 2.0 | 1.6 | 140 | 670 | 7.8 |
| | basin | 20-25 | 58 | 35 | 30 | 156 | 193 | 2 | 20 | 233 | 0.45 | 3.0 | 1220 | 2.0 | 0.3 | 2.1 | 1.9 | 134 | 640 | 7.7 |
| | | 30-35 | 60 | 37 | 32 | 138 | 196 | 2 | 16 | 241 | 0.40 | 2.8 | 1250 | 2.3 | 0.4 | 2.0 | 1.6 | 135 | 640 | 7.4 |
| | | 50-55 | 61 | 53 | 32 | 150 | 194 | 2 | 16 | 241 | 0.40 | 2.8 | 1220 | 2.3 | 0.4 | 2.0 | 1.5 | 135 | 630 | 7.3 |
| | | 70-75 | 62 | 37 | 32 | 145 | 197 | 2 | 20 | 240 | 0.37 | 2.6 | 1190 | 2.3 | 0.4 | 2.0 | 1.5 | 131 | 650 | 7.2 |
| 3. | Centre of | 0-5 | 66 | 46 | 35 | 181 | 216 | 2 | 26 | 253 | 0.48 | 3.5 | 750 | 2.7 | 0.4 | 1.9 | 1.3 | 147 | 730 | 8.9 |
| | south basin | 10-15 | 67 | 42 | 37 | 175 | 206 | 3 | 20 | 254 | 0.45 | 3.4 | 1140 | 2.3 | 0.5 | 1.8 | 1.2 | 155 | 730 | 8.4 |
| | | 20-25 | 66 | 39 | 35 | 181 | 204 | 2 | 14 | 251 | 0.48 | 3.5 | 970 | 2.3 | 0.5 | 1.9 | 1.3 | 147 | 700 | 8.6 |
| | | 30-35 | 64 | 38 | 33 | 158 | 194 | 2 | 16 | 235 | 0.46 | 3.2 | 670 | 2.2 | 0.5 | 1.8 | 1.4 | 156 | 680 | 8.1 |
| | | 50-55 | 68 | 39 | 38 | 182 | 206 | 2 | 18 | 259 | 0.51 | 3.7 | 1050 | 2.3 | 0.5 | 2.0 | 1.4 | 160 | 730 | 9.2 |
| | | 70-75 | 65 | 38 | 35 | 165 | 203 | 2 | 18 | 254 | 0.46 | 3.4 | 1100 | 2.3 | 0.4 | 1.8 | 1.2 | 148 | 690 | 8.4 |
| 4. | North of centre | 0-5 | 61 | 46 | 37 | 184 | 224 | 2 | 20 | 260 | 0.47 | 2.3 | 1390 | 2.9 | 0.5 | 1.6 | 1.2 | 155 | 600 | 8.8 |
| | of south | 10-15 | 60 | 42 | 35 | 166 | 219 | 2 | 32 | 245 | 0.45 | 2.2 | 1280 | 3.0 | 0.4 | 1.6 | 1.2 | 155 | 630 | 8.8 |
| | basin | 20-25 | 64 | 45 | 37 | 187 | 230 | 2 | 24 | 276 | 0.47 | 3.2 | 1390 | 3.1 | 0.4 | 1.7 | 1.2 | 160 | 610 | 8.9 |
| | | 30-35 | 65 | 39 | 37 | 172 | 225 | 2 | 20 | 267 | 0.47 | 3.2 | 1420 | 2.8 | 0.4 | 1.6 | 1.2 | 159 | 600 | 8.9 |
| | | 50-55 | 64 | 39 | 36 | 164 | 219 | 2 | 8 | 259 | 0.49 | 3.1 | 1370 | 2.9 | 0.4 | 1.6 | 1.2 | 158 | 590 | 8.9 |
| | | 70-75 | 64 | 41 | 37 | 178 | 222 | 2 | 12 | 270 | 0.48 | 3.3 | 1380 | 3.0 | 0.4 | 1.7 | 1.2 | 159 | 610 | 9.2 |
| 5. | Traverse Bay - | 0-5 | 43 | 29 | 27 | 118 | 169 | 2 | 26 | 141 | 0.34 | 1.8 | 580 | 1.9 | 0.7 | 2.9 | 4.3 | 179 | 490 | 6.0 |
| | near Winnipeg | 10-15 | 43 | 28 | 27 | 126 | 157 | 2 | 14 | 143 | 0.32 | 1.7 | 600 | 1.7 | 0.7 | 2.7 | 4.0 | 184 | 490 | 5.9 |
| | River | 20-25 | 40 | 23 | 25 | 116 | 143 | 2 | 12 | 115 | 0.28 | 1.5 | 520 | 1.4 | 0.8 | 3.2 | 5.1 | 191 | 460 | 5.5 |
| | | 30-35 | 38 | 25 | 25 | 109 | 138 | 1 | 8 | 112 | 0.28 | 1.5 | 510 | 1.4 | 0.8 | 3.0 | 4.8 | 194 | 450 | 5.6 |
| | | 50-55 | 39 | 22 | 25 | 112 | 136 | 1 | 26 | 109 | 0.27 | 1.4 | 490 | 1.3 | 0.7 | 3.4 | 5.5 | 187 | 440 | 5.1 |
| | | 70-75 | 36 | 17 | 22 | 97 | 134 | 1 | 18 | 97 | 0.28 | 1.4 | 390 | 1.1 | 0.8 | 3.7 | 6.0 | 197 | 450 | 5.3 |
| 6. | North End of | 0-5 | 61 | 42 | 36 | 177 | 216 | 2 | 26 | 244 | 0.45 | 3.2 | 1240 | 2.7 | 0.4 | 1.6 | 1.3 | 159 | 620 | 8.6 |
| | south basin | 10-15 | 66 | 44 | 38 | 189 | 229 | 2 | 34 | 259 | 0.49 | 3.4 | 1580 | 2.9 | 0.5 | 1.7 | 1.3 | 171 | 640 | 8.9 |
| | | 20-25 | 63 | 40 | 36 | 185 | 208 | 2 | 30 | 241 | 0.48 | 3.2 | 1410 | 2.7 | 0.5 | 1.6 | 1.3 | 169 | 590 | 8.5 |
| | | 30-35 | 62 | 37 | 36 | 178 | 205 | 2 | 22 | 248 | 0.48 | 3.1 | 1380 | 2.7 | 0.5 | 1.6 | 1.3 | 174 | 600 | 8.6 |
| | | 50-55 | 66 | 39 | 39 | 186 | 216 | 2 | 18 | 251 | 0.51 | 3.3 | 1380 | 2.7 | 0.5 | 1.7 | 1.4 | 178 | 600 | 8.9 |
| | | 70-75 | 61 | 38 | 35 | 170 | 208 | 2 | 30 | 236 | 0.47 | 3.0 | 1170 | 2.6 | 0.5 | 1.6 | 1.4 | 177 | 590 | 8.4 |
| 7. | South End of | 0-5 | 58 | 37 | 30 | 141 | 154 | 2 | 16 | 151 | 0.36 | 2.1 | 780 | 2.1 | 1.3 | 1.1 | 1.8 | 211 | 580 | 7.2 |
| | narrows | 10-15 | 52 | 44 | 29 | 134 | 152 | 2 | 14 | 154 | 0.37 | 2.1 | 880 | 2.0 | 1.3 | 1.1 | 1.8 | 211 | 580 | 7.2 |
| | | 20-25 | 59 | 41 | 30 | 144 | 161 | 2 | 12 | 171 | 0.40 | 2.3 | 970 | 2.1 | 1.1 | 1.2 | 1.7 | 209 | 580 | 7.4 |
| | | 30-35 | 57 | 44 | 31 | 143 | 165 | 2 | 22 | 171 | 0.41 | 2.3 | 870 | 2.1 | 1.3 | 1.3 | 1.8 | 211 | 590 | 7.9 |
| | | 50-55 | 59 | 40 | 30 | 148 | 166 | 2 | 20 | 172 | 0.42 | 2.4 | 900 | 2.3 | 1.2 | 1.3 | 1.7 | 210 | 570 | 7.7 |
| | | 70-75 | 52 | 40 | 29 | 138 | 160 | 2 | 20 | 167 | 0.40 | 2.2 | 830 | 2.3 | 1.3 | 1.2 | 1.7 | 211 | 590 | 7.9 |
| 8. | North End of | 0-5 | 61 | 50 | 32 | 159 | 183 | 2 | 16 | 179 | 0.41 | 2.4 | 640 | 2.4 | 1.0 | 1.4 | 1.6 | 207 | 580 | 7.4 |
| | narrows | 10-15 | 61 | 46 | 32 | 171 | 193 | 2 | 14 | 200 | 0.45 | 2.8 | 710 | 2.5 | 0.9 | 1.6 | 1.6 | 202 | 580 | 7.5 |
| | | 20-25 | 64 | 47 | 36 | 172 | 199 | 2 | 20 | 214 | 0.49 | 2.9 | 640 | 2.6 | 0.9 | 1.6 | 1.5 | 203 | 590 | 7.8 |
| | | 30-35 | 66 | 47 | 36 | 173 | 198 | 2 | 12 | 213 | 0.48 | 2.8 | 670 | 2.7 | 0.9 | 1.5 | 1.6 | 205 | 590 | 8.6 |
| | | 50-55 | 61 | 40 | 37 | 167 | 188 | 2 | 20 | 201 | 0.46 | 2.8 | 620 | 2.4 | 1.0 | 1.5 | 1.6 | 207 | 590 | 8.6 |
| | | 70-75 | 57 | 42 | 31 | 156 | 190 | 2 | 28 | 190 | 0.45 | 2.5 | 600 | 2.6 | 1.3 | 1.4 | 1.7 | 211 | 610 | 8.3 |
| 9. | Centre of | 0-5 | 66 | 46 | 35 | 176 | 208 | 3 | 20 | 187 | 0.45 | 3.1 | 580 | 2.5 | 0.8 | 2.0 | 1.9 | 196 | 620 | 8.6 |
| | north basin | 10-15 | 76 | 53 | 41 | 189 | 216 | 2 | 24 | 207 | 0.47 | 3.2 | 750 | 2.6 | 0.8 | 2.0 | 1.7 | 198 | 640 | 8.7 |
| | | 20-25 | 73 | 53 | 39 | 186 | 220 | 3 | 16 | 203 | 0.45 | 3.2 | 830 | 2.5 | 0.7 | 2.0 | 1.5 | 198 | 630 | 8.5 |
| | | 30-35 | 74 | 48 | 37 | 197 | 222 | 2 | 16 | 207 | 0.47 | 3.4 | 810 | 2.5 | 0.7 | 2.1 | 1.7 | 196 | 620 | 8.7 |
| | | 50-55 | 73 | 53 | 39 | 196 | 224 | 2 | 16 | 211 | 0.49 | 3.3 | 800 | 2.7 | 0.8 | 2.1 | 1.7 | 201 | 630 | 9.0 |
| | | 70-75 | 67 | 50 | 35 | 194 | 211 | 2 | 26 | 199 | 0.43 | 3.0 | 770 | 2.4 | 0.7 | 2.0 | 1.6 | 199 | 600 | 8.2 |
| 10. | West side of | 0-5 | 64 | 42 | 32 | 163 | 204 | 2 | 20 | 172 | 0.41 | 3.1 | 730 | 2.1 | 0.6 | 2.4 | 2.5 | 183 | 590 | 7.9 |
| | north basin - | 10-15 | 67 | 41 | 35 | 176 | 214 | 2 | 10 | 181 | 0.42 | 3.1 | 970 | 2.4 | 0.7 | 2.5 | 2.5 | 190 | 620 | 8.0 |
| | near Saskatch- | 20-25 | 66 | 42 | 35 | 170 | 202 | 2 | 8 | 177 | 0.41 | 3.1 | 1280 | 2.1 | 0.6 | 2.6 | 2.8 | 181 | 610 | 7.6 |
| | ewan River | 30-35 | 70 | 43 | 37 | 181 | 211 | 3 | 16 | 187 | 0.45 | 3.2 | 1490 | 2.5 | 0.7 | 2.7 | 3.2 | 192 | 700 | 8.1 |
| | | 50-55 | 68 | 44 | 36 | 174 | 205 | 2 | 6 | 187 | 0.46 | 3.3 | 1480 | 2.4 | 0.7 | 2.6 | 2.9 | 190 | 670 | 8.2 |
| | | 70-75 | 72 | 45 | 38 | 179 | 215 | 2 | 18 | 191 | 0.49 | 3.4 | 1190 | 2.6 | 0.7 | 2.7 | 3.0 | 193 | 640 | 8.6 |
| 11. | North End of | 0-5 | 72 | 50 | 39 | 185 | 224 | 2 | 22 | 199 | 0.43 | 3.2 | 1360 | 2.7 | 0.8 | 1.7 | 1.5 | 198 | 740 | 7.9 |
| | north basin | 10-15 | 76 | 48 | 39 | 195 | 231 | 2 | 18 | 210 | 0.47 | 3.5 | 1080 | 2.9 | 0.8 | 1.9 | 1.5 | 198 | 710 | 8.5 |
| | | 20-25 | 58 | 36 | 27 | 145 | 164 | 2 | 20 | 153 | 0.35 | 3.3 | 670 | 2.0 | 0.5 | 1.4 | 1.1 | 161 | 510 | 6.1 |
| | | 30-35 | 78 | 50 | 40 | 186 | 218 | 2 | 28 | 211 | 0.51 | 3.5 | 860 | 2.7 | 0.8 | 2.1 | 1.6 | 199 | 670 | 9.0 |
| | | 50-55 | 81 | 50 | 43 | 190 | 223 | 2 | 26 | 223 | 0.50 | 3.5 | 940 | 2.7 | 0.9 | 2.1 | 1.5 | 199 | 670 | 9.1 |
| | | 70-75 | 75 | 49 | 40 | 191 | 223 | 2 | 16 | 216 | 0.53 | 3.6 | 990 | 2.9 | 0.8 | 2.1 | 1.7 | 198 | 660 | 9.0 |

We include this raw data here because of our general criticism that such multielement information is scarce in the literature. Using the location maps and raw data tables (Figure 1, Tables 2 and 3; Figure 3, Table 4), a multitude of single element distribution diagrams can be constructed. Although much information can be extracted from such diagrams (e.g. Hg in the Great Lakes (Thomas, 1972; 1973) we do not present these here.

The sample techniques employed at other sites (Sudbury, Red Lake, Bear-Slave area-tundra lakes) varied from the use of extension soil augers to Eckman dredges and Phleger corers. Detailed descriptions can be found in Allan et al. (1973) and Allan and Timperley (1975). The raw data used here can be located in Allan and Timperley, 1975 and in the maps accompanying Allan et al. (1973).

The data sets, in $\mu$ moles/g dry weight were used in linear regression analysis to obtain correlation coeffficients for all possible pairs of elements. Where significant, the

Table 4. Mercury and arsenic in sediment cores from the south basin of Lake Winnipeg.

| Depth in Core | Core Number | | | | | | | | | Core Number | | | | | | | | |
|---|---|---|---|---|---|---|---|---|---|---|---|---|---|---|---|---|---|---|
| | 1 | 2 | 3 | 4 | 5 | 6 | 7 | 8 | 9 | 1 | 2 | 3 | 4 | 5 | 6 | 7 | 8 | 9 |
| cm | Hg in ppb | | | | | | | | | As in ppm | | | | | | | | |
| 0-2 | 94 | 96 | 106 | 150 | 120 | 94 | 54 | 80 | 120 | 2 | 1 | 5 | 5 | 6 | 9 | 4 | 2 | 5 |
| 2-4 | 225 | 120 | 106 | 94 | 285 | 94 | 80 | 80 | 106 | 1 | 7 | 3 | 3 | 6 | 7 | 5 | 4 | 3 |
| 4-6 | 150 | 150 | 106 | 94 | 150 | 106 | 94 | 54 | 68 | 3 | 6 | 3 | 3 | 1 | 5 | 2 | 4 | 4 |
| 6-8 | 195 | 150 | 120 | 100 | 165 | 94 | 68 | 54 | 54 | 7 | 3 | 5 | 3 | 1 | 4 | 7 | 4 | 7 |
| 8-10 | 165 | 150 | 106 | 94 | 120 | 94 | 106 | 42 | 54 | 7 | 5 | 7 | 3 | 1 | 5 | 8 | 2 | 5 |
| 10-14 | 180 | 135 | 120 | 61 | 113 | 76 | 87 | 35 | 54 | 7 | 1 | 8 | 1 | 1 | 6 | 4 | 2 | 2 |
| 14-18 | 150 | 135 | 100 | 70 | 158 | 61 | 61 | 28 | 48 | 8 | 1 | 5 | 1 | 2 | 6 | 5 | 1 | 3 |
| 18-22 | 61 | 135 | 74 | 87 | 94 | 54 | 67 | 28 | 48 | 7 | 4 | 2 | 1 | 2 | 6 | 5 | 1 | 2 |
| 22-30 | 60 | 110 | 77 | 95 | 111 | 57 | - | 33 | 45 | 6 | 3 | 2 | 4 | 2 | 6 | - | 1 | 1 |
| 30-40 | 68 | 61 | 63 | 80 | 87 | 68 | - | 28 | 42 | 5 | 2 | 2 | 3 | 3 | 4 | - | 1 | 1 |
| 40-50 | 72 | 68 | 83 | 61 | 71 | 68 | - | 28 | 45 | 2 | 3 | 2 | 2 | 3 | 2 | - | 2 | 1 |
| 50-60 | 54 | 61 | 74 | 54 | 109 | 56 | - | 28 | 42 | 4 | 4 | 2 | 4 | 3 | 2 | - | 1 | 1 |
| 60-70 | 68 | 80 | 73 | 68 | 74 | 56 | - | - | 28 | 5 | 5 | 3 | 4 | 3 | 1 | - | - | 1 |

1. The top of core 1 may well have been disturbed by dredging.
Because of the much higher water flows from the Winnipeg River, Traverse Bay where cores 7, 8 and 9 were collected may be an area of erosion rather than deposition.
1 ppm As is the detection limit, all values shown as 1 are less than 1 ppm.

slopes of the linear relationships between selected element pairs (RAV) were used for comparison between data sets.

## Results and Discussion

Although we are aware of the valuable information supplied by single element concentrations alone relative to geochemical cycles and anthropogenic sources of elements in drainage sediments, our proposal is that a more general and geographically independent index of element geochemistry or contamination should be developed. This index must necessarily involve more than the concentration of one element. It should permit comparisons to be made between diverse geological and limnological environments. A few workers have made previous efforts along similar lines. These include Cowgill and Hutchinson, 1966; Thomas, 1972; Bruland *et al.*, 1974; Thomas and Jacquet, 1976; and Kemp *et al.*, 1976.

The literature indicates that geochemists lean towards the use of ratios rather than concentrations alone, but studies so far have not involved data from widely separated sites of different geological and limnological characteristics so as to determine geochemical normalcy or contamination indexes in a more continental or global sense.

Element Concentrations Distribution patterns of element concentrations for heavy metals and nutrients in dredged and cored sediments of lakes provide valuable clues as to the existing or changing geochemical status of a drainage basin. The predominance of study sites in the literature are ones of serious known or suspected contamination. This has been usually established by determination of concentrations of elements in dredged sediments taken at sucessive distances from the suspected source or by analysis of sectioned cores at critical locations. However, direct comparison of degree of contamination between different sites has remained a problem. Examples of the site specific type of studies reported in the literature include mercury in Lake Ontario (Thomas, 1972) Lake Huron (Thomas, 1973), Lake Geneva (Vernet and Thomas, 1972), Lake Erie (Wolery and Walters, 1974), Lake Maggiore (Damiani and Thomas, 1974), Lake St. Clair (Thomas et al., 1975), and lakes in the English Lake District (Aston et al., 1973); arsenic in Lake Washington (Shapiro et al., 1971; Crecelius, 1975); lead in Lake Washington (Crecelius and Piper, 1973), Lake Balkhash (Lubchenko and Turovskii, 1973); heavy metal suites in Lake Michigan (Shimp et al., 1970 and 1971), in Lake Erie (Wolery et al., 1974; Kemp et al., 1976), in Lake Constance and the Tegernsee (Forstner and Muller, 1974), in Wisconsin lakes (Iskandar and Kenney, 1974), in the Baltic Sea (Erlenkeuser et al., 1974; Hallberg, 1974), in various estuaries (Muller and Forstner, 1974; Ali et al., 1975; Harding and Brown, 1975), and in off-shore basins (Chow et al., 1973; Bruland et al., 1974).

The type of distinct heavy metal contamination detected in many of the above studies does not occur in the bottom sediments of Lake Winnipeg (Tables 2,3 and 4). The cored sections provided good surface samples, as the micro-oxidised zone was distinct in all cores. The general similarity in concentrations between the dredged samples and surface cored samples substantiates this conclusion. There was also a strong similarity between the concentrations of many elements in suspended sediments from the Red River and its main tributary the Assiniboine River and those in the bottom sediments of the south basin of Lake Winnipeg (Tables 2 and 5)

Table 5. Concentrations of elements in suspended sediments of the Red River.

| Sample[1] Location | Ni ppm | Cu ppm | Co ppm | Cr ppm | Zn ppm | Cd ppm | Pb ppm | As ppm | V ppm | Ti ppm |
|---|---|---|---|---|---|---|---|---|---|---|
| Red River 1. | 48 | 34 | 30 | 143 | 198 | 2.0 | 20 | 5 | 224 | 0.37 |
| Assiniboine River 1. | 45 | 45 | 29 | 130 | 193 | 1.9 | 20 | 6 | 202 | 0.35 |
| Red River 2. | 43 | 32 | 26 | 130 | 184 | 1.8 | 10 | 7 | 199 | 0.35 |

| Sample Location | Al % | Fe % | Mn ppm | Be ppm | Na % | Mg % | Ca % | Sr ppm | P ppm |
|---|---|---|---|---|---|---|---|---|---|
| Red River 1. | 7.2 | 2.5 | 1029 | 1.9 | 0.3 | 2.0 | 3.4 | 136 | 592 |
| Assiniboine River 1. | 6.6 | 2.4 | 1017 | 2.0 | 0.4 | 2.1 | 3.9 | 151 | 568 |
| Red River 2. | 6.8 | 2.3 | 1010 | 1.8 | 0.3 | 2.2 | 3.9 | 148 | 579 |

1. The Assiniboine River is the main tributary of the Red River. The confluence of the rivers is in the centre of Winnipeg. Winnipeg is circled by a perimeter road, which crosses the Red River south and north of the city, and the Assinboine to the west of the city. Sample Red River 1 comes from the south perimeter crossing. Sample Assinboine 1 comes from the west perimeter crossing. Sample Red River 2 comes from Lockport bewteen Winnipeg and Lake Winnipeg. The Red River flows from south to north and the Assinboine from west to east. The Assiniboine enters the Red between samples Red River 1 and 2.

For most elements determined her, Lake Winnipeg sediments have not yet recorded any significant impact of agricultural, industrial or urban-domestic activities in the drainage basin. A significant exception to this general statement is for mercury (Table 4). Mercury contamination of fish in Lake Winnipeg resulted in a two year closure of the fishing industry on the lake. The source of the mercury is thought to be industrial activities in northwestern Ontario. Mercury would reach Lake Winnipeg via the Winnipeg River, which flows into Traverse Bay (Figures 1 and 2). Hg would be expected to accumulate in the surface sediments of the Bay. Results of sediment analyses (Figure 2 and Table 4) indicates an increase Hg concentration in sediments derived from the Red River. A first analysis (Table 4) did not reveal high concentrations in Traverse Bay sediments. A second analysis by a second laboratory did not change the mean results for the sediments of Red River origin (150 ppb vs 148 ppb mean Hg). However, reanalysis of Traverse Bay sediments resulted in two concentrations of 520 and 512 ppb respectively. These isolated values may be the result of Hg associated with randomly distributed wood fibers seen in some of the Traverse Bay sediments during sample collection.

In addition to the above there were geochemical differences between the sediments of the north and south basin. These differences are discussed below in the section on Relative Atomic Variation. Some of the concentrations (Zn and V) were higher than recorded in the literature (Bruland et al.,1974; Kemp et al., 1976) for contaminated sites. This is partly a function of analytical methods, and emphasises the problem of comparison of data between sites of different geology and hydrology.

## Relative Atomic Variation for Elements in Lake Winnipeg

Linear regression analyses were performed on the Lake Winnipeg data set[1] (Table 2) to determine correlation coefficients for all possible pairs of elements Highly significant correlations, both positive and negative, existed for a large number of element combinations. The heavy metals,(Ni, Cu, Co, Cr, Zn, V, Ti, Fe, Mn) exhibited a high degree of positive correlation. These elements are probably present in the heavy mineral fraction of the sediment in minerals such as spinels, ilmenite, rutile and beryl. It is also likely that this suite includes elements found in amorphous iron-manganese-aluminum coatings of minerals. In the south basin, a second suite was Na, Sr and K[2]. This most surely reflects the coarse grained constituents of the sediment, and in particular the feldspar component. One element, Pb, remained unrelated to the others. High concentrations of Pb were occasionally found (Table 2) and may indicate contamination. Its unique lack of correlation with other elements may indicate that its chemistry is dominated by different parameters,

1. From this point on all concentrations which are given in the Tables in ppm or % were converted to $\mu$. moles/gram for calculation of correlation coefficients and Relative Atomic Variation.
2. The K data is not included in Table 2. It will appear in a forthcoming publication by Brunskill et al. (in prep.).

perhaps the sulphur cycle.

In addition, correlation matrices for other Canadian lakes were calculated. In general the patterns were similar to those for the Lake Winnipeg sediments.

Where significant correlations existed e.g. Cr/Fe, the slopes of the linear regression were calculated (Table 6).

Table 6. Relative Atomic Variation (RAV)[1] of elements in Lake Winnipeg sediments.

| | | RAV (Trace Element/Fe) | | | | | | |
|---|---|---|---|---|---|---|---|---|
| Surface Sediments | n | Ni | Cu | Co | Cr | V | P | Zn |
| South Basin | 21 | 10 | 8 | 5 | 38 | 73 | 167 | 35 |
| North Basin | 21 | 12 | 12 | 4 | 38 | 55 | 168 | 28 |

| | | RAV (Trace Element/Ti) | | | | | | |
|---|---|---|---|---|---|---|---|---|
| Surface Sediments | n | Ni | Cu | Co | Cr | V | P | Zn |
| South Basin | 21 | 82 | 66 | 42 | 305 | 559 | 1270 | 276 |
| North Basin | 21 | 111 | 107 | 45 | 343 | 484 | 1390 | 249 |

Although it is evident from Table 2 that the concentrations of chromium in the sediments of the north basin are in general higher than those in the south basin (North y=1.17 u moles Cr/g, South y=0.76 u moles Cr/g), the slope of the Cr/Fe relationship is very similar for both basins. Thus within these sample suites, a similar geochemical balance between concentrations of the two elements is maintained over a wide range of concentrations for both elements. Other highly correlated pairs of elements can be similarily used to determine Relative Atomic Variation Table 6.

The major elements, except under rare circumstances, are unlikely to have enhanced concentrations related to anthropogenic activities. The conservative major elements that might be used are Fe, Al, Si, Ti, Na, K or Ca in carbonate-rich sediments. The relatively high concentrations of these elements even if enhanced by cultural activities makes them less sensitive to change.

For Lake Winnipeg, we present RAV values for several elements that may be enhanced in lake sediments because of industrial, agricultural, or urban-domestic developments relative to Fe and Ti, which we considered to be conservative elements in this well aerated lake. These results are based on the dredged samples alone, because we have already established (by comparison of dredge and core analysis) that the surface samples have similar concentrations to older (70-90 cm deep) sediments (Table 3). The results (Table 6) indicate that Cu and perhaps Ni are relatively enriched in north basin relative to Fe or Ti compared to south basin sediments. Vanadium is enriched relative to Fe and Ti in the south basin compared to the north basin sediments. In California, higher V/Al ratios in the sediments of the Santa Barbara Basin were attributed to natural oil seeps containing vanadium porphyrins (Bruland et al., 1974). A similar explanation for the V in sediments of the Red River (Table 5) and the south basin (Table 2) may be the recent discovery of some 2 gallons of oil per ton of Pleistocene clays in the Red River valley (Russell, Univ. Man., Pers.Comm.) Cr, Co and P were slightly enhanced relative to Ti in the north basin compared to the south basin. Zn appeared to be enriched in south basin sediments. This may be due to geology, and may be related to Zn rich shales or dolomites in the Red River basin.

Relative Atomic Variation for Elements in Sediments from Other Canadian Sites Compared to Lake Winnipeg

Published data which might allow RAV comparisons to be made is extremely scarce. Analyses are often confined to one element, or to a suite of trace elements. However, a more common failing, and one of which we were also guilty when we examined out data files, is the tendency to analyse for only trace metals in pollution studies or for only major elements in geological studies. If at one site we could obtain RAV for trace element/Si we discovered that we did not have Si analyses for other sites. Thus from the vast numbers of analyses that exist, the comparable suites were rare indeed. Even for trace element/trace element combinations, where the situation was better, it was often the case that the analytical suite varied because of the different objectives of the studies. We were also reluctant to use trace element/trace element combinations because of a general lack of information on detailed geochemical cycles of trace elements in the surficial environment in general and surficial sediments in particular.

Trace Element /Conservative Element RAV

No complete data set from one large lake such as Lake Winnipeg has been published. It was however, assumed that RAV could be calculated by using data from a set of small lakes all within one small physiographic and geological area. Thus, data from lakes in the Red Lake, Sudbury and Great Bear Lake-Great Slave Lake areas (Table 1) were used. Calculations of RAV (Table 7) for Ni/Fe, Zn/Fe and Cu/Fe for these sites were determined for comparison with sediment geochemical data

1. RAV = the numerically expressed slope ($x10^{-4}$) of the linear regression of each trace element against Fe or Ti. All regressions are significant at the 95% level. "n" refers to the number of sediment samples analysed. RAV can also be thought of as the "average ratio".

from the north and south basins of Lake Winnipeg.

Table 7. Relative Atomic Variation (RAV) of elements in sediments from various lakes and lake areas in Canada.
Trace Element/Major Element Combinations[1].

| Sub-Site[1] | n | Ni/Fe r | Ni/Fe RAVx$10^{-4}$ | Zn/Fe r | Zn/Fe RAVx$10^{-4}$ | Cu/Fe r | Cu/Fe RAVx$10^{-4}$ |
|---|---|---|---|---|---|---|---|
| North Basin | 21 | 0.771 | 12 | 0.769 | 28 | 0.869 | 12 |
| South Basin | 21 | 0.871 | 10 | 0.891 | 35 | 0.866 | 8 |
| Red Lake | 29 | 0.617 | 5 | 0.436 | 8 | 0.340 | ns |
| Snib Lake | 16 | 0.584 | 32 | 0.549 | 48 | 0.311 | ns |
| Sully Lake | 20 | 0.329 | ns | 0.864 | 32 | 0.528 | 14 |
| Gullrock Lake | 20 | 0.914 | 23 | 0.914 | 26 | 0.933 | 15 |
| Sudbury Lakes | 22 | 0.316 | ns | 0.534 | 25 | 0.320 | ns |
| Agricola | 21 | 0.521 | 10 | 0.375 | ns | 0.171 | ns |
| Bear | 25 | 0.867 | 9 | 0.364 | ns | 0.308 | ns |
| Hepburn | 26 | 0.742 | 9 | 0.782 | 54 | 0.690 | 12 |
| Granite | 26 | 0.725 | 12 | 0.814 | 20 | 0.479 | 9 |
| Gneiss | 26 | 0.716 | 14 | 0.736 | 21 | 0.555 | 14 |
| Volcanic | 20 | 0.780 | 11 | 0.582 | 12 | 0.891 | 11 |

For sediments from these other sites, high correlations existed between Ni and Fe. The exceptions were the Sudbury area and Sully Lake. The former correlation may be lacking because of variable and massive airborne input of Ni relative to Fe (Figure 3) from the Sudbury smelters.

The latter may be a result of extremely low and uniform Ni concentrations in Sully Lake sediments. The Sully Lake watershed is in acid volcanic and granitic bedrock. The high RAV for Snib lake is a trend towards the Sudbury example. However, in this case the enhanced Ni is due to drilling of basic bedrock beneath the lake with associated contamination of the lake sediments (Allan and Timperley, 1975).

Whereas the Sudbury and Red Lake areas (excluding Red Lake itself) have sediments which are commonly referred to as gyttja, the Bear-Slave lakes and Lake Winnipeg have sediments which are of inorganic silt and clay composition. It thus seems that at geographically widely separated sites, where the sediment is primarily of inorganic composition, that Ni/Fe RAV values are generally consistent, in spite of diverse geological and limnological environments. We must remember that the Ni and Fe concentrations alone are quite different at these different locations. It is the average ratio Ni/Fe of a set of samples that is similar (Table 7). This similarity may mean that the normal geochemical balance of these elements is maintained at all sites.

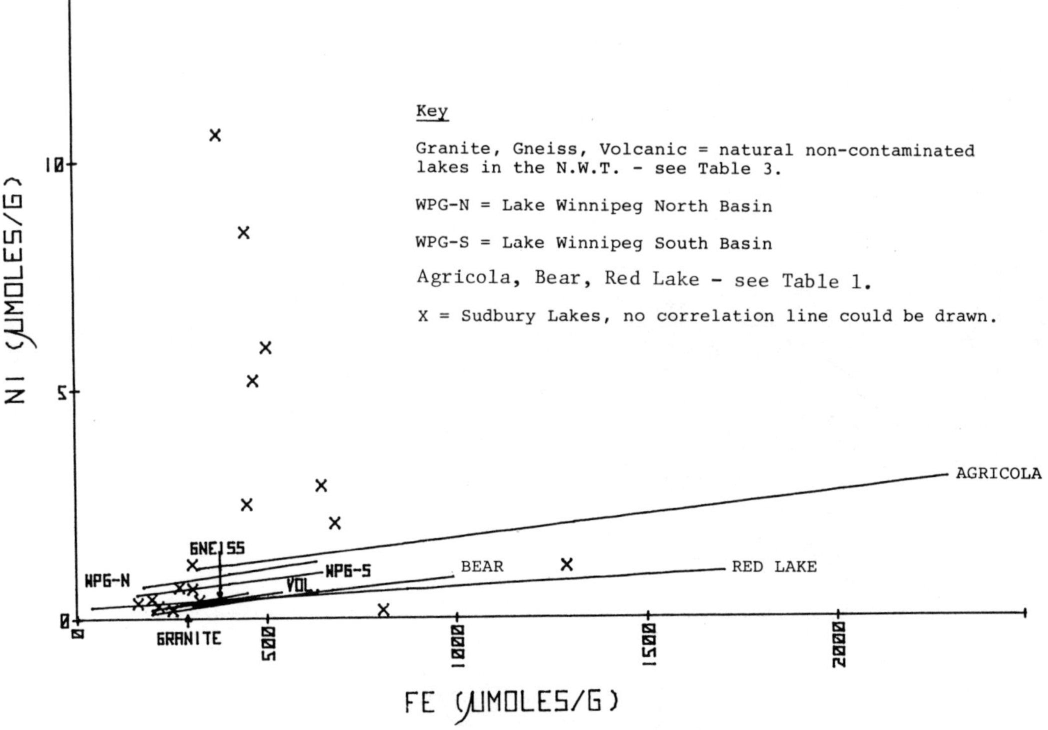


Figure 3. Linear Correlation Plots for Ni/Fe in Lake Sediments from Various Lakes and Lake Areas in Canada.

1. For more detailed site descriptions see Table 1. "ns" not significant. "r" is the correlation coefficient. "n" is the number of samples.

Graphically (Figure 3), the reason for the lack of a Ni/Fe correlation at Sudbury is immediately obvious and the similarity of slopes of the other sites is also clear.

Conservative element data other than for iron was only available for Lake Winnipeg and the Bear-Slave sites. RAV computations were done for trace elements/Ti (Table 6), and also for the relatively conservative elements Al and Si (not shown here). Al was significantly related to many trace elements in this diverse set of sediment samples, but Si was correlated poorly with only a few other elements. As for Ni/Fe, these RAV values indicate a great deal of geochemical similarity between these diverse sites, considering that the resulting relative concentrations in sediments are a function of source, weathering, transport, deposition and diagenesis.

Zn/Fe RAV values are given in Table 7. In this calculation only two sub-sites have non-significant correlations. These were the Agricola and Bear, both of which are natural geological anomalies for Zn in lake sediments, in the sense that their lake sediments have higher concentrations than those found in surrounding lakes. The lack of correlation for Zn/Fe at these sites is seen in Figure 4 where in both cases a wide range of concentrations of Zn occurs relative to a small range of Fe concentrations. All of the remaining correlations are very high for both gyttja and inorganic sediment. The corresponding Zn/Fe RAV values are not as consistent as for Ni/Fe, but are still surprisingly similar considering the variables involved - including sediment type, geology, limnology, geographic location, sediment collection methodology, and analytical techniques. For Zn/Ti only the Lake Winnipeg sediments were significantly correlated. The Zn/Ti RAV's are given in Table 6. The Zn/Al RAV values were $9 \times 10^{-4}$ and $5 \times 10^{-4}$ for the south and north basins of Lake Winnipeg respectively. Again, this showed a relatively higher input of Zn to the south as opposed to the north basin.

At five of the sub-sites, the Cu/Fe correlations was not significant (Table 7). In three of these cases (Red Lake, Snib Lake and Sudbury Lakes) copper contamination (in excess of Fe supply) from exploration or refining activities is known or suspected. At the other two sites, Agricola and Bear, elevated Cu concentrations (relative to Fe) in lake sediments have been attributed to geological anomalies, (Allan and Cameron, 1973).

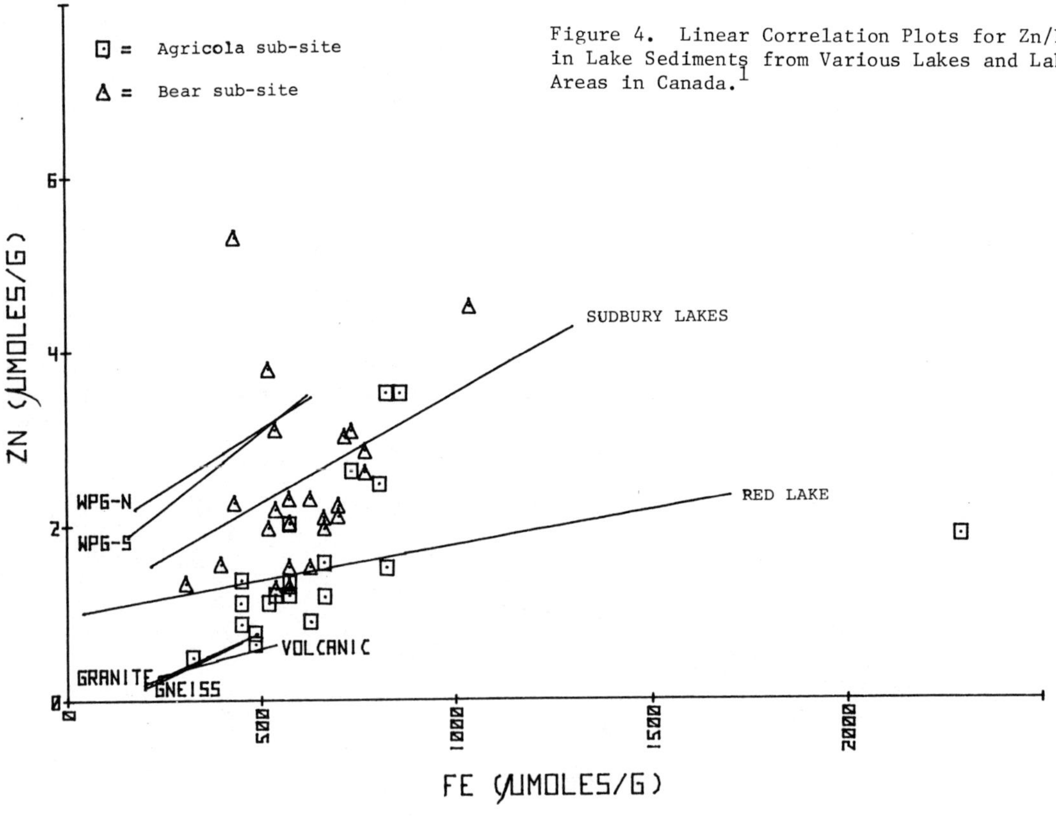


Figure 4. Linear Correlation Plots for Zn/Fe in Lake Sediments from Various Lakes and Lake Areas in Canada.[1]

1. For key see Figure 3 and Table 1.

Again, with these five exceptions, the consistency of Cu/Fe RAV values was striking. Pb/conservative element correlations were sought but not found. RAV values are likely most meaningful if the element pair has related geochemical behavior. There were cases in the Lake Winnipeg sediment data, however, where correlations were high (e.g., for P and Ti), but where no geochemical relationship for this high correlation could be reasonably proposed. In such cases, the correlations may be an artifact of both elements being correlated to some other sediment parameter (such as clay content or cation exchange capacity). For further investigations of RAV, parameters such as sulphide, organic, and carbonate fractions of the sediment should be incorporated into the correlation matrices.

Trace Element/Trace Element RAV  Our argument for the use of element ratios and element correlations has been based initially on the use of trace element/major (conservative) element pairs. We also propose that the use of trace element/trace element RAV may provide clues to internal geochemical balance of trace metals in lake sediments. As an example, we have calculated RAV values for Ni/Cu, Zn/Cu and As/Cu (Table 8).

Table 8. Relative Atomic Variation (RAV) of elements in sediments from various lakes and lake areas in Canada.
Trace Element/Trace Element Combinations

| | | Ni/Cu | | Zn/Cu | | As/Cu | |
|---|---|---|---|---|---|---|---|
| Sub Site[1] | n | $r^2$ | RAVx$10^{-4}$ | r | RAVx$10^{-4}$ | r | RAVx$10^{-4}$ |
| North Basin | 21 | 0.8 | 10 | 0.9 | 23 | - | - |
| South Basin | 21 | 0.9 | 11 | 0.9 | 38 | - | - |
| Red Lake | 29 | 0.7 | 9 | 0.8 | 22 | 0.7 | 45 |
| Sudbury Lakes | 22 | 0.9 | 19 | 0.9 | 1 | -0.2 | ns[3] |
| Agricola | 21 | 0.7 | 7 | 0.9 | 11 | 0.6 | 15 |
| Bear | 25 | 0.4 | 3 | 0.5 | 23 | 0.5 | 3 |
| Granite | 26 | 0.7 | 6 | 0.7 | 9 | 0.7 | 12 |
| Gneiss | 26 | 0.7 | 5 | 0.7 | 8 | 0.6 | 7 |
| Volcanic | 20 | 0.8 | 9 | 0.7 | 12 | 0.6 | 20 |

1. For more detailed descriptions see Table 3.
2. r = simple correlation coefficient rounded to first decimal place.
3. ns = not significant.

Correlations for Ni/Cu were usually high and more consistently so than for the trace element/major element pairs. The Ni/Cu RAV values for all sites fall within the range 3-19x$10^{-4}$. The lower values were for the sub-Arctic Bear-Slave sub-sites, and higher values were found in the temperate Red Lake and Lake Winnipeg sub-sites. This is probably a function of source material composition. The major exception was for the Sudbury lakes. The high RAV indicates a higher Ni input (relative to Cu) than at the other sites. Although both elements are added aerially from smelting activities, the smelting process results in deposition at rates which are different from normal weathering and geochemical cycles. The Bear-Slave sub-sites show minimal variation in RAV values, despite distinct differences in element concentration ranges between geologically anomalous and background areas.

Zn/Cu RAV (Table 8) are less consistent perhaps due to a wider range of possible geologic sources for Zn at the various study sites. As for Ni/Cu, there is a tendency for the Zn/Cu RAV values to be higher at the temperate sub-sites. The exception is again the Sudbury lakes sub-site. The low RAV indicates the unusually low Zn relative to Cu input from the smelting activities. A second facet of the Zn/Cu RAV values is the distinction between the Agricola and Bear sub-sites. Both are geological anomalies for Zn in the sense that lake sediment Zn concentrations in both areas are higher than occur in lake sediments from surrounding areas (Allan and Cameron, 1973). There has been speculation that these anomalies have different geological sources. Agricola has been shown to be an area of sulphide occurrences. It appears from the RAV values that the Bear anomly is in fact different and thus may not reflect a sulphide source. It may thus be feasible to use RAV values to differentiate trace element anomalies of sulphide relative to silicate source. There is unfortunately insufficient data presently available to expand this theory.

The calculation of As/Cu RAV was limited to very few sites. A high RAV at Red Lake can be attributed to the excess input of As relative to Cu. This excess As originates from gold mining and processing activities at the east end of Red Lake. The Agricola sub-site is a geological anomaly for As (Allan et al., 1974). However, the RAV value of 15x$10^{-4}$ (Table 8) falls within the values for the Bear-Slave background areas. Again, in the case of a natural geologic anomaly, the concentrations for each element are anomalous individually, but the RAV based on their relative variation is not. No significant correlation existed at the Sudbury site because of the very wide range of Cu concentrations that occurred over a relative narrow range of low As concentrations.

## Conclusions

Lake Winnipeg sediments are apparently little influenced in their metal concentration by the present level of development in the lake drainage basin. There does not exist in Lake Winnipeg sediments the logarithmic decrease of metals with depth that is experienced for example in some of the Laurentian Great Lakes of North America or some of the contaminated Alpine lakes of Europe such as Lake Constance. In this sense, Lake Winnipeg sediments actually constitute background RAV values, excluding the variations dealt with in the text for such metals as Hg, Pb and P.
RAV values cannot be calculated for other large lakes (such as the Laurentian Great

1. For more detailed site descriptions see Table 1; r = simple correlation coefficient rounded to the first decimal place; n = number of samples; ns = not significant.

Lakes, Lake Washington, Lake Constance, Lake Geneva, Lake Balkash and others) because data for these is often limited to one metal such as Hg or Pb, or when multielement data is presented, to diagramatic representations from which absolute values cannot be easily extracted.

Despite the above limitations, the Canadian data base we were able to accumulate, particularily that on Lake Winnipeg sediments, has made it possible for us to make certain propositions, to test some hypotheses on relative element abundances, and to draw some general conclusions.

(1) Element concentrations alone in lake and river sediments are at best only of value in detecting cultural or natural anomalies when the concentrations are very high and readily related to point sources such as nearby smelters or weathering ore deposits.

(2) A second level of anomaly or contamination detection should rely on more exact definition of geochemical imbalance than can be detected by concentrations alone. It is proposed that a more sensitive index is the determination of relative atomic variation (RAV) of metal pairs.

(3) Calculation of correlation coefficients for metal pairs shows that significant correlations normally exist for many trace/major and trace/trace element pairs. These correlations appear to disappear or change in magnitude at sites of cultural contamination and natural anomalies.

(4) When significant correlations exist (i.e., when two elements are geochemically in proportion over a range of concentration changes, slopes of linear regression lines can be calculated. This slope is referred to as the relative atomic variation or RAV.

(5) At many diverse limnological sites widely separated within Canada, there appears to be generally similar RAV values for several element pairs despite great variation in absolute concentrations due to geological source materials.

(6) The sensitivity of the RAV value in detecting elemental proportions depends on the pairs of elements selected for correlation. For example, for phosphorus in Lake Winnipeg, it appears that titanium is a more sensitive predictor of phosphorus than iron. However, it should be noted that a high correlation and distinct RAV value do not necessarily mean that the two elements are mineralogically associated.

(7) Further work with sulphide and organic sediment components is expected to increase the power to correlate elements which were found to have low correlations with elements determined in the Lake Winnipeg suite (e.g., lead).

(8) Although no data is presented here for rivers, the determination of RAV values for river bottom sediments is considered to be of greater significance than for lake sediments is often quite complicated or may even be impracticable. River sediment metal concentrations are often measured only on dredged sediments. With surface sediment surface sediments. With surface sediment sample suites it can be somewhat naive to base conclusions relative to cultural contamination on concentrations alone. Such absolute comparisons are relatively flimsy support for legal action when concerns between provinces, states or countries are at stake.

(9) The RAV values for a particular lake basin, whole lake, or for several smaller adjacent lakes provide an index of the geochemical normalcy of the drainage basin. In this context, RAV values may be of significance in terms of global geochemical balances and cycles of elements and lead to conclusions relevant to the determination of global loadings of metals to oceans.

Acknowledgements

We wish to acknowledge the help of many people from the Freshwater Institute and the crew of vessel Bradbury, who were involved in the collection of sediments in 1968-71. We also thank S. Withers who supervised the collection of the 1975 samples. All of the statistical work was done by S.E.M. Elliott. Much of the integration of the data was done by Messrs. B. Graham and M. Roy of the Freshwater Institute. The final manuscript was typed by S. Bannerman.

References

Ali, S.A., G.M. Gross & J.R.L. Kishpaugh, 1975. Cluster analysis of marine sediments and waste deposits in New York Bight. Environ. Geol. Vol. 1, p.143-148.

Allan, R.J., E.M. Cameron & C.C. Durham, 1973. Reconnaissance geochemistry using lake sediments of a 36,000 square mile area of the northwestern Canadian Shield. Geol. Surv. Canada Paper 72-50, 70p.

Allan, R.J., & E.M. Cameron,1973. U, Zn, Pb, Mn, Fe, Cu, Ni, K content of lake sediments; Bear-Slave Operation. Geol. Surv. Canada Maps 9-15, 1972.

Allan, R.J., E.M. Cameron & I.R. Jonasson, 1974. Mercury and arsenic levels in lake sediments from the Canadian Shield. Proc. 1st. Internat. Mercury Cong., Vol. 2, p. 93-119.

Allan, R.J. & M.H. Timperley, 1975. Prospecting by use of lake sediments in areas of industrial heavy metal contamination in Prospecting in Areas of Glaciated Terrain, Ed. M. Jones, I.M.M. Special Pub., 87-111.

Aston, S.R., D. Bruty, R. Chester & R.C. Padgham, 1973. Mercury in lake sediments: a possible indicator of technological growth.Nature, 241, p.450-451.

Bruland, K.W., K. Bertine, M. Koide & E.D. Goldberg, 1974. History of metal pollution in southern California costal zone. Environ. Sci. and Tech., Vol. 8, No. 5, p. 425-432.

Brunskill, G.J., 1973. Rates of supply of nitrogen and phosphorus to Lake Winnipeg, Manitoba, Canada. Verh. Internat. Verein. Limnol., 18, p.1755-1759.

Brunskill, G.J., S.E.M. Elliot & P. Campbell, 197-. Speculations on the future of Lake Winnipeg, with specific reference to eutrophication. (in press).

Cowgill, U.M. & G.E. Hutchinson, 1966. A general account of the basin and the chemistry and mineralogy of the sediment cores in History of the Laguna de Petenxil, Mem. Conn. Acad. Arts and Sci., Vol. 17, p.2-62.

Chow, T.J, K.W. Bruland, K. Bertine, E.D. Goldberg, 1973. Lead pollution : records in southern California coastal sediments. Science, 181, p.551-552.

Crecelius, E.A. & D.Z. Piper, 1973. Particulate lead contamination records in sedimentary cores from Lake Washington, Seattle. Environ. Sci. and Tech., Vol. 7, No. 11, p.1053-1055.

Crecelius, E.A., 1975. The geochemical cycle of arsenic in Lake Washington and its relation to other elements. Limnol. and Ocean., Vol. 20, No. 3, p.441-451.

Damiani, V. & R.L. Thomas, 1974. Mercury in the sediments of the Pallanza Basin. Nature, Vol. 251, No. 5477, p.696-697.

Elson, J.A., 1967. Geology of Glacial Lake Agassiz in Life, Land and Water (W. Mayer-Oakes, Ed.) U. Man. Press, Winn., p.36-95.

Erlenkkeuser, H., E. Suess & H. Willkomm, 1974. Industrialization affects heavy metals and carbon isotope concentrations in recent Baltic Sea sediments. Geochim. Cosmochim. Acta., 38, p.823-842.

Förstner, U. & G. Muller, 1974. Accumulations of heavy metals in dated sediment cores from Lake Constance and Tegersee. Techermaks Mineralogica and Petrographic News, Vol. 21, p.145-163.

Hallberg, R.O., 1974. Paleoredox conditions in the eastern Gotland Basin during the recent centuries. Merentutkimuslait Julk/ Havsforskningsinst. Skr. No. 238, p.3-16.

Harding, S.C. & H.S. Brown, 1975. Distribution of selected trace elements in sediments of Pamlico River Estuary, North Carolina, Environ. Geol., Vol. 1, p.181-191.

Iskandar, I.K. & D.R. Kenney, 1974. Concentration of heavy metals in sediment cores from selected Wisconsin Lakes. Environ. Sci. and Tech., Vol. 8, p.165-170.

Kemp, A.L.W., R.L. Thomas, C.I. Dell & U.M. Jacquet, 1976. Cultural impact on the geochemistry of sediments in Lake Erie. Jour. Fish. Res. Bd., Vol. 33, No. 3, p.440-462.

Kushnir, D.W., 1971. Sediments in the south basin of Lake Winnipeg, M.Sc. Thesis, Univ. Man., 92p.

Lubchenko, I. Yu. & D.S. Turovskii, 1974. Distribution of lead in the surface layer and middle of bottom sediments of Lake Balkhash. Dolk. Acad. Nauk. SSSR, Vol. 214, No. 5, p.1182-1185.

Muller, G. & U. Förstner, 1975. Heavy metals in sediments of the Rhine and Elbe estuaries : mobilization or mixing effect. Environ. Geol., Vol. 1, p.33-39.

Shapiro, J., W.T. Edmondson & D.E. Allison, 1971. Changes in the chemical composition of sediments of Lake Washington, 1958-1970. Limnol. and Ocean., 16, p.437-452.

Shimp, N.F., H.V. Leland & W.A. White, 1970. Distribution of major. minor and trace elements in unconsolidated sediments from southern Lake Michigan. Ill. Geol. Surv. Envir. Geol. Notes, 32, 19p.

Shimp, N.F., J.A. Schleicher, R.R. Ruch, D.G. Heck & H.V. Leland, 1971. Trace element and organic carbon accumulation in the most recent sediments of southern Lake Michigan. Ill. Geol. Surv. Envir. Geol. Notes, 41, 25p.

St. John, B.E., E.C. Carmack, R.J. Daley, C.B. Gray & C.H. Pharo, 1976. The limnology of Kamloops Lake, B.C. Inland Waters Directorate Unpublished Report, 167p.

Thomas, R.L., 1972. The distribution of mercury in the sediments of Lake Ontario. Can. Jour. Earth Sci., Vol. 9, p.66-84.

Thomas, R.L., 1973. The distribution of mercury in the sediments of Lake Huron, Can. Jour. Earth Sci., Vol. 10, p.194-204.

Thomas, R.L. & J.M. Jacquet, 1976. Mercury in the surficial sediments of Lake Erie. Jour. Fish. Res. Bd., Vol. 33, No. 3, p.404-412.

Thomas, R.L., J.M. Jacquet & A. Murdoch,1976. Sedimentation processes & associated changes in surface sediment trace metal concentrations in Lake St. Clair, 1970-1974. Proc. Int. Conf. on Heavy Metals in the Envir.,(in press).

Vernet, J.P. & R.L. Thomas, 1972. The occurrence & distribution on mercury in the sediments of the Petit Lac (West Lake Geneva). Ecologae Geol.Helv.,Vol.65, p.307-316.

Walters, Jr., L.J., T.J. Wolery & R.D. Myers, 1974. Occurrence of As,Cd,Co,Cr,Cu,Fe,Hg,Ni, Sb & Zn in Lake Erie sediments. Proc. 17th. Conf. Great Lakes Res., (in press).

Wolery, T.J. & L.J. Walters, Jr. 1974. Pollutant mercury and sedimentation in the western basin of Lake Erie. Proc. 17th. Conf. Great Lakes Res., (in press).

Key words: Lake Winnipeg
Sediments
Geochemistry
Relative atomic variation (RAV)

# CHEMICAL CHARACTERISTICS OF LAKE LUGANO SEDIMENTS*)

*G. Premazzi and O. Ravera*

Biology Group – D.G. Research, Science and Education C.C.E.
Joint Research Centre, 21020 Ispra (Italy)**)

## Summary

Cores from the central area of the eight basins of Lake Lugano, at 1 cm intervals were analyzed for organic matter, total phosphorus, nitrogen, carbon, calcium, iron and manganese. The pattern of Fe and Mn deposition was not related to that of total P. In most of the cores, a high positive correlation coefficient between total P and organic C, and between nitrogen concentration and organic C has been calculated. Organic C, total P and N concentration decreased sharply from the surface down to 7 - 12 cm in the majority of the cores. This decrease may be due to the combined effects of the degradation processes of the sediment and the increase of the eutrophication of this lake during the last 15 years.

## Introduction

Sediments are the result of the accumulation of allochthonous and autochthonous material partly degraded during and partly subsequent to its sedimentation. From the study of chemical, physical and biological characteristics of successive layers of the sediment core, the most relevant changes occurring within the lake and its watershed can be recorded.

Lake Lugano is an interesting subject for this kind of research and, particularly, for studying the evolution of this ecosystem towards a highly eutrophicated stage. This research, under a contract between the J.R.C. - Euratom and the Government of Canton Ticino, is part of a series of limnological studies started on this lake from the beginning of 1972 and still in progress. The principal aim of the present study is to follow the trophic evolution of Lake Lugano through the chemical characteristics of its sediments. The most important characteristics of Lake Lugano are listed in Table 1 and more information on the ecology of this lake is reported in : Baldi et al, 1949; Gygax and Nydegger, 1964; Vollenweider, 1965; Jaag and Märki, 1970; and in our reports 1972, 1973, 1974 and 1975.

**) Paper presented at the "Symposium on Interactions between Sediments and Water", Amsterdam, 6 – 10 September 1976*

***) Contribution No. 1348 of Biology Program, Directorate General XII of the Commission of the European Communities*

**Table 1 : Physical and chemical characteristics of Lake Lugano**

| | |
|---|---|
| Northern latitude | 45° 51° – 46° 02′ |
| Eastern longitude | 8° 52′ – 9° 07′ |
| Altitude (m) | +271 |
| Surface area ($Km^2$) | 48.9 |
| Maximum depth (m) | 320 |
| Mean depth (m) | 130 |
| Volume ($Km^3$) | 6.5 |
| Cathment area (including the lake surface) ($Km^2$) | 615 |
| Trophic stage | eutrophic |
| Alkalinity (meq/1)(1972-74) | 1.30 – 2.38 |
| Conductivity ($\mu S$ at 18°C) | 136 - 226 |
| $0_2$ in hypolinmetic water | depletion in summer ($0_2$ consumption: 34 g/$m^2$ month) |

## Methods

Surface sediment cores were collected with a "Jenkin sampler" in February - March 1973 from 8 basins of Lake Lugano. Sampling depths varied from a minimum of -50 meters to a maximum of -310 meters. The cores, soon after collection, were cut into slices 1 cm thick by a procedure described in detail by Premazzi (1970). Each slice was dried at 105°C for 24 hours and then homogenized by grinding. Each slice was subdivided in the following 4 aliquots:

1) for ash content
2) for analyses of Ca, Fe and Mn
3) for total phosphorus
4) for total C and N

## Results and conclusions

All data are the mean value of two replicate determinations and are expressed on an oven-dried 105°C weight basis. The mean concentration of total organic C, N, P, Fe, Mn, Ca and organic matter are listed in Table 2. The stratification of sediments from the central area of the basins, is probably not affected by the activities of bottom organisms, since the density of benthonic populations in this lake is rather low (Ravera, 1973).

Table 2 : Lake Lugano surface sediment data (mg/g)

a: average values for upper layer (5 cm) b: average values for lower layer

| Lake basin | organic matter a | organic matter b | organic C a | organic C b | total N a | total N b | total P a | total P b | total Fe a | total Fe b | total Mn a | total Mn b | total Ca a | total Ca b |
|---|---|---|---|---|---|---|---|---|---|---|---|---|---|---|
| Morcote | 155 | 111 | 74.6 | 55.6 | 9.0 | 6.5 | 2.0 | 1.3 | 24.2 | 25.9 | 0.6 | 0.6 | 138 | 87 |
| Melide | 125 | 131 | 58.5 | 61.1 | 5.9 | 5.0 | 2.0 | 1.8 | 25.7 | 29.6 | 3.2 | 4.5 | 120 | 98 |
| P. Tresa | 138 | 85 | 89.4 | 28.4 | 11.3 | 5.4 | 4.5 | 2.4 | 31.7 | 36.8 | 1.4 | 0.8 | 105 | 49 |
| Lugano | 118 | 98 | 43.1 | 35.1 | 4.1 | 4.0 | 1.9 | 1.0 | 45.1 | 43.7 | 0.5 | 0.6 | 32 | 20 |
| Capolago | 111 | 57 | 57.4 | 22.4 | 5.6 | 4.0 | 1.4 | 1.1 | 22.9 | 26.5 | 0.3 | 0.3 | 107 | 84 |
| Figino | 103 | 91 | 45.8 | 43.6 | 4.0 | 4.0 | 2.1 | 1.7 | 41.1 | 42.2 | 1.2 | 1.2 | 61 | 39 |
| Gandria | 93 | 76 | 36.5 | 26.4 | 4.7 | 1.0 | 1.1 | 1.5 | 36.2 | 42.8 | 0.5 | 1.1 | 92 | 60 |
| Agno | 72 | 60 | 29.4 | 19.7 | 4.0 | 4.0 | 2.5 | 1.8 | 47.1 | 51.7 | 1.1 | 1.2 | 20 | 18 |
| $\bar{X} \pm S_x$ | 114±26 | 89±25 | 54±20 | 37±16 | 6.1±2.7 | 4.6±.9 | 2.2±1.0 | 1.6±.5 | 34±10 | 37±9 | 1.1±.9 | 1.3±1.3 | 84±42 | 57±31 |

A C/N ratio of 9.2 (range 7.4 - 11.5), a C/P ratio of 27.1 (range 11.8 - 41.0) and a N/P ratio of 3.0 (range 1.6 - 4.5) was calculated for all basins and mud samples.

As to Fe and Mn, relevant differences between the different basins have been observed (Fig.1): the sediment of the Melide basin has a very low Fe/Mn ratio (5 - 10 : 1); in Figino, Agno, Morcote and Ponte Tresa basins the ratio Fe/Mn ranged from 40 to 50. Otherwise for Lugano and Capolago basins, a higher Fe/Mn ratio has been calculated (80).

For percent of water, loss on ignition, concentrations of organic C, total N, total P and the ratios C/N and N/P the sediments of Lake Lugano are very similar to those of other lakes (Table 3). From the results of statistical analysis the following conclusions may be drawn:

1) there is a direct correlation between total P and organic C (r = 0.66). This seems to support the hypothesis that organic P and calcium phosphate together represent an important fraction of total P in the lake sediments. In fact, analyses of $PO_4$-P dissolved in the interstitial water of recent sediments showed that orthophosphate never exceed 0.5% of the total P (Sorensen, J., 1976);
2) because no correlation between total P and Fe has been found, it seems that P precipitation is not influenced by Fe. Bortleson (1974) found that the Fe, Mn and P concentration profiles in Weber Lake are not related, he concludes that the Fe is probably associated with clay minerals which would provide little opportunity for a coprecipitation of P and Fe.

It is probable that the deepest area of the lake (Gandria and Lugano) with a continuous $O_2$ depletion has no precipitation of phosphates. Other parts of the lake have oxygenation of the deepest waters at least for some months and periods with precipitation of phosphate will follow periods with release of phosphates from the sediments.

An interesting point is the evolution pattern of some parameters in cores from the Melide and Gandria basins. From 15 - 17 cm down a simultaneous decrease of Ca, organic C and N, and an increase of P, Fe and Mn was observed.

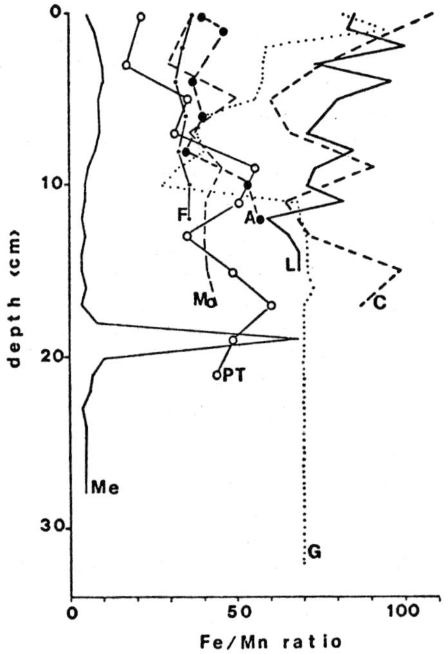


Fig.1 _ Variation with depth of the ratio of Fe to Mn in the sediments of Lake Lugano

Table 3 : Comparative surface sediment chemistry for Lake Ceresio and other lakes.
One standard deviation is given for values from group of lakes or different basins of same lake

| LAKE | water content (%w.w.) | loss on ignition (mg/g d.w.) | organic C (mg/g d.w.) | total N (mg/g d.w.) | total P (mg/g d.w.) | org. C/N | tot. C/P | N/P |
|---|---|---|---|---|---|---|---|---|
| Ceresio, Italy-Switz. | 78 ± 8 | 114 ± 26 | 54 ± 2 | 6.1 ± 2.7 | 2.2 ± 1.0 | 9.2 ± 1.5 | 43.5 ± 18 | 3.0 ± 1.1 |
| Maggiore, Italy-Switz. | – | 105 ± 17 | 49 ± 5 | 4.8 ± 1.5 | 2.5 ± 1.0 | 10.7 ± 2.6 | 25.7 ± 8.6 | 2.1 ± 0.7 |
| Comabbio, Italy | ~ 90 | 288 ± 5 | 152 ± 5 | 17.6 ± 0.6 | 2.6 ± 0.3 | 8.6 ± 0.5 | 60.6 ± 6.9 | 6.8 ± 1.0 |
| Varese, Italy | – | 157 ± 29 | 63 ± 11 | 9.5 ± 2.4 | 1.7 ± 0.7 | 6.9 ± 1.4 | 96.5 ± 20.5 | 5.8 ± 1.3 |
| Monate, Italy | – | 239 ± 12 | 111 ± 6 | 11.8 ± 1.1 | 2.3 ± 0.5 | 9.5 ± 0.5 | 49.7 ± 7.6 | 5.2 ± 0.8 |
| Orta, Italy | – | 146 ± 55 | 80 ± 18 | 12.3 ± 5.7 | 3.3 ± 2.1 | 7.9 ± 3.4 | 28.4 ± 9.8 | 3.6 ± 1.1 |
| Zürichsee, Switz. | 78 | – | 100 | 3.5g | 2.0 | 28.4h | – | – |
| Windemere (S.basin),Engl. | ~ 90 | 200 | 104 ± 25 | 6.0g | 2.2 ± 0.3 | 11.0h | 17.0 | – |
| Esthwaite, Engl. | – | 280 | 105 ± 12 | 11.0g | 2.5 | 12.0h | 75.0 | – |
| Ennerdale, Engl. | – | ~150 | 70 | 6.4g | ~ 1.3 | 11.0h | 63.0 | – |
| Mikolajki, Poland | 95 | 110 | 60 | 7.0g | – | 8.6h | - | – |
| Kolno, Poland | 88 | 580 | – | 5.0g | – | – | – | – |
| Kinnereth, Israel | 50–295dw | – | ~ 40 | 1.0–2.4g | 1.0–4.5 | 21.0h | 36.0 | 1.7 |
| Erie, Canada | 72 | 350 | 22.6 | 3.3 | 0.8 | 8.5 | – | 4.1 |
| Ontario, Canada | 80 | – | 33 | 4.3 | 1.9 | 8.2 | – | 2.3 |
| ELA, Canada | 94 ± 3 | 440 ± 140i | 200 ± 7 | 21.0 ± 8.0 | 2.2 ± 0.5 | 10.3 ± 1.7 | 103 ± 45 | 10.2 ± 3.7 |
| Mendota, USA | 90 | – | 85 ± 8 | 8.0 ± 11.6g | 1.4 ± 0.2 | 7.8 ± 12.4h | – | -- |
| Bantam, USA | 90 | 10 – 290 | – | 1.0–14.6g | 0.4–2.7 | 12–20h | – | – |
| Wisconsin lakes, USA | 65–90 | 460i | 200 ± 100 | 18.0 ± 6.8g | 1.2 ± 0.9 | 11.0h | – | – |
| Washington, USA | 81 | 107 | 50 | 7.0 | 1.0–5.0 | 5.7 | – | 1.4–7.0 |

*i) at 900°C* *g) Kjeldahl N* *h) C: Kjeldahl N*

According to Serruya, 1971, we may suppose that at the time of deposition of these elements the environment was more oxidative, Mn and Fe precipitated as oxides; higher amounts of P were also incorporated to the muds as iron phosphate. The contemporaneous decrease of organic C, N and Ca could indicate a more modest productivity or a higher rate of mineralization;

3) A close relationship between total N and organic C is statistically supported (r = 0.95). This shows that the greatest part of N is in organic form.

The actual quantity of C, N and P stored in the sediments partly depends on the oxidative capacity of the water. This parameter is difficult to estimate but one can compute the quantity of C withdrawn annually into the sediments from the open-waters. On the basis of Cs-137 distribution in the cores (Ravera and Premazzi, 1972) we have estimated that (on the average of 14 years) the organic carbon accumulated into the sediment is 101 g · $m^{-2}$ $yr^{-1}$.

The high concentration of organic C in the sediments may be the effect of two factors:

1) the increase of the production of organic matter in the water caused by the eutrophication process;
2) the $O_2$ depletion (occurring for some months of the year near the lake bottom) prevents the degradation of the organic compounds. On the other hand, the degradation of organic compounds is, obviously, more complete in the deeper part of the sediments than in the upper part.

According to the Cs-137 distribution in the thickness of the sediments and supposing a constant sedimentation rate, it has been calculated (Table 4) that:

a) the organic C content has decreased about 50% over the last 15 years;

Table 4 : Changes in the ratio of organic matter, total N, total P, total Fe, total Mn and total Ca in the last 15 years in the Lake Lugano sediments

| Lake basin | organic matter | total N | total P | total Fe | total Mn | total Ca | sedimentation rate cm/year* |
|---|---|---|---|---|---|---|---|
| Morcote | 2.0 | 2.2 | 1.5 | – | – | 3.9 | 1.3 |
| Melide | 1.2 | 1.6 | 1.5 | 0.8 | 0.6 | 1.5 | 1.0 |
| P. Tresa | 1.2 | 1.7 | 2.5 | 0.9 | 1.8 | 2.8 | 1.5 |
| Lugano | 2.0 | 1.0 | 1.0 | 1.0 | 0.7 | 1.1 | 1.0 |
| Figino | 1.2 | 1.3 | 1.2 | 1.0 | 1.0 | 3.2 | 0.8 |
| Capolago | 1.6 | 2.5 | 1.7 | 1.1 | 1.0 | 1.3 | 1.0 |
| Gandria | 1.6 | 1.3 | 1.6 | 0.6 | 0.4 | 2.3 | 1.0 |
| Agno | 1.2 | 1.0 | 1.3 | 0.9 | 1.3 | 2.6 | 1.0 |
| $\bar{X} \pm S_x$ | 1.50±0.35 | 1.58±0.54 | 1.58±0.45 | 0.90±0.16 | 0.97±0.47 | 2.34±0.98 | 1.08±0.22 |

** Data unpublished (Premazzi, G. 1975)*

b) the P and N content have decreased about 60%;

c) Fe and Mn seem not to be influenced either by variations of trophic state or by the mineralization;

d) Ca has markly increased, probably as the result of more intense photosynthesis over the last years.

### References


Baldi, E., Pirocchi Livia & Tonolli, V, 1949. Relazione preliminare sulle ricerche idrobiologiche condotte sul Lago di Lugano (1946 - 1947). Ispettorato Federale Svizzero per la Pesca, Berna, Gennaio 1949

Bortleson, G.C. & Lee, C.F., 1974. P, Fe and Mn distribution in sediment cores at six Wisconsin lakes. Limnol. Oceanogr. **19**, 794-801

Gygax F. & Nydegger, P., 1964. Contributo alle indagini sulle attuali condizioni del Ceresio. Tipografia Grafica Bellinzona, S.A., pp. 1-36

Jaag, O. & Märki, E., 1970. Recent developments and present-day situation in relation to the lake of Lugano and its tributaries. in: "L'eutrophisation dans les grands lacs et les retenues d'eau", Uppsala Symposium, OECD, Paris

Premazzi, G., 1970. Ricerche sulle relazioni tra la quantità di sostanza organica del plancton e la quantità di sostanza organica accumulata nei sedimenti. Thesis, Univ. Milan

Ravera, O. & Premazzi, G., 1972. A method to study the history of any persistent pollution in a lake by the concentration of Cs-137 from fall-out in sediments. Proc. Int. Symp. Radioecology applied to the protection of man and his environmental, **1**, 703-722

Ravera, O., 1973. Studio sull'eutrofizzazione del Lago di Lugano. Rapporto sulle ricerche condotte nel corso del 1972. Ed. O. Ravera - EUR/C-IS/152/73i

Ravera O., 1974. Studio sull'eutrofizzazione del Lago di Lugano. Rapporto sulle ricerche condotte nel corso del 1973. Ed. O. Ravera - EUR/C-IS/116/74i

Ravera, O., 1975. Studio sull'eutrofizzazione del Lago di Lugano. Rapporto sulle ricerche condotte nel corso del 1974. Ed. O. Ravera - EUR/C-IS/551/75i

Serruya, C., 1972. Lake Kinneret: The nutrient chemistry of the sediments. Limnol. Oceanogr. **16**, 510-521

Soerensen, J., 1976. Observations on interstitial water in Lake Lugano sediments. Int. Report EUR, 1699

Vollenweider, R.A., 1965. Materiali ed idee per una idrochimica delle acque insubriche. Mem.Ist.Ital.Idrobiol.**19**, 213-286



**Key - words**

Sediment
Eutrophication
Chemistry
Radioecology
Lake Lugano

COMPOSITION OF WATER AND SEDIMENTS IN LAKE MJØSA, SOUTH NORWAY, IN RELATION TO WEATHERING PROCESSES

J.O. Englund[1], P. Jørgensen[2], E. Roaldset[2], and P. Aagaard[2*]

[1]Dept. of Geology, Agricultural University of Norway, 1432 ÅS-NLH, Norway

[2]Dept. of Geology, University of Oslo, Box 1047 Blindern, Oslo 3, Norway

Summary

From the largest lake in Norway, Lake Mjøsa, and its drainage area, different types of water samples were studied. Water chemistry is discussed in relation to mineral stability and weathering processes. The conclusions are that podzol weathering of tills and weathering of bedrock are the major processes causing the release of cations to the lake.

Introduction

Mjøsa is the largest lake in Norway with the following hydrological characteristics (Table 1).

Table 1. Hydrological characteristics of Lake Mjøsa

| | | |
|---|---|---|
| Surface area | 362 | $km^2$ |
| Water volume | 56.2 | $km^3$ |
| Maximum depth | 443 | m |
| Drainage area | 15 514 | $km^2$ |
| Annual inflow | 9.3 | $km^3$ |
| Theoretical renewal time | 6 | years |

The Lågen river is one of the main river systems in Norway, about 200 km long with a drainage area of about 11 500 $km^2$ of which only 2% is cultivated. The area is sparsely populated with little industrial activity.

The lake is situated in an area with various bedrock types (Fig. 1), viz. Precambrian gneisses and granites, late Precambrian and Cambro-Silurian sedimentary rocks and Permian plutonics.

North of Mjøsa the Lågen river drains a large area as far north as the Dovre and Jotunheimen mountain areas. The rocks here are mainly late Precambrian and Cambro-Silurian sedimentary rocks, which become metamorphosed towards the northwest. The late Precambrian ones are dominated by conglomerates, sandstones, shales and limestones (Englund, 1972).

The Cambro-Silurian sedimentary rocks consist mostly of shales and limestones (Skjeseth, 1963; Bjørlykke, 1974). These rocks weather easily and thus support relatively densely populated, rich, agricultural land.

All rocks within the present area are impervious, except for the porous Brumunddal Sandstone (Englund & Jørgensen, 1975). Water is therefore mostly concentrated in openings along faults, joints, other fractures and solution openings.

The soils within the drainage area are mainly tills formed in the course of the Weichselian glaciation. The tills consist of short transported material and thus reflect the composition of the underlying bedrock (Haldorsen, in press).

Water samples from fifteen rivers were collected monthly, and occasionally more frequently, throughout the two years 1974-75 by the Norwegian Institute of Water Research (NIVA).

The sampling sites are located near the river mouths, but upstream beyond the influence of Lake Mjøsa.

The ground water samples are from the eastern and northern areas around the lake. They were sampled regularly during the years 1971-75 and analyzed by the Institute of Public Health, Oslo.

Sediments in Lake Mjøsa

Seventy-four sediment samples from the upper 5 cm of bottom sediments were collected at a number of localities for grain-size analyses and mineralogical studies. The samples were divided into five major groups on the basis of their mechanical composition. The regional distribution of these groups is illustrated in Fig. 2.

The major part of the lake bottom is covered with silty clays, and only in shallow, nearshore areas do coarser sediments occur.

We observe (year 1973) that approximately 19 500 tons of suspended inorganic material is annually brought into this lake (92% comes with the river Lågen). About 11 700 tons left

*Present address: Dept. of Geology and Geophysics, Univ. of California, Berkely, Calif., U.S.A.

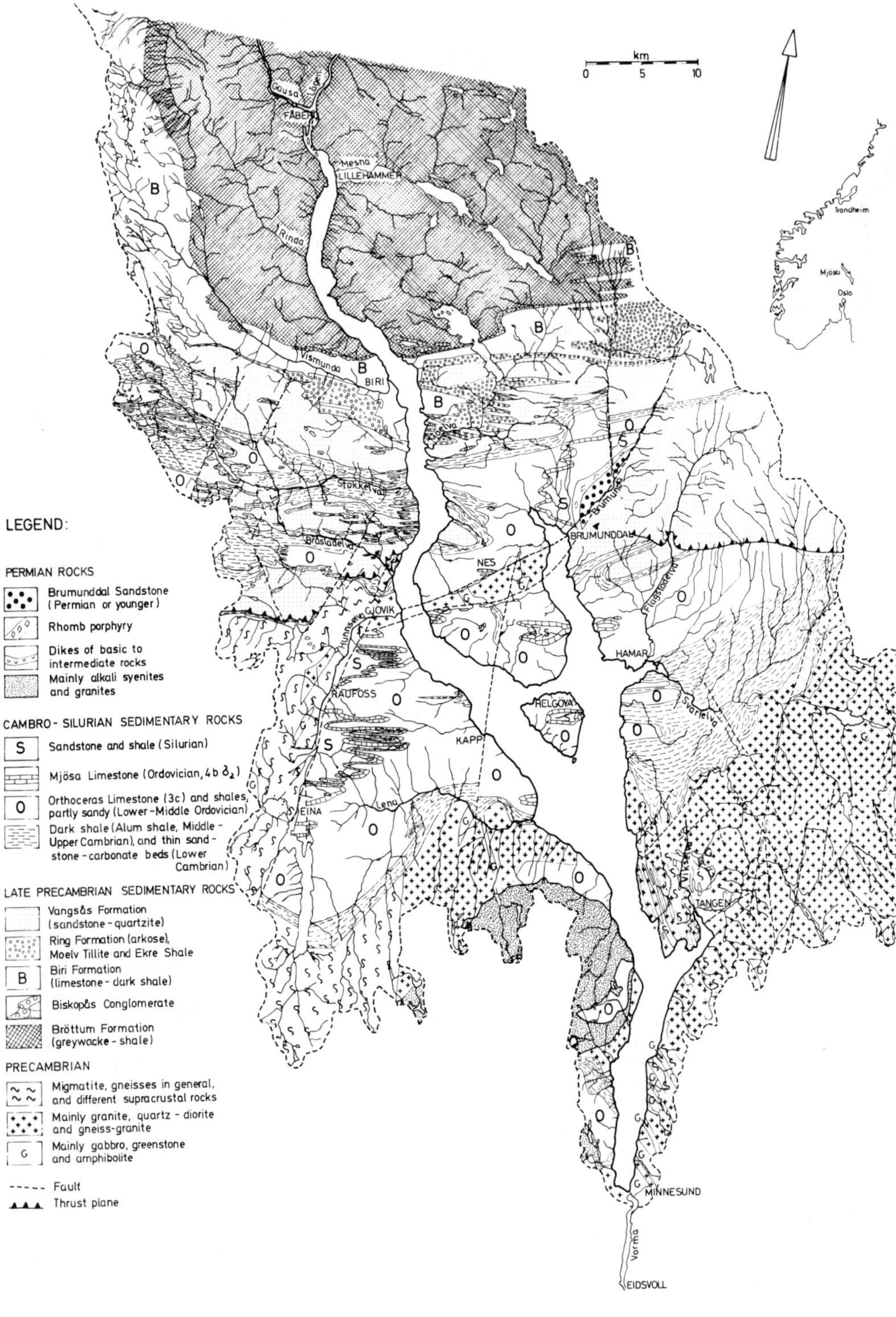


Fig. 1. Simplified geological map of the southern drainage area of Lake Mjøsa.

Mjøsa with the outlet river Vorma this year. The remaining 7 800 tons of inorganic material were deposited in the lake.

With a sediment density of 2 for a normally consolidated clay and a homogeneous distribution of the material over the whole lake bottom, we calculate an average sedimentation rate of 0.01 mm/year, which is very low.

The mineralogical composition of the clay fraction in the bottom sediments was calculated on the basis of semi-quantitative X-ray diffraction and chemical analyses. The minerals were identified according to the criteria given by Brown (1961). The phyllosilicates, illite and chlorite, constitute from 50 to 70% of the clay fraction (Fig. 3). Beside quartz and two types of feldspar (sodic plagioclase and K-feldspar) are found in considerable amounts. In addition, small amounts of mixed layer minerals and smectite are present.

The high illite and chlorite contents demonstrate the immaturity of these sediments and show a close relationship to the illite and chlorite rich tills and their underlying rocks (Haldorsen, in press).

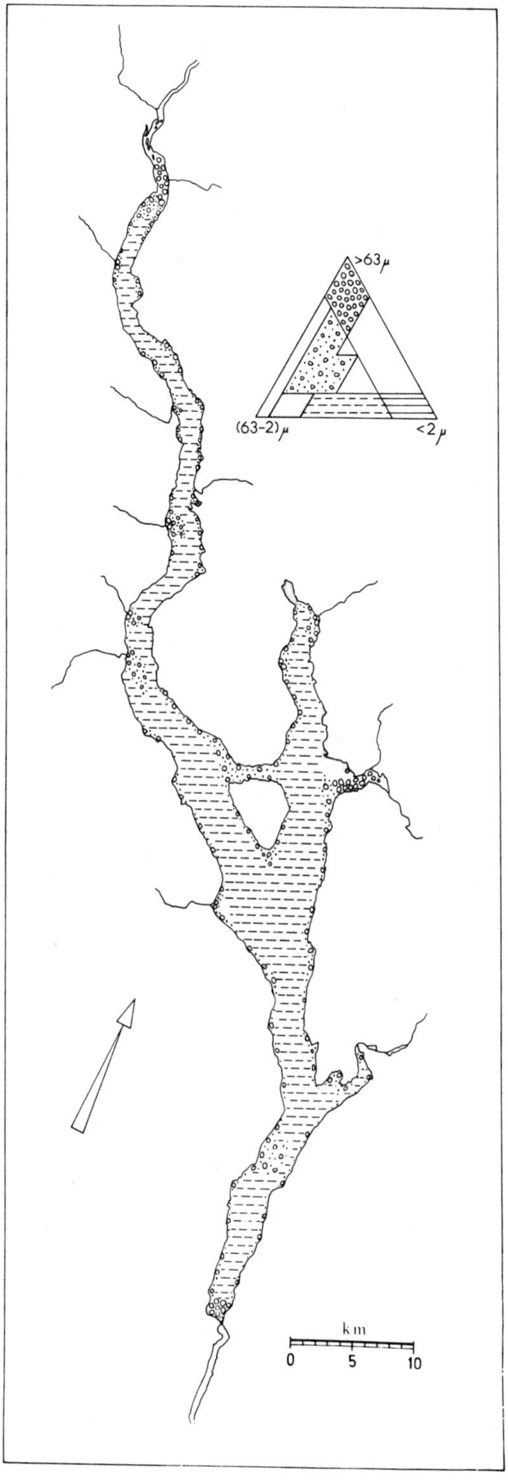


Fig. 2. Grain size distribution of Mjøsa bottom sediments.

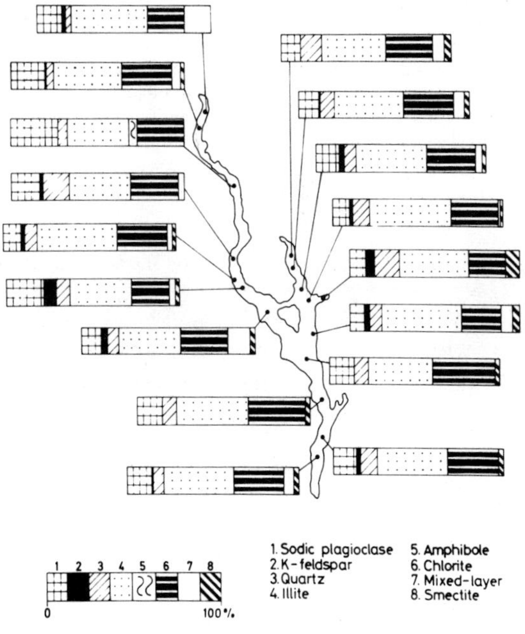


Fig. 3. Mineralogical composition of the clay fraction (< 2 μm) of Mjøsa bottom sediments.

## Water chemistry and mineral stability

The main purpose of this paper is to discuss water-mineral interactions. The considerations are therefore based on the $K^+$, $Na^+$, $Mg^{++}$, $Ca^{++}$, $Si^{4+}$ ($SiO_2$), $H^+$ (pH) and $Cl^-$ contents in precipitation, river and lake water and some types of ground water (Fig.4).

The mean total concentration of the abovementioned constituents in precipitation at Kise near Mjøsa is 1.5 mg/l for the period 1966-73. Ground water may contain up to 180 mg/l, while river water contains between 8 and 50 mg/l.

pH values are shown below, and they increase remarkably from the precipitation to ground and river water.

As normally found, the amounts of dissolved cations in river water are highest in the winter period (November-March) due to a larger inflow of ground water (Fig. 4).

These data were used to calculate the amounts of cations released from the different drainage areas (tons/km$^2$ year), which is presented in Fig. 5 together with flow rates for the various rivers.

When we remember the regional distribution of different rock types, we notice a clear correlation between bedrock type and water composition (Fig. 5 and Table 2). This means that Cambro-Silurian calcareous rocks give the highest concentrations, while the lowest values are obtained for rivers draining sandstone areas and Precambrian gneisses and granites.

The most important process responsible for the liberation of these extra cations to the rivers is podsol weathering of the tills and underlying bedrock.

If the sediments in Mjøsa reflect the composition of the fine fractions in the tills, the most probable proton weathering processes are the following (idealized):

$$NaAlSi_3O_8 + H^+ \rightarrow Na^+ + Al(OH)_3 + H_4SiO_4$$

albite

$$KAlSi_3O_8 + H^+ \rightarrow K^+ + Al(OH)_3 + H_4SiO_4$$

microcline

$$CaAl_2Si_2O_8 + 2H^+ + 6H_2O \rightarrow Ca^{++} + 2Al(OH)_3 + 2H_4SiO_4$$

anorthite

Weathering of calcite:

$$CaCO_3 + H^+ \rightarrow Ca^{++} + HCO_3^-$$

The stability of these minerals in relation to water composition is frequently illustrated by plotting water composition in theoretical calculated stability diagrams (Helgeson et al., 1973).

It is clearly seen (Fig. 6) that albite is unstable in contact with precipitation, lake water and river water as well as ground water. This underlines the importance of albite weathering. No kaolinite has so far been found in weathering profiles. This is probably due to a rapid release of cations during weathering, while authigenic growth of kaolinite is a much slower process.

Table 2. Cations released from some drainage areas.

| Drainage area | $H^+$ supplied by precipitation (kg ekv/year km$^2$) | Na,K,Mg,Ca released by weathering and other processes (kg ekv/year km$^2$) | Na,K,Mg,Ca $-H^+-Cl^-$ (kg ekv/year km$^2$) |
|---|---|---|---|
| Lågen | 10 | 137 | 120 |
| Mjøsa district average (excl. Lågen) | 15 | 390 | 318 |
| Hunnselva (polluted) | 15 | 882 | 588 |
| Flagstadelva | 15 | 658 | 519 |
| Vikselva | 15 | 205 | 181 |
| Vorma | 12 | 285 | 246 |

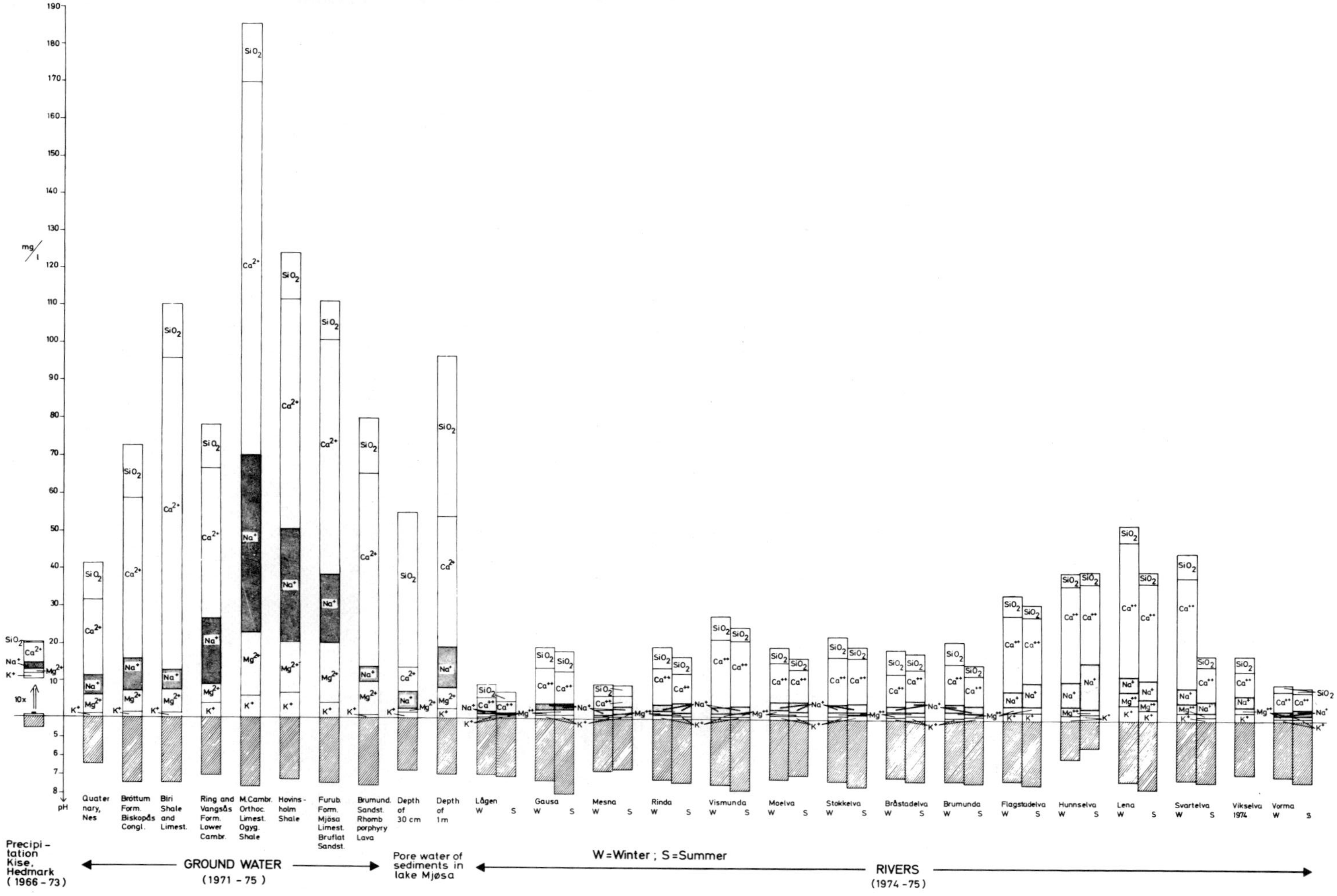


Fig. 4. Average chemical composition of different water types in the Lake Mjøsa district.

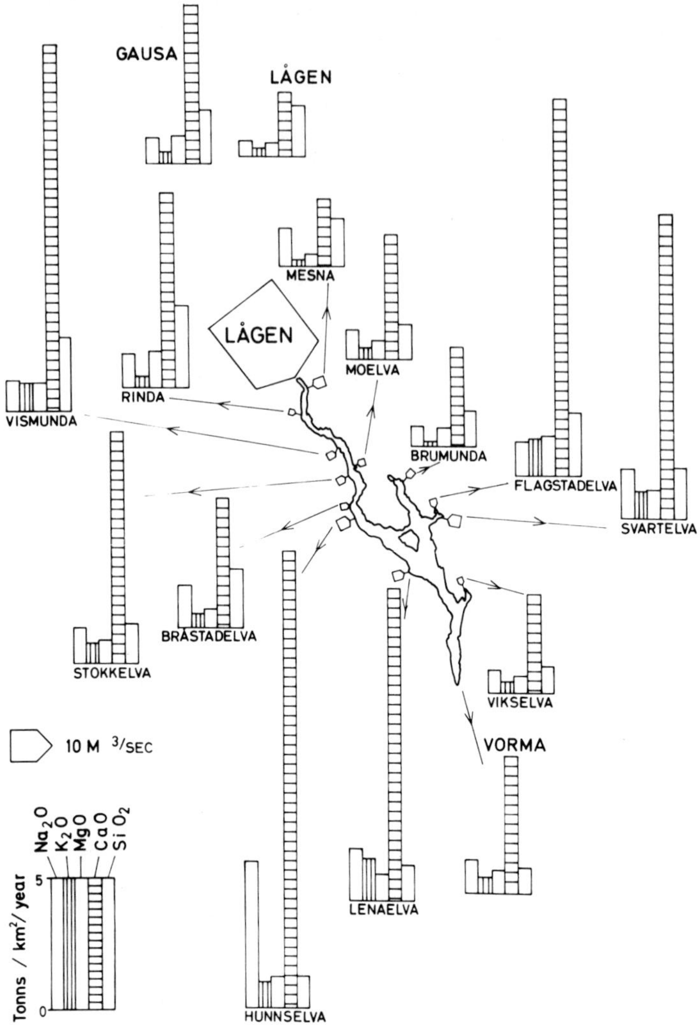


Fig. 5. Amounts of different ions entering and leaving Lake Mjøsa in one year.

Table 3. Amounts of different ions entering and leaving Lake Mjøsa during one year (tons/year).

| | $Na^+$ | $K^+$ | $Mg^{++}$ | $Ca^{++}$ | $H_4SiO_4$ | $H^+$ |
|---|---|---|---|---|---|---|
| Lågen | 7400 | 3760 | 4150 | 23800 | 35600 | 0.75 |
| Other rivers | 5850 | 2980 | 2800 | 27900 | 12300 | 0.35 |
| Precipitation into Mjøsa | 30 | 30 | 20 | 150 | - | 2.50 |
| SUM ions entering | 13280 | 6770 | 6970 | 51850 | 47900 | 3.60 |
| SUM ions leaving with Vorma | 13050 | 7300 | 7300 | 53100 | 24000 | 0.75 |

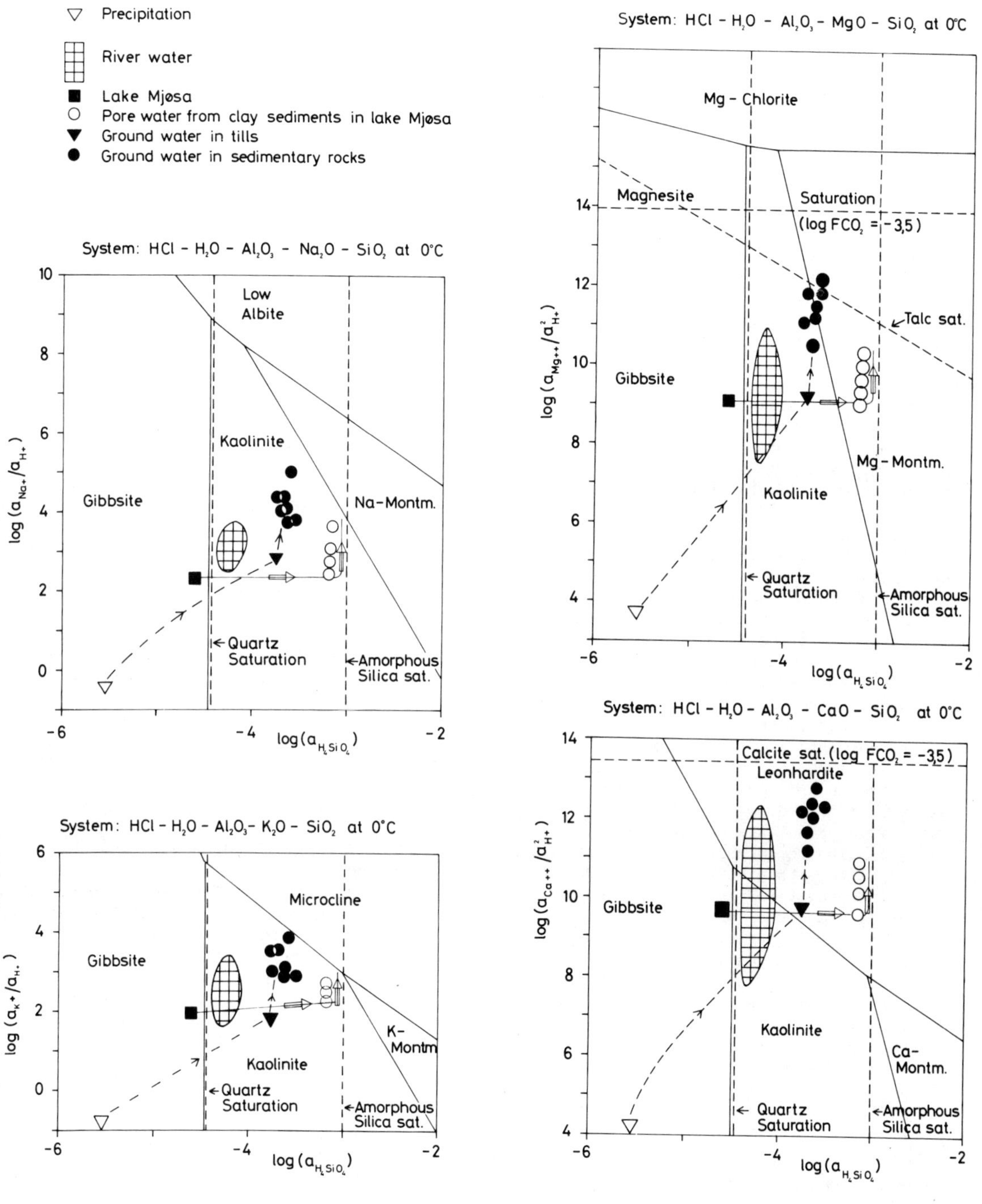


Fig. 6. Stability digrams after Helgeson et al. (1973).

Stability diagrams obtained by replacing sodium with either potassium, magnesium or calcium show that the given reactions must be expected. In the presented stability diagrams (Fig. 6) two main trends in the development of water are shown. Starting with the precipitation it is seen that the water increases its concentration of $H_4SiO_4$ and cations, as it percolates down to the ground water level in tills and bedrock.

River water is composed of rain water and ground water. It is noticeable that Lake Mjøsa contains less silica than all the rivers entering this lake. This is due to the production of diatoms and algae (Berge, 1973; 1974). Incorporation of diatoms skeletons in the lake sediments explains why the pore water contains even more silica than the ground water.

It should be pointed out that in some parts of Lake Mjøsa biogene silica is an important part of the sediments, while clastic material is a minor part. This is further illustrated in Table 3.

Comparing the amounts of ions entering Lake Mjøsa with the amounts leaving it, we observe no significant change for sodium, potassium, magnesium and calcium, while we observe a 50% reduction for the amount of silica. This is, as previously mentioned, related to formation of diatoms in the lake.

We have also considered the amounts of cations supplied by diffusion out of the bottom sediments. Less than 1/4% of the $Ca^{++}$ brought into the lake comes from the sediments within the lake basin.

It is concluded that "subaerial" weathering processes are far more efficient than diffusion processes.

We also observe that only a small part of the protons entering Lake Mjøsa are leaving it. This may be due to proton-consuming weathering and biological processes within the lake basin. Since the weathering processes (see above) are proton-consuming, it is interesting to compare the amounts of protons supplied with precipitation with the equivalent amounts of ions released from weathering processes (Table 2).

From this it is seen that less than 10% of the protons needed for the weathering processes are supplied with the precipitation. The remaining amounts of protons needed must come from the breakdown of humic acids and carbon dioxide-water reactions. Bicarbonate is the major anion in the river and ground water.

The river Lågen transports annually about 315 tons aluminum and 576 tons iron. From this we conclude:

(1) More than 98% of the alumina,
(2) " " 80% " " silica, and
(3) most of the iron

are retained in the soil profiles, where they are found as authigenic reaction products.

## Acknowledgements

We want to thank The Norwegian Institute for Water Research (NIVA), The Institute of Public Health (SIFF), and The Norwegian Institute for Air Research (NILU) for all help received during this work.

## References

Berge, F., 1973 and 1974. The eutrophication of Lake Mjøsa, Norway. Reports 1-4 to the Norwegian Institute for Water Research.

Bjørlykke, K., 1974. Depositional History and Geochemical Composition of Lower Palaeozoic Epicontinental Sediments from the Oslo Region. Norges geol. unders. 305: 81 pp.

Brown, G., 1961. X-ray Identification and Crystal Structures of Clay Minerals. The Mineralogical Society, London.

Englund, J.O., 1972. Sedimentological and Structural Investigations of the Hedmark Group in the Tretten-Øyer-Fåberg District, Gudbrandsdalen. Norges geol. unders. 276: 59 pp.

Englund, J.O. & Jørgensen, P., 1975: Weathering and hydrogeology of the Brumunddal Sandstone, Southern Norway. Nordic Hydrology 6:43-62.

Haldorsen, S., 1976. The petrography of till - a study from Ringsaker, South-eastern Norway. Norges geol. unders. (in press).

Helgeson, H.C., Brown, T.H. and Leeper, R.H., 1973. Handbook of Theorctical Activity Diagrams. Freeman, Cooper and Company.

Skjeseth, S., 1963. Contribution to the geology of the Mjøsa districts and the classical sparagmite area in Southern Norway. Norges geol. unders. 220: 126 pp.

Key words

Sediment composition
Clay mineralogy
Water chemistry
Mineral stability
Weathering

# ORGANIC COMPOUNDS AS INDICATORS OF ALLOCHTHONOUS AND AUTOCHTHONOUS INPUT TO LAKE SEDIMENTS

P.A. Cranwell

Freshwater Biological Association, Ambleside, Cumbria, Great Britain

## Summary

The contribution of allochthonous and autochthonous material to lake sediments can be recognised by analysis of the solvent-soluble organic fraction of sediments and by comparisons with the corresponding extract of soils and aquatic detritus. Separation of the extract gives several classes, differing in functional group, in which the abundance and percentage composition of each class may be related to the source of the organic material. Sediments from two lakes differing in trophic status illustrate the influence of source material on the composition of the organic fraction.

## Introduction

Organic matter in lake sediments is derived from primary production within the aquatic ecosystem (autochthonous sources) and also from terrestrial biota (allochthonous sources) by transport of leached and eroded material into the lake (Likens, 1972). Organic material which survives degradation in the water column and in a shallow zone at the sediment surface is incorporated into the sediment and undergoes a slower process of diagenesis.

Diagenesis is minimal in sediments deposited since the last glaciation so that the carbon skeleton of many sedimentary constituents is unchanged. Therefore compounds which occur only in a specific group of organisms may be preserved in the sediments and act as specific indicators of sedimentary source organisms. Alternatively, a class of organic compounds which occurs widely in biological material may differ in composition according to its source. In some compound classes obtained from the wax fraction of sediments, the distribution characteristic of the parent material is preserved. Thus a change in the nature of allochthonous input, deduced from pollen analysis of the sediment, was correlated with a change in the chain-length distribution pattern of the n-alkanes (Cranwell, 1973b) while input from autochthonous and allochthonous sources was recognised from the n-alkanoic acid distribution pattern (Cranwell, 1974). In this paper components of the wax fraction obtained from sediments of two contrasting lakes are compared and correlated with either autochthonous or allochthonous source material.

## Experimental

### Sites studied

The sediments were obtained from two contrasting lakes, Cam Loch in NW Scotland and Crose Mere in the English Midlands. The valleys of NW Scotland, like those of the Lake District, were formed during the last glaciation, consequently the lakes lie in steep-sided valleys and have vigorous streams flowing through, while rainfall is high. Cam Loch (area: 2.6 $km^2$, maximum depth 37 m) has a drainage area (41.4 $km^2$) of complex geology, containing Torridonian and Cambrian sedimentary rocks, Lewisian gneiss and outcrops of Durness limestone, and high rainfall (229cm/year on Cul Mor). Local vegetation consists of heather, deer grass, sphagnum bog and some birch woods.

The meres of the Shropshire-Cheshire plain lie in shallow hollows in the thick mantle of glacial drift deposited towards the end of the Pleistocene period. Crose Mere (area: 0.15 $km^2$; maximum depth 9.2 m) is maintained and fed mainly by the sub-surface flow of groundwater and has a catchment area of 2.1 $km^2$; rainfall in the region is low (70-80 cm/year). The mere becomes thermally stratified during the summer and the hypolimnion rapidly becomes deoxygenated. The seasonal phytoplankton succession shows a spring chlorophyll maximum due to diatoms with a later maximum due to Ceratium hirundinella (Reynolds, 1973). Thus Crose Mere is an eutrophic lake in which allochthonous input is expected to be relatively low, while Cam Loch is now oligotrophic.

### Sampling and separation techniques

Sediment samples (Table 1) were obtained using pneumatic coring apparatus (Mackereth, 1958, 1969).

The procedure for extraction of the sediment and saponification of the extract, to obtain the neutral and acidic fractions, has been described (Cranwell, 1973b). Separation of the neutral fraction, to obtain the hydrocarbon, ketone and alcohol components, followed the method used for algal detritus (Cranwell, 1976a). The acidic fraction was esterified and the esters separated by chromatographic methods as described previously (Cranwell, 1973a). Each separated component was analysed by gas-liquid chromatography using flame ionisation detection. Quantitative analyses were obtained by

Table 1. Analysis of sediment samples.

| Source | Sample no. | Depth (cm) | Analysis % C | H | N | Organic C % | Yield of wax extract (% of org. C) |
|---|---|---|---|---|---|---|---|
| Cam Loch | 1 | 150-175 | 12.5 | 2.1 | 0.9 | 12.5 | 12.8 |
| | 2 | 350-400 | 12.4 | 1.7 | 1.0 | 12.4 | 9.7 |
| | 3 | 525-575 | 11.0 | 1.8 | 1.4 | 11.0 | 8.2 |
| | 4[1] | 464-487 | 7.9 | 1.8 | 1.3 | 7.9 | 7.6 |
| | 5[1] | 519-550 | 1.8 | 0.7 | 0.2 | 1.6 | 12.5 |
| Crose Mere | 1 | 0-12 | 18.5 | 2.6 | 1.3 | 10.8 | 24.0 |
| | 2 | 25-40 | 18.8 | 1.3 | 1.4 | 11.6 | 12.0 |

[1]Sections 4 and 5 from core with smaller depth of post-glacial deposit. Age of sediment sections increases with numerical order of samples

measurement of GLC peak area.

## Discussion

### Recognition of allochthonous and autochthonous indicators

The standard scheme of analysis (outlined above) was used to separate the components of the wax extract from sediments taken from a number of lakes in the English Lake District. The lakes have been studied for a number of years and the sediments provide a record of past changes (Macan, 1970). Similar analyses were made on terrestrial organic matter from a variety of environments (peat, forest humus) to characterise allochthonous input. This data can be extended by referring to earlier work on soil lipids (Morrison, 1969) and work on the composition of higher plant waxes (Eglinton & Hamilton, 1967). To obtain evidence of an autochthonous contribution the scheme of analysis was applied to natural algal populations and to the detritus obtained by allowing microbial attack on the algae to proceed in the laboratory under aerobic and anaerobic conditions. There is a considerable additional literature on algal constituents, particularly fatty acids (Erwin, 1973), sterols (Goodwin, 1973) and also hydrocarbons (Gelpi et al., 1970).

### *n*-Alkanes

In Cam Loch, the $C_{31}$ *n*-alkane is dominant in the two uppermost horizons and $C_{27}$ is dominant in sections 3 and 4 (Fig. 1). The former pattern is characteristic of sediments derived from acidic peat, while the latter pattern resembles the distribution in forest soil-derived sediments (Cranwell, 1973b). The Cam Loch profile has been $^{14}C$-dated and it is known, from study of an adjacent catchment (Pennington et al., 1972), that the change in alkane distribution pattern between sections 2 and 3 corresponds with the major period of deforestation. These distribution patterns show a high carbon preference index (CPI) and chain-length characteristic of terrestrial vegetation and therefore of allochthonous material. The lowest section of the profile, however, while showing abundant $C_{29}$ and $C_{31}$ alkanes as in other sections, has a very low CPI in the $C_{17}$-$C_{26}$ region. Studies of algal decay showed that a distribution with low CPI was obtained during microbial attack on Ceratium hirundinella (Cranwell,

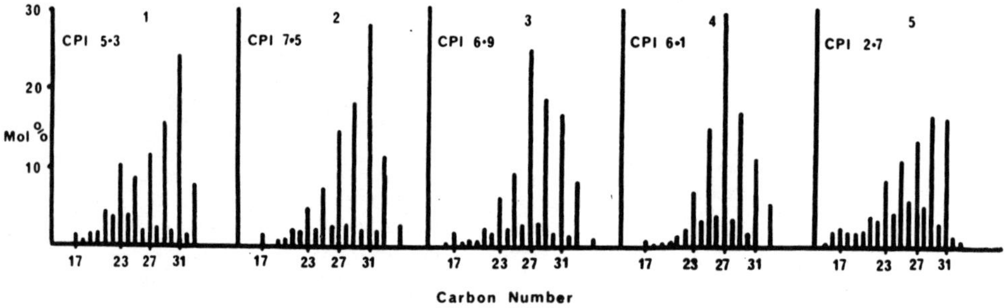


Fig. 1. Percentage composition of *n*-alkanes in Cam Loch sediment. (Samples are numbered as in Table 1)

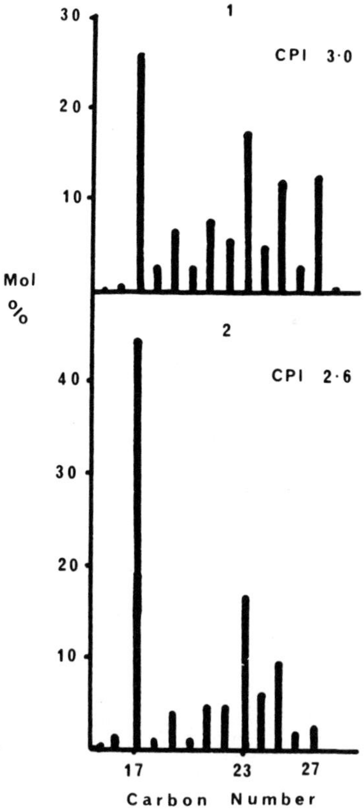


Fig. 2. Percentage composition of n-alkanes in Crose Mere sediment. (Sediment sections numbered as in Table 1)

1976a). Since the low CPI does not occur in samples of terrestrial detritus, the distribution pattern in the lowest horizon of Cam Loch may result from intense microbial activity within the lake at the time of deposition.

In Crose Mere sediments, $C_{17}$ is dominant, while the remaining alkanes have a low CPI and do not extend above $C_{27}$ (Fig. 2), thus showing the absence of a significant allochthonous input. The n-$C_{17}$ alkane is dominant in many blue green algae (Gelpi et al., 1970) and has also been found in algal mats (Cardoso et al., 1976), algal detritus (Cranwell, 1976a), and bacteria (Han et al., 1968).

Monocarboxylic acids

Sedimentary fatty acids were separated into n- and branched/cyclic alkanoic, and alkenoic components. Analysis of these acids in sediments of lakes differing in trophic status showed that eutrophic lakes contained a higher proportion of branched/cyclic and alkenoic acids than oligotrophic lakes (Cranwell, 1974).

a) n-Alkanoic acids

Previous results showed that a bimodal chain-length distribution pattern was associated with eutrophic lakes and a unimodal pattern with oligotrophic lakes. In Cam Loch the two uppermost horizons show the latter distribution while the lowest two show the pattern associated with productive lakes (Fig. 3). Section 3, in which the distribution pattern is intermediate in type, may represent the period during which the trophic status decreased. The contemporary spread of pine at an adjacent site, deduced from pollen analysis, provides evidence of nutrient depletion within the drainage basin (Pennington et al., 1972) and a consequent decrease in trophic status of the lake.

Algae contain alkanoic acids in the $C_{14}$-$C_{20}$ region, with n-$C_{16}$ predominant. Analysis of algal detritus revealed a similar pattern to that of the parent material (Cranwell, 1976a), indicating that the maximum at $C_{16}$ in sediments of productive lakes was associated with autochthonous input.

The distribution in Crose Mere sediments (Fig. 4) shows a higher autochthonous component and correspondingly less significant allochthonous input of higher molecular

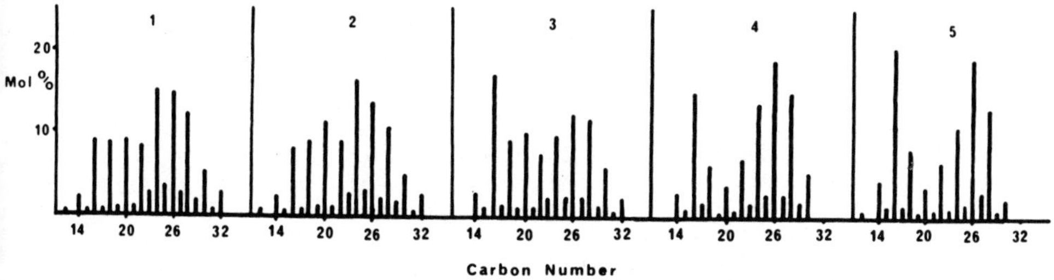


Fig. 3. Percentage composition of n-alkanoic acids in Cam Loch sediment. (Sediment sections numbered as in Table 1)

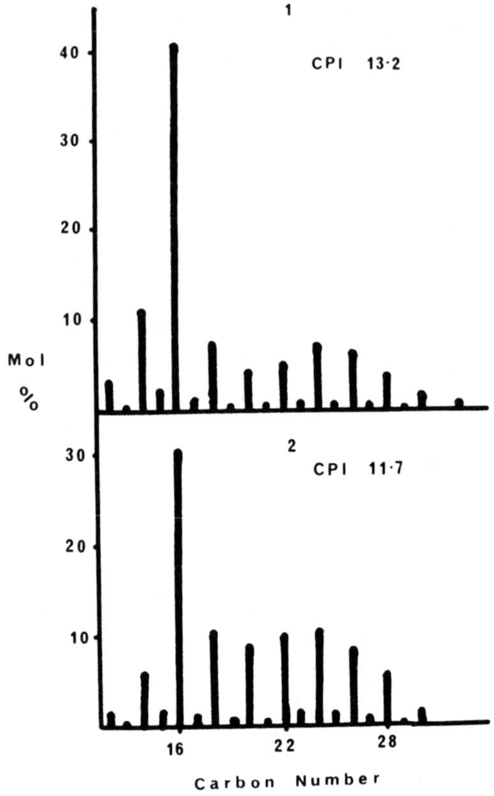


Fig. 4. Percentage composition of n-alkanoic acids in Crose Mere sediment. (Sediment sections numbered as in Table 1)

weight acids, consistent with the eutrophic nature of this lake.

b) Alkenoic acids

Small amounts of mono-unsaturated relative to saturated acids were found in soil organic matter and oligotrophic lake sediments, but a larger proportion was found in productive lakes (Cranwell, 1974). These observations suggest that the unsaturated acids in productive lakes are mainly derived from autochthonous sources. Algae are rich in mono- and polyunsaturated acids and bacteria also contain monounsaturated acids (Erwin, 1973), also anaerobic conditions, which occur in these productive lakes (Macan, 1970), may aid preservation of these acids.

The top three sections of the Cam Loch profile contain a small proportion of monoenoic acids (Table 2), consistent with the mainly allochthonous origin of the sediment deduced from the distribution of n-alkanoic acids. The larger proportion of monounsaturated acids in sections 4 and 5 suggests some autochthonous input. Independent evidence of oxygen depletion when section 5 was being deposited has been obtained from the iron and manganese content (Cranwell, 1974). Both sections of the Crose Mere profile contain a large amount of mono- and polyunsaturated acids, consistent with the large autochthonous input inferred from the n-alkanoic acids and with rapid deoxygenation during summer stratification (Reynolds, 1973).

Table 2. Percentage composition of monocarboxylic acids from lake sediments.

| Sediment source & sample | Abundance (% of total) alkanoic n | br/cyclic[1] | alkenoic |
|---|---|---|---|
| Cam Loch 1 | 85 | 11 (11.5) | 4 |
| 2 | 84 | 13 (13.4) | 3 |
| 3 | 85 | 12 (12.4) | 3 |
| 4 | 75 | 13 (14.8) | 12 |
| 5 | 70 | 18 (20.5) | 12 |
| Crose Mere 1 | 49 | 15 (21.6) | 36 |
| 2 | 52 | 13 (20.0) | 35 |

[1] Figures in parenthesis show % of total alkanoic acids

c) Branched/cyclic alkanoic acids

The higher proportion of these acids in the saturated acids of productive compared with unproductive lakes (Cranwell 1974) cannot be attributed solely to algal constituents, since many of the low molecular weight compounds (those below $C_{20}$), such as the iso, anteiso, 10-methyl-branched acids and those containing cyclopropyl rings, occur mainly in bacteria and aquatic fungi (Cranwell, 1973a). Also present among low molecular weight compounds are the isoprenoids 4,8,12-trimethyltridecanoic, pristanic and phytanic acids derived in part from chlorophyll (Brooks & Maxwell, 1974). The unexpected presence of phytenic acid in the 'saturated' acid fraction obtained from a lake sediment was reported by Boon et al. (1975). Acids derived from chlorophyll precursors may be of terrestrial or aquatic origin, since chlorophyll pigments occur widely, but the high concentration of pigments in sediments of productive lakes was correlated with a high algal crop (Gorham et al., 1974), indicating that the isoprenoid acids in eutrophic lakes were mainly autochthonous. The small amount of branched/cyclic alkanoic acids in soils also indicated that the acids found in sediments were mainly formed within the lake. Support for this hypothesis is provided by the similarity in qualitative composition of sedimentary branched/cyclic acids and those isolated from algal detritus (Cranwell, 1976a, b). The

Table 3. Composition of branched/cyclic alkanoic acids.

| Sediment | | Percentage as function of Equivalent Chain Length (ECL) | | | % of 17βH,21βH-$C_{32}$ acid |
|---|---|---|---|---|---|
| source | sample no. | below $C_{20}$ | $C_{28}$-$C_{32}$ (group i) | $C_{35}$-$C_{38}$ (group ii) | |
| Cam Loch | 1 | 44 | 11 | 38 | 27 |
| | 2 | 41 | 15 | 37 | 27 |
| | 3 | 39 | 20 | 32 | 25 |
| | 4 | 33 | 12 | 45 | 35 |
| | 5 | 63 | 5 | 28 | 25 |
| Crose Mere | 1 | 75 | 1 | 17 | 12 |
| | 2 | 66 | 3 | 23 | 18 |

increase in relative abundance during algal decay provides further evidence of a microbial origin.

The presence in sediments of a small percentage of branched/cyclic alkanoic acids with equivalent chain length (ECL) greater than $C_{20}$ was reported earlier (Cranwell, 1974). However it is now known that major components with very long GLC retention times had not been detected. In the Cam Loch profile these complex esters constitute 37-67% of the total fraction (Table 3), while in Crose Mere amounts of these complex esters were lower, especially in the surficial sediment. There are two major groups of acids (Table 3), based on molecular size and structural similarities:-

i) acids eluting between n-$C_{28}$ and n-$C_{32}$ (on OV-17 phase)

ii) acids eluting between n-$C_{35}$ and n-$C_{38}$ on the same GLC columns

The major component of group ii and the most abundant of all branched/cyclic acids in these sediments was identified by GC/MS as a $C_{32}$ acid, 17βH,21βH-bishomohopanoic acid. One of the two possible diastereoisomers with this structure was found in a wide variety of recent and ancient sediments (Van Dorsselaer et al., 1974) and in an algal mat (Cardoso et al., 1976). Two minor components of this group, with 31 and 33 carbon atoms, appear to be homologues of the $C_{32}$ acid while small amounts of the $C_{31}$ and $C_{32}$ 17αH,21βH isomeric hopane acids may also be present, as indicated by the relative intensity of the two main fragment ions characteristic of the hopane system (Ensminger et al., 1974). The origin of these compounds is uncertain, though possible triterpene precursors exist in both higher plants and lower organisms such as prokaryotes (Ensminger et al., 1974). No hopanoid acids were detected during short-term algal decay (Cranwell, 1976b) but their presence in an algal mat (Cardoso et al., 1976) suggests that these compounds, being less easily degraded than acyclic branched-chain acids, may accumulate during microbial breakdown of organic matter. Analysis of the Crose Mere sediments supports this hypothesis since the $C_{32}$ 17βH,21βH-hopanoic acid constitutes a larger proportion in the lower, older horizon. The greater relative stability of group ii acids is also reflected in their abundance in the sediment, decreasing from 320 ppm organic C in section 1 to 150 ppm in section 2 whereas the other branched/cyclic acids decrease from 1600 to 450 ppm. Elemental analyses of these two sections and analyses of the soluble organic fractions reported here indicate little overall change in sediment composition and source material, respectively, and thus seem to exclude factors other than diagenesis from influencing the abundance and composition of the branched/cyclic acids.

The acids within group i are believed to be derived from allochthonous sources since they have not been found in algal detritus and are only trace constituents of Crose Mere sediments. In the Cam Loch profile these acids are most abundant in horizons in which erosion increased as deforestation occurred. GC/MS analysis enabled some structural features to be deduced and showed the presence of minor components unresolved by GLC. The four main components show fragment ions characteristic of polycyclic terpenoid compounds (m/e 177,191,205) but lack strong peaks associated with ester loss (m/e 74,87, 88,101), the latter feature being also observed with the hopane derivatives discussed above. In the Cam Loch profile changes in terrestrial vegetation influence the relative abundance of these components suggesting that the latter may be characteristic of distinct taxonomic groups.

## Sterols

The major sterols in the leafy material of higher plants, stigmasterol and sitosterol are not abundant in algae (Goodwin, 1973), which probably provide a considerable sterol input to sediments of productive lakes. Analysis of soils and sediments showed that

Table 4. Composition of sterol fractions from sediments and soils.

| Sterol | Range of % abundance values | | | | | |
|---|---|---|---|---|---|---|
| | soils[1] | sediment survey | | cam loch | | crose mere |
| | | oligotr[2] | eutr[3] | 1-3 | 4-5 | 1-2 |
| Cholesta-5,22-dien-3β-ol | | 1-2 | 3-4 | 2 | 2-3 | 4-5 |
| Cholesterol + 5α-stanol | 1-4 | 9-15 | 18-22 | 11-14 | 25 | 31-32 |
| 24-Methylcholesta-5,22-dien-3β-ol | 1-3 | 2-4 | 5-9 | 2-5 | 6-8 | 8-10 |
| 24-Methylcholesterol+ 5α-stanol | 6-13 | 7-11 | 8-9 | 7-9 | 5-8 | 8-11 |
| 24-Ethylcholesta-5,22-dien-3β-ol | 4-15 | 7-12 | 9 | 5-8 | 6 | 3-6 |
| 24-Ethylcholesterol+ 5α-stanol | 66-72 | 51-69 | 41-47 | 53-60 | 28-42 | 23-27 |

Notes [1] Range of values for 4 soil samples.
[2] " " " " 11 sections from 5 oligotrphic lakes.
[3] " " " " 2 " in an eutrophic lake.

the composition of the sterol fraction was an indicator of input from terrestrial and aquatic source material. Reduction of sterols to stanols has been shown to occur in sediments (Gaskell & Eglinton, 1975) but the major reduction product, the 5α-stanol, is not easily resolved from the parent sterol by GLC, so that the peak area reflects the original sterol input. The survey (Table 4) showed very small amounts of cholesterol in soils, in which the $C_{29}$ sterol was dominant, but somewhat larger amounts of $C_{27}$ $\Delta^5$-sterol and related stanol in oligotrophic lake sediments, and an even larger proportion in sediments of a productive lake, with a reduction in the $C_{29}$ component.

In the Cam Loch profile the sterol distribution of the upper three horizons is typical of oligotrophic lakes previously studied, and in the lowest two sections the distribution resembles that found in a modern productive lake (Table 4). The $C_{27}$ sterol is even more abundant in Crose Mere, consistent with the high autochthonous contribution deduced from the n-alkanoic acids. Cholesterol was dominant in the algal species subject to decay (Cranwell, 1976a, b) and was also dominant in both phytoplankton and surficial sediment from Rostherne Mere (Gaskell & Eglinton, 1975). Cholesterol is not abundant in the major algal groups apart from Rhodophyta (Goodwin, 1973), which are absent in Crose Mere, so that zooplankton may be a significant source of cholesterol, especially in Crose Mere where high zooplankton populations have been observed (Reynolds, personal communication).

Ketones

A homologous series of methyl n-alkyl ketones, with $C_{27}$ or $C_{29}$ as major component and a high odd/even CPI, occurs in lake sediments from Cam Loch and was also found in soils and peat (Morrison, 1969). The similarity in composition of soils and sediments suggests that sedimentary ketones are allochthonous in origin. They may be formed by microbial action on n-alkanes or carboxylic acids (Morrison, 1969). No extended series of ketones was detected in algal detritus, in which 6,10,14-trimethylpentadecan-2-one was the major ketone. This isoprenoid ketone was abundant in Crose Mere sediment which also contained some $C_{17}$-$C_{30}$ ketones but here, unlike Cam Loch, the chain length distribution did not correspond with that of the n-alkanes in the same sediment.

Alcohols

Cam Loch sediments contain a homologous series of n-alkanols, from $C_{16}$-$C_{32}$, with a high even/odd CPI as in alcohols of higher plant waxes (Eglinton & Hamilton, 1967), $C_{28}$ most abundant in sections 1-4 and $C_{26}$ dominant in section 5. No long-chain n-alcohols were found in the algal materials, but in Crose Mere n-alkanols were more abundant than anticipated for mainly autochthonous input and the $C_{26}$ constituent was unusually strong. The abundance and composition of the alkanols in Crose Mere may reflect a contribution from the abundant zooplankton.

αω-Dicarboxylic acids and ω-Hydroxyacids

Cam Loch sediments contain series of αω-dicarboxylic and ω-hydroxy acids having an even number of C atoms in the range $C_{16}$-$C_{28}$ in which the $C_{22}$ acid is dominant in both series. A parallel chain-length distribution between these two classes, previously reported by Eglinton et al. (1968) was interpreted as evidence that the diacids were in part derived from ω-hydroxyacids, which occur in the cutin of higher plants. ω-Hydroxyacids and dicarboxylic acids above $C_{19}$ have not been reported in algae and were not detected in algal detritus. In Crose Mere, dicarboxylic acids were less abundant than in Cam Loch and the $C_{16}$ acid was more prominent, however ω-hydroxyacids were not detected. Thus the ω-hydroxyacids in sediments probably represent allochthonous input but αω-dicarboxylic acids may be partly derived from autochthonous sources.

## Conclusions

Analysis of organic compound classes from sediments, soils and algal detritus facilitates the recognition of allochthonous and autochthonous input to sediments.

In Cam Loch, the predominantly allochthonous input, revealed by the n-alkane, n-alkanoic acid, sterol, ketone, αω-dicarboxylic acid and ω-hydroxyacid components, is consistent with environmental features (climate and relief) favouring erosion of the drainage basin. The older horizons show a greater autochthonous input in the early post-glacial period when leaching had not depleted the soil nutrients.

The dominance of autochthonous sources of organic matter in the surface sediments of Crose Mere, indicated by these analyses, is consistent with the high trophic status and the absence of features causing intense soil erosion.

## Acknowledgements

I thank Professor G. Eglinton, Organic Geochemistry Unit, Bristol University, for the use of the computer/GC/MS instrument, and Drs W. Tutin, E.Y. Haworth and C.S. Reynolds for useful discussions.

## References

Boon, J.J., W.I.C. Rijpstra, J.W. deLeeuw & P.A. Schenk, 1975. Phytenic acid in sediments. Nature (London), 258:414-416.

Brooks, P.W. & J.R. Maxwell, 1974. Early stage fate of phytol in a recently-deposited lacustrine sediment. In: B. Tissot & F. Bienner (Ed.): Advances in Organic Geochemistry 1973. Editions Technip, Paris. p. 977-991.

Cardoso, J., P.W. Brooks, G. Eglinton, R. Goodfellow, J.R. Maxwell & R.P. Philp, 1976. Lipids of recently-deposited algal mats at Laguna Mormona, Baja California. In: J.O. Nriagu (Ed.): Environmental Biogeochemistry. Vol. 1. Ann Arbor Science, Ann Arbor. p. 149-174.

Cranwell, P.A., 1973a. Branched-chain and cyclopropanoid acids in a recent sediment. Chem. Geol., 11:307-313.

Cranwell, P.A., 1973b. Chain-length distribution of n-alkanes from lake sediments in relation to post-glacial environmental change. Freshwater Biol., 3:259-265.

Cranwell, P.A., 1974. Monocarboxylic acids in lake sediments: indicators, derived from terrestrial and aquatic biota, of paleoenvironmental trophic levels. Chem. Geol., 14:1-14.

Cranwell, P.A., 1976a. Decomposition of aquatic biota and sediment formation: organic compounds in detritus resulting from microbial attack on the alga Ceratium hirundinella. Freshwater Biol., 6:41-48.

Cranwell, P.A., 1976b. Decomposition of aquatic biota and sediment formation: Lipid components of two blue-green algal species and of detritus resulting from microbial attack. Freshwater Biol., 6: in press.

Eglinton, G. & R.J. Hamilton, 1967. Leaf epicuticular waxes. Science (Wash. DC), 156:1322-1334.

Eglinton, G., D.H. Hunneman & K. Douraghi-Zadeh, 1968. Gas chromatographic-mass spectrometric studies of long-chain hydroxy acids-II. The hydroxy acids and fatty acids of a 5000-year-old lacustrine sediment. Tetrahedron, 24:5929-5941.

Ensminger, A., A.Van Dorsselaer, C. Spyckerelle, P. Albrecht & G. Ourisson, 1974. Pentacyclic triterpenes of the hopane type as ubiquitous geochemical markers: origin and significance. In: B. Tissot, & F. Bienner (Ed.): Advances in Organic Geochemistry 1973. Editions Technip, Paris. p. 245-260.

Erwin, J.A., 1973. Comparative biochemistry of fatty acids in eukaryotic microorganisms. In: J.A. Erwin (Ed.): Lipids and Biomembranes of Eukaryotic Microorganisms. Academic Press, London p. 41-143.

Gaskell, S.J. & G. Eglinton, 1975. Rapid hydrogenation of sterols in a contemporary lacustrine sediment. Nature (London), 254:209-211.

Gelpi, E., H. Schneider, J. Mann & J. Oro, 1970. Hydrocarbons of geochemical significance in microscopic algae. Phytochemistry, 9:603-612.

Goodwin, T.W., 1973. Comparative biochemistry of sterols in eukaryotic microorganisms. In: J.A. Erwin (Ed.): Lipids and Biomembranes of Eukaryotic Microorganisms. Academic Press, London. p. 1-41.

Gorham, E., J.W.G. Lund, J.E. Sanger & W.E. Dean 1974. Some relationships between algal standing crop, water chemistry, and sediment chemistry in the English lakes. Limnol. Oceanogr. 19:601-617.

Han, J., E.D. McCarthy, W. van Hoeven, M. Calvin & W.H. Bradley, 1968. Organic geochemical studies II. A preliminary report on the distribution of aliphatic hydrocarbons in algae, in bacteria, and in a recent lake sediment. Proc. Nat. Acad. Sci., USA, 59:29-33.

Likens, G.E., 1972. Eutrophication and aquatic ecosystems. In: Nutrients and Eutrophication. Special Symposium, Am. Soc. Limnol. Oceanogr., Lawrence. p. 3-13.

Macan, T.T., 1970. Biological studies of the English lakes. Longman, London, 260pp.

Mackereth, F.J.H. 1958. A portable core sampler for lake deposits. Limnol. Oceanogr. 3:181-191.

Mackereth, F.J.H., 1969. A short core sampler for subaqueous deposits. Limnol. Oceanogr., 14:145-151.
Morrison, R.I., 1969. Soil lipids. In: G. Eglinton & M.T.J. Murphy (Ed.): Organic Geochemistry. Springer-Verlag, Berlin. p. 558-575.
Pennington, W., E.Y. Haworth, J.P. Lishman & A.P. Bonny, 1972. Lake sediments in northern Scotland. Phil. Trans. R. Soc. (B), 264:191-294.
Reynolds, C.S., 1973. The phytoplankton of Crose Mere, Shropshire. Br. phycol. J., 8:153-162.
Van Dorssalaer, A., A. Ensminger, C. Spyckerelle, M. Dastillung, O. Sieskind, P. Arpino, P. Albrecht, G. Ourisson, P.W. Brooks, S.J. Gaskell, B.J. Kimble, R.P. Philp, J.R. Maxwell & G. Eglinton, 1974. Degraded and extended hopane derivatives ($C_{27}$ to $C_{35}$) as ubiquitous geochemical markers. Tetrahedron Lett. 1349-1352.

Key words: organic
indicators
sediments
source material

THE RELATIONSHIP BETWEEN LIPIDS FROM FONTINALIS ANTIPYRETICA, ITS DETRITUS AND THE UNDERLYING SEDIMENT: THE FATE OF WAXESTERS AND STEROLESTERS

Jan W. de Leeuw[1], W. Irene C. Rijpstra, Jaap J. Boon, Frits de Lange and P.A. Schenck

Delft University of Technology, Department of Chemistry and Chemical Engineering, Organic Geochemistry Unit, Delft, The Netherlands

Summary

Sterolesters of Fontinalis antipyretica are found to be rapidly hydrolysed during decay. The resulting sterols are modified into the corresponding 5α(H)- and 5β(H)-stanols and 4-methyl-5α(H)-stanols in the sediment. Isoprenoid waxesters are thought to be indicators for processes of hydrolysis and resynthesis of waxesters in detritus and sediment.

Introduction

Detailed structural analysis of lipid material from standing crop, detritus and sediment in the Lonnekermeer was performed with the object to investigate processes leading to fossil sedimentary organic matter in a fresh water environment.

The Lonnekermeer is at present a small lake, situated on private property near Hengelo (The Netherlands); it started 50 years ago as a sandpit. It was left to itself and kept free of waste discharge since that time. The lake recieves rain and well water.

The periodic abundance of springmoss - Fontinalis antipyretica - in this lake opened the opportunity to investigate the intact plant material and the decaying detritus of Fontinalis antipyretica in relation to sedimentary organic matter of older date. Comparison of the molecular structures identified in these stages indicate that several lipid classes and individual compounds underwent reactions in the process of sedimentation. We have restricted ourselves mainly to the discussion of the fate of waxesters and sterolesters. The unique structural characteristics of the building blocks of these compound classes provide clues to mechanisms of transformation of lipid material in the lake ecosystem.

Experimental

Samples and sampling

Fontinalis antipyretica was sampled in the Lonnekermeer, which is situated near Hengelo coordinates: 6° 51' 13" E, 52° 17' 00" N). Decaying Fontinalis antipyretica, in this paper further on called detritus, was found above the organic rich sediment layer. It could be sampled by hand picking. The sediment of the lake consisted of an organic layer of 8 cm thick laying on coarse sand. The sediment was sampled by insertion of a 30 cm glass core. The core was taken in the centre of the lake and consisted of 8 cm organic layer and 20 cm sand. The sediment sample investigated was taken from this core and consists of 7 cm sandlayer and 5 cm organic layer. This sample is referred to as the sediment. The upper 3 cm organic rich sediment present in the core was not investigated. Immediately after sampling, the samples were transferred to large glass jars kept in dry ice.

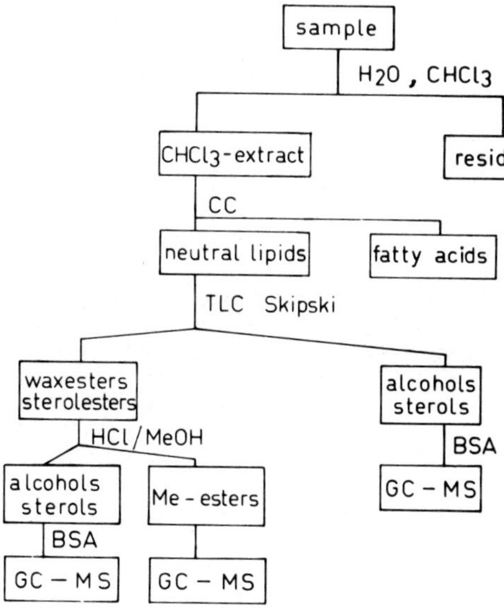


Figure 1. Flow scheme of analytical procedure

1. Address: de Vries van Heystplantsoen 2, Delft, The Netherlands

Table 1. Inventory of lipid classes

| | F.antipyretica | Detritus | Sediment |
|---|---|---|---|
| Hydrocarbons | + | + | + |
| Waxesters | +) $C_{26}-C_{60}$ [1] | + $C_{28}-C_{54}$ | + $C_{40}-C_{51}$ |
| Sterolesters | +) | - | - |
| Triglycerides | + | + | - [2] |
| Hexahydrofarnesylacetone | + | + | |
| Free fatty acids | + | + | + |
| Free fatty alcohols | + | + | + |
| Diglycerides | + | - | - |
| Free Sterols | - | + | - |
| Free Stanols | - | - | + |
| Free 4-methylstanols | - | - | + |

[1] determined by Field Desorption Mass Spectrometry
[2] not yet determined

Extraction, isolation and identification

In the simplified extraction scheme we have only included those fractions pertinent to the alcohol, sterol, waxester and sterolester isolation and identification. Each of the samples was blended in bidistilled water using a Waring blendor at high speed. Coarser plant remains were separated from the pureed samples and the watery mass was extracted by shaking with chloroform using a separatory funnel. The sandy part of the sediment was moreover extracted with chloroform by ultrasonication. Each of the chloroform layers was evaporated using a rotatory evaporator and the obtained extracts were weighed. In this way 450 mg of Fontinalis antipyretica, 1850 mg detritus and 300 mg sediment extract were obtained. These extracts were surveyed for lipid classes by thin layer chromatography (Skipski et al., 1965). Data are presented in table 1.

Neutral lipids were separated from lipid material with free carboxylic functions by $K^+$-impregnated silicic acid column chromatography (McCarty & Duthie, 1962). Waxesters + Sterolesters, fatty alcohols + sterols or stanols were isolated from the neutral lipid fraction by preparative thick layer chromatography (Skipski et al., 1965).

The molecular weight range of each waxester and sterolester fraction was determined by Field Desorption Mass Spectrometry. The fractions were also transesterified by 5% HCl-methanol (Gellerman et al., 1975).

Hydrolysis products were separated into hydrocarbons, fatty acid methyl esters and phytylmethylethers, fatty alcohols and sterols by thin layer chromatography (Skipski et al., 1965)

Identification of individual organic compounds was performed by capillary gas chromatography-mass spectrometry using the Varian - MAT 111 instrument. The gas chromatograph was equipped with a 30 m stainless steel open tubular column (I.D. 0.25 mm) coated with OV - 101. Helium was used as carrier gas. The mass spectrometer was used under standard conditions at 70 eV.

The fatty acid methyl esters, phytadienes and phytylmethylethers were analysed as such, alcohols, sterols and stanols were derivatised by B.S.A. (Bistrimethylsilylacetamide, Merck Chem. Comp.) to the corresponding trimethylsilylethers. Cholesteryllaurate (Sigma Chem. Comp.) and Cetylpalmitate (Applied Science Inc.) were used as authentic standards.

Synthesis of the 4-methylcholestan-3-ol

The synthesis of this reference compound was started from purified cholesterol (Merck) which was converted into cholestenone via Oppenauer oxidation (Vogel, 1961). Methylation of cholestenone was performed with $CH_3I$ in base according to Atwater (1960). Catalytic reduction of the double bond with 5% Pd/C/$H_2$ in ethanol yielded the corresponding 4-methylcholestanone. The 4-methylcholestan-3β-ol was obtained by reduction of the ketone with $NaBH_4$ in methanol.

The mass spectrum of the 4-methylcholestan-3β-trimethylsilylether was obtained by capillary GCMS (Figure 2)

Results

Table 1 represents an inventory of the classes of lipid compounds present in the samples investigated. Their identification is based on qualitative thin layer chromatography, Field Desorption Mass Spectrometry and results of the ultimate analysis of individual compounds by GCMS after preparative layer chromatography.

From this survey of lipid species it was decided to study the wax- and sterolesters in detail. Acid catalyzed transesterification of these fractions yielded fatty acid methyl esters, fatty alcohols, sterols, phytadienes and phytylmethylethers. Phytadienes and phytylmethylethers are artifacts of phytol in this procedure (Gellerman et al., (1975), Rijpstra et al., (1976). Major hydrolysis products of the wax- and sterolester fraction together with data on free fatty alcohols, sterols, stanols and stenols are summarized in Table 2. Major compounds in this table are defined as compounds with a relative abundance of more than 50% of the highest peak in the gaschromatogram. All compounds were identified from mass spectral data by gas chromatography mass spectrometry after appropiate chromatographic separations and derivatisations. Interpretation of mass spectra was based on literature data (McCloskey (1970), Brooks et al., (1969), Boon et al., (1975).

The identified stanols need a more extensive discussion. Preparative thin layer chromatography of the neutral lipids of the sediment extract yielded two bands with a retention distance comparable to that of sterols. After GCMS analysis the upper band consisted of a mixture of 5β(H)-stanols and 4-methylstanols, while the lower band consisted of 5α(H)-stanols. This difference in chromatographic behaviour is probably related to the stereochemistry of the A and B rings in the steroid skeleton.

Figure 3 represents the mass spectra of the sedimentary 5α(H)-β-sitostanol, 5β(H)-β-sitostanol and 4-methyl-β-sitostanol as trimethylsilylether.From literature data it is known that the 5α(H) and 5β(H) configuration in these derivatives can be judged from differences due to the loss of the sidechain and D-ring cleavage. (Brooks et al., 1968, Henderson et al., 1972). 5α(H)-β-sitostanol therefore shows a strong m/e 215 ($M^+$-90-side chain-42) and m/e 305 ($M^+$-side chain-42). The 5β(H)-β-sitostanol spectrum does not show m/e 305, but the m/e 257 ($M^+$-90-side chain) is of the same abundance as the m/e 215.

The spectrum of the authentic 4-methyl-cholestan-3β-trimethylsilylether (Figure 2) shows m/e 229 ($M^+$-90-side chain-42) and m/e 271 ($M^+$-90-side chain) reminiscent of a 5β(H) configuration. The mass spectrum in Figure 3C presents m/e 319 and m/e 229 as characteristic fragment ions together with a low intensity of m/e 217. This would point to 5α(H) configuration of the 4-methylstanols present in the sediment. The stereochemistry of the 4-methyl group in 4-methylstanols is thought to have no evident influence on the formation of the fragment ions (Brooks et al., 1968, Gustafsson et al., 1966).

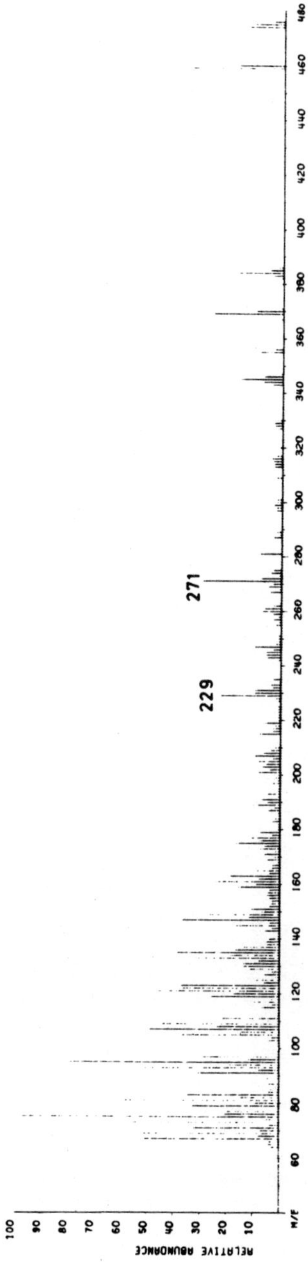


Figure 2. Mass spectrum of the authentic 4-methyl-cholestan-3β-trimethylsilylether.

Table 2. Inventory of free major alcohols, sterols and stanols, and major hydrolysis products of the waxester, sterolester fraction.

| | |
|---|---|
| **Fontinalis antipyretica** | |
| Free fatty alcohols | nC16:0, nC16:1, nC18:0, isopr.C20:1, nC20:0, nC22:0, nC24:0, nC26:0 |
| Bound fatty alcohols | nC14:0, iC15:0, nC16:0, nC16:1, nC18:0, nC18:1, nC18:2 |
| Bound sterols | cholesterol ($\Delta^5$C27:1)(A), campesterol ($\Delta^5$C28:1)(B), stigmasterol ($\Delta^5$, $\Delta^{22}$C29:2)(C), β-sitosterol ($\Delta^5$C29:1)(D) |
| Bound fatty acids | nC16:0, nC16:1, nC18:0, nC18:1, nC18:2, isopr.C20:1, nC20:0, nC20:1, nC20:2, nC20:3, nC22:0 |
| Phytadienes, Phytyl-methylethers | present (indicative of phytol (isopr. C20:1) in waxesters) |
| **Detritus** | |
| Free fatty alcohols | nC14:0, nC15:0, nC16:0 |
| Bound fatty alcohols | nC14:0, nC15:0, nC16:0, nC16:1, nC18:0, nC18:1, nC22:0, nC24:0 |
| Bound fatty acids | nC14:0, nC16:0, nC16:1, nC18:0, nC18:1, nC18:2, isopr.C20:1, nC22:0, nC24:0 |
| Free sterols | cholesterol(A), campesterol(B), stigmasterol(C), β-sitosterol(D) |
| **Sediment** | |
| Free fatty alcohols | nC12:0, nC14:0, iC15:0, ai15:0, nC16:0, nC22:0, nC24:0, nC26:0, nC28:0 |
| Free 5α(H)stanols | cholestanol(A), campestanol(B), stigmastenol ($\Delta^{22}$C29:1)($C^*$), β-sitostanol(D) |
| Free 5β(H)stanols | cholestanol(A), campestanol(B), stigmastenol ($\Delta^{22}$C29:1)($C^*$), β-sitostanol(D) |
| Free 4-methyl-5α(H)-stanols | -cholestanol(A?), -campestanol(B), -stigmastenol ($\Delta^{22}$C29:1)($C^*$), -β-sitostanol(D) |
| Bound fatty alcohols | nC15:0, nC16:0, nC20:0, nC22:0 |
| Bound fatty acids | nC16:0, isopr.C20:0, nC18:0, isopr.C20:1, nC20:0, nC22:0, nC23:0, nC24:0 |

Alcohol : isopr.C20:1 = phytol
Fatty acid : isopr.C20:1 = phytenic acid; isopr.C20:0 = phytanic acid
The highest peak in each compound class is underlined.
The lettercodes (in brackets) correspond with those in Figure 4.

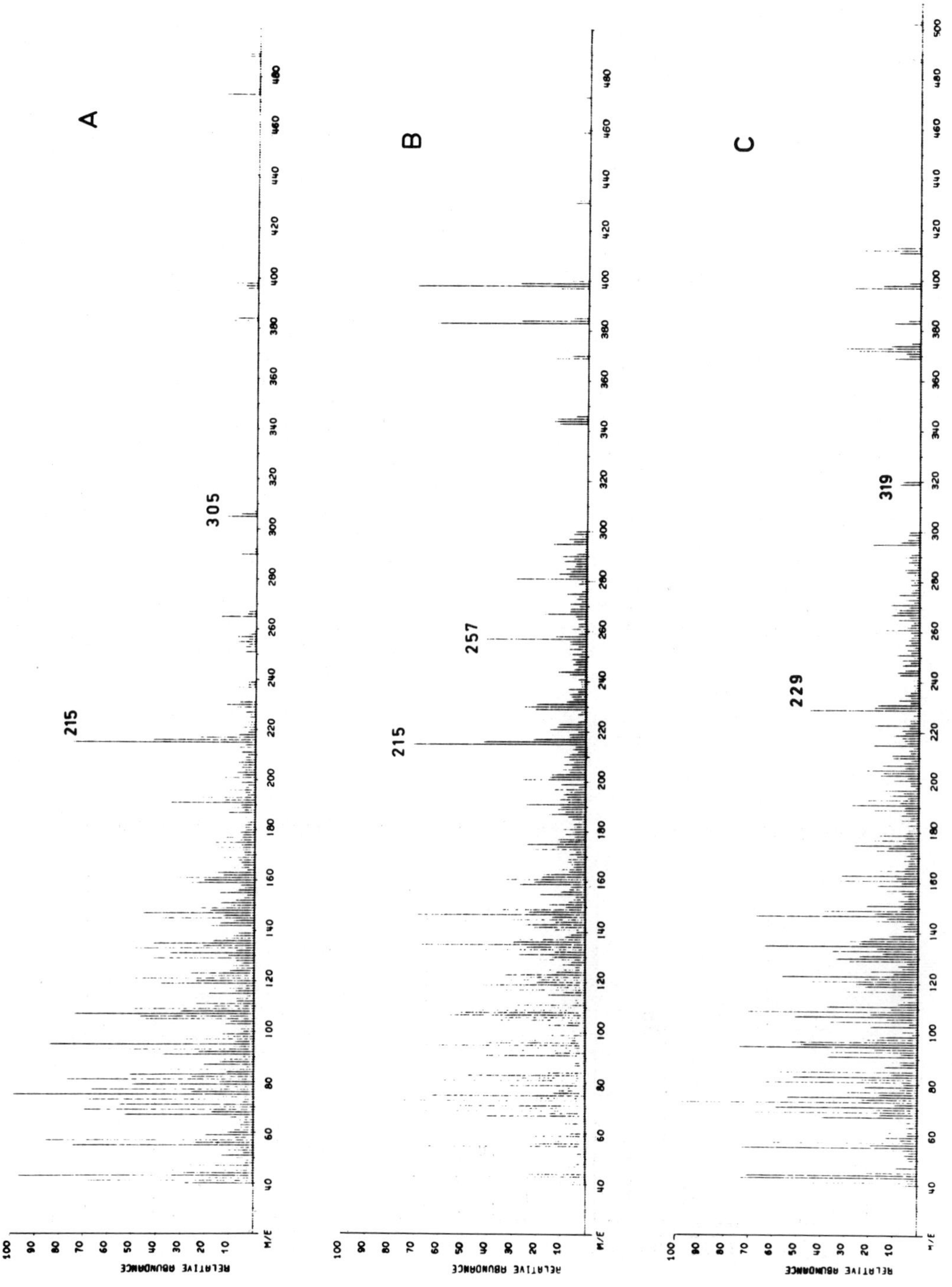


Figure 3. Mass spectra of sedimentary steroidal alcohols as trimethylsilylethers.
A=5α(H)-β-sitostanol, B=5β(H)-β-sitostanol, C=4-methyl-5α(H)-β-sitostanol.

## Discussion

The importance of hydrolytic processes carried out by microorganisms on the more complex neutral lipid species present in the three stages of organic matter transformation in the lake is pointed out by the inventory of lipid classes (Table 1).

The change in the sterolesters is dramatic. The sterolesters of Fontinalis antipyretica are rapidly hydrolysed as is reflected in the character and distribution of free sterols in the detritus.

The $\Delta^5$ double bond of the free sterols is reduced in the sediment stage and yields 5α(H)- and 5β(H)-stanols. The relative abundance of the individual steroid skeletons is retained (Figure 4), which means that there seems to be no significant introduction of other steroidal compounds in the sediment stage.

The process of production of stanols from sterols has been observed before by Gaskell & Eglinton (1975) in the productive Rostherne mere. $^{14}$C labelled cholesterol was introduced into the lake bottom and yielded more 5α(H)- than 5β(H)-cholestanols. No data are available with regard to the ratio of 5α(H)- and 5β(H)-stanols in the Lonnekermeer sediment. The formation of these isomers however is evidently a result of the properties of the microbial enzymes involved in this hydrogenation process. The occurrence of 4-methylstanols in a lake sediment has not been reported before. The relative distribution of these compounds (Figure 4), their co-occurrence with the 5α(H)- and 5β(H)-stanols, their mode of occurrence and the character of the steroidal skeletons indicate that these compounds are also formed from the sterols present in the detritus or maybe even from the sedimentary stanols. No microorganisms are known as yet which are able to modify sterols or stanols into 4-methylstanols. An incubation experiment with 4-$^{14}$C-cholesterol is currently in progress to verify the hypothesis of the formation of 4-methylstanols from sterols in the sediment.

The waxesters are found in the three stages in the lake ecosystem. It is of interest to note the variability in the range of molecular weights in the waxester fractions. This variability together with the hydrolysis data indicates that several as yet not understood processes take place. On the one hand we suppose that resynthesis of waxesters may take place on the other hand complete hydrolysis and introduction of waxesters from another source in the sediment stage is a possibility. Details of the exact processes remain to be elucidated. However one should be careful with assumptions that waxesters in sediments represent waxesters directly derived from plants or organisms in the lake.

Hydrolytic processes are not only indicated by the fate of the sterolester and part of the waxesterfraction. Triglycerides and diglycerides, which can be produced also from a variety of complex polar lipids, disappear in the detritus stage, while preliminary evidence point to increased amounts

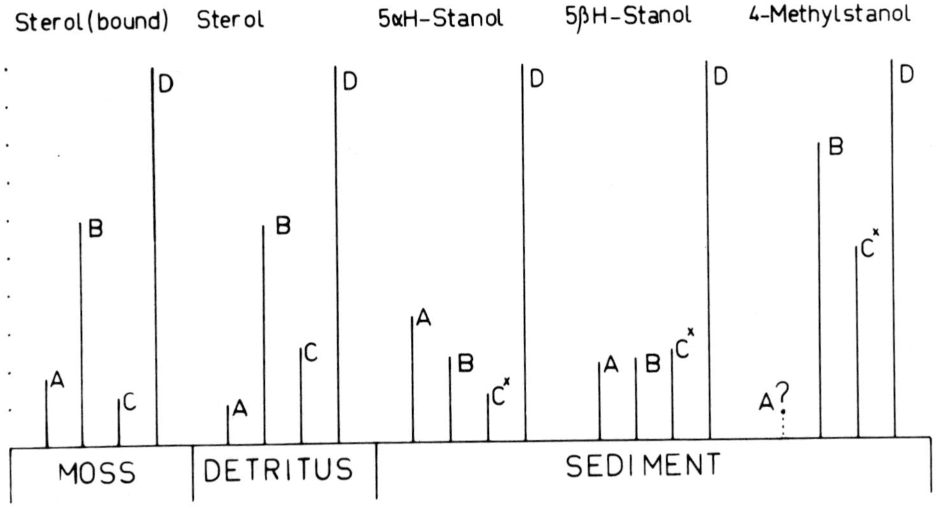


Figure 4. Relative abundance of steroidal alcohols in Fontinalis antipyretica (moss), detritus and sediment. The letter codes in the compound classes correspond with those in Table 2.

of free fatty acids. This phenomenon is in accordance with data of Poltz (1972).

The reduction of double bonds in the sediment stage is not only pointed out by the stanols but also by the free fatty alcohols and the bound fatty alcohols and acids (compare Table 2). The only unsaturated compound still present in the sediment is phytenic acid. The explanation for the origin of this compound (Boon et al., 1975) must be reconsidered. Phytylphytenate is an abundant waxester in F. antipyretica and other mosses (Gellerman et al., 1975) but no free phytenic and phytanic acid is present in the moss. The phytol in the free alcohol fraction of the moss is very probably derived from chlorophyl derivatives.

In the detritus no phytylphytenate is found, but phytenates are present. In the free fatty acid fraction of the detritus phytenic and phytanic acid are major acids. In the sediment phytenates are present, but no phytylphytenate. Only phytanic acid is present in the free acid fraction of the sediment.

The structural rigidity of the ester region in the phytylphytenate molecule warrants a good access for an hydrolytic enzyme. This could be the reason that phytylphytenate is rapidly hydrolysed in the detritus stage into phytenic acid and phytol.

Earlier studies have demonstrated that U-$^{14}$C-phytol, introduced into the Lonnekermeerdetritus, is converted for the greater part into phytanic acid (de Leeuw et al., 1975) The presence of phytenates in detritus and sediment therefore is seen as an indication of resynthesis of waxesters in these stages of organic matter transformation (Figure 5).

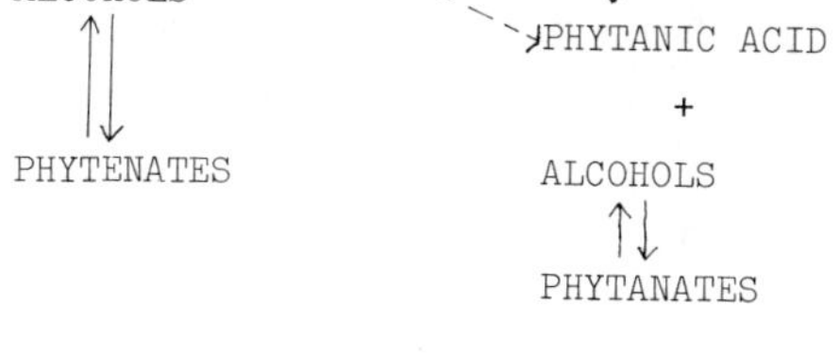


Figure 5. Proposed early diagenetic pathway for waxesters concluded from the fate of isoprenoid waxesters.

This hypothesis will be studied by introduction of radiolabelled phytylphytenate into the detritus.

## Literature

Atwater, N.W., 1960. 4-Substituted steriods. J. Am. Chem. Soc. 82:2847-2854.

Boon, J.J., W.I.C. Rijpstra, J.W. de Leeuw & P.A. Schenck, 1975. Phytenic acid in sediments. Nature 258:414-416.

Brooks, C.J.W., E.C. Horning & J.S. Young, 1968. Characterization of sterols by gas chromatography - mass spectrometry of the trimethylsilylethers. Lipids 3:319-402.

Gaskill, S.J. & G. Eglinton, 1975. Rapid Hydrogenation of sterols in a contemporary lacustrine sediment. Nature 254:209-211.

Gellerman, J.L., W.H. Anderson & H. Schlenk, 1975. Synthesis and analysis of phytol and phytenoyl waxesters. Lipids 10:656-661.

Gustafsson, B.E., J.-Å. Gustafsson & J. Sjövall, 1966. Intestinal and fecal sterols in Germfree and Conventional rats. Acta Chem. Scand. 20:1827-1835.

Henderson, W., W.E. Reed & G. Steel, 1972. The origin and incorporation of organic molecules in sediments as elucidated by studies of the sedimentary sequence from a residual pleistocene lake. In: H.R. v. Gaertner & H. Wehner (Ed.): Advances in Organic Geochemistry Pergamon Press, Oxford. p. 335-352.

de Leeuw, J.W., B.R. Simoneit, J.J. Boon, W.I.C. Rijpstra, F. de Lange, J.C.W. v.d. Leeden, V.A. Correia, A.L. Burlingame & P.A. Schenck. Phytol derived compounds in the geosphere. Paper presented at the 7th International Congres on Organic Geochemistry. Madrid, 1975.

McCarty, R.D. & A.H. Duthie, 1962. A rapid quantitative method for the separation of free fatty acids from other lipids. J. Lipid Res. 3:117-120.

McCloskey, J.A., 1970. Mass spectrometry of fatty acid derivatives. In: F.D. Gunstone (Ed.): Topics in Lipid chemistry vol 1. Logos Press, London. p. 369-441.

Poltz, J., 1972. Untersuchungen über das Vorkommen und den Abbau von Fetten und Fettsäuren in Seen. Arch. Hydrobiol. Acta. Suppl 40:315-399.

Rijpstra, W.I.C., J.W. de Leeuw & P.A. Schenck 1976. The action of borontrifluoride - methanol on isoprenoid alcohols. Geochim. Cosmochim. Acta. 40:1289-1290.

Skipski, V.P., A.F. Smolowne, R.C. Sullivan & M. Barclay, 1965. Separation of lipid classes by thin layer chromatography. Biochim. Biophys. Acta. 106:386-396.

Vogel, A.I., 1961. A textbook of practical organic chemistry. Third edition. Longmans. London. p. 887 en 888.

Key words: Sterolester
Waxesters
4-Methylstanols
Mass spectrometry
Fontinalis antipyretica

# PRODUCTION, SEDIMENTATION, AND ACCUMULATION OF PARTICULATE CARBON AND NITROGEN IN A SHELTERED SUBALPINE LAKE

Bruce L. Kimmel and Charles R. Goldman

Division of Environmental Studies, University of California, Davis 95616 USA

## Summary

Particulate carbon and nitrogen production, sedimentation, and accumulation rates for the pelagic region of Castle Lake, California were compared for the purpose of assessing the relative importance of internal nutrient cycling and sedimentation losses in an unproductive, highly-sheltered lake. Mineralization efficiency was found to be high in both the epilimnion and upper hypolimnion. Of the areal particulate carbon and nitrogen production, only 27% of the carbon and 19% of the nitrogen reached the sediment surface and 14% and 6%, respectively, accumulated in permanent deposits. For Castle Lake, which has both low productivity and a shallow epilimnion, trophogenic zone depth provides a better index of the fate of the organic matter supply than does mixed-layer depth.

## Introduction

Of the processes involved in nutrient cycling and energy flow in stratified lakes, perhaps the most fundamental is the gravitational sinking of "suspended" particulate matter through the water column. The sedimentation of organic particles from the trophogenic zone provides an essential link between planktonic and benthic communities (Jónasson 1964; Hargrave 1973, 1975) and constitutes a net vertical displacement of energy and nutrients, which is of major importance to lake metabolism (Saunders 1972, Wetzel et al. 1972). Thomas (1950, 1955), Kleerekoper (1953), and Ohle (1956, 1962) recognized that, as a result of mineralization processes in the water column, only a fraction of the particulate production reaches the sediment-water interface. By comparing the quantity and quality of sedimented material collected at various depths in the water column and at the sediment surface, numerous investigators have concluded that most mineralization occurs within the epilimnion with little additional decomposition in the hypolimnion (e.g., Kleerekoper 1953, Ohle 1962, Lawacz 1969, Kajak et al. 1972, Wetzel et al. 1972, Bloesch 1974). Direct measurements of sediment oxygen uptake show that sedimented materials undergo considerable further degradation at the sediment-water interface (Hayes and MacAulay 1959, Hargrave 1971, and Neame 1975). Generally, only the most refractory materials accumulate as permanent sediment deposits (Pamatmat 1971).

The proportion of the organic matter supply which sediments to the lake bottom has been related to both lake morphometry and the mixing characteristics of the water column. Ohle (1956) showed that in six north German lakes the fraction of the production mineralized within the water column was directly related to mean depth. Larkin (1964) suggested that the organic matter supply to benthic communities in some Canadian lakes was dependent on the extent of mineralization in the euphotic zone and was inversely related to the depth of mixing. Hargrave (1973, 1975) correlated the fate of organic carbon with productivity and mixed-layer depth, and concluded that in oligotrophic waters primary production is less efficiently utilized and a greater proportion of it sediments, than in more productive systems.

This study was conducted as part of an ongoing investigation of internal nutrient cycling and phytoplankton productivity in Castle Lake. Our objective was to determine, in regard to carbon and nitrogen, the proportions of the particulate production mineralized within the water column, deposited at the sediment-water interface, and accumulated as bottom sediment. These data provide a basis for comparing the relative importance of organic matter mineralization and sedimentation in a sheltered, unproductive lake with those systems previously investigated (Hargrave 1975).

## The Study Area

Castle Lake (Fig. 1) is a small, highly sheltered, subalpine lake lying in a granitic cirque basin at an elevation of 1706 m in the Klamath Mountains of northwestern California. The watershed area (80.9 ha) is only four times that of the lake surface (20.1 ha). The lake is dimictic with well-developed thermal stratification during the short growing season, usually early June to mid-October. Due to the sheltered nature of the lake, the epilimnion is very shallow (ca. 5-6 m) relative to the maximum depth (35 m). The lake surface freezes in late

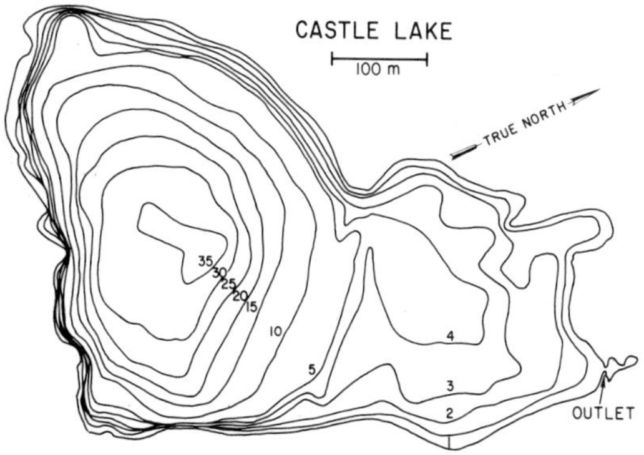


Fig. 1. Morphometric map of Castle Lake, California. Depth contours are in meters.

November or early December, and ice and snow cover usually persists until late May or early June.

Castle Lake is oligo-mesotrophic with a mean phytoplankton productivity of approximately 70 mg C $m^{-2}hr^{-1}$ (3.5 mg C $m^{-3}hr^{-1}$) in the euphotic zone during the growing season. Because the water is highly transparent (Secchi depth = 12-14 m in summer), much of the hypolimnion lies within the euphotic zone. The summer vertical distribution of phytoplankton productivity is bimodal with peaks at about 5 and 20 m. High dissolved oxygen levels are present throughout the year except in the deep hypolimnion where oxygen depletion below 25 m is common during prolonged periods of stratification. Vertical profiles typical of midsummer conditions are shown in Fig. 2.

Allochthonous inputs to Castle Lake are minimal. Inflow occurs primarily in the spring from snowmelt and results in a water turnover rate of less than 0.001 $day^{-1}$ by late June (Goldman 1961). Castle Creek, which cuts through the recessional moraine underlying the shallow end of the lake, is the only surface outlet. Summer precipitation is almost completely absent. The University of California Limnological Laboratory is the only permanent facility on the lake.

## Methods

All samples were collected in the Castle Lake deep basin in water of greater than 30 m depth. Phytoplankton productivity was measured in situ with the $^{14}C$ uptake method (Steemann Nielsen 1952, as modified by Goldman 1963) at 5-day intervals during the growing season and approximately monthly thereafter. Water samples for "suspended" seston analysis, collected at about 20-day intervals, were filtered through 100-nm nylon mesh to remove zooplankton. Particulate matter was then collected by low vacuum pressure (<100 mm Hg) filtration through precombusted (500 °C) 25-mm glass-fiber filters (Whatman GF/C). Samples were stored at -20 °C until analysis.

Sedimenting particulate matter was collected at 10, 20, and 30 m with messenger-closed sediment traps. Sediment trap design, operation, and suspension are described by Kimmel et al. (1977). Sedimented material was homogenized by continuous stirring, subsampled, collected by filtration, and stored as described above. Dry weight (105 °C) was determined by weighing samples collected on preweighed filters to the nearest microgram with a Cahn Model G electrobalance. Collection periods varied, but usually were of about 20-days duration throughout the growing season. Unusual ice conditions prevented retrieval of sediment traps in the early winter of 1975, so the final collection interval was three months (18 October to 14 January). No attempt was made to prevent degradation of sedimented material during collection periods.

Duplicate sediment cores were obtained in July 1976 with a 1-m gravity corer. A stratified box sampler (Milbrink 1968) was used at the same location to collect samples at 1-cm intervals near the sediment-water interface. Cores were immediately extruded, sectioned at 2- or 3-cm intervals, and stored at -20 °C until analysis. Sediment dry weight was determined with a Mettler H31 analytical balance (0.1 mg accuracy) after oven-drying samples overnight at 105 °C.

Carbon and nitrogen analysis was by combustion gas chromatography with a Hewlett-Packard 185 B elemental analyzer. All

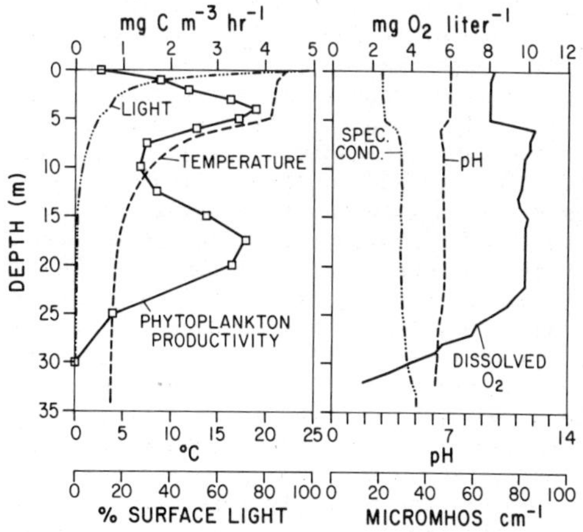


Fig. 2. Vertical profiles of light penetration, temperature, specific conductivity, pH, dissolved oxygen and phytoplankton productivity in the deep basin of Castle Lake, 14 August 1975. All measurements except phytoplankton productivity were made at 1-meter intervals.

filtered samples were processed as described by Sharp (1974). Approximately 1-mg subsamples of sediment were weighed to the nearest 0.1 μg on a Cahn electrobalance and then combusted by the usual procedures. Since acid treatment has shown that carbonates form a negligible fraction (<0.1% by weight) of the total particulate carbon in Castle Lake samples, total carbon was interpreted as organic carbon. All samples were subsampled to permit replicate determinations of dry weight, particulate carbon, and particulate nitrogen; and results presented here are averages of at least triplicate determinations. For "suspended" seston, sedimented seston, and bottom sediment samples, coefficients of variations for replicate subsamples averaged less than 10% and were usually about 5%. The coefficient of variation among means for triplicate sediment trap samples was 13% for dry weight, 16% for particulate carbon, and 19% for particulate nitrogen.

Sedimentation within discrete water column strata was estimated by subtracting the amount of material collected at a shallower depth from that collected below; e.g., sedimentation within the 10-20 m stratum = (sedimentation to 20 m) - (sedimentation to 10 m). Similarly, production within a stratum was estimated as, for example, the difference between areal productivity to 20 m and that to 10 m. The extent of mineralization is reflected by the difference between production and sedimentation. Particulate nitrogen production was taken as one-sixth of that estimated for carbon (Vallentyne 1974, Redfield et al. 1963, Strickland 1960). Analysis of periphyton indicated an average C:N ratio (by weight) of 6.3±0.32 ($\bar{x}$±SD) for Castle Lake algae during the growing season, so this approximation seems reasonable. Since in situ productivity measurements have consistently shown that little production occurs during the winter period of ice and snow cover, production and sedimentation measured from late May to mid-January were assumed to approximate that occurring annually. It was further assumed that phytoplankton production, as measured by the $^{14}C$ method, and the subsequent secondary production derived from it constitute the organic matter supply in the pelagic water column and that allochthonous and littoral contributions are negligible. Although these assumptions may be challenged on numerous points, their requirements are probably more nearly met in Castle Lake than in most previously studied systems.

## Results and discussion

The amount of material collected in sediment traps consistently increased with depth except during spring and fall circulation periods when sedimentation within 10-m strata was approximately equal. Corresponding

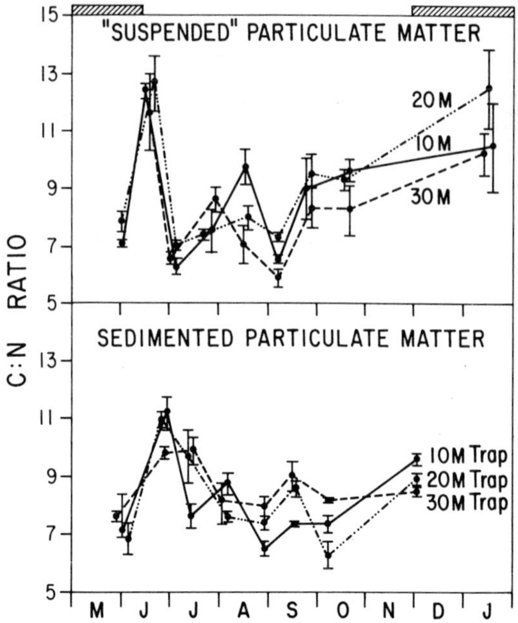


Fig. 3. Carbon:nitrogen ratios (by weight) of "suspended" and sedimented particulate matter at 10, 20, and 30 m in Castle Lake, 1975. Vertical bars represent means ±1 SD. Shaded horizontal bars at top indicate periods of snow and ice cover.

increases in C:N ratios suggest that resuspension of recently sedimented material may have occurred at these times (Fig. 3). Pollen from Abies-Pinus stands in the surrounding basin probably also contributed to the elevated C:N ratios observed in late June. C:N ratios of "suspended" and sedimented particulate matter corresponded closely throughout the growing season in regard to both their magnitude and fluctuations with time. This indicates that the composition of material collected in the sediment traps accurately reflects that of the particulate matter settling through the water column. Although the C:N ratios of "suspended" and sedimented particulate matter fluctuated seasonally, consistent changes with depth were not evident.

Production and sedimentation rates for particulate carbon and nitrogen are presented in Fig. 4 for successive 10-m strata and for the entire water column. Although 0-10 m productivity increased throughout most of the growing season, the amount of particulate matter sedimenting to 10 m was greatest in early summer and decreased thereafter. This pattern reflects increased mineralization efficiency with the higher epilimnetic water temperatures of mid- and late summer. Sedimentation in the 10-20 m stratum followed a similar pattern with production declining rapidly after mid-August. Production was low in the deep hypolimnion (20-30 m), but the proportion of the particulate production sedimented was much greater than in overlying strata.

Production and sedimentation data are summarized in Table 1. Because the 0-10 m stratum included the mixed layer, the metalimnion and part of the upper hypolimnion, its production value was adjusted to account for the morphometric concentration of particulate matter settling through the metalimnion (Hutchinson 1957, Ohle 1962, Wetzel et al. 1972). Since the thermocline occurs at about 5 m during the growing season and the lake surface area is approximately twice that of the 5 m contour, we doubled the measured areal production for the 0-10 m stratum. Both corrected and uncorrected values are given in Table 1.

Comparison of production and sedimentation in water column strata indicates that mineralization occurs with almost equal efficiency in the mixed layer and the oxygenated portion of the hypolimnion. Only in the deep hypolimnion, where oxygen is depleted during stratification periods, is the extent of mineralization reduced substantially (Table 1). These results initially appear to contrast with those of numerous investigators who have concluded that the hypolimnion is relatively unimportant in regard to the mineralization of sedimenting particulate matter. However, most previous conclusions have been derived from studies of more productive systems in which temperature and oxygen conditions favorable for rapid oxidative decomposition existed only in the epilimnion. In Castle Lake, a significant portion of the trophogenic zone lies in the hypolimnion, and oxygen sufficient for high mineralization rates is present to at least 25 m (Fig. 2). The lack of a consistent increase in C:N ratios of "suspended" and sedimented seston with depth beneath the mixed layer (Fig. 3) reflects the extensive vertical overlap of assimilation, production, and decomposition processes in the water

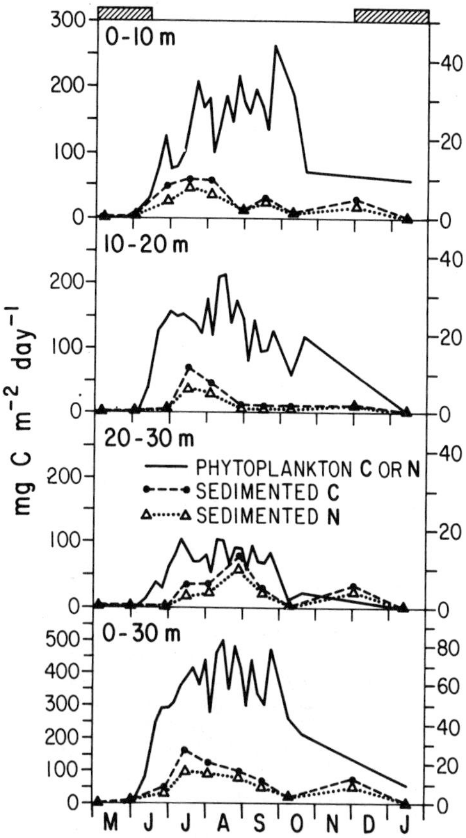


Fig. 4. Production and sedimentation of particulate carbon and nitrogen within discrete water column strata (0-10, 10-20, 20-30 m) and the entire water column (0-30 m) in Castle Lake, 1975. Sedimentation data are plotted at the mid-points of collection intervals. Shaded horizontal bars at the top indicate periods of snow and ice cover. Note the change of scale for the 0-30 m water column.

Table 1. Production and sedimentation rates (g m$^{-2}$yr$^{-1}$) for particulate carbon and nitrogen in the Castle Lake water column, 1975. Values in parentheses represent approximate corrections for the morphometric concentration of particulate matter settling through the metalimnion.

| Water Column Stratum | Part. C Prod. | Part. C Sed. | % Sed. | Part. N Prod. | Part. N Sed. | % Sed. |
|---|---|---|---|---|---|---|
| 0-10 m | 25.59(~50) | 6.19 | 24.2(~12) | 4.26(~8) | 0.70 | 16.4(~8) |
| 10-20 m | 21.92 | 3.60 | 16.4 | 3.66 | 0.42 | 11.5 |
| 20-30 m | 9.33 | 5.42 | 58.1 | 1.55 | 0.64 | 41.3 |
| 0-30 m | 56.86 | 15.11 | 26.6 | 9.48 | 1.76 | 18.6 |

Table 2. Average carbon and nitrogen composition (as percent of dry weight) of particulate matter collected in sediment traps throughout the growing season and of recent bottom sediments. Results are expressed as means ±1 SD.

| | % Carbon | % Nitrogen |
|---|---|---|
| Sediment Traps | | |
| 10 m | 24.7 ± 6.95 | 3.08 ± 1.21 |
| 20 m | 25.5 ± 5.14 | 3.14 ± 0.81 |
| 30 m | 24.8 ±11.00 | 2.94 ± 1.38 |
| Bottom Sediment | | |
| upper 3 cm | 12.7 ± 0.97 | 0.95 ± 0.02 |
| upper 40 cm | 12.1 ± 1.45 | 0.91 ± 0.11 |

Table 3. Comparison of annual production, sedimentation, and accumulation rates for particulate carbon and nitrogen in the Castle Lake water column, 1975.

| | $g\ m^{-2}yr^{-1}$ | % of Prod. | % of Sed. |
|---|---|---|---|
| Carbon | | | |
| Production | 56.86 | 100 | - |
| Sedimentation | 15.11 | 26.6 | 100 |
| Accumulation | 7.75 | 13.6 | 51.3 |
| Nitrogen | | | |
| Production | 9.48 | 100 | - |
| Sedimentation | 1.76 | 18.6 | 100 |
| Accumulation | 0.57 | 6.0 | 32.3 |

column, rather than low mineralization activity.

Recent Castle Lake sediments were found to be remarkably uniform in respect to carbon and nitrogen content. Composition of the upper 3 cm of sediment collected with the stratified box sampler was approximately the same as the most recent 40 cm of sediment obtained by coring (Table 2). Carbon and nitrogen release by long-term post-depositional processes appears to be insignificant relative to that occurring prior to permanent sediment accumulation. Decomposition at the sediment surface was estimated by comparing the average C and N composition of the uppermost 3 cm of sediment with that of sedimented material collected at 30 m (Table 2). It appears that an additional 49% of the carbon and 68% of the nitrogen which reached the sediment-water interface is mineralized between the time of initial deposition and permanent accumulation. However, due to the relatively small proportion of material deposited at the sediment surface, this represents only about 13% of both the annual C and N production. Estimated annual rates of particulate C and N accumulation in permanent deposits, based on amounts sedimented to 30 m, are compared to those for production and sedimentation in Table 3.

Results of previous studies had indicated that the magnitude of nutrient loss by sedimentation in Castle Lake would far exceed that actually observed (Table 3). Phytoplankton loss rates estimated by Jassby and Goldman (1974) suggested that a major fraction of the annual phytoplankton production in Castle Lake directly enters the "detritus pathway" (Odum 1962, 1963; Wetzel et al. 1972) by physiological cell death unrelated to grazing or sinking. Since aquatic bacteria in low nutrient environments may assimilate more dissolved nutrients from the surrounding water than they release by mineralizing organic substrates (Johannes 1968), it appeared likely that bacterial attachment to sinking particulate detritus (Paerl and Goldman 1972, Paerl 1973) could enhance vertical nutrient displacement by seston sedimentation in Castle Lake. Additionally, because Castle Lake is highly-sheltered and a substantial fraction of its phytoplankton production occurs in the relatively non-turbulent water below the metalimnion, significant nutrient loss by sedimentation of living phytoplankton and particulate detritus was expected. Gessner (1948) observed a rapid, progressive loss of phytoplankton cells by passive sinking in Königsee, a similarly-sheltered Bavarian lake having a shallow mixed layer and a deep trophogenic zone. Both the productivity and mixed-layer depth are low in Castle Lake and, according to Hargrave's (1973, 1975) observations, most of the primary production should sediment rather than decompose within the water column.

However, our results indicate that internal cycling of carbon and nitrogen in the Castle Lake water column in quite efficient. Mineralization of sedimenting organic matter is efficient not only in the epilimnion, but also in the oxygenated portion of the hypolimnion (Table 1). Consequently, most of the particulate carbon and nitrogen production is mineralized within the water column rather than at the sediment surface (Table 3) despite low phytoplankton productivity, a shallow mixed layer, and an extensive hypolimnion. The sediments do not appear to be an especially large sink for carbon and nitrogen, and their relatively high organic content (Table 2) is probably best explained by (i.) the proximity of the hypolimnetic production peak to the sediment surface and (ii.) low levels of turbulence, temperature, and dissolved oxygen in the deep hypolimnion. Refractory humic substances may also contribute to the organic content of bottom sediments in Castle Lake (Hargrave 1975). Our nitrogen results

(Table 3) are very similar to those of Toyoda et al. (1968) who found that nitrogen loss by sedimentation in Lake Biwa was about 5% of the particulate nitrogen supply. As expected for nitrogen-poor systems (Strickland 1970), nitrogen appears to be cycled with particular efficiency and very little is lost to the sediments.

Using data from several freshwater and marine systems, Hargrave (1975) observed a negative correlation between mixed-layer depth and the proportion of the carbon supply sedimented. However, his analysis did not include data from an unproductive system having both a very shallow mixed layer and a high maximum depth : mixed-layer depth ratio. Our study provides data for that case and indicates that considerably less of the primary production reaches the sediment surface (ca. 27%) than would be expected (ca. 59%) on the basis of the mixed-layer depth relationship. Since methods of production and sedimentation measurement were essentially the same in our study as those cited by Hargrave, it is unlikely that technical differences can adequately account for this discrepancy. The high mineralization efficiency in the oxygenated portion of the hypolimnion provides a better explanation for the relatively small fraction of the primary production reaching the sediments in Castle Lake. More commonly, the trophogenic zone is confined to the epilimnion and, therefore, the thermocline depth adequately reflects the extent of mineralization. However, for Castle Lake and similar systems in which the euphotic layer extends well into the hypolimnion, the depth of the trophogenic zone probably provides a more accurate index of the proportion of the carbon supply sedimented than does mixed-layer depth.

Our results also reemphasize the sensitivity of sheltered unproductive mountain lakes to disturbance by man. Such systems, which are often particularly attractive to recreationists and cottage-builders, are usually unproductive by virtue of both their morphometry and low rates of external nutrient input (Ohle 1956, Wetzel and Allen 1970, and many others). In lakes having a significant fraction of the trophogenic zone within the hypolimnion, even a slight increase in nutrient loading is likely to stimulate epilimnetic algal production and result in reductions of light penetration, trophogenic zone depth, and volume of the oxic portion of the hypolimnion. Increased hypolimnetic anoxia would not only reduce mineralization efficiency and accelerate sediment accumulation, but also would increase internal nutrient loading by expanding the sediment area exposed to anoxic conditions. Such events may be expected to produce rapid shifts toward eutrophy, particularly in lakes of relatively small volume.

As recently emphasized by Wetzel (1975), Hargrave (1975) and Hargrave et al. (1976), investigations which attempt to compare production and sedimentation rates require broad assumptions, and this study is no exception. Uncertainties are involved in primary production measurements, the qualitative and quantitative relationships of sediment trap collections to the material actually settling through the water column, the magnitude of allochthonous and chemosynthetic contributions to the particulate pool and the sediments, and the extent of sediment resuspension and redeposition. Therefore, such studies can provide only indirect approximations rather than highly quantitative determinations. However, with the limitations in mind, these comparisons can provide valuable insight to the processes of nutrient cycling, nutrient loading and lake metabolism as long recognized by Ohle (1956, 1962) and recently reemphasized by Vollenweider (1976).

Acknowledgements

Castle Lake limnological studies are supported by National Science Foundation grant DEB 76-19524 to C. R. Goldman. University of California Jastro research fellowship funds, intercampus research funds, and research travel funds all contributed to this project. R. P. Axler, J. A. Coil, R. M. Gersberg, P. A. Neame, L. J. Paulson, and G. W. Redfield provided valuable assistance at various stages of this work. O. Holm-Hansen, R. W. Eppley, and E. Renger permitted and assisted our use of the Marine Food Chain Research Group CHN analyzer, Scripps Institution of Oceanography. Recent conversations with B. T. Hargrave, J. Bloesch, R. M. Davis, H. W. Paerl, and P. A. Neame were particularly stimulating and helpful. L. Kimmel, B. Jost, E. de Amezaga, and D. Payne aided in manuscript preparation.

References

Bloesch, J., 1974. Sedimentation und Phosphorhaushalt im Vierwaldstätter (Horwer Bucht) und im Rotsee. Schweiz. Z. Hydrol. 36:71-186.

Gessner, F. 1948. The vertical distribution of phytoplankton and the thermocline. Ecology 29:386-389.

Goldman, C.R., 1961. The contribution of alder trees (Alnus tenuifolia) to the primary productivity of Castle Lake, California. Ecology 42:282-288.

Goldman, C.R., 1963. The measurement of primary productivity and limiting factors in freshwater with carbon-14. In: M.S. Doty (Ed.): Proceedings of the Conference on primary productivity measurement, marine and freshwater, Univ. Hawaii, Aug. 21-Sept. 6, 1961. U.S. Atomic Energy Comm., Div. Tech. Inform. Rep. TID - 7633. p. 103-113.

Hargrave, B.T., 1971. A comparison of sediment oxygen uptake, hypolimnetic oxygen deficit and primary production in Lake Esrom, Denmark. Verh. Int. Verein. Limnol. 18:134-139.
Hargrave, B.T., 1973. Coupling carbon flow through some pelagic and benthic communities. J. Fish. Res. Bd. Canada 30:1317-1326.
Hargrave, B.T., 1975. The importance of total and mixed-layer depth in the supply of organic material to bottom communities. Symp. Biol. Hung. 15:157-165.
Hargrave, B.T., G.A. Phillips and S. Taguchi. 1976. Sedimentation measurements in Bedford Basin, 1973-1974. Environment Canada, Fish. and Mar. Service Tech. Rep. 608. 129p.
Hayes, F.R. and M.A. MacAulay, 1959. Lake water and sediment, 5. Oxygen consumed in water over sediment cores. Limnol. Oceanogr. 3:291-298.
Hutchinson, G.E., 1957. A treatise on limnology, vol. 1. Wiley, New York.
Jassby, A.D., and C.R. Goldman, 1974. Loss rates from a lake phytoplankton community. Limnol. Oceanogr. 19:618-627.
Johannes, R.E., 1968. Nutrient regeneration in lakes and oceans. Adv. Microbiol. Sea 1:203-213.
Jónasson, P.M., 1964. The relationships between primary production and production of profundal bottom invertebrates in a Danish eutrophic lake. Verh. Int. Verein. Limnol. 15:471-479..
Kajak, Z., A. Hillbricht-Ilkowska and E. Pieczyńska, 1972. The production processes in several Polish lakes. In: Z. Kajak and A. Hillbricht-Ilkowska (Ed.): Productivity problems of fresh waters. IBP-UNESCO Symposium, Poland, 1970. p. 129-147.
Kimmel, B.L., R.P. Axler and C.R. Goldman, 1977. A closing, replicate-sample sediment trap. Limnol. Oceanogr., In Press.
Kleerekoper, H., 1953. The mineralization of plankton. J. Fish. Res. Bd. Canada 10: 283-291.
Larkin, P.A., 1964. Canadian lakes. Verh. Int. Verein. Limnol. 15:76-90.
Lawacz, W., 1969. The characteristics of sinking materials and the formation of bottom deposits in a eutrophic lake. Mitt. Int. Verein. Limnol. 17:319-331.
Milbrink, G., 1968. A microstratification sampler for mud and water. Oikos 19: 105-110.
Neame, P.A., 1975. Oxygen uptake of sediments in Castle Lake, California. Verh. Int. Verein. Limnol. 19:792-799.
Odum, E.P., 1962. Relationships between structure and function in the ecosystem. Japan. J. Ecol. 12:108-118.
Odum, E.P., 1963. Primary and secondary energy flow in relation to ecosystem structure. Proc. 16th Int. Cong. Zool. 4:336-338.
Ohle, W., 1956. Bioactivity, production, and energy utilization of lakes. Limnol. Oceanogr. 1:139-149.
Ohle, W., 1962. Der Stoffhaushalt der Seen als Grundlage einer allgemeinen Stoffwechseldynamik der Gewässer. Kiel. Meeresforsch. 18:107-120.
Paerl, H.W., 1973. Detritus in Lake Tahoe: structural modification by attached microflora. Science 180:496-498.
Paerl, H.W. and C.R. Goldman, 1972. Stimulation of heterotrophic and autotrophic activities of a planktonic microbial community by siltation at Lake Tahoe, California. Mem. Ist. Ital. Idrobiol. 29 Suppl.:129-147.
Pamatmat, M.M., 1971. Oxygen consumption by the seabed, 4. Shipboard and laboratory experiments. Limnol. Oceanogr. 16:536-550.
Redfield, A.C., B.H. Ketchum, and F.A. Richards, 1963. The influence of organisms on the composition of seawater. In: M.N. Hill (Ed.): The sea, vol. 2. Interscience. p. 26-77.
Saunders, G.W., 1972. Summary of the general conclusions of the symposium. Mem. Ist. Ital. Idrobiol. 29 Suppl.:533-540.
Sharp, J.H., 1974. Improved analysis for "particulate" organic carbon and nitrogen from seawater. Limnol. Oceanogr. 14:799-801.
Steeman Nielsen, E., 1952. The use of radioactive carbon ($^{14}C$) for measuring organic production in the sea. J. du Conseil. 18: 117-140.
Strickland, J.D.H., 1960. Measuring the production of marine phytoplankton. Fish. Res. Bd. Canada Bull. 122. 172 p.
Strickland, J.D.H., 1970. Introduction, recycling of organic matter. In: J.H. Steele (Ed.): Marine food chains. Univ. Calif. Press, Berkeley. p. 3-5.
Thomas, E.A., 1950. Beitrag zur Methodik der Produktionsforschung in Seen. Schweiz. Z. Hydrol. 12:25-37.
Thomas, E.A., 1955. Stoffhaushalt und Sedimentation im oligotrophen Aegerisee und im eutrophen Pfäffiker-und Greifensee. Mem. Ist. Ital. Idrobiol. 8 Suppl.:357-465.
Toyoda, Y., S. Horie, and Y. Saijo, 1968. Studies on the sedimentation of Lake Biwa from the viewpoint of lake metabolism. Mitt. Int. Verein. Limnol. 14:243-255.
Vallentyne, J.R., 1974. The algal bowl-lakes and man. Misc. Spec. Publ. 22, Dept. of Environ., Ottawa. 185 p.
Vollenweider, R.A., 1976. Rotsee, a source, not a sink for phosphorus? A comment to and a plea for nutrient balance studies. Schweiz. Z. Hydrol. 38:29-34.
Wetzel, R.G., 1975. Limnology. Saunders, Philadelphia.
Wetzel, R.G., and H.L. Allen, 1970. Functions and interactions of dissolved organic matter and the littoral zone in lake metabolism and eutrophication. In: Z. Kajak and A. Hillbricht-Ilkowska (Eds.): Productivity problems of freshwaters. IBP-UNESCO Symposium Poland, 1970. p. 129-147.

Wetzel, R.G., P.H. Rich, M.C. Miller, and H.L. Allen, 1972. Metabolism of dissolved and particulate detrital carbon in a temperate hard-water lake. Mem. Ist. Ital. Idrobiol. 29 Suppl.:185-243.

Key words: lake metabolism
nutrient cycling
sedimentation
mineralization
carbon, nitrogen

# ON THE STRUCTURE OF SOLUBLE ORGANIC SUBSTANCES IN SEDIMENTS OF LAKE PLUßSEE

H.-H. Stabel and U. Münster

Max-Planck-Institut für Limnologie, Dept. Allgemeine Limnologie, Plön, Federal Republic of Germany

## Abstract

The molecular weight pattern of dissolved organic substances from interstitial waters of Lake Plußsee sediments equalled that of the water column. According to the pH of the medium different types of organic molecules could be distinguished in sediment pore water samples by Sephadex-G-15-filtration. Under slightly acid conditions a peak of oligomeric compounds (800 - 900 Daltons) was liberated. This was confirmed by extraction of mud with inorganic buffers and by original samples of various pH. It is suggested that considerable amounts of organic carbon are liberated from the sediments into the adjoining water layers due to the strong concentration gradient of soluble organic carbon and the pH variations at the mud water interface.

## Introduction

Studies have been made on the function of soluble organic molecules in river or lake waters, but of those concerning substances in the interstitial there is little known. Mobility and diffusibility enable dissolved organic molecules to participate in metabolic pathways within the mud and to join in the organic carbon metabolism in the hypolimnion by penetrating through the mud water interface. Cations, e. g. $Ca^{++}$, precipitate part of the substances within the water column. Sinking down again they add to the solid phase. In spite of their significance for several limnetic pathways not much data concerning amount and quality of dissolved organic substances in sediment pore waters is available, hitherto.

Only a few authors measured interstitial soluble organic carbon (Kemp, 1969; Ishiwatari, 1971; Kemp & Wong, 1974). The molecular constitution as well as the molecular weight distribution pattern of pore water organic matter from different types of lake sediments still remains unknown. Studying structure and function of soluble organic material in different aquatic habitats we investigated interstitial waters in order to compare the chemical composition of their soluble molecules to that of the corresponding lake waters.

## Material and methods

The basin of eutrophic Lake Plußsee (Plön, Northern Germany) has a maximum depth of 28 m and 135 000 $m^2$ surface area and holds 1.2 x $10^6$ $m^3$ water. Allochthonous material from forests on the surrounding slopes strongly influences the metabolism of the lake. The chemical composition of the sediments therefore is dominated by leaf and wood debris. Although more than 55 % of the dry weight was found to be organic matter, Ungemach (1960) pointed out that Plußsee sediments cannot be classified as typical Dy. Some characteristics like the small degree of humification (only 1 - 4 % of the ignition loss was extractable with 0.1 n NaOH) and the high alkalinity of the lake water (always exceeding 1.55 mVal) can be assigned to Gyttja. Inorganic constituents of the interstitial water were measured by Ohle (1964).

Mud samples were taken with a Züllig corer. Once we took a 10 m core with a Livingstone piston corer. Mud samples were pressed through membrane filters or centrifuged (10 000 r.p.m.). In the supernatant pH was measured immediately and compared to the pH of water in contact with the mud and lake water samples.

Dissolved organic carbon was determined by measuring the chemical oxygen demand (COD) with a Technicon AutoAnalyzer as described by Wagner (1966).

Aliquots of the mud were extracted with definite volumes of Soerensen buffers (pH 5.25 to pH 8.0). In the range from pH 9.25 to pH 10.75 boric acid/NaOH buffers were used. Buffers in the range from pH 8.0 to pH 9.25 contain halogenids and interfere with the carbon determination method. We separated the pool of organic substances by Sephadex-G-15 gel filtration (Stabel, 1976). The columns were equilibrated and eluted with 0.05 molar buffers corresponding to the pH measured in situ or used for extraction.

## Results

A typical stratification of the pH in Lake Plußsee is shown in Fig. 1. The highest deviations between sediment pore water and supernatant lake water were found at the mud water interface at 14 m depth, where the steepest slope of the basin is located (Fig. 1, curve a and b). All through the investigated period similar distribution patterns were typical.

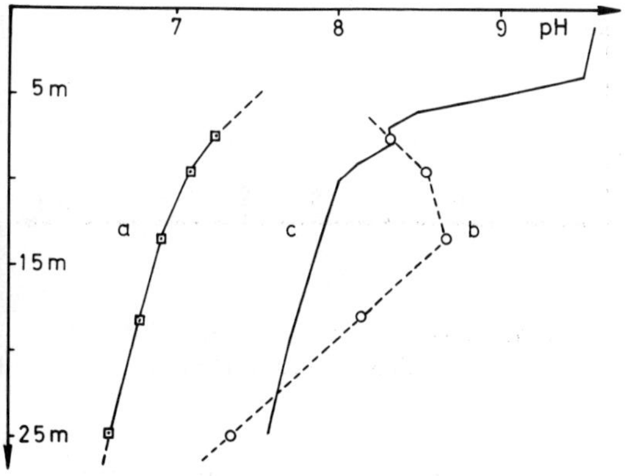


Fig. 1. pH-values in Lake Plußsee basin: a) interstitial; b) adjoining layers; and c) lake water.

Like many inorganic constituents soluble organic carbon is enriched in the interstitial. Fig. 2 shows the concentrations of soluble organic carbon at five sampling stations of different depths. Concentrations within the water column of the profundal vary from 6 to 9 mg $C_{org}$/l while those in the interstitial (squares) and in the contact water layers (open circles) are much higher (often exceeding 30 mg $C_{org}$/l).

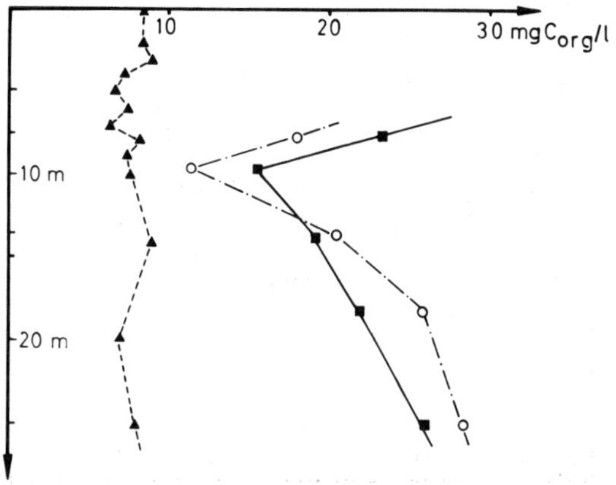


Fig. 2. Profiles of the concentration of dissolved organic carbon: Triangles are due to the lake water, while open circles denote the pore water and squares refer to the adjoining water layers.

This distribution of organic substances was found during the whole period of investigation. From the curves in Fig. 2 it can be concluded, that organic molecules from the pore water strongly influence the chemical composition in adjoining water layers. It seems that organic molecules are liberated into the lake water and that pH plays a part in this process.

Fig. 3 shows the pH oscillations of both interstitial and contact water during the period of investigation. The curves do not correlate, the maximum of deviation occurring in early August, when anaerobic decomposition of a precipitated algal bloom acidifies the mud water contact zone.

The pH in the topmost sediment layers is alterated by internal biological processes as well as by external effects.

These variations seem to influence the liberation of dissolved organic compounds. In order to study this effect we extracted aliquots of mud (1 - 2 g freshweight) with 25 ml of inorganic buffers (24 h, 4° C, $N_2$-atmosphere). The organic carbon content was measured after centrifugation and filtration.

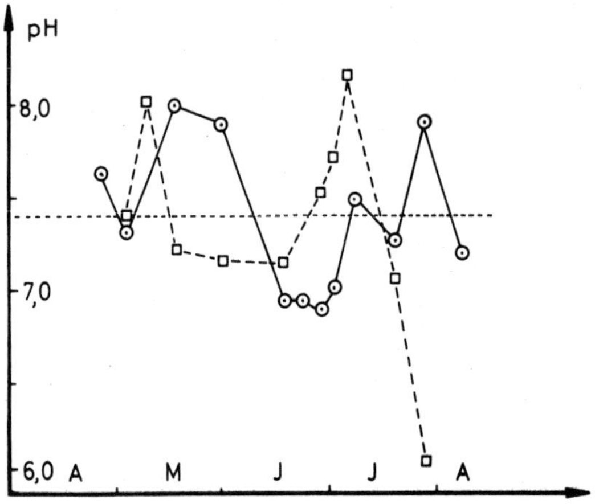


Fig. 3. pH-oscillations at the mud water interface: squares are due to the adjoining water layers while the circles refer to the interstitial.

Obviously the amount of extractable organic matter is closely correlated with increasing pH in the range from pH 7.4 to pH 10.3 (Fig. 4). There is a minimum of $C_{org}$ at pH 7.4, and with decreasing pH higher concentrations are liberated. The significant shape of the curve indicates a discontinuous effect of pH on the liberation of organic compounds of different chemical quality. It suggests that in addition to an increasing amount of organic acids liberated with increasing pH a pool of substances is redissolved under weakly acid conditions.

We separated the set of molecules from each wing of the curve in Fig. 4 with Sephadex-G-15 columns equilibrated and eluted with inorganic buffers corresponding to the pH measured in situ or used for extraction. Untreated interstitial water was thus compared to the solutions gained by extraction.

The elution pattern of the organic molecules obtained under alkaline conditions is recorded in Fig. 5. The set of molecules split into two fractions: the major part was eluted at void volume, indicating high molecular weights, the minor consisted of small molecules.

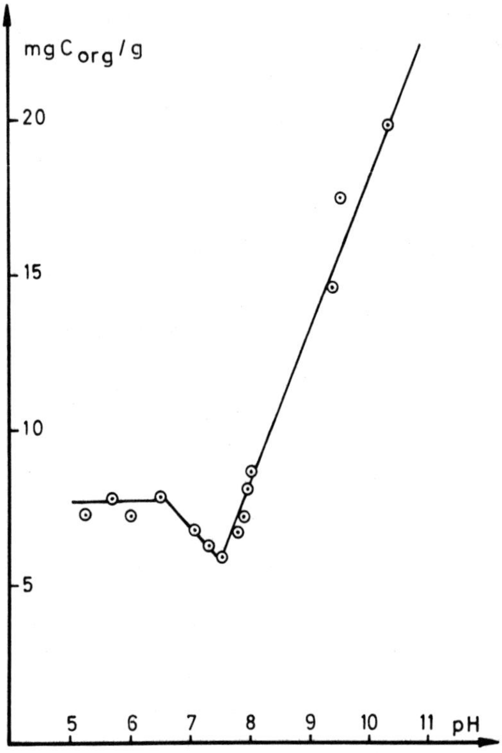


Fig. 4. Extraction curve of soluble organic carbon from mud samples.

Equal results were obtained by separating any natural interstitial water of pH above 7.4 from Lake Plußsee. A similar distribution pattern was observed in the lake water, too, and was ascribed to small, deep lakes strongly influenced by allochthonous organic matter (Stabel, 1976). Disregarding a concentration gradient, this similarity strongly suggests a continuum of lake water and interstitial concerning dissolved organic molecules. This conclusion is confirmed by the elution diagram of pore water from 4 m depth of a mud core.

Occasionally the interstitial becomes slightly acid, leading to considerable changes in the elution curve. The share of macro- and monomolecules is diminished by an additional peak of oligomers, which is obviously liberated by weak acids. The macromolecular fractions proved to be stabile towards acid treatment in experiments within a range from pH 5.25 to pH 7.0. Hence the oligomers do not derive from macromolecular complexes.

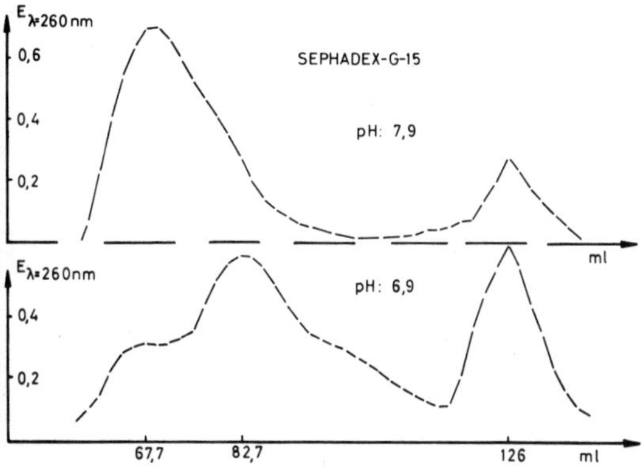


Fig. 5. Sephadex-G-15 elution pattern of two interstitial water samples different in pH.

Since there were no significant differences in the absorption spectra of all fractions molecular weight determinations were done as a means of chemical characterization. The fraction eluted at total volume represented substances from 100 to 200 Daltons, e. g. monomolecules. The oligomeric compounds were rechromatographed on a calibrated Sephadex-G-15 column. Several oligosaccharides and peptides were used for calibration (Stabel, 1976).

The molweight of the interstitial substances liberated under acid conditions appeared to be 800 - 900 Daltons. Since the macromolecules were eluted at void volume on a Sephadex-G-50 column their molweights must exceed 30 000 Daltons.

## Discussion

Apparently dissolved organic molecules penetrate from the sediment into the hypolimnion by diffusion and methane convulsion. Concerning molecular weight distribution both lake water and interstitial organic compounds form a continuum, suggesting that interstitial constituents determine to a high degree the molecular composition of the lake water. Allochthonous organic matter which forms the main part of the sediment in Lake Plußsee is to some extent slowly redissolved by chemical or bacterial degradation and diffuses within the mud layers.

Our results show, that Lake Plußsee interstitial soluble organic matter is made up to three different fractions of molecules, depending on the pH of the medium. Extraction with alkaline buffers liberates increasing amounts of organic carbon. Slightly acid conditions on the other hand liberate an additional set of molecules of low molecular weight. This distribution pattern differs very much from those obtained by Kemp & Wong (1974) from the Great Lakes sediments. They always found more than 70 % of the organic matter to have molecular weights smaller than 700 Daltons. We think, that different compositions of interstitial dissolved organic matter are due to different types of lake sediments and lake metabolism.

## Acknowledgement

We wish to thank Prof. J. Overbeck for his kind interest in this work and for helpful criticism. The Deutsche Forschungsgemeinschaft supported this study by giving a grant to H. H. St.

References

Ishiwatari, R., 1971. Molecular-weight distribution of humic acids from lake and marine sediments. Geochem. J. 5: 121 - 132.

Kemp, A. L. W., 1969. Organic matter in the sediments of Lake Ontario and Erie. Proc. 12th Conf. Great Lakes Res. p. 237 - 249.

Kemp, A. L. W. & H. K. T. Wong, 1974. Molecular-weight distribution of humic substances from Lakes Ontario and Erie sediments. Chem. Geol. 14: 15 - 22.

Ohle, W., 1964. Interstitiallösungen der Sedimente, Nährstoffgehalt des Wassers und Primärproduktion des Phytoplanktons in Seen. Helgol. Wiss. Meeresunters. 10: 411 - 429.

Stabel, H. H., 1976. Gebundene Kohlenhydrate als stabile Komponenten im Schöhsee und in Scenedesmus-Kulturen. Thesis, Univers. Kiel.

Ungemach, H., 1960. Sedimentchemismus und seine Beziehungen zum Stoffhaushalt in 40 europäischen Seen. Thesis, Univers. Kiel.

Wagner, R., 1966. Bestimmung der chemischen Oxidierbarkeit nach der Dichromat-Methode (COD) mit Hilfe des kontinuierlichen Digestors. Technicon Symposia (II): 133 - 143.



Key words: dissolved $C_{org}$
interstitial
molecular weight
pH-dependence

STRATIGRAPHY OF DIATOMS AND CHLOROPHYLL DERIVATIVES IN THE UPPERMOST SEDIMENT OF A SWISS LAKE

E. Staub

Swiss Federal Institute of Technology Zurich, Institute for Aquatic Sciences and Water Pollution Control, 8600 Dübendorf, Switzerland

Summary

Sediment cores of 30 cm length were taken from a deep mesotrophic Swiss lake and analysed in 0.5 - 1.0 cm intervals. The stratigraphy of six dominant diatom groups, of chlorophyll and chlorophyll degradation products, of organic and inorganic carbon and of calcium was analysed. The chlorophylls were investigated by a new chromatographic method: separation by high-pressure liquid chromatography.

The diatom community shows sharp changes with sediment depth. These changes were dated with time markers found in the sediment from which an annually sedimentation rate of 1.8 mm/year was also calculated. The diatom succession gives evidence for eutrophication and is supported by other parameters showing a similar trend.

Introduction

Diatom analyses from lake sediments are used to estimate qualitative and quantitative changes in diatom communities. In addition, chlorophyll analyses in the sediment can give a picture of the primary productivity of a lake in the past. Individual components of chlorophylls and their degradation products can give estimates about taxonomic composition of former phytoplancton populations, relative contributions of allochthonous and autochthonous organic matter, changes of the chemical environment in the sediment and estimations of trophic development of a lake through time (Daley, 1973).

Lake of Lucerne, which is changing from a mesotrophic to an eutrophic state, provides ideal conditions for finding states of transition in the sediment.

Materials and methods

Undisturbed sediment cores with a length of 30 cm were taken with the sediment sampler of Ambühl & Bührer (1975) which was modified by replacing the sampling tube by two half-tubes and in cutting the sediment core into two half-cores by the preparation technique described by Staub (1974). Thus the perceptable warves can be determining for the subsequent fractionation and subsampling from the cut core. All subsamples were dried by lyophilization immediately after collection. They were taken in 0.5 - 1.0 cm intervals.

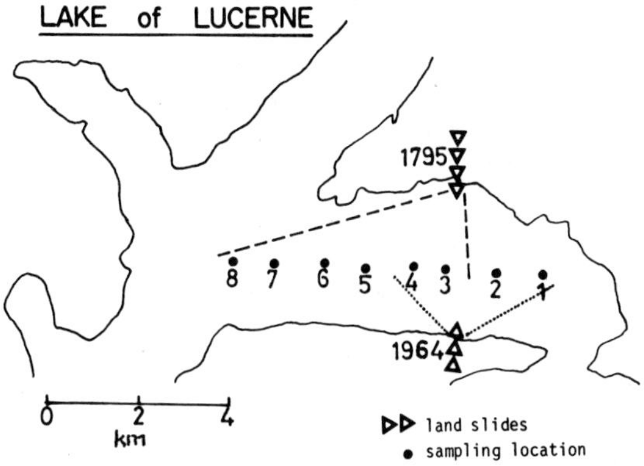


Fig.1. Lake of Lucerne with sampling locations and land slides

Sediment cores were taken from Lake of Lucerne, a mesotrophic lake in Switzerland with a water depth of 100 to 140 m in the sampling area (Fig. 1). Samples for the chlorophyll analyses were taken from core 3, all other analyses were made with samples from core 5.

Chlorophylls and chlorophyll degradation products were studied using high-pressure liquid chromatography (HPLC), a new method where sediment extracts can be readily separated and then detected by visible absorption (Staub & Giger, in press). An example of the HPLC-method is shown in Fig. 2. The retention times of the chlorophylls were comparable with the $R_f$ values of chromatogramms with reversed-phase thin-layer chromatography by Daley (1973). A preliminary quantitation was achieved with the extinction coefficients by Mackinney (1941).

Results and discussion

Sedimentation rate

Two sriking layers in the sediment profile, a yellow and a grey one, could be related to historically dated land slides from 1795 and 1964. An annually sedimentation rate of 1.8 mm was calculated using the depth of these time markers.

The sedimented material of these land sli-

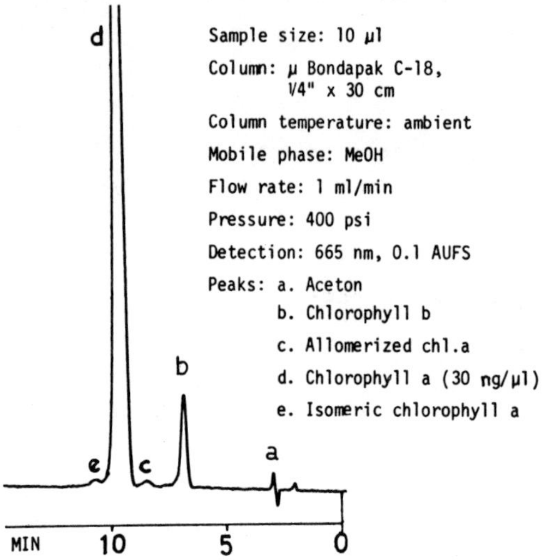


Fig.2. HPLC-separation of a green algae extract

des shows a deviation to the right, interpretable as effect of the Coriolis acceleration. Thus we found the time marker from 1964 in the cores 1 to 4 and the time marker from 1795 in the cores 3 to 8 (Fig. 1).

Diatoms

The diatom frustules did not show a marked corrosion effect with increasing depth in the sediment. So the fine frustule of Asterionella formosa was found as well in 1 as in 40 cm depth.

The dominant species found in the sediment were:

Cyclotella catenata BRUNNTH.
Cyclotella bodanica var. lemanensis O. MEULL.
Cyclotella ocellata PANTOCS.
Cyclotella stelligera CL. et GRUN.
Cyclotella operculata (AG.) KG.
Stephanodiscus Hantzschii GRUN.
Stephanodiscus alpinus HUSTEDT
Stephanodiscus astrea (E.) GRUN.
Tabellaria fenestrata (LYNGB.) KG.
Asterionella formosa HASSALL

The percentage frequency of the individual species is shown in Fig. 3 (Cyclotella spp. includes C. bodanica, C. stelligera and C. operculata; other diatoms includes all other diatoms).

The relative composition of the diatom flora remains the same in the lower part of the sediment (deeper than 5 cm) and shows sudden changes in the 0 to 5 cm region, e.g. the appearance of Tabellaria fenestrata and Stephanodiscus Hantzschii. These algae are typical for eutrophic lakes (Cholnoky, 1968) and

Fig.3. Percentage frequency of individual diatoms in the sediment core

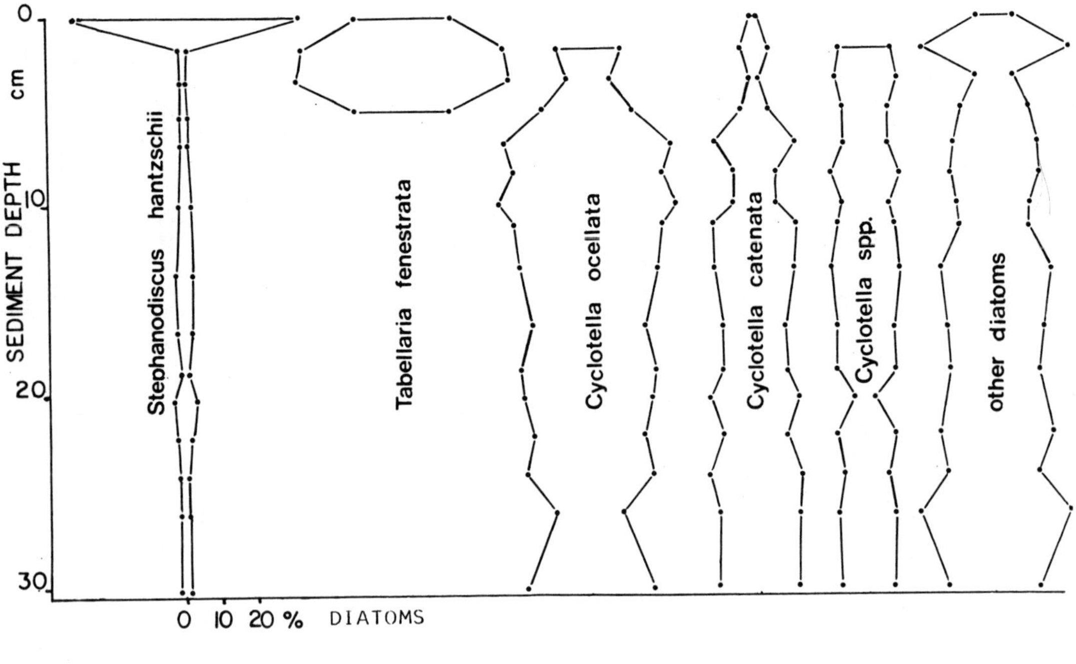

appear with progress of eutrophication, a fact which is also noticed in other Swiss lakes (Nipkow, 1927). The abundant appearance of these two species is dated with the time markers:

Tabellaria fenestrata ca. 1940
Stephanodiscus Hantzschii ca. 1964

Concerning the physiological demands of St. Hantzschii, it is interesting to note that the phosphate concentration was 18 µg/l in the sampling area of the lake in 1964 (Ambühl, 1975).

The unchanged frequency of four dominant diatom groups in the lower part of the sediment reflects constant environmental conditions for the diatoms.

Autochthonous sediment was diluted by a fluctuating amount of allochthonous material. Therefore it was difficult to get information about the increasing total diatom deposition. But the increasing biological activiy of the lake is shown in Fig. 4 and 5. The concentrations of organic carbon and calciumcarbonate increase together with St. Hantzschii and Tab. fenestrata. This is a second indication for progressing eutrophication.

Chlorophylls

A HPLC analysed sediment extract with the different chlorophyll fractions is shown in Fig. 6. The peak A is caused by the extraction solvent (1 g sediment was extracted with 5 ml acetone). Native chlorophylls reaching the sediment surface are degraded within the first 10 cm sediment depth (Fig. 7 A). Pheophytines and chlorophyllides are two of the degradation products which are formed (Fig. 7 B and C). The comparison of the degradation

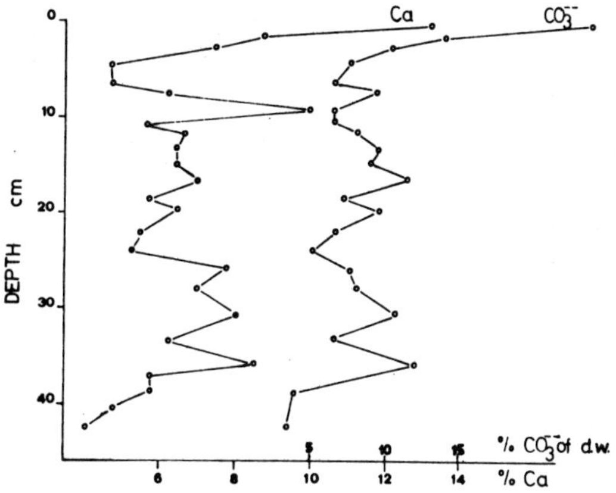


Fig.4. Distribution of calcium and carbonate in the sediment core

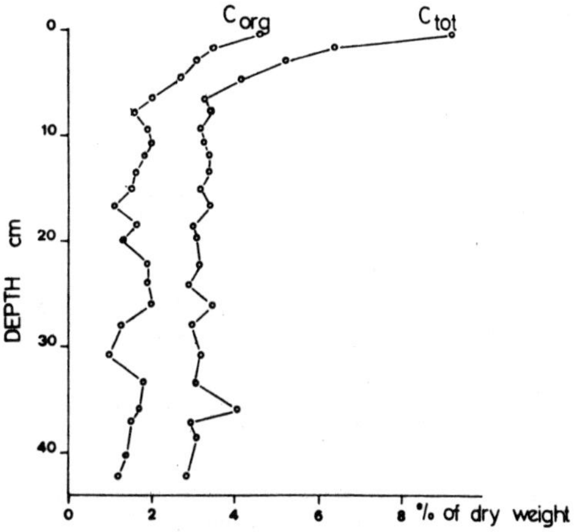


Fig.5. Distribution of organic and total carbon in the sediment core

Fig.6. Sediment extract separated by HPLC

Sample size: 100 µl
Column: µ Bondapak C-18,
1/4" x 30 cm
Column temperature: ambient
Mobile phase: methyl alcohol
Flow rate: 2 ml/min
Pressure: 400 psi
Detection: 665 nm, 0.05 AUFS

Peaks:
A Aceton
B Chlorophyllides
C Allomerized chlorophyll a
D Chlorophyll a
E Isomeric chlorophyll a
F Pheophytin a
G Allomerized pheophytin a
H Not identified

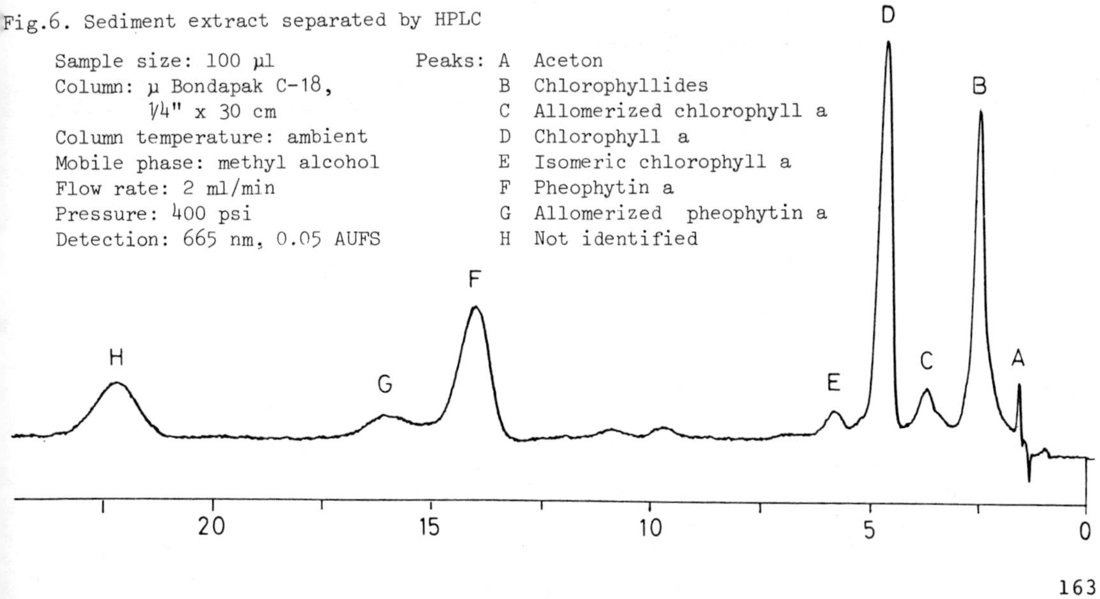

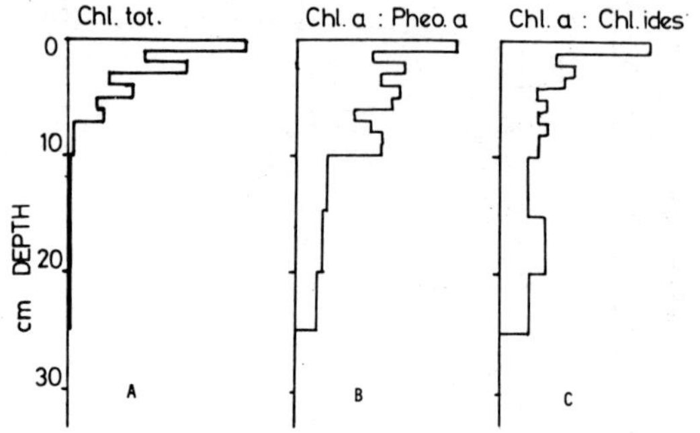


Fig.7. A: Distribution of chlorophylls (incl. degradation products) in the sediment core; B and C: Ratio of chlorophyll a to its degradation products in the sediment core

curves of chlorophyll and C org shows a faster decrease for the C org curve if the undecomposable background of the C org fraction is not considered (Fig. 8). A possible explanation for this fact is a slower degradation velocity of the chlorophylls. Further explanations could be:

1) a changed phytoplankton/zooplankton ratio because of an increased grazing effect by the zooplankton,

2) a changed chlorophyll/biomass ratio because of an altered phytoplankton community of the lake, and

3) a changed allochthonous/autochthonous ratio of organic matter contributed to the sediment.

The allomerization (oxidation) of chlorophylls takes place only in the presence of oxygen either during settling through the water body or in the sediment. Therefore the allomerized chlorophyll fraction (peak C in Fig. 6) is of special interest. It is expected to be comparable with results of redox measurements.

Further investigations on chlorophylls separated by HPLC will give more information on the subject.

## References


Ambühl H., 1975. Forschung am Vierwaldstättersee und die jüngste Entwicklung schweizerischer Seen. GWA 55: 5-16

Ambühl H. & H. Bührer, 1975. Zur Technik der Entnahme ungestörter Grossproben von Seesedimenten: ein verbessertes Bohrlot. Schweiz. Z. Hydrol. 37: 175-186

Cholnoky B.J., 1968. Die Oekologie der Diatomeen. Verl. v. J. Cramer

Daley R.J., 1973. Experimental characterization of lacustrine chlorophyll diagenesis. I. Physiological and environmental effects. Arch. Hydrobiol. 72: 277-304


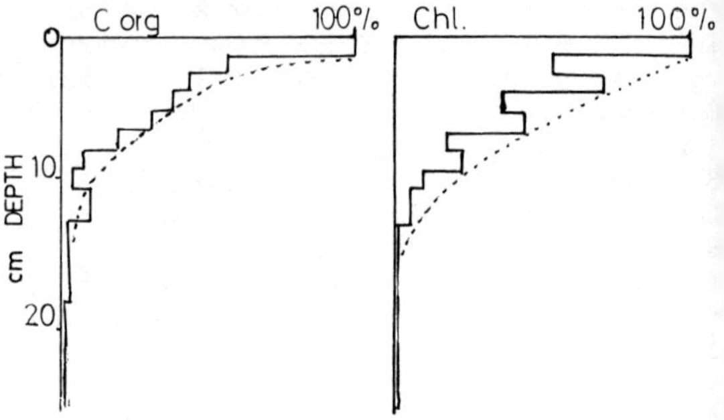


Fig.8. Degradation curves of C org and chloro rophyll


Daley R.J., C.B.J. Gray & S.R. Brown, 1973. Reversed-phase thin-layer chromatography of chlorophyll derivatives. J. Chromatogr. 76: 175-183

Mackinney G., 1941. Absorption of light by chlorophyll solutions. J. biol. Chem. 140: 315-311

Nipkow F.H., 1927. Ueber das Verhalten der Skelette planktischer Kieselalgen. Diss. ETH Zurich

Staub E., 1974. Biologisch-stratigraphische Untersuchungen an jungen Sedimenten des Vierwaldstättersees. Diplomarbeit EAWAG/ETH Zurich (unpublished)

Staub E. & W. Giger, in preparation. A quantitative method for determining algal and sedimentary chlorophyll derivatives by HPLC separation



Key words: sedimentation rate
organic components
memory-effect of sediment
eutrophication and sediment
methods

# CHANGES IN THE PURINE AND PYRIMIDINE CONCENTRATIONS AND ORGANIC CARBON CONTENTS IN LAKE SEDIMENTS

W. van der Velden and Alan W. Schwartz

Department of Exobiology, University of Nijmegen, Nijmegen, The Netherlands

Summary

The vertical distribution of purines and pyrimidines and total organic carbon was studied in a variety of recently deposited lake sediments, including the Great Lakes of North America and the Bodensee (Lake Constance).

Steep increases in both organic carbon and purine and pyrimidine concentrations were observed in the top layers of sediments of lakes subject to cultural eutrophication. The distribution patterns of the investigated parameters indicate a correlation with increased rates of loading with organic material and suggest also a higher microbial turnover in sediments of lakes with serious recent changes in productivity.

A zooplankton sample and a core from the same location of Lake Ontario were studied in greater detail, including total organic nitrogen and amino-acid distributions.

Organic carbon, nitrogen and purines and pyrimidines decreased uniformly, with a sharp change in the rate of degradation around 20 cm, suggesting a shift from microbial to chemical diagenesis.

The similarity of the amino-acid distributions in sediments and zooplankton samples has been attributed to a low microbial turnover. However, the lack of any apparent correlation between the purine and pyrimidine content of the sediment and the investigated zooplankton sample indicates a high bacterial activity.

The purines and pyrimidines decreased more rapidly with depth than the total organic C or the amino-acids. Pyrimidines (especially uracil) degraded at a faster rate than the purines.

These observations suggest the existence of selective breakdown mechanisms operating in the organic matter in lake sediments.

References

van der Velden, W. and A.W. Schwartz, 1976. Nucleic acid base contents as indicators of biological activity in sediments. In: J.O. Nriagu (Ed.): Environmental Biogeochemistry. Ann Arbor Science, Ann Arbor. p. 175-183.

van der Velden, W. & A.W. Schwartz. Purines, pyrimidines and organic carbon in lake sediments - a comparison of sediments from lakes of varying degrees of eutrophication. Chemical Geology, in press (expected to appear Jan.-Feb. 1977).

Dungworth, G., M. Thijssen, J. Zuurveld, W. van der Velden & A.W. Schwartz. Distribution of amino acids, amino sugars, purines and pyrimidines in a Lake Ontario sediment core. Chemical Geology, in press (expected to appear March-April 1977).

Key words: Purines
Pyrimidines
Organic carbon
Lake sediments
Eutrophication

Theme 2 - Sediment formation and rate of sedimentation

ACCUMULATION OF SEDIMENTARY ORGANIC MATTER AT THE BASE OF STEEP BOTTOM GRADIENTS

B. T. Hargrave[1] and L. Kamp Nielsen[2]

Summary

Changes in sediment organic carbon and nitrogen with increasing depth in a marine bay and a eutrophic lake are inversely related to the gradient of bottom slope. Maximum values occur at the base of steep slopes between 15 and 20 m in both basins. Accumulation could be due to downslope transport caused by erosion from currents which meet the bottom at these depths.

Introduction

Post-depositional movement of sediment, through resuspension and transport, are common features of many aquatic environments which occur when high levels of current or turbulence reach the bottom. This occurs in shallow littoral and marine coastal areas as well as in the deep sea where horizontal velocities reach or exceed several centimeters per second. Erosion and horizontal transport of material occurs to regions where turbulent velocities are less than those required to sustain particle suspension. The effect is to physically distribute sedimenting and sedimentary material into a gradient of decreasing particle size and density with increasing distance from shore or with depth (Allen, 1970). Depositional zones should exist in lake, estuarine and marine sediments, therefore, which correspond to areas where velocity thresholds fall below critical values for particle suspension.

Depositional conditions which cause sediment accumulation may be characterized by organic content or by particle size. The amount of organic matter in sediments is usually inversely related to the median grain size (Hargrave, 1972a; Loring & Nota, 1973). This may be due to increased surface area per unit weight for adsorption of organic matter as particle size decreases. It may equally well reflect diagenic loss or formation of organic matter in situ in sediments at rates dependent on particle size. Also, high rates of deposition of coarse-grained inorganic material may lower the concentration of organic matter accumulated in sediments by simple dilution. Although measures of gross organic composition in sediments give no information on their origin or rates of formation (Vallentyne, 1969), differences in spatial distribution of organic content can indicate regions where accumulation occurs.

Observations summarized here compare concentrations of organic matter at short distances along downhill slopes in a marine and lake basin. Low concentrations along upper slope faces are thought to reflect the constant erosion of material through water motion with deposition occurring at deeper depths where turbulence is reduced.

Methods

Sediment cores were collected from Esrum Sø, Zealand, Denmark, in June and from Bedford Basin, Nova Scotia, Canada in September 1975. Both bodies of water have been extensively studied in the past in terms of phytoplankton production (Jónasson, 1972; Platt, 1975) and sedimentation (Lastein, 1976; Hargrave et al., 1976). Cores (25 $cm^2$) from each area were collected over a few hours by sampling along transects normal to shore (Fig. 1) using a Kajak corer (Brinkhurst et al., 1969).

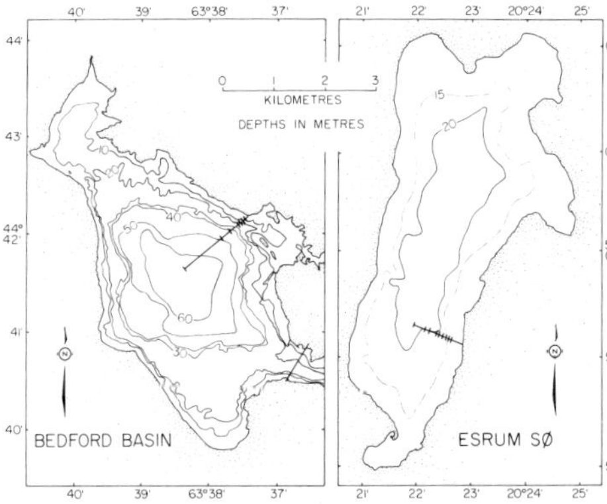


Fig. 1 - Sampling transects crossing depth contours from littoral to central basin areas in Bedford Basin and Esrum Sø. Total surface area of both basins is 17 $km^2$.

Positions of stations were pre-selected from bathymetry maps to cross closely spaced depth contours. Replicate cores were

[1] Bedford Institute of Oceanography, Dartmouth, Nova Scotia, Canada
[2] Freshwater Biological Laboratory, University of Copenhagen, Hillerød, Denmark

collected at some stations for analysis of within station variation. Cores held in dim light at 5°C were extruded and sectioned (1 cm intervals) within 24 hr of collection. Sediment from Bedford Basin was suspended in 25 ml of distilled water and filtered onto glass fiber filters to remove salts. Samples were dried (60°C for 24 hr), pulverized and stored in glassine envelopes in a dessicator containing Silica gel. No attempt was made to enumerate fauna but macrofauna (molluscs, amphipods, polychaetes and Chironomus larvae) present in some samples were removed when observed either before or after drying.

Total organic matter (weight loss on ignition by combustion at 550°C for 1 hr) and organic carbon and nitrogen (using a Perkin-Elmer 240 Elemental CHN Analyzer) determinations were made for each sample. Carbonates in sediment were removed by treating 50 mg samples with 2 ml of 4N HCL. A Metler HT 20 (accurate to 0.05 mg) or a Cahn electrobalance (accurate to 0.005 mg) was used for weight determinations. Samples weighed between 5 and 10 mg and thus weighing errors could account for up to 1% of measured organic content. Coefficients of variation (standard deviation x 100/mean) of subsamples (10) from one homogenized sample amounted to 3% for organic matter, 8% for organic carbon and 4% for nitrogen.

## Results

### Within station variation in sediment organic content

Samples collected from replicate (2-4) cores taken at one station showed that between core variation could be considerable and in excess of values expected from analytical error alone (Table 1).

Table 1. Coefficient of variation [($\sigma/\bar{X}$.100)] of percent sediment dry weight as organic matter, carbon and nitrogen in surface (0-5 mm) sediment for sediment cores taken within one sampling depth at different locations in Bedford Basin and Esrum Sø.

| | | | Coefficient of variation as percent | | |
|---|---|---|---|---|---|
| | Depth (m) | # of cores | Organic matter | Organic carbon | Nitrogen |
| Bedford | 9 | 4 | 13.1 | 9.0 | 16.2 |
| Basin | 18 | 4 | 8.0 | 10.6 | 19.0 |
| | 70 | 3 | 10.5 | - | 18.5 |
| Esrum | 15 | 2 | 17.1 | 1.7 | 0 |
| Sø | 19.5-20 | 4 | 4.1 | 3.6 | 28.7 |

Nitrogen content was the most variable of the three measures of organic content.

### Bottom slope and sediment organic content

The carbon and nitrogen content of surface (upper 5 mm) sediment increased over the upper slope portion of the transect in Bedford Basin (Fig. 2).

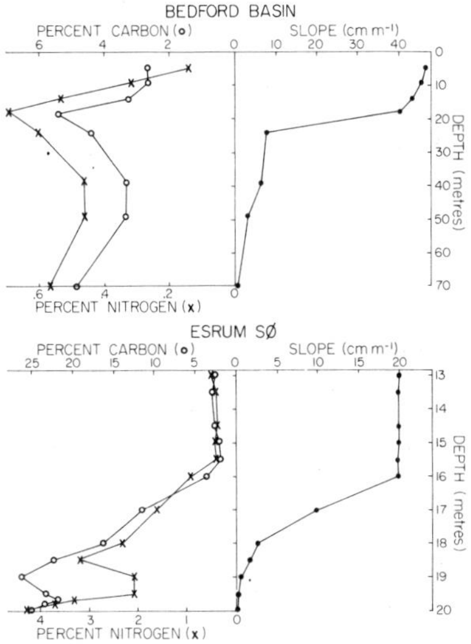


Fig. 2 - Organic carbon and nitrogen content, as a percentage of dry weight, in surface (0-5 mm) sediment of cores and bottom slope gradient (calculated as mean of the depth difference between adjacent sites divided by the horizontal distance at various depths) along transects in Bedford Basin and Esrum Sø.

Bottom slope was high (1:2) and only decreased slightly in this region. Maximum carbon and nitrogen content occurred below 18 m, corresponding to a decrease in slope (to 1:15). Concentrations of carbon and nitrogen in sediments at deeper depths (to a maximum of 70 m) where bottom slope decreased further (to 1:50) were intermediate between values for upper slope deposits and maximum amounts associated with the change in bottom gradient at 18 m.

Percentage content of organic carbon and nitrogen in Esrum Sø sediments showed no significant variation between the five mid-slope sampling sites where bottom gradient was steep and constant (1:5) over a short (60 m) distance. Between 16 and 18 m depth, however, the gradient decreased abruptly to 1:45 and carbon and nitrogen content both increased by an order of magnitude. At depths greater than 18 m variation in organic content, exceeding analytical error, demonstrated the presence of high and variable concentrations of organic matter as bottom gradient decreased to values greater than 1:100 near the center of the lake.

Organic content of surface sediment from areas of steep gradient was similar in both basins. Organic carbon and nitrogen at the base of the slope and in deeper water in Esrum Sø, however, increased by an order of magnitude. No similar increase existed in Bedford Basin.

Organic carbon and nitrogen content in surface sediment samples from both areas was directly related to total organic content but the proportion of organic matter as carbon and nitrogen differed (Table 2).

Table 2. Comparison of organic carbon and nitrogen content expressed as percentage of total organic matter in surface sediment from different depths along transects in Bedford Basin and Esrum Sø. Parentheses indicate standard deviation of adjacent mean values determined from at least three different cores. All values are means of replicate determinations of organic carbon and nitrogen content combined with single determinations of weight loss on ignition.

| | | Percentage of organic matter | | |
|---|---|---|---|---|
| | Depth (m) | As carbon | As nitrogen | C:N |
| Bedford Basin | 4 | 37.1 | 1.9 | 19.5 |
| | 9 | 37.6 | 4.1 | 9.2 |
| | 14 | 29.3 | 4.9 | 6.0 |
| | 18 | 38.2 | 4.7 | 8.1 |
| | 24 | 35.5 | 3.9 | 9.1 |
| | 70 | 30.0 | 3.5 | 8.6 |
| Esrum Sø | 13-15 | 30.3 (7.2) | 4.3 (1.2) | 7.0 |
| | 16-18 | 45.1 (12.1) | 6.1 (1.6) | 7.4 |
| | 18-20 | 59.5 (3.3) | 7.1 (2.2) | 8.4 |

Carbon and nitrogen tended to account for similar but variable fractions of organic matter with increasing depth in Bedford Basin. Organic matter in Esrum Sø sediments below 18 m, however, contained approximately twice the amount of carbon and nitrogen as that in deposits above 15 m.

The relative proportions of organic carbon and nitrogen in surface sediment also changed with depth in Esrum Sø. The ratio of C:N (on average) increased slightly with increasing depth (Table 2). Average values, however, do not indicate that between 19 and 20 m depth near the center of the lake basin, both high (12.8, 11.9) and low (6.9, 6.3, 5.9) C:N ratios occurred in closely spaced samples. Little change in bottom slope existed between the sampling sites and total sediment organic content was not significantly different (40.6% ± 1.5). Apparently heterogeneous conditions exist in the profundal zone of the lake which cause differences in the quality of organic matter (proportion of carbon:nitrogen) but not the total quantity. Carbon:nitrogen ratios in surface sediments from Bedford Basin were more variable than those in Esrum Sø. Minimum values (6-7), however, occurred midway down the slope in regions of the steepest gradient in both basins (Fig. 2).

Profiles of sedimentary organic matter

Approximately uniform concentrations of organic carbon and nitrogen occurred in the upper 10 cm of sediment cores taken along transects in Bedford Basin and Esrum Sø in regions of steepest bottom gradient (Fig. 3)

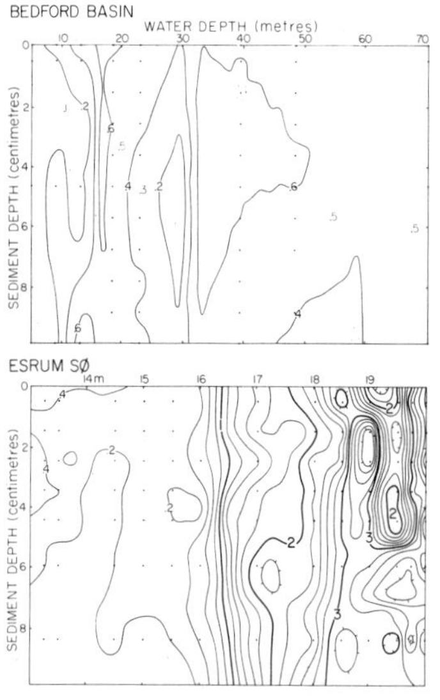


Fig. 3 - Contour diagrams of the vertical distribution of nitrogen as percent dry weight in sediments along sampling transects in Bedford Basin and Esrum Sø. Dots indicate measurements from which contour lines were derived. Lines drawn in areas where derived. Lines drawn in areas where data points do not exist result from interpolated values from polynomial curves generated by the plotting program (General Purpose Contouring Program, Calcomp Corp. using a PDP8 computer to drive a 563 plotter). Contours of organic carbon with sediment depth along the transects showed similar homogeneous vertical distributions between 15 and 18 m.

Gradients of decreasing organic content with sediment depth were present in cores taken between 4 and 14 m in Bedford Basin and at 13 - 13.5 m in Esrum Sø. Profiles in samples from deeper water in both areas indicated variable amounts of total organic matter, carbon and nitrogen and no clear trend with depth.

Coefficients of variation of the three measures of organic content also indicate that gradients were greatest in cores taken at depths shallower than 18 m in Bedford Basin and 16 m in Esrum Sø (Table 3).

Table 3. Coefficients of variation of vertical profiles of organic content (as percent dry weight) in sediment cores taken from various depths along transects normal to shore in Bedford Basin and Esrum Sø. Each value derived from at least seven determinations per core with samples taken at 1-2 cm intervals from the upper 10 cm of undisturbed sediment cores.

| | Water depth (m) | Total organic matter | Organic carbon | Nitrogen |
|---|---|---|---|---|
| | 4 | 24.3 | 63.5 | 24.3 |
| | 9 | 31.5 | 26.9 | 38.4 |
| | 14 | 31.4 | 84.2 | 46.5 |
| Bedford Basin | 18 | 7.0 | 7.0 | 14.1 |
| | 24 | 12.0 | 13.3 | 23.2 |
| | 39 | 11.0 | 11.5 | 11.7 |
| | 49 | 10.5 | 13.2 | 15.2 |
| | 70 | 14.5 | 13.8 | 11.0 |
| | 13.0 | 30.5 | 54.0 | 61.6 |
| | 13.5 | 33.8 | 58.5 | 57.8 |
| | 14.5 | 20.5 | 41.9 | 41.9 |
| | 15.0' | 32.6 | 29.5 | 31.9 |
| | 15.0" | 19.1 | 31.5 | 27.6 |
| Esrum Sø | 16.0* | 8.9 | 11.7 | 12.0 |
| | 17.0 | 3.6 | 13.1 | 12.5 |
| | 18.0 | 5.7 | 7.1 | 6.3 |
| | 18.5 | 3.7 | 10.6 | 11.8 |
| | 19.0 | 1.0 | 5.6 | 19.6 |
| | 19.5 | 3.4 | 6.3 | 27.1 |
| | 19.7' | 2.9 | 4.5 | 10.3 |
| | 19.7" | 1.7 | 4.7 | 9.7 |

(') (") replicate cores from same site
* core sectioned at three depths (0-5, 5-7, 7-10 cm)

Tendency towards homogeneous vertical distributions occurred in cores from 18 m and 19 m in both basins. Variation in total organic matter and carbon in cores from deeper depths was more or less similar, although significant differences ($p < 0.05$) in nitrogen content occurred at certain depths. Vertical differences in organic carbon and nitrogen also tended to be greater than differences in total organic content.

## Discussion

Differences in organic content in sediments along steep bottom gradients can often be related to erosional features caused by water movement. Turbulence in these areas continually transports fine grained material from the sediment surface into deeper, more quiescent areas. We have no data on near-bottom currents in Bedford Basin or Esrum Sø, but the existence of erosional zones between 16 and 18 m is suggested by parallel changes in bottom slope and sedimentary organic content in both basins. Regions of high slope gradient (1:2 to 1:5) along both transects were characterized by similar amounts of organic carbon (2 to 3%) and nitrogen (.2 to .4%) (Fig. 2). Changes in bottom gradient (1:15 to 1:50) between 16 and 18 m, however, were associated with increased organic matter which was an order of magnitude greater in Esrum Sø than in Bedford Basin.

Differences in levels of accumulated organic matter deposited with depth in Bedford Basin and Esrum Sø may be related to the bathymetry of the two basins. Although both water bodies have identical surface area, the morphometry of Bedford Basin provides a large deep central basin into which erosion may deposit sediment (Fig. 1). Depths > 18 m occupy 11 $km^2$ (73% of the total surface area) in Bedford Basin but only 7 $km^2$ (41% of the total) in Esrum Sø. Thus, a longer residence time for suspended organic matter and more extensive mineralization before permanent deposition may occur in Bedford Basin. In addition, seasonal oxygen depletion below 15 m is almost complete in Esrum Sø (Jónasson, 1972; Hargrave, 1972b) thus inhibiting aerobic oxidative degradation of deposited material. Dissolved oxygen below 20 m in Bedford Basin is depleted during certain periods of reduced circulation but minimal values (20-40% saturation) only occur in the deepest layers (Hargrave et al., 1976).

More extensive decomposition of sediments in Bedford Basin may also be indicated by low carbon (30-35%) and nitrogen (3.5-4.7%) content of deposited organic matter relative to that in Esrum Sø (C - 60%, N - 7%) (Table 2). These differences may reflect the origin of sedimented material. In Bedford Basin sediments may be derived more from terrestrial sources (river runoff) than in Esrum Sø where products from phytoplankton and macrophyte production are the primary source of deposited sediments. Despite this, enrichment of sediment organic matter with carbon and nitrogen with depth occurs in Esrum Sø suggesting accumulation of less degraded material below 18 m. No such accumulation occurs with depth in Bedford Basin.

Changes to C:N ratios of organic matter in sediments might also be used to index differences in composition of deposits from different depths. Although difficulties exist in interpreting these ratios in samples of mixed living and non-living organic material (Banse, 1974), patterns of change with depth might be useful in indicating accumulation of nitrogen rich sediments. In our data, low ratios (6-7) occurred in regions of steep slope in both basins and in certain locations in the profundal zone in Esrum Sø. Apparently, localized pockets of nitrogen rich (relative to organic carbon) deposits exist. Even closely spaced samples contain significantly different organic content both in quality and quantity (Table 1). Similar enrichment on slope faces in both basins, however, suggests that sediments in these zones contain freshly derived material from aquatic production as opposed to terrestrially derived material which would be expected to have a higher C:N ratio (Stephens et al., 1967).

The horizontal distribution of total amounts and ratios of organic carbon: nitrogen in both Bedford Basin and Esrum Sø imply that slopes are sites of enhanced flux of organic matter in comparison to that which occurs in deeper water where bottom gradients are low. Differences in profiles of organic carbon and nitrogen support this view. If water movement and supply of particulate and dissolved material to the sediment surface are higher in these areas than in regions of low bottom gradient, both an increase in organic concentration and a more homogeneous distribution with depth, as observed (Fig. 3, Table 3), might be expected. Further sampling with more closely spaced cores in localized areas would quantify this variation.

No attempt was made to estimate macrofauna biomass in cores collected during the present study but physical stirring through burrowing could mix sediments to at least 10 cm depth (Guinasso and Schink, 1975). Physical mixing through resuspension in small lakes can redistribute 1-2 cm of sediment (Davis, 1973) and this can probably occur to a much greater depth in shallow-water marine sediments. Lastein (1976) observed resuspension of surface sediment in central Esrum Sø during fall overturn which would tend to homogenize the depth distribution of sediment organic matter. Relatively homogeneous vertical concentrations of organic content in cores taken at 18-19 m depth in both Bedford Basin and Esrum Sø, however, indicate that the greatest physical redistribution may occur at these depths. Enhancement of sedimentation or erosion, metabolic activity and nutrient release might be expected in these zones.

The expression of sediment organic matter on the basis of its organic carbon and nitrogen content (Table 2) also demonstrates enrichment at certain depths along both transects. This accumulation is consistent with the view that shallow-water sediments are continually eroded and carried to deeper areas in a basin (Wetzel et al., 1972; Thomas et al., 1976) with permanent deposition dependent on topological characteristics of bottom gradients. It is well known that steep-sided basins are subject to mass slumping. Turbidity currents can be generated in regions of bottom slopes susceptible to slumping (Ludlam, 1974). Even in areas of more gentle gradients, echo-sounder traces usually show thicker sediments in depressions. Wieckowski (1969) attributed spatial differences in sediment thickness within a lake basin to such gravity induced motion. Recently sedimented material was thought to form a semi-fluid suspension which slides down into all bottom depressions.

Benthic ecologists have recognized the importance of depressions and gradients of bottom slope in controlling characteristics of bottom communities. Distinctive foraminifera assemblages in areas of relief in the deep sea may reflect segregation on the basis of density differences which occurs through scouring and movement of less dense tests into depressions (Berger and Piper, 1972). Eggleton (1952) attributed the non-even distribution of benthic macrofauna in many lakes to distinctive groups present in bottom depressions. Deposits in such areas are usually dominated by less dense fine-grained particulate matter. They may contain enhanced numbers of certain macrofauna species and they may also be sites for concentrated fish feeding activity The importance of these areas in cycling nutrients and organic matter, either naturally produced or man-added, may be far more important than their limited area in relation to the total area of a particular basin suggests.

## References

Allen, J.R.L., 1970. Physical processes of sedimentation. Georg. Allen & Unwin Ltd., London. 248 pp.

Banse, K., 1974. On the interpretation of data for the carbon-to-nitrogen ratio of phytoplankton. Limnol. Oceanogr. 19: 695-699.

Berger, W.H. & D.J.W. Piper, 1972. Planktonic foraminifera: differential settling, dissolution, and redeposition. Limnol. Oceanogr. 17: 275-287.

Brinkhurst, R.O., K.E. Chua & E. Batoosingh, 1969. Modifications in sampling procedures as applied to studies on the bacteria and tubificid oligochaetes inhabiting aquatic sediments. J. Fish. Res. Board Can. 26: 2581-2593.
Davis, M.B., 1973. Redeposition of pollen grains in lake sediment. Limnol. Oceanogr. 18: 44-52.
Eggleton, F.E., 1952. Dynamics of inter-depression benthic communities. Trans. Am. Micros. Soc. 71: 189-228.
Guinasso, N.L., Jr. & D.R. Schink, 1975. Quantitative estimates of biological mixing rates in abyssal sediments. J. Geophys. Res. 80: 3032-3043.
Hargrave, B.T., 1972a. Aerobic decomposition of sediment and detritus as a function of particle surface area and organic content. Limnol. Oceanogr. 17: 583-596.
Hargrave, B.T., 1972b. A comparison of sediment oxygen uptake, hypolimnetic oxygen deficit and primary production in Lake Esrom, Denmark. Int. Ver. Theor. Angew. Limnol. 18: 134-139.
Hargrave, B.T., G.A. Phillips & S. Taguchi, 1976. Sedimentation measurements in Bedford Basin, 1973-1974. FMS Tech. Rept. No. 608. 129 pp.
Jónasson, P.M., 1972. Ecology and production of the profundal benthos. Oikos 14 (Suppl.): 1-148.
Lastein, E., 1976. Recent sedimentation and resuspension of organic matter in eutrophic Lake Esrom, Denmark. Oikos 27: 44-49.
Loring, D.H. & D.J.G. Nota, 1973. Morphology and sediments of the Gulf of St. Lawrence. Fish. Res. Bd. Canada Bull. 182 pp.
Ludlam, S.D., 1974. Fayetteville Green Lake, New York 6. The role of turbidity currents in lake sedimentation. Limnol. Oceanogr. 19: 656-664.
Platt, T., 1975. Analysis of the importance of spatial and temporal heterogeneity in the estimation of annual production by phytoplankton in a small, enriched, marine basin. J. exp. mar. Biol. Ecol. 18: 99-103.
Stephens, K., R.W. Sheldon & T.R. Parsons, 1967. Seasonal variations in the availability of food for benthos in a coastal environment. Ecol. 48: 852-855.
Thomas, R.L., J.-M. Jaquet, A.L.W. Kemp & C.F.M. Lewis, 1976. Surficial sediments of Lake Erie. J. Fish. Res. Bd. Canada 33: 385-403.
Vallentyne, J.R., 1969. Sedimentary organic matter and paleolimnology. Int. Ver. Theor. Angew. Limnol. Mitt. 17: 104-110.
Wetzel, R.G., P.H. Rich, M.C. Miller & H.L. Allen, 1972. Metabolism of dissolved and particulate detrital carbon in a temperate hard-water lake. Mem. Ist. Ital. Idrobiol. 29: 185-243.
Wieckowski, K., 1969. Investigations of bottom deposition in lakes of N.E. Poland. Int. Ver. Theor. Angew. Limnol. Mitt. 17: 332-342.

Key words: Sediments
Organic matter
Accumulation
Bottom slope

# SEDIMENTATION–RATE IN A BASIN (AGNO) OF LAKE LUGANO*

*O. Ravera & Marita Viola*

Biology Group, Ispra - D.G. Research Science and Education
C.E.C. - EURATOM, Joint Research Centre, 21020 Ispra (Italy)

**Summary**

Sedimentation rate and its seasonal distribution with depth in the basin of Agno (Lake Lugano) from July 1975 to July 1976 has been studied. Sedimentation traps, consisting of two sections (upper for collecting tripton and lower for recovery of attached material), were suspended at 4 depths (-5m; -15m; -40m; -73m). All data was corrected for the attached material, which may attain the 50% of the dry weight of the total sample. High values of sedimentation rate have been measured from spring to autumn and the lowest in the winter. Relationships between the pattern of sedimentation rate and those of temperature, rainfall and Secchi disk values were observed. During sedimentation from -5m to -73m tripton lost about the 30% of its dry weight. The mean sedimentation rate at -73m (2m above the lake bottom) amounts to 635 $g.m^{-2}.d^{-1}$ on dry weight basis. This value is roughly in agreement with that calculated on the $^{137}Cs$ distribution in the recent sediments of the same environment.

**Introduction**

In spite of the important results on the metabolism of lakes obtained by studying the sedimentation of tripton (e.g. Serruya et al., 1974), tripton is not commonly listed among the more important limnological parameters. A suspicion that tripton traps give biased results is quoted by Kirchner (1975) as the main reason why such traps have not been more widely used in sedimentation studies.

Sediment trap is the unique method of collecting tripton (e.g. Edmonson & Winberg, 1970; Kirchner, 1975) and has been used successfully by several authors (e.g. Scott & Miner, 1936; Rossolimo, 1937; Thomas, 1950; Grim, 1950; Kleerekoper, 1952; Jarnefelt, 1955; Tutin, 1955; Ohle, 1962; Bombowna, 1962; Stepánek, 1963; Zeitzschel, 1965; Brunskill, 1969; Bloesch, 1974). The more recent collectors prevent the greatest part of the causes of error (e.g. White and Wetzel, 1973; Fuhs, 1973) and some authors (e.g. Davis, 1967; Pennington, 1974; Rigler et al., 1974) demonstrated, by means of different approaches, that applying the method of the trap, actual values of sedimentation rate of tripton may be obtained.

Since July 1967 the effluent to the Agno basin of Lake Lugano, which accounts for 65% of the total volume of the lakes effluents, has received tertiary treatment. In 1975 it seemed, therefore, appropriate to add a research study on sedimentation rates in the Agno basin to the other studies already in progress. The sedimentation rate of tripton and its seasonal distribution in relation to depth has been studied in the Agno basin from July 1975 to July 1976. The results obtained from this investigation will provide a basis for comparison which results to be obtained in the coming years and thus evaluate the probable influence of the treatment process on the sedimentation rate.

**Method**

The general characteristics of Lake Lugano, which is an environment heavily eutrophicated, have been reported in another paper (Premazzi and Ravera, 1976). The basin of Agno, which is in the North-Western arm of this lake, attains a maximum depth of 87 m and receives about 3.2 $m^3$/sec totally from its main tributaries: River Vedeggio and Agnuzzo (Fig. 1).

Tripton has been collected from 9th July 1975 to 1st July 1976 by means of traps at the depth of 5 and 73 meters. From 16th October 1975 to 1st July 1976, using the same method, tripton from two other depths (-15m and -40m) has been collected. These depths were chosen to obtain information about tripton in the epilimnic layer (-5m), the metalimnic (-15m), at the middl of the hypolimnion (-41m) and at 2 meters above the mud surface (-73m). The exposition time of 17 collections, distributed in different months, ranged, in relation with the season from 6 to 29 days (average value = 13 days) for a total of 220 days.

The perspex trap used by us (Fig. 2) has been built taking into account the criticism reported in literature and particularly the model designed by Fuhs (1973). At the end of the exposition, the material was agitated and subsamples were taken for weight at 105°C and 450°C, chlorophyll, carbon and nitrogen analysis. Composition of each sample was examined by microscope. In this paper only the data on weight at 105°C and 450°C are reported.

**Results and discussion**

To test the reproducibility of the results, all colle

* *Contribution No. 1353 of the Biology Programme, Directorate General XII of the Commission of the European Communities.*

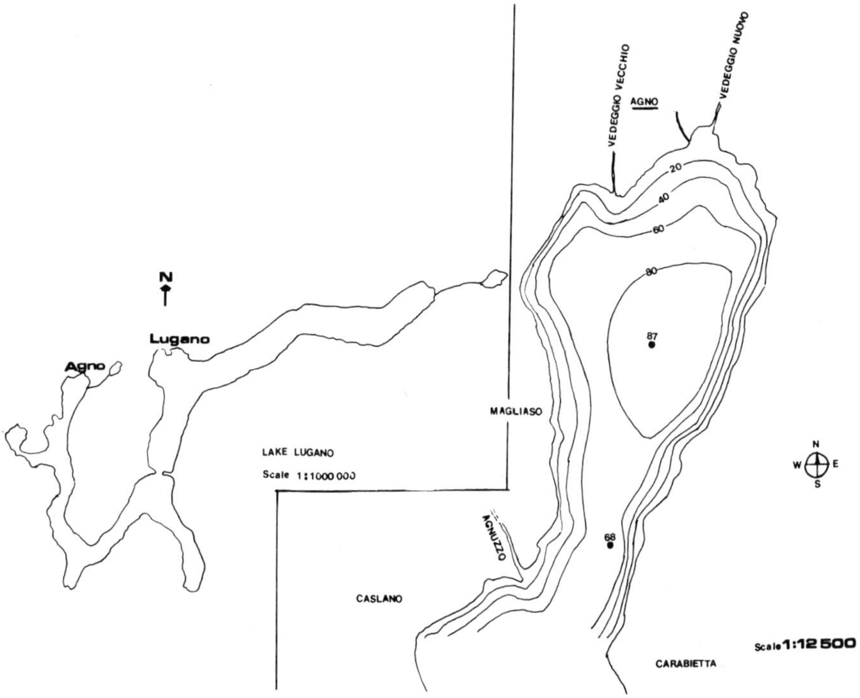


*Fig. 1 : Lake Lugano and basin of Agno*

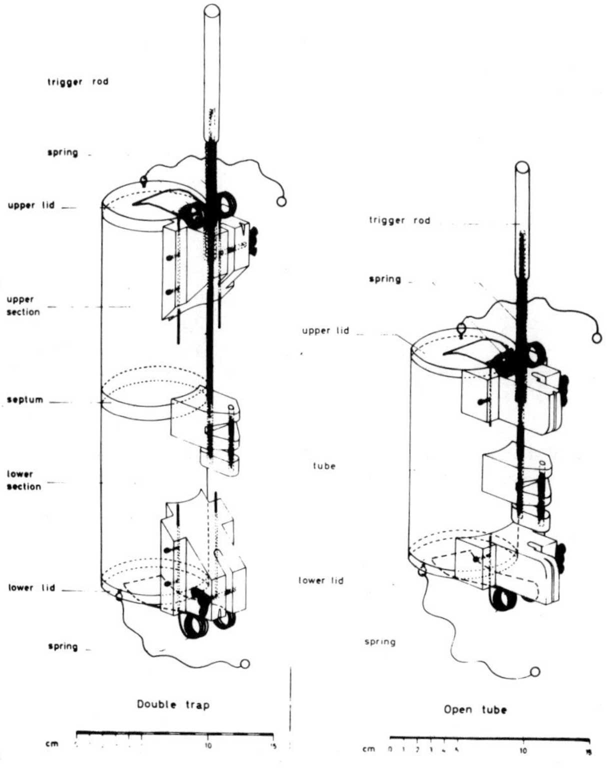


*Fig. 2 : Sediment trap*

tions were replicated from traps suspended at the same depth by means of two buoys (buoy 1 and 2) placed at the distance of about 20 meters. For each collection the differences between the data from the two series are not relevant and even more low between averages calculated on all collections (Table 1).

This means that there is no systematic difference between data from the two series of collections and, then, the results obtained are, probably, reliable for the whole central area of the basin of Agno. High values of standard deviation are, obviously, caused by seasonal variations of the sedimentation-rate.

The amount of material (consisting principally of algae and fungi) attached to the inner wall of the trap may be quantitatively relevant, even for a short exposition time, particularly in eutrophicated lakes. As a consequence, using a simple trap, the sedimentation rate may be overestimated because, together with tripton, attached material is recovered (White et al., 1973; Fuhs 1973).

For this reason, double traps have been adopted and the data corrected by subtracting from the values obtained for the upper section of the trap that of the lower one (Fig. 2).

In our collections the attached material may represent, on the weight basis, 50% of the total sample. Our values on the attached material are higher than those reported by White et al. (1973) for an oligotrophic lake (Lawrence Lake Michigan). Our high value could be the result of conditions particularly favorable to the growth of micro-organisms, existing inside the lower section and, then, the corrected values should result in an underestimation. Indeed, inside the lower section, closed on top, turbulence is reduced to a minimum.

To ascertain if this suspicion was true, four experiments were carried out in duplicate. The amount of material collected by means of a perspex tube open at two ends (suspened at -10m) has been compared with that recovered from the lower section of traps settled at -5m and -15m. The length and diameter of the open tube were the same as those of the lower section of the trap and the device of closure identical. Since the lower section of the trap was closed above by a septum its inner surface resulted 12% greater than that of the open tube.

If all collections are considered, the differences between the material collected by means of the open tube and that of the lower section of the traps seems to be random, but the average calculated for two lower sections is greater than that for the open tube (Table 2). The difference between the average is due, at least partly, to the larger surface of the lower section. As a consequence, the effect of turbulence on the attached material does not seem very important and, then, the values obtained for the lower section may be used to correct those of the upper one. The high standard deviation is due to the seasonal variations.

**Table 1 : Mean and standard deviation of the sedimentation rate ($g.m^{-2}.d^{-1}$) calculated on the weight basis at 105°C and 450°C from upper and lower section of buoy 1 and 2. Data for 5 and 73 meters depth concern collections from 9th July 1975 to 1st July 1976, those for 15 and 40 meters depth collections from 16th October 1975 to 1st July 1976.**

| Depth in meters | Upper section | | | | Lower section | | | |
|---|---|---|---|---|---|---|---|---|
| | buoy 1 | | buoy 2 | | buoy 1 | | buoy 2 | |
| | 105°C | 450°C | 105°C | 450°C | 105°C | 450°C | 105°C | 450°C |
| 5 | 4.6 ± 3.2 | 3.4 ± 3.0 | 4.3 ± 3.5 | 3.4 ± 3.2 | 2.0 ± 0.9 | 1.5 ± 0.5 | 2.1 ± 1.3 | 1.6 ± 1.1 |
| 15 | 3.6 ± 3.2 | 2.8 ± 3.1 | 4.7 ± 3.9 | 3.6 ± 3.5 | 2.0 ± 0.8 | 1.4 ± 0.7 | 2.3 ± 1.2 | 1.6 ± 0.9 |
| 40 | 3.6 ± 3.6 | 2.9 ± 3.2 | 4.8 ± 4.4 | 3.6 ± 3.6 | 1.9 ± 0.6 | 1.4 ± 0.4 | 2.2 ± 1.1 | 1.5 ± 0.8 |
| 73 | 3.6 ± 1.9 | 2.6 ± 1.7 | 4.7 ± 3.1 | 3.3 ± 3.2 | 2.1 ± 0.8 | 1.5 ± 0.6 | 2.6 ± 1.9 | 1.8 ± 1.2 |

**Table 2 : Amount of material attached to the inner wall of the lower section of traps suspended at 5 meters and 15 meters depth and that of open tube at 10 meters depth.**

| Weight mg/collector.day | 105°C | | | 450°C | | |
|---|---|---|---|---|---|---|
| Depth in meters | 5 | 10 | 15 | 5 | 10 | 15 |
| Feb. 24/March 22, 1976 | 2.55 | 2.49 | 2.33 | 1.83 | 1.79 | 1.65 |
| April 5/April 15, 1976 | 7.14 | 4.76 | 6.22 | 5.17 | 3.53 | 4.38 |
| April 29/May 10, 1976 | 8.15 | 7.21 | 7.00 | 5.82 | 5.35 | 5.21 |
| May 10/May 19, 1976 | 5.69 | 4.86 | 6.21 | 1.66 | 1.24 | 1.76 |
| $\overline{X} \pm s$ | 5.88 ± 2.44 | 4.83 ± 1.93 | 5.44 ± 2.11 | 3.62 ± 2.18 | 2.98 ± 1.86 | 3.25 ± 1.82 |

The seasonal variation of sedimentation rate with depth, calculated on the mean values of two series of collections and corrected for the attached material, are schematized in Fig. 3.

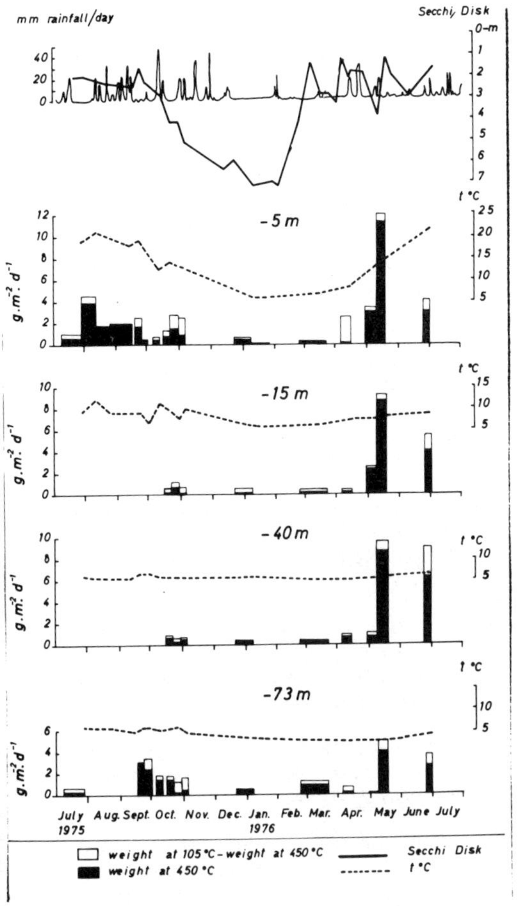


*Fig. 3 : Seasonal variations of rainfall, water transparency, temperature and tripton*

It is probable that several factors control, directly or indirectly, the seasonal pattern of sedimentation rate. For example, an aliquot of tripton is allochtonous, consisting of material carried from the watershed into the lake. Indeed, there is a certain relation between the sedimentation rate and the seasonal distribution of rainfall. Dead phytoplankton seems also to be a relevant part of tripton, whereas zooplankton and its pellets are far less important. Because the phytoplankton growth follows the annual cycle of the radiation and temperature of the epilimnetic layer the relation between these parameters and the abundance of tripton is easy to predict.

High values of sedimentation rate have been measured from spring (the maximum value falls in May) to autumn and the lowest in winter. The inverse relation between the pattern of Secchi disk and that of sedimentation rate is, evidently, the effect of the control exerted by the suspended material on the water transparency.

Since the seasonal pattern of the sedimentation rate is roughly the same at different depths, we may conclude that it is (at least for the central area of the basin) controlled by epilimnetic variations and no factor (e.g. turbidity, currents) disturbs, significantly, the sedimentation of tripton.

The difference between the sedimentation rate at -5m and at -73m varies with the season. In winter the rather homogeneous distribution of tripton with depth is probably due to: 1) low plankton production; 2) very scarce rainfall and strongly reduced run-off from the watershed; 3) uniform water temperature (and consequent water density) from the surface to the lake bottom. The higher values at -73m than at -5m in September—October may be the effect of the decrease of phytoplankton growth in the epilimnion and the accumulation in the deeper layer of material produced in the elilimnion during the preceeding weeks. In addition, it is evident that in May the relevant sedimentation rate at all the depths investigated but particularly in the epilimnetic layer, reflects the high primary production of this month.

On the basis of weight at 105°C the sedimentation rate at -5m ranges from 0.1 to 12 $g.m^{-2}.d^{-1}$ (average = 2.51) and at -73m the range is between 0 to 4.9 $g.m^{-2}.d^{-1}$ (average = 1.74). The sedimentation rate, calculated from the weight at 450°C, at -5m varies between 0 and 11.4 $g.m^{-2}.d^{-1}$ (average = 1.92) and at -73m ranges between 0 and 4.1 $g.m^{-2}.d^{-1}$ (average = 1.25).

The higher percentage of ignition loss at -73m (28%) compared to that calculated for -5m (23%) seems to testify an increase of organic material during the sedimentation of tripton. This result is to be considered absurd if the value of the loss for ignition at 450°C is assumed to be that of organic matter. Because this equation is not always true, the lower value of the ignition loss at -5m than at -73m could be explained by this hypothesis. Some organic compounds are resistent at 450°C when they are at -5m, but during sedimentation they are degradated to simpler compounds which are volatilized at the same temperature.

If the loss of weight of tripton for mineralization is not considered, the amount of material collected at -73m should be greater than that recovered at -5m, for the addition of material produced below this depth. Because the mean sedimentation rate at -5m (2.51 $g.m^{-2}.d^{-1}$) is higher than that at -73m (1.74 g), it is probable that the weight reduction caused by the mineralization of tripton, during the sedimentation from -5 to -73m, is such to overcome the amount of material produced below the 5 meter depth.

## Conclusions

From the results obtained the importance of tripton in the trophy of lakes is evident, for two main reasons: 1) it enriches the water with various substances during its sedimentation (about 30% of its dry weight is lost from -5m to -73m); 2) the amount of material accumulated yearly on the bottom of lake is relevant.

The mean sedimentation rate, calculated at -73m (that is 2 m above the sediment surface) attains 635.1 $g.m^{-2}.d^{-1}$ on the dry weight basis.

From the water content in the first centimeters of sediment core, collected at -75m from the basin of Agno, the thickness of surface sediment corresponding to 635 gram per unit surface has been calculated. This value is about 1.5 cm. The annual increase of the sediment thickness, calculated on the centrifuged volume (Bachofen, 1960) is 1.8 mm, value lower than that (2.5 mm) calculated with the same method by Bloesch (1974) for the eutrophic Rotsee. On the distribution of Cs-137 from fall-out in the thickness of sediment cores collected from the same depth and basin the mean sedimentation rate has been estimated about 1 $cm.yr^{-1}$ (Premazzi & Ravera, in press). The discrepancy between this value and that calculated from tripton collections is probably due to different intervals of time considered. Indeed, our value (1.5 cm) is referred to the material just sedimented, that is with a high content of water, whereas that sedimented by Cs-137 method (1 cm) concerns sediment deposited during the last 14 years and, then, with a lower water content and partly degradated. This agreement with Cs-137 method and the trap technique in evaluating the sedimentation rate has been noted even by Pennington (1974).

#### References


Bachofen, R., 1960, "Stoffhaushalt und Sedimentation im Baldegger- und Hallwilersee", Diss. (Juris-Verlag Zürich)

Bloesch, J., 1974, "Sedimentation und Phosphorhaushalt im Vierwaldstättersee (Horwer Bucht) und im Rotsee", Schweiz. Z. Hydrol, 36, 71-186

Bombowna, M., 1962, "Sedimentieren von Sinkstoffen im Staubecken-Goczalkowice", Acta Hydrobiol., 4, 69-118

Brunskill, G.J., 1969, "Fayetteville Green Lake, New York", Limnol. Oceanogr., 14, 830-847

Davis, M.B., 1967, "Pollen deposition in lakes as measured by sedimentation traps", Bull. Geol. Soc. Amer., 78, 849-858

Edmonson, W.T. & Winberg, G.G., 1971, "A manual on methods for the assessment of secondary productivity in freshwater", I.B.P. Handbook No. 17, Blackwell Scientific Publications, Oxford, p. 1-358

Fuhs, G.W., 1973, "Improved device for the collection of sedimenting matter", Limnol. Oceanogr., 18, 989-993

Grim, J., 1950, "Versuche zur Ermittlung des Produktionskoeffizienten einiger Planktophyten in einem flachen See", Z. Biol., 69, 147-174

Järnefelt, H., 1955, "Ueber die Sedimentation des Seston", Int. Rev. Theor. Angew. Limnol. Verh., 12, 144-148

Kirchner, W.B., 1975, "An evaluation of sediment trap methodology", Limnol. Oceanogr., 20, 657-660

Kleerekoper, H., 1952, "A new apparatus for the study of sedimentation in lakes", Can. J. Zool., 30, 185-190

Ohle, W., 1962, "Stoffhaushalt der Seen als Grundlage einer allgemeinen Stoffwechseldynamik der Gewässer", Kiel Meeresforsch., 18, 107-120

Pennington, W., 1974, "Seston and sediment formation in five Lake District lakes", J. Ecol., 62, 215-251

Premazzi, G. & Ravera, O., "Chemical characteristics of Lake Lugano sediments", (In press)

Rigler, F.H., MacCallum, M.E. & Roff, J.C., 1974, "Production of zooplankton in Char Lake", J. Fish. Res. Bd. Canada, 31, 637-646

Rossolimo, N.N., 1957, "Materialy k poznaniju sedimentacii ozernych otlozenij", Tr. Limnol. Stancii v. Kosine, 21, 5-20

Scott, W. & Miner, D.H., 1936, "Sedimentation in Winona Lake and Tippecanoe Lake, Kosciusko County, Indiana, July 31, 1930 to July 30, 1935", Proc. Ind. Acad. Sci., 45, 275-286

Serruya, C., Edelstein, M., Pollingher, U. & Serruya, S., 1974, "Lake Kinnereth sediments: nutrition composition of the pore water and mud water exchange", Limnol. Oceanogr., 19, 489-508

Stepánek, M., 1963, "Sampler of sedimenting plankton", Sb. Vys. Sk. Chem. Technol. Praza, Oddlil Fak. Technol. Paliv. Vody, 7, 171-173

Thomas, E.A., 1950, "Beitrag zur Methodik der Produktionsforschung in Seen", Schweiz. Hydrol., 12, 25-37

Tutin, W., 1955, "Preliminary observations on a year's cycle of sedimentation in Windermere, England", Mem. 1st. Ital. Idrobiol., 8 (suppl), 467-484

White, W.S. & Wetzel, R.G., 1973, "A modified sedimentation trap", Limnol. Oceanogr., 18, 986-988

Zeitzschel, B., 1965, "Zur Sedimentation von Seston, eine produktionsbiologische Untersuchung von Sinkstoffen und Sedimenten der westlichen und mittleren Ostsee", Kul. Meeresforsch., 21, 55-90



#### Key words

Sedimentation-rate
Sediment trap
Eutrophication
Insubrian lake

# SESTON AND SEDIMENTATION IN FARMOOR RESERVOIR, GREAT BRITAIN

D. Johnson

Water Research Centre, Medmenham, Great Britain.

## Summary

Observations on the composition and quantities of material caught in a seston trap sited in a hardwater, eutrophic, pumped storage reservoir showed that the material sedimenting to the bottom of the reservoir was of autochthonous origin. A recording thermogravimetric balance was used to analyze dried samples for organic, carbonate and silica residues. The calcium budget for the reservoir showed the trap to be overestimating the amount of calcium carbonate deposition. This was associated with periods of high winds disturbing the existing sediments and was confirmed by observations of empty and broken diatom frustules.

## Introduction

Farmoor Reservoir is a hard water, eutrophic, pumped storage reservoir situated by the River Thames upstream of the City of Oxford. The reservoir has been studied by the Water Research Centre since it was first filled in April 1965. Results of the study have been given by Youngman (1975) who showed inter alia that during periods of thermal stratification the hypolimnion became deoxygenated to less than ten percent of saturation. Chemical analyses of discrete depth samples taken before, during, and after stratification showed increased levels of phosphate, silicate and iron in the hypolimnion. Calculations have also shown the reservoir to be retaining a number of nutrients. Crowe (1974) has reported an accumulation of about 2.5 cm of sediment in the reservoir over a seven year period. With this background it was decided to investigate the composition and rate of accumulation of the sediments in the reservoir.

## Methods

### Collection of samples

During 1972 sedimenting material was caught in an inverted bell-jar 0.15 m diameter and 0.4 m high suspended from a surface buoy sited at the centre of the reservoir. The length of the support rope was adjusted to maintain the top of the trap 1 m above the bottom of the reservoir. Whenever possible the trap was raised at weekly intervals to remove the material within. This was usually found as an undisturbed layer at the bottom of the trap. At times of high algal concentrations in the surface waters the trap was filled with filtered water before being returned. On return to the laboratory the sample was mixed, and sub-samples taken for the following analyses:

### Chlorophyll content

The concentrations of chlorophyll a and phaeo-pigment in the samples were estimated by filtration through a Whatman GF/C glass paper filter. The pigments were extracted into 90% aqueous acetone and calculated in accordance with the spectrophotometric equations of Lorenzen (1967).

### Algal counts

The estimation of algae in the samples was carried out using a nannoplankton counting chamber (Youngman 1971). Cells with and without contents were counted separately, as were broken diatom frustules.

### Calcium

Dissolved calcium concentrations in the river and reservoir were determined using the British Standards Institution method (BSI, 1967), but omitting the potassium cyanide.

### Dry weight

The samples were settled and the supernatant decanted. The remaining material was dried at 85$^{o}$C under an air flow, and then weighed.

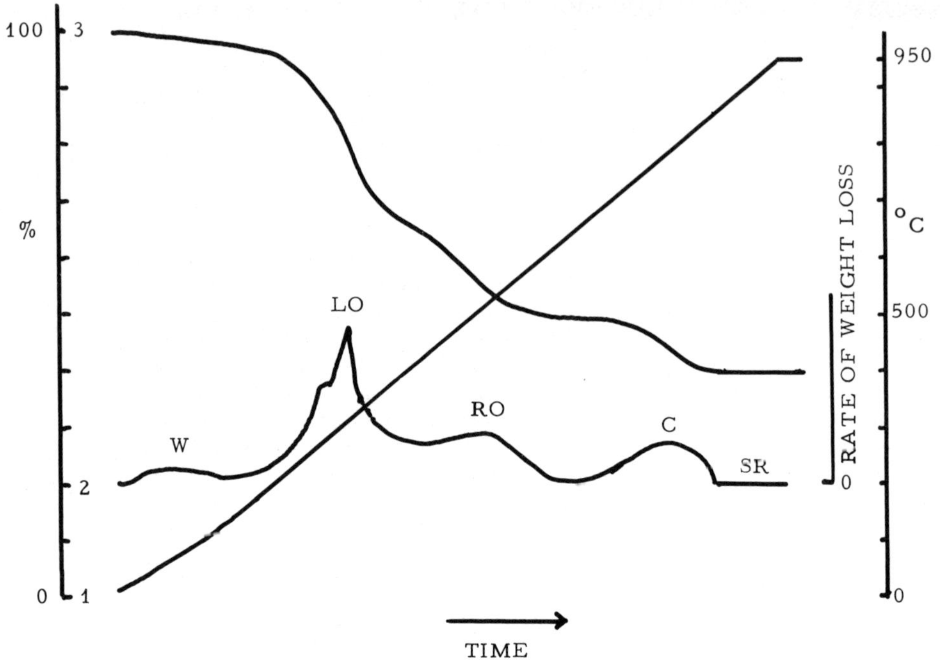


| | |
|---|---|
| 1 | Temperature |
| 2 | Differential weight loss |
| 3 | Loss in weight as % starting weight |
| W | Water |
| LO | Labile organic material |
| RO | Residual organic material |
| C | Carbonate |
| SR | Silica residue |

Figure 1. Typical output from Thermogravimetric Balance for Seston Sample.

Organic, Carbonate and Residue content

Sub-samples of the dried material were ground to a fine powder and analysed on a Stanton Redcroft TG 750 Thermobalance. This was programmed to heat from room temperature to 950°C at a rate of 20°C.min$^{-1}$ under a dried air flow of 0.01 l.min$^{-1}$. Temperature, percentage of starting weight and differential weight loss were recorded simultaneously on a chart recorder. The weight losses between 105°C and 650°C, and 650°C and 950°C were regarded as representing organic content and carbon dioxide from calcium carbonate respectively. The residue was that weight remaining at 950°C, less its calcium oxide content and was regarded as consisting mainly of silica.

Results

Figure 1 shows a typical plot from the thermobalance. The shape of the differential weight loss curve can be used to identify the various components of the seston, these were as follows:-

- W - a loss of water absorbed during storage.
- LO - a peak stepped at 280°C and 310°C representing the initial breakdown of organic material. The first part arising from the loss of chemically bound water giving rise to a conjugated carbon chain. The second part resulting from the loss of volatile aromatic molecules originally in the sample and also produced during the bound water loss.
- RO - a broad peak due to the oxidation of the organic carbon residue produced at the first peaks.
- C - the loss of carbon dioxide from calcium carbonate.
- SR - the remaining ash, when corrected for calcium oxide content, assumed to be silica.

The quantities of each of the three major components, organic material, carbonate and residue in each of the weekly samples were estimated and converted to daily rates per unit area (g.m$^{-2}$.d$^{-1}$).

The numbers of frustules of the different diatoms present in each sample were calculated. Stephanodiscus astraea (Ehr.) Grün was by far the most abundant diatom present, other diatoms contributing usually less than five percent of total diatom cell volume. These sub-dominant diatoms were expressed as S. astraea units in a proportion dependent on their cell volume, one S. astraea unit corresponding to a cell volume of 13 000 $\mu^3$. The quantity of total chlorophyll pigment in the caught material was calculated and converted to a daily rate per unit area.

Discussion

Konrad et al (1970) have shown that organic carbon can be determined in the presence of carbonate in a micro-combustion furnace if the maximum temperature is maintained between 600-650°C as virtually all the organic carbon is lost by 650°C. They showed by differential thermal analysis of a number of calcium carbonate minerals that the breakdown of the minerals began between 640°C and 680°C and was complete by 950°C. Natural calcite decomposes at 825°C while magnesium carbonate, which is also insoluble, decomposes at 350°C which is within the region designated as the loss of weight due to organic combustion. Analyses of water from the River Thames and Farmoor Reservoir show a 10-25% decrease in calcium in the reservoir, while there was no significant loss of magnesium. The contribution of magnesium carbonate to the weight loss between 105°C and 650°C can therefore be ignored in this situation although it may be significant in other bodies of water.

The following linear regressions and correlation coefficients were calculated for the relationships between

(i) S. astraea units and silica residue

$y = 1.05\,x + 0.505 \quad r = 0.974 \quad \ldots(1)$

where y = silica residue g.m$^{-2}$.d$^{-1}$

x = S. astraea $10^8$ units.m$^{-2}$.d$^{-1}$

(ii) Chlorophyll pigment content and organic content

$y = 64.5\,x + 0.467 \quad r = 0.935 \quad \ldots(2)$

where y = organic content g.m$^{-2}$.d$^{-1}$

x = chlorophyll pigments g.m$^{-2}$.d$^{-1}$

The regression equation (1) gives the silica content of S. astraea as 10.5 mg.$10^6$ cells$^{-1}$ which compares favourably with the value of 9.9 mg.$10^6$ cells$^{-1}$ calculated for the 1972 spring population.

The comparison between organic content and chlorophyll pigments, equation (2), gives chlorophyll as 1.6% of the organic content. This agrees with the quoted figures of 0.3-2% of the dry weight (Rabinowitch 1945). These values indicate that the material caught by the trap is almost entirely a result of phytoplankton activity within the reservoir. The embankments of the reservoir are bare concrete and support no macrophyte growth, the only allochthanous inputs being from the river and aeolian input. Stephanodiscus hantzschii Grün was present both in the reservoir and the river but it was not possible to differentiate between those from different sources. However the allochthanous material appears largely to be deposited in the channels, which were designed for this purpose, leading from the river to the pumping station, since Crowe (1974) has reported that considerable quantities of material have had to be removed from therein. In order to assess the efficiency of the trap a calcium budget was calculated for the reservoir. This showed a retention of 64 tonnes of calcium between April and December and the observed quantity caught by the trap was 130 tonnes, this discrepancy was thought to be a result of wind induced resuspension of previously settled deposits. The calcium budget was recalculated on a weekly basis and the periods of overestimation compared with the numbers of empty and broken S. astraea frustules. On 66% of the occasions when the trap was overestimating the calcium deposition S. astraea frustules were recorded thereby giving evidence of resuspension of previously sedimented material. The S. astraea found in the plankton samples were usually entire and with cell contents.

It is accepted that the trap used does not always quantitatively measure the rate of sedimentation within the reservoir, the largest errors probably being due to wind induced resuspension of previously settled material. These observations confirm the conclusion of Kirchner (1975) that the vertical sided trap is less realistic in shallow unstratified bodies of water. During 1972 the reservoir was only thermally stratified for a period of about three weeks. The results however do show that the material caught was almost entirely of reservoir phytoplankton origin.

In view of this and the quantities of sediment removed from the inlet channels it would appear that the channels are working efficiently. This situation could change as the demand for more water and hence throughput of the channels increases.

## Conclusions

A vertical sided sedimentation trap used in a shallow reservoir overestimated the quantities of seston reaching the sediment surface. This was probably a result of wind induced recirculation currents disturbing previously deposited material. The results did show that the origin of the sediment was almost entirely from within the reservoir.

## References

British Standards Institution, 1967. Method of testing water used in industry; part 4 - aluminium, calcium, magnesium, fluoride; metric units. London, BSI.

Crowe, P.J., 1974. The Farmoor Source Works in Operation. J. Instn Wat. Engrs 28 : 101-124.

Kirchner, W.B., 1975. An evaluation of sediment trap methodology. Limnol. Oceanogr. 20 : 657-660.

Konrad, J.G., G. Chesters & D.R. Keeney 1970. Determination of organic- and carbonate-carbon in freshwater lake sediments by a micro-combustion procedure J. Thermal Anal. 2 : 199-208.

Lorenzen, G.J., 1967. Determination of chlorophyll and pheopigments : Spectrophotometric equations. Limnol Oceanogr 12 : 343-346.

Youngman, R.E., 1971. Algal monitoring of water supply reservoirs and rivers. Water Research Association, TM 63, Medmenham, The Association.

Youngman, R.E., 1975. Observations on Farmoor, a eutrophic reservoir in the upper Thames Valley, during 1965-1973. In: The effects of storage on water quality. The Water Research Centre, Medmenham.

Key words Eutrophic; Reservoir; Sedimentation; Seston.

# A MULTIVARIATE PREDICTION EQUATION FOR RATE OF SEDIMENTATION IN NORTH INDIAN RESERVOIRS

V.B. Lal & S. Banerji

Council of Scientific and Industrial Research, New Delhi, India

## Summary

Data of sedimentation surveys on 11 major Indian reservoirs distributed all over the country, are available. Out of these 5 reservoirs, viz. Bhakra, Maithon, Panchet, Mata Tila and Mayurakshi, lying in the sub-Himalayan belt and the Indo-Gangetic plain have been selected and their data analysed in search of a general law that could give order and meaning to the data and be utilised for the purpose of estimation and prediction of silt-index i.e. annual rate of sediment yield per unit area of the catchment, at least for reservoirs in North India. Taking silt-index as a function of capacity-inflow ratio and the 'inflow-index' i.e. average annual inflow per unit area of the catchment, a multivariate prediction equation has been drived. Unfortunately, there are no data that can be used at this stage as a test-set for the prediction equation, however, the equation will mark a considerable improvement in the estimation of sedimentation rates of reservoirs for purposes of design of river valley projects. Presently a silt index of 375 to 430 $m^3/km^2$ is being adopted in the design, though in many projects, its value has been found to be many times higher.

## Introduction

Data on Sedimentation of reservoirs in India have been scarce. During the III Plan period (1961-66), the Central Board of Irrigation and Power included systematic surveys of reservoirs as an item under their programme of Fundamental and Basic Research, and Sedimentation Surveys on 11 major reservoirs, built in different parts of the country were taken up. Results of these surveys over the years are now available.

In an earlier paper (Banerji & Lal, 1971), the authors examined these data. Observing that certain parameters like the capacity - inflow ratio, (C/I), capcity-watershed ratio, (C/W), and the inflow-index (I/W), would influence significantly the silt-index, i.e. the annual sediment yield per unit area of watershed, they indicate that since the selected Indian reservoirs fall under widely different hydro-climatic regions, the data obtained could not be lumped together for drawing any conclusion about the effect of a factor/parameter on the silt-index or about 'the interrelationship between different factors/parameters'.

The present paper is a continuation of the earlier effort and presents the results of the authors' search for a general law that could be evolved from the available data-set, to give an orderly meaning to observations so painstakingly collected, and provide a tool with which engineers/designers can have better estimates of the expected rate of sedimentation in a reservoir, which has a crucial bearing on the 'life' of the reservoir and is an important economic factor in the project design.

## Nature of the Data available

The elevan reservoirs, on which results of sedimentation surveys are available, are as follows:-

1. Bhakra (Panjab)
2. Panchet (Bihar)
3. Maithon (Bihar)
4. Mayurakshi (West Bengal)
5. Mata Tila (Uttar Pradesh)
6. Tungabhadra (Karnataka)
7. Nizam Sagar (AP)
8. Sivaji Sagar (Maharashtra)
9. Hirakud (Orissa)
10. Lower Bhavani (Tamil Nadu)
11. Gandhi Sagar (Madhya Pradesh)

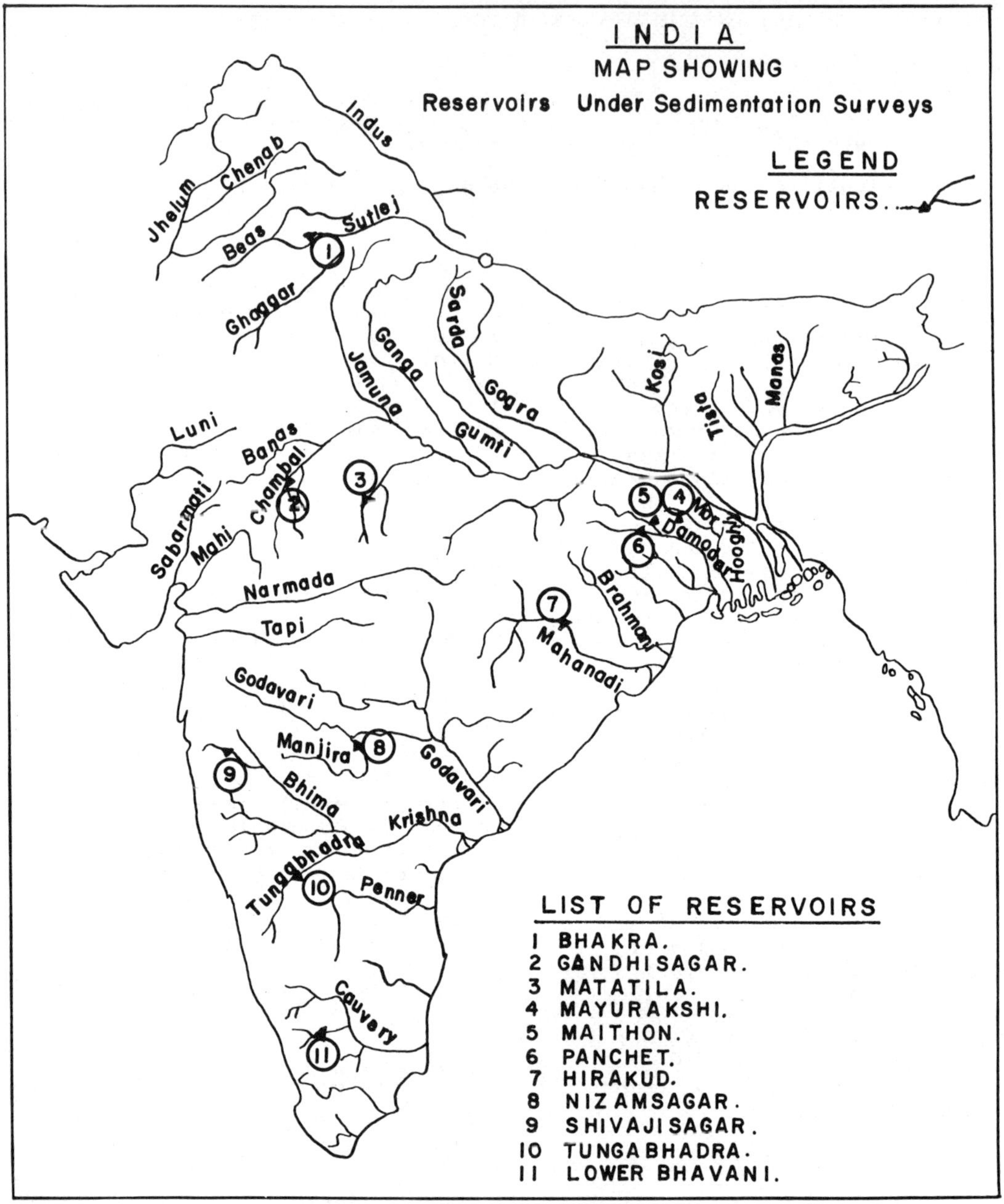
INDIA
MAP SHOWING
Reservoirs Under Sedimentation Surveys
LEGEND
RESERVOIRS.
Indus
Chenab
Jhelum
Beas
Sutlej
Ghaggar
Sarda
Ganga
Jamuna
Gogra
Gumti
Kosi
Tista
Manas
Luni
Banas
Chambal
Sabarmati
Mahi
Damodar
Hooghly
Brahmani
Narmada
Tapi
Mahanadi
Godavari
Manjira
Bhima
Krishna
Tungabhadra
Penner
Cauvery
LIST OF RESERVOIRS
1 BHAKRA.
2 GANDHISAGAR.
3 MATATILA.
4 MAYURAKSHI.
5 MAITHON.
6 PANCHET.
7 HIRAKUD.
8 NIZAMSAGAR.
9 SHIVAJISAGAR.
10 TUNGABHADRA.
11 LOWER BHAVANI.

The type of survey carried out in these reservoirs is not the same. Reservoirs at Serial 1,2,3,5,8,10 are covered by range surveys; reservoirs at serial 4,6 by contour surveys; reservoirs at serial 9,11 by inflow-outflow surveys; and the reservoir of serial 7 by a combination of range - contour surveys. Results from different types of surveys are not very satisfactory for mutual comparison. This is apart from the difficulty that the reservoirs are located in regions widely different in climate, geology, nature and pace of human activity and such other relevant factors.

The way in which the grain-size analysis of the sediment is being reported is also not uniform. Some surveys report size-distribution of the suspended sediment, while some report it for the deposited sediment. Also, some report the size-ranges, as percentages under 'clay and fine silt, medium silt- and coarse silt', some mention these categories as 'Clay, Silt and Sand', and some, as 'Clay and fine silt, coarse silt and fine sand, sand, and gravel'. Actual grain-size analysis curves were not readily available.

A similar difficulty is felt in the reporting of the density of the deposited sediment. No density observations are available for some, density has been assumed in some cases, some report dry density, while for others it is difficult to make out if the figure reported is dry density. Thus at the present stage of our information, it was difficult to include a representative grain-size, or the dry density, as parameters for the prediction equation.

## Selection of Reservoirs and Parameters for Co-relation:

Five reservoirs, lying in the sub-Himalyan belt and the Indo-Gangetic plain have been selected for the study viz. Bhakra, Maithon, Panchet, Mata Tila and Mayurakshi. Out of the choice available for a homogeneous grouping of the surveyed reservoirs according to hydro-climatology, geology and human activity, this set would perhaps present the least divergence. According to the Borland and Miller classification indicating the distribution of the deposited sediment at different reservoir depths, Bhakra falls in type-III(Hill), Maithon, Panchet and Mata Tila in type-II (Flood-Plain-Hill)and Mayurakshi in type-I(lake). From the point of view of operational characteristics, all these reservoirs fall in the 'normally moderate to considerable reservoir draw down' category.

The silt index, S,i.e. the annual sediment-yield per unit area of the watershed, cu. meters per sq. km. per year, is visualised as a function of (a) Capacity-inflow ratio, C/I, where C is the original capacity of the reservoir and I the average annual volume of inflow into the reservoir, and (b) the inflow-index, I/W, which is the average annual inflow per unit area of the watershed-1,000 cu. meters per sq. km. As the reservoirs under consideration fall in the same operational category, C/I will also reflect the trap efficiency of the reservoirs to a certain degree. The inflow-index is, of course, a most relevant factor for the transportation of the eroded sediment in the watershed to the reservoir, incorporating the composite effect of a number basin characteristics like the average annual precipitation, evapotranspiration, infiltration and catchment cover and geomorphologic characteristics like the basin slope and the drainage network. Thus:

$$S = f\ (C/I\ ,\ I/W\ )$$

Let the form of the prediction equation for S be as follows:-

$$S = K\ (C\ /I)^{m}\ .\ \ (I/W)^{n}$$

where K is a constant parameter, and m,n are linear regression coefficients when logarithmic transformation is applied to the variables.

$$\text{Or}\ \log S = \log K + m \operatorname{Log}\ (C/I) + n\ \log\ (I/W) \quad - (2)$$

which is a linear regression equation in three variables of the type:-

$$X = B_1 + B_2\ X_2 + B_3\ X_3 \quad - (3)$$

where $B_1$ is the intercept; $B_2$ is the multiple regression coefficient of $X_1$ on $X_2$ keeping $X_3$ constant, and $B_3$ is the multiple regression coefficient of $X_1$ on $X_3$ when $X_2$ is constant.

The values of coefficients $B_1, B_2, B_3$ are obtained by the solution of the

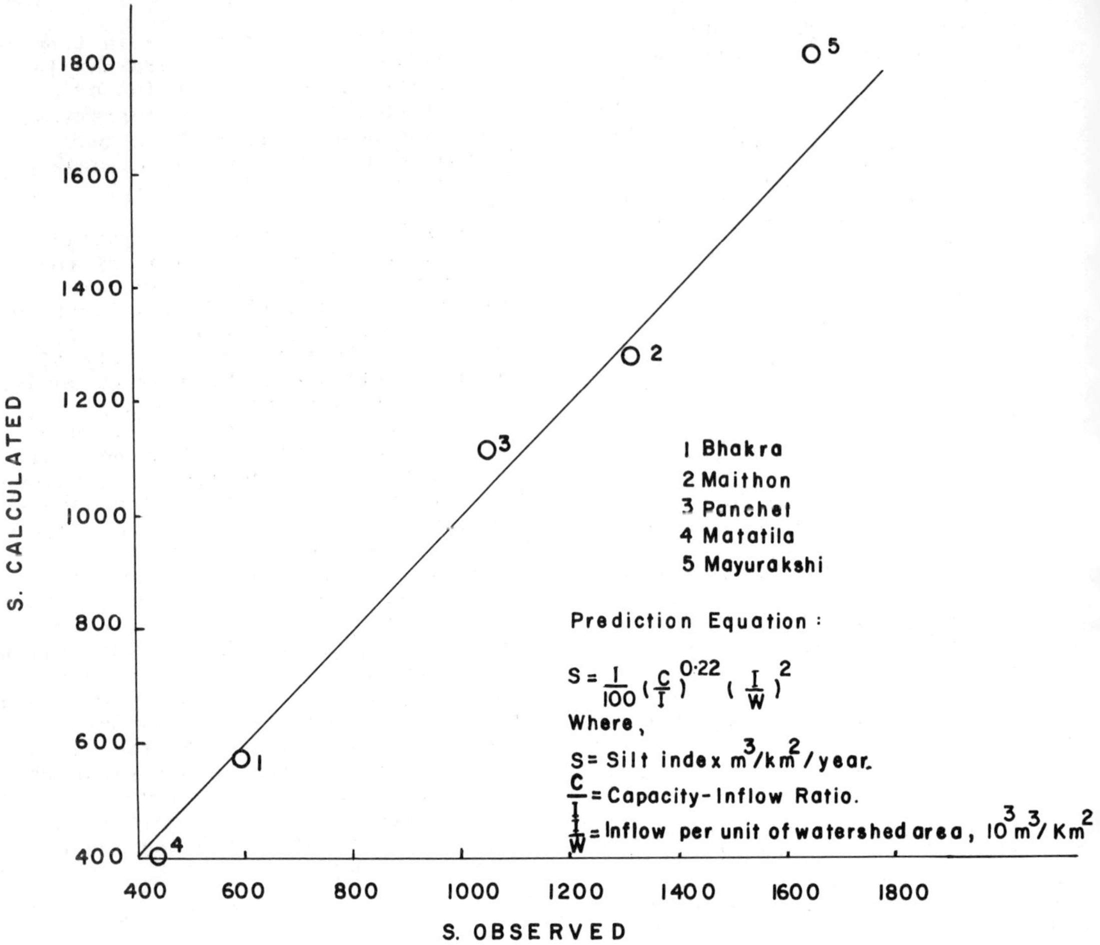

1800
1600
1400
1200
1000
800
600
400
S. CALCULATED
400
600
800
1000
1200
1400
1600
1800
S. OBSERVED
1 Bhakra
2 Maithon
3 Panchet
4 Matatila
5 Mayurakshi
Prediction Equation :
$S = \frac{1}{100} \left(\frac{C}{I}\right)^{0.22} \left(\frac{I}{W}\right)^{2}$
Where,
S = Silt index $m^3/km^2$/year.
$\frac{C}{I}$ = Capacity-Inflow Ratio.
$\frac{I}{W}$ = Inflow per unit of watershed area, $10^3 m^3/Km^2$

resulting three linear equations derived on the principle of least squares (V.Vevjevich, 1964).

Relevant data from reservoir surveys are given in Table-1. Solving these equations, we get

$$B_1 = -2 \; ; \; B_2 = 0.22 \; ; \; B_3 = 2$$

The prediction equation comes out to be $S = \frac{1}{100} (C/I)^{0.22} (I/W)^2$ (10)

Fig.I gives a plot of the S -Values as observed from sedimentation surveys against the S-values worked back from equation (10. The variation in the two values is greatest in the case of Mayurakshi (1650 observed Vs1810 calculated back i.e. .9.7%); it is of interest to note that in Mayurakshi the survey was a contour survey while in all others, it was the 'range-survey'

## Conclusion:

Unfortunately, there are no other reservoirs surveyed for sedimentation in this area, which could provide an independent 'test-set' for the prediction equation. It is realized that the equation will gain in 'Universality' of application if a term representing the sediment - production in the catchment could be incorporated in it. However, it is believed that it will give a better estimate of expected rate of sedimentation in reservoirs, particularly in the Northern India, than is being presently done. The river-valley projects in the country are designed taking an estimated siltation rate of 75 to 90 Ac. ft. per 100 sq. miles (375 to 430) $m^3/Km^2$), for major catchments over 2600 sq. km. However, actual sedimentation rates in many cases have been many times higher, and equation (10) will prove helpful in more correctly estimating the **sedimentation rates in reservoirs.**

## References


Banerji, S & V.B.Lal; 1971/1974. Silting of reservoirs: Indian data and the needed direction of efforts. Proc. of the Indian National Science Academy, Vol. 40. Part-A, Nos. 5&6, P.356-365, New Delhi

Murthy, B.N. 1968. Capacity Survey of Storage Reservoirs, Central Board of Irrigation and Power, Publication No.89, New Delhi.

Murthy, B.N. 1974. A Note on Estimated and Actual Sediment Load into Reservoir, Damodar Valley Corporation, Maithon Dam.

Murty, B.N. 1975. Reservoir Sedimentation in the country. Damodar Valley Corporation(Hydraulic Data) Maithon Dam.(Unpublished report).

Yevjevich, V; Satistical and Probability Analysis of Hydrological Data Part-II. Regression and Correlation Analysis. In : Chow, V.T.(ed.) 1964. Hand book of Applied Hydrology, Mc Graw Hill Publishers, New York, P-8-60.

Bhakra Management Board, 1975, Sedimentation Studies of Bhakra Reservoir 1973-74 (Un published Report).

Damodar Valley Corporation, 1973. Sedimentation Survey of Maithon Reservoirs, Annual Review - Maithon (Unpublished Report

Damodar Valley Corporation, 1975. The Fourth Sedimentation Survey of Panchet Hill Reservoir, Maithon Dam (Unpublished Report).

Irrigation Research Institute, U.P., 1973. Sedimentation Studies at Mata Tila Reservoir, Roorkee (Unpublished Report).

River Research Institute, West Bengal, 1974. Sedimentation Studies of Mayurakshi Reservoir, 1954-70, Calcutta, (Unpublished Report).


------


Key words: Indian reservoirs
Sedimentation
Silt index
Prediction equation

# TABLE - I

| Sl. No. | Name of Reservoir | Catchment area in $Km^2$ | Capacity $\frac{C}{\text{Inflow}}$ $\overline{I}$ | Inflow/Water-shed, $\frac{I}{W}$ $10^3 m^3/Km^2$ | Silt Index, $Sm^3/Km^2/year$ | S.Worked back from the Eqn. |
|---|---|---|---|---|---|---|
| 1 | 2 | 3 | 4 | 5 | 6 | 7 |
| 1. | Bhakra | 57000 | 0.7 | 247.8 | 596.6 | 576 (3.0) |
| 2. | Maithon | 6290 | 0.6 | 372.6 | 1310.0 | 1280(2.3) |
| 3. | Panchet | 10950 | 0.36 | 375.8 | 1049.0 | 1116(6.4) |
| 4. | Mata Tila | 20750 | 0.2 | 238.2 | 438.5 | 405 (8.3) |
| 5. | Mayurakshi | 1860 | 0.70 | 442.0 | 1650.0 | 1810(9.7) |

N.B:-

1. : Figures for Panchet reflect results upto the III Survey, before the Tenughat dam 120 km. upstream of Panchet, came into operation.

2. : All reservoirs belong to the 'normally moderate to considerable draw-down' operational category.

3. : Reservoir at Serial 5 is covered by contour-survey; all others have 'range-surveys'.

4. : Figures with brackets in col. 7, represent percentage variation from col. 6.

LAKES YAXHA AND SACNAB, PETEN, GUATEMALA: PLANKTONIC FOSSILS AND SEDIMENT FOCUSING

E. S. Deevey, Hague Vaughan, and G. B. Deevey

Florida State Museum, University of Florida, Gainesville, Florida, U. S. A.

Summary

Modern sedimentation rates, needed for budgeting present-day throughflow of nutrients in 2 lacustrine systems formerly influenced to different degrees by Maya urbanization, were measured in sediment traps and checked against plankton studies giving estimated rates of production of carbonaceous sediment and of 3 types of planktogenic fossils. The conical or hyperconical basins, overdeepened by solution in karsted limestone along fault scarps, are conducive to strong sediment focusing, apparently demonstrated in the deeper lake (Yaxha, $z_{max}$27 m, $\bar{z}$ 7.4 m) by enhanced accumulation in the deeper of 2 sets of traps. The other lake (Sacnab, $z_{max}$13 m, $\bar{z}$ 6.5 m) is shallower, and traps were recovered from it at only one (deep) station.

The tropical limnoclimate also favors strong diagenesis, to which the alga Botryococcus, the cladoceran Eubosmina tubicen, and the pelagic ostracod Cypria petenensis are differentially subject in the 2 lakes. Microfossil concentrations per unit organic matter are similar between trap and uppermost core sediments and between lakes, the minor dif-

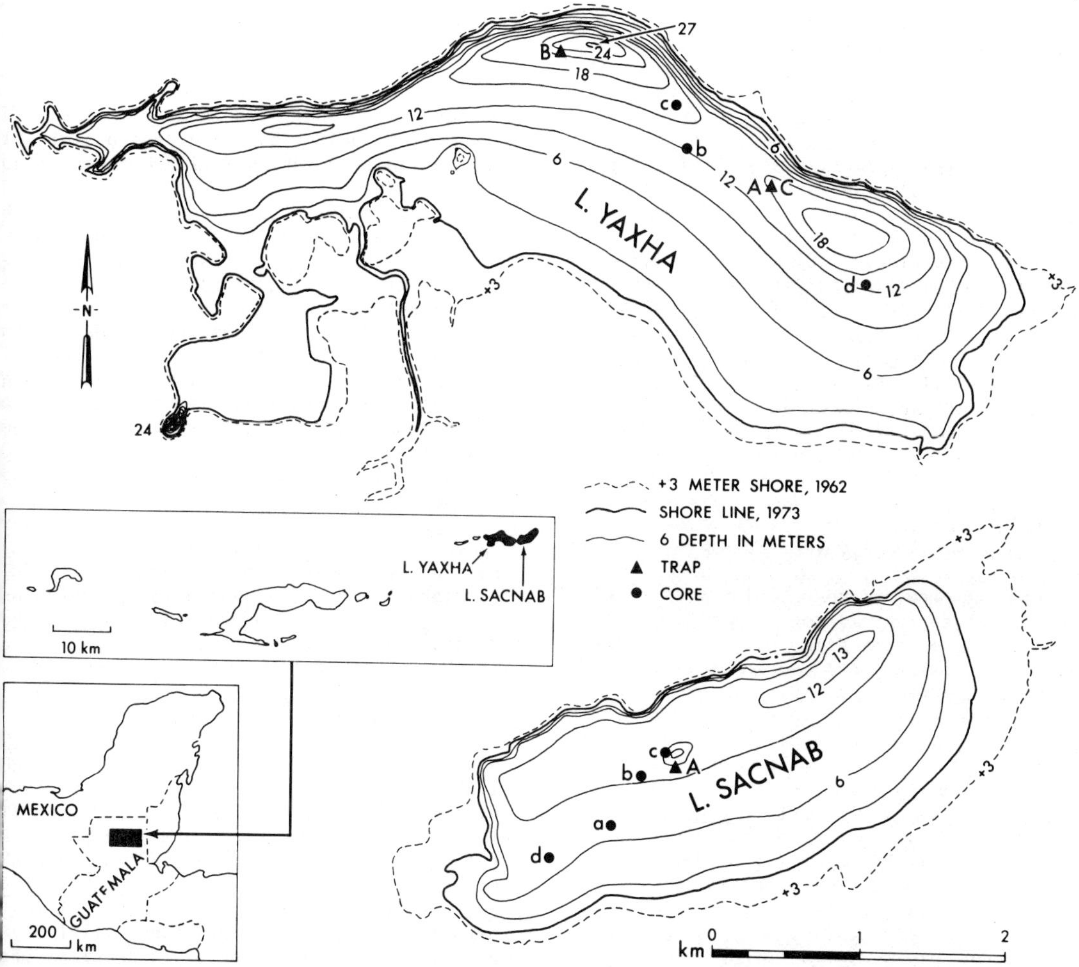


'ig. 1. Bathymetric maps of Lakes Yaxha and Sacnab, locating traps and short cores. Bathymetry done by H. K. Brooks.

ferences suggesting that loss of Botryococcus is more nearly complete in the deeper sediments of Yaxha, while Bosmina may be preferentially lost in Sacnab. These results are insecure, because horizontal transport and diagenesis can work against each other; relative enrichment of Bosmina in Yaxha sediments, implied by core-to-core differences and by relative scarcity of Bosmina in Yaxha plankton, probably reflects both differential transport in Yaxha and relative depletion of Bosmina in Sacnab.

More serious ambiguities arise when inputs, supposedly measured in traps, are divided by standing crops to give apparent turnover rates; except perhaps for Botryococcus in Yaxha, all turnover rates are impossibly high, and a 5-fold enhancement of inputs in Sacnab, though directly observed in the traps, is entirely unsupported by measurements of $O_2$ production and of zooplankton populations. Evidently all trap accumulations, and especially those in Sacnab, are grossly distorted by horizontal transport. Rejecting such accumulation rates, and pending restudy of the problem by other methods, we compute from internally consistent concentrations of carbon and microfossils in core sediments that fresh sediment forms in both lakes at similar rates, $\gtrsim$ 1 $cm^3$ per year (10 $mm.cm^{-2}.yr^{-1}$).

## Introduction

Similar and adjacent lakes, Yaxha and Sacnab, 17° 04' N Lat, 89° 23' W Long, alt 155 m, were chosen for a paired-comparison study of effects of millennium-long agroengineering and urbanization in the forested Classic Maya heartland. In the combined basin of 70 $km^2$ (arbitrarily defined by the 200-m contour) within which population densities were archaeologically estimated (Rice, 1977a, 1977b; Rice & Rice, ms) the 2 lakes (at high 1962 levels) occupy 8.9 + 5.2 $km^2$ or 20% of the area. The Yaxha sector was more than twice as heavily settled as the Sacnab sector (84.6 vs 40.8 mounds per $km^2$, maximum nonurban population 322 vs 129 persons per $km^2$). In addition, the Yaxha sector contained a large Classic (200 B.C.-A.D. 900) and a smaller Postclassic (ca. A.D. 1300) urban center (Bullard, 1970; Hellmuth, 1972). The region is densely forested and essentially uninhabited today.

Borings, apparently complete but no older than ca. 3000 yr (Yaxha) or ca. 8000 yr (Sacnab) were made in 1973. Radiocarbon dates are not yet available. Provisional pollen and microfossil stratigraphy was established by Habib Yezdani (Queensland Institute of Technology, Brisbane) and is being refined for both lakes by H. V. (pollen) and Mark Brenner (animals). About one-third of the chemical analyses of the paleolimnologic record are complete. Their quantitative interpretation requires evaluation of rates of sedimentation. Recent rates, as measured in sediment traps and short cores, are considered in this paper.

Our studies have been supported by the National Science Foundation (BMS 72-01859) and have been assisted by many friends and public agencies in Guatemala and Florida. A grant (no. 6932) to E.S.D. from the American Philosophical Society was of assistance in completing field work during hydrologic year 1973-74.

## Comparative basin morphology

Most lake basins in the karsted and fault-wrinkled Tertiary limestone of central Peten are asymmetric: steep N shores are fault scarps having relief of 20-50 m, with trenches, evidently overdeepened by solution at their feet. Despite strongly seasonal rainfall of 1500-2500 mm, surface drainage in the lake district is interior, and any outlets are subterranean. Yaxha (maximum depth 27 m) and Sacnab (maximum depth 13 m) basins are sealed by several m of unfossiliferous clay, deposition of which was sharply accelerated by Maya land clearance and urban construction. Recent sediments are more organic, both in sediment traps and in cores, with 20-26% (Yaxha) or 40-50% (Sacnab loss on ignition. Bathymetric maps are show in Figure 1.

This morphometry, in Yaxha at least, shoul be conducive to strong focusing of sediment into trenches. Several S-N fathometer trace are convex upward, and Figure 2, comparing hypsometric properties of Yaxha and Sacnab with 4 of Lehman's most divergent models of basin shape (Lehman, 1975), confirms the hyperconical (flaring) form of Yaxha's basin In the shallower Lake Sacnab, although the form becomes conical only as the indistinct trench is approached, significant nonuniformity of sedimentation accumulation might als be expected.

## Production and sedimentation

Sediment traps were quadruple sets of open Lucite tubes, 11.34 $cm^2$ in area, 50 cm high, suspended 1 m off bottom for 67-268 days. Several sets were lost. Table 1 shows accumulation in Yaxha to be nonuniform, organic and bulk sediment being concentrated 3.5-3.8 times in traps at the deeper station. As th only Sacnab traps recovered were near the trench, even greater accumulation there is not demonstrably due to sediment focusing.

Computing mean organic sedimentation in each lake as mg $C.cm^{-2}.yr^{-1}$, for comparison with phytoplankton production (Table 2), magnifies the excessive accumulation as measured in Sacnab. Production of $O_2$ in the 2 lakes is very similar, probably being a little higher in Yaxha. Even maximal figures (by the diurnal-curve method, considered less reliable because applied on a single date) fall short of the measured sediment production, and fail by a large margin to account for organic deposition in Sacnab.

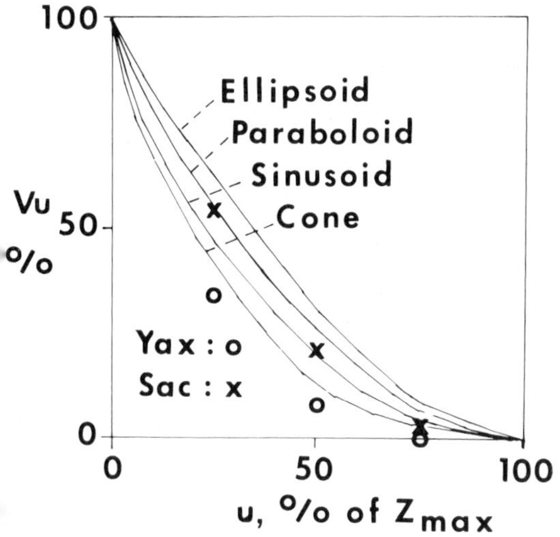


Fig. 2. Morphometry of Lakes Yaxha and Sacnab, compared to 4 theoretical models devised by J.T. Lehman (1975). $V_u$ is volume below depth (u), as % of total volume; depth u is expressed as % of maximum depth $z_{max}$.

Table 1. Organic and bulk deposition in sediment traps, $mg.cm^{-2}.yr^{-1}$.

| Station | | Organic | Total |
|---|---|---|---|
| Yaxha A/C | 16 m | 66.8 ± 31 | 273.2 ± 148 |
| B | 24 m | 233.7 ± 83 | 1032.9 ± 348 |
| Sacnab A | 13 m | 815.6 ± 412 | 1819.2 ± 858 |

Table 2. Production and sedimentation.

| | Daily rate, mg $C.cm^{-2}.day^{-1}$ | Annual rate, mg $C.cm^{-2}.yr^{-1}$ |
|---|---|---|
| Dark-and-light bottles: | | |
| Yaxha | 0.0261 | 9.53 |
| Sacnab | 0.0223 | 8.14 |
| Diurnal $O_2$ curve: | | |
| Yaxha | 0.099 | 36.16 |
| Sacnab | 0.145 | 52.94 |
| Organic inputs, sediment traps: | | |
| Yaxha | | 51.00 |
| Sacnab | | 326.25 |

In contrast to accumulation in traps, fresh sediment at the tops of short cores is notably homogeneous, both with respect to depth and from lake to lake. Organic content of a wet ml of core sediment ranges from 8.2 to 10.8 mg C per $cm^3$, averaging a little higher (10.18 mg C) in 3 cores from Yaxha than in 4 cores from Sacnab (8.73 mg $C.cm^{-3}$). If similar sediment is formed in both lakes, estimates of carbon under 1 $cm^2$ of mud-water interface, in short cores only (8-11 mg C per $cm^3$), can be divided by production of carbon in dark-and-light bottles (8-10 mg $C.cm^{-2}.yr^{-1}$). Formation of 1 $cm^3$ of sediment, or 10 mm per $cm^2$, then requires ca. 1 yr. Faster sedimentation is possible if diurnal-curve data are used, and is likely if pondweeds and/or allochthonous sources contribute carbon to sediments.

Station-to-station differences in accumulation, observed only in traps, are possibly important only in Yaxha; the Sacnab anomaly may need a different explanation. One possible cause--substantially higher proportions of pondweed to planktogenic detritus in the shallower lake--is not confirmed by microscopic examination. Hoping to explain the anomaly, and at the same time to estimate the magnitude of net horizontal transport of sediments in Yaxha, we looked more systematically at production and deposition of fossilizable plankton.

## Plankton composition

Nine complete vertical series of Schindler-trap samples, spaced through hydrologic year 1973-74, have been counted by G.B.D. (larger animals) and E.S.D. (algae and rotifers). Five series were taken in Yaxha; as sampling was missed in Sacnab in August 1973, when a dry-season minimum occurred in Yaxha, August samples have been omitted in computing annual means. Qualitative seasonal changes are unimportant. Quantitative differences between the lakes (Table 3) are also minor; however, Sacnab's relative richness in colonial algae, all with gelatinous sheaths, in diatom-poor plankton dominated by Botryococcus, may account for the initially higher organic content of Sacnab sediments.

For this study we concentrate attention on 3 fossilizing species. The alga Botryococcus (cf. B. Braunii) dominates the phytoplankton throughout the year, and is also conspicuous in sediments. The cladoceran Bosmina is Eubosmina tubicen (Brehm), first reported (Brehm, 1939) from Lake Peten Itza, now known (Deevey & Deevey, 1971) to be common in the American tropics. Most unusual is the pelagic, apparently endemic ostracod Cypria petenensis (Brehm), redescribed by Ferguson et al. (1964). Valves of this animal have not been seen as fossils, but large body fragments and appendages are common. Estimates of natality, mortality, and instar number, intended to provide independent measures of rates of production of fossil Bosmina and Cypria, are being undertaken by John Cooley (Canada Centre for Inland Waters, Burlington), but are not yet available.

In (arithmetic) mean standing populations, Cypria is more abundant than Bosmina, and Yaxha plankton has more of both than does Sacnab plankton. Cypria is more abundant than Bosmina at all depths, and relatively more so in deeper water, of which Yaxha (mean depth 7.4 m) has a larger volume than Sacnab (mean depth 6.5 m). Seasonal occurrence of Cypria is also more uniform. As a result, Bosmina populations weighted by water volume and averaged over the year (Table 3) are reversed in rank, Sacnab having more Bosmina under unit area than Yaxha.

Table 3. Composition of plankton, Yaxha and Sacnab.

| | Yaxha | Sacnab |
|---|---|---|
| Arithmetic mean, no. per liter: | | |
| Botryococcus | 671.66 | 725.35 |
| colonial algae | 635.93 | 1488.12 |
| Staurastrum | 6.27 | 100.47 |
| Pediastrum | 2.29 | 2.32 |
| Keratella | 21.47 | 70.40 |
| Ptygura | --- | 62.16 |
| Filinia | 1.68 | 0.93 |
| Bosmina | 3.42 | 2.93 |
| Cypria | 7.89 | 5.32 |
| calanoid copepods | 33.32 | 21.34 |
| cyclopoid copepods | 35.36 | 40.20 |
| nauplii (GBD) | 101.44 | 69.82 |
| Weighted mean crops, per $cm^2$: | | |
| Bosmina | 0.6480 | 2.0044 |
| Cypria | 4.9899 | 2.7611 |
| Botryococcus | 552.231 | 465.408 |

Sediment traps: microfossil inputs

Volumes of fresh sediment taken for microfossil counting, and processed with hot 10% KOH, are adjusted and diluted to yield single-drop (19.1 μl) preparations containing several tens of identifiable algae and cladoceran carapaces and 10 or more ostracod fragments. For core sediments the usual volume is 0.455 ml, but contents of traps were suspended in appropriate amounts of water before subsamples of ca. 10% were withdrawn. In Table 4 mean abundances of the commonest microfossils are arranged by depth of station, 6 traps at Yaxha station A/C, 4 at station B, and 4 in Sacnab being averaged without regard to time and duration of exposure. Standard deviations of such means (not shown in Table 4, but given for core samples in Table 5), though smaller than those of plankton counts, are rather large, the coefficient of variation being 35-50%.

Differences within Yaxha are those expected if sediment is being focused into the trench. Not only are inputs of microfossils higher at the deeper station, but concentrations per unit organic matter, for all types except the statistically unreliable Chaoborus, are also higher there. Comparing Sacnab inputs with maximal inputs in Yaxha, Table 1 shows bulk and organic sediment to accumulate faster by 1.76 and 3.5 times, respectively. Inputs of microfossils are even higher, by 5.65 times for Bosmina, 8.72 times for Cypria, and 5.11 times for Botryococcus. Although the Sacnab traps were near the maximum depth, differences between lakes seem too large to be explained by sediment focusing.

Table 4. Abundance of microfossils in sediment traps.

| Station: | YAX A/C | YAX B | SAC |
|---|---|---|---|
| Depth: | 16 m | 24 m | 13 m |
| Number per $mg_{org}$: | | | |
| Bosmina (1/2's) | 55.82 | 95.85 | 202.93 |
| Cypria | 11.89 | 17.70 | 58.02 |
| Chydorids (1/2's) | 40.10 | 63.99 | 248.88 |
| Botryococcus | 176.36 | 269.59 | 497.06 |
| Chaoborus | 9.89 | 5.65 | 11.62 |
| Rates of settlement, per $cm^2$.yr: | | | |
| Bosmina | 1637 | 11504 | 65013 |
| Cypria | 624 | 4261 | 37142 |
| Botryococcus | 10044 | 61470 | 313925 |

Just as production of $O_2$ fails to explain enhanced production of organic sediment in Sacnab (Table 2), relative abundance of Cypria and Botryococcus in mean plankton crops (Table 3) suggests no source for enrichment of these types in Sacnab sediments; while alive, both types are commoner in Yaxha. After weighting, however, Sacnab shows a larger mean Bosmina crop than Yaxha. If more Bosmina than Cypria fossils are moulted skins, as is likely for any cladoceran as compared with an ostracod, we can explain why Bosmina fossils are invariably commoner than Cypria fossils. Excess production of Bosmina moults will not explain relative enrichment in Sacnab, however, for the excess of Bosmina fossils, relative to Cypria, is actually greater in Yaxha

(Tables 5, 6).

Excessive accumulation in Sacnab traps has been shown for 3 types of planktogenic fossils as well as for bulk and organic sediment; but Yaxha traps also accumulate excessive numbers of fossils. Production of fossils was thought to have been directly measured, after estimation of plankton crops, by calculating rates at which crops of each fossilizing plankter are replaced. Such rates should be higher for algae than for crustaceans. Essential data are given in lower portions of Tables 3 and 4. In Sacnab, the number of Bosmina under 1 $cm^2$ turns over 32,429 times per year; Cypria replaces itself 13,437 times per year; Botryococcus turns over 674 times. Minimum figures for Yaxha are only a little less grotesque: 2526 Bosmina crops, 125 Cypria crops, and 18 Botryococcus crops theoretically replace each other in one year. Only the last of these estimates can be regarded as reasonable, others in both lakes being grossly distorted.

Our sediment traps differed in no important way from traps used successfully in many temperate lakes. Whether because of horizontal movement and focusing of sediment in these peculiar basins, or for some other reason, inputs measured in traps are plainly not to be trusted in these lakes.

Table 5. Abundance of microfossils in cores, per $mg_{org}$.

| | Yaxha | Sacnab |
|---|---|---|
| No. of samples | 9 | 16 |
| Bosmina (1/2's) | 287.28 | 118.06 |
| σ | ± 261.72 | ± 43.56 |
| Cypria | 22.09 | 43.06 |
| σ | ± 13.89 | ± 23.47 |
| Chydorids (1/2's) | 139.10 | 120.77 |
| σ | ± 92.50 | ± 49.52 |
| Botryococcus | 214.55 | 238.18 |
| σ | ± 114.36 | ± 77.30 |
| Chaoborus | 7.80 | 9.30 |
| σ | ± 6.01 | ± 5.22 |

## Core sediments: microfossils

Microfossils were counted at several levels in 3 short cores at the mud-water interface in Yaxha and 4 in Sacnab. Wet, dry, and ignited weights on all trap and core sediments, by H.V., show fresh Sacnab muds to be appreciably more organic than Yaxha muds. In Sacnab cores, but not in Yaxha cores, loss on ignition and water content diminish significantly downward, materials below ca. 50 cm becoming very similar to Yaxha sediments, which lose some water but little or no organic matter with age. Aware of probable diagenetic changes in Sacnab, we confined this study to microfossil analyses above 20 cm, in sediments at most a few years old. No diagenetic changes were detected in the microfossils themselves: analysis of variance shows that some core-to-core differences, e.g. for Bosmina and Botryococcus in both lakes, are probably significant, but depth variance within cores is only slightly larger than residual variance. Error terms are uncomfortably large, as shown by standard deviations in Table 5, but do not affect the important differences between lakes.

Ignoring core-to-core differences and considering mean abundance per unit organic matter only, planktogenic microfossils are very similar in composition in the 2 lakes (Table 5), and the assemblage in core sediments is also very similar to that in trap sediments (Table 4). The main differences concern Bosmina, the commonest microfossil in Yaxha cores. Not only are its half-carapaces more concentrated in Yaxha than in Sacnab core samples, reversing the order in trap samples, but Bosmina is enriched in Yaxha cores relative both to Cypria and to Botryococcus. Bosmina:Cypria ratios vary from core to core, as discussed in the next section. When the ratio of Botryococcus to "total herbivores" is computed, by combining half-carapaces of Bosmina and chydorids as whole exoskeletons and adding major fragments of Cypria, the Botryococcus:herbivore ratio per $mg_{org}$ is 0.966 in Yaxha cores and 1.466 in Sacnab cores. The difference is unchanged when ratios are expressed per unit dry sediment or per ml of fresh sediment, and we judge it to be significant.

Phytoplankton:zooplankton ratios are usually low in thanatocoenoses, and ratios of ca. 100 to 1, comparable to ratios in living plankton (Table 3), are not expected among fossils. In Yaxha cores, however, the alga:herbivore ratio as computed above is ca. one-third lower than in Sacnab cores, whereas in trap sediments (Table 4) it is at least one-third higher (2.95 at Yaxha station A/C, 2.76 at station B, 1.75 in Sacnab). The relative rankings imply substantial loss of Botryococcus during fossilization, incipient in the comparatively shallow traps in Sacnab, observable in the deeper of the 2 trap stations in Yaxha, and more nearly complete in the Yaxha cores, all of which were deeper than any Sacnab cores.

## Diagenesis: Bosmina/Cypria ratios

In living plankton, after weighted mean crops are computed, Sacnab has more Bosmina than Yaxha, and several times more Bosmina relative to Cypria (Table 6). In trap and core sediments the relation between lakes is reversed: although counts of Bosmina half-carapaces (Tables 4 and 5) are divided by 2 to obtain numbers of whole animals, Bosmina: Cypria ratios are higher, occasionally much higher, in Yaxha. In contrast to temperate lakes, where loss of Bosmina exoskeletons has not been suspected, tropical limnoclimates may favor postdepositional losses both of Bosmina and of ostracods; but microscopic examination has yielded no evidence of enzymatic attack on animal remains, other than Cypria valves. Relative enrichment of fossil Bosmina in Yaxha sediments was initially attributed to winnowing as in Linsley Pond (Deevey, 1942; Vallentyne & Swabey, 1955). As winnowing separates lighter anterior from heavier posterior half-carapaces, but has not done so in Yaxha or Sacnab, apparent enrichment probably reflects relatively intense loss of Bosmina in sediments of shallower Lake Sacnab.

Table 6. Bosmina:Cypria ratios in plankton and in sediments of traps and cores.

| | | No. Bosmina | B/C ratio |
|---|---|---|---|
| Plankton, weighted mean crops per $cm^2$: | | | |
| Yaxha | | 0.648 | 0.1299 |
| Sacnab | | 2.0044 | 0.7259 |
| Trap sediments, per $mg_{org}$: | | | |
| Yaxha | 16 m | 27.94 | 2.348 |
| Yaxha | 24 m | 47.93 | 2.708 |
| Sacnab | | 101.47 | 1.749 |
| Core sediments, per $mg_{org}$: | | | |
| Yaxha | b, 12.5 m | 24.97 | 2.23 |
| | c, 16.5 m | 141.82 | 5.69 |
| | d, 12.5 m | 264.15 | 8.76 |
| Sacnab | a, 8.0 m | 52.85 | 1.32 |
| | b, 11.0 m | 58.14 | 1.05 |
| | c, 12.2 m | 56.27 | 1.24 |
| | d, 7.5 m | 31.74 | 1.45 |

Taking data from Yaxha trap station A/C as probably least affected by horizontal transport, Bosmina:Cypria ratios in apparent turnover rates are related to Bosmina:Cypria ratios in Yaxha plankton as 20.21 to 1. A similar ratio (20.79 to 1) applies to the deeper Yaxha trap station B. Gross distortion of apparent turnover rates does not affect these figures, for high Bosmina: Cypria ratios (18 or 20 to 1), relative to the ratio in plankton, are also found in Yaxha trap sediments when normalized per unit of organic matter. In Sacnab, however, fossil Bosmina is only 2.4 times as abundant, relative to Cypria in plankton. The ratio is the same, whether computed from apparent turnover rates or on the basis of unit organic matter in trap sediments. Even if Bosmina is concentrated by winnowing in Yaxha, and/or if Cypria is also subject to diagenetic loss, Bosmina fossils are evidently being lost from Sacnab.

In core sediments, Bosmina:Cypria ratios are also much higher in Yaxha than in Sacnab, and support the same inferences as do trap sediments. Using mean abundances per $mg_{org}$ in cores (Table 5), and again dividing half-carapaces by 2, Bosmina:Cypria ratios are 6.05 in Yaxha and 1.37 in Sacnab. Nearly identical values are found when microfossil data are expressed per wet ml or per mg of bulk sediment. Comparing core with trap sediments, the lake-to-lake disparity, or relative enrichment of Bosmina over Cypria in Yaxha, appears to be smaller in cores than in traps. Unfortunately, such a comparison, suggesting diagenetic loss of Bosmina in Yaxha, is invalidated by the high core-to-core variance in Yaxha as compared to Sacnab. In Table 6, where Bosmina:Cypria ratios are given for individual cores, numerical values of the ratio are highest in cores where Bosmina is most abundant. If real, the positive correlation between relative and absolute abundance of Bosmina in cores agrees with enhanced abundance in the deeper Yaxha traps, and may perhaps be attributed to winnowing.

Considering relative abundance in sediments only, without regard to planktonic abundance or to inputs in traps, the Bosmina:Cypria ratio in Sacnab falls from 1.75 in traps to a homogeneous mean of 1.37 in cores. In Yaxha, where Bosmina is least abundant (least concentrated by winnowing?) at trap station A/C and in core b, the minimal Bosmina:Cypria ratio also falls from 2.35 in trap sediments to 2.23 in core sediments. Although the latter difference obviously lacks statistical significance, core sediments must be a little older than trap sediments and seem in both lakes to have lost more Bosmina than Cypria.

## Conclusions: sedimentation rates

Modern sedimentation rates, the primary objective of this study, were to have been directly observed in traps, checked against rates of production of planktogenic fossils, and compared to net accumulation rates of older sediments in the age-range of $C^{14}$. Gross anomalies, visible when apparent turnover rates are computed from ratios of inputs to standing crops, prove that inputs have not been correctly measured in the traps. One explanation, focusing of sediment into deeper parts of these unusual basins, is supported by relative concentrations of microfossils at the deeper of 2 trap stations in Yaxha.

In Sacnab, where only 1 trap station provides no control over horizontal transport of sediment, remarkably high rates of accumulation, observed in traps, are confirmed and amplified by microfossil counts. Sacnab is shallower than Yaxha, and postdepositional loss of organic matter occurs, as it does not in Yaxha, in the upper 50 cm of core sediments. Substantial loss of the alga Botryococcus, relative to remains of Cladocera and ostracods, occurs in both lakes, and we infer that this loss is less complete in the less altered trap and uppermost core sediments of Sacnab. If Bosmina carapaces are also subject to diagenetic loss, this loss should also be less complete in Sacnab. In fact, relative to the proportion of Cypria in plankton and in sediments, Bosmina either suffers greater loss during fossilization in Sacnab or is somehow made superabundant in Yaxha. Anomalously high sedimentation rates in Sacnab are therefore not simply explicable as the result of less intense diagenesis; horizontal transport must also be important.

Local and lake-to-lake differences in sedimentation rate can plainly not be deduced from such data, and must now be approached by other methods, e.g. by microstratigraphy of radioisotopes or paleomagnetic properties in cores to be collected in 1977. Another objective of the 1973-74 study, estimation of the mean production of Bosmina fossils and of the mean instar number (number of exoskeletons per individual lifetime), must also be temporarily set aside. If we discard all measurements from sediment traps, however, microfossil counts in core sediments nevertheless permit order-of-magnitude estimates of rates of accumulation.

Yaxha core b is least enriched in Bosmina of the 3 cores, and most similar to Sacnab cores in Bosmina:Cypria ratio. It contains 662 whole Bosmina carapaces per wet ml. If the instar number is 10, and the Bosmina crop of 0.648 animals per $cm^2$ replaces itself 10 times per year, 10.2 yr are required to produce the number in 1 ml. But Bosmina fossils appear to be greatly enriched in Yaxha sediments, relative to Sacnab sediments and relative to Cypria fossils. If we derive a "Yaxha Bosmina-enrichment factor" of 46.8 from the excess input in Yaxha traps compared with Sacnab traps, the otherwise conservative assumptions yield a sedimentation rate of 4.6 $cm^3$ per year (and per $cm^2$).

For Cypria in the same Yaxha core, containing 303 fossils per wet ml, we can guess the instar number to be 5, and the crop of 4.99 animals per $cm^2$ to replace itself 10 times per year. But these fossils are not necessarily whole animals, and if there are 2 per animal per instar, the sedimentation rate is 1.65 $cm^3$ (16.5 mm) per year. For Botryococcus, assuming that the crop turns over 10 times per year (or 20 times but with 50% diagenetic loss), 2825 fossils per wet ml may settle at a bulk rate of 1.95 $cm^3$ per year and per $cm^2$.

Applying the same assumptions to the mean concentrations in 4 Sacnab cores (except that Bosmina is not assumed to be enriched), calculated sedimentation rates are slower, by 2-28 times, than those calculated for Yaxha. Unenriched Bosmina here gives the most aberrant figure, 6.24 years per ml or 1.6 $mm.yr^{-1}$. Cypria yields 1.49 $mm.yr^{-1}$ as the bulk rate, while the estimate from Botryococcus is most similar to that for Yaxha, 9.6 mm or slightly less than 1 $cm^3$ per year and per $cm^2$. If Bosmina (and Cypria) are more depleted by diagenetic loss in Sacnab than in Yaxha, the principal disparities are reduced. All figures are then of the order of 1 $cm^3$ per year and per $cm^2$, the highest estimate, for Bosmina in Yaxha, being the most uncertain because of probable distortions by sediment focusing.

Where probable errors of microfossil counting are large, and there are serious uncertainties as to magnitude of 2 dominant processes (diagenesis and sediment focusing may cancel each other to different degrees for particular microfossils in each lake), no confidence can be felt in such estimates of sedimentation rate. They are at least not in conflict with the deduction from carbon production (Table 2), that sediment forms in both lakes at similar rates, probably 1 $cm^3.yr^{-1}$ or higher. Data from sediment traps play no part in these conclusions, and are rejected except for the dim light they throw on differential diagenesis.

## References


Brehm, V., 1939. La fauna microscópica del Lago Petén, Guatemala. Escuela Nac. Cien. Biol., Anales 1:173-203.

Bullard, W. R., 1970. Topoxte, a Postclassic Maya site in Peten, Guatemala. In: Bullard, W. R. (Ed.): Monographs and papers in Maya archaeology. Peabody Mus. Archaeol. & Ethnol., Harvard Univ. Papers 61:245-307.

Deevey, E. S., 1942. Studies on Connecticut lake sediments. III. The biostratonomy of Linsley Pond. Amer. J. Sci. 240:233-264, 313-338.

Deevey, E. S. & G. B. Deevey, 1971. The American species of Eubosmina Seligo (Crustacea, Cladocera). Limnol. & Oceanogr. 16:201-218.

Ferguson, E., G. E. Hutchinson & C. E. Goulden, 1964. Cypria petenensis, a new name for the ostracod Cypria pelagica Brehm 1932. Peabody Mus. Nat. Hist., Yale Univ., Postilla 80:1-4.

Hellmuth, N. M., 1972. Excavations begin at Maya site in Guatemala. Archaeology 25: 148-149.

Lehman, J. T., 1975. Reconstructing the rate of accumulation of lake sediment:

the effect of sediment focusing. Quatern. Res. 5:541-550.
Rice, D. S., 1977a. Population growth and subsistence alternatives in a tropical lacustrine environment. Internat. Cong. Americanists, Paris 1976, Proc., in press.
Rice, D. S., 1977b. Middle Preclassic Maya settlement in the central Maya lowlands. Subm. to J. Field Archaeol.
Rice, D. S. & P. M. Rice, ms. The ceramic and non-ceramic artifacts of Yaxha-Sacnab, El Peten, Guatemala.
Vallentyne, J. R. & Y. S. Swabey, 1955. A reinvestigation of the history of Lower Linsley Pond, Connecticut. Amer. J. Sci. 253:313-340.

Key words: Sedimentation rate
Planktonic fossils
Nutrient budgets
Digenesis

THE USE OF DIATOM MICROFOSSILS IN INTERPRETING THE DISTRIBUTION OF PORE WATER SOLUTES IN SMALL MOMELA LAKE (TANZANIA)

R. E. Hecky

Fisheries and Environment Canada, Freshwater Institute, Winnipeg, Manitoba

Summary

Analysis of aquatic microfossils can provide essential information to improve interpretation of the chemical stratigraphy of sediments and interstitial waters. Because they are immobile once deposited and their information content is less reactive with other sediment parameters, they are independent indicators of what environmental conditions were when the sediments were deposited. They can demonstrate whether environmental conditions are in flux or whether they have been constant. This information can aid decisions as to whether observed concentration gradients are the result of steady state phenomena or transient conditions. The utility of diatom microfossils for this purpose is demonstrated with an example from an East African lake where changing salinity conditions in the lake and highly porous sediments have combined to produce very dynamic gradients in interstitial salts.

Introduction

Diffusion of solutes in sediments is demonstrable and quantifiable experimentally in the laboratory (Duursma and Hoede, 1967) or in the natural environment (Schindler et al., this volume). Therefore, when concentration gradients are observed in sediments it is possible to estimate instantaneous flux within sediments or between sediments and overlying water. To project instantaneous rates into the future, in order to predict their effect on the sediment-water system, a decision is required as to whether the observed gradient is the result of a steady-state or transient phenomenon. Usually an assumption is made because, except in the experimental situation, the history of observed concentration gradients is unknown, i.e. what the initial concentration differences were, the causes of the differences, and the time at which the differences were initiated. All too often, the gradients are viewed as static as if they did not change with time. This occurs because of the infrequency of observation, usually only once. This static view favors the choice of the steady-state assumption.

To interpret observed concentration gradients in sediments properly, their history should be defined as well as possible. An independent estimator of initial solute concentration in pore waters is required to define the history of the gradient and to demonstrate the effectiveness of diffusion in redistributing solutes within and out of the sedimentary column. The ideal recorder of initial conditions must be sensitive to the conditions, immobile in deposited sediment and relatively non-reactive with pore waters. Among the components of sediments, only microfossils have the potential to fulfill these criteria. In this paper, the utility of diatom microfossils in elucidating the history of concentration gradients and demonstrating the effectiveness of diffusion in redistributing solutes will be explored with an example from an East African lake.

Although diatom microfossils have, in the past, been recognized as potential repositories of information on chemical concentration of depositing waters, the information has been considered too qualitative to be useful in analytical studies. However, in East African alkaline, saline lakes the simplified diatom flora appears to be determined strictly by dissolved salt concentration with dominant species occurring within definable chemical ranges (Hecky and Kilham, 1973). Analysis of fossil diatom assemblages in an 8.4 m sediment core from Small Momela Lake allowed Hecky (1971) to reconstruct with some quantitative assurance the history of chemical concentrations in the lake.

Methods

The core was taken in one meter or shorter sections using a modified Livingstone sampler. The cores were shipped to Duke University (North Carolina, U.S.A.) where the aluminum sample tubes were milled open. Samples for water content, organic matter, carbonate content, and diatom analysis were taken every 10 cm. Water content was determined by weight loss at 110°C, and organic content of dry material was estimated by loss of weight on ignition at 550°C. Carbonate content was determined manometrically after adding 25% HCl to a known weight of dried sample. Interstitial water was obtained by squeezing sediment samples taken every 20 cm using an apparatus similar to that described by Reeburgh (1967). Methods of diatom preparation and counting and the analytical methods used for the chemical analyses of water are given in Hecky and Kilham (1973). Bulk density of the sediment was determined from the volume and weight gain of a 40 ml graduated centrifuge tube containing water when approximately 15 ml of sediment were added isothermally.

Small Momela Lake

Small Momela Lake is a small (59 ha) relatively shallow lake ($\bar{z}$ = 4 m and $z_{max}$ = 10 m) located in Northern Tanzania (3°14' S. by 36°54' E.). The mean annual air temperature is approximately 20°C, and the lake is holomictic throughout the year. Sodium and bicarbonate-carbonate account for 80% of the total concentration of cations and anions, on a meq $l^{-1}$ basis, in the modern lake (Table 1). The lake is presently fed by direct runoff, a small stream and a small spring while water leaves the lake via a stream. All identifiable inflows are more dilute than the lake.

Table 1. Major ion concentrations (meq $l^{-1}$) in Small Momela Lake and its tributaries.

| | Lake | Spring | Stream |
|---|---|---|---|
| Na | 54 | 18 | 39 |
| K | 5 | 1 | 3 |
| Ca | 0.04 | 0.05 | 0.14 |
| Mg | 0.15 | 0.04 | 0.17 |
| $SO_4$ | 6 | 4 | 2 |
| Cl | 3 | 1 | 3 |
| $HCO_3$ + $CO_3$ | 46 | 13 | 37 |
| F | 5.8 | 3.2 | 3.2 |

Sediment Composition

From 0-7.25 m the sediment is a uniform gyttja with the exception of a coarse sand layer at 2.25 m. From 7.25 m to 8.00 m there is a transition to fine volcanic ash which constitutes the lahar on which the lake sets. The core represents a complete stratigraphic section and an additional 0.4 m of the substratum was recovered.

Water, organic material and diatom frustules are the main components of the lacustrine sediments. Weathered volcanic ash, calcium carbonate, and authigenic zeolites (Hecky, 1971) account for less than 10% of the dry sediments. The low bulk density (average 1.1 g $cc^{-1}$) and high water content throughout the core (Fig. 1) except for the sand layer (2.25 m) and bottom of the core results from the loosely packed mixture of diatoms and organic material and the hydrophilic nature of the organic matter. Although the lake waters are in equilibrium with a calcium carbonate phase (Table 1), most of the carbonate deposition occurs before the water enters the lake. Therefore the carbonate content of the sediments is low (Fig. 1). The time scale in Fig. 1 is based on carbon dating of organic material and is corrected for historical atmospheric variation of the $C^{14}/C^{12}$ ratio (Stuiver, 1967). Details on the radiocarbon dates can be found in Hecky (1971).

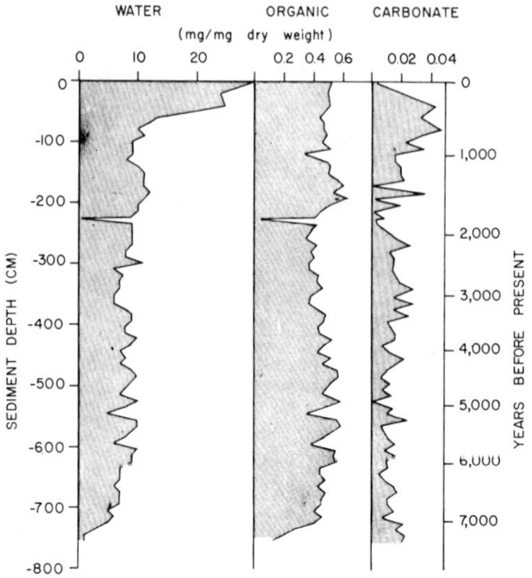


Figure 1. Vertical distribution of water, organic material and carbonate in Small Momela Lake sediments.

Diatoms

Diatom valves are the major inorganic component of the dry sediment. A diagram of the relative frequencies of the diatom species plotted against depth is given in Hecky and Kilham (1973). Four species Cyclotella meneghiniana Kutz., Coscinodiscus rudolfi Bachmann, Navicula elkab O. Mull., and Nitzchia frustulum (Kutz.) Grun., attained dominance during the existence of the lake. These species have been shown to have preferred ranges of ionic concentration in East African lakes (vide Fig. 4 of Hecky and Kilham, 1973 which relates species occurrence to total alkalinity). C. meneghiniana is found in more dilute waters, mean alkalinity of 15 meq $l^{-1}$, while N. frustulum prefers waters of greater than 80 meq $l^{-1}$, mean of 150 meq $l^{-1}$. C. rudolfi and N. elkab occupy intermediate concentration ranges on the alkalinity gradient.

Because these species appear to sort themselves along a concentration gradient and tentative ranges and means of preferred concentrations are available, it is possible to interpret the succession of assemblages in the Small Momela Lake core in terms of approximate salt concentration (Fig. 2). This is done by assigning a single concentration value, generally the midpoint of the preferred range of chemical concentration for the dominant species as reported by Hecky and Kilham (1973), to each fossil diatom assemblage. An exception to this scheme of interpretation for the depth interval -60 to

-210 cm is made in order to illustrate a case discussed below. The concentration shown for this interval in Fig. 2 is 110 meq $l^{-1}$ which is the maximum concentration observed in the interstitial waters. A concentration of 110 meq $l^{-1}$ is near the low end of the preferred range of N. frustulum the dominant diatom in this sediment interval. The midpoint of the range is 150 meq $l^{-1}$. The uncertainty of the most appropriate concentration for this time interval is implied by the question mark in Fig. 2.

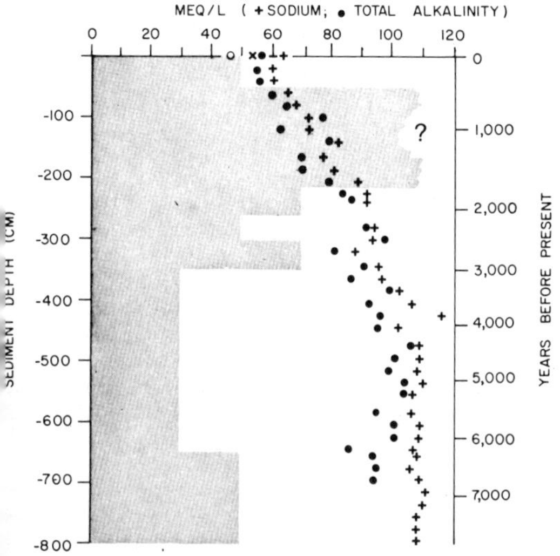


Figure 2. Vertical distribution of sodium and total alkalinity ($HCO_3^- + CO_3^{--}$) in the interstitial waters of Small Momela Lake sediments. Present lake concentrations of sodium (X) and total alkalinity (O) are shown on abscissa. Shaded area represents probable salinity history of the lake based on interpretation of diatoms in sediments.

## Interstitial Waters

The interstitial water concentrations of sodium and total alkalinity are shown in Fig. 2. As in the lake, these two ions accounted for over 80% of the total concentration of major ions in all interstitial water samples (complete analysis in Hecky 1971). Concentrations of these two ions are highly correlated in the core (r = 0.92) so that the description of one profile (sodium) applies to the other (total alkalinity). From 0 to -40 cm sodium concentrations are nearly identical. These near-surface sediments had very high water content (Fig. 1) and were fluid. The homogeneous interstitial sodium concentrations in this interval may have resulted from mixing in the core tube during transport. Such mixing would account for the interstitial sodium concentrations (Fig. 2). Below -40 cm, ion concentrations monotonically increase with sediment depth (considering likely sampling and analytical errors) to more than double the concentration of the modern lake at -450 cm. Between -450 cm and the bottom of the core concentrations are essentially constant. If the slope of the gradient between -40 cm and -140 cm were extended to the surface of the sediment, it would intersect the concentration of the modern lake. It is likely that the in situ gradient does extend to the surface, and this is assumed in the discussion. Changes in sediment composition at -225 cm and -725 cm have no visible effect on the local gradients of sodium.

## Discussion

The salient feature of Fig. 2 is the complete lack of agreement between measured interstitial ion concentrations, and the original ion concentrations of the waters depositing the sediments as reconstructed from fossil diatom assemblages. The reconstructed salinity history and its climatic implications are correlative with other East African lakes during the past 7,000 years (Richardson and Richardson, 1972) and the known recent geological history of the region. The East African climate was becoming drier from 5,000 to 2,000 yr B.P., and Small Momela Lake responded by becoming more saline. Approximately 1,900 yr B.P. a catastrophic flood (which deposited the sand layer in Small Momela Lake) resulted in a river capture which took away the lake's major tributary (Hecky, 1971) leaving it with only the local inflows it has now. The lake consequently entered its most saline phase, and it was a closed basin lake until 500 yr B.P. when it once again became an open drainage lake under a more humid climate. As the salinity history of Small Momela Lake given in Fig. 2 is consistent with both the ecology of the microfossils in the sediments and the regional climatic and geological history, explanation for the interstitial sodium profiles must be found.

Ion concentrations in interstitial waters can be affected by a variety of processes. Five processes or conditions which might produce the observed profile were considered: 1) microbial diagenesis, 2) compaction, 3) pore water mineral equilibria, 4) ground water percolation, and 5) diffusion. A number of microbial processes under both oxic and anoxic conditions can cause an increase in the alkalinity as diagenesis proceeds (Berner, 1971). However, none of these would concomitantly increase sodium in these well buffered alkaline solutions. Salt filtering caused by compaction of clays under high pressures deriving from deep burial can produce profiles qualitatively similar to the observed (Degens and Chilingar, 1967), but the required conditions of high clay content and high overburden pressure are not present in the Small Momela core.

The observed concentration profile in Fig. 2 cannot be the result of equilibrium reactions with a sodium carbonate mineral deposited at an earlier stage of the lake's history as the concentrations are well below saturation for even the most insoluble sodium carbonate mineral (Garrels and Christ, 1965), and no such minerals are observed or would be expected in the lower portions of the core. Saline groundwater percolating along an aquifer and up through the sediments might produce a similar profile. However, ground waters in the immediate vicinity are less saline than the lake (Table 1), and sediment composition which determines hydraulic conductivity is uniform between -400 and -500 cm where the observed concentrations decline from the constant values below -500 cm.

Diffusion appears to be the most likely explanation for the observed gradient. From the interpretation of the diatom assemblages, the lake is proposed to have had a variable salinity history which would have set up significant concentration differences between the sediments and the lake. The high porosity, 90%, (calculated from water content and density measurements) and high mean annual temperature 20°C, of Small Momela Lake sediments will maximize the diffusion coefficient. Relations between temperature, porosity, and D given by Manheim (1970) indicate that D should approach $2 \times 10^{-5} cm^2 sec^{-1}$ in Small Momela Lake sediments. Direct in situ measurements of D in lake sediments of similarly high organic and water content estimated a D of $1\text{-}2 \times 10^{-5} cm^2 sec^{-1}$ (Schindler et al., this volume). This value of D should be near the maximum attainable for molecular diffusion in interstitial waters as it is equivalent to D's observed in dilute, ionic solutions at similar temperatures (Berner, 1971).

If diffusion produced the observed profile, then it should be possible to demonstrate that a) enough sodium could enter the sediments during the lake's saline phase to raise the sodium concentrations to the values observed in the deepest sediments and b) that enough sodium could subsequently be removed from the sediments when the lake became dilute to produce the gradient observed in the upper sediments. The reconstruction of the actual shape of the gradient through time would be mathematically complex or impossible even if the exact salinity history of the lake were known: but, by making simplifying assumptions, the feasibility of diffusion fulfilling conditions a) and b) can be evaluated. Mathematically the simplest situation to consider is for a constant concentration difference and a fixed sediment depth. For the purpose of this analysis these boundary conditions are adequate even though they are artificial. The appropriate formula for these boundary conditions is given by Duursma and Hoede (1967),

$$1)\quad \frac{C_{(x,t)}}{C_o} = \text{erfc}\ \frac{x}{2\sqrt{Dt}}$$

where $C_o$ = initial difference in concentration between sediment and overlying water
$C_{(x,t)}$ = the concentration difference at depth, x, at time, t
D = diffusion coefficient
erfc is the error function.

In the solution of this equation for the cases to be considered, x, D, and t are considered to be known within satisfactory limits from the information available from analyses of the core or from the literature. The greatest uncertainty is in $C_o$ and the suitability of the boundary conditions.

To see if adequate sodium could reach the deepest sediments during the lake's most saline phase, 1800 yr to 600 yr B.P., the lake was assumed to have a mean concentration of 110 meq $l^{-1}$ for this period of time. This value is the maximum observed in the interstitial water. The value is consistent with the dominant diatom species in the sediments deposited during this time period, but the value is near the low end of the preferred concentration range of the species as mentioned above. During this period of time the lake was in a closed basin, and the sodium concentration of the lake should have been continuously increasing. Therefore the boundary condition is unrealistic, but the assumption of a mean concentration lower than what likely occurred should minimize transport of sodium into the sediment using equation 1). If adequate sodium is transported for the minimum condition then the feasibility of condition a) would be established. Fig. 3 shows that using the most likely value for the diffusion coefficient, $2 \times 10^{-5} cm^{-2} sec^{-1}$, sufficient sodium would enter the deepest sediments to yield concentrations of greater than 92 meq $l^{-1}$ which compares well with the observed concentration at the deepest depth, 110 meq $l^{-1}$. If a higher $C_o$ were assumed, there would be more than adequate sodium transported. Fig. 3 also shows that D values significantly less than $2 \times 10^{-5} cm^2 sec^{-1}$ would not be adequate.

Condition b), whether adequate sodium would be removed from the sediments when the lake became diluted, was tested in a similar fashion. The observed concentration gradient has nearly constant values below -450 cm depth. The lake is considered to have nearly constant concentration when it has open drainage. $C_o$ in equation 1) was determined by this difference. These boundary equations imply that 500 yr B.P. the interstitial sodium concentrations were uniform at 110 meq $l^{-1}$. The expected concentration

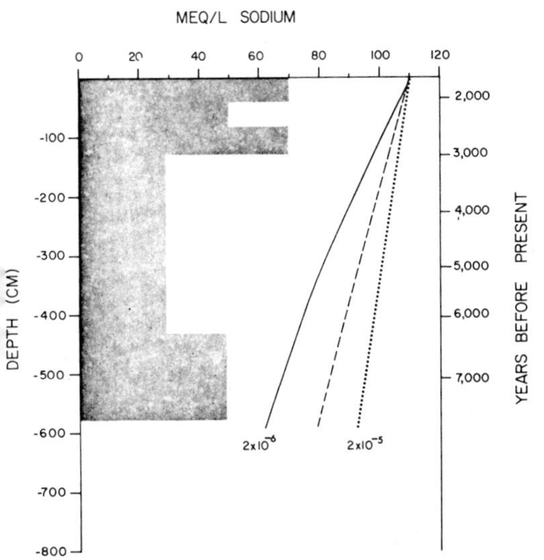


Fig. 3. Sodium distribution which could result in sediment depths existing 1,800 yr B.P. from the assumed sodium concentrations existing in the lake then and different diffusion coefficients acting over 2,600 yr. Shaded area as in Fig. 4. Solid line = $2x10^{-6}cm^2sec^{-1}$, dashed line = $6x10^{-6}cm^2sec^{-1}$, dotted line = $2.5x10^{-5}cm^2sec^{-1}$.

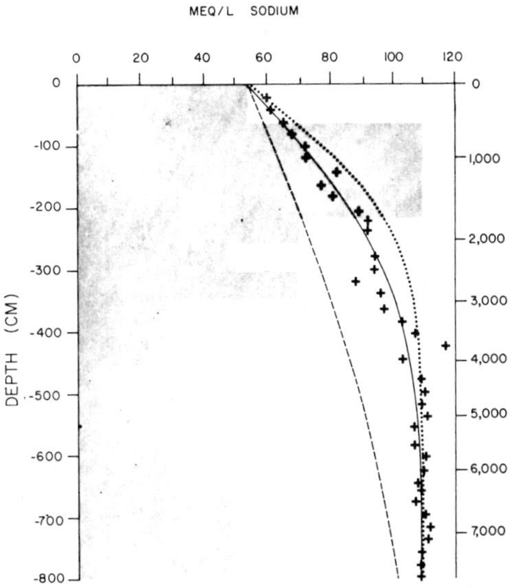


Fig. 4. Vertical distribution of sodium in interstitial waters (+) and reconstructed history (shaded) as in Fig. 2. Lines represent different distributions of sodium expected with different values for the diffusion coefficient: $D = 2x10^{-5}cm^2sec^{-1}$ (dashed), $D = 2x10^{-6}cm^2sec^{-1}$ (solid), and $D = 1x10^{-6}cm^2sec^{-1}$ (dotted).

profile after 500 years of these conditions is given by Fig. 4. If these conditions were true, sodium concentrations in interstitial waters should be much less than observed. This means that the initial concentration difference between interstitial waters in sediment and overlying waters was much greater than assumed. The microfossil assemblage in the sediments deposited during the lake's most saline phase would also imply that the initial concentration of interstitial sodium in the sediments 500 yr B.P. was greater than 110 meq $l^{-1}$.

The qualitative evolution of the salinity gradient in the sediments through the history of the lake is most probably as follows. During the early history of the lake when it had open drainage and was relatively dilute, sodium concentrations in the interstitial waters were nearly uniform because there was little variation in the lake's sodium concentration. Around 3,000 yr B.P., in response to a more arid climate and higher evaporation rates, the lake's sodium concentration began to increase and sodium began to be transported into the sediment by diffusion. The loss of the lake's major tributary 1,900 yr B.P. created a closed basin lake which became increasingly more saline until the lake once more overflowed at its effluent 500 yr B.P. During this saline stage, sodium was transported into the sediments at an increasing rate. The lower sediments did not have time to come to equilibrium with the maximum concentration the lake achieved.

Therefore, when the lake became more dilute a cusped concentration gradient was established, i.e. a maximum in interstitial sodium concentration would have been sandwiched between more dilute interstitial waters above and below the sediments deposited during the most saline phase. This cusped gradient would be degraded by diffusive transport along two concentration gradients until the maximum concentration fell to the concentration of the lower sediments at which time diffusion of sodium would proceed along only one gradient, up to the lake, as presently observed.

Only a strict interpretation of the dominant diatom species present in microfossil assemblages would directly indicate that the sodium concentration of the lake ever exceeded 110 meq $l^{-1}$. The diatom microfossils are also the only indication that the lake was ever more dilute than 54 meq $l^{-1}$. It would not be possible to interpret the shape of the observed gradient, understand its causation, nor appreciate its transience without the information the diatom microfossils yield. Because of the potential capability of microfossils to detect transient environmental conditions, they can be a useful tool in chemical studies of interstitial waters, or sediments in general, where the recent history of the aquatic system is unknown.

Acknowledgments

The work in Tanzania and at Duke University was supported by National Science Foundation Grant GB-8328X to Dr. D. A. Livingstone and an NSF predoctoral fellowship to myself. Dr. R. Hesslein provided assistance with computational procedures and much valuable advice on diffusion processes. Dr. G. J. Brunskill critically read an early draft, and his comments altered the presentation significantly. S. J. Guildford improved both the grammar and the clarity by reading many drafts.

Key words: Diffusion
Micro-fossils
Sediments
Lake
Paleolimnology

References

Berner, R. A. 1971. Principles of chemical sedimentology. McGraw-Hill, New York.

Degens, E. T. and G. V. Chilinger. 1967. Diagenesis of subsurface waters. In: G. Larsen and G. V. Chilingar (Ed.): Developments in Sedimentology 8. p. 477-502.

Duursma, E. K. and C. Hoede. 1967. Theoretical, experimental and field studies concerning molecular diffusion of radioisotopes in sediments and suspended solid particles of the sea. Part A: Theories and mathematical calculations. Neth. J. Sea Res. 3: 423-457.

Garrels, R. M. and C. L. Christ. 1965. Solutions, minerals and equilibria. Harper, New York.

Hecky, R. E. 1971. The paleolimnology of the alkaline, saline lakes on the Mt. Meru lahar. Ph.D. thesis. Duke University, Durham 209 p.

Hecky, R. E. and P. Kilham. 1973. Diatoms in alkaline, saline lakes: ecology and geochemical implications. Limnol. Oceanogr. 18: 53-71.

Manheim, F. T. 1970. The diffusion of ions in unconsolidated sediments. Earth Planet. Sci. Letters 9: 307-309.

Reeburgh, W. S. 1967. An improved interstitial water sampler. Limnol. Oceanogr. 12: 163-165.

Richardson, J. L. and A. E. Richardson. 1972. History of an African rift lake and its climatic implications. Ecol. Monogr. 42: 499-534.

Schindler, D. W., R. Hesslein, and G. Kipphut. In press. Interactions between sediments and overlying waters in an experimentally-eutrophied Precambrian Shield lake. Proc. Symp. on Sediment Water Interactions, Sept. 1976. Amsterdam. This vol., page 235.

Stuiver, M. 1967. Origin and extent of atmospheric $C^{14}$ variations during the past 10,000 years. In: International Atomic Energy Agency, Proceedings of the IAEA Symposium on Radioactive Dating and Methods of Low-level Counting, March 2-10, 1967. Monaco. IAEA Vienna. p. 27-40.

# A RAPID METHOD FOR DETERMINING RECENT DEPOSITION RATES OF FRESHWATER SEDIMENTS[1]

Jerry C. Ritchie and J. Roger McHenry

USDA Sedimentation Laboratory, Oxford, Mississippi 38655, USA

## Summary

Measuring deposition of recent freshwater sediments deposited in the last 20 years is difficult and the usual survey methods requires several years to obtain reliable results. Methods are available for dating older sediment by using carbon-14 or other natural radioactivity that are quick and reliable, and can be made from a single core sample. Nuclear test have released into the atmosphere several radionuclides that have returned to the earth's surface as fallout. The rate of fallout can be related to the number and size of nuclear detonations in a time period. One of these fallout radionuclides, cesium-137, is tightly adsorbed on the smaller-sized soil particles and any subsequent release of this Cs-137 from the soil particles by natural chemical process is very limited. Thus, Cs-137 acts as an effective tracer that can be used to study soil erosion and sediment deposition. Since the major input of Cs-137 into the atmosphere, soil, and sediment occurred in 1959 and 1963 and these fallout deposits are effective tracers for identifying sediment layers deposited in those years. Extensive studies of the distribution of Cs-137 in sediment cores collected from small lakes and reservoirs (18 to 4000 ha), the upper Mississippi River, and estuaries along the United States coast showed that Cs-137 can be used to determine the sediment layers deposited in 1963 and often those deposited in 1959 with results that are comparable with that determined using classical survey techniques. With the survey technique the same area must be surveyed several times to obtain the data needed, but with the Cs-137 technique (like the C-14 technique) only one sediment core is collected, divided into vertical increments and analyzes to determine which increment has the highest Cs-137 concentration. Since Cs-137 is strongly adsorbed by the finer soil particles, this technique does not work as effectively in sediment with a high percentage of sand. The Cs-137 sediment dating technique will furnish reliable estimates of the amount of sediment deposited since 1963 and since 1959 for any areas where fine sediment accumulates. This paper will show typical results that were obtained using the Cs-137 tracer technique in several different freshwater environments.

## Introduction

Deposited sediments affect many properties of freshwater lakes. They can adsorb or release phosphorus (Syers, Harris, and Armstrong, 1973), nitrogen (Keeny, 1973), carbon (Kerr, et al., 1973), heavy metals (Jacobs and Keeney, 1974), and other chemicals which affect the chemical properties of a waterbody. Deposited sediments fill reservoirs causing loss of capacity for flood control, electric power production, irrigation, or other purposes for which the reservoir was built. A study of 968 reservoirs in the United States showed an average 1.77% of their storage capacity was lost each year to sediment. The median loss was 0.72% per year (Dendy, 1968). Sediment deposition in freshwater lakes is a major world problem, yet there is no simple method to determine deposition rates for recent sediments--those deposited in the past 10 to 20 years and those having the greatest effect on the freshwater system.

Present techniques for measuring sediment deposition rates require the surveying of the sediment surface for several years until enough data is accumulated to accurately calculate sediment deposition rates. This paper describes a method for estimating the amount of sediment deposited since 1963-1964 and possibly since 1958 by measuring the concentration of radioactive fallout Cs-137 in depth increments of a sediment core collected from a waterbody. This method, using Cs-137, makes it possible to estimate sediment deposition rates on lakes and reservoirs where no previous measurements of sediment deposition have been made or where survey data are insufficient or unavailable.

## Background

Cesium-137 is a manmade radionuclide produced during nuclear fission either in a nuclear reactor or the detonation of a nuclear device. The input of strontium-90 as well as Cs-137 into the atmosphere and its subsequent deposition on the earth's surface has varied with time (Fig. 1)

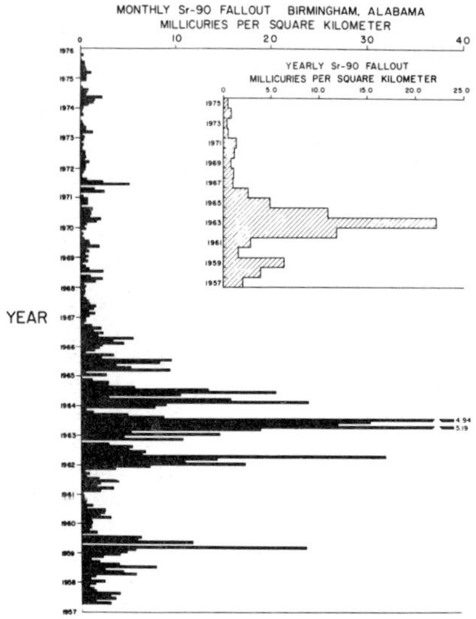


Figure 1. Strontium-90 fallout at Birmingham, Alabama. (Cesium-137 to Strontium-90 ratio is approximately 1.5)

(Hardy, 1976; Lindblom, 1967; Linden and Gustafsson, 1967), but the total deposition has been uniform over regional areas (List, et al., 1965). Once Cs-137 comes in contact with the surface soil, it is strongly adsorbed on the smaller-sized particles and organic matter, and further movement by natural chemical processes is limited (Tamura, 1964; Davis, 1963; Klechkouskii, et al., 1973). Subsequent movement within a watershed from place of deposition would be by physical processes, like erosion (Rogowski and Tamura, 1970; Ritchie, et al., 1974). Cesium-137 deposited on a lake surface quickly becomes adsorbed by suspended material in the water column and is deposited on the lake bottom. Once adsorbed to the suspended or deposited sediment, it desorbs very slowly (Duursma and Bosch, 1970; Duursma and Gross, 1971; Lomenick and Tamura, 1965). Cesium-137 also has a strong gamma emission (0.662 mev) making it easy to detect with a minimum of sample preparation.

To use Cs-137 to estimate sediment deposition rates, the following assumptions are made.

1. Cesium-137 deposition in a sediment column is related to the annual input of Cs-137 from the atmosphere to the watershed.

2. Once Cs-137 is adsorbed on a sediment or soil particle, very little chemical exchange or movement of the Cs-137 atom occurs.

3. Physical mixing of the sediment column by wave action or by burrowing organism is at a minimum so that little or no redistribution of Cs-137 occurs in the sediment column.

Studies have shown that assumptions 1 (Ritchie, et al., 1973; Pennington, et al., 1973; Robbin and Edgington, 1975) and 2 (Duursma and Bosch, 1970; Duursma and Gross, 1971; Lomenick and Tamura, 1965) are valid and can be made with confidence in most freshwater lakes. There have been a few studies of the effects that biological and physical mixing of sediment has on the use of Cs-137 to estimate deposition rates (Bailey, et al., 1975). However, if there has been any physical mixing of the sediment distribution of Cs-137 in the profile with depth will be more uniform, and this sedimen core could not be used to determine deposition rates. For these reasons, we feel that Cs-137 can be a very useful tracer to estimate the deposition rates of sediments in lakes and reservoirs.

## Methods and materials

To use the Cs-137 technique, a sediment core sample is taken of the deposited sediment profile and divided into depth increments (e.g. 1, 2, 5, 10 cm) that would give the needed definition of Cs-137 in the profile. Multiple cores may be taken from the same location and composited to give a sufficiently large sample for gamma-ray analysis.

Once the sample has been collected, it is dried and passed through a screen to give a uniform aggregate size. Since Cs-137 has a strong gamma-ray, no further sample preparation is needed. The sample is analyzed using a thallium activated sodium-iodide crystal or a lithium drifted germanium solid state crystal as a detector and a multiple channel pulse height analyzer. The complex gamma-ray spectra are then reduced to give the Cs-137 concentration in the sample.

## Example of results

In 1970, samples of deposited sediment were collected from two small (<10 ha) flood control reservoirs. One reservoir was in Iowa, in the eastern part of the grasslands, and the other was in the desert of Arizona. Both reservoirs were dry at the time of sampling. The reservoir in Iowa had filled with sediment to the top of the outlet pipe so that no water storage capacity remained. The reservoir in Arizona was designed to retain water from a runoff event for only five days. Sediment deposition in both reservoirs had been measured by standard surveying techniques. The reservoir in Iowa had been surveyed 11 times since it was constructed in 1954, and the reservoir in

Arizona had been surveyed four times since it was constructed in 1962. Sediment samples were collected on or near survey points so they could be compared with the survey data.

A comparison of the depth distribution of Cs-137 in the sediment profiles with the surveyed sediment depth (Fig. 2) showed that

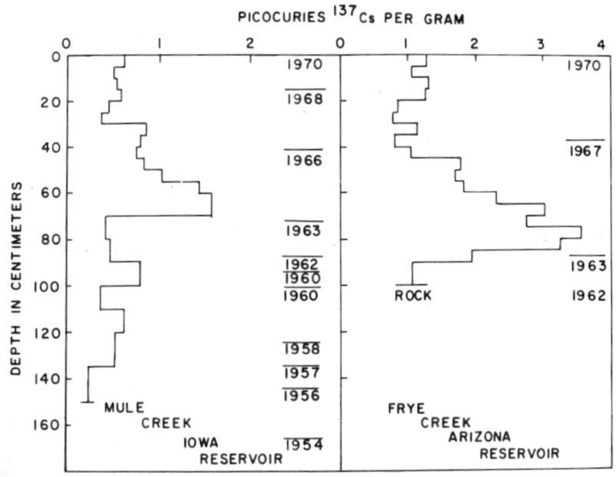


Figure 2. Vertical distribution of cesium-137 in sediment profiles collected from Mule Creek Reservoir in Iowa and Frye Creek Reservoir in Arizona. The line above the dates at the right side indicates the surveyed sediment depth at that time.

the maximum Cs-137 concentration in the sediment profile was associated with the sediment layer deposited in 1963-1964, when fallout of radioactivity from the atmosphere was maximum (Fig. 1). A calculation of sediment deposition rates (Table 1) showed that at both Mule Creek and Frye Creek, the Cs-137 technique underestimated the sediment deposition rate by 0.4 cm/yr as compared with that calculated from the conventional survey data.

Table 1. Comparison of sediment deposition calculated from survey records and estimated by the Cs-137 technique.

| Reservoir | Deposition rate Sediment survey | Cs-137 survey |
|---|---|---|
| | (cm/yr) | |
| Mule Creek, IA | 10.4 | 10.0 |
| Frye Creek, AZ | 12.5 | 12.1 |
| Tippah, MS | 2.5 | 2.8 |
| Horn's Ferry, MS | 4.9 | 4.7 |
| Murphy, MS | 1.3 | 1.5 |
| Smith, MS | 1.8 | 2.0 |

In 1971, samples of sediment deposited in the Tallahatchie River Valley in northern Mississippi were collected along the river bottom where spring flooding occurs. These were areas of sediment deposition and were also near survey monuments established in 1936. The vertical distribution of Cs-137 in the sediment profiles again showed a peak of concentration (Fig. 3). Assuming the

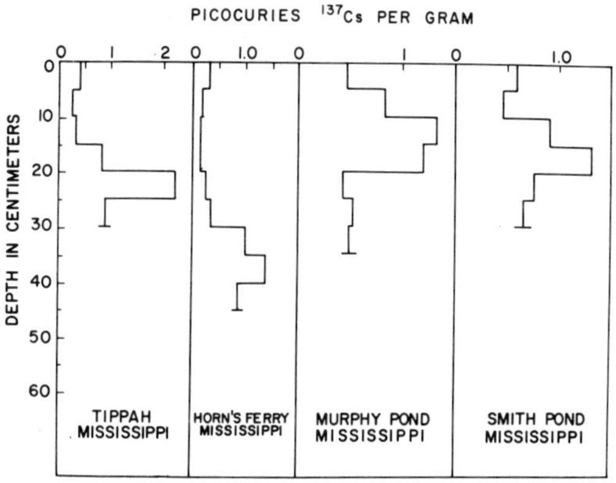


Figure 3. Vertical distribution of cesium-137 in sediment profiles taken from four different locations in northern Mississippi.

high concentrations of Cs-137 were associated with the 1963 peak fallout level, sediment deposition rates were estimated and compared with those calculated from survey records (Table 1). In this study of valley sediment deposition, the Cs-137 technique estimated the sediment deposition rates to be 0.3 cm/yr greater than that calculated from the survey records.

In another study of two small reservoirs (<5 ha) in north Mississippi, the vertical distribution of Cs-137 in the sediment was related to atmospheric input of Cs-137 (Fig. 3). Sediment deposition rates as estimated from the Cs-137 distribution in the deposited sediment were greater than those calculated from survey records by 0.2 cm/yr.

In these studies, where it was possible to compare sediment deposition rates calculated from survey records with those estimated by the Cs-137 technique, the sediment deposition rates were within 10% of each other. Agreement might have been better if samples had been collected at less than 5 cm increments. Other sediment profiles from these and other reservoirs have given similar results (Ritchie, et al., 1970; 1972; 1973; 1975). Other researchers have used the Cs-137 technique to estimate sediment deposition rates in France (Krishnaswawy, et al., 1971), Italy (Ravera and Premazzi, 1971), England (Pennington, et al., 1973; Pennington, 1974), Israel (Stiller and Assaf, 1973), Russia (Makhoi'ko, et al., 1975), and the U. S. (Robbins and Edgington, 1975; Plato and Jacobson, 1976) with similar results.

On the upper Mississippi River, where backwater lakes exist due to the presence of a series of locks and dams, the Cs-137 technique is being used to estimate sediment deposition rates in these lakes. Few other records are available so that the Cs-137 technique (Fig. 4) provided a unique measurement of the sediment being deposited in these

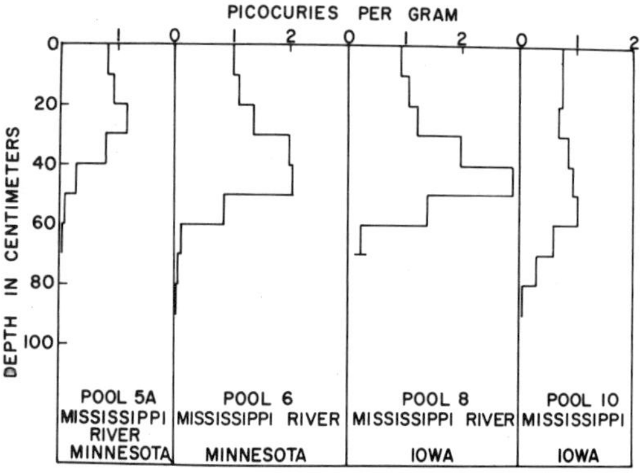


Figure 4. Vertical distribution of cesium-137 in sediment profiles collected in the backwater lakes of the Upper Mississippi River.

lakes. This study has shown that areas of sediment deposition in the backwater lakes are extensive where commercial and sports fishing are important sources of income and relaxation for the people of the area.

## Conclusions

Cesium-137 is a useful and unique tool for studying the amount and rate of sediment deposited since 1963-1964. The Cs-137 technique is unique in that it provides an accurate estimates of the sediment deposited from a single sampling with no resampling or later resurvey necessary. In using the Cs-137 technique, one must consider whether any mixing of the sediment has occurred. Also, as Cs-137 is adsorbed by the finer-size particles, one should know something about the particle-size distribution, so that the presence or absence of Cs-137 in layers of high clay or high sand will not be misinterpreted. The Cs-137 technique offers both researchers and operations personnel a unique tool to better understand the sediment deposition in lakes and reservoirs. We have shown that the Cs-137 technique can provide information on sediment deposition rates when no other method is available.

## Acknowledgement

This paper is a contribution of the Sedimentation Laboratory, USDA-ARS, Alabama-North Mississippi Area, Southern Region. Paul Hawks, Angela Gill, Jonathan May, and Kathleen Roush helped with the field and laboratory analyses. Robert Wilson helped with the computer analyses of the data, and Donna Bethany helped prepare the manuscript.

## References

Bailey, J.H., C.T. Hess, S.A. Norton, & R.B. Davis, 1975. Cs-137 distribution in profundal sediments of Maine lakes. Abstracts of papers for the 38th Annual Meeting of the American Society of Limnology and Oceanography. p. 4. Abstract.

Davis, J.J., 1963. Cesium and its relationship to potassium in ecology. p. 539-556. IN: V. Schultz and A.W. Klements, Jr. (eds.), Radioecology. Reinhold, New York.

Dendy, F.E., 1968. Sedimentation in the Nation's reservoirs. J. Soil and Water Conservation 23:135-137.

Duursma, E.K. & C.J. Bosch, 1970. Theoretical experimental and field studies concerning diffusion of radioisotopes in sediments and suspended particles of the sea. Part B. Methods and Experiments Netherlands. J. Sea Res. 4:395-469.

Duursma, E.K. & M.G. Gross, 1971. Marine sediments and radioactivities. IN: Radioactivity in Marine Environment. Nat. Acad. Sci., Washington, D. C. pp. 147-160.

Hardy, E.P., Jr., 1976. Health and Safety Laboratory Environmental Quarterly and Appendix, March 1 through June 1, 1976. Health and Safety Laboratory Report, HASL-306.

Jacobs, L.W. & D.R. Keeney, 1974. Methyl mercury formation in mercury-treated river sediment during in situ equilibrium. J. Environmental Quality 3:121-126.

Keeney, D.R., 1973. The nitrogen cycle in sediment-water systems. J. Environmental Quality 2:15-29.

Kerr, P.D., D.L. Brockway, D.F. Paris, & S.E. Craven, 1973. Carbon cycles in sediment-water systems. J. Environmental Quality 2:46-52.

Klechkouskii, V.M., G.G. Polikarpov, & R.M. Aleksakhin, 1973. Radioecology. John Wiley New York.

Krishnaswamy, S., D. Lal, J.M. Martin, & M. Meybeck, 1971. Geochronology of lake sediment. Earth and Plan. Sci. Lett. 11:407-414.

Lindblom, G., 1967. Measurements of radionuclides in the environment and in the human body. p. 53-57. IN: B. Aberg and F. P. Hungate (eds.), Radioecological Concentration Processes. Pergamon Press, Oxford, England.

Linden, K. & M. Gustafsson, 1967. Relationships and seasonal variations of Cs-137 in lichen, reindeer and man in northern Sweden 1961-1965. p. 193-208. IN: B. Aberg and F. P. Hungate (eds), Radioecological Concentration Processes. Pergamon Press, Oxford, England.

List, R.J., L. Maihta, L.T. Alexander, J.S. Allen, M.W. Meyer, V.T. Valassis, & E.P. Hardy, Jr., 1965. Strontium-90 on the earth's surface. III. IN: A. W. Klements, Jr. (ed.), Radioactive fallout from nuclear weapons test. U. S. Atomic Energy Commission Report CONF-765.
Lomenick, T.F. & T. Tamura, 1965. Naturally occurring fixation of cesium-137 on sediments of lacustrine orgin. Soil Sci. Soc. Am. Proc. 29:383-386.
Makoni'ko, K.P., Ts.I. Bobounikova, A.S. Auramenko, A.V. Dibtseva, & A.A. Volokintin, 1975. Vertical distribution of Sr-90 and Cs-137 in bottom deposits of lakes and rivers. Ekologiya 3:90-93.
Pennington, W., 1974. Seston and sediment formation in five district lakes. J. Ecol. 62:215-251.
Pennington, W., R.S. Cambray, & E.M. Fisher, 1973. Observations on lake sediments using fallout Cs-137 as a tracer. Nature 242: 324-326.
Plato, P.A. & A.P. Jacobson, 1976. Cesium-137 in Lake Michigan sediments: Areal distribution and correlation with other manmade material. Environ. Pollut. 10:19-34.
Ravera, O. & G. Premazzi, 1971. A method to study the history of any persistant pollution in a lake by the concentration Cs-137 from fallout. IN: Radioecology applied to the protection of man and his environment.
Ritchie, J.C., J.R. McHenry, & A.C. Gill, 1972. Dating recent reservoir sediments using radioactive fallout. The ASB Bull. 19:95.
Ritchie, J.C., J.R. McHenry, & A.C. Gill, 1973. Dating recent reservoir sediments. Limnol. Oceanogr. 18:254-263.
Ritchie, J.C., J.R. McHenry, A.C. Gill, & P.H. Hawks, 1970. The use of fallout cesium-137 as a tracer of sediment movement and deposition. Proceedings Mississippi Water Resources Conference 1970:149-163.
Ritchie, J.C., P.H. Hawks, & J.R. McHenry, 1975. Deposition rates in valleys determined using fallout cesium-137. Geol. Soc. Amer. Bull. 86:1128-1130.
Ritchie, J.C., J.A. Spraberry, & J.R. McHenry, 1974. Estimating soil erosion from the redistribution of fallout Cs-137. Soil Sci. Soc. Am. Proc. 38:137-139.
Robbins, J.A. & D.N. Edgington, 1975. Determination of recent sedimentation rates in Lake Michigan using Pb-210 and Cs-137. Geochemica et Cosmochima Acta 39:285-300.
Rogowski, A.S. & T. Tamura, 1970. Environmental mobility of cesium-137. Radiation Botany 10:35-45.
Stiller, M. & G. Assaf, 1973. Sedimentation and transport of particles in Lake Kinneret traced by Cs-137. Int. Ass. Hydro. Sci. Publ. No. 109 p. 383-396.
Syers, J.K., R.F. Harris, & D.E. Armstrong, 1973. Phosphate chemistry in lake sediments. J. Environmental Quality 2:1-14.
Tamura, T., 1964. Selective sorption reaction of cesium with mineral soil. Nuclear Safety 5:262-268.

Key words: Cesium-137
Radioactivity
Fallout
Sediment Accumulation
Erosion

Theme 3 - Role of sediments in element cycles

# THE INFLUENCE OF SEDIMENTS UPON NUTRIENT EXCHANGES IN TROPICAL LAKES

A.B. Viner

D.S.I.R. Freshwater Section, P.O. Box 415, Taupo, New Zealand[1]

## Summary

A description relevant to nutrient transfers is given for the sediments of a shallow equatorial lake. Compared to the water column the sediments are much less important for nutrient recycling. This is thought to be an outcome of the climatic regime, but the functional difference between the two zones is obscured by the movements of primary producers between them.

The lake shows features of tropical limnology to an extreme degree and may be used as a reference basis for comparing conditions in other waters.

## General considerations

It might perhaps be useful to review some of the well known features of tropical ecology which are of relevance to this topic. Attention will be largely restricted to African limnology since it is in this continent that the greatest number and variety of tropical waters exist. Seasonal changes of solar radiation, and so of temperature, are very much reduced compared to higher latitudes. Consequently organic productivity in aquatic systems is continuous.

Because the radiant energy input is also high the metabolic rates and turnover of elemental material are high too. This results in very rapid responses to those seasonal and similar duration environmental changes which do take place. There is thus a very small lag-phase of biotic adaptation to changing circumstances.

Such conditions are conducive to utilisation of important limiting primary nutrient material to the maximum that these are supplied, with proportionately little of them passing through a lake to the outflow due to lack of organisms to make use of them. This contrasts somewhat to the temperate situation in which seasonal flushes of nutrient inputs are not precisely in phase with maximal standing stock, so that a proportion of annual nutrient input passes out of the lake unused.

Although high metabolic rates mean high gross production, to maintain this production the net production need only to balance the emigration due to outflow and to permanent sedimentation of organic material, which, depending on the circumstances, may be small compared to the gross flux of material.

A steady-state system is thus approached. For small water bodies where there is complete mixing of the water mass, the dynamics can be paralleled to a laboratory continuous culture. In such a system the residence time of an atom in any one compartment within a lake will be short.

This pattern allows for potential nutrient demands of the primary producers to be calculated from the relative proportions of the elements in the seston, when these are compared to metabolic rate as measured by gross photosynthesis or respiration. The different turnover rates of one element to another, which undoubtedly must occur in principle, are not of importance in this context.

Depending upon their morphometry, shelter from local climatic changes and position relative to the equator, tropical lakes will deviate in varying amounts from the above pattern, but generally between the tropics the continuous climatic temperatures makes for lakes to be either polymictic, and even diurnally mixing, or, if they are sufficiently deep relative to surface disturbance, to have long-term or permanent thermoclines which separate the sediments from the trophogenic zones. In the latter type there may well be a degree of seasonal or aseasonal erosion of the thermocline, largely due to wind disturbance rather than by the solar input and concomitant density changes normally associated with the temperate lake. (A good review on this aspect is given by Beadle, 1974. See also Talling, 1969). Circulation to the permanent thermocline is functionally the same as circulation to a solid sediment surface, since the physicochemical conditions there are chemically essentially similar to the conditions in the superficial zone of the sediment. The

1. Ex Royal Society African Freshwater Biological Team, which was organised by the United Kingdom for the International Biological Programme

shallower regions which are within the mixed depth of very large lakes with permanent stratification will, of course, function similarly to a shallow permanently mixed lake.

The issue is how much the nutrient exchange with the sediment (or upper zone of the hypolimnion) contributes to the total nutrient flux under these circumstances.

## The importance of sediments in Lake George from mass budget data

Detailed descriptions of work upon these sediments and their relationship to the lake's ecosystem are to be found in Ganf, 1974 a, b, c; Ganf & Viner, 1973; Viner, 1975 a, b, c, d, 1976; Viner & Smith 1973.

Lake George (Uganda) is a particularly good example of a stable ecosystem in the sense outlined above. It is sufficiently shallow (c. 2.4m) to mix diurnally, and climatic and hydrological features combine to give a marked constancy of limnological environment. The diurnal limnological changes are in fact more marked than the seasonal ones, and Ganf (1974c) has discussed how this cycle duration is shorter than what one might expect to be the generation time of the phytoplankton, so that a single diurnal cycle is not going to be observable as a significant growth response. The usefulness of this lake is that it is possible to view it as being at the end of a sequence of limnological types, and as a base against which such variants may be contrasted.

A budget could be made to compare the relative amounts of the limiting nutrients of nitrogen and phosphorus. Inorganically combined N and P do not normally appear in the water, so that for budget purposes fluctuations in dissolved concentrations of these elements need not be taken into account. The quantity of material lost by emigration from the ecosystem must be balanced by the inputs in a fairly direct manner, and these portions of the total material may be compared to the quantity of standing stock so as to see their relative importance (c.f. Viner, 1977).

Using the means of the ratios of elements in the water and the water replacement rate, 25 000 tonnes of carbon, 3400 tonnes of nitrogen and 220 tonnes of phosphorus are flushed from the lake per year.

Sediment cores were analysed for organic material and other chemical components and $^{14}C$ dating showed a current organic deposition rate of 15 – 20 $g/m^2/yr$; equivalent to 2500 tonnes of carbon, 235 tonnes of nitrogen and 18.7 tonnes of phosphorus per year, if the highest deposition rate is used. The values for total inorganic input which should balance this organic emigration were calculated from inputs by Ganf & Viner (1973) in a more approximate manner using fewer analyses, and gave 2800 tonnes of N and 320 for P. This estimate for nitrogen is 23% lower, and phosphorus is 34% higher, than the export values quoted here, but the differences are of no importance when considering the orders of magnitude involved in the comparison.

Measured gross photosynthetic production for a typical day; 5.8 g C $m^{-2}day^{-1}$ (Ganf & Horne, 1975) may be multiplied up to give a figure of 529 250 tonnes which is passed through the plankton community during a year. The proportion of this which is net production in the form of total emigration is seen to be small (Table 1). The ratio of gross to this net carbon is about 19, while the ratio of gross carbon to sediment carbon alone is 212 which means that the sediment is of complete insignificance in the total budget since it must be well within the errors of calculations and chemical analyses. (Ganf & Horne op.cit., on the basis of their physiological experiments, actually calculate a net production which, when multiplied up in the above manner, yields a value of about one tenth of the export in Table 1, but their estimation is probably subject to the errors which they themselves discuss.)

Table 1. Mass budgets, as tonnes per year, for C, N and P in L.George, from seston data.

| | | C | N | P |
|---|---|---|---|---|
| Export | Effluent | 25000 | 3400 | 220 |
| | Sediment | 2500 | 235 | 18.7 |
| Total Export | | 27500 | 3635 | 239 |
| Gross metabolised | | 529250 | 71325* | 4649* |
| Gross metab. / Total export | | 19.3 | 19.6 | 19.5 |
| Gross metab. / Sediment | | 212 | 303 | 248 |

*Potential rates, assuming mean rates of uptake are similar to the mean rate of carbon metabolism.

Were the flushing rate of water to exceed the turnover rate of elements, such that some of the nutrients could remain unused, the nutrients from sediments would still not become an important component of the budget because these too would be flushed from the lake at a proportionate rate.

## Sediment metabolism

Whatever might be precipitated must therefore be almost entirely decomposed and recycled, in which case sediment decomposition is very efficient and important, or, no significant amount of material ever remains at the bottom of the lake sufficiently long to contribute to the processes there. Investigation from different viewpoints suggested a very inefficient sediment metabolism and that it contributed little to the overall lake metabolism.

The muds are disturbed by turbulence to well below their surface, but different observations gave different depth for the penetration.

By using a fairly simple model of the probable wave mechanics at the sediment-water interface together with measured features of climate, a calculation suggested that mechanical disturbance could be common to about 10 cm, with frequency of disturbance increasing for depths above this.

However, sediment core interstitial water analyses for dissolved $NH_4$-N and $PO_4$-P found that almost nothing of these radicles existed in the uppermost 5 cm, but that there was a sharp increase of concentration from 6 cm to about 15-25 cm, with a certain amount of variation between cores (Figure 1). Below this level the concentration increases were slight. (Similar profiles are discussed for other lake sediments in this symposium.) On this criterion one might believe that common disturbance was only to 5 - 6 cm, and infrequently to about 25 cm. Or perhaps the burrowing fauna exhibited their effect on the profile below 6 cm.

Figure 1. Profiles of interstitial $NH_4$-N and $PO_4$-P in the L. George sediments. The increases in concentration often begin at slightly lower depths than those in this particular example (Redrawn from Ganf & Viner, 1973).

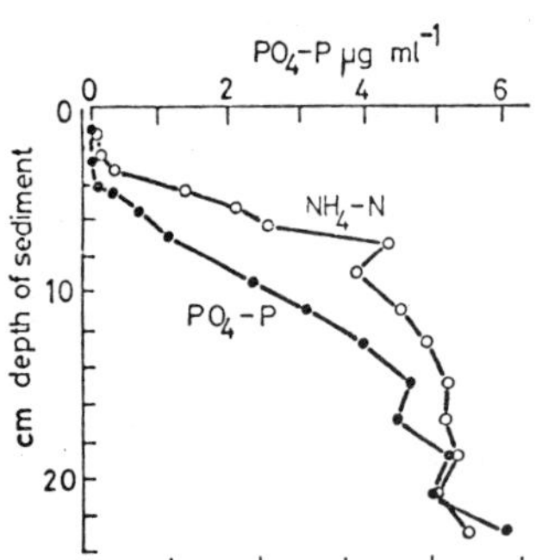


The rates at which oxygen is consumed by, and carbon dioxide released from, sediment is an indication of the metabolic activity therein and so of the extent of stimulation of this activity by circulating water. For L.George, a gradient of increasing oxygen uptake with increasing depth was found, as for similar studies upon sewage sludge and natural sediments (c.f. Edwards & Rolley, 1965). A peak in uptake was found at about 15-18 cm, and a decrease below this (Figure 2). There was evidence shown, by enhanced oxygen uptake upon the introduction of sterilised mud, that a significant factor in the metabolic profile was a limiting substrate, such that the sediment at the oxygen uptake peak region had potentially more bacterial substrate than that above. Limitations of the experimental technique meant that one could not be sure of this interpretation because radical alteration of the sediment may have been caused by sterilisation.

An alternative explanation could be that the bacterial population was metabolically differentiated with depth. Laboratory experiments of liberation of $CO_2$ by samples from different sediment depths, measured in an oxygen or nitrogen environment, indicated that this differentiation could well have existed because the bacterial performance changed progressively to more efficient aerobic metabolism during the course of switching successively to aerobic from anaerobic conditions and vice versa, as if the population was becoming better adjusted to an aerobic environment, either in its population structure or in preferred aerobic metabolism.

Figure 2. Oxygen uptake rates in L.George sediments. Each profile is for a different site. Location of sites are shown in the inset map of the lake, for which the curve and map symbols correspond. (From Viner, 1975a).

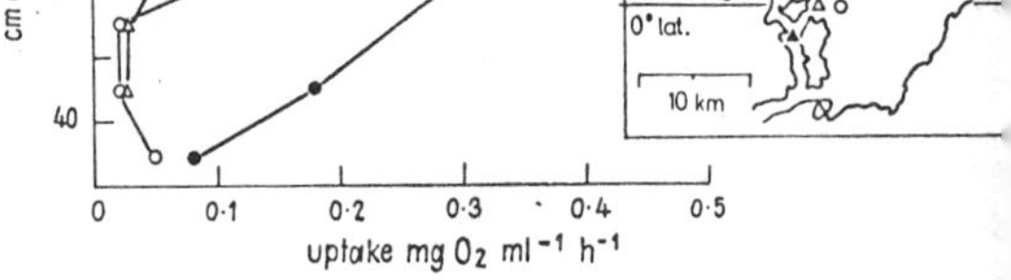

It is evident that the degree of turbulence of water into the mud would control both the thoroughness of organic degradation and the aerobic/anaerobic adjustments of the decomposer population in such a way that there could well be rather less of readily decomposable intermediate breakdown products of organic material near the sediment surface where there would be the most abundant influx of oxygen. Further evidence of the relationship between degree of mixing and vertical sediment characteristics was seen in the shift upward of the peak of oxygen consumption in cores taken from sites in the lake more sheltered from wind stress nearer the shore (Figure 1).

The decrease in metabolism below the peak encompasses a zone where extremely infrequent oxygenation, and the exhaustion over a long time of aerobically decomposable substrates, has reduced the metabolic abilities of the sediment. Deeper layers of sediment have been shown to be more resistant to microbial breakdown than those nearer the surface in other water bodies (Allen <u>et al</u>, 1953).

To quantify this in terms of chemical components supplied to the water becomes very much more problematical. Presence or absence of the bacterially optimal oxygenation state in the mud will depend upon the amount of transported oxygen rather than the actual mixing. If the amount transported is small relative to the amount of very reduced organic material present then clearly proportionately little of the reduced organic mass will be oxidised. For Lake George the rapidity of redox potential movements to low values after mixing (complete recovery within 12 hours) demonstrated an extremely great oxygen demand consonant with the high organic percentage in the sediment (about 70%). Evidently sediment decomposition is inefficient.

An attempt in the laboratory to quantify the minimum output of $NH_4$-N and $PO_4$-P from a completely static mud surface demonstrated that non-biotic factors (diffusion gradients, desorption, etc.) rather than decomposition rates were predominant in the dynamics of these nutrients. For an anaerobic system they both rapidly increased in the overlying water during the first three days, in responce to their amounts in the mud interstitial water. Thereafter increase virtually ceased. Phosphate in an aerobic system showed no significant release at all, conforming to conventional understanding of phosphate dynamics. The $NH_4$-N release was recorded as oxidised forms of nitrogen, but since nitrification was considered to be abnormal under the lake conditions, this result was thought to be an experimental artifact. Decomposition was thus considered to be of limited effectiveness.

The normal sediment profile of these nutrients would indicate that the effective (sharpest) diffusion gradient was actually within the mud and not at an undisturbed sediment/water interface. This location will retard the rate of nutrient transfer relative to the rate across the mud surface. Molecular and hydrological diffusion rates are influenced, to an order of magnitude (Klotz, 1975), by the presence and nature of solid particles. In a laboratory experiment with undisturbed cores into which had been injected a layer of $^{14}C$ bicarbonate, no significant movement of this tracer up the core could be detected after two weeks.

Also, diffusion transfer of phosphate will be much influenced by its capacity to be adsorbed (Gupta & Greenkorn, 1974). For L.George mud, such an undersaturation with respect of $PO_4$-P adsorption was found, to the extent that, in the upper 6 cm, 540 mg was taken up across 1 $m^2$. If this is typical for the whole lake sediment area, then 135 tonnes are capable of being taken up, which is equivalent to about half the annual seston export.

As there is no, or very little, $NH_4$-N or $PO_4$-P in the top 6 cm the combined effect of diffusion and decomposition releases must be at a slower rate than that of the flushing away of the nutrients by turbulence, and, since the main release mechanism appears to be abiotic, the sediment contribution of nutrients must be limited by the diffusion phenomena. Such indirect observations support the notion that the sediments provide only a small part of the nutrient economy of the water column.

There is, however, another dimension to consider. The L.George sediments contain large quantities of occluded algae down to 20 cm of depth but especially to 6 cm. They are viable in the sense that they photosynthesise, and their viability decreases with increasing depth of mud; the uppermost responding very similarly to the algae of the water column. These data and their resulting complication of the measured rates of algal biomass increase in the lake have been discussed by Ganf (1974 b, c), who shows how very greatly the amount of material available to photosynthesis can be changed if hydrological disturbance of the sediment permits.

The sediment is thus functioning as a reservoir of primary producing material which in such a position is able to metabolise relatively slowly. The algal respiratory rate is approximately 1/8 - 1/5 of the remaining material, probably because lack of sunlight does not initiate a fast turn over rate.

The time any alga can remain a potential contributor to this reservoir will depend upon how long it can remain viable. Scattered information from other studies show that algae are capable of preserving a viable state in an apparently unfavourable environment for at least some months. Most of such information concerns the diatom Melosira (Lund, 1955) and the idea has been evoked to explain changes in phytoplankton in L.Victoria (Talling, 1966). But it seems that similar resuspension of deeply sited algae of a different and unidentified sort can occur in L. Tanganyika (G.W. Coulter, pers.comm.). For the Cyanophyte algae there are numerous examples of vertical changes of peak distribution within the water column, especially for those movements performed by species capable of being buoyed by gas vacuoles, (Fogg et al 1973 p.105-107, 261-268), but there does not seem to be information about their long term survival in non-optimal conditions. It is known that many blue-green algae can metabolise heterotrophically in the dark, which in L.George would aid their longevity while in the sediment.

When algae are accumulated in the upper layer of mud, or in the upper hypolimnion in the sense applied here, they are in closest proximity to the highest concentration of their own products of respiration. Uptake of these products can be carried out as long as there has been sufficient photosynthesis immediately prior to their occlusion in the mud. Hence to sustain a population nutrient uptake and photosynthesis need not be in phase (Viner, 1973). Nutrient recycling will thus continue in the upper part of the sediment in the same way as it does in the water column above, and under these circumstances recycling can be very efficient.

## Flood plain phenomena

While the conditions in L.George can make a useful reference point for conditions in other African lakes, it is necessary to say that a somewhat contrasting regime of sediment-water interaction of great importance exists in the tropics. That is, the seasonal or periodic flood-plain phenomenon with its large flushes of productivity, e.g. Kainji Lake (Nigeria), the Kafue Gorge (Zambia) and Nymbe ya Mungu (Tanzania) lake flood-plains. This is normally a riverine rather than lacustrine phenomenon, and to that extent we are not concerned with it here, but there are notable examples which do have a lacustrine relevance, e.g. L.Chilwa (Malawi) which periodically dries up and is refilled (A.J. McLachlan, et al 1972) and also the sediment exposure at the edge of man-made lakes due to seasonal drawdown as exemplified by L.Kariba (Zambia-Rhodesia) (S.M. McLachlan, 1970). In both these examples the element of time is eviden[t]. Because of the very rapid re-establishment o[f] a community that is possible under the tropical conditions, the exposed sediment acquir[es] terrestrial primary producers and consumers which sustain the movement of chemical nutrients throughout the dry stage, and whose organic products provide decomposer substrates upon their subsequent re-inundation. This in turn permits a rapid return of the aquatic community. An analogy is the seasonal draining and fallowing of fish ponds but in the tropics this cycle need only take a few weeks.

## Conclusions

The extreme pattern for biological processes in a tropical lake is of continuous and constant production with high turnover rate of elemental material and sustained low concentrations of dissolved nutrients. Most lakes will deviate from this pattern in some degree depending on local condition and proximity to the equator.

L.George (Uganda) comes very close to the extreme pattern, and its nutrient dynamics can be viewed from the standpoint of a continuous culture. On this basis it is conceptually possible to calculate the nutrient demands of the primary producers from the ratios of elements present in their standing crop.

Complications for this approach would be from recycling processes in the sediments, but it is seen from simple mass budgeting that permanent organic deposition is a very small compartment for element accumulation.

Also, physiological measurements strongly indicated a very inefficient decomposition of what was precipitated. Penetration of turbulence was seen to be very important for sediment features, but it was insufficient for efficient decomposition, though it could flush out some dissolved nutrients. The rat[e] at which nutrients could increase in this disturbed zone was seen to be more dependent upon the diffusion rate from lower sediment than by organic decomposition, but the diffusion rate must be retarded by being into mud rather than water and, for $PO_4$-P, by adsorption phenomena as well.

Recycling through sediment in this situation is of minor importance compared to water column recycling. It is considered that the lack of marked seasonality in the sediment environment inhibits the development of bacterial diversity and so of decomposition there. Parallel to this, continuous optimal temperatures in the oxygenated water column makes high decomposition rates here an inevitability. Because so much viable phytoplankton is temporarily located in the upper sediment, and mineralisation due to their own respiration occurs, it is not possible to distinguish the vertical extent of water column recycling. An analogous pattern is seen in deep, permanently stratified, tropical waters where the hypolimnion can function similarly to the upper layer of mud of shallow polymictic lakes.

A very important mud-water relationship in the tropics is the flood-plain phenomenon, in which terrestrial and aquatic communities replace each other successively and mineral recycling may thereby be maintained.

## References


Allen, L.A., J. Grindley & E. Brooks, 1953. Some chemical and bacterial characteristics of bottom deposits from lakes and reservoirs. J. Hyg., Camb. 51, 185-94.

Beadle, L.C., 1974. The inland waters of Africa. London. Longman.

Edwards, R.W. and H.L.J. Rolley, 1965. Oxygen consumption of river muds. J. Ecol. 53, 1-19.

Fogg, G.E., W.D.P. Stewart, P. Fay & A.E. Walsby, 1973. The blue-green algae. London, New York. Academic Press.

Ganf, G.G., 1974a. Diurnal mixing and the vertical distribution of phytoplankton in a shallow equatorial lake (Lake George, Uganda). J. Ecol. 62, 641-59.

Ganf, G.G., 1974. Incident solar irradiance and underwater light penetration as factors controlling the chlorophyll <u>a</u> content of a shallow equatorial lake (Lake George, Uganda). J. Ecol. 62, 593-609.

Ganf, G.G., 1974c. Phytoplankton biomass and distribution in a shallow eutrophic lake (Lake George, Uganda). Oecologia (Berl.) 16, 9-29.

Ganf, G.G. & A.J. Horne, 1975. Diurnal stratification, photosynthesis and nitrogen fixation in a shallow, equatorial lake (Lake George, Uganda). Freshwat. Biol. 5, 13-39.

Ganf, G.G. & A.B. Viner, 1973. Ecological stability in a shallow equatorial lake (Lake George, Uganda). Proc. R. Soc. Lond. B. 184, 321-46.

Gupta, S.P. & R.A. Greenkorn, 1974. Determination of dispersion and non-linear adsorption parameters for flow in porous media. Water Resour. Res. 10, 839-46.

Klotz, D. 1975. Physical fundamentals of the mixing of solutions of pollutants and sewage in porous media. Wat.Res.9, 783-90.

Lund, J.W.G., 1955. Further observations on the seasonal cycle of the plankton diatom <u>Melosira italica</u> ssp. <u>subarctica</u> Müll. J. Ecol. 43, 91-102.

McLachlan, A.J., P.R. Morgan, C. Howard-Williams, S.M. McLachlan & D. Bourn, 1972. Aspects of the recovery of a saline African lake following a dry period. Arch Hydrobiol. 70, 325-340.

McLachlan, S.M., 1970. The influence of lake level fluctuations and the thermocline on water chemistry in two gradually shelving areas in Lake Kariba, Central Africa. Arch. Hydrobiol. 66, 499-510.

Talling, J.F., 1966. The annual cycle of stratification and phytoplankton growth in Lake Victoria (East Africa). Int. Rev. ges Hydrobiol. 51, 545-621.

Talling, J.F., 1969. The incidence of vertical mixing, and some biological and chemical consequences, in tropical African lakes. Verh.Int.Ver.Limnol.17, 998-1012.

Viner, A.B. 1973. Responses of a mixed phytoplankton population to enrichments of ammonia and phosphate, and some associated ecological implications. Proc. R. Soc. Lond. B. 183, 351-70.

Viner, A.B., 1975a. The sediments of Lake George, Uganda. I. Redox potentials, oxygen consumption and carbon-dioxide output. Arch Hydrobiol. 76, 181-97.

Viner, A.B., 1975b. The sediments of Lake George, Uganda. II. Release of ammonia and phosphate from an undisturbed mud surface. Arch. Hydrobiol. 76, 368-78.

Viner, A.B., 1975c. The sediments of Lake George, Uganda. III. The uptake of phosphate. Arch Hydrobiol. 76, 393-410.

Viner, A.B., 1975d. Non-biological factors affecting phosphate recycling in the water of a tropical eutrophic lake. Verh.Int.Ver. Limnol. 19, 1404-15.

Viner, A.B., 1976. The sediments of Lake George, Uganda. IV. Vertical distribution of chemical characteristics in relation to ecological history and nutrient recycling. Arch. Hydrobiol. 78, (in press).

Viner, A.B., 1977. Relationships of nitrogen and phosphorus to a tropical phytoplankton population. Hydrobiologia 51, (in press).

Viner, A.B. & I.R. Smith, 1973. Geographical, historical and physical aspects of Lake George. Proc. R. Soc. Lond. B 184, 235-70.



Key words: lakes
Africa
sediments
nutrients
recycling

# BIOTURBATION AND EXCHANGE OF CHEMICALS IN THE MUD-WATER INTERFACE

T. Petr

Department of Zoology, Monash University,
Clayton, Victoria 3168, Australia

Summary

Macrobenthos disturbs the normal equilibrium state of the sediment-water interface, and the microgradients of physico-chemical properties in the sediment. The vertical burrowing of aquatic invertebrates has been implicated in the movement of interstitial water, dissolved gases, and reduced substances. Bioturbation increases the influx of water to the sediments, which must equal the volume of water containing solutes flushed from sediments. The role of benthic fauna in circulating materials from the sediments is of chemical, pollutional, as well as ecological importance. Also, benthos plays a role in production, transport and modification of sediments, and in their stratification.

## Introduction

The regeneration of nutrients in the sediments is a two stage process: the first one which includes bacterial activity and physico-chemical reactions, determines the concentration of nutrients in the pore water; the second one governs their release through the mud-water interface (Serruya et al.1974).

The rate and character of mineralization of aquatic matter in bottom sediments depend on the composition of substances reaching the bottom, and on the character and depth of the water body. Physico-chemical conditions of the overlying water and in sediments themselves have a significant influence on the mineralization of organic matter in sediments. The wind and the surface area of a shallow lake will determine the actual size of resuspension and the dynamics of the interface. The frequent stirring of the sediment may even cause a remarkable lack of zoobenthos (Lastein 1976). However, in relatively deep lakes, water movements will cause mixing of only a few mm of surface sediments. If their effect were more intensive, the subtle time-depth changes in stratigraphy of fossils in cores would not exist (Stockner & Lund 1970).

The concentration of dissolved chemical components in interstitial water is considerably higher than in overlying water (Glass & Podolski 1975). In two Canadian lakes the lake floor contributed 15-20% of investigated cations to the total input to the water. However, Mortimer (1971) stated that as long as the oxygen concentration at the sediment surface does not fall below 1 or 2 mg $l^{-1}$,typical deep water sediments in the American Great Lakes will probably be found to exert a measurable but quantitatively unimportant influence on the chemistry of the overlying water.

The exchange across the interface is regulated by a variety of mechanisms. Among the most important are: mineral-water equilibria, sorption processes, notably ion exchange, redox interactions dependent on oxygen supply, and the activities of organisms. The exchange rates depend on the local diffusion coefficient as well as on environmental control of inorganic and organic (enzymatic) reactions (Mortimer 1971). The process of release of dissolved substances takes place by actual dissolution of particles, as well as by metabolic activity of organisms. The latter is responsible for the decomposition of organic matter and its mineralization, a process of release of its inorganic components and their return back into circulation. Products of bacterial reactions can bring about further changes in sediment chemistry. Wood (1975) suggested that of several mechanisms responsible for the movement of material across the mud-water interface, ordinary diffusion is probably the least important, while turbulent exchange, release of gases from microbial activity, and the movement of certain biota are of considerable importance. While this may be so in shallow water deposits, it would appear that molecular diffusion would be of considerable importance in deep waters, where the other mechanisms will be of low intensity (Berner 1976).

The character of components liberated from the substrate by zoobenthos in the process called bioturbation will depend primarily on the presence of mineralized and dissolved organic components, which in turn will depend on the stage of decomposition of organic matter.

## Definition of bioturbation

Bioturbation is stirring of sediments by the activity of burrowing benthonic organisms. The burrowing activity of large decapods, certain fish and diving birds should possibly be included here. However, the animals usually understood to be involved in bioturbation are invertebrates such as oligochaete worms, insect larvae, crustaceans and bivalve molluscs. A number of other organisms probably participates as well, but as their activity is less conspicuous than that of the above, they are usually neglected. This concerns also the meiofauna.

The impact of bioturbation is a function of the body size of the animal, its activity, and its depth penetration. The bioturbation results in (a) pumping the enriched interstitial water out of sediment, and bringing in water mostly richer in oxygen, (b) active transport of particulate material to the surface and into deeper layers (c) discarding faecal pellets onto the surface.

Most of the studies have dealt with the following aspects of bioturbation: (a) the stratigraphy of sediments (b) biological interrelationships between (i) individual macrozoobenthos species (ii) macrofauna and microorganisms of the sediment (c) chemical and physical changes in the interstitial water and overlying water (d) removal of nutrients and decomposition processes in treatment systems (e) release of heavy metals from sediments.

Four major groups of organisms, but only few species in each of them, have been studied for bioturbation. Crustacea: Pontoporeia affinis, Marzolf 1966, Milbrink 1973; Ostracoda, Berg 1938; Mollusca: Pisidium, Berg 1938, Edmondson 1975, Jónasson 1972; Anodonta, Jernelöv 1970; Oligochaeta: Alsterberg 1925, Berg 1938, Brinkhurst & Kennedy 1965, Chua & Brinkhurst 1973, Brinkhurst, Chua & Kaushik 1972, Kleckner 1967, Davis 1974a, 1974b, Hargrave 1972b, Poddubnaya 1961, 1965, Ravera 1955, Rybak 1969, Schumacher 1963, Stockner & Lund 1970, Wagner 1968, Wavre & Brunkhurst 1971, Wood 1965: Insecta: Chironomidae: Brundin 1951, Edwards 1958, Ganapati 1949, Jónasson 1970, 1972, Jónasson & Thorhauge 1972, McLachlan & McLachlan 1976. Milbrink 1969, Rybak 1969, Tessenow 1964, Walshe 1950; Hexagenia limbata (Ephemeroptera): Hunt 1953.

## Vertical distribution of macrofauna in lake sediments and its effect on stratigraphy

In the vertical distribution of benthic macrofauna the nature of food and the amount of oxygen in the habitat will be of considerable importance.

Chironomids have been found 1 m deep in lake deposits, but their density at depths more than 15 cm is generally insignificant (Berg 1938). Stockton & Lund (1970) found two chironomid larvae at 55 and 65 cm in Esthwaite Lake in the anoxic zone. Generally, however, they tended to be distributed within the top 5 cm in both Esthwaite and Windermere.

Davis (1974a) summarizing the literature data available on tubificid vertical distribution averaged the information and pointed out that 97% of the worms are above 16 cm with 42% in the upper 4 cm, 32% 4-8 cm, 18% 8-12 cm, 5% 12-16 cm, and then sporadically to 40 cm. In his experimental containers burrows occurred rarely below 40 cm. Poddubnaya & Sorokin (1961) noted that the worms did not feed below 20 cm. Kleckner (1967) concluded from laboratory experiments in which chironomid larvae and tubifics mixed glass microbeads into the surface sediment that generally only the top 3 cm were disturbed to a significant degree. Brinkhurst & Kennedy (1965) established the maximum abundance of tubificids in 4-5 cm depth, and Stockner & Lund (1970) stated that burrowing invertebrates will disturb stratigraphy only in the surface immediately surrounding them and to a depth equal to the length of their burrows. They believed that invertebrate mixing does not destroy the stratigraphy but it merely blurs it. Davis (1974a) found most intensive feeding in about 3-4 cm below the surface. From field observations and laboratory experiments he concluded that at least a few decades of deposit are affected by tubificid burrowing activities. The stratigraphy was disturbed most intensively in a depth less than 12 cm but slight disturbance of layering was evident down to 20 cm. Davis concluded that a given vertical extent of bioturbation is more important if sediment accumulation is slow, because more years of deposit are involved. Continuous homogenization of sediments by ingestion and defaecation as observed by Poddubnaya (1961), followed by their reworking in coprophagy, with further erosion by bottom water currents will lead to destruction of any lamination. Low densities of benthos in deep sediments will cause less disturbance and this will be reflected in a good lamination of sediments, such as observed in some English lakes (Stockner & Lund 1970).

## Penetrability, compaction and water content of the sediments

Loose sediments and a high content of interstitial water favour vertical penetration by burrowing animals. Particle size of the substrate is of obvious importance for the burrower, and for example, Hexagenia limbata of North American lakes and rivers does not burrow into sand as the tube there does not keep shape. This mayfly prefers for its burrowing a substrate with 66.5 to 84.0% water content (Hunt 1953). Berg (1938) suggested that the compactness of the sediments sets the limit to depth penetration of burrowing forms. In a marine littoral dominated by bivalve deposit feeders, the water content in the uppermost sediment was 60% by weight, at stations with fewer deposit feeders the water content was between 30% and 50% (Rhoads & Young 1970). This biological control of compaction was confirmed by an experiment, in which removal of bottom organisms from muds resulted in a decrease of approximately 10% in the water content. The compaction of reworked sediments fluctuated with water temperature, reflecting seasonal changes in metabolic activity of benthic organisms (Rhoads & Young 1970). In another set of experiments, lower compaction of mud and higher water content resulted in an increased uptake of lithium chloride by deep sediments under the impact of the burrowing and respiratory activity of Chironomus riparius larvae (Edwards & Rolley 1965). This suggests the importance of a less compact substrate for the increased import of dissolved chemical compounds and gases, e.g. oxygen.

Some authors consider organic content in sediments to be more responsible for depth penetration by burrowing animals than the water content (e.g. Stockner & Lund 1970). Finally the anatomy of the burrowing organisms is also of importance: in Rybinsk reservoir the strong musculature of Limnodrilus newaensis (Oligochaeta) is reflected in the occurrence of this worm in a larger variety of substrates than Limnodrilus hoffmeisteri with weaker musculature. In the same water body the latter species prefers clay with plant detritus (Poddubnaya 1965).

## Transport of sediments

The burrowing and feeding activities of zoobenthos results in transport of particles mostly to the surface. To follow the movement of individual particles Davis (1974a) used acridin orange for staining pollen grains to follow their movement under UV illumination which records the fluorescence. Poddubnaya & Sorokin (1961) used $^{14}C$ labelled algae layered at various depths in aquarium experiments, and Wood (1975) applied rhodamine B as a tracer substance.

Limnodrilus spp. under experimental conditions transfer 17 ml of sediments worm $^{-1}$yr$^{-1}$ (Davis 1974). Summarizing the literature data, Davis gives a range of 0 to 83 ml worm $^{-1}$yr$^{-1}$. There is a difference with species and with temperature. A rise in temperature in the laboratory speeds up the feeding activity of oligochaetes, resulting in faster reworking of sediments and in turn speeding up biochemical processes. For Limnodrilus udekemianus and L. hoffmeisteri, the mean value of volume of sediment defaecated at 21-22°C was 59 and 61 ml worm $^{-1}$yr$^{-1}$ respectively (Poddubnaya 1961). A rise in temperature is also reflected in an increase in the pumping of water through their burrows; in experiments, the pumping rate of the oligochaetes L. hoffmeisteri and Potamothrix vejdovskyi over a temperature of 10°C (from 10°C to 20°C) increased 5 times (Wood 1975).

The sediment turnover by two species of oligochaete was followed in another set of experiments in which L. hoffmeisteri and L. udekemianus were placed on a 5 cm layer of sand covering silt. In two days almost all individuals reached the silt, and in two weeks they transferred silt on to the sand, which became overlayed by a 5 cm layer of faecal pelogen (Poddubnaya 1961). In marine environments the amount of time required by species populations to rework completely all the sediment in a given area of the depth accessible to individuals varies from a low of once every 15 years to a high of once every 10 weeks. Most rates fall in the range of 1 to 5 years (Nichols 1974). Poddubnaya's experiments indicate that much higher rates can be expected in freshwater environments. Her experimental data support Davis' (op.cit.) conclusion that burrowing benthic macrofauna must exert considerable disturbance on the stratigraphy of sediments.

The intensity of the transfer also depends on the composition of the soil: the higher the concentration of utilizable matter in the sediment, the slower the passage of food through the gut, or in other words the more food available in the substrate the slower the turnover rate. Nichols (1974) noted that the highest organic carbon content coincided with the lowest sediment turnover rate by polychaetes. In Lake Maggiore oligochaetes were feeding most slowly on muds with very high nutritive value (Ravera 1955). Poddubnaya (1961) found oligochaetes feeding fastest on peat soils, more slowly on grey soils, and on the transitional sediment they fed even more slowly. If allowed to completely utilize the nutritive content of the substrate the worms in peat stopped feeding in 19 days, in the transitional type of soil in 24 days, and in the grey silt-sand in 32 days. The peat was utilized almost completely; when the worms stopped feeding, only large particles of peat and a few sand grains were left over. In Lake Maggiore the different amounts of mucus admixture to faeces among three species of oligochaetes resulted in different dissipation rates of these pellets, with Peloscolex ferox pellets dissipating faster than those of Bythonomus lemani and Psammoryctes barbatus (Ravera 1955). Psammoryctes produced a larger quantity of faecal pellets per unit of time than the other two species. Ravera also found that different species have different metabolic levels which will result in differential excretion rates.

Davis (1974a) found the downward transport of sediment by oligochaetes to represent only 14-19% of the upward transport.

It has been suggested that nematodes are more efficient detritus feeders than the larger macrobenthos (McIntyre 1964). Although no data are yet available, one could speculate that their ability to transport particles over larger distances is probably slight.

## Bacterial feeding

In sediments, bacterial densities show a steep gradient, with highest concentrations in the first 10 cm depth (Hayes & Anthony 1959). Assuming that the highest density of microbial feeders such as oligochaetes is there where food is abundant, and that other factors such as oxygen supply are favourable, high bioturbation may be expected.

Pontoporeia affinis is a burrowing amphipod dominating benthic fauna in the deeper parts of Michigan, Huron and Superior Lakes. It is also a common benthic animal in Swedish lakes where it is considered along with chironomid larvae as the most mobile benthic invertebrate (Milbrink 1973). Marzolf (1966) found that the distribution of P. affinis does not show any significant correlation with depth of sampling, particle size distribution or percentage loss on ignition. Positive correlation was found only with the number of bacteria in the sample. P.affinis selects particles whose surfaces were enriched with organic matter, or which were allowed to "condition", assumedly by the growth of bacteria on the substrate surfaces. Marzolf's investigations indicated that the bacteria in the sediments may serve as the source of organic nutrients and energy to this important benthic form. The observation that animals can utilize bacteria as food is not new. Alsterberg (1925) was one of the first scientists to observe a remarkable reduction in number of bacteria by freshwater oligochaetes. Fenchel (1970) and Hargrave (1970) stressed that it is not the detritus which supplies the energy, but rather the associated microbiota. Preference experiments with marine meiofauna indicate that most forms will select sands having an organic coating. This coating is usually bacterial in origin. Chua & Brinkhurst (1973) described selective bacteria feeding by tubificids; after one week only one bacterial species survived passage through the worm gut, a different species in each instance: Acromonas sp. in Peloscolex multisetosus, Micrococcus sp. in Limnodrilus hoffmeisteri, Pseudomonas sp. in Tubifex tubifex.

Coull (1973) noted that the species of bacteria and not their numbers per unit is the factor controlling the attractiveness of a particular unit. Benthic fauna recognize it by "tactile chemical sense", i.e. direct contact with the bacteria.

Poddubnaya (1961) kept tubificids on the same substrate for a long period of time and defined three periods in their activity: (1) worms are feeding predominantly on organic particles, rich in food; (2) worms are feeding actively on less valuable particles, and (3) worms stop feeding, as the substrate consists almost entirely of mineral particles. Experimental findings by Wagner (1968) support point two: tubificids were more active in sandy substrate than in silt, as the former contained less nutritive substances. Poddubnaya's experiments may indicate that in fresh water the recolonization of the substrate by bacteria might be slower than in a brackish environment where the bacterial community on faeces of amphipods is restored within 4 days after defaecation (Fenchel 1970). It could also mean that when the amount of substrate is limited, as in Poddubnaya's experiments, the large sized faeces and clasts produced remove these from rapid circulation, as erosional forces are absent and the size is above the limit to be taken up by the oligochaetes.

## Trophic interactions

Jónasson & Thorhauge (1972) found in Lake Esrom that a large population of Chironomus anthracinus lowered recruitment of oligochaetes. Davis (1974a) found that the 180μ fraction which represents almost entirely chironomid faecal pellets, had 33 ± 10% more organic matter than the finer portion of the sediment. This fraction was considered to be too big to be ingested by examined tubificid species, and Davis concluded that the presence of chironomids resulted in a diminished food supply for tubificids which was reflected in the low organic matter content in tubificid faeces. McLachlan & McLachlan (1976) established that apart from a significant increase in the amount of fines (i.e. particles less than 105μ in diameter) as a result of chironomid larval activity, the larvae simultaneously are producing faecal pellets and silt, both of which are aggregates greater than 250μ in diameter. In laboratory conditions tubificids started gradually consuming larger particles as they were running out of smaller ones. But even under starvation, particles >125μ formed only a small percentage of food taken (Wagner 1968).

A tubificid species interaction was found by Brinkhurst, Chua & Kaushik (1972) in experiments with tubificids of Canadian lakes. The presence of Peloscolex multisetosus increased the weight gain of Tubifex tubifex and Limnodrilus hoffmeisteri by a factor of 2-3 x that found in pure culture. There was no decrease in percentage organic matter, percentage nitrogen or caloric content in faeces; the values were actually higher than those present in the mud taken as food. This was considered to be a result of selective feeding on bacteria. Wavre & Brinkhurst (1971) demonstrated a reduction in the number of heterotrophic aerobic bacteria of about 72% from the number present in mud offered as food.

By partitioning nutritional resources three or more species of oligochaetes can coexist by sharing a seemingly identical sediment as a food source. By their selective feeding on different bacteria within the same sediment they affect the composition of this microflora, as well as being affected by it (Chua & Brinkhurst 1973). This will help maintain a higher species diversity of the zoobenthos. Trophic interactions will be reflected indirectly in the intensity of bioturbation through their effect on the population densities, species diversity and feeding activity of macrozoobenthos.

Trophic specialization of the sulfide community to bacterial food was clearly established in ciliates of which 9 out of 11 dominant species fed on bacteria only (three predominantly on sulfur bacteria (Fenchel & Riedl 1970). Studies on feeding interactions within the sulfide biome are yet to be initiated.

## The effect of bioturbation on physical and chemical factors

After disturbance sediments without macrofauna will reach an equilibrium state within two weeks (Hargrave 1975). The equilibrium can be easily disturbed by bioturbation. Bioturbation influences both the interstitial water and that overlying the sediments. Diffusion into and out of sediments is facilitated by burrows.

Visually, pumping of oxygenated water into the sediment may be detected by orange coloration of tubes (ferric hydroxide) situated within black ferrous sulfide mud (Gorham 1958). The import of oxygen appears to be of the utmost significance for the majority of physico-chemical processes connected with the release or fixation of dissolved substances from and within the sediments.

The rate of water pumping out of sediments was measured at 20°C in the highly polluted Keating Channel in Toronto Harbour, with a density of oligochaetes 194 000 $m^{-2}$ $hour^{-1}$ (Brinkhurst 1970). This may be considered as one of the highest values for this type of system, as densities of benthic macrofauna in less polluted waters are considerably lower. The pumping activity will also be correspondingly lower at lower temperatures. The release of ions from mud into the overlying water by pumping activity is not dissimilar from that due to erosion of bottom sediments by water.

In marine environments reworking of muddy sediments by deposit-feeding organisms may result in an unstable bottom whose properties are not suitable for suspension feeding. Rhoads and Young (1970) call this trophic group amensalism.

### (i) Turbidity

Under experimental conditions in sand filters, along with a number of chemical changes, chironomid larvae increased the turbidity in the overlying water as a result of their burrowing activity (Ganapati 1949). McLachlan & McLachlan (1976) observed a conductivity and turbidity increase in Lake Chilwa (Malawi and Lake Lady Burns(England A greater turbidity above the burrowed sediment may be also a result of differentia erosion: faecal pellets and mud clasts are preferentially eroded and suspended at water current velocities greater than 4 cm $sec^{-1}$. The increase in interface roughness also lowers the critical erosion velocity (Rhoads & Young 1970).

### (ii) pH

The irrigation activity of tubificids reduced the pH in the upper 1-2 cm, and an increase was recorded below 2-3 cm (Davis 1974b). The introduction of chironomid larvae to settled activated sludge increased the pH throughout the mud profile, with higher increase (from 6.85 to 7.73) recorded at a depth of 6 cm (Edwards 1958). The caus of the smaller change in pH in surface sediments still needs to be clarified. According to Edwards the increase in pH of the deeper mud was due to removal of acid metabolic products due to irrigation activit of midge larvae. The changes of pH influenc numerous chemical reactions, and in conjunction with other factors e.g. $E_h$, dissolved $O_2$ concentration, they will result in fixation in or release of chemical compounds to the mud-water interface.

### (iii) Oxygen and redox potential

The activity of oligochaetes and their vertical movement in the sediment is directl related to oxygen conditions near the sediment interface. During summer stratificatic the hypolimnetic water may become rapidly depleted of oxygen and feeding activity is minimal; at that time in Lake Easthwaite th tubificids spend a great portion of their time at the mud-water interface (Stockner & Lund 1970). With the restoration of aerobic conditions tubificids are actively feeding.

In the laboratory the larvae of Chironomou plumosus in well-aerated water spend about 50% of the time on respiratory irrigation, 35% on filter feeding, and the remainder on rest. With decrease in oxygen concentration less time is occupied by filter feeding and immobility and more by respiratory irrigatio At less than 10% $O_2$ saturation the larvae no longer feed (Walshe 1950). Further decrease in $O_2$ accompanied by increase in $CO_2$ and decrease in pH, and an increase in alkalinit causes lengthening of the irrigation period. Final immobility takes place at less than 3% $O_2$ saturation as a result of excessive accumulation of anaerobic products.

Brundin (1951) has shown that with decreasing body size chironomid larvae are less capable of managing under low oxygen concentrations. Jónasson (1970) concluded that th ability of the larva to break through a micr stratification of the bottom layers increases with increasing size.

The mayfly Hexagenia limbata larvae have lowest oxygen consumption in the most suitable substrate. Larger larvae produce a significantly greater current, and hence could increase the oxygen concentration in the mud.

The microbial activity can be greatly enhanced by improved import of oxygen. The increase in oxygen uptake by bottom muds due to bioturbation is known, for example, from experiments by Zvetkova (1972), who noted that under experimental conditions the rate of biochemical oxidation of organic matter in sediments increased 1.5 to 2 times in the presence of tubificids. A considerably larger increase has been recorded for the marine littoral environment, where the rate of energy degradation of the bacteria living in the seagrass Spartina was increased about 25 times when they were experimentally supplied with oxygen (Teal & Kanwisher 1961). However this was not due to the bioturbation which was not tested in this experiment.

Mass transport of oxygen into the sediments by bioturbation will be of considerable importance where suitable organisms exist in large densities. A concentration of 50 000 tubificids $m^{-2}$ resulted in an extension of the oxidized zone to at least 15 cm deep (Schumacher 1963), and a high concentration of chironomid larvae in Lake Esrom more than doubled the depth of oxygen penetration (Hargrave 1975). The relative value of chemical uptake of sediments increased after the emergence of chironomids from Lake Esrom muds.

Sediments with lowest $E_h$ have the greatest oxygen uptake (Mortimer 1941, 1942). Hallberg (1973) suggested that a sediment having a redox potential value lower than zero should not necessarily be considered as being anoxic. Oxygen does not control $E_h$ potential but rather the strongly reducing substances are responsible for lowering $E_h$ and they do not occur in the presence of oxygen. However, reducing organic substances such as humic acids from Sphagnum with $E_h$ +300 to +400 mV, can coexist with oxygen (Visser 1964).

Burrowing activities of benthic fauna such as Chironomus riparius in sludge (Edwards 1958) and Ilyodrilus hammoniensis in Lake Esrom (Hargrave 1972) resulted in an increase in the $E_h$ in surface sediments. In a north American lake the presence of tubificids increased the $E_h$ in 2 and 3 cm depth and the oxidized layer (>200 mV) was deepened by 0.3 to 1.6 mm (Davis 1974b).

In experiments with Lake Chilwa mud, the burrowing activity of chironomid larvae resulted in a decrease in $E_h$ from +168 mV to -130 mV in the surface mud layer, and in a decrease from -313 to -319 mV in $E_h$ of the bottom layer (McLachlan & McLachlan 1976).

To get a better understanding of the impact of bioturbation on the chemicals exchange between sediments and the overlying mud, we have to briefly summarize the oxidation-reduction processes taking place at different $E_h$. It has already been mentioned above that the level of $E_h$ is correlated with the oxygen concentration in the interstitial and overlying water. At +400 mV and more there is 4-10 mg $O_2$ $l^{-1}$, at +360 to +400 mV about 4 mg $O_2$ $l^{-1}$, at +300 mV about 0.3 mg $O_2$-$l^{-1}$, and at +200 mV 0.1 or less mg $O_2$ $l^{-1}$. The change in $E_h$ is reflected in chemical reductions, and it is well known that at < +350 mV $NO_3^-$ is reduced to $NH_4^+$ (Edwards 1958), at<+300 mV $Fe^{+++}$ is reduced to $Fe^{++}$ (Mortimer 1941), with maximum solubility of Fe at +80 to -100 mV (Serruya 1971), at < +100 mV of sulphur takes place (Baas-Becking et al. 1960), and that strongly reduced sediments have $E_h$ of 0 to -250 mV. While Hallberg (1973) recorded oxygen conditions at <-400 mV, Hargrave (1972) and Mortimer (1941) found no oxygen at +100 mV.

Some of the most important reactions directly resulting from the bacterial decomposition of organic matter are the removal of dissolved oxygen, reduction of sulphate, nitrate, manganese, iron and some other elements and compounds, and production of $CO_2$, $CH_4$, $NH_4^+$, $PO_4^{3-}$ and $H_2S$. The intensity will depend on the content of organic matter and oxygen availability. Products of bacterial reactions can bring about further changes in sediment chemistry. The concentrations of Na, Ca, Mg, K, Fe, Mn, and possibly $SiO_2$ in pore waters are affected by the microbiological decomposition of organic matter as well as by the concentrations of oxygen, $HCO_3^-$, $SO_4^{2-}$, $NO_3^-$, $N_2$, $H_2S$, $PO_4^{3-}$, $CO_2$ and $CH_4$ (Berner 1976). Bacterial reactions often seem to dominate the chemistry of pore water (Manheim & Sayles 1974). Thus the major function of bioturbation, apart from providing feeding and respiratory currents, will be in the release of the dissolved compounds to the overlying water.

### (iv) Nitrogen

Viner (1975) found that under experimental conditions a rapid initial increase in ammonia and phosphate in overlying water from Lake George (Uganda) is due to non-biogenic simple diffusion from the large reservoir of the interstitial water of the mud, but subsequent slow release is biogenic. In another set of experiments by Ganapati (1949), Chironomus riparius larvae increased the concentration of the free $NH_4^+$ in the overlying water as a result of their activity in sand filters of the Madras Water Works. Also Edwards (1958) found an increase in $NH_4^+$ in the overlying water where chironomid larvae were introduced into the sludge. The result was also a decrease in oxidized N in mud. A dentrification results in a significant loss of inorganic N produced in mud by dentrifying bacteria. Excretion of $NH_4^+$ by chironomid larvae was not significant.

(v) Phosphorus

In sediments, P can exist in a considerable number of compounds. Under the effect of bioturbation $E_h$ sensitive P will be released from muds, but the bioturbation will probably not affect $E_h$-indifferent P such as that adsorbed on the clay fraction. The amount of P released will also be determined by the strength of the adsorption of P on the substrate. In some experiments, the large difference in P release between untreated sediments was assumed to be due to an aging effect making adsorbed P more strongly bound by rearrangement of the phosphate ions on the surface (Kuo & Lotse 1974). The rate of release of P by bioturbation then will be determined by the physico-chemical situation.

At pH 8 and higher precipitation of phosphate is enhanced, (Helfrich & Kevern 1973). There is a high adsorption to clays at low pH: at pH <6.5 release into water of two orthophosphates, $H_2PO_4$ and $HPO_4$ takes place. The main result of these desorption-precipitation reactions is a transformation of adsorbed phosphate into hydroxyapatite, which is practically insoluble in alkaline waters (Andersen 1975). It will therefore be of importance to know at what pH and $E_h$ bioturbation takes place in order to assess the solubility of P compounds which might be flushed out of sediments. In the laboratory addition of sulphate reducing bacteria seems to enhance the rate of oxygen decrease in the water, but it results in an immediate release of phosphate (Schippel et al 1973).

Lake Kinneret sediments exposed to $H_2S$ released 17 times more $PO_4$-P from mud than those in aerobic conditions (Serruya et al. 1974). This indicates that bioturbation bringing oxygen into sediments would subsequently flush out only a small amount of P due to its precipitation in an oxidized environment. On the other hand, macrofauna which can be active in deoxygenated environment such as that present in muds of Lake Kinneret (Por & Masry 1968) might enhance the release of P and other dissolved substances into the overlying water.

At pH 6.1 Chironomus plumosus larvae increased the release of P from the slow sand filters of the Madras Water Works (Ganapati 1949). In the marine environment of Puget Sound, input of P from interstitial into the overlying water was suggested by Nichols (1974) to be due solely to the activity of the polychaete Pectinaria californiensis.

In experiments by Davis et al. (1975) tubificid worms reduced the proportion of phosphate in the uppermost interstitial water. This may have resulted from the increase in $E_h$ caused by worms, leading to the precipitation of $PO_4^{3-}$ with ferric ions or to entrapment in flocks. The effect of the worms on the proportion of phosphate to non phosphate P was at least in part mediated by their effect on $E_h$ of the sediment. However the impact of this on the exchange of P between sediment and overlying water will vary with the proportion of redox-sensitive P. When most of the P is redox-insensitive the effect of worms on its mobility would be due primarily to their effect of increasing the rate of physical contact between sediment and water (Davis et al. 1975).

(vi) Silicon dioxide

In the laboratory, addition of chironomid larvae to sediments resulted in a decrease in $SiO_2$ in interstitial water. A higher larval density led to lower $SiO_2$ concentrations and to its removal from deeper sediment than a lower larval density. As a result of bioturbation the increase in silica in the overlying water was 4 to 5 times greater than in control experiments without larvae (Tessenow 1964). Tessenow suggested (but not tested) that apart from chironomids, other organisms such as tubificids, ostracods, copepods, ciliates and bacteria might increase the rate of flux of silica at the mud-water interface.

(vii) Iron, calcium, magnesium

In an experimental set up with sediment, Edwards (1958) recorded a black layer of ferrous sulphide in a depth of 6 cm with an $E_h$ of -120 to -203 mV. The addition of chironomid larvae resulted in a drift of the $E_h$ isopleth deeper into the mud, and emergence of the adults resulted in an upward drift of the isopleth. McLachlan & McLachlan (1976) found that bottom fauna also promotes release of $Ca^{2+}$ and $Mg^{2+}$ from sediments.

(viii) Methylmercury

When tubificid worms or Anodonta mussels are present in sediments, deposits down to 2.5 or 9 cm respectively are activated and release methylmercury which is formed there by microbial activity (Jernelöv 1970). In estuarine environments resuspension of bottom sediments by benthic organisms is considered to partly account for the decrease of mercury with time in contaminated estuarine sediments (Bothner 1973, in Nichols 1974). As the inorganic mercury, as well as methylmercury, has a high affinity for organic substances (Jernelöv 1975), it will tend to accumulate in areas of high productivity with high concentrations of benthic organisms. Bioturbation in such productive (eutrophic) areas can be expected to be more intensive than in oligotrophic areas, and it will result in higher methylmercury concentration in the overlying water.

## Sulphide system

In most situations, as soon as the organic material has been buried more than a few mm by either animal movement or further deposition on top, it is in an anaerobic environment.

Recently, a complex anaerobic ecosystem was found to have world-wide distribution and was called sulphide system due to the presence of $H_2S$ (Fenchel & Riedl 1970). Its upper boundary is definable by a redox-potential discontinuity (RPD), a layer where oxidizing processes become displaced by reducing processes. The lower limit of the system is defined by termination of all biological activities.

If the RPD is at or near the surface then the macro- and meiofauna of the oxidized biome is drastically reduced or completely wiped out, or restricted to those few species that can survive anaerobically. However, there is a specialized living community present in the sulphide system including Monera, Protista, Fungi, and Metazoa such as Turbellaria, Gnathostomulida, Nematoda, Gastrotricha, Rotifera, Oligochaeta and some Polychaeta. The presence of eggs and embryos suggests that a number of metazoan undergo all the life cycle in the sulphide system, and many species, i.e. ciliates, appear to have a definite distributional pattern (Fenchel 1969). Considering the immense trophic importance of anaerobic processes in the aquatic environment it is desirable to pay more attention to the interstitial fauna of the sulphide ecosystem.

The reduced materials from the sulphide system will be transported slowly upward by diffusion in a similar way to that in which oxygen moves down. To what extent the sulphide community affects this transfer is not known. It is known, however, that the Nematoda in this system are sensitive to small changes in substrate composition and loci of nematode activity may occur within an apparently uniform and homogenous substrate. The migration of nematodes to loci containing viable bacteria is known, as well as the fact that these bacteria serve them as food (Wilt & Smith 1970). The effect of nematodes on the transfer of chemicals within and out of the sediment, apart from that due to removal of nematodes from the substrate by predation is unknown. There is little evidence to suggest any important interaction between meiofauna of the sulphide system and organisms in the oxidized upper layer. While the biomass and turnover appear to be high (Muus 1967, Gerlach 1971), at present it would appear that much of the biomass will stay within the sulphide system from which the decomposing products reach the oxidized layer only by slow diffusion. A number of invertebrates are known to "dive" into the highly reduced black layer of the sulphide system. The number of these divers, the frequency of their visits, as well as the rapidity with which they will return back into the oxidized system, may possibly have some impact on the chemistry of the sulphide and oxidized systems and on the rate of the transfer of chemicals between them.

## Recommendations

1) The community approach in bioturbation studies is important as is evident from species interaction studies. More data is required concerning the selective feeding of macrobenthos on bacteria, on trophic interactions of bacterial feeders, and on bacteria stripping as a means of energy source and chemical transfer.

2) The significance of meiobenthos for the exchange of chemical substances in the sediment and at the sediment-water interface is unknown. Meiobenthos is known to often reach densities exceeding those of macrobenthos. The role of the sulphide biome metazoans in the transfer of chemicals from this system into the oxidized layer is poorly known.

3) While some information is available on the pumping rate of chemicals into overlying water, the rate of dispersion and further transformations of these chemicals need to be closely examined.

4) Methods for substrate analysis should aim at preservation of organic matter and bacteria on mineral particles, and at preservation of faecal pellets, as these prove to be of crucial importance for biological and chemical processes in the mud-water interface.

5) As organic detritus represents all types of biogenic particulate and subparticulate material in various stages of decomposition, including dissolved vitamins, amino acids and other organic end products of decay whose return into circulation via overlying water will be speeded up by bioturbation, close attention to various forms of DOM is desirable.

## References

Alsterberg, G., 1925. Die Nahrungszirkulation einiger Binnenseetypen. Arch. Hydrobiol. 15:291-338.

Andersen, J.M., 1975. Influence of pH on release of phosphorus from lake sediments. Arch. Hydrobiol. 76:411-419.

Berg, K., 1938. Studies on the bottom animals of Esrom Lake. K. Danske vidensk. Selsk., Naturv. Math. Avd. 9,8. 255 p.

Baas Becking, L.G.M., I.R. Kaplan & D. Moore, 1960. Limits of the natural environment in terms of pH and oxidation-reduction potentials. J. Geol. 68:243-284.

Berner, R.A., 1976. The benthic boundary from the viewpoint of a geochemist. In: I. N. McCave (Ed.): The Benthic Boundary Layer. Plenum Press, N.York & London. p. 33-55.
Bothner, M.H., 1973. Mercury: some aspects of its marine geochemistry in Puget Sound, Washington. Ph.D. thesis, Univ.Wash., Seattle. 126 p.
Brinkhurst, R.O., 1970. Distribution and abundance of tubificid (Oligochaeta) species in Toronto Harbour, Lake Ontario. J. Fish. Res. Bd. Can. 27:1961-1969.
Brinkhurst, R.O., K.E. Chua & N.K. Kaushik, 1972. Interspecific interactions and selective feeding by tubificid oligochaetes. Limnol. Oceanogr. 17:122-133.
Brinkhurst, R.O. & Kennedy, C.R., 1965. Studies on the biology of the Tubificidae (Annelida, Oligochaeta) in a polluted stream. J. anim. Ecol. 34:429-444.
Brundin, L., 1951. The relation of $O_2$ microstratification at the mud surface to the ecology of the profundal bottom fauna. Report Inst. Freshw. Res. Drottningholm No. 32: 32-42.
Chua, T.E. & R.O. Brinkhurst, 1973. Bacteria as potential nutritional resources for three sympatric species of tubificid oligochaetes. In: H. Stevenson & R.R. Colwell (Eds.): Estuarine Microbial Ecology. Univ. S. Carolina Press. p.115-125.
Coull, B.G., 1973. Estuarine meiofauna: a review: trophic relationships and microbial interactions. In: H.Stevenson & R.R. Colwell (Eds.): Estuarine Microbial Ecology. Univ. S. Carolina Press. p.499-512.
Davis, R.B., 1974a. Stratigraphic effects on tubificids in profundal lake sediments. Limnol. Oceanogr. 19:466-488.
Davis, R.B., 1974b. Tubificids alter profiles of redox potential and pH in profundal lake sediment. Limnol. Oceanogr. 19:342-346.
Davis, R.B., D.L. Thurlow & F.E. Brewster, 1975. Effects of burrowing tubificid worms on the exchange of phosphorus between lake sediment and overlying water. Verh. Internat. Verein. Limnol. 19:382-394.
Edwards, R.W., 1958. The effect of larvae of _Chironomus riparius_ Meigen on the redox potentials of settled activated sludge. Ann. appl. Biol. 46:457-464.
Edwards, R.W. & H.L.J. Rolley, 1965. Oxygen consumption of river muds. J. Ecol. 53:1-19.
Edmondson, W.T., 1975. Microstratification of Lake Washington sediments. Verh. Internat. Verein. Limnol. 19:770-775.
Fenchel, T., 1969. The ecology of marine microbenthos. IV. Structure and function of the benthic ecosystem, its chemical and physical factors and the microfauna communities with special reference to the ciliated Protozoa. Ophelia 6:1-182.
Fenchel, T., 1970. Studies on the decomposition of organic detritus derived from the turtle grass _Thalassia testudinum_. Limnol. Oceanogr. 15:14-20.
Fenchel, T.M. & R.J. Riedl, 1970. The sulphide system: a new biotic community underneath the oxidized layer of marine sand bottoms. Marine Biol. 7:255-268.
Ganapati, S.V., 1949. The role of the blood-worm, _Chironomus plumosus_, in accounting for the presence of phosphates and excessive free ammonia in the filtrates from the slow sand filters of the Madras Water Works. Zool. Soc. India J. 6:41-43.
Gerlach, S.A., 1971. On the importance of marine meiofauna for benthos communities. Oecologia (Berl.) 6:176-190.
Glass, G.E. & J.E. Podolski, 1975. Interstitial water components and exchange across the water sediment interface of western Lake Superior. Verh. Internat. Verein. Limnol. 19:405-420.
Gorham, E., 1958. Observations on the formation and breakdown of the oxidized microzone at the mud surface in lakes. Limnol. Oceanogr. 3:291-298.
Hallberg, R.O., 1973. The microbiological C-N-S cycles in sediments and their effect on the ecology of the sediment-water interface. Oikos Suppl. 15:51-62.
Hargrave, B.T., 1970. The effect of a deposit-feeding amphipod on the metabolism of benthic microflora. Limnol. Oceanogr. 15:21-30.
Hargrave, B.T., 1972. Oxidation-reduction potentials, oxygen concentration and oxygen uptake of profundal sediment in a eutrophic lake. Oikos 23:167-177.
Hargrave, B.T., 1975. Stability in structure and function of the mud-water interface. Verh. Internat. Verein. Limnol. 19: 1073-1079.
Hayes, F.R. & E.H. Anthony, 1959. Lake water and sediment. VI. The standing crop of bacteria in lake sediments and its place in the classification of lakes. Limnol. Oceanogr. 4:299-315.
Helfrich, L.A. & N.R. Kevern, 1973. Availability of phosphorus-32, adsorbed on clay particles, to a green algae. The Michigan Academician. 6:71-81.
Hunt, B.P., 1953. The life history and economic importance of a burrowing mayfly, _Hexagenia limbata_, in southern Michigan lakes. Bull. Inst. Fish. Res. No.4. 151 p
Jónasson, P.M., 1970. Bottom fauna and eutrophication. In: Eutrophication: Causes, Consequences, Correctives. Nat. Acad. Sci, Washington, D.C. p. 274-305.
Jónasson, P.M., 1972. Ecology and productio of the profundal benthos in relation to phytoplankton in Lake Esrom. Oikos suppl. 14:1-148.

Jónasson, P.M. & F. Thorhauge, 1972. Life cycle of Potamothrix hammoniensis (Tubificidae) in the profundal of a eutrophic lake. Oikos 23: 151-158.
Jernelöv, A., 1970. Release of methyl-mercury from sediments with layers containing inorganic mercury at different depths. Limnol. Oceanogr. 15:958-960.
Jernelöv, A., 1975. Cycling of mercury in the environment. In: M.J. Chadwick & G.T. Goodman (Eds.): The ecology of resource degradation and renewal. The 15th Symp. Brit. Ecol. Soc. 10-12 July 1973. Blackwell, p. 49-55.
Kleckner, J.F., 1967. The role of bottom fauna in mixing lake sediments. M.S.thesis Univ.Wash., Seattle. 61 p.
Kuo, S. & E.G. Lotse, 1974. Kinetics of phosphate adsorption and desorption by lake sediments. Soil Sci. Soc. of America Procs. 38:50-54.
Lastein, E., 1976. Recent sedimentation and resuspension of organic matter in eutrophic Lake Esrom, Denmark. Oikos 27:44-49.
Manheim, F.T. & F.L. Sayles, 1974. Composition and origin of interstitial waters of marine sediments based on deep sea drill cores. In: E.D. Goldberg (Ed.): The Sea. Wiley-Interscience, N.Y., p. 527-568.
Marzolf, G.R., 1966. The trophic position of bacteria and their relation to the distribution of invertebrates. In: Organism-substrate relationship in streams. The Pymatuning Symposia in Ecology, Spec.Publ. No.4, p. 131-135.
McIntyre, A.D., 1964. Meiobenthos of sub-littoral muds. J. Mar. Biol. Ass. U.K. 44:665-674.
McLachlan, A.J. & S.M. McLachlan, 1976. Development of the mud habitat during the filling of two new lakes. Freshwater Biol. 6:59-67.
Milbrink, G., 1969. Microgradients at the mud-water interface. Report Inst. Freshw. Res. Drottningholm No. 49:129-148.
Milbrink, G., 1973. On the vertical distribution of oligochaetes in lake sediments. Report Inst. Freshw. Res. Drottningholm No. 53: 34-50.
Mortimer, C.H., 1941 & 1942. The exchange of dissolved substances between mud and water in lakes. J. Ecol. 29:280-329 & 30:147-201.
Mortimer, C.H., 1971. Chemical exchanges between sediments and water in the Great Lakes - speculations on probably regulatory mechanisms. Limnol. Oceanogr. 16:387-404.
Muus, B., 1967, The fauna of Danish estuaries and lagoons. Distribution and ecology of dominating species in the shallow reaches of the mesohaline zone. Medd. Danmarks Fisk. Havunders.(n.ser.) 5: 316 p.
Nichols, F.H., 1974. Sediment turnover by a deposit-feeding polychaete. Limnol. Oceanogr. 19:945-950.
Poddubnaya, T.L., 1961. Materialy po pitaniyu massovykh vidov tubificid Rybinskogo vodokhranilishcha. (Concerning the feeding of the high density species of tubificids in Rybinsk Reservoir. In Russian). Trudy Inst. Biol. Vodokranilishch AN SSSR. 4(7):219-231.
Poddubnaya, T.L., 1965. (On the tubificid fauna of the Rybinsk reservoir. In Russian). Ekol. Biol. Presnov. Bespozv., Trudy Inst. Biol. Vnutr. Vod. Vyp. 8(11): 20-36.
Poddubnaya, T.L. & Yu.I. Sorokin, 1961. Glubina sloya optimalnogo pitania tubificid v svyazi s ikh peremeshcheniami v grunte. (The depth of the layer of optimal feeding of tubificids in connection with their movements in the sediment. In Russian). Izv. Inst. Biol. Vodokhr. AN SSSR 10:14-17.
Por, F.D. & D. Masry, 1968. Survival of a nematode and an oligochaete species in the anaerobic benthal of Lake Tiberias. Oikos 19:388-391.
Ravera, O., 1955. Amount of mud displaced by some freshwater oligochaeta in relation to the depth. Mem. Ist. Ital. Idrobiol., Suppl. 8:247-264.
Rhoads, D.C. & D.K. Young, 1970. The influence of deposit-feeding organisms on sediment stability and community trophic structure. J. Marine Res. 28:150-178.
Rybak, J.E., 1969. Bottom sediments of the lakes of various trophic types. Ekologia Polska, 17(ser.A):611-662.
Schippel, F.A., R.O.Hallberg & S. Oden, 1973. Phosphate exchange at the sediment-water interface. Oikos Suppl. 15:64-67.
Schumacher, A., 1963. Quantitative Aspekte zwischen Stärke der Tubificidenbesiedlung und Schichtdicke der Oxydationszone in den Süsswasserwatten der Unterelbe. Arch. Fischwiss. 14:88-51.
Serruya, C., 1971. Lake Kinneret: the nutrient chemistry of the sediments. Limnol. Oceanogr. 16:510-521.
Serruya, C., M. Edelstein, U. Pollingher, & S. Serruya., 1974. Lake Kinneret sediments: nutrient composition of the pore water exchanges. Limnol. Oceanogr.19:489-508.
Stockner, J.G. & J. Lund, 1970. Live algae in postglacial lake deposits. Limnol. Oceanogr. 15:41-58.
Teal, J.M. & J. Kanwisher, 1961. Gas exchange in a Georgia salt marsh. Limnol. Oceanogr. 6:388-399.
Tessenow, U., 1964. Experimentaluntersuchungen zur Kieselsäurerückführung aus dem Schlamm der Seen durch Chironomiden-larven (Plumosus-Gruppe). Arch.Hydrobiol. 60:497-504.
Viner, A.B., 1975. The sediments of Lake George (Uganda). II. Release of ammonia and phosphate from an undisturbed mud surface. Arch. Hydrobiol. 76:368-378.

Visser, S.A., 1964. Oxidation-reduction potentials and capillary activities of humic acids. Nature (Lond.) 204:581.
Wagner, G., 1968. Zur Beziehung zwischen der Besiedlungsdichte von Tubificiden und dem Nahrungsangebot im Sediment. Int. Rev. Gesamten Hydrobiol. 53:715-721.
Walshe, B.M., 1950. The function of haemoglobin in Chironomus plumosus under natural conditions. J. Exp. Biol. 27: 73-95.
Wavre, M. & R.O. Brinkhurst, 1971. Interactions between some tubificid oligochaetes and bacteria found in the sediments of Toronto Harbour, Ontario. J. Fish Res. Bd. Can. 28:335-341.
Wilt, G.R. & R.E. Smith, 1970. Studies on the interactions of aquatic bacteria and aquatic nematodes. Auburn Univ. Bull. 701 Auburn, Ala.: Water Resources Res. Inst.
Wood, L.W., 1975. Role of oligochaetes in the circulation of water and solutes across the mud-water interface. Verh. Internat. Verein. Limnol. 19:1530-1533.
Zvetkova, L.J., 1972. Oligochaeta in the oxygen balance of reservoirs. The aquatic Oligochaeta. Trans. Hydrobiol. Soc. USSR 17:118-125.

Key words: Bioturbation
Sediments
Macro benthos
Chemical exchange
Nutrients

# SEDIMENTS AS A NUTRIENT SOURCE IN SHALLOW POLLUTED LAKES

S.-O. Ryding and C. Forsberg

The National Swedish Environment Protection Board
Algal Assay Laboratory, Institute of
Physiological Botany, Box 540, S-751 21
Uppsala, Sweden.

## Summary

By thoroughly made budget calculations estimates of the net flux of N and P to and from the sediments are illustrated for three lakes. The water quality in the lakes showed a characteristic pattern with peaks of $PO_4$-P, high pH-values, and great amounts of organic substances in the middle of the summer.

Substantial amounts of P were released during the summer. The N budgets showed a variable pattern concerning gaining or losing N. By summarizing input-output data on a yearly basis, the magnitude of the sediment release compared with other sources could be illustrated. In all three lakes great amounts of P originated from the sediments, from 22 to over 400 percent of the total external loading per year. The external input of N was greater than contributions from the sediments. Considerable amounts of N were lost due to denitrification. The net loss of N ranged from 35 to 60 percent of the annual external loading.

A comparison of the maximum net release of P observed under natural aerobic conditions in the lakes and gross release rates obtained in the laboratory, showed large differences. In one lake the maximum net release value, 47 mg P/m$^2$ · d, far exceeded the "laboratory release" under anaerobic conditions, 18 mg P/m$^2$ · d.

In one lake the impact of sediments on the overlying water could be demonstrated in a somewhat different way. The water quality and the force and duration of the lengthwise winds in the lake were correlated, indicating that mixing plays a dominant role for the transport of nutrients from the sediments in shallow lakes.

## Introduction

The ability of sediments to absorb nutrients is well documented (see e.g. Shukla et al., 1970). However, the significance of lake sediments as an internal source of lake fertilization is still poorly understood. There is little doubt that in shallow parts of a lake, turbulent mixing frequently extends to the bottom emphasizing the possibility of sediments to act as a major nutrient source. External sources of nutrients may be largely eliminated through modern methods of wastewater treatment or through diversion techniques. Internal loading, however, is more difficult to control. The uptake and release of nutrients by sediments is a complicated function of interacting physical, chemical, and biological processes (Syers et al., 1973).

The N-release from the sediments is controlled to a large extent by biological processes (Kamp-Nielsen, 1974) and is partly a function of the N-content in the sediments and the temperature (Jacobsen & Jørgensen, 1975). The hydrodynamic conditions (Lee, 1970) as well as the oxygen content (Austin & Lee, 1973) and the amount of organic matter (Kamp-Nielsen, 1974) are of great importance. A large pool of the sediment-N can be a source of N to the overlying waters (Byrnes et al., 1972). The N-cycle in the sediment-water systems has been reviewed by Keeney (1973).

The P-release from the sediments is in many lakes primarily regulated by the oxygen conditions (Mortimer 1941, 1942; Fillos & Swanson, 1975). The P-release is also a function of the P-content in the sediments and interstitial water (Stumm & Leckie, 1971; Wildung & Schmidt, 1973; Sanville et al., 1974) and is enlarged by an increase in pH (McPhearson et al., 1958; Graneli, 1974) and organic substances (Tessenow, 1972; Hallberg et al., 1973). A low P-content in the overlying water, (Kamp-Nielsen, 1974) increased turbulence (Sullivan, 1967; Lee, 1970) and high water temperatures (Wilding & Schmidt 1973; Kamp-Nielsen, 1975) also stimulate the release rates. Radioactive tracer techniques with $P^{32}$ have revealed that exchangeable inorganic-P represents a large pool of sediment inorganic-P (Li et al., 1972, 1973).

A comprehensive review of the factors involved in the transfer of materials between water and sediments has been presented by Lee (1970).

In 1972, the National Swedish Environment Protection Board started an investigation of the recovery of 20 polluted lakes after nutrient diversion (Forsberg

Table 1. Background data for the lakes.

| | | L. Glaningen | L. Ramsjön | L. Ryssbysjön | L. Uttran |
|---|---|---|---|---|---|
| Catchment area | $km^2$ | 34.7 | 12.3 | 99.1 | 17.1 |
| Forests/agriculture/lakes/ densely populated areas | % | 95/1/0/4 | 80/13/0/7 | 72/12/5/11 | 36/20/42/2 |
| Surface area | $km^2$ | 0.75 | 0.39 | 2.73 | 2.78 |
| Mean depth | m | 1.5 | 1.8 | 1.8 | 5.7 |
| Hydraulic retention time | year | 0.08 | 0.20 | 0.13 | 4.4 |
| Immediately adjacent land | % of the catchment area | 22 | 20 | 10 | 100 |
| annual N-load | $g/m^2$ | 2.4 | 2.7 | 0.9 | 2.9 |
| annual P-load | $g/m^2$ | 0.03 | 0.05 | 0.01 | 0.09 |

et al., 1975). Special interest is devoted to the role of the sediments as a nutrient source, particularly for N and P. In this paper, results from 1974 are presented for four of these lakes. All are shallow and have been heavily polluted by domestic wastewaters.

## Methods

Water samples were taken in the inlet and outlet once a week the year around simultaneously with observations of the water flow. Automatic sampling of sewage effluent in proportion to the flow was carried out at the treatment plants (Elfving et al., 1975). Lake water sampling (surface water, 0-2 m) was very frequent, between 3-5 times a week. All samples were preserved by deep-freezing. More details about the sampling and analysis performed will be presented elsewhere (Forsberg et al., in prep.)

Net gain and net loss of N and P have been calculated according to the following equation: Net gain, net loss = output -input + $\Delta$ lake water. If the calculations ended up with a positive difference it was interpreted as a net gain; if negative a net loss.

The annual net release of N from the sediments has been estimated by multiplying the annual P-release with a factor describing a mean N/P ratio in the suspended solids in the water and in superficial sediments.

## Background data

Table 1 summarizes some background data concerning morphometrical and hydrological characteristics as well as some data describing the catchment area and loading from the immediately adjacent land. Precipitation on the lake surface add about 1 g $N/m^2 \cdot yr$ and 0.02 g $P/m^2 \cdot yr$ (Ahl & Odén, 1975). The lakes have for several decades been recipient lakes for municipal wastewater discharges. They are all in poor condition and could be classified as highly eutrophic. During recent years the wastewater treatment has gradually been improved. In 1973, chemical treatment of wastewater started at two of the lakes, L. Ramsjön, and L. Ryssbysjön, yielding a reduced input of P. In the two other lakes L. Glaningen, and L. Uttran, wastewater was totally diverted in early 1974.

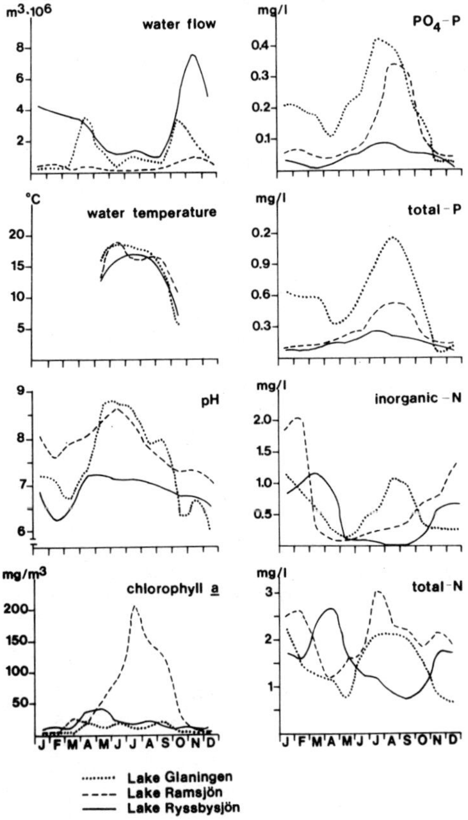


Fig. 1. L. Glaningen, L. Ramsjön and L. Ryssbysjön 1974. Water flow and surface water quality. Monthly average values.

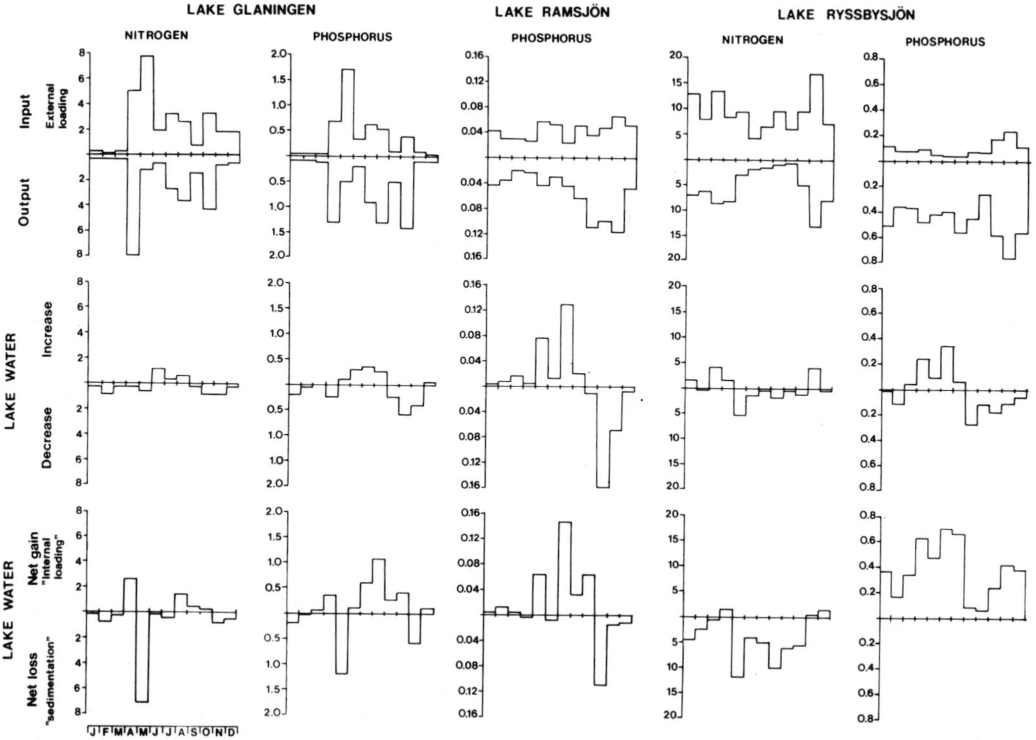


Fig. 2. L. Glaningen, L. Ramsjön and L. Ryssbysjön 1974. Input to and output from the lakes, increase and decrease in the lake water and net gain and net loss for N and P. Monthly average values.

## Results

L. Glaningen, L. Ramsjön and L. Ryssbysjön

Water Quality

Water flow through the lakes and some physical-chemical data from the lakes expressed as monthly mean values are presented in fig. 1. The winter conditions in L. Ryssbysjön and L. Uttran are represented by data for the water quality of the outlet.

Several parameters showed great variations with a similar pattern throughout the year. Maximum water flow was observed in spring and autumn, highest water temperature in June - July, and high pH-values during the vegetation period. The peak of $PO_4$-P and total-P is a characteristic feature for these shallow polluted lakes. The samples from L. Glaningen gave the lowest chlorophyll content. The highest concentration of inorganic-N in this lake was obtained in the middle of the summer. L. Ramsjön had a pronounced peak in chlorophyll _a_ while the inorganic-N was low during the summer.

Loading

The transport of N and P to and from the lakes as well as the increase and decrease in lake water and the net gain and net loss data based on monthly calculations are presented in fig. 2. No data for N from L. Ramsjön is available due to incomplete N-analysis of the sewage effluent.

The transport of N and P to and from L. Glaningen is quite irregularly distributed throughout the year. The input to L. Ramsjön and L. Ryssbysjön is about the same for all months, while the output from these lakes was somewhat higher in late autumn. The P-content in the lake waters showed an increase in late spring - early summer and a decrease in autumn. The same pattern could be illustrated for N in L. Glaningen.

A net gain considering P is equal to a net sediment release; a net loss as net sedimentation. Considering N, the picture becomes somewhat more complicated. Here, $N_2$-fixation as a N-source and denitification as a N-sink must be considered. The net gain and net loss data was somewhat different for the lakes. In L. Glaningen a marked loss of N and P occured in May, while a net gain could be observed mainly for P in the summertime. L. Ramsjön showed a considerable sediment release of P in the summer and a high sedimentation in late autumn. In L. Ryssbysjön, there was a net loss for N and a net gain for P almost throughout the entire year. In fig. 3, the fluxes of N and P are summarized on a yearly basis. In all lakes, there was a considerable net release of P from the sediments. In L. Ryssbysjön, the net release of P was as high as four times the external loading, and in L. Glaningen and L. Ramsjön it was 20-30%. A net outflow of P was observed in all the lakes. The N-situation was somewhat different. Here, the external input was larger than the output. The net loss of N, denitrification - $N_2$-fixation, (arrows in the upper part of the N-boxes) was high in all lakes, from 35-60% of the external input.

L. Uttran

A comprehensive nutrient budget as shown above. cannot be calculated for L. Uttran. This depends on difficulties in measuring the nutrients reaching the lake by many small brooks. However, the role of sediments as nutrient source in this lake can be demonstrated indirectly. After a more total diversion of nutrients in January 1974 the lake water during the summer of 1974 showed very poor quality compared with the situation in 1972 and in 1973. The only reason for this - an increase of the chlorophyll content by a factor of four, seemed to be diffusion or additions of nutrients from the sediments. In order to study if water quality varied parallel to weather and winds, water chemical data was plotted against different directions and forces of the winds (fig. 4). When th wind was blowing lengthwise in the lake (S and SW), an increase in $NH_4$-N and total-P could be observed. Here, the wind conditions obviously regulate, to a great extent, the internal loading and nutrient supply in the water.

## Discussion

Many lakes have, by increased nutrient concentrations, reached levels of eutrophication that make them unacceptable - both for recreational activities and for municipal and industrial supply sources. This increase in nutrient concentration may be caused by wastewater discharge and urban or agricultural stormwater runoff (external sources).

The importance of sediments as a source of N and P for the overlying waters (internal loading) has been stressed by many authors (see e.g. Gahler, 1969; Mckee et al., 1970 a,b). However, because of the complexity of the systems involved, littl success has been achieved by experiments in predicting the magnitude of N and P uptake or release from sediments. In shal low, homomictic lakes with short hydrauli residence times, thoroughly made budget calculations in short intervals (shown

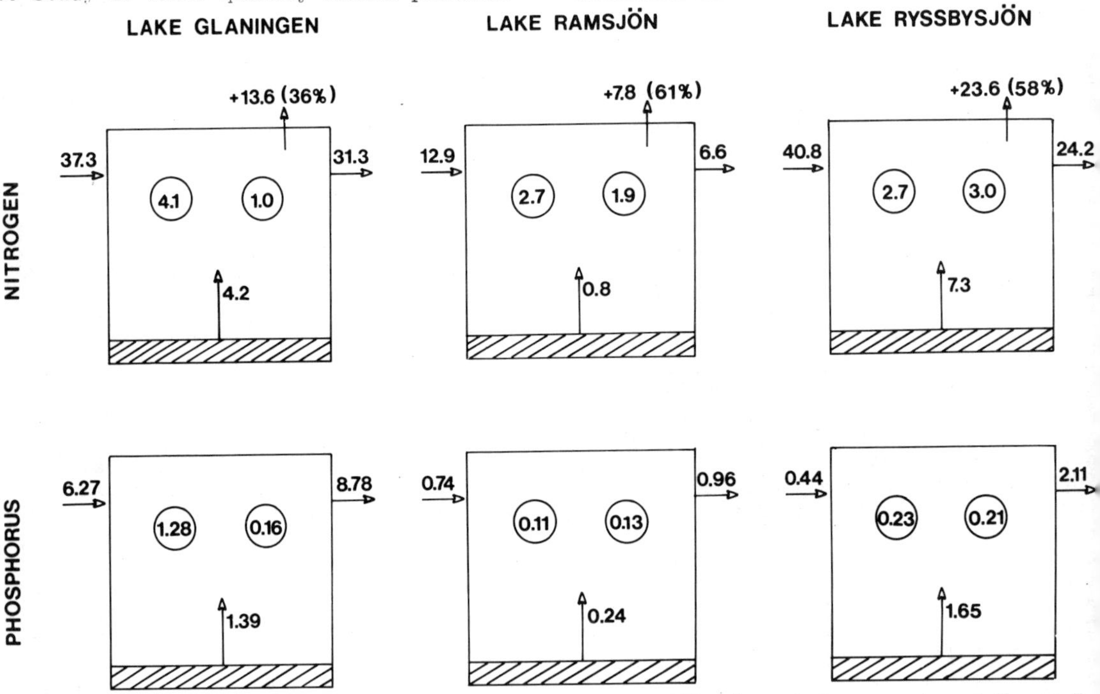


Fig. 3. L. Glaningen, L. Ramsjön and L. Ryssbysjön 1974. Annual transports of N and P ($g/m^2$ . year). The arrows in the upper part of the N-boxes indicate the difference between denitrification and $N_2$-fixation and its proportion to the external N-input (within brackets). The encircled figures show the total content of N and P in the lake water ($g/m^2$) at the beginning and at the end of the year.

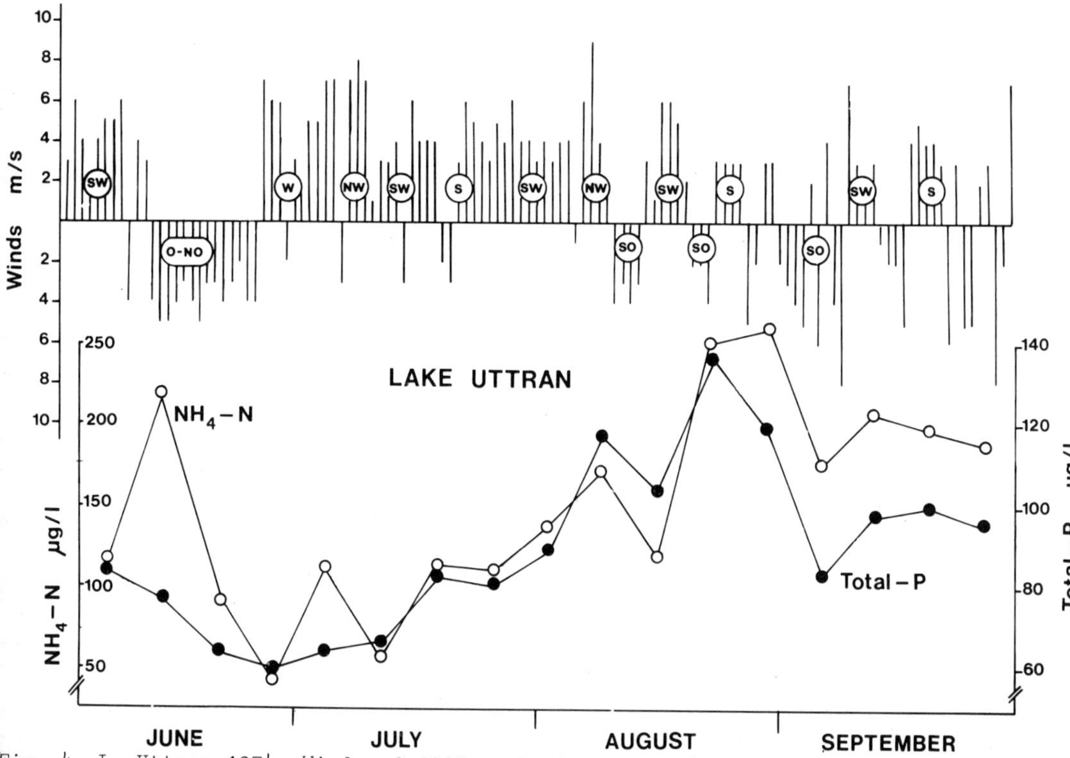


Fig. 4. L. Uttran 1974. Winds of different directions (daily average values) and the variations in $NH_4$-N and total-P (weekly average values) in surface water (0-2 m).

above) yield a good idea of the net flux of nutrients. A similar approach has been used in other lake budget studies (Bengtsson, 1975; Malueg et al., 1975).

Biological, physical, and chemical processes are involved in the transfer of materials between water and sediments - although the interrelationship among the various processes makes them difficult, if not impossible, to evaluate separately. By comparing net gain - net loss data with other physical-chemical conditions in the lake, one can indicate the conditions prevailing when a high sediment release is observed. In L. Glaningen, L. Ramsjön, and L. Ryssbysjön, there was low flushing, a high water temperature, high pH-values, and high concentrations of organic substances during that period.

Properly identifying the active mechanisms regulating the nutrient release from sediments by budget calculations is not possible. Here, laboratory experiments can be of further help. However, the conditions prevailing in the laboratory are not the natural ones. Therefore, an attempt to translate laboratory experiments should be made with caution. The problems can be illustrated by comparing results from field studies (budget calculations) and laboratory experiments (performed by N. Edberg, unpublished). The maximum net P-release observed under natural aerobic conditions far exceeded the aerobic gross release obtained in the laboratory (Table 2). The maximum net release in L. Glaningen was even higher than the "laboratory" release under anaerobic conditions. Other ways to study the role of the sediments as a nutrient source is

Table 2. Maximum net release of P observed by field measurements (Lake) and gross release of P obtained in laboratory tests (tube experiments) under aerobic and anaerobic conditions (mg P/$m^2$ · d).

| | | L. Glaningen | L. Ramsjön | L. Ryssbysjön |
|---|---|---|---|---|
| Lake | Max. net release (aerobic) | 47 | 13 | 9 |
| Laboratory | Gross release (aerobic) | 2 | 0.3 | 0.7 |
| | Gross release (anaerobic) | 18 | 20 | 20 |

. Data from Edberg, unpublished

by culturing algae in a vessel with sediments (Golterman et al., 1969; Fitzgerald, 1970; Sagher et al., 1975) or by making a model approach (Jacobsen & Jørgensen, 1975; Jørgensen et al., 1975; Kamp-Nielsen 1975).

Mixing by wave actions, currents, burrowing animals, and gas ebullition, is of much more importance for the nutrient release from sediments than is diffusion (Lee, 1970). The degree of mixing is also dependent upon the sharpness of the sediment-water interface, sediment density, and water depth. That mixing could be of great importance to water quality in shallow lakes, was shown above for L. Uttran. The variations in water quality in this lake during the last 10 years may be explained from wind conditions. At L. Uttran one of the first Swedish biological-chemical treatment plants started to operate in February 1968. L. Uttran showed an unexpectedly good water quality already during the summer of 1968. The retention time of the water is 4-5 years. In spite of a "total" diversion of sewage in January, 1974, water quality deteoriated during the summer of 1974 and 1975 (Forsberg et al., 1976). Factors other than the external loading seemingly played the dominant role. Looking at the wind conditions and especially at winds from W and SW (L. Uttran is located in a W - SW direction) it was obvious that the force and duration o winds increased during 1974 and 1975 (fig 5). The calm conditions of 1968 can be th explanation for the very good water quality observed during this year.

A characteristic feature for shallow polluted lakes is the peak values of P in the middle of the summer caused by an extensive leakage from the sediment. Simila fluctuations have been reported from othe shallow lakes (Hutchinson, 1957; Andersso 1975; Sonzogni et al., 1975). In some of them, however, the increase in P was explained by decomposition of algae. In shallow polluted lakes, the high productive period can last throughout the entire vegetation period - due to the continous enrichment of nutrients from the bottoms. In deeper, stratified, eutrophic lakes, ther is often a minimum of nutrients and algae in the middle of the summer (Rodhe, 1966).

Of great importance when studying the impact of sediments in the recovery processes in polluted lakes is to know the magnitude of the nutrient release in rela tion to other sources. The net P-release from the sediments in shallow polluted lakes can be of a considerable amount whe compared with the external loading. Looking at the input-output figures it is ob-

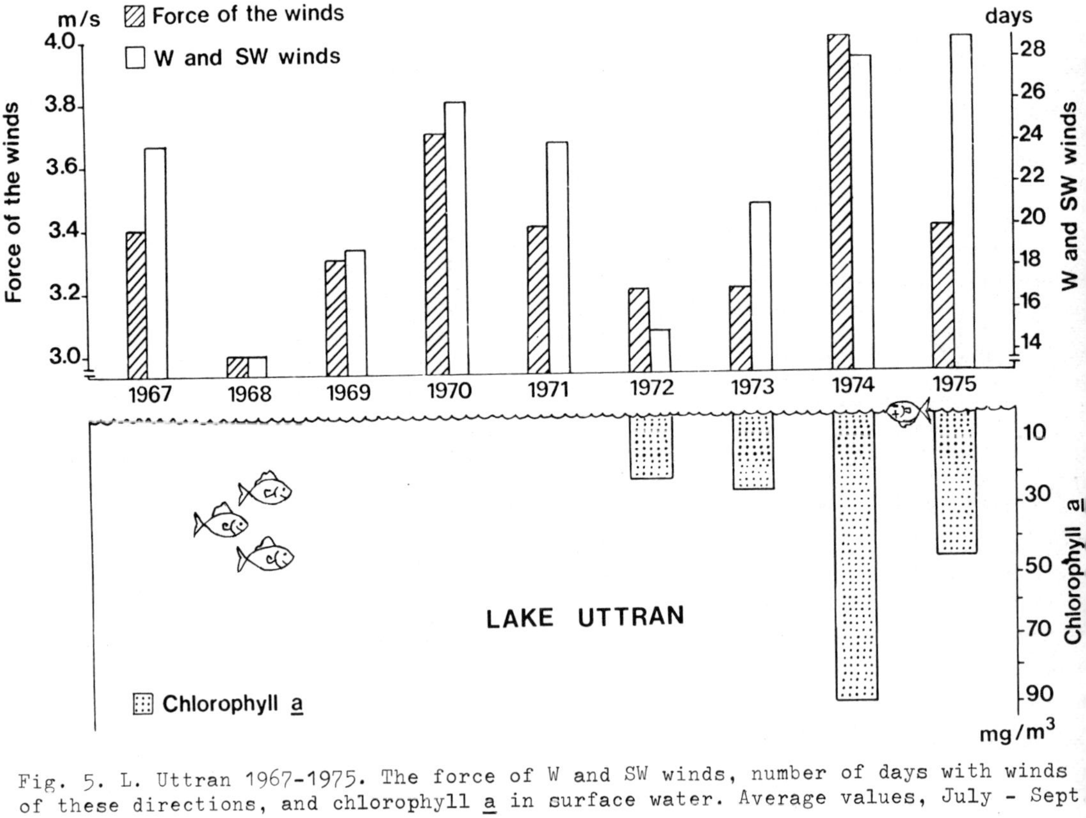


Fig. 5. L. Uttran 1967-1975. The force of W and SW winds, number of days with winds of these directions, and chlorophyll a in surface water. Average values, July - Sept

rious that in the equilibruim in the system, input-lake water-sediments are being markedly disturbed when the P-input diminished after advanced wastewater treatment or sewage diversion. A net outflow of P was observed in all lakes. On the N side, the external input was larger than the output. The net N-release from the sediment seemed to play a minor role. The net loss of N (denitrification - $N_2$-fixation) ranged from 35-60% of the external input. In other eutropicated lakes similar figures have been reported: L. Norrviken 3-30 g $N/m^2$ · yr, 41-61% of the external N-input (Ahlgren, 1973); L. Mälaren 6.4 g $N/m^2$ · yr, 57% (Ahl, 1973); and L. Mendota 0.7 g $N/m^2$ · yr, 11% (Brezonik & Lee, 1968).

If a great deal of the P-release from the sediments is recycled back to the ecosystem each season through sedimentation, a lowered P-input to the lake will have no significant effect on water quality, at least not in the short run. Furthermore, the high P-content in the water in many lakes makes N the most limiting nutrient (Claesson & Ryding, 1975). Attempts to decrease the N input to these type of lakes and thereby improve water quality must be considered carefully since inorganic-N concentration is often low during the high-productive period and a significant N-reduction takes place within the lakes themselves by denitrification.

The present study will emphasize the use of thoroughly made budget calculations to study the magnitude of the sediment release compared with input from other sources. These data are of great importance for succesful predictions of the effects of lake renewal proposals such as nutrient reduction or diversion, flushing, or dredging.

## References

Ahl, T., 1973. Mälarens belastning och vattenkvalitet. Scripta Limnologica Upsaliensia. 332.

Ahl, T. & S. Odén, 1975. Närsaltskällor - en översikt. "Eutrophication". 10th Nordic Conf. on Water Research. Nordforsk 1975:1, p. 99-134.

Ahlgren, I., 1973. Limnologiska studier av sjön Norrviken. III. Avlastningens effekter. Scripta Limnologica Upsaliensia. 333.

Andersson, G., 1975. Restaureringseffekter i sjön Trummen. Limn. Inst., Lund.

Austin, E.R. & G.F. Lee, 1973. Nitrogen release from lake sediments. JWPCF 45: 870-879.

Bengtsson, L., 1975. Phosphorus release from a highly eutrophic lake sediment. Verh. Int. Verein. Limnol. 19: 1107-1116.

Brezonik, P.L. & G.F. Lee, 1968. Denitrification as a nitrogen sink in Lake Mendota. Environ. Sci. Techn. 2: 120-125.

Byrnes, B.H., D.R. Keeney & D.A. Graetz, 1972. Release of ammonium-N from sediments to waters. Proc. 15th Conf. Great Lakes Research. 249-259

Claesson, A. & S.-O. Ryding, 1975. Nitrogen - a growth limiting nutrient in eutrophic lakes. Proc. Conf. on Nitrogen as a Water Pollutant (IAWPR). 2. Copenhagen. August 18-20.

Elfving, E., Å. Forsberg & C. Forsberg, 1975. Minitest method for monitoring effluent quality. JWPCF 47: 720-726.

Fillos, J. & W.R. Swanson, 1975. The release rate of nutrients from river and lake sediments. JWPCF 47: 1032-1042.

Fitzgerald, G.P., 1970. Aerobic lake muds for the removal of phosphorus from lake waters. Limn. Ocean. 15: 550-555.

Forsberg, C., S.-O. Ryding & A. Claesson, 1975. Recovery of polluted lakes. A Swedish research program on the effects of advanced wastewater treatment and sewage diversion. Wat. Res. 9: 51-59.

Forsberg, C., S.-O. Ryding & B. Forsberg, 1976. Naturvårdsverkets RR-undersökning. Vad händer i Uttran? Vatten 32: 285-294.

Forsberg et al. Recovery of polluted lakes. Background data and methods. (in prep.)

Gahler, A.R., 1969. Sediment-water nutrient interchange. Proc. Eutrophication. Biostimulation Workshop, Berkeley, Calif., p. 243-257.

Golterman, H.L., C.C. Bakels & J. Jakobs-Mögelin, 1969. Availability of mud phosphates for the growth of algae. Verh. Internat. Verein. Limnol. 17: 467-479.

Graneli, W., 1974. Fosforomsättning - sedimentens roll. "Eutrophication". 10th Nordic Conf. on Water Research. Nordforsk. 1975: 1, p. 213-222.

Hallberg, R.O., L.-E. Bågander, A.-G. Engwall, H. Lindström, S. Odén & F.A. Schippel, 1973. The chemical microbiological dynamics of the sediment-water interface. Contrib. Askö Lab., 2: 1-117.

Hutchinson, G.E., 1957. A Treatise in Limnology. I. John Wiley and Sons, Inc., New York, N.Y.

Jacobsen, O.S. & S.E. Jørgensen, 1975. A submodel for nitrogen release. Ecol. Modelling 1: 147-151.

Jørgensen, S.E., L. Kamp-Nielsen & O.S. Jacobsen, 1975. A submodel for anaerobic mud-water exchange of phosphate. Ecol. Modelling, 1: 133-146.

Kamp-Nielsen, L., 1974. Mud-water exchange of phosphate and other ions in undisturbed sediment cores and factors affecting the exchange rates. Arch. Hydrobiol. 73: 218-237.

Kamp-Nielsen, L., 1975. A kinetic approach to the aerobic sediment-water exchange of phosphorus in Lake Esrom. Ecol.

Modelling, 1: 153-160.
Keeney, D.R., 1973. The nitrogen cycle in sediment-water systems. J. Environ. Quality 2: 15-28.
Lee, G.F., 1970. Factors affecting the transfer of material between water and sediments. Eutrophication Information Program. Univ. Wisconsin, Madison, Literature Review No 1.
Li, W.C., D.E. Armstrong, J.D.H. Williams, R.F. Harris & J.K. Syers, 1972. Rate and extent of inorganic phosphate exchange in lake sediments. Soil Sci. Soc. Amer. Proc. 36: 279-285.
Li, W.C., D.E. Armstrong & R.F. Harris, 1973. Measurement of exchangeable inorganic phosphate in lake sediments. Environ. Sci. & Techn. 7: 454-456.
MacPhearsson, L.B., N.R. Sinclair & F.R. Hayes, 1958. Lake water and sediment III. The effect of pH on the partition of inorganic phosphate between water and oxidized mud or its ash. Limnol. Ocean. 3: 318-326.
Malueg, K.W., D.P. Larsen, D.W. Schults & H.T. Mercier, 1975. A six-year water, phosphorus and nitrogen budget for Shagawa Lake, Minnesota. J. Environ. Qual. 4: 236-242.
McKee, G.D., L.P. Parrish, C.R. Hirth, K.M. Mackenthum & L.E. Keup, 1970 a. Sediment-water nutrient relationships. Part I. Water and sewage works, 117: 203-206.
McKee, G.D., L.P. Parrish, C.R. Hirth, K.M. Mackenthum & L.E. Keup, 1970 b. Sediment-water nutrient relationships. Part II. Water and Sewage Works, 117: 246-249.
Mortimer, C.H., 1941, 1942. The exchange of dissolved substances between mud and water in lakes. J. Ecol. 29: 280-329; 30: 147-201.
Rodhe, W., 1966. Hur reagerar recipienten? Presented at the IVL-Conf., Stockholm, Dec., 6th, 1966. Limn. Inst., Uppsala.
Sagher, A., R.F. Harris & D.E. Armstrong, 1975. Availability of sediment phosphorus to microorganisms. Water Resources Center, Univ. of Wisconsin, Madison, Technical Report 75-01.
Sanville, W.D., C.F. Powers & A.R. Gahler, 1974. Sediments and sediment-water nutrient interchange in Upper Klamath Lake, Oregon. EPA 660/3-74-015.
Shukla, S.S., J.K. Syers, D.H. Williams, D.E. Armstrong & R.F. Harris, 1970. Sorption of inorganic phosphate by lake sediments. Soil Sci. Soc. Amer. Proc. 35: 244-249.
Sonzogni, W.C., G.P. Fitzgerald & G.F. Lee, 1975. Effects of wastewaters diversion on the lower Madison lakes. - JWPCF 47: 535-542.
Stumm, W. & J.O. Leckie, 1971. Phosphate exchange with sediments, its role in the productivity of surface waters. Proc. 5th Int. Wat. Poll. Res. Conf. Pergamon Ltd.
Sullivan, W.T., 1967. Chemical composition of the mud-water interface zone, with the description of an interface sampling device. Proc 10th Conf. Great Lakes Res., p. 340-403.
Syers, J.K., R.F. Harris & D.E. Armstrong 1973. Phosphate chemistry in lake sediments. J. Environ. Quality 2: 1-14.
Tessenow, U., 1972. Lösungs,- Diffusions- und Sorptionsprozesse in der Oberschicht von Seesedimenten. I. Ein Langzeitexperiment unter aeroben und anaeroben Bedingungen in Fliessgleichgewicht. Arch Hydrobiol. Suppl., 38: 353-398.
Wildung, R.E. & R.L. Schmidt, 1973. Phosphorus release from lake sediments. EPA-R3-73-024.

Key words: nitrogen
phosphorus
release rates
denitrification
recovery

# INTERACTIONS BETWEEN SEDIMENTS AND OVERLYING WATERS IN AN EXPERIMENTALLY EUTROPHIED PRE-CAMBRIAN SHIELD LAKE

D. W. Schindler, R. Hesslein and G. Kipphut

Experimental Lakes Project, Freshwater Institute, 501 University Crescent, Winnipeg, Canada
and Lamont-Doherty Geological Observatory, Palisades, New York

## Summary

Several recently-developed methods were used to study the role of sediments in the nutrient economy of a small Precambrian Shield lake which has been artificially fertilized for several years with phosphorus and nitrogen. Measurement include gross and net nutrient fluxes, estimated by employing a combination of mass-balance techniques, whole-lake additions of isotopic tracers and recent dating techniques.

Input-output mass balances revealed that 81 and 61 percent of phosphorus and nitrogen, respectively, added over 6 years were deposited in sediments, while 1 and 11 percent appeared in the water column. Direct measurements of phosphorus and nitrogen in sediment cores agreed well with mass-balance calculations of sedimentation.

Sediment "peepers" (racks of small vials spaced at 1 cm intervals, filled with distilled water and covered with dialysis membrane) were inserted into sediments, and left to equilibrate. Analysis of the contents of these allowed rapid estimation of nutrient concentrations in pore waters, while eliminating the problems associated with extraction of pore water under anoxic conditions. Coefficients of diffusion were estimated simultaneously by tagging water of the entire lake with tritium and measuring its rate of penetration into pore waters. Values so obtained averaged $2x10^{-5}$ $cm^2$/second. Coupled with observed nutrient gradients in pore waters, the following conclusions were reached:

Pore water gradients of phosphorus were absent or weak, and concentrations were low. Sediments did not release significant amounts of phosphorus to overlying waters under either oxic or anoxic conditions. This result was confirmed by mass-balance results. Almost all of the phosphorus in sediments was bound to the particulate fraction.

Steep gradients of $NH_4$-N in pore waters indicated that over 50% of summer epilimnetic total nitrogen could potentially be replaced from littoral sediments each day. On the other hand, nutrients released from the hypolimnion or hypolimnion sediments were not important in the maintenance of epilimnetic algal blooms in midsummer, because vertical eddy diffusion in the region of the thermocline was less than $10^{-4}$ $cm^2$/sec. Studies of horizontal and vertical diffusion in the water column revealed that horizontal, rather than vertical water movements were chiefly responsible for the resulting distribution of nitrogen and carbon in the water column.

Lead-210 and cesium-137 from fallout, artificially added radium-226, and the added nutrients were used to measure recent sedimentation rates and mixing depths. The sedimentation rate averaged 1.0 mm/yr. Recently-deposited material was found mixed to a depth of 5 to 10 cm. The possible importance of this mixing as a "buffer" delaying the alteration of lake trophic conditions when external sources of nutrients are changed is discussed. The half-time for establishing a new equilibrium between the sediment mixed layer and overlying water is estimated to be 35 years.

## Introduction

Although numerous thorough studies of sediment-water interactions in freshwater lakes have been done, no author has related the role of sediments to the overall nutrient economy of a lake, particularly in the critical years when a lake is being rapidly eutrophied. Lake 227, in the Experimental Lakes Area, northwestern Ontario, provided an ideal site for such an investigation. The lake, originally oligotrophic, was rendered eutrophic by known additions of phosphorus and nitrogen fertilizer from 1969 onwards. The total amount of both nutrients entering the lake from fertilizer plus natural sources is known to within ±5 percent or better. Likewise, continuous measurements of outflow volume and weekly chemical analysis have allowed precise calculation of amounts leaving the lake (Schindler et al., 1971, 1973; Schindler, 1975 and unpublished data). Nitrogen fixation and denitrification have been measured, (R. Flett, unpublished data) and are known to be insignificant during the period 1969 to early 1975, the time period discussed in this paper. There are no significant inputs or losses of either element via groundwater. These facts allowed accurate mass-balances to be calculated for phosphorus and nitrogen, and thus provided precise controls on other sediment-water calculations and experiments.

## Description of the lake

Lake 227 lies in a basin of Precambrian Shield granites and gneisses. Thin rocky and gravelly non-calcareous overburden covers parts of the drainage basin. The lake is a headwater, receiving only direct drainage from the surrounding slopes, which are virgin boreal forest, primarily *Pinus banksiana* and *Picea mariana*. There are no inhabitants in the drainage basin. The climate is extreme and continental. The lake is covered by ice of 0.5 to 1 m depth for 6 months of the year, with snow of 25 to 50 cm. overlying the ice for about 5 months.

Spring overturn is typically incomplete, due to the rapid onset of warm weather in May. In autumn, circulation is usually complete, except in an occasional, exceptional year when the lake cools and freezes rapidly. As a result, the hypolimnion is usually anoxic for 10 to 11 months of the year. Some general limnological characteristics of the lake are shown in Table 1.

Table 1. Lake 227 Limnological characteristics.

| | |
|---|---|
| Surface area, ha. | 5.00 |
| Terrestrial drainage area, ha. | 26.8 |
| $\bar{z}$, m. | 4.4 |
| Water renewal time, years | 1.4 to 2.9 |
| Mean annual chlorophyll in epilimnion, μg/l* | 44 |
| Maximum epilimnion chlorophyll, μg/l* | 198 |
| Mean annual total phosphorus, μg/l* | 38 |
| Average phosphorus input, $g/m^2/yr$ | 0.55 |
| Average nitrogen input, $g/m^2/yr$ | 7.65 |

* Data for 1973

The lake has extremely soft water (conductivity 21 μmhos/cm in 1968, when experiments began). Waters are slightly humic (15-20 Pt. Units). Near-surface sediments in the lake have a water content of over 95%, and dried sediment contains over 50% organic matter. Sediments are black to brown, extremely flocculent, and have no visible laminar structure.

## Methods

### Whole-lake mass-balance calculations

Net sedimentation or return from sediments (S) was calculated for the whole lake:

$$S = I - E - \Delta M$$

where I, E and ΔM are the input, output and change in mass of nutrient in the waters of lake, respectively, for the time period considered. S will be positive when the net flux is toward sediments and negative when from the sediments to overlying water. Such calculations were done both annually, to look for long-term trends as the lake continued to be fertilized, and for shorter periods to observe any seasonal trends. From December through early April of each year, when the lake is ice-covered, both I and E are known to be zero; therefore ΔM is a direct measure of S.

### Phosphorus and nitrogen content of sediments

Short (15-30cm) cores were taken at several sites and depths in each lake, using a piston equipped plexiglass core tube operated by a SCUBA diver. Core sites were selected to be representative of broad regions of the lake bottom, and great care was taken to obtain undisturbed mud-water interfaces. Cores were kept at *in situ* temperature until sectioning at cm intervals, usually within a few hours after collection. Dry weights were obtained by drying sediments at 60°C, then desiccating for several days before weighing. Phosphorus was determined by ignition and acid digestion (Andersen, 1976). Nitrogen was determined on a Carlo erba nitrogen analyser.

### Nutrient gradients in interstitial waters

Pore water "peepers" (Hesslein, 1976) were made by filling small (2ml) autoanalyser vials with distilled water, covering them with millipore 0.45μ membrane, and inserting racks of the vials at 1 cm intervals in the sediments. Peepers were left in place for three weeks before collection. Contents were analysed for molybdate-reactive phosphorus and ammonia by auto analyser (Stainton et al. 1974).

### Diffusion coefficients

Diffusion rates in the lake waters and sediments of lake 227 were measured by adding tritiated water to the lake, then measuring the rate of penetration of the tritiated water into sediments by liquid scintillation counting the contents of the peepers describe above (Hesslein, 1976). Gross fluxes of nutrients from the sediments were calculated by multiplying maximum nutrient gradients by these diffusion coefficients.

### Deposition rates and sediment mixing

Recent deposition rates for lake 227 were determined by lead-210 dating of several core (Kipphut, 1976; Bruland, 1974; Robbins & Edgington, 1975). Radium-226, added to the lake in 1970, and fallout cesium-137 were also measured, giving some evidence for sediment mixing.

## Results

### Accumulation of phosphorus and nitrogen in sediments of lake 227

The long-term average sedimentation rates for phosphorus and nitrogen were 81 and 61% of inputs, respectively (Tables 2 & 3). This represents an input to sediments of 2.46 $gP/m^2$ and 40.67 $gN/m^2$ during the period of fertilization, 1969-1974.

Several lead-210 profiles from the lake revealed an average sedimentation rate of approximately 1.0 mm/year, regardless of the depth in the basin from which a core was taken (Kipphut, 1976). Based on the above known sedimentation of phosphorus and nitrogen, in six years the concentrations of P and N should be 410gP and 6780gN per $m^3$ of wet sediment, or 20 mgP and 330 mg N per g of dry sediment if all sedimented material remained at the mud surface.

Actual measurements revealed that such concentrations were not approached in surface sediments. Instead, the "excess" phosphorus was commonly mixed into the upper 9 to 10 cm, representing several decades of sediment accumulation (Figure 1). The distribution of nitrogen is similar. Depth-profiles of fallout cesium-137 and experimentally-introduced radium-226 also indicated significant mixing of the upper sediment layers (Figure 2).

Integration of the excess (fertilizer) phosphorus and nitrogen measured in the sediments accounted for 85 ± 29 percent of the net sedimented phosphorus and 79 ± 14 percent of the net sedimented nitrogen calculated from input-output mass balances.

Table 2. Lake 227 phosphorus budget, 1970-1974 average.

| | $gP/m^2/yr$ | | |
|---|---|---|---|
| | Average | Range | % of $I_p$ |
| Input ($I_p$) | 0.551 | 0.523 to 0.582 | 100 |
| Outflow (Ep) | 0.095 | 0.044 to 0.175 | 17 |
| Increase in $\Sigma P$ ($\Delta M_p$) | 0.008 | -0.063 to 0.047 | 1 |
| Sedimentation (Sp) = $I_p - E_p - \Delta M_p$ | 0.448 | 0.315 to 0.556 | 81 |

Table 3. Lake 227 nitrogen budget, 1970-1974.

| | $gN/m^2/yr$ | | |
|---|---|---|---|
| | Average | Range | % of $I_N$ |
| Input ($I_N$) | 7.65 | 7.23 to 7.90 | 100 |
| Outflow ($E_N$) | 2.15 | 1.34 to 3.39 | 28 |
| Increase in $\Sigma N$ in water ($\Delta M_N$) | 0.86 | -2.10 to 3.40 | 11 |
| Sedimentation ($S_N$) = $I_N - E_N - \Delta M_N$ | 4.63 | 1.11 to 7.64 | 61 |

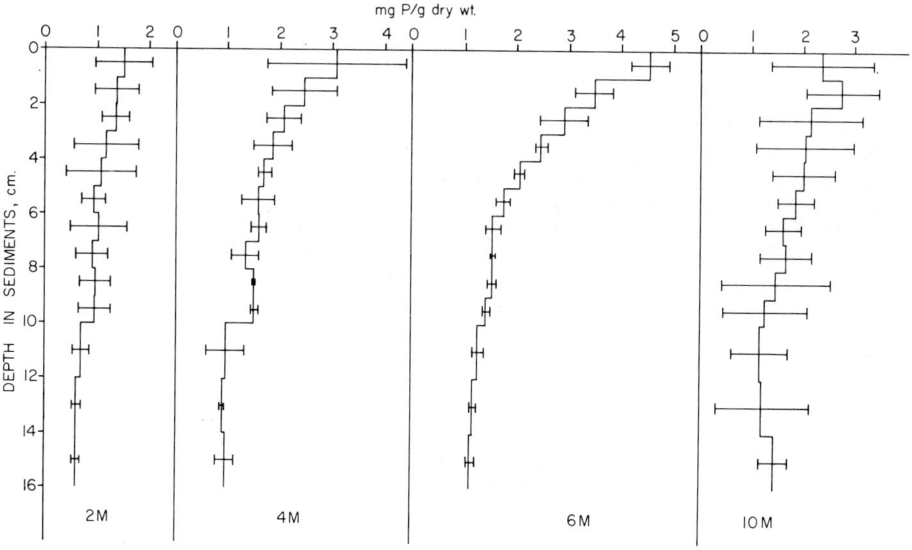


Figure 1. Representative total phosphorus profiles from lake 227 sediments, May 1975. The horizontal bars are 95% confidence limits based on three cores at each depth. Cores were taken along three transects in different parts of the lake.

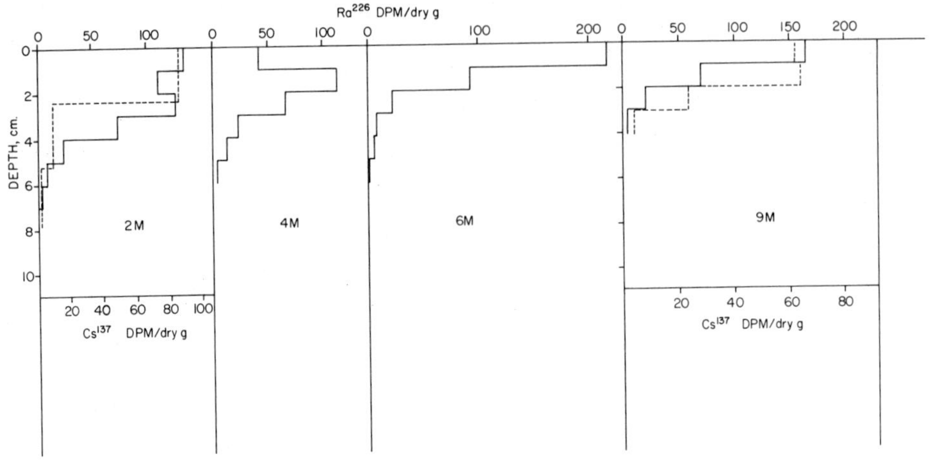


Figure 2. Representative profiles in lake 227 sediments of fallout cesium-137 (dotted line) and radium-226 (solid line) added to the lake in 1969.

Flux of phosphorus and nitrogen from pore water

Despite the observed increase in the total phosphorus content of sediments resulting from fertilization, the concentration of molybdate-reactive phosphorus in pore water was less than 1 μg/l, as in oligotrophic lakes of the area. The lack of any gradient in concentration regardless of whether sediment surfaces were oxic or anoxic, made it seem unlikely that return of phosphorus from sediments was of great significance in lake 227. This fact was confirmed by in situ radiotracer experiments under both oxic and anoxic conditions (Schindler and Lean, 1974).

Ammonium-nitrogen, on the other hand, had a strong gradient in pore water (Figure 3). Profiles from unfertilized Lake 302S, which has similar sediments, had only about 10% the gradient of ammonium in pore water, suggesting that the gradient in lake 227 has developed in response to fertilization. Potential returns of nitrogen from sediments are discussed below.

Diffusion rates in sediments

Three profiles of tritium vs. depth in epilimnion sediments (for example, Figure 4) indicated an average diffusion coefficient $2x10^{-5}$ $cm^2$/second. The observed maximum gradient in pore-water ammonium was 0.84 mg N/cm. Using traditional calculations for gross sediment flux, we find that 145 mgN/$m^2$/day could diffuse from epilimnion sediments. This represented a supply of 27 mgN/$m^3$/day to epilimnion water, a substantial part of the total supply to the epilimnion (Table 4).

Assuming that diffusion in Lake 302 sediments would take place at a similar rate,

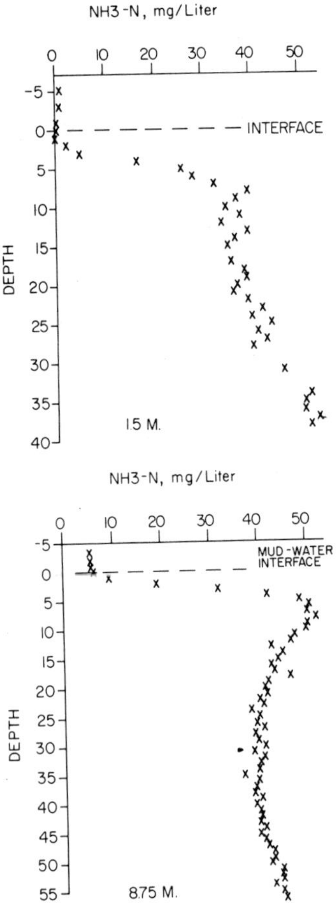


Figure 3. The gradient of pore-water ammonium nitrogen at two depths in lake 227. Depth is in cm.

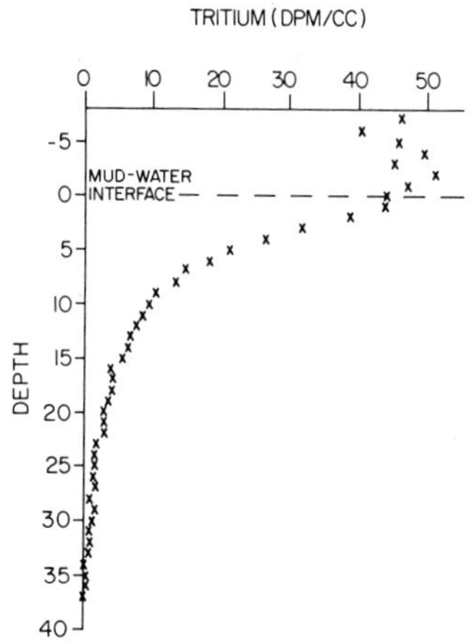


Figure 4. A typical example of the distribution of tritium with depth in lake 227 sediments. Depth is in cm.

Table 4. Typical input and turnover of nitrogen in the epilimnion of Lake 227, midsummer.

| | $Mg/M^3$/Day | % of [ΣN]/ day | % of total inputs |
|---|---|---|---|
| Fertilizer | 22 | 2.2 | 43 |
| Return from sediments | 27 | 2.7 | 53 |
| Natural sources | 2 | 0.2 | 4 |
| Return from hypolimnion | <0.06 | 0.0 | 0.0 |
| Total | 51 | 5.1 | |

$[\Sigma\bar{N}]$ = 1000 $Mg/m^3$
Renewal time = 20 days

which is likely because of the similarity in sediment appearance and water content, the much smaller gradient in ammonium observed there would allow a flux of only 3 $mgN/m^2$/day. It thus appears that the nitrogen supply from sediments has increased rapidly as the result of fertilization.

Seasonal changes in net sediment-water exchange

If inflows, outflows and changes in mass of phosphorus or nitrogen in the water column are measured precisely several times during the year, the net exchange between water and sediment at different seasons may be calculated.

Even in late summer and late winter, when 50 to 100 percent of the sediment surface was in contact with anoxic water, there was no net return of phosphorus from sediments to overlying water (Figure 5). Throughout most of the year net exchange between sediments and water was low. Almost all of the sedimentation occurred during late summer and early fall, when midsummer phytoplankton blooms were deteriorating.

The seasonal pattern for nitrogen was slightly different from that for phosphorus. During snowmelt and spring overturn, there was a significant net return of nitrogen, possibly due to mixing and resuspension of littoral sediments with their high concentrations of pore water ammonium. (Figure 6).

Some resuspension of sediments probably also occurred during fall overturn, although because of the high sedimentation of algal material at that time, the net flux was slightly to rather than from sediments. (Compare July and August-September maxima sedimentation of phosphorus and nitrogen.)

The above general pattern has been observed for six years. There was some year-to-year variation in the magnitude of nitrogen return in spring and fall, due to the amount and timing of spring melt, and to the duration of spring and fall overturn. The high sedimentation in late summer and fall was also displaced a few weeks in timing in different years.

## Discussion

Net and gross exchange between sediments and water

Although gradients of ammonia in sediment pore water indicated that there should be substantial return of nitrogen from pore water, annual and seasonal mass balances for the lake indicate that the net flux of nitrogen was almost always to, rather than from the sediments. The gravitational settling of seston and possibly also interception of dissolved nitrogen returning from sediments by algae or bacteria at the sediment surface may be responsible. The important point is that lake 227 sediments are not presently an important net source of either phosphorus or nitrogen. The nutrient balance of the lake is determined almost entirely by external sources of the elements and by water renewal.

While it is possible that this result is due to the short period during which the lake has been fertilized, or to the fact that allochthonous organic materials rather than mineral particles dominate lake 227 sediments, there is some evidence that at least for phosphorus this result may not be unusual. Lee (this volume) found that there was no net annual income of phosphorus from the sediments of Lake Mendota, a calcareous basin which has been eutrophic for many years. Likewise, Ahlgren (this volume) found that

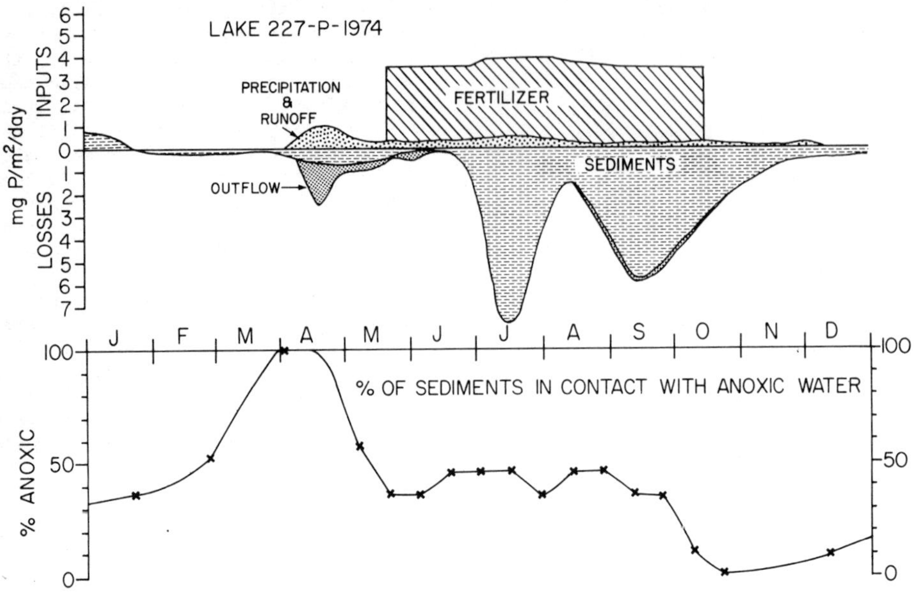


Figure 5. Upper panel: Typical seasonal inputs and losses of phosphorus for the water of lake 227. Lower panel: The percent of bottom sediments overlain by anoxic water. The lake is ice covered from January through April and November-December.

while there was some seasonal return of phosphorus from sediments, the decrease in phosphorus concentration over several years in highly eutrophic Lake Norrviken after diversion of sewage could be explained by water dilution alone. While high gross rates of flux of phosphorus from mineral sediments have been demonstrated for Greifensee (Emerson, 1976; Imboden and Emerson, this volume) and Lake Erie (Burns and Ross, 1972) these figures must be compared to input, outflow and sedimentation before their importance can be assessed.

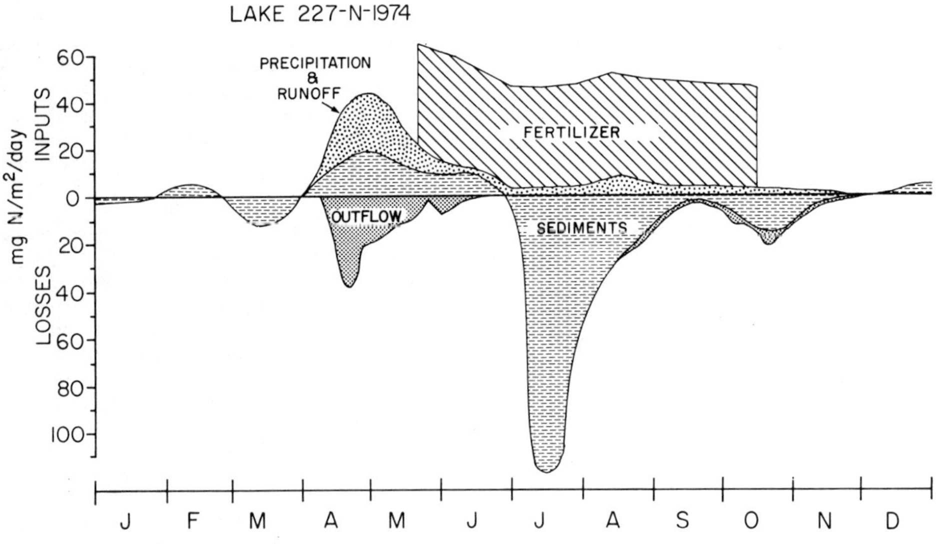


Figure 6. Typical seasonal inputs and losses of nitrogen from the water of lake 227.

## The effects of anoxia on phosphorus exchange

The classical view is that phosphorus is always liberated from the sediments of anoxic hypolimnions (see Mortimer, 1941-2; Wetzel, 1975). This is simply not true in lake 227. Annual and seasonal mass-balance studies, interstitial water gradients and isotope studies all indicate the absence of any change in phosphorus exchange in response to oxygen concentration. Even under winter ice, when sedimentation of particulate phosphorus is nearly zero, there is no increase in phosphorus in the lake under anoxic conditions (Figure 5). It is obvious that the results of Mortimer's very convincing study have been extrapolated too widely by limnologists, and that more comparative work on this subject is needed.

There appear to be no detectable iron-phosphorus minerals in lake 227 sediments (Coey, Schindler and Weber, 1974). Isotopic tracer studies indicate that the initial binding of phosphorus at the sediment surface may be by bacterial uptake. Labelled $PO_4$ added to deep hypolimnion water was rapidly removed by seston (Levine and Schindler, unpublished data). Most of this phosphorus was retained in the seston fraction. The little which was excreted was almost entirely bound in the form of refractory, high molecular weight compounds. These compounds appeared to bind readily to sediments, and binding did not appear to be affected significantly by the presence or absence of oxygen.

## Hydrodynamics and sediment-water interactions

Several features of water circulation must be taken into account when considering the production of summer algal blooms. The epilimnion of the lake is in full, rapid circulation (>100 $cm^2/sec$ expressed as an apparent diffusion coefficient), so that materials released from epilimnion sediments would be rapidly and thoroughly mixed throughout the mixed layer.

On the other hand, nutrients below the thermocline are unavailable to epilimnion algae during the period of stratification, because the low coefficient of vertical eddy diffusion in the region of the thermocline ($10^{-4} cm^2/sec$) restricts movement of nutrients even when gradients in concentration are steep (Table 4).

Hypolimnion circulation, too, is more rapid in the horizontal than in the vertical direction (5 $cm^2/sec$ as opposed to $10^{-3} cm^2/sec$; Quay, 1976), so that the concentration of substances in hypolimnion water is closely correlated with the area of sediment in contact with the volume of water sampled (Hesslein, 1976).

## Modelling sediment characteristics

Berger and Heath (1968) developed a simple mathematical model to investigate the effects of sediment mixing on the distribution of substances deposited at the sediment-water interface. Their model may be applied to an investigation of the post-fertilization nutrient distributions in the sediments of lake 227.

The model assumes that material added at the sediment-water interface is rapidly and uniformly mixed throughout a thickness of sediment, L. The change in concentration (C) of an element X in this mixed layer may be expressed as:

$$1. \quad \frac{dC}{dt} = \frac{F-SC}{L}$$

where F is the downward flux of element X from the water column, and S is the sediment deposition rate. At steady state:

$$2. \quad C = F/S$$

If we assume that F, S and L remain constant during an interval of time, t, the concentration of element X in the mixed layer at any time before steady state is reached will be given by:

$$3. \quad C = Ke^{-\frac{S}{L}t} + \frac{F}{S}$$

where K is a constant. Letting $C = C_o$ when $t = 0$ defines $K = (C_o - \frac{F}{S})$:

$$4. \quad C = (C_o - \frac{F}{S})e^{-\frac{S}{L}t} + \frac{F}{S}$$

The half time for reaching steady state can be calculated from equation 4:

$$5. \quad t_{\frac{1}{2}} = \frac{-\ln\frac{1}{2}}{S/L} = \frac{.693}{S/L}$$

Equation 4 can be used to predict the surficial sediment phosphorus concentrations at any time during the fertilization experiment. During the period 1969-1975 F averaged 0.045 $mgP/cm^2/yr$, $S = 0.10$ cm/yr, and $L =$ 5 cm. $C_o$ was approximately 0.048 $mgP/cm^3$. The equation predicts the surface phosphorus concentration in 1975 should have been 0.09 $mgP/cm^3$. This value agrees fairly well with observed concentrations. If surface samples shown in figure 1 are corrected for water content, phosphorus concentrations ranging from 0.08 to 0.24 $mgP/cm^3$ are obtained. As mentioned above, if sediment mixing were not occurring, all of the phosphorus added during fertilization would be found in the upper 0.6 cm of the sediments, and much higher surface concentrations would be expected. Equation 2 indicates that the surficial phosphorus concentration expected in the absence of mixing would be $C = \frac{.045}{.1} = 0.45$ mgP/cm$^3$. In lake 227 this value should be approached with a half time of about 35 years.

Figures 1 and 2 indicate that the upper 5 cm of sediment are not uniformly mixed with respect to phosphorus, cesium-137, or radium-226, as required by our model. This is undoubtedly because mixing in lake 227 sediments operates with a time constant on the order of several years. This would result in higher surficial concentrations than predicted by the model. The model remains useful, however, in explaining the first order features of the phosphorus distribution.

Work in other aquatic ecosystems suggest that the upper layers of sediment are often more thoroughly mixed than in Lake 227. Both benthic organisms (Rhoads, 1973; Davis, 1974; Pemberton et al., 1976) and physical processes (Davis, 1973) have been found to be important.

The implications of this sediment buffering zone may be important for eutrophication and recovery of lakes. Our model indicates that a lake would have to be heavily loaded for several decades before sediments would become saturated with nutrients, ie. sediments would remain oligotrophic long after the overlying waters are highly eutrophic.

Conversely, if external nutrient sources to a eutrophic lake which has been subjected to many years of high loading are suddenly reduced, the sediments will require an equally long response time, delaying the recovery of the lake.

The duration of eutrophication (and thus sedimentation of phosphorus-rich material) and the depth of the mixed layer in sediments may thus be important parameters to consider in predicting lake eutrophication and recovery. It is probably not a coincidence that among recovery experiments to date, lakes which were eutrophic for only a few years recovered nearly immediately (Schindler, 1975; Michalski and Conroy, 1973), while those eutrophied for only a few decades recovered somewhat slower (Larsen, 1975; Edmondson, 1972) and those with a long history of eutrophication show even slower recovery (Ahlgren, this volume; Bjork, 1972). These implications are discussed further by Schindler (1976).

References

Ahlgren, I., 1977. Role of sediments in the process of recovery of a eutrophicated lake. (This volume, page 372).

Anderson, J. M., 1976. An ignition method for determination of total phosphorus in lake sediments. Water Research 10: 329-331.

Berger, W. H., and G. R. Heath, 1968. Vertical mixing in pelagic sediments. J. Mar Res. 26: 134-143.

Bjork, S., 1972. Swedish lake restoration program gets results. Ambiol: 153-165.

Bruland, K. W., 1974. Pb-210 geochronology in the coastal marine environment. Ph.D. dissertation, Univ. California, San Diego. 106 pp.

Burns, N. M, & C. Ross, 1972. Oxygen-nutrient relationships within the central basin of Lake Erie. In: N. M. Burns & C. Ross (Ed.): Project Hypo. Canada Centre for Inland Waters, Paper No. 6, Burlington Ontario. pp. 85-119.

Coey, J. M. D., D. W. Schindler & F. Weber, 1974. Iron compounds in lake sediments. Can. J. Earth Sci. 11: 1489-1493.

Davis, M. B., 1973. Redeposition of pollen grains in lake sediments. Limnol. Oceanogr. 18: 44-52.

Davis, R. B., 1974. Stratigraphic effects of tubificids on profundal lake sediment. Limnol. Oceanogr. 19: 466-488.

Edmondson, W. T., 1972. The present condition of Lake Washington. Verh. Internat. Verein. Limnol. 18: 284-291.

Emerson, S., 1976. Early diagenesis in anaerobic lake sediments: chemical equilibria in interstitial waters. Geochim. Cosmochim. Acta 40: 925-934.

Hesslein, R., 1976. An in situ sampler for close interval pore water studies. Limnol. Oceanogr. (in press).

Imboden, D. M. & S. Emerson, 1977. Study of transport processes through the sediment-water interface in lakes using natural radon-222. (This volume).

Kipphut, G., 1976. Ph.D. thesis, Columbia University (In prep.).

Larsen, D. P., K. W. Malueg, D. W. Schults & R. M. Brice, 1975. Response of eutrophic Shagawa Lake, Minnesota, U.S.A. to point source phosphorus reduction. Verh. Internat. Verein. Limnol. 19: 884-892.

Lee, G. F., 1977. Significance of sediment phosphorus in lake fertilization. (This volume)

Michalski, M. F. P & N. Conroy, 1973. The "oligotrophication" of Little Otter Lake, Parry Sound District. Proc. 16th. Conf. Great Lakes Res., pp. 934-938.

Mortimer, C. H., 1941-42. The exchange of dissolved substances between mud and water in lakes. J. Ecol. 29: 280-289;30: 147-201.

Pemberton, G. S., M. J. Risk & D. E. Buckley 1976. Supershrimp: deep bioturbation in the Strait of Canso, Nova Scotia. Science 192: 790-791.

Quay, P., 1976. Ph.D. thesis, Columbia University. (In prep.).

Rhoads, D. C., 1973. The influence of deposit-feeding benthos on water turbidity and nutrient recycling. Amer. J. Sci. 273: 1-22.

Robbins, J. A., & D. N. Edgington, 1975. Determination of recent sedimentation rates in Lake Michigan using Pb-210 and Cs-137. Geochim. Cosmochim. Acta. 39: 285-304.
Schindler, D. W., 1975. Whole-lake eutrophication experiments with phosphorus nitrogen and carbon. Verh. Internat. Limnol. 19: 3221-3231.
Schindler, D. W., 1976. Biogeochemical evolution of phosphorus limitation in nutrient-enriched lakes of the Precambrian Shield. In: J. O. Nriagu (Ed.): Environmental Biogeochemistry. Volume 2. Metal transfers and ecological mass-balances. Ann Arbor Science Publishers, Inc., Ann Arbor, Mich. p. 647-664.
Schindler, D. W., F. A. J. Armstrong, S. K. Holmgren & G. J. Brunskill, 1971. Eutrophication of lake 227, Experimental Lakes Area, northwestern Ontario, by addition of phosphate and nitrate. J. Fish. Res. Board Can. 28: 1763-1782.
Schindler, D. W., H. Kling, R. V. Schmidt, J. Prokopowich, V. E. Frost, R. A. Reid & M. Capel, 1973. Eutrophication of lake 227 by addition of phosphate and nitrate: the second, third and fourth years of enrichment, 1970, 1971 & 1972. J. Fish. Res. Board Can. 30: 1415-1440.
Schindler, D. W., & D. R. S. Lean., 1974. Biological and chemical mechanisms in the eutrophication of freshwater lakes. Ann. New York Acad. Sci. 250: 129-135.
Stainton, M. P., M. J. Capel & F. A. J. Armstrong, 1974. The chemical analysis of freshwater. Misc. Special Publ. No. 25, Canada Dept. of Environment, Freshwater Institute, Winnipeg. 129 pp.
Wetzel, R. G., 1975. Limnology. W. B. Saunders Co., N.Y. 743 pp.

Note added in proof at the request of the editor.

The sediments of lake 227 are extremely rich in organic matter (over 50-60%) due to input of allochthonous organic matter from the surrounding forests and due to a lack of dilution by clastic materials.

The inorganic portion of sediments was composed almost entirely of oxides of Si, Al and Fe, primarily in the minerals quartz, plagioclase and K-feldspar. Nearly 50% of the inorganic fraction of sediments is of clay size ($< 2\ \mu$). Kaolinite, illite and chlorite were also abundant in this small particle size. More detailed information on the inorganic sediments of lake 227 and other ELA lakes is given by Brunskill et al (1971).

Brunskill, G.J., D. Povoledo, B.W. Graham & M.P. Stainton, 1971. Chemistry of surface sediments of sixteen lakes in the Experimental Lakes Area, northwestern Ontario. J. Fish. Res. Board Can. 28: 277-294

Key words: Experimental Lake Area
Sediments
Eutrophication
Phosphate
Mass balance

A PRELIMINARY REPORT: NUTRIENTS AND METALS TRANSPORTED BY SEDIMENTS WITHIN THE GENESEE RIVER WATERSHED, NEW YORK, U.S.A.*

Michael M. Reddy

Division of Laboratories and Research
New York State Department of Health, Albany, N.Y., U.S.A.

Summary

Representative preliminary results from a program instituted to determine the sediment distribution and transport of nutrients and metals in the Genesee River basin are discussed. Concentration ranges for total and fractional analyses of Genesee River watershed sediments are consistent with data obtained by several workers for similar sediments. Total P is higher in main tributaries than in the Genesee River. Calcium carbonate appears to be involved in sediment P mineralization. Average atomic ratios for C/N/P and Fe/Mn/Zn in sediments collected from the Genesee River and two of its main tributaries before and after the 1976 spring runoff show consistent values, indicating that they may be related to long-term geological and cultural trends within the watershed. Organic C/N and organic C/P in sediment are directly related to sediment organic C. These relationships may be helpful in identifying the source of excess nutrients in sediments.

Introduction

The purpose of this paper is to present a partial approach to, and some preliminary results from, a study of the chemical interactions of nutrients and metals being transported by sediments. The system under study is the watershed of the Genesee River, a tributary to Lake Ontario. The watershed covers 2400 square miles in New York State, U.S.A.

Theoretical background

The major factors which affect the composition of transported sediments in unpolluted watersheds are weathering processes, including erosion, and the composition of source rocks (Self, 1975). Sediments also reflect the presence of natural or man-made point and non-point sources of nutrients and metals. Systematic investigation of sediment will show the extent and distribution of contamination and will aid in the identification of natural and man-made pollution sources (Müller & Förstner, 1975).

Recent research indicates that the major sources of fluvial sediment in North American streams and rivers are agricultural and urban runoff and stream bank erosion; for example, the annual suspended load of the Maumee River, Ohio, has been reported to be largely surficial in origin (Wall & Wilding, 1976). Stall (1973) reported that most fluvial sediment in the U.S.A. originates from agricultural land.

Biological processes may also have a dominant effect on sediment composition. In Buzzards Bay, Massachusetts, for example, the characteristics of sediments, benthic animals and bottom waters appear to be interrelated (Driscoll, 1975). Other examples of this interrelation have been reviewed by Rhoads (1974).

The way that sediments interact with nutrients and metals in the water column and their rates of transport within the watershed depend on their type. The PLUARG Quality Control Handbook for Pilot Watershed Studies (International Reference Group on Great Lakes Pollution from Land Use Activities, 1976) defines two types: sediment and suspended sediment. Each of these has distinctive physical properties, e.g., particle-size distribution, density and surface area.

The amount of trace metal carried by suspended sediment in a river is strongly dependent on the chemical form of the element (Gibbs, 1973). Within a sediment an element may exist in several different chemical forms. The specific chemical form in turn regulates its reactivity and bioavailability. In the case of sediment P a small fraction may be readily available for biouptake in the form of ion-exchangeable

*This study was carried out as part of the Task C work of the Pollution from Land Use Activities Reference Group, International Joint Commission for the Great Lakes and was funded through the United States Environmental Protection Agency and the State of New York.

Table 1. Bottom sediment and suspended bottom sediment sample collection, preservation and pretreatment.

| | Sediment sample type | |
|---|---|---|
| | Bottom sediment | Resuspended sediment |
| Sample collection | Collected midstream where possible and wet sieved immediately, on site, through a 2-mm polyethylene sieve using river water. | Bottom sediment suspended by washing approximately 2 kg of river water. The resulting suspension is allowed to settle for 30 min. Settled sediment is removed and the supernatant is placed in a 2-l plastic bottle. |
| Sample preservation | Wet sediment samples are frozen immediately on site and are kept frozen until analysis. | A 2-l sample of suspended sediment is frozen immediately on site and kept frozen until analysis. |
| Sample pretreatment | Frozen sediment samples are thawed prior to analysis, dried at 110°C, crushed and sieved through a 100-mesh all-plastic sieve.* | Frozen suspended sediment samples are thawed and centrifuged (10,000 rpm for 20 min.) to reduce sample volume from 2 to 0.1 l. This suspension is freeze-dried. |

* A comparison of analytical results obtained for wet sediments, dried unsieved sediments and dried crushed and sieved sediments showed minor differences. However, the analytical precision was significantly poorer for unsieved samples.

phosphate ion. A much greater fraction may be in a biologically unavailable form such as apatite. Thus, information limited to the total nutrient and metal contents of sediments and water is insufficient to characterize the distribution and ecologic effects (O'Shea & Mancy, 1976).

Study organization

Field work for the Genesee River watershed study was performed at 28 stations on three consecutive days each during each season of the year. Major emphasis is placed on conducting sampling during periods of high flow. Two mobile field laboratories are used. Sample collection and pretreatment procedures have been adapted from techniques employed by the U. S. Geological Survey (Malo, 1975). Contamination is avoided by using plastic or plastic-coated equipment.

All sediment samples are analyzed for total nutrients and metals. In addition, for several elements the chemical species are determined by fractional analysis of phases: (1) an adsorbed phase consisting of ions or complexes adsorbed to sediments; (2) an incorporated metal or phosphate phase formed by physical, chemical or biological incorporation of phosphorus or trace metals in a sediment organic matrix, metal oxide or metal hydroxide matrix; and (3) a crystalline phase consisting of P or trace metals which are part of the crystal lattice of sediment minerals.

Representative results are presented here for total and fractional analyses of bottom and suspended sediments collected between June 1975 and March 1976. A complete report on the transport of nutrients and trace metals by sediments in the basin is in preparation.

## Experimental

Methods used to collect and treat bottom and suspended sediment samples are outlined in Table 1.

Total analysis

Details of experimental procedures for total analysis of bottom and suspended sediments, including information about the effect of total sediment digestion procedures on the precision and accuracy of metal analyses have been presented elsewhere (Krishnamurty et al., 1975, 1976). Since silicate minerals are not solubilized in these procedures the results are considered only as an estimate of the total available elements in sediments.

Total P analysis of a National Bureau of Standards sample of orchard leaves (certified value 2100 ± 100 μg/g) was 2038 μg/g with a relative standard deviation (RSD) of 5% for N=21. Precision for total P analysis of three actual sediments used as quality control check samples was RSD 12%, N=12; 16%, N=11; 13%, N=11 (means 412, 487 and 758 µg P/g).

### Fractional analysis

Procedures for fractional analysis were evolved from methods used for soils and lacustrine sediments. In particular, techniques with high within-run and between-run precision were chosen for P and trace metal fractionation. An extraction procedure employing NaOH (with NaCl) to determine nonoccluded P and subsequently HCl to determine occluded P was adapted from the work of Williams et al. (1971 a,b). Work performed in this laboratory (unpublished data) and elsewhere (Sagher, 1974) indicates that the NaOH extractable P is a measure of bioavailable sediment P. Hydrochloric acid extractable P reflects apatite and P incorporated in iron oxides.

Methods used to measure extractable metals were chosen to identify the elemental association between P and hydrous metal oxides in sediments. These extraction procedures and those for P are semiquantitative. Schwertmann (1973) emphasized that the results of such procedures are best considered as a measure of the relative amount of a phase or, more generally, of an element's reactivity in a sediment under carefully controlled conditions.

A hydroxylamine extraction procedure developed by Chau (1972) for specific extraction of Mn from sediments under mildly reducing acidic conditions was used to examine the Mn-P relationship in sediments. Subsequent extraction with ammonium oxalate-oxalic acid solution removes amorphous iron oxides and organic Fe, and the P associated with these constituents (Schwertmann, 1973). Precision for replicate P fractional analysis on actual sediment samples used for quality control checks shows a RSD of about 5% for all extraction procedures for samples containing more than 100 µg extractable P per gram. Precision for metal extraction analysis is similar to that for P.

## Results and Discussion

### Total Analysis

Total sediment content for a number of elements in bottom sediment has been determined for samples obtained from 25 stations throughout the watershed. The ranges were (µg/g): Al, 2,800-11,000; Cr, 10-79; Cu, 11-41; Fe, 2,350-26,000; Mn, 180-1,035; Ni, 10-87; Pb, 10-320; Zn, 35-210; and (%): total C, 0.31-8.26; organic C, 0.09-5.69; total N, 0.011-0.63; and total P, 0.033-0.096.

Analysis of bottom sediments collected in Lake Ontario at the mouth of the Genesee River (12 sampling stations) from June 1972 to May 1973 (Carlson, 1975) gave ranges for selected results of (µg/g): Cu, 6-50; Mn, 25-800; Ni, 10-450; Pb, 5-80; Zn, 25-350; and (%): organic C, 0.10-3; total N, 0.025-0.26; total P, 0.01-0.1. Lower values in lake sediments for Mn may result from solubilization under reducing conditions. High values for Zn and Ni in the lake sediments reflect the presence of point-source inputs adjacent to the lake (Carlson, 1975). The upper limit of the ranges for Pb, organic C and total N are higher in the watershed sediments. Total P and Cu analytical ranges are quite similar for the fluvial and lacustrine sediments. Total sediment P of .0825% in a sample collected in the Genesee River on 9 October 1973 (Bannerman et al. 1975) is consistent with the result range found in this study.

### Fractional analyses

The sediment contents extracted by selective reagents have been determined for bottom sediment samples from 25 stations within the watershed. Ranges for P fractional analysis were (µg/g): after NaOH extraction, 11-410; after subsequent HCl extraction, 187-731; after hydroxylamine extraction, 6-313; and after subsequent ammonium oxalate extraction, 71-405. Values within these ranges were found in sediments collected in the Genesee River on 9 October 1973 (Bannerman, 1975): after NaOH extraction, 100 µg/g, and after HCl extraction (following NaOH and citrate-dithionite-bicarbonate extractions) 328 µg/g.

Sediments from the Rochester Basin, the central basin of Lake Ontario, contain a significantly higher minimum level of NaOH extractable P (129 µg/g, N=9; Bannerman et al. 1975) than bottom sediments from the Genesee River watershed (11 µg/g, N=30). This difference between lake and river sediments reflects the general increase in total P content of lake sediments with increasing depth (Delfino et al., 1969) and results from deposition of fine-textured materials in low-energy areas of the lake. Suspended sediment in the Genesee River watershed has a higher P content than bottom sediments (unpublished data) indicating that P accumulates in the smaller components of basin sediments. Recent research has indicated the importance of P transport by suspended sediment (Gill et al., 1976; Johnson et al., 1976).

Bottom sediment Fe and Mn ranges were µg/g): Fe after hydroxylamine extraction, 33-150; after subsequent ammonium oxalate extraction, 800-5,700; Mn after hydroxylamine extraction, 41-385; and after subsequent ammonium oxalate extraction, 15-133. Higher

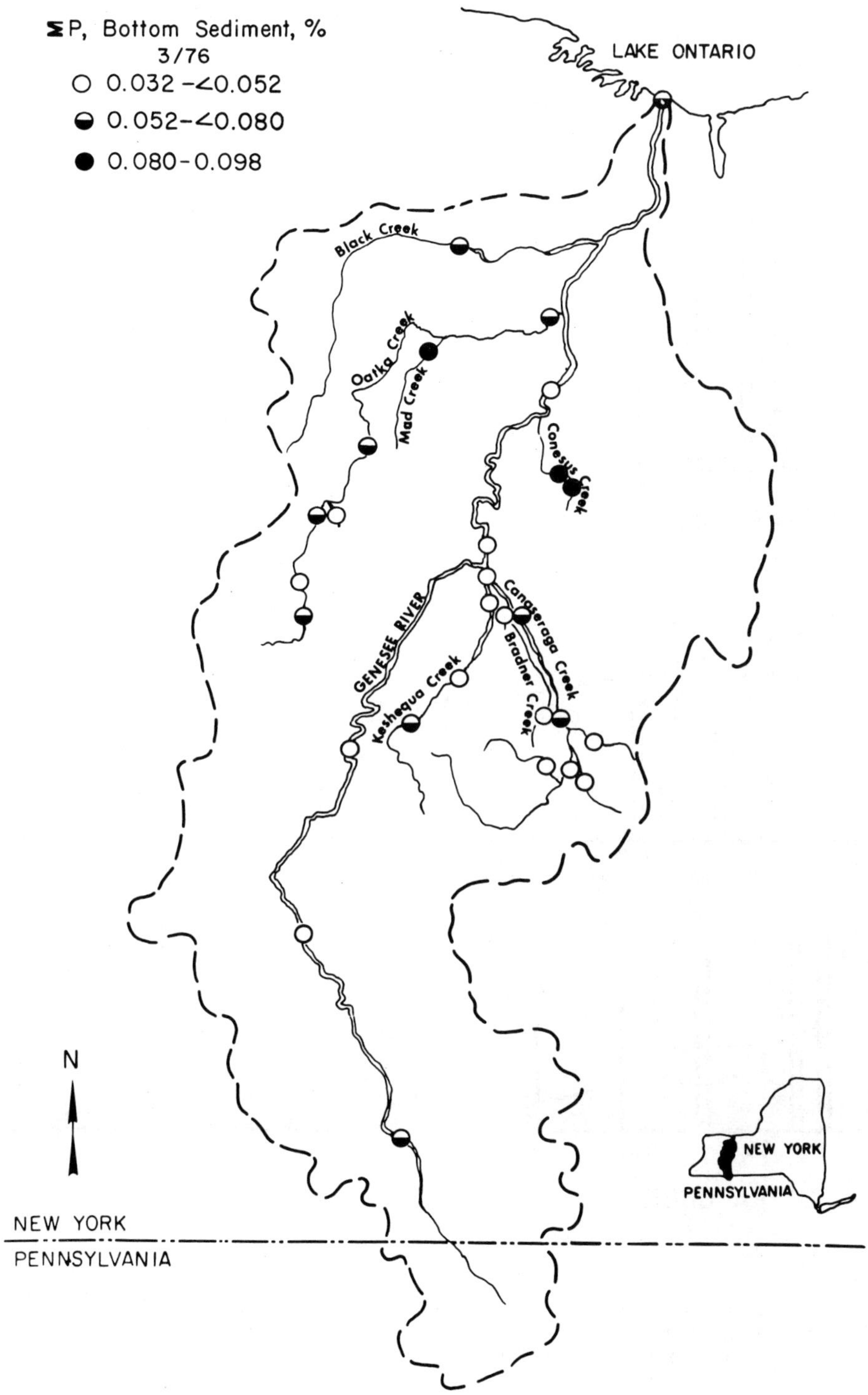


Fig. 1. Areal distribution diagram of total P in bottom sediments of the Genesee River watershed, March 1976.

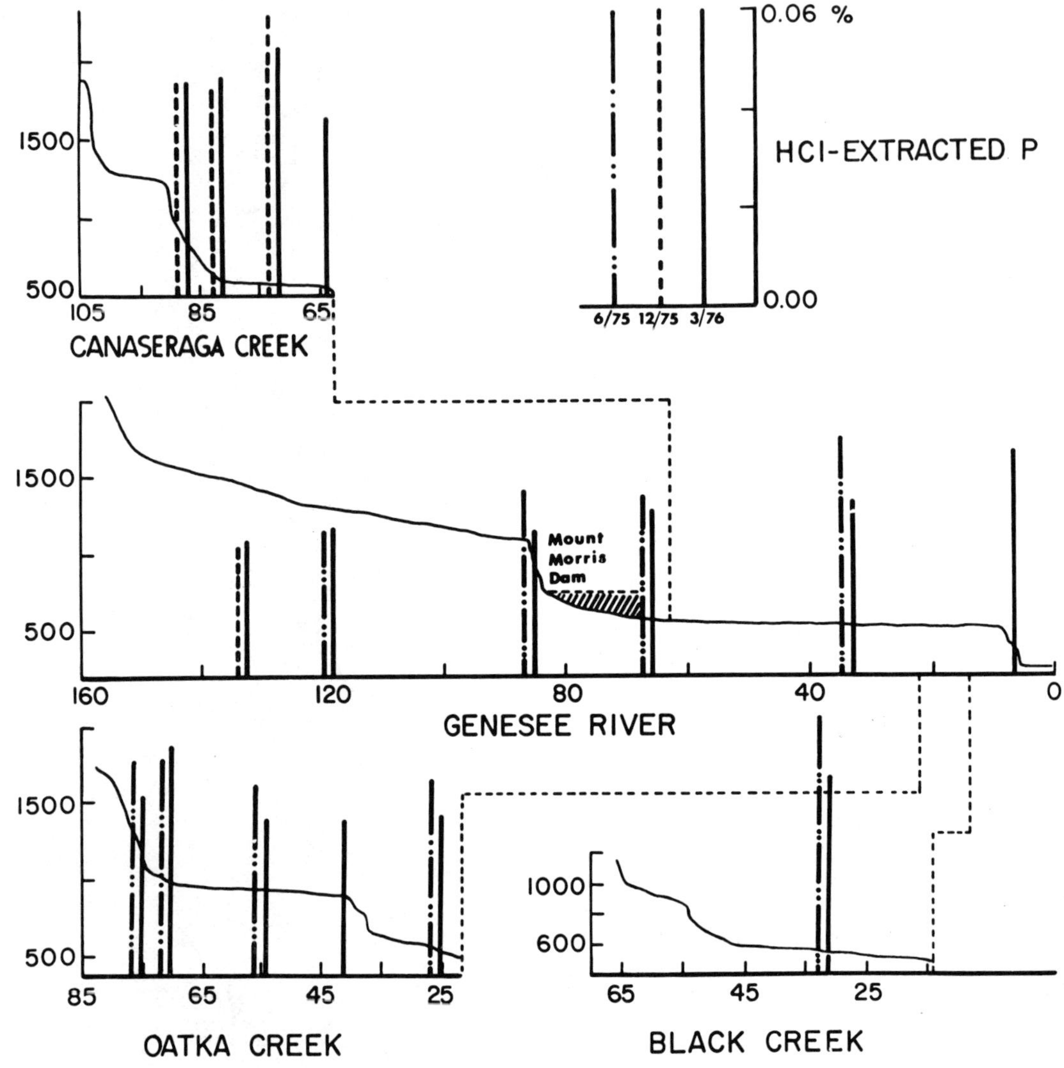


Fig. 2. HCl-extracted P in bottom sediments from the Genesee River and its major tributaries. Type of bar indicates date of sample collection (scale on upper right-hand corner). The horizontal scale indicates distance (miles) from the mouth of the Genesee River; the vertical scale indicates elevation above mean sea level (feet).

Table 2. Average atomic ratios of major and minor nutrients in sediment.

| Sediment type | Sampling date | N | C/N/P | N | Fe/Mn/Zn |
|---|---|---|---|---|---|
| Bottom | Dec. 1975 | 6 | 54/5/1 | 10 | 278/9/1 |
| | March 1976 | 19 | 53/3/1 | 28 | 511/9/1 |
| Suspended | Dec. 1975 | 5 | 79/9/1 | 4 | 333/8/1 |
| | March 1976 | 10 | 53/5/1 | 13 | 388/9/1 |

ranges were obtained for 10 lake sediment samples from central and western U.S.A. (extractable Fe, 3,700-11,800 μg/g; extractable Mn, 120-660 μg/g; Jenne et al., 1974), probably mainly because of differences in sample types and analytical procedure. For these lake sediments oxides of both Fe and Mn were dissolved with a strong reducing agent, sodium dithionite. In a comparison of Mn extraction from stream sediments using dithionite and hydroxylamine procedures, the dithionite extracted 10-20% more of the total Mn present (Chao & Anderson, 1974).

Areal distribution of sediment P

Areal distribution diagrams for sediment nutrients and metals in the Genesee basin show two major types of distributions, the first characteristic of nutrients, the second characteristic of the major mineral components.

A plot of total P, a representative nutrient, in bottom sediment samples collected at 28 stations during March 1976 (Fig. 1) shows higher levels in tributaries than in the Genesee River - a pattern which we found characteristic of nutrients. In particular, tributaries flowing through agricultural land (e.g., Conesus and Oatka Creeks) show the highest P content in the basin. A preliminary thermodynamic analysis for such a tributary, Oatka Creek, shows that the water column is saturated with calcium carbonate. This compound may be instrumental in forming carbonate apatite and binding P to sediments (Leckie & Stumm, 1970).

The lower P contents of the Genesee River sediments may arise by dilution of tributary sediment (relatively high in P) with low-P river sediment. A careful mass balance of sediment and P in the Genesee basin should indicate the magnitude of sediment dilution. Dilution of contaminated sediment has been proposed by Müller and Förster (1975) to explain the decreased content of pollutants in the lower Rhine River.

The irregular sediment P content for eastern tributaries in the central basin may be related to the greater sediment heterogeneity for streams in subbasins with high gradients (Self, 1975). High P content at the southernmost station on the Genesee River appears to be due to a sewage treatment plant outfall at Wellsville, New York (Carlson, 1975).

Hydrochloric acid-extractable P

Hydrochloric acid-extractable P is a major component of Genesee River basin sediments. The gross distribution pattern exhibited by total P (higher in tributary than in river sediments) is also exhibited by the HCl-extractable fraction (Fig. 2).

Two major tributaries, Canaseraga Creek and Oatka Creek, do not show wide spatial or temporal variation in this fraction. The Genesee River, on the other hand, shows a steady increase as it flows toward the lake. This increase appears to be a function, not only of increased inputs of P, but also of increased mineralization of the sediment phosphate, which occurs by the formation of carbonate apatite and probably increases as the basin geology becomes dominated by carbonate minerals. It is interesting that calcium carbonate saturation in the Genesee River follows a pattern similar to that of the HCl-extractable P of the sediments (unpublished data). This fact points again to a major role of calcium carbonate in P mineralization and immobilization in the watershed (Leckie & Stumm, 1970).

Hydrochloric acid-extractable phosphorus decreased at several sites in the basin after the spring runoff (Fig. 2). This reduction may be associated with the early snow-melt runoff during 1976 and is being investigated further.

Elemental ratios for sediment components

The atomic ratios of major and minor nutrients in natural waters and sediments are often used as indicators of the sources and nature of nutrient inputs. In addition, elemental ratios may give an indication of the suitability of a sediment for biological transformation and its potential as a source of nutrients when transported to a lake (Porter et al., 1975). (Table 2)

In the Genesee River and two major tributaries, Oatka and Canaseraga Creeks, the atomic ratios for the major nutrients in sediment (organic C, total N and total P) and for minor nutrients (total extractable Fe, Mn and Zn) have been determined for two moderate-flow periods, one preceding and one following the 1976 runoff. The overall consistency of these ratios (Table 2) suggests that they are not affected by short-term variations in basin development and utilization.

Organic C to N ratio

More detailed information about sediment nutrient content and distribution may be obtained by examining the relationship between the atomic ratio of organic C to N (C/N) and the organic C content. Bacterial action in stream environments results in the accumulation of refractory N-depleted organic material. Wetzel (1975) has therefore suggested that high C/N of dissolved organic matter in lakes indicates allochthonous input, while low C/N indicates autochthonous input. He further points out that C/N is a function of total organic C; lakes with high total C (arising principally from allochthonous inputs) generally show high C/N.

Seven sediment samples collected on March 15, 1976 from five sampling sites on a 100-mile section of the river from Wellsville upstream to Avon, New York, contained 1.6 - 8.2 mg C/g and showed C/N ratios of 7 to 25. The least-squares equation for the relationship between C/N and C was: $C/N=1.51\ (\pm0.93)\ C+9.62(\pm4.63)$, where organic C is given in mg/g and the values in parentheses are standard errors. The correlation coefficient was 0.59. Scatter associated with the data may be due to additional sediment inputs of high N organic material from municipal and agricultural wastes and to N losses from sediments by denitrification (Keeney, 1973; Iserimah et al., 1976). An eighth sample (22 mgC/g, C/N=15) collected at Wellsville, New York, at the same time did not follow this relationship and may have been contaminated by a nearby sewage treatment plant outfall.

This equation indicates that the relation between C/N and total organic C in lakes may also apply to river and stream sediments.

As limiting cases of natural organic substances with high and low N contents we chose commercial peat (low N) and processed sewage sludge sold as Millorganite, a high organic N fertilizer. Each of these products was analyzed by the procedures developed for sediments. The peat had 216 mg C/gm and C/N=108, while the dried sewage sludge had 375 mg C/gm and C/N= 6. While these limiting cases do not follow the relationship obtained for the Genesee River sediments, they do illustrate the range of C/N values which might be encountered in an area of point or nonpoint source pollution. The C/N for these limiting cases are more extreme than those cited by Wetzel (1975): allochthonous organic matter, C/N = 50/1; autochthonous organic matter, C/N = 12/1. Carbon-nitrogen ratios may be especially useful in identifying sediments containing excess nutrients because of contamination by domestic or agricultural wastes.

Organic C to P ratio

The seven sediments from the Genesee River which showed a linear relationship between C/N and organic C also show a linear relationship between the C to P atomic ratio (C/P) and organic C. The least-squares equation obtained was $C/P = 4.72\ (\pm0.39)\ C + 2.0\ (\pm1.72)$. The values in parentheses are standard errors; the correlation coefficient was 0.98. Ranges for these seven samples for C and C/P were 1.6-8.2 mg C/g and 9-41. As with C/N, the eight sediment sample appeared to be contaminated by sewage and did not follow this equation (C = 22 mg C/g, C/P=59).

Sediments with high organic C in the Genesee River probably arise principally from inputs of low-P allochthonous carbonaceous material. The high correlation coefficient may result from the conservative nature of P in natural waters and sediments (Johnson et al., 1976).

Conclusion

The sediment sampling and analysis program developed for this study is useful for characterization of sediment nutrients and metals in an intermediate-sized river basin. In particular, relationships between the major nutrients (organic C, N and P) have been confirmed which may be helpful in identifying the source of excess nutrients in sediments. Further fractional analyses of sediments and examination of chemical equilibria in the sediments and water column are expected to contribute information on the distribution, transformation and transport of nutrients and metals in this watershed.

Acknowledgment

The author wishes to express his appreciation to G. W. Fuhs for his valuable suggestions and discussions during the course of this work; K. V. Krishnamurty and E. Shpirt for assistance in the development of the Genesee River watershed sediment analysis program; and to K. Wang, C. Baggenstoss, N. Sopchak, C. Snyder and the staff of the Analytical Chemistry Section of this Division for technical support.

## References


Bannerman, R.T., D.E. Armstrong, R.F. Harris & G.C. Holdren, 1975. Phosphorus uptake and release by Lake Ontario sediments, EPA-660/3-75-006. USEPA, Corvallis, Oregon 97330.

Carlson, G.A., 1975. Personnal communication.

Chao, T.T., 1972. Selective dissolution of manganese from soils and sediments with acidified hydroxylamine hydrochloride. Soil Sci. Am. Proc. 36:764-768.

Chao, T.T. & B.J. Anderson, 1974. The scavenging of silver by manganese and iron oxides in stream sediments collected from two drainage areas of Colorado. Chemical Geology 14:159-166.

Delfino, J.J., G.C. Bortleson & G.F. Lee, 1969. Distribution of Mn, Fe, P, Mg, K, Na, and Ca in the surface sediments of Lake Mendota, Wisconsin. Environ. Sci. Technol. 3:1189-1192.

Driscoll, E., 1975. Sediment-animal-water interaction, Buzzards Bay, Mass. J. Marine Research 33:275-302.

Fuhs, G.W., S.D. Demmerle, E. Canelli & M. Chen, 1972. Characterization of phosphorus-limited plankton algae (with reflections on the limiting-nutrient concept). In: G.E. Likens, (Ed.): Nutrients and eutrophication: the limiting nutrient controversy. Special Symposium, Amer. Soc. Limnol. Oceanogr. 1:113-133.

Gibbs, R.J., 1973. Mechanism of trace metal transport in rivers. Science 180:71-73.

Gill, A.E., J.R. McHenry & J.C. Ritchie, 1976. Efficiency of nitrogen, carbon and phosphorus retention by small agricultural reservoirs. J. Environ. Qual. 5:310-315.

International Reference Group on Great Lakes Pollution from Land Use Activities, 1976. Quality Control Handbook for Pilot Watershed Studies. Revised June 1976. International Joint Commission, Regional Office, Windsor, Ontario, Canada. p. 2.221-2.222.

Isirimah, N.O., D.R. Keeney & E.H. Dettmann, 1976. Nitrogen cycling in Lake Wingra. J. Environ. Qual. 5:182-188.

Jenne, E.A., J.W. Ball & C. Simpson, 1974. Determination of trace metals in sodium dithionite-citrate extracts of soils and sediments by atomic absorption. J. Environ. Quality 3:281-287.

Johnson, A.H., D.R. Bouldin, E.A. Goyette & A.M. Hedges, 1976. Phosphorus loss by stream transport from a rural watershed: Quantities, processes and sources. J. Environ. Qual. 5:148-157.

Keeney, D.R., 1973. The nitrogen cycle in sediment-water systems. J. Environ. Qual. 2:15-29.

Krishnamurty, K.V. & M.M. Reddy, 1975. The chemical analysis of water and sediments in the Genesee River Watershed Study. On file in library of International Joint Commission, Regional Office, Windsor, Ontario, Canada.

Krishnamurty, K.V., E. Shpirt & M.M. Reddy, 1976. Trace metal extraction of soils and sediments by nitric acid-hydrogen peroxide. Atomic Absorption Newsletter 15:68-70.

Leckie, J. & W. Stumm, 1970. Phosphate precipitation. In: E.F. Gloyna & W.W. Eckenfelder (Eds.): Water quality improvement by physical and chemical processes. Univ. of Texas Press, Austin, Texas. p. 237-249.

Malo, B., 1975. Personal communication.

Müller, G. & U. Förstner, 1975. Heavy metals in sediments of the Rhine and Elbe Estuaries: Mobilization or mixing effect? Environmental Geology 1:33-39.

O'Shea, T.A. & K.H. Mancy, 1976. Characterization of trace metal-organic interactions by anodic stripping voltammetry. Anal. Chem. 48:1603-1607.

Porter, K.S., D.A. Lauer, J. Messinger & D.R. Bouldin, 1975. Flows of nitrogen and phosphorus on land. In: K.S. Porter (Ed.): Nitrogen and phosphorus (Food Production, waste and the environment). Ann Arbor Science Publishers, Ann Arbor, Michigan. p. 121-165.

Rhoads, D.C., 1974. Organism-sediment relations on the muddy sea floor. Oceanogr. Mar. Bio. Ann. Rev. 12:263-300.

Sagher, A., 1974. Microbial availability of phosphorus in lake sediments. M.S. thesis, Department of Soil Science, University of Wisconsin, Madison.

Schwertmann, U., 1973. Use of oxalate for Fe extraction from soils. Can. J. Soil Sci. 53:244-246.

Self, R.P., 1975. Petrologic changes in fluvial sediments in the Rio Nautla basin Veracruz, Mexico. J. Sed. Pet. 45:140-149.

Stall, J.B., 1972. Effects of sediment on water quality. J. Environ. Quality 1:353-360.

Wall, G.J. & L.P. Wilding, 1976. Mineralogy and related parameters of fluvial suspended sediment in northwestern Ohio. J. Environ. Qual. 5:168-173.

Williams, J.D.H., J.K. Syers, R.F. Harris & D.E. Armstrong, 1971a. Fractionation of inorganic phosphate in calcareous lake sediments. Soil Sci. Soc. Amer. Proc. 35:250-255.

Williams, J.D.H., J.K. Syers, D.E. Armstrong & R.F. Harris, 1971b. Characterization of inorganic phosphate in noncalcareous lake sediments. Soil Sci. Soc. Amer. Proc. 35:556-561.

Wetzel, R.G., 1975. Limnology. W.B. Saunders Co, Philadelphia, Pa. p. 208-212.



Key words: fluvial sediment
sediment transport
nutrients
metals
North American Great Lakes

# PRELIMINARY RESULTS FROM A RECIRCULATING EXPERIMENTAL SYSTEM: OBSERVATIONS ON INTERACTION BETWEEN CHALK STREAM WATER AND INORGANIC SEDIMENT

M. Ladle, J. H. Baker, H. Casey & I. S. Farr

Freshwater Biological Association, River Laboratory, Wareham, Dorset, England

## Summary

Some effects of water circulation and sediment on water chemistry in a large circulating channel are described. The system is designed to permit investigation of the biologically related processes taking place in a simulated chalk stream ecosystem. Water pumped from a borehole in a chalk aquifer is circulated in a sectional, glass-reinforced plastic tank (53 m in length, having a trapezoidal section 1 m in width at the base) by means of an Archimedian screw pump.

The water is similar in character and origins to that forming the major proportion of the discharge in local chalk streams. Initially, changes in pH, alkalinity, calcium, magnesium, sodium, potassium, reactive phosphate, silicate, nitrate, dissolved and total organic carbon and bacterial numbers were monitored. Circulation trials were carried out in the dark, both in the presence and absence of a gravel substratum, to test the stability of the water chemistry.

Preliminary results indicate that the presence of a gravel substratum comparable to that of chalk streams exerts an important stabilising influence on the levels of dissolved elements in the flowing water. Major interactions between the inorganic sediment and water appear to affect the concentrations of calcium, reactive phosphate and silicate, whereas magnesium, sodium and potassium seem to be affected to a much lesser extent. In addition there are substantial differences in the levels and population dynamics of the suspended, heterotrophic bacteria in the presence and absence of the gravel substratum.

## Introduction

Major difficulties arise in the study of stream ecosystems because of unpredictable changes in flow and nutrient levels. Nevertheless seasonal patterns of change of some dissolved chemical species in chalk streams have been ascribed to primary production cycles (Casey & Newton, 1973; Casey, 1976). However, both seasonal and progressive downstream changes are difficult to relate quantitatively to autochthonous biological activities because of the short and seasonally variable retention time of the water ( a few days). In addition, changes are often difficult to detect against background fluctuations due to agricultural, urban and storm inputs.

Spates and river management practices can dramatically affect stream bed configuration as well as sediment, plant and faunal distributions. These result in abrupt changes in biomass and development of populations, which can raise difficulties in the study of natural stream biota.

To facilitate studies of the relationships between hydrological, chemical and biological processes in chalk streams, a circulating channel has been constructed to simulate steady flow stream conditions. The channel has known inputs and outputs and can retain a body of water for any desired period of time, allowing experimental control or protection from many of the factors causing unpredictable changes in natural streams.

Smaller systems incorporating pumped or paddled recirculation of water have been used to study particular aspects of processes in flowing ecosystems (Kevern & Ball, 1965; Lauff & Cummins, 1964; Von Lutz & Keller, 1975; McIntyre et al., 1964; Odum & Hoskin, 1957) and large, through-flow troughs have been in use in Zurich for many years for the investigation of self-purification following pollution (Wuhrmann, 1964; Wuhrmann et al., 1966). The present channel will be the first attempt to build up a more or less complete ecosystem in a channel having a recirculating facility.

## Description of the apparatus

The channel is at the Waterston outstation (National Grid Reference SY 742953) of the Freshwater Biological Association's River Laboratory. It consists of a "race-track" shaped tank (Figure 1) composed of fibreglass reinforced plastic sections which are bolted together. The joints are sealed with a bituminous compound and coated internally with fibreglass resin. The channel is 53 m in length and has a trapezoidal cross section of similar dimensions to many small natural streams. The size of the channel was determined by the demand for a model which could contain sufficiently large populations of plants and animals and areas of sediments to permit destructive sampling with minimal disturbance to the system as a whole. Water is circulated by an Archimedian screw pump which has been modified by the replacement of the original oil-filled lower bearing by a water-lubricated nylon bearing, thereby avoiding oil seepages into the channel. The channel and pump are partially buried in the gravel of a disused watercress bed which supports the structure. Groundwater in the gravel is at a constant temperature of 10°C and this helps to stabilise the temperature of the circulating water.

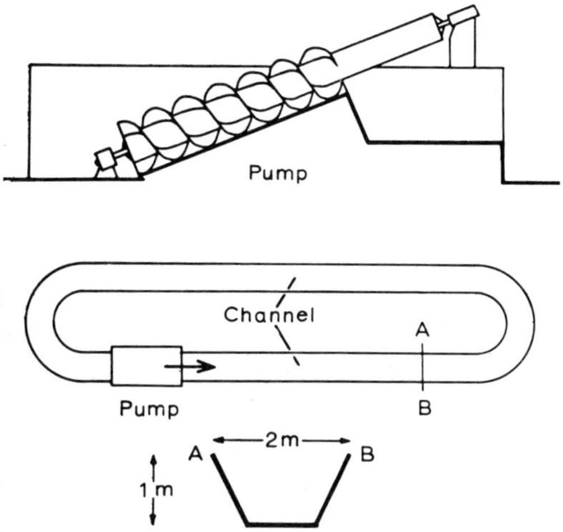


Figure 1. Circulating channel. Elevation of screw pump, plan and cross section of channel.

Water supply to the channel is from a chalk aquifer borehole, which had an original depth of 54.8 m. The chemical composition of the input water is almost constant (Table 1) and is similar to that of many springs and boreholes which form the major sources of chalk stream discharge throughout the year (Casey, 1969; Casey & Newton, 1973; Westlake et al., 1972).

The incorporation of an overflow notch and the use of a controlled, continuously pumped inflow provides a facility for the exchange of water which is necessary to prevent nutrient depletion in the periods between prolonged retention experiments. Water level in the channel is maintained at approximately 0.7 m above the base.

The channel has been used both with borehole water alone, and in the presence of a gravel substratum of similar particle size and nature to the materials forming the bed of natural chalk streams. The channel, without gravel, holds between 60 and 63 $m^3$ of water and after placement of gravel 50 $m^3$.

The screw pump was adjusted to develop a mean velocity of 0.2 m $sec^{-1}$ in the centre of the channel in the absence of gravel. This increased to 0.4 m $sec^{-1}$ after addition of the substratum. Temperature and oxygen are recorded continuously. Oxygen throughout the water body and within the gravel remains at 100% saturation relative to air. Sodium chloride dilution tests show that the volume of water above and within the gravel (calculated from channel geometry and pore space in the gravel) was totally mixed within a period of less than two hours. Oxygen saturation of the borehole supply is 70%. In the tests reported in this paper, light and allochthonous materials were excluded by covering the channel with a sheet of black polythene.

Table 1. Chemical composition of inflow water and range of values observed in channel water.

| | Waterston Borehole | Waterston Channel (without gravel) | Waterston Channel (with gravel) |
|---|---|---|---|
| | range | range | range |
| pH | 7.2 -7.8 | 7.4 -8.6 | 7.9 -8.6 |
| Alkalinity (m.e.$l^{-1}$) | 4.44-4.80 | 1.86-4.60 | 3.30-4.50 |
| Calcium (mg $l^{-1}$) | 97.4-103.0 | 42.4-100.0 | 80.0-100.0 |
| Magnesium (mg $l^{-1}$) | 2.00-2.30 | 2.00-2.20 | 1.90-2.20 |
| Sodium (mg $l^{-1}$) | 8.4-9.1 | 8.5-8.8 | 8.6-9.2 |
| Potassium (mg $l^{-1}$) | 1.0-1.4 | 1.0-1.2 | 0.8-1.2 |
| Reactive phosphate (µg $l^{-1}$ $PO_4P$) | 30.8-49.4 | 0 -39.4 | 9.5-31.0 |
| Nitrate (mg $l^{-1}$ $NO_3N$) | 4.76-5.57 | 4.46-5.06 | 4.40-4.70 |
| Silicate (mg $l^{-1}$ Si) | 3.80-5.45 | 4.30-4.62 | 2.12-4.38 |
| T.O.C. (mg $l^{-1}$) | 0.4-0.6 | 0.4-0.7 | - - |

Aspects of the study made repeated checks on chemical results highly desirable. However, it was impracticable to use the large channel for this purpose and hence a small (2 m) laboratory channel with paddled circulation was set up. There was good qualitative agreement between processes taking place in the two systems.

## Results

In an initial test of the stability of the water chemistry, borehole water was added to the large channel, without substratum, in the dark. In this experiment temperature fluctuated little (8.0 to 10.3°C) and was well within the range recorded for chalk streams (Westlake et al., 1972). Total organic carbon in the water showed no appreciable change (Table 1). Only a small number of invertebrates gained access to the system.

The most striking changes were in the chemistry of the water. With the water stationary (not circulating) for three days there were only very small changes in chemical composition. When pumped circulation started the pH increased from 7.9 to 8.6, and the calcium concentration (Figure 2) and alkalinity decreased rapidly. There was visible precipitation of calcium carbonate on the walls and floor of the channel.

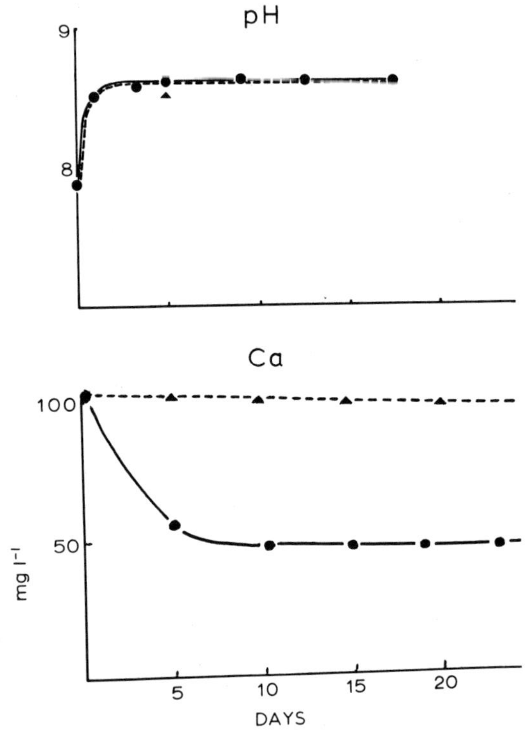


Figure 2. Effect of gravel substratum on circulating water. Changes in pH and calcium concentration with ▲——▲ and without ●—● gravel.

Loss of dissolved inorganic phosphate was associated with the precipitation of calcium carbonate until phosphate was no longer detectable (< 1μg $l^{-1}$) (Figure 3) by the standard method (Murphy & Riley, 1962) after eleven days. After calcium precipitation had ceased, phosphate was added in the form of potassium dihydrogen orthophosphate in sufficient quantity to increase the concentration to approximately its value at the start of the experiment. The phosphate remained in solution. The experiment was repeated with fresh borehole water introduced into the channel which was still coated with precipitated calcite. Calcium and phosphate losses from solution were the same on each occasion. Calcium precipitation and phosphate loss could be arrested or reversed by bubbling in gaseous carbon dioxide.

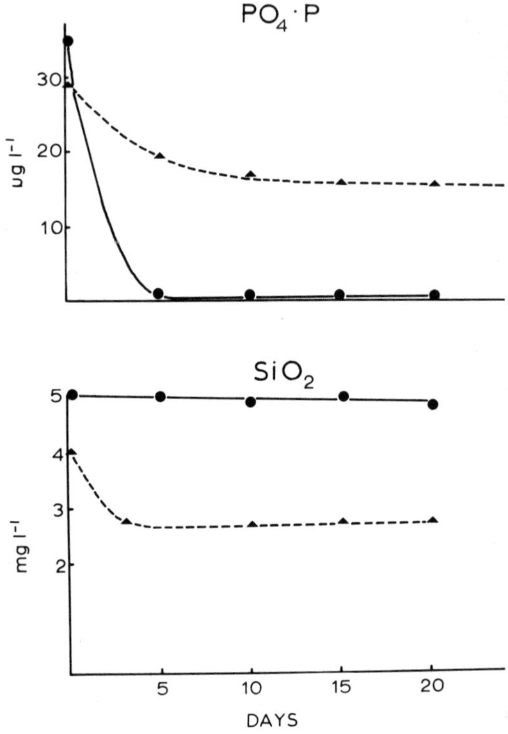


Figure 3. Effect of gravel substratum on circulating water. Changes in reactive phosphate and silicate with ▲——▲ and without ●——● gravel.

The concentration of suspended heterotrophic bacteria was determined daily by spread plate counts on casein/peptone/starch medium (Jones, 1970), incubated at 20°C. The bacteria increased logarithmically for the first nine to twelve days in consecutive experiments, with a doubling time of 36 and 49 hours respectively (Figure 4). The channel therefore acted as a large batch culture in a way which would be most unusual in any natural water body. The reason for the increased doubling time in the second experiment is not known, but it does not seem to be due to the water temperature, because the latter remained close to 10°C throughout both experiments.

After exponential increase, the popuations in both cases showed signs of oscillating reminiscent of classical predator prey theory (Krebs, 1972). As expected, faecal coliforms were not detected by standard tests (Anon, 1969) performed daily. The final bacterial population was of the same order of magnitude as that found in natural chalk streams (Baker & Farr, in press).

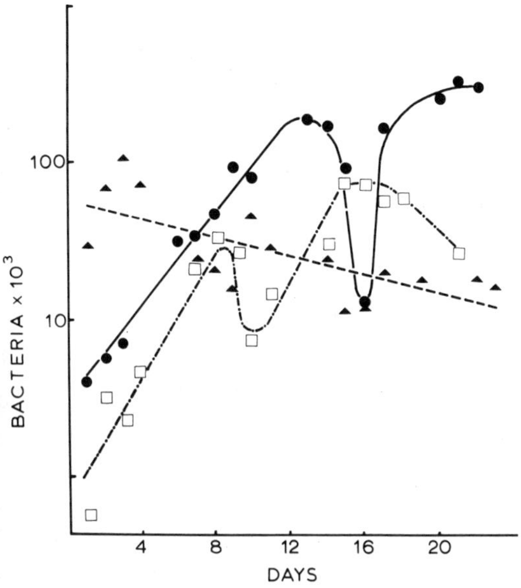


Figure 4. The variation in plate counts of bacteria per ml plotted against time from filling the channel. The linear dashed line refers to the experiment with a gravel substratum. The curvilinear lines refer to successive experiments without gravel.

Chlorophyll a could not be detected either in suspension or in the precipitate on the sides and bottom of the channel (Marker, pers. comm.).

The experiment was repeated following a complete change of water and the addition of commercially washed "reject" flint gravel (nominal 40 mm). The depth of gravel added varied from 0.5 m immediately downstream of the pump to 0.3 m at the other end of the channel, giving a gradient approximately the same as in natural chalk streams (1/250).

Although the gravel had been washed, a small amount of fine material adhering to its surface clouded the water during its placement in the channel. Repeated exchanges of water during circulation reduced suspended solids concentration to a level below 0.5 mg $l^{-1}$. X-ray diffraction of the suspended material taken from the gravel surface and collected in sediment traps showed it to be predominantly clay with 12-30% of fine quartz. The clay is a mixture of Illite and Kaolinite in a ratio of approximately 3.5:1.0, its organic carbon content is less than 1%.

Initially when pumped recirculation was started, gaseous carbon dioxide was bubbled into the water to maintain the calcium at a high solution concentration. Progressive reduction in the amount of carbon dioxide supplied had only a slight effect on the calcium levels in solution. Finally, after 21 days, the carbon dioxide was switched off and the pH rose to 8.6 within two days, in a similar fashion to the experiment without the gravel. Turbulence appeared to be just as great in the presence as in the absence of gravel, however the calcium concentration showed only a very gradual decline (Figure 5). The reactive phosphate concentration which had fallen from an initial value of 31.0 μg $l^{-1}$ to 9.5 μg $l^{-1}$, rose to 12.0 μg $l^{-1}$ when the carbon dioxide was switched off, and stayed at that level (Figure 5). Potassium dihydrogen orthophosphate was added to the channel after another seven days to raise the concentration by 12 μg $l^{-1}$ to 24 μg $l^{-1}$. Thereafter the reactive phosphate levels declined to about 14 μg $l^{-1}$ in three days. A second phosphate spike added eight days after the first, behaved in a similar fashion but the final concentration remaining in solution was 16.0 μg $l^{-1}$ (Figure 5). No effect on calcium concentration or pH directly associated with these phosphate additions was observed, but subsequent addition of carbon dioxide reversed the slight decline in calcium concentration mentioned above. This carbon dioxide addition reduced the pH from 8.6 to between 8.1 and 8.3 and caused the reactive phosphate concentration to decrease from 16.0 μg $l^{-1}$ to 11.0 μg $l^{-1}$ (Figure 5).

After repeated exchanges of water the channel was again circulated this time without the carbon dioxide supply. A sharp rise in pH to 8.6 indicated that an equivalent carbon dioxide loss to that in the absence of gravel occurred. Calcium concentrations showed a gradual, though small, decline. Reactive phosphate decreased to a plateau of approximately 16.0 μg $l^{-1}$ and silicate concentration fell from 4.0 mg $l^{-1}$ to a plateau of 2.7 mg $l^{-1}$ over three days (Figure 3).

Suspended bacteria which had increased exponentially in the absence of gravel, showed no increase after an initial high level had been reached (Figure 4) in the presence of the substratum. Subsequently the population declined slowly but significantly (t value for the significance of slope was 3.95) and had not levelled out by the end of the experiment.

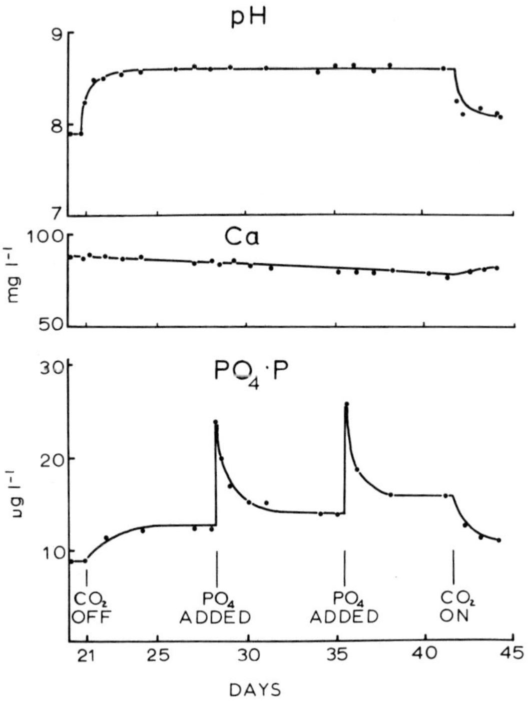


Figure 5. Effects of supply in $CO_2$ and additional inorganic phosphate on water chemistry.

In order to check the results of chemical changes taking place in the channel, the smaller (2 m) system was set up in the dark, containing only borehole water. A rapid decrease of dissolved calcium and reactive phosphate took place (Figure 6). As with the large channel this could be arrested or reversed by supplying gaseous carbon dioxide. In a series of replicate experiments, the addition of washed gravel similar to that used in the large system resulted in substantial stabilisation of calcium, whilst phosphate initially fell then gradually increased to a plateau above 10 µg $l^{-1}$ after seven to eight days.

Gravel scrubbed and thoroughly washed in tap water behaved in a similar manner to commercially washed materials. However, experiments with gravel extracted with Olsen's reagent, which is used to remove available phosphate from calcareous soils, reduced the reactive phosphate concentration initially to 5 or 6 µg $l^{-1}$ followed by a gradual increase to approximately 10 µg $l^{-1}$, and also resulted in calcium losses from solution to a level intermediate between those achieved with and without commercially washed gravel (Figure 6).

The effect of clays purchased commercially added to the small channel with clean glass marbles as support was similar to that of gravel washed in Olsen's reagent.

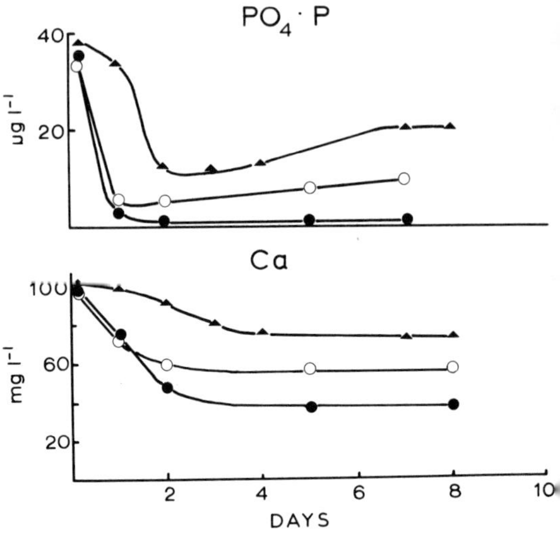


Figure 6. 2 m channel, effects of gravel substratum on circulating water. Changes in calcium and reactive phosphate concentrations without gravel ●—● , with gravel ▲—▲ and with gravel washed in Olsen's reagent ○—○

## Discussion

Although a model of this type is unlikely to resemble closely any natural stream, the extension of time scales by prolonged retention of water and the increased accuracy of measurements made on a single body of water should make it possible to detect and estimate the magnitudes of many chemical, physical and biological processes.

In chalk streams with sources of water comparable to that of the channel, calcium concentration and alkalinity vary to some extent both seasonally and with distance from the source. The loss of carbon dioxide from source water when exposed to violent mixing in the screw pump, leads to greater changes in pH, calcium and alkalinity in the channel. In nature equilibration with the atmosphere is likely to be slower and biological carbon dioxide outputs probably oppose the establishment of conditions suitable for the precipitation of calcium carbonate. Nevertheless, considerable loss of calcium from the water has been observed in nature and there is seasonal variation at any one site related to retention time (Casey, pers. comm.).

Decreases in soluble phosphate and reactive silicate concentrations have sometimes been observed in chalk streams (Casey, 1969; Casey & Newton, 1973). It is difficult to ascribe such changes to any particular cause although uptake by stream biota is a major factor. For example, large fluctuations in benthic diatom populations sometimes, but not always, affect silicate concentrations (Marker, in press). Variations in reactive silicate and phosphate concentrations in source water have been observed and these are apparently unrelated to discharge (Casey, 1976).

The present results suggest that for silicate an equilibrium between clay complexes and soluble forms could be operating. The stabilisation of soluble reactive phosphate concentration which took place after the addition of gravel and the response of the concentration to phosphate spikes and pH changes associated with carbon dioxide supply, suggest that the phosphate is competing (possibly with $OH^-$ ions) for anion exchange sites on the clays. In nature this effect may be masked by biological mineralisation of organic phosphate, uptake of phosphate as a nutrient and also by the large and unpredictable inputs from agriculture and sewage treatment.

The most obvious effect of the gravel on the suspended bacterial population was the rapid establishment of a comparatively large population (Figure 4) presumably washed off the gravel. This population is likely to differ qualitatively from those populations which partly accounts for the subsequent steady decline in numbers. If the decline was due to predatory activity, then one might have expected to observe oscillations similar to those beginning to appear in the earlier experiments, but there was no evidence of these.

## Acknowledgements

This work is part of the Department of the Environment (DoE) contract number DGR/480/33. We wish to thank DoE for permission to publish this paper. We are also grateful to the Institute of Geological Sciences for clay analyses and to our colleagues at the Freshwater Biological Association's River Laboratory for advice and encouragement.

## References

Anon, 1969. The bacterial examination of water supplies. Report No. 71. HMSO.

Baker, J. H. & Farr, I. S., In press. Origins, characterisation and dynamics of suspended bacteria in two chalk streams. Arch. Hydrobiol.

Casey, H., 1969. The chemical composition of some southern English chalk streams and its relation to discharge. Yb. Ass. River Auth. 1969: 100-113.

Casey, H., 1976. The changing chemical composition of chalk streams and rivers. Freshwater Biological Association Annual Report: 28-35.

Casey, H. & Newton, P. V. R., 1973. The chemical composition and flow of the River Frome and its main tributaries. Freshwat. Biol. 3: 317-333.

Jones, J. G., 1970. Studies on freshwater bacteria: effect of medium composition and method on estimates of bacterial population. J. Appl. Bacteriol. 33 (4): 679-686.

Kevern, N. R. & Ball, R. C., 1965. Primary productivity and energy relationships in artificial streams. Limnol. Oceanogr. 10: 74-87.

Krebs, C. J., 1972. Ecology: The experimental analysis of distribution and abundance. Harper and Rowe. 694 pp.

Lauff, G. H. & Cummins, K. W., 1964. A model stream for studies in lotic ecology. Ecology 45 (1): 188-191.

McIntyre, G. D., Garrison, R. L., Phinney, H. K. & Warren, C. E., 1964. Primary production in laboratory streams. Limnol. Oceanogr. 9: 92-102.

Murphy, J. & Riley, J. P., 1962. A modified single solution method for the determination of phosphate in natural waters. Anal. Chim. Acta 27: 31-36.

Odum, H. T. & Hoskin, C. M., 1957. Metabolism of a laboratory stream microcosm. Publ. Inst. Mar. Sci., Texas 4: 115-133.

Von Lutz, I. & Keller, A., 1975. Das Rheodrom, eine neue Fliesswasserversuchanlage. Schweiz. Z. Hydrol. 31 (1): 187-192.

Westlake, D. F., Casey, H., Dawson, F. H., Ladle, M., Mann, R. H. K. & Marker, A. F. H., 1972. The chalk stream ecosystem. Proc. IBP/UNESCO Symposium on productivity problems of freshwaters. Kazimierz Dolny 1970. (Eds. Z. Kajak & A. Hillbricht-Ilkowska). PWN (Polish Scientific Publishers), Warszawa-Krakow. p. 615-635.

Wuhrmann, K., 1964. River bacteriology and the role of bacteria in self-purification of rivers. In: H. Heukelekion & N. C. Dondero (Eds.): Principles and applications in aquatic microbiology. Wiley, New York. p. 167-188.

Wuhrmann, K., Ruchti, J. & Eichenberger, E., 1966. Quantitative experiments on self-purification with pure organic compounds. 3rd Int. Conf. Wat. Pollut. Res. Munich 1966: 1-17.

Key words: chalk stream
water chemistry
inorganic sediment
recirculating system
gravel substratum

# DECOMPOSITION OF MACROPHYTES AND THE METABOLISM OF ORGANIC MATTER IN SEDIMENTS

Gordon L. Godshalk and Robert G. Wetzel

W. K. Kellogg Biological Station, Michigan State University,
Hickory Corners, Michigan, U.S.A.

## Summary

The littoral zone and its autochthonous productivity have a significant impact on the metabolism of lake ecosystems. The inputs of littoral organic matter through decomposition are often greater than those from planktonic and allochthonous sources. Decomposition of macrophytes is slower than that of planktonic material because of resistant structural tissue. Decomposition of particulate (POM) and dissolved organic matter (DOM) of six species of aquatic angiosperms were monitored through time under various conditions of controlled oxygen and temperature. Four fates of decomposing macrophytic tissue were evaluated: conversion of POM to $CO_2$, to labile DOM, and to refractory DOM, and sedimentation of undegraded particles. Temperature was instrumental in controlling conversion of POM to DOM, while oxygen concentration controlled rates of conversion of DOM to $CO_2$.

The experimental results are significant in relation to conditions present at various times of the year in all zones of typical temperate dimictic lakes. Under all conditions, macrophytic tissue is converted to DOM. If conditions are anoxic, relatively little DOM is degraded, but in aerated water the conversion is nearly complete. As expected, processing of both POM and DOM occurs faster at high than at low temperatures. Semiannual lake circulation exposes organic matter to oxygenated water and resuspends POM. The resultant impact to the lake ecosystem is that even though a large amount of organic matter becomes available to the detrital food chain in autumn in the form of dead macrophytes, environmental conditions constrain decomposition such that inputs of reduced carbon to the system are spread out over the entire year. This continued input of organic matter from regulated decomposition of macrophytes enables mesotrophic lakes to process nearly all autochthonous production with little permanent sedimentation of undegraded POM. As lakes become more productive, controlled decomposition is unable to process the more abundant resistant macrophytic tissue; permanent sedimentation rates then increase.

## Introduction

A majority of the lakes of the world are small, i.e. less than 50 ha in surface area. In these lakes, where a relatively greater amount of littoral zone occurs per total lake volume than in the more rare truly large lakes of the world, the littoral vegetation plays an important role in the total metabolism of the lake ecosystem (Wetzel, 1975). Photosynthetic production by aquatic macrophytes colonizing the littoral zone often dominates total autochthonous primary productivity.

Coupled with high littoral productivity is the impact on the lake after the growing season when most aquatic angiosperms senesce and collapse to the sediment-water interface. The organic matter resulting from decomposition of higher plant material may have a significant stabilizing influence on the lake, particularly in regard to carbon metabolism, between growing seasons (Wetzel, et al., 1972; Wetzel, 1975; Rich and Wetzel, 1977).

The typically oversimplified account of decomposition in lakes involves only the fate of planktonic plants and animals (e.g. Kuznetsov, 1975). Data are available that show that much of this planktonic material is decomposed while still in the water column, and usually still within the warmer oxygenated waters of the epilimnion (Ohle, 1962; Wetzel, 1975). Phytoplankton, which contains very little structural organic matter, is relatively easily decomposed. The ultimate fate of this material is primarily rapid conversion to $CO_2$ with a small amount of the more refractory compounds remaining in particulate and dissolved form. A small fraction of the particulate component remains insoluble sufficiently long to reach the pelagial sediments. Macrophytic plant material, on the other hand, is characterized by having substantial structural material (cellulose, lignin), occurring in very large particle sizes relative to plankton, and having much lower surface area to volume ratios of the particles. All of these characteristics make macrophytic tissue relatively resistant to decomposition.

A series of controlled experiments was conducted to evaluate the rates of decomposition of littoral macrophytes and the ultimate fate of these plants and their dissolved and particulate organic fractions in the lake ecosystem. Senescent plants of five fresh-water and one marine species of aquatic angiosperms were collected, washed free of sediments, and lyophilized. Replicated samples of dried plant material were put in one-liter flasks of synthetic lake water (Wetzel Media 5) or sea water, inoculated with natural littoral sediment, and incubated up to 180 days under anaerobic and aerated conditions at 10°C and at 25°C. The various conditions of oxygen concentration were selected to simulate three situations under which naturally occurring decomposition proceeds: (1) completely aerobic conditions, as in wave-swept, turbulent littoral regions; (2) completely anaerobic conditions, as in many quiescent littoral regions where plant material compacts at the sediment-water interface under continuously strongly reducing conditions; and (3) aerobic to anaerobic conditions (i.e. synthetic lake water was initially oxidizing but allowed to become anoxic as a result of decomposition).

Flasks were sacrificed after 2, 4, 10, 24, 50, 90, and 180 days of decomposition and the contents of the flasks filtered (0.22 μm Millipore GS membrane filters). Dissolved organic compounds (< 0.2 μm) were fractionated by Amicon membrane ultrafiltration into several molecular weight categories (<30000 MW; <10000 MW; <1000 MW, and <500 MW). Each fraction was examined for total dissolved organic carbon, ultraviolet absorbance, and fluorescence; unfractionated filtrate was analyzed for dissolved carbon and nitrogen content, and by infrared spectrophotometry. The particulate material recovered was lyophilized, weighed to determine organic weight loss during decomposition, and analyzed for content of ash, total non-structural carbohydrates, hemicellulose, cellulose, lignin, and particulate carbon and nitrogen. In addition, microbial activity of identical fresh material at each time interval was assessed by determining both dehydrogenase activity and ATP content. Present discussion of the dynamics of decomposition is based on the disappearance of particulate plant material and the quantity and quality of dissolved organic matter produced. Typical data of weight loss and concentrations of dissolved organic carbon during decomposition are given in Figures 1 and 2 for four experimental conditions. A detailed treatment of the decomposition of dissolved and particulate fractions of the six species will appear elsewhere (Godshalk and Wetzel, in preparation). Discussion here summarizes the observed experimental data and attempts to place them in an ecological perspective in relation to in situ conditions.

## Results and discussion

It was observed in these experiments that there were four alternative fates for decomposing macrophytic tissue: (a) Complete and efficient decomposition of organic materials to $CO_2$ and water; little or no particulate material would remain and there would not be a large concentration of dissolved organic matter in the media. Less complete decay would involve the loss of particulate plant material and production and accumulation of dissolved organic matter with relatively small amounts of the dissolved organic compounds being oxidized. This pool of dissolved organic matter can be further subdivided into (b) relatively refractory, large molecular weight compounds and into (c) relatively labile, small molecular weight compounds. (d) The final possible fate of plant material in lakes is to decompose little; some or nearly all of the plant tissue, depending upon the species and its morphology, may not be converted significantly to any other form and is permanently interred in the sediments.

The composite results of our experiments are summarized in Figure 3. The influence of temperature and oxygen conditions on rates and completeness of degradation are elaborated into the four general types; complete decomposition to $CO_2$, conversion to refractory and/or labile dissolved organic matter, or sedimentation of macrophytic particulate matter in relatively unaltered form. The most conspicuous features of the results are direct proportionalities of the (a) decomposition of particulate organic matter to dissolved organic matter with temperature, and (b) decomposition of dissolved organic matter to $CO_2$ with oxygen concentration. While both of these findings were anticipated from general knowledge of controls of decomposition of organic matter, they have not been evaluated previously for aquatic angiosperms. The ecological implications are of greater interest.

Decomposition of macrophytic tissue is proceeding under one or more of these conditions at all times in the different zones of typical dimictic lakes of the temperate regions. For example, warm aerobic conditions occur in the epilimnion and at the sediment-water interface of the littoral zone during summer. The hypolimnion is always cold and usually anoxic near the sediment-water interface except for the two periods of seasonal mixing of the entire water column at which time the entire water mass is aerated, but is still cold.

How these conditions occur seasonally in a typical temperate, dimictic, mesotrophic lake

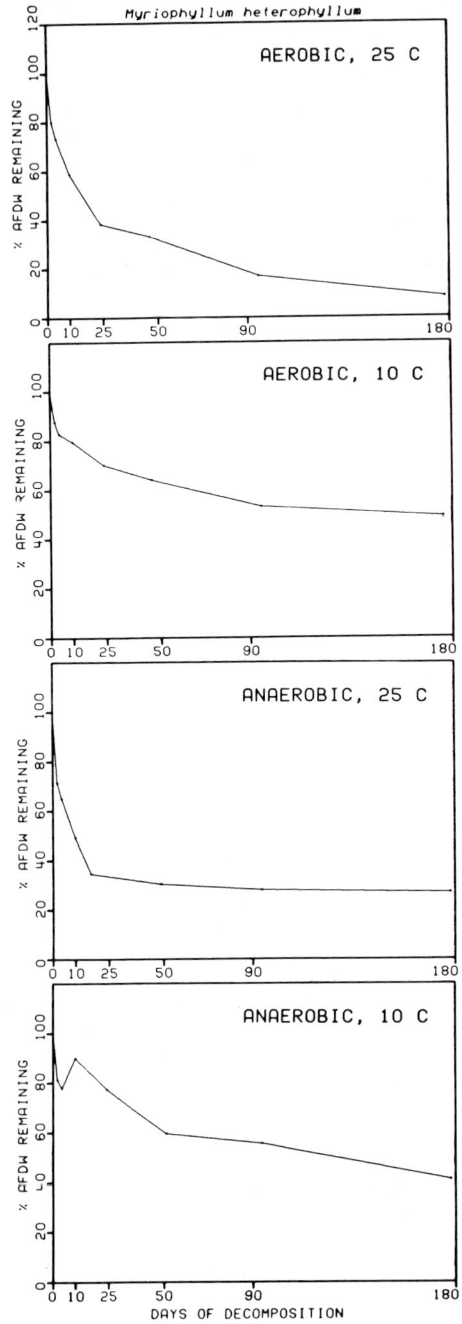


Figure 1. Ash-free dry weight remaining, as per cent of initial, during laboratory decomposition of Myriophyllum heterophyllum under four conditions.

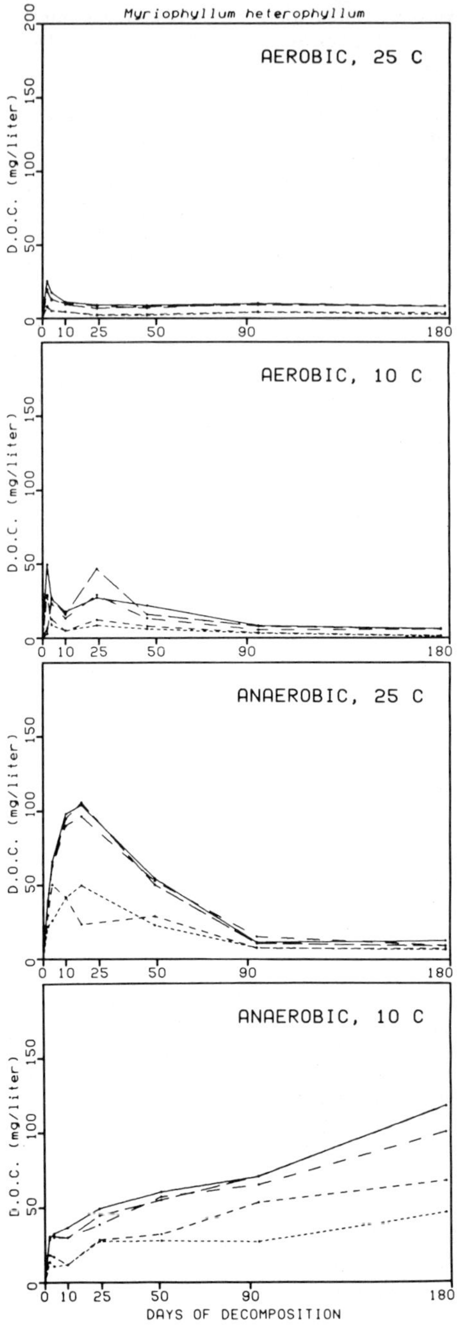


Figure 2. Concentration (mg/l) of dissolved organic carbon in various molecular weight fractions of media during laboratory decomposition of Myriophyllum heterophyllum under four conditions.

——— < 0.2 μm
— — — < 30000 MW
— — — < 10000 MW
- - - - < 1000 MW
· - · - · < 500 MW

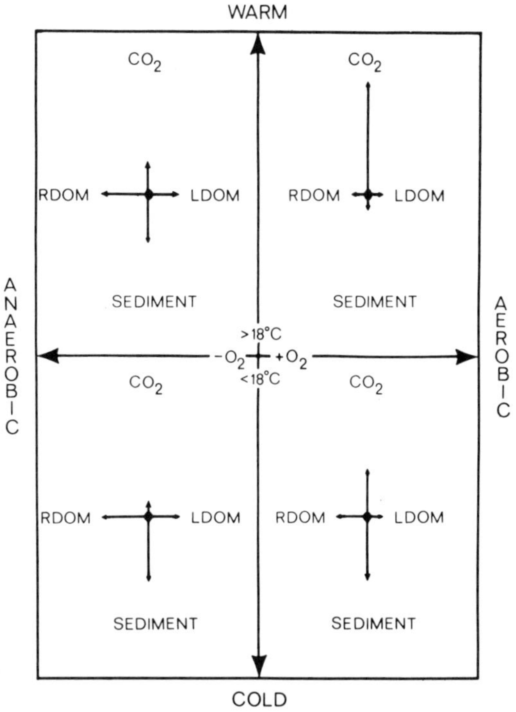


Figure 3. Relative importance, with respect to rates and total accumulation, of the four possible fates of macrophytic tissue observed during laboratory decomposition under various conditions of temperature and oxygen concentration. See text for further explanation.

is depicted in Figure 4a-4f, along with some important aspects of carbon metabolism in the ecosystem. In late summer and early fall (Fig. 4a), the growing season for most annual macrophytes ends, and they begin to senesce. The actual beginning of decomposition is difficult to define but certainly some plant parts remain viable while other parts begin to lose their integrity, leach dissolved organic matter, and begin to be colonized by microflora. While the release extracellularly of dissolved organic matter from actively growing submergents increases in the fall (Hough and Wetzel, 1975), the production of dissolved organic matter by autolysis of the plant tissue in the autumn is a major input of photosynthetically reduced carbon to the detrital food chain (cf. definitions of Wetzel, 1975; Rich and Wetzel, 1977). As the plants collapse to the sediment-water interface, they are rapidly colonized by bacteria and fungi (Oláh, 1972; Aliverdieva-Gamidova, 1969). If this happens while the water of the littoral zone has not yet cooled greatly and contains high dissolved oxygen concentrations, decomposition will

Figure 4. General diagrammatic representation of seasonal conditions of carbon metabolism of particulate and dissolved organic carbon of macrophytic origin in a typical temperate, dimictic lake of intermediate productivity. POM = particulate organic matter; DOM = dissolved organic matter; LDOM = labile DOM; RDOM = resistant DOM.

(a) Late summer, early autumn: high input of freshly senescent macrophytes with much autolytic release of DOM; oxygen abundant in epilimnion with rapid, efficient conversion of DOM to $CO_2$, some particle formation from DOM as a result of complexation of humic materials with other humics and/or calcium carbonate as well as incorporation of DOM as microbial biomass; oxygen deficient in hypolimnion with accumulation of DOM and sedimentation of POM from littoral zone.
(b) Autumnal circulation: usually longer, more complete aeration of total water volume than during spring circulation; relatively efficient conversion of DOM to $CO_2$ under cold, aerobic conditions; resuspension and net movement of fine POM from littoral to pelagic zones and from pelagic sediments while coarse POM remains in littoral sediments.
(c) Winter stratification under ice cover: littoral decomposition slow with accumulation of DOM and little conversion to $CO_2$; slow conversion of pelagial DOM to $CO_2$ under aerobic conditions.
(d) Spring circulation: cold, aerobic conditions in most of volume; resuspension of fine POM of littoral and pelagial sediments with net movement to pelagial zone; slow decomposition of DOM from winter.
(e) Late spring, early summer stratification: rapid, efficient utilization of DOM of past macrophytic sources, decomposing phytoplankton, and DOM excreted from actively growing plants in aerobic conditions; reduced in colder hypolimnion.
(f) Summer stratification: highest temperatures; rapid, efficient conversion of POM to DOM to $CO_2$ in oxic epilimnion; slow decomposition and accumulation in anaerobic, cold hypolimnetic waters.

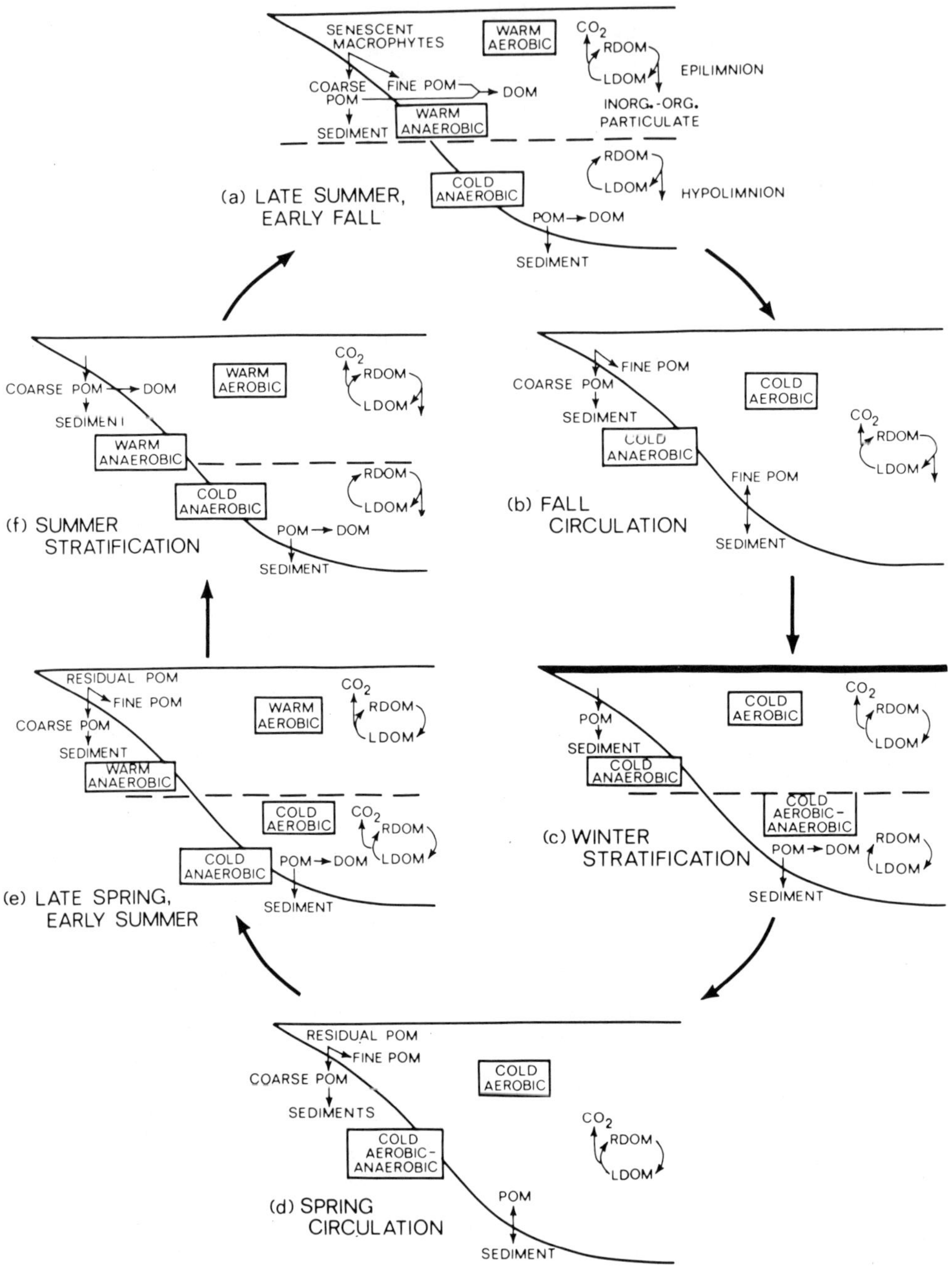
SENESCENT MACROPHYTES
WARM AEROBIC
$CO_2$
RDOM
LDOM
EPILIMNION
COARSE POM
FINE POM
DOM
INORG.-ORG. PARTICULATE
SEDIMENT
WARM ANAEROBIC
RDOM
LDOM
HYPOLIMNION
COLD ANAEROBIC
(a) LATE SUMMER, EARLY FALL
POM → DOM
SEDIMENT
COARSE POM → DOM
WARM AEROBIC
$CO_2$
RDOM
LDOM
SEDIMENT
WARM ANAEROBIC
COLD ANAEROBIC
RDOM
LDOM
(f) SUMMER STRATIFICATION
POM → DOM
SEDIMENT
FINE POM
COARSE POM
COLD AEROBIC
SEDIMENT
$CO_2$
RDOM
LDOM
COLD ANAEROBIC
FINE POM
(b) FALL CIRCULATION
SEDIMENT
RESIDUAL POM
FINE POM
COARSE POM
WARM AEROBIC
$CO_2$
RDOM
LDOM
SEDIMENT
WARM ANAEROBIC
COLD AEROBIC
$CO_2$
RDOM
LDOM
COLD ANAEROBIC
POM → DOM
(e) LATE SPRING, EARLY SUMMER
SEDIMENT
POM
COLD AEROBIC
$CO_2$
RDOM
LDOM
SEDIMENT
COLD ANAEROBIC
COLD AEROBIC-ANAEROBIC
(c) WINTER STRATIFICATION
POM → DOM
RDOM
LDOM
SEDIMENT
RESIDUAL POM
FINE POM
COARSE POM
COLD AEROBIC
SEDIMENTS
$CO_2$
RDOM
LDOM
COLD AEROBIC-ANAEROBIC
POM
(d) SPRING CIRCULATION
SEDIMENT

Figure 4.

result in the rapid conversion of particulate organic matter to dissolved organic matter and of dissolved organic matter to $CO_2$. However, in most small temperate lakes the temperature of the water will already be significantly lowered at this time and, from the results of our laboratory studies, slow degradation of macrophytic particulate organic matter can be predicted. Meanwhile, in the colder hypolimnion of the lake, which commonly has, at least to some degree, become anoxic from decomposition of organic materials associated with the bottom sediments, an accumulation of dissolved organic matter has been taking place. Data from deep-water samples of Lawrence Lake, Michigan, for example, at this time of the year, show that there is a drastic increase in the absorbance of ultraviolet light and in fluorescence activity, implying an accumulation of refractory dissolved organic compounds of plant origin which are known to have these properties (Otsuki and Wetzel, 1973, 1974; Wetzel and Otsuki, 1974). Lack of oxygen in the hypolimnetic waters prevents rapid conversion of this dissolved organic matter to $CO_2$.

At fall turnover (Fig. 4b), mixing of the total lake volume accomplishes two things. First, the entire water body is aerated, which allows the dissolved organic matter that had been accumulating as a result of anaerobic decomposition to be further oxidized to $CO_2$ and water. Secondly, recently sedimented surficial particulate organic matter is resuspended. Particles that had previously been confined to the anaerobic conditions of the sediment-water interface are brought into contact, at least temporarily, with oxygenated water. While decomposition is slow because of low water temperatures, further conversion of particulate organic matter to $CO_2$ is made possible.

During the cold winter stratification (Fig. 4c), decomposition continues both in the water and at the sediments of all depths. Depending on the types and amounts of organic matter present at the sediment-water interface, and the degree of oxygenation accomplished by fall circulation, the sediment surface often rapidly becomes anoxic at increasing water depths. Most particulate organic matter lies at the sediment surface without being appreciably degraded; what decomposition does occur results in the production of dissolved organic matter, with conversion to $CO_2$ being dependent on available dissolved oxygen.

Spring turnover (Fig. 4d) again causes aeration of the water mass and resuspension of fine particulate organic matter from the sediments. At this time of the year, however, the water is constantly warming and consequently, as long as dissolved oxygen is present, microbial conversion of the dissolved organic matter which accumulated over winter and of the particulate organic matter brought up from the sediments will be accelerated and be more complete than under winter conditions.

As the epilimnetic waters and littoral sediments begin to warm, more of the heavier, coarser, non-resuspended material may be degraded (Fig. 4e). As the rates of decomposition increase coincident with higher temperatures, oxygen demand increases and, by midsummer (Fig. 4f), littoral sediments are commonly anaerobic, and complete oxidation of organic matter is greatly reduced. Dissolved organic matter is continually generated from the macrophytic detritus at these warmer temperatures and as it moves to oxygenated waters of the pelagial zone a greater conversion to $CO_2$ can result. The hypolimnion, however, is in large part sealed off physically from oxygen renewal and becomes increasingly anaerobic, allowing an accumulation of dissolved organic matter that is being degraded at rates less than those of inputs.

The implications of this variety of conditions under which decomposition occurs in a lake are great. Essentially, the input of organic matter into the detrital food chain occurs constantly over an entire year but decomposition is pulsed. Influxes of metabolizable carbon into the system can be largely decomposed in the fall when many macrophytes senesce in the littoral zone. Lower water temperatures in autumn prevent rapid conversion of particulate organic materials to dissolved forms. Constantly low temperatures at the sediment-water interface where particulate matter comes to rest cause this material to be metabolized gradually. Periodic oxygenation of the entire lake allows accumulated labile and especially refractory dissolved organic matter produced at the sediment-water interface to be converted to $CO_2$. Furthermore, the effect of seasonal mixing is to bring particulate material out of conditions of incomplete decomposition for further degradation while suspended periodically in oxic waters. Coarse, heavy particles are not resuspended and therefore are not removed from the littoral zone, where they are more effectively decomposed under warmer and more frequently oxygenated conditions. Thus only the most refractory particulate organic materials will be permanently buried in the sediments of the basin.

As lakes become more productive, there is frequently a tendency for massive development of emergent macrophytes (cf. Wetzel, 1975). Tissue of these plants is much more resistant to decomposition than that of submersed and floating-leaved hydrophytes. The gradual systematic decomposition of this particulate detrital carbon under many lake conditions is overwhelmed by the excessive quantities of resistant carbon compounds and generation of anaerobic conditions in the decomposing littoral sediments as well as in the hypolimnion. Accumulation rates of sediment then accelerate rapidly.

Key words: decomposition
macrophytes
organic matter
temperature
oxygen

Acknowledgments

We greatly appreciate the technical assistance of J. S. Sonnad, S. Morrison, and A. J. Johnson. The analytical work was supported by subventions from the National Science Foundation (Grants GB-40311, BMS-75-20322, and OEC 74-24356 A01) and the U. S. Energy Research and Development Administration (EY-76-S-02-1599, COO-1599-112). Contribution No. 312, W. K. Kellogg Biological Station of Michigan State University.

Literature Cited

Aliverdieva-Gamidova, L. A. 1969. Microbiological processes in Lake Mekhteb. Mikrobiologiya 38:1096-1100.

Hough, R. A. and R. G. Wetzel. 1975. The release of dissolved organic carbon from submersed aquatic macrophytes: diel, seasonal, and community relationships. Verh. Int. Ver. Limnol. 19:939-948.

Kuznetsov, S. I. 1975. The microflora of lakes and its geochemical activity. Univ. Texas Press, Austin. 503 pp.

Oláh, J. 1972. Leaching, colonization and stabilization during detritus formation. Mem. Ist. Ital. Idrobiol. 29(Suppl.): 105-127.

Ohle, W. 1962. Der Stoffhaushalt der Seen als Grundlage einer allgemeinen Stoffwechseldynamik der Gewässer. Kieler Meeresforschungen 18:107-120.

Otsuki, A. and R. G. Wetzel. 1973. Interaction of yellow organic acids with calcium carbonate. Limnol. Oceanogr. 18:490-493.

Otsuki, A. and R. G. Wetzel. 1974. Release of dissolved organic matter by autolysis of a submersed macrophyte, *Scirpus subterminalis*. Limnol. Oceanogr. 19:842-845.

Rich, P. H. and R. G. Wetzel. 1977. Detritus in the lake ecosystem. Amer. Naturalist (In press).

Wetzel, R. G. 1975. Limnology. W. B. Saunders Co., Philadelphia. 743 pp.

Wetzel, R. G. and A. Otsuki. 1974. Allochthonous organic carbon of a marl lake. Arch. Hydrobiol. 73:31-56.

Wetzel, R. G., P. H. Rich, M. C. Miller, and H. L. Allen. 1972. Metabolism of dissolved and particulate detrital carbon in a temperate hard-water lake. Mem. Ist. Ital. Idrobiol. 29(Suppl.):185-243.

# THE ROLE OF SEDIMENTS WITH REGARD TO LAKE OXYGEN CONSUMPTION

N. Edberg

Institute of Limnology, Uppsala, Sweden

## Abstract

Investigations have been performed during winter periods in order to estimate the amount of oxygen consumed by sediments and hypolimnetic water. The four lakes investigated are deep enough not to be influenced by photosynthetic oxygen production in bottom-near water. The gas exchange with the atmosphere is prevented by ice cover.

The intention was to compare the oxygen depletion in lakes with rates of oxygen consumption found in laboratory investigations.

Water samples were taken at a depth of 10 m and sediment cores from the deep parts in the lakes at the start of the winter period. The oxygen consumption of lake water was measured in BOD bottles (60 ml) during 20-40 days. The rate constant, $k_1$-value ($day^{-1}$), for a first-order oxygen-depletion curve was calculated. The rate of oxygen uptake by sediments, $k_2$-value ($g\ O_2 \cdot m^{-2} d^{-1}$), was measured for cores in the laboratory. It has been suggested from these experiments that there is a linear relationship between oxygen concentration and rate of oxygen uptake by sediments. The distribution in water column of sediment oxygen uptake was observed in the lakes by measuring the concentration decrease at different depths. As long as there was logarithmic declining oxygen concentration decrease from bottom and upwards the sediment was proposed to influence the concentration. In the calculations the $k_4$-value (m) is the limit of influence from sediment and $k_3$-value ($m^{-1}$) is the constant in the logarithmic oxygen decrease. A comparison of oxygen concentration above sediment surface, calculated from laboratory rate constants, with the observations in the lakes showed that calculated and in-situ observed oxygen concentration agree very well in three of the four lakes as shown in the table.

## Reference

Edberg, N., 1976. Oxygen consumption of sediment and water in certain selected lakes. Vatten 32:2-12.

Key words: oxygen consumption
sediment
water
hypolimnion

Table 1. Constants used in calculations of hypolimnetic oxygen concentration and oxygen measured concentrations in the water after a period of time.

| Lake | Year | Time interval (days) | $k_1$ ($day^{-1}$) | $k_2$ ($g\ O_2 \cdot m^{-2} day^{-1}$) | $k_3$ ($m^{-1}$) | $k_4$ (m) | Oxygen concentration at sediment surface ($mg\ O_2 \cdot l^{-1}$) Measured | Calculated |
|---|---|---|---|---|---|---|---|---|
| Uttran | 72-73 | 60 | 0.00177 | 0.50 | 0.029 | 4 | 10 -2.15 | 10 -2.28 |
| Uttran | 72-73 | 98 | " | " | " | " | 10 -0.3 | 10 -0.7 |
| Uttran | 73-74 | 74 | " | " | 0.0108 | " | 9 -0.4 | 9 -0.8 |
| Uttran | 73-74 | 117 | " | " | " | " | 11- -3 x | 11- -2 |
| Sillen | 73-74 | 71 | 0.000933 | 0.34 | 0.166 | 5 | 8.3-0.6 | 8.3-0.35 |
| Molkomsjön | 72-73 | 77 | 0.000984 | 0.17 | 0.040 | 4 | 9 -4 | 9 -3.82 |
| Molkomsjön | 73-74 | 79 | " | " | " | " | 9 -2.8 | 9 -3.71 |
| Molkomsjön | 73-74 | 102 | " | " | " | " | 9 -1.4 | 9 -2.5 |
| Kyrkviken | 72-73 | 75 | 0.000971 | 0.47 | 0.064 | 4 | 9.8-5.4 | 9.8-1.47 |
| Kyrkviken | 73-74 | 101 | " | " | " | " | 9.6-4.3 | 9.6-0.16 |
| Uttran | 75-76 | 105 | 0.00297 | 0.50 | 0.028 | 3 | 10- -0.6 x | 10- -2.9 |
| Molkomsjön | 75-76 | 84 | 0.00264 | 0.17 | 0.032 | " | 14.8-5.6 | 14.8-3.9 |

x Analysed as $H_2S$

EFFECT OF SEDIMENTS ON THE QUALITY OF THE OVERLYING WATER

J. Fillos

The City College, Department of Civil Engineering, New York, N. Y., U.S.A.

## Summary

Lake sediments and simulated sludge bottoms were used in laboratory experiments to measure the oxygen uptake rate and the phosphorus release rate to the overlying water under different environmental conditions. The electrode-potential profile across the water-sediment interface was also measured indicating the presence of an aerobic layer at the sediment surface only a few millimeters deep.

## Introduction

The accumulation of sludge in the bottoms of many lakes, estuaries, and river stretches is a result of continuous discharges of settleable and colloidal solids by wastewater discharges and agricultural or urban drainage. The stabilization of these sludge bottoms, usually referred to as benthal decomposition, involves both aerobic and anaerobic processes. These processes can affect the overlying water in two basic ways. First, dissolved oxygen (DO) is removed from the overlying water by bacterial respiration in the aerobic layer and by the immediate oxygen demand of the reduced substances emanating from the deeper anaerobic layers. Second benthal deposits can also release organic substances and various nutrients into the overlying water. These release processes can cause or aggravate the familiar lake problems of nuisance growths of planktonic and attached plants, offensive odors, turbidity, etc., all of which limit the usefulness of an inlet water in terms of commercial or recreational activities.

In view of the important effect sediments can have on water quality, a long term study was initiated to measure the oxygen uptake rate and the release rate of phosphorus from lake sediments to the overlying water under various environmental conditions. Realizing that the water-sediment interface is always saturated with water and, except at shallow depths, is subjected to little variations in temperature or light, the most important variable and the one used in this investigation is dissolved oxygen. Also in the investigation both simulated sludge and lake sediments were used. The simulated sludge used was a mixture of raw primary wastewater sludge solids and fine sand (Fillos & Molof, 1972). The lake sediments used were obtained from Lake Mohegan, a partially eutrophic lake and one of the largest in Westchester County, New York. Lake Mohegan has a surface area of approximately 421,000 m$^2$, an average depth of 4.3 m and a volume of about 1.8 x $10^6$ m$^3$. The sediments were collected from the Lake using a Peterson dredge.

## Experimental Apparatus

The effect of sediments on water quality was measured using both continuous flow and batch systems. The continuous flow system was described in detail in earlier papers (Fillos & Swanson, 1975). Essentially the system consisted of five air-tight reactors submerged in a controlled temperature bath. Each reactor was made from a 15-cm ID plexiglass cylinder, 31 cm long. Figure 1 (a) shows the cross section of a typical reactor used in this investigation. The cover plate

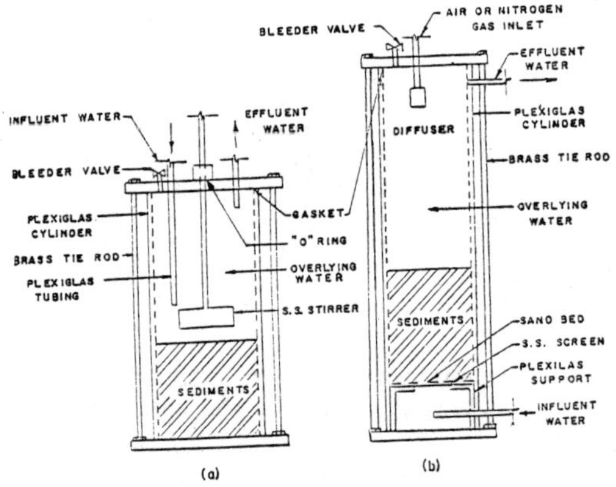


Fig. 1. Representative reactors

contained an influent and an effluent tube with two additional plug points through which samples of sediments could be obtained from within the reactors and through which the accumulated gases could be released. The water flow rate into each reactor was controlled individually by a peristaltic pump, while the oxygen concentration in the influent water was controlled by alternately aerating the influent water with either air or nitrogen gas. The effluent from all the reactors was intermittently monitored for different substances. Mechanical mixing of the overlying water was provided in each reactor at a rate sufficient for completely mixed conditions in the overlying water without disturbing the bottom sediments.

The batch system consisted of round bottom glass tubes with plastic screw caps having a total volume of 33 ml. This system was used to observe the maximum phosphate concentration in the overlying water when in contact with the lake sediments and the adsorptive capacity of undisturbed sediments.

In an effort to measure the effect of ground-water seepage through sediments on the release rate of phosphates, a special reactor was built as shown in Figure 1 (b). The reactor consisted of a cylindrical plexiglass tube 11.5 cm ID, 49 cm long, with a false bottom at about 4 cm above the actual bottom of the reactor. The false bottom consisted of a stainless steel wire mesh kept in place by a plexiglass platform. The false bottom supported the column of sediments, about 14 cm deep, through which ground water flowed. The reactor was covered and a gas diffuser was used to control the oxygen concentration in the overlying water by changing the purging gas from nitrogen to air. The ground-water influent velocity was maintained at $9 \times 10^{-5}$ cm/sec. This flow rate is common in silty soils and sediments (DeWiest, 1965).

## Experimental Procedure

Initially the sediments were stored in a number of glass jars and refrigerated at 4°C until they were used in the experiments. Before transferring the sediments to the reactors, they were thoroughly mixed, allowed to reach room temperature, and analyzed for total solids, volatile solids, phosphorus, and iron. The Lake Mohegan Sediments were about 6% dry solids of which 44.3% was volatile solids. Total phosphorus was 1.19 mg/g dry solids while total iron was 42.5 mg/g dry solids.

New York City tap water was the influent water to each reactor in all experimental runs. The tap water was first brought to 20°C, the temperature at which most of the experiments were conducted, then aerated and passed through an activated carbon column before it flowed to the reactors. The overlying water was mixed continuously so that it had properties identical with those of the effluent water. Thus the oxygen uptake rate and the release rate of nutrients was calculated by performing a mass balance for each reactor. All analyses were performed according to Standard Methods (1971).

## Experimental Results

### Effect of Dissolved Oxygen on the Overlying Water on Oxygen Uptake Rate.

The oxygen uptake rate of sediments, measured using the simulated sludge bottoms, is a result of the oxygen demand of the aerobic microorganisms in the mud surface and of the immediate chemical demand of anaerobic decomposition end products emanating from the deeper sludge layers. The effect of oxygen concentration in the overall oxygen uptake rate was measured at various DO levels ranging from 7.7 to 0.15 mg/l. The desired DO levels were maintained until steady-state conditions with respect to oxygen uptake rate were observed. The data for the 7.6 cm sludge depth reactors are summarized in Figure 2. The 12.7 cm reactors

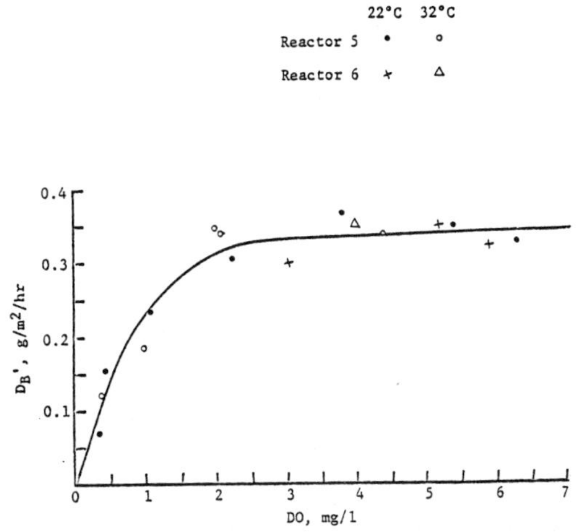


Fig. 2. Oxygen uptake rate of 5 inch sediments as a function of DO

showed similar results. Figure 2 indicates that the oxygen uptake rate of benthal deposits changes very little with increasing oxygen concentration in the range of 2 to 7.5 mg/l. At values below 2 mg/l, the oxygen uptake rate decreases dramatically with decreasing DO in the overlying water.

### Electrode Potential Profile Across The Water-Sediment Interface

To obtain a quantitative measurement of the thickness of the aerobic layer, the vertical electrode potential profile across the water-sediment interface was measured. With a multipoint electrode having 15 different exposed platinum electrode surfaces at predetermined locations, the vertical electrode potential profile was measured and is shown in Figure 3. The profile indicates that aerobic conditions may exist only at the mud surface, while immediately underneath anaerobic conditions exist. At 3 cm below the mud-water interface, the potential reached its minimum value of $E_c$ = -450 mV,

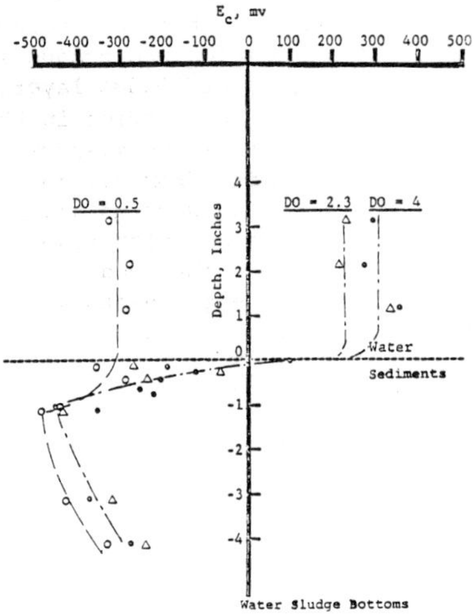


Fig. 3. Vertical distribution of electrode potential in benthal deposits.

with conditions becoming less reductive below this depth. (Fillos & Molof, 1972).

During the measurement of the electrode potentials, the oxygen concentration in the overlying water was varied. Varying the DO in the overlying water caused no appreciable changes in the potential profile within the sediments. In contrast, the change of potential in the overlying water was pronounced as it decreased from an $E_C$ = +230 mV at a DO of 2.3 mg/l to an $E_C$ = -300 mV at a DO of 0.5 mg/l. This large change cannot be explained by the decrease in DO alone. Other factors, such as the lack of poise in the overlying water along with the release of reduced substances from the benthal deposits into the overlying water, probably caused the ultimate change in the electrode potential.

## Dissolved Phosphate in Equilibrium with Sediments

Lake sediments contain significant quantities of phosphorus. This phosphorus exists in both organic and inorganic forms. The organic forms include nucleic acids, nucleotides, phospholipids, and a variety of polyphosphate esters(Syers et al,1973). The inorganic forms may include calcium phosphate, iron phosphate complexes, phosphates adsorbed on clays, and orthophosphate dissolved in the interstitial water (Frink, 1969; Lee 1970). The concentration of orthophosphate in the interstitial water of sediments is controlled by a variety of complex reactions that include the interaction of the ground-water recharge system, mineralogical solution and precipitation reactions, biological activity, and the degree of interaction between the sediments and the overlying water (Sanville, 1974).

The interaction of the phosphorus in the Lake Mohegan sediments with the overlying water was first considered in terms of the equilibrium concentration of dissolved inorganic phosphorus in the overlying water when in contact with undisturbed lake sediments under aerobic and anaerobic conditions.

Equal quantities of lake sediments were brought in contact with equal volumes of water in glass test tubes. The sediments throughout the contact period remained undisturbed simulating the sediment water interface in lakes under stratification. The phosphate and oxygen concentration in the overlying water of each test tube were initially different, varying from zero to 10mg/l for phosphorus and saturation to zero for dissolved oxygen. All anaerobic tubes were sealed with screw caps and melted wax throughout the contact time to prevent any exchange with the atmosphere. The temperature was maintained at 20°C for both the aerobic and anaerobic tubes.

The experimental data obtained from the aerobic and anaerobic tubes after a contact time of 41 days and 69 days are presented in Figure 4. The data show that under aerobic

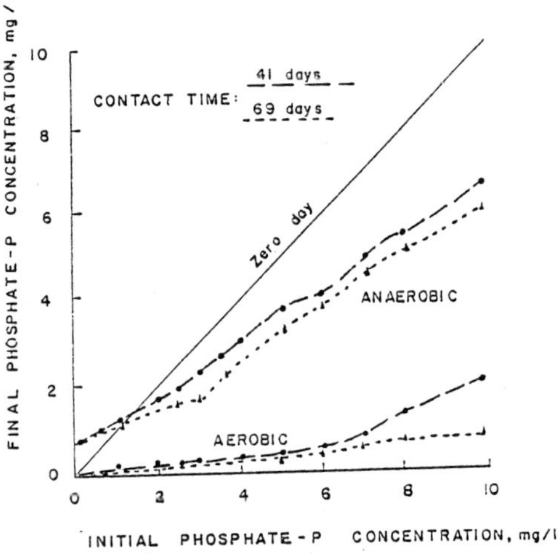


Fig. 4. Phosphate concentration in overlying water in contact with Lake Mohegan sediments.

conditions, the sorption of dissolved phosphate on sediments was high throughout the phosphate range of the experiment reaching a value of 900 mg/m$^2$ for an equilibrium phosphorus concentration of 1 mg/l. Figure 4, therefore indicates that under aerobic conditions, Lake Mohegan sediments will always act as a "sink" for phosphorus. When conditions were anaerobic, the sediments acted as a source until the phosphate concentration in the overlying water reached a value between 1 mg/l and 1.4 mg/l. Above this phosphate concentration, the sediments acted as a "sink" causing a decrease in the phosphate concentration of the overlying water. The data therefore suggest that in lakes that stratify, the sediments will act as a phosphate source only when the hypolimnion becomes anaerobic and in the absence of external phosphorus sources, the maximum expected phosphate concentration in the hypolimnion will be between 1 mg/l and 1.4 mg/l (Fillos & Biswas, 1976).

The adsorption capacity of sediments under aerobic conditions has been attributed to a thin aerobic layer several millimeters thick on top of the sediments. This layer is present as long as the overlying conditions are aerobic. Mortimer, (1971), suggested that the adsorption capacity is largely due to the presence of colloidal ferric complexes. Clays and other substances also contribute to the sorption capacity of sediments. When conditions become anaerobic the ferric complexes are reduced, the sorption capacity is abolished resulting in a freer exchange of dissolved substances between the sediments and the overlying water. Figure 4 tends to support the aforementioned, though under anaerobic conditions at high phosphate concentrations it also shows a large sorption capacity. Patrick and Khalid, (1974), working with different soils also reported large sorption capacity under anaerobic conditions. An explanation for this phenomenon may be that though ferric oxyhydroxide is apparently capable of binding orthophosphate ions more firmly than the ferrous form, the probably greater surface area of the gel-like ferrous compounds results in high phosphate sorption when phosphate concentration in the water phase is high. (Patrick & Khalid, 1974).

## Release Rate of Phosphorus from Lake Sediments

The release rate of phosphorus from Lake Mohegan sediments was measured in several experimental runs using the continuous flow system. In all experimental runs the sediment depth was about 13 cm. This depth is sufficient to give results indicative of the natural system. After some initial experments, the effects of dissolved oxygen and inorganic phosphorus concentration in the overlying water on the release rate of phosphorus from lake sediments were considered.

Fillos and Molof, (1972), showed that the release rate of phosphorus from simulated sludges increases with decreasing oxygen concentration in the overlying water reaching a maximum value under anaerobic conditions. Lake Mohegan sediments showed a zero release rate under aerobic conditions. Under anaerobic conditions the release rate was approximately 3 mg/day/m$^2$. Concurrent to the release rate of phosphorus, the release rate of iron was measured to be 6 mg/day/m$^2$.

Recently, Vollenweider, (1968), proposed permissible and critical loading rates for nitrogen and phosphorus as a function of lake mean depth. The dangerous loading rate for a lake of 5 m in depth is 0.36 mg/day/m$^2$, This rate is about one-tenth of the release rate from Lake Mohegan sediments under anaerobic conditions. It is therefore important to consider the effect of sediments on the overlying water in any lake management scheme. Maintaining aerobic conditions in the water sediment interface not only limits phosphorus release, but in addition, phosphorus from the overlying water will be "sorbed" by the sediments.

In addition to disolved oxygen, the phosphate concentration was increased to 0.025 mg/l. Figure 5 shows that under aerobic

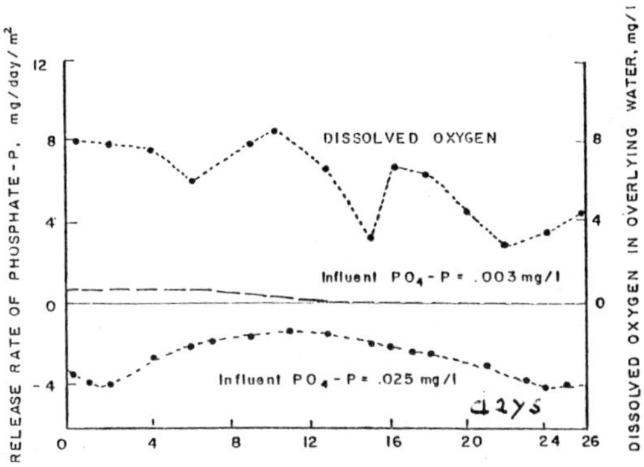


Fig. 5: Effect of phosphate concentration in overlying water on release rate of phosphates from sediments.

conditions the phosphate up-take rate by the lake sediment was approximately 2.5 mg/day/m$^2$ when the influent phosphate concentration was 0.025 mg/l. At an influent phosphate concentration of 0.003 mg/l, this value dropped to zero. Since the phosphate concentration in the lake water is usually much higher than 0.003 mg/l the sediments of Lake Mohegan will act as a phosphate "sink" as long as the bottom waters are aerobic.

When a similar experiment was conducted under anaerobic conditions, the sediments acted as a source of nutrients as shown in Figure 6. It is indicated that under anaerobic condition the sediments will act as a source with an increase phosphate concentration in the overlying water. The release rate could eventually decrease to zero when the overlying water approaches 11 mg/l as suggested by Figure 5. Figure 6 also shows that initially when conditions were aerobic the sediments acted as a sink. Beginning on the eighth day when conditions in the overlying water were made anaerobic the sediments began to release phosphates along with iron. Sridharan and Lee (1974) having made

Fig. 6. Effect of phosphate in overlying water on release rates of phosphate from sediments.

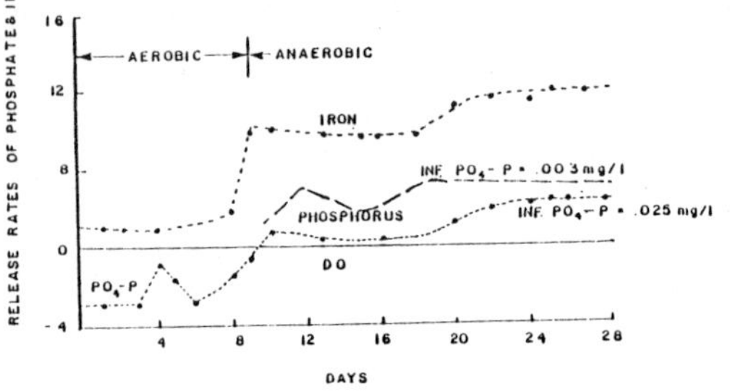


a similar observation with lake sediments, suggested an association of phosphorus with iron species. They also suggested that the high release rate of phosphorus results from chemical reactions and is most likely abiotic.

The Effect of Biochemical Activity on Release Rate of Phosphorus from Sediments

To study the effect of biochemical activity on the release rate of phosphorus, dihydrostreptomycin sulfate was used to arrest bacterial activity. Streptomycin sulfate is active against Gram positive and Gram negative bacteria (Cutting, 1962). In the experiment, five grams of the antibiotic were thoroughly mixed with the sediments before they were transferred to one of the reactors. A concentration of 100mg/l of the antibiotic was also maintained in the influent water to arrest biochemical activity in the overlying water within the reactor. Another reactor was set up without any antibiotic. These two reactors were identical in all respects except that one contained the dihydrostreptomycin sulfate. Figure 7 shows the results of phosphate and

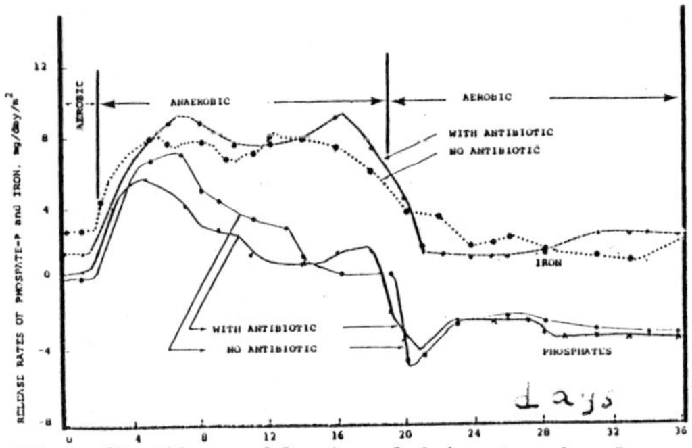


Fig. 7. The effect of biochemical activity on release rates of nutrients from sediments.

iron release rate as conditions were varied from anaerobic to aerobic. Throughout the experimental run, no significant differences in the release rate of the two nutrients was noted between the two reactors. This suggests that bacterial activity played a small part in the short term release rate of phosphorus and iron from the sediments. The release reaction was probably governed by a desorption process.

Groundwater-Flow Through the Sediments

Groundwater seepage into lakes, where large accumulation of sediments exist, may appreciably increase the transfer of phosphorus to the overlying water through convection. In the reactor where this was simulated, the overlying water conditions could be maintained either aerobic or anaerobic by aerating the overlying water or purging it with nitrogen gas. In each case, the gas flow rates were controlled to prevent any disturbances of the lake sediments. Figure 8 shows the changes in nutrient

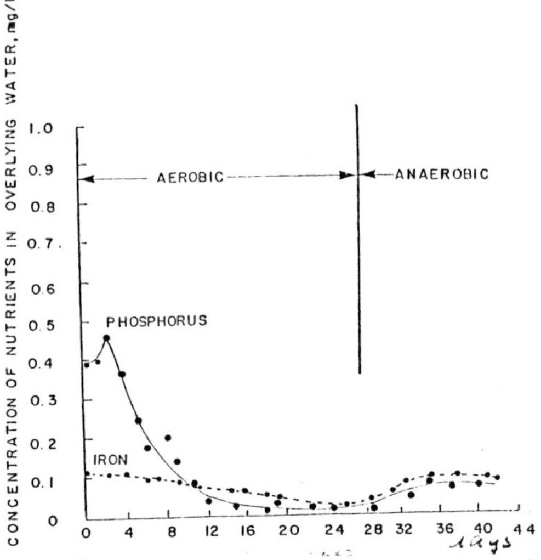


Fig. 8. Influence of ground-water flow on release of nutrients.

concentrations in the overlying water under aerobic and anaerobic conditions as affected by water flow through the lake sediments. During these two phases, the concentration of both iron and phosphates stabilized sufficiently to allow determination of the release rate of these two nutrients. The average release rates for iron and phosphorus were 2.9 mg/day/$m^2$ and 2.0 mg/day/$m^2$, respectively, under aerobic conditions, and 4 mg/day/$m^2$ and 11.3 mg/day/$m^2$, respectively, under anaerobic conditions. These values are higher than the release rates previously measured because of physical transport provided by groundwater-flow through the sediments.

Conclusions

In sediments, phosphorus may exist as calcium phosphate, organic phosphorus, orthophosphate dissolved in the interstitial water, aluminum and iron phosphate compounds, and as phosphates adsorbed on silicates. In the natural purification of these sediments, phosphate release from organic forms may occur through dephosphorylation by the phosphatase enzymes of the various microorganisms involved. In the degradation of organic matter, organic and inorganic acids are also formed, which may effectively dissolve inorganic phosphate compounds. A number of organic acids may also act as chelating agents, complexing calcium, iron, manganese, and aluminum. Such reactions result in additional solubilization of phosphorus. The resulting mineralized phosphorus may be released to the overlying water or retained by the sediments. Though the exact retention mechanism of phosphorus by the sediments is not known, it is believed that the aerobic layer of sediment plays a major role. When such an aerobic layer exists, it possesses a certain capacity for adsorption, by which many of the nutrients including phosphorus are retained by the sediments.

The existence of the aerobic layer in sediments depends on the oxygen concentration in the overlying water. By using the vertical distribution of the electrode potential in the sediments, Figure 3 showed that the aerobic layer may not be more than a few millimeters deep. Below this depth, conditions are anaerobic. Figure 3 also showed that the electrode potentials in the anaerobic layer are not affected by the level of oxygen concentration in the overlying water. This suggests that changes in the oxygen concentration of the overlying water will only affect conditions within the top few millimeters of sediments. Therefore, a decrease of oxygen concentration in the overlying water down to zero will eliminate the aerobic layer, destroy the adsorption capacity of the sediments, and result in an immediate nutrient release. Such a series of events seems to have occurred in this experiment.

The oxygen uptake rate of benthal deposits as measured in this experiment includes the respiration requirements of biological actvities in the sediments and the immediate oxygen demand of reduced substances, such as ferrous and sulfide ions, emanating from the lower anaerobic layers. The overall data relating DO changes from 0.15 to 7.7 mg/l to oxygen uptake rates showed a critical oxygen concentration of about 2 mg/l, above which the oxygen uptake rate was independent of oxygen concentration. A similar observation was made by Baity (1938), using a continuous system with sludge bottoms of 0.5-cm depth.

## References


Baity, H.G., 1938. Some Factors Affecting the Aerobic Decomposition of Sewage Sludge Deposits. Sew. Works Jour. 10:539-554.

Cutting, W.C.,1962. Handbook of Pharmacology Second Edition, Appleton-Century-Croffs, Div. of Meredith Publishing Co., N.Y.

DeWiest, R.J., 1965. Geohydrology. John Wiley & Sons, Inc., New York, N.Y.

Fillos, J. & H. Biswas, 1976. Phosphate Re - lease and Sorption by Lake Mohegan Sediments. Jour. Enviromental Engr. Div. ASCE 102:239-249.

Fillos, J. & A.H. Molof, 1972. The Effects of Benthal Deposits on Oxygen and Nutrient Economy of Flowing Water. Jour. Water Poll. Control Fed. 44:644-662.

Fillos, J. & W.R. Swanson,1975. The Release rate of Nutrients from River and Lake Sediments. Jour. Water Poll. Control Fed. 47:1032-1041.

Frink, C.R., 1969. Fractionation of Phosphorus - Analytical Evaluation. Soil Sci. Soc. of Amer. Proc. 33:326-328.

Lee, G.F., 1970. Factors Affecting the Transfer of Material Between Water and Sediments. Occasional Pap. 1, Eutrophication Information Program, Univ. of Wisconsin, Madison.

Mortimer, C.H., 1971. Chemical Exchanges Between Sediments and Water in Great Lakes--Speculation on a probably Regulatory Mechanism. Limnol. Oceanogr. 16:386-397.

Patrick, W.H. Jr.& R.A. Khalid, 1974. Phosphate Release and Sorption by Soils and Sediments: Effect of Aerobic and Anaerobic Conditions. Science. 186:53-56.

Sanville, W.D., 1974. Sediments and Sediment-Water Nutrient Interchange in Upper Klamath Lake, Oregon. Environmental Protection Agency 660/3-74-015.

Sridharan, N. &G.F. Lee, 1974. Phosphorus Studies in Lower Green Bay, Lake Michigan. Jour. Water Poll. Control Fed. 46:684-696.

Standard Methods for the Examination of Water and Wastewater. 1971. Amer. Pub. Health Assn., Washington, D.C.

Syers, J.K., R.F. Harris & D.E. Armstrong, 1973. Phosphate Chemistry in Lake Sediments. Jour. of Environmental Qual. 2:1-14.

Vollenweider, R.A., 1968. Scientific Fundamentals of the Eutrophication of Lakes and Flowing Waters with Particular Reference to Nitrogen and Phosphorus as Factor in Eutrophication. O.E.C.D. Report. DAS/CIS/68:27, Organization of Economic Co-operation and Development, Paris.



Key words: lakes
sediments
phosphorus
water quality
electrode-potential

# RELATIONSHIP BETWEEN RATE OF OXYGEN CONSUMPTION AND CONTENT OF ADENOSINETRIPHOSPHATE (ATP) IN LAKE SEDIMENTS

B. Ulén

Institute of Limnology, Uppsala, Sweden

## Abstract

The ATP content of sediments was determined by extraction with sulphuric acid and EDTA followed by a bioluminiscent reaction with luciferin-luciferase in a scintillator. Oxygen consumption of aerated sediments was determined according to the BOD-dilution method. The oxygen consumption of sediments of different types decreased with time and decreasing ATP content. Sterile sediments inoculated with untreated sediment resulted in increased oxygen uptakes along with increasing amounts of ATP. A good correlation existed when the same sediment was investigated. ATP values are discussed as means for measurements of biomass and activity.

## Introduction

ATP is the central molecule that transfers energy in living material. Holm-Hansen & Booth (1966) regarded it as a biomass indicator and introduced it as such in the study of marine organisms. Thereafter ATP has often been used for this purpose and also in studies of freshwater organisms. However, it cannot only be described as a measure of biomass. Because of its fundamental role in the processes of the cell it has also been used to study the response of the cell to its environment. For example, ATP has been used as an indicator of nutrient bioassay by Brezonik, Brown & Fox (1975) and for the study of toxicity by Brezonik & Patterson (1972).

ATP has also been considered as a measure of activity by, for instance, Hast & Rosswall (1973). Their opinion is based on relationships between oxygen respiration and ATP content in samples of soils, bacteria and fungi.

Some authors regard the concentration of ATP as an intermediate between a biomass indicator and an activity measurement. For instance, Ausmus (1971) considered ATP as a biomass indicator that is dependent on metabolic activity. Patterson, Brezonik & Putnam (1969) point out that ATP endogenously indicates the density of a population but together with substrate it represents the oxidative biomass.

## Material and methods

This paper describes two experiments which measure the relationship between oxygen consumption and the ATP content of sediments. In the first experiment four different sediments were adapted aerobically to 20°C for several days and thereafter used for measurements of biological oxygen demand and ATP. In the second experiment, sediment was dried at 80°C and thereafter wetted again. In this way the amount of bacteria was substantially reduced. This sediment was innoculated with fresh sediment and a biomass increase could be followed.

In both experiments the oxygen demand of the sediment was measured in darkness in bottles filled with water. The bottles were shaken in an end-over-end rotator. The chemical oxygen consumption was measured on mercurydichloride-treated sediment in the same way. This value was subtracted from the total oxygen demand to obtain the biological oxygen demand. ATP was measured according to Karl & LaRoch (1975) by adding a bacteria-ATP standard to the sediment. Adenosinediphosphate (ADP) and adenosinemonophosphate (AMP) was measured according to Laake (1974). Attempts were also made to measure bacterial biomass by use of the staining technique with acridine orange and counting in the fluoroscence microscope. However, there was a large variation in the biomass values in parallel measurements and the results have not yet been evaluated.

## Results and discussion

Fig. 1 shows the oxygen consumption and ATP concentration plotted against time in the adapted sediments. There was a good correlation between ATP and respiration with a coefficient of correlation that varied between 0.73 and 0.97.

These correlations are illustrated in Figure 2. With the same sediment-water system a close relationship is evident but there are variations between sediments from different lakes and also in sediments from different localities in the same lake. In the second experiment one part of the dried se-

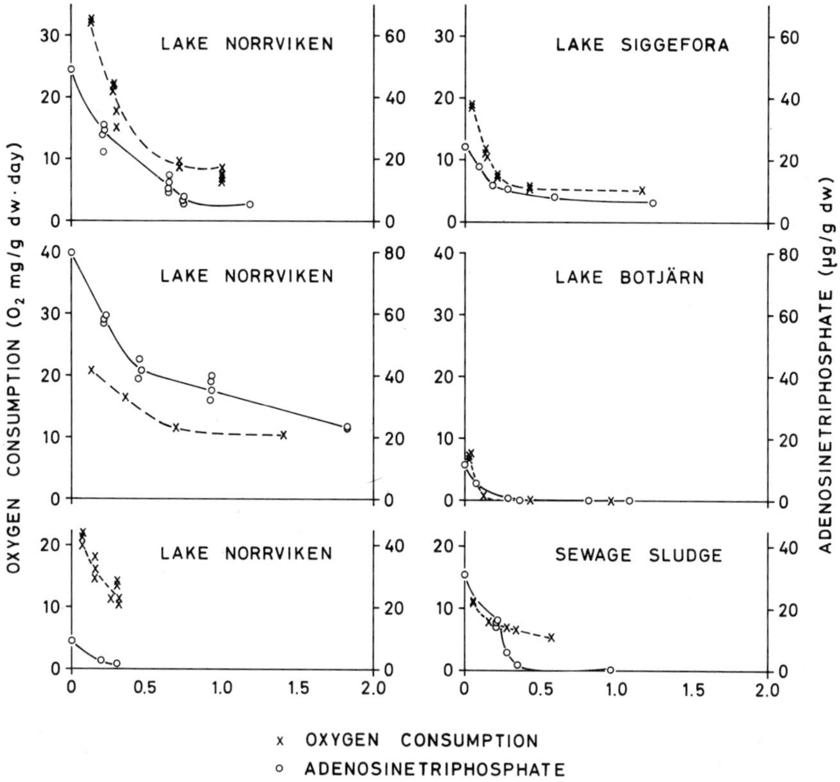


Figure 1. Oxygen consumption and ATP-content versus time of different sediments.

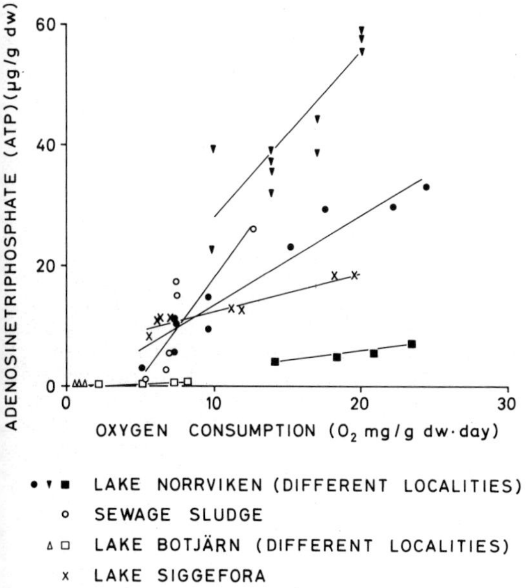


Figure 2. ATP-content versus oxygen consumption of different sediments.

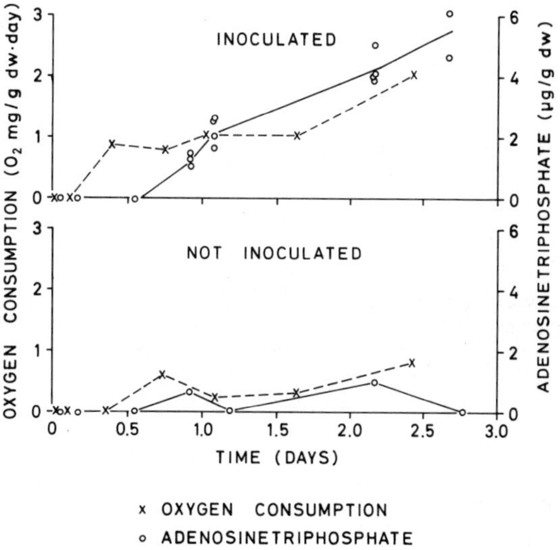


Figure 3. Oxygen consumption and ATP-content of inoculated sediment.

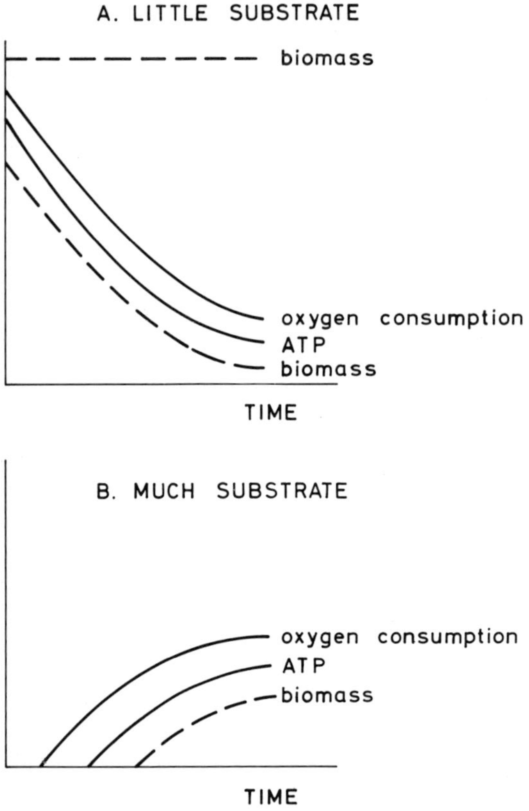


Fig. 4. Principal figure describing the effect of little or much substrate on oxygen consumption and ATP content.

diment was inoculated but not the other. The result is shown in Figure 3. Respiration and ATP increased with time and showed good correlation.

The principle of the experiments is shown in Figure 4. The upper half represents sediments that only contain a small amount of substrate. It is known that only a small part of the organic content in the sediment is easily degraded and will be consumed. If the biomass is constant (upper dashed line) the ATP should be a true measure of activity. With declining biomass (lower dashed line) the ATP concentration could be dependent on both activity and biomass. The lower half represents the sediment that was heated and afterwards inoculated. This sediment had high content of substrate and a low amount of bacteria that ought to have increased. An increasing concentration of ATP is mainly dependent on an increase of biomass.

Laake (1975) held the opinion that the ATP/ADP ratio might give valuable information on the activity level and energetic

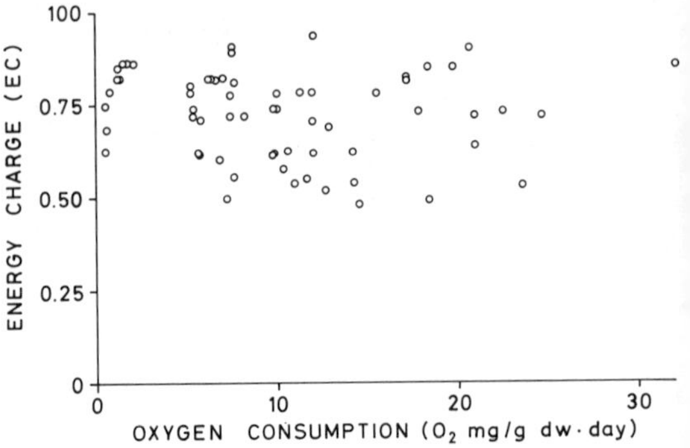


Fig. 5. Energy charge and oxygen consumption of different sediments.

status of organisms living in the sediment. In Figure 5 an energy charge is used that is the ratio

$$\frac{ATP + 1/2\ ADP}{ATP+ADP+AMP}$$

which shows the amount of the adenylate pool that is phosphorolysised. The energy charge is dependent on the balance of energy in the cell and it has been shown to change at the same time as the activity of single enzymes changes (Atkinson, 1968). Figure 5 illustrates the energy charge as a function of oxygen consumption. The energy charge was always high, which showed that the cells were in an active state. It varied between 0.48-0.96 and had no relationship with oxygen consumption. This irregularity is difficult to explain. One explanation could be that different groups of bacteria in the population might have had very different ratios of ATP:ADP:AMP.

## Conclusion

These experiments have not provided a direct answer to the question whether ATP could be regarded as a biomass and/or activity indicator. In all the experiments there was, however, a fairly close relationship between ATP and oxygen consumption for the separate runs. The results indicate that measurements of ATP could replace measurements of oxygen consumption. This is useful, since ATP measurements are independent of experimental conditions such as stirring etc. However, more data on this question is obviously needed.

## References

Atkinson, D., 1968. The energy charge of the adenylate pool as a regulatory parameter. Interaction with feedback modifiers. Biochemistry 7:4030-4034.

Ausmus, B., 1971. Decomposition studies:

mass and metabolic activity of microflora in litter and soil. Eastern Deciduous Forest Biome Memo Report No. 71:66, US-IBP, Oak Ridge National Laboratory.

Brezonik, P.L., F.X. Browne & J.L. Fox, 1975. Application of ATP to plankton biomass and bioassay studies. Water Research 9:155-162.

Brezonik, P.L. & J.W. Patterson, 1972. Activated sludge ATP: effects of environmental stress. J. Sanit. Engrg. Div. Amer. Soc. Civil. Engrs. 97:813-824.

Hast, R. & T. Rosswall, 1973. Activity of soil microorganisms as measured by respiration, ATP and dehydrogenase. Technical report No. 16, Ecol. Res. Comm. (Stockholm) 57-71.

Holm-Hansen, O. & C.R. Booth, 1966. The measurement of adenosinetriphosphate in the ocean and its ecological significance. Limnol. Oceanogr. 11:510-519.

Karl, D.M. & P. LaRoch, 1975. Adenosinetriphosphate measurements in soil and marine sediments. J. Fish. Res. Board Can. 32: 599-607.

Laake, M., 1974. Determination of ATP and interpretation of the results. NIVA-report XB-06, Norwegian Institute for Water Research, Oslo.

Laakc, M., 1975. Determination of ATP in aquatic sediments. In: N. Edberg & A. Wilander (Ed.): Biologisk omsättning i sediment. SNV PM 694:17-44.

Levin, G.V., J.R. Scrot & W.C. Hess, 1975. Methodology for application of adenosinetriphosphate determination in wastewater treatment. Environ. Sci. Technol. 9: 961-965.

Patterson, J.W., P.L. Brezonik & H.D. Putnam, 1969. Sludge activity parameters and their application to toxicity measurements and activated sludge. Proceedings of the Industrial Waste Conference. Purdue University Engineering Bulletin. Purdue University, Engineering Extension Series No. 135 pp. 127-154.

Key words: Adenosinetriphosphate
Oxygen consumption
Sediment
Biomass
Activity

# SEDIMENT OXYGEN UPTAKE IN SOUTH SWEDISH LAKES

Wilhelm Granéli

Institute of Limnology, Fack, S-220 03 Lund 3, Sweden

## Summary

Sediment oxygen demand was measured monthly or bimonthly in five lakes of different trophic state in South Sweden. Plexiglass tubes with undisturbed sediment cores were incubated in the laboratory at the in situ temperature. The relationship between oxygen saturation in the water and sediment oxygen uptake was linear. In addition to oxygen availability, temperature seemed to regulate the seasonal variations in sediment oxygen demand. Once the influence of oxygen saturation on oxygen uptake was eliminated there was a good correlation between temperature and oxygen consumption (correlation coefficient after logarithmic transformation 0.76-0.94 excluding one lake). $Q_{10}$ values calculated from the regression lines were 2.0-3.0 for the interval 5-10$^{o}$C and 1.3-1.6 for the interval 15-20$^{o}$C. In two of the lakes the oxygen uptake of undisturbed sediment cores sampled in late autumn and the oxygen uptake of sieved and restratified sediment (sampled with an Ekman dredge) were investigated at temperatures from 5 to 20$^{o}$C. At the same temperatures, the oxygen uptake of these samples was similar to that of the monthly samples for each lake, indicating a strong "buffering capacity" for sediment respiration. Thus freshly sedimented organic matter seemed to be of minor importance in determining the seasonal variation in sediment oxygen uptake. This was confirmed by studies of sediments from two cylindrical plastic enclosures in one of the lakes. Despite a large difference in plankton biomass in the two enclosures, sediment oxygen uptake was similar when correction was made for macrobenthic respiration. Although the variation among the five lakes in yearly pelagic primary production was considerable (20-575 g $C \cdot m^{-2} \cdot year^{-1}$), the yearly oxygen consumption of the sediments showed a much more restricted range (120-235 l $O_2 \cdot m^{-2} \cdot year^{-1}$). There was a positive correlation between yearly sediment oxygen demand and the quotient between primary production and mixed-layer depth of the form proposed by Hargrave (1973). Measurement of sediment growth rate in three of the lakes indicated that decomposition of organic matter in the sediment does not increase in proportion to pelagic primary production. The sediment growth rate is directly related to the primary production, however, which explains the rapid filling of shallow eutrophic lakes.

## Reference

Hargrave, B.T. 1973. Coupling carbon flow through some pelagic and benthic communities. J. Fish. Res. Board Can. 30:1317-1326.

Key words: sediment
oxygen
uptake
production
temperature

MODELLING THE TEMPORAL VARIATION IN SEDIMENTARY PHOSPHORUS FRACTIONS

ars Kamp-Nielsen

reshwater Biological Laboratory, Hillerød, Denmark

ummary

The temporal variation in sedimen-ation and sedimentary phosphorus ractions has been studied in L. Es-om. The sedimentation of phosphorus aried between 1.1 $10^{-3}$ g P $m^{-1}$ $d^{-1}$ nd 53.3 $10^{-3}$ g P $m^{-2}$ $d^{-1}$. The amount f exchangeable phosphorus in the se-iment varied between 1.4 and 4.2 g P $^{-2}$ and the ammonium chloride extract-ble phosphate fluctuates between .02 and 0.18 g P $m^{-2}$.

A model based on previous experi-ents describing the variations in he sedimentary phosphorus fractions s functions of sedimentation, phos-hate concentrations in the overlying ater, temperature and oxygen condi-ions has been compared with the field bservations.

ntroduction

A model describing the anaerobic xchange of phosphorus across the se-iment-water interface has been deve-oped by Jørgensen et al. (1975). his model considers the sedimenta-ion of phosphorus and the phosphate oncentration gradient across the se-iment-water interface as driving va-iables. This form of the submodel as been used in the total modelling f L. Glumsø (Jørgensen, 1976). How-ver, the exchange model has only een validated in laboratory experi-ents (Jørgensen et al., 1975), where o sedimentation is occurring and it escribes the total release of phos-horus from the sediment by simple iffusion only.

In dimictic lakes distinct diffe-ences between reduced and oxidized ediment surface are observed and hese shifts have to be considered in he modelling of stratified lakes Kamp-Nielsen, 1975 a, b). In this tudy the measured rates of phospho-us sedimentation, bottom temperature, hosphate concentration above the ediment and the ratio exchangeable hosphorus/non-exchangeable phospho-us are used as driving variables and he variations in the state variables exchangeable sedimentary phosphorus and interstitial phosphate - will be compared with measured values.

Previously, the kinetics of the liberation of phosphate from the sediment in L. Esrom have been studied and the net exchange described by the sum of a zero order temperature dependent biological degradation, and two first order transport processes (Kamp-Nielsen, 1975 a). These equations do not give a satisfactory response to a reduced external loading of phosphorus due to the zero order process. Using the latter equation the liberation of phosphate will continue at almost the same rate after reduced sedimentation following a reduced external loading of phosphorus, which is not in accordance with our previous predictions (Jørgensen et al., 1975).

This study is a part of the "Joint Lake Research Project" concerned with the modelling of eutrophication in lakes. One of the major ideas of the project is that the total model can be separated into submodels, each of which can be physically isolated and experimentally validated e.g. phytoplankton growth, grazing by zooplankton and sediment-water exchange of nutrients.

Materials and methods

L. Esrom is situated in the northern part of Zealand. It has an area of 17.3 $km^2$ and the average depth is 12.3 m (Jónasson, 1972).

Sedimentation during 1975 was measured by sediment traps exposed for periods of 2 weeks 1 m above the bottom at 20 m depth (Kamp-Nielsen, 1975 b, Lastein, 1976).

The analysis for sedimentary phosphorus was carried out on 4 parallel cores collected at 20 m depth twice a month. The cores were split up into intervals: 0-1 cm, 2-3 cm, 3-5 cm and 5-10 cm. The amount of exchangeable phosphorus in each depth interval was calculated by subtracting the total phosphorus in the stabilized sediment - 5-10 cm depth - from the total amount of phosphorus in each of the overlying slices. Using the content of minerogen matter (density = 2.6) and the content of organic matter (density = 1.0) the amount of exchangeable phosphorus in each slice and per $m^{-2}$ could be calculated (Jørgensen et al., 1975).

The total phosphorus content in the sediment and in the tripton was determined by an ignition method (Møller Andersen, 1976). The phosphate concentration in the bottom water and the extracted phosphate from the sediment was determined by the method described by Murphy & Riley (1962).

The mobile phosphate - represented by the interstitial and sorbed phosphate - was measured by the ammonium-chloride extraction method (Williams et al., 1971). This procedure is considered as the most reproducible measurement of mobile phosphate in calcareous sediments (Kamp-Nielsen, 1974). The model was programmed in CSMP and executed at NEUCC at the Technical University, Copenhagen using an IBM 370/165.

## Results

The temporal variations in temperature, oxygen concentrations and phosphate concentrations at 20 m depth appear on Fig. 1. The lake is stratified from May 15th to Oct. 1st and anoxic bottom water exists from July 15th until Oct. 1st. Beyond the stagnation period, the surface sediment is assumed to be anaerobic in the periods indicated on Fig. 1. The considerations leading to this assumption will be discussed later. The seasonal variations in gross primary production, sedimentation and resuspension of phosphorus and the ratio: exchangeable phosphorus/non-exchangeable phosphorus in the sedimentating phosphorus are shown on Fig. 2.

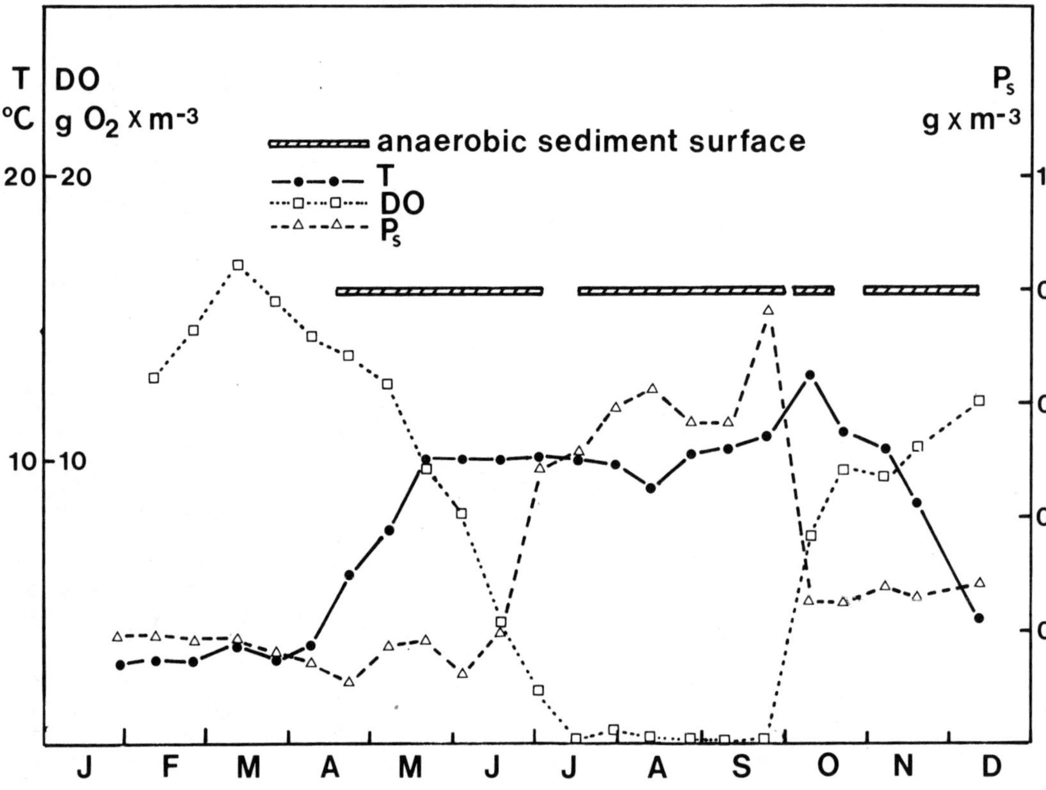


Fig. 1. The seasonal variations in temperature (T), dissolved oxygen (DO) and phosphate ($P_S$) at 20 m depth in L. Esrom. The periods with anoxic sediment surface are indicated with hatched bars.

The bimodal pattern of sedimentation and its relation to primary production has previously been discussed by Kamp-Nielsen (1975 b) and Lastein (1976). A correction for resuspension occurring prior to and during the autumn overturn is made according to Lastein (1976) and indicated on Fig. 2.

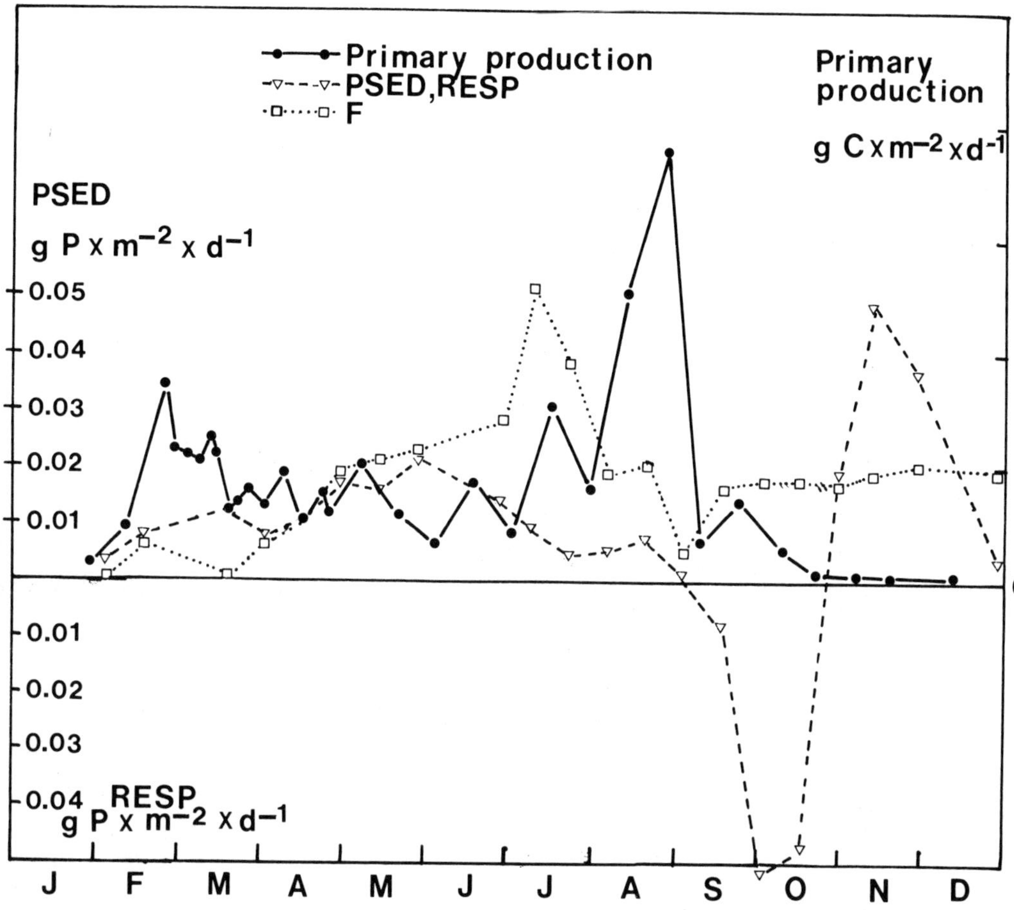


Fig. 2. The seasonal variations in gross primary production, sedimentation of phosphorus (PSED), resuspension of phosphorus (RESP) and the ratio: F = exchangeable phosphorus/non-exchangeable phosphorus in L. Esrom. The resuspension is indicated as negative sedimentation.

Until now it has not been possible to distinguish analytically between the exchangeable and the non-exchangeable phosphorus in the tripton. The ratio: exchangeable/non- exchangeable phosphorus in the tripton has therefore been calculated assuming the non-exchangeable fraction to be a constant proportion of the dry matter and equal to the concentration of phosphorus in the stabilized part of the sediment (5-10 cm depth). The seasonal variation apparently shows the same trend as the seasonal variation in zooplankton biomass cf. Jónasson (1972) and Bosselmann (1974). An explanation might be that a substantial part of the sedimentating phosphorus is associated with zooplankton fecal pellets. Most of the phytoplankton, decomposed in the water body, is mineralized through the grazing by zooplankton (Jónasson, 1972). The mineralized nitrogen is easily dissolved in the form of ammonia from the excretion products of the zooplankton, whereas phosphorus can be adsorbed to the surface of the fecal pellets in the alkaline surface water and decomposed in the more acidic bottom water. This assumption is supported by the observation of low nitrogen/phosphorus ratio in the tripton during the maximum in zooplankton biomass.

The annual rates of primary production and sedimentation are presented in Table 1.

Table 1. Annual rates of primary production and sedimentation in L. Esrom

| Rate | gram x $m^{-2}$ x $yr^{-1}$ |
|---|---|
| Gross primary production of carbon | 260 |
| Gross sedimentation of dry matter | 1036 |
| - - - - - (corrected for resuspension) | 793 |
| - - - organic matter | 385 |
| - - - - - (corrected for resuspension) | 294 |
| - - - phosphorus | 6.12 |
| - - - - (corrected for resuspension) | 4.58 |

Development of sediment model

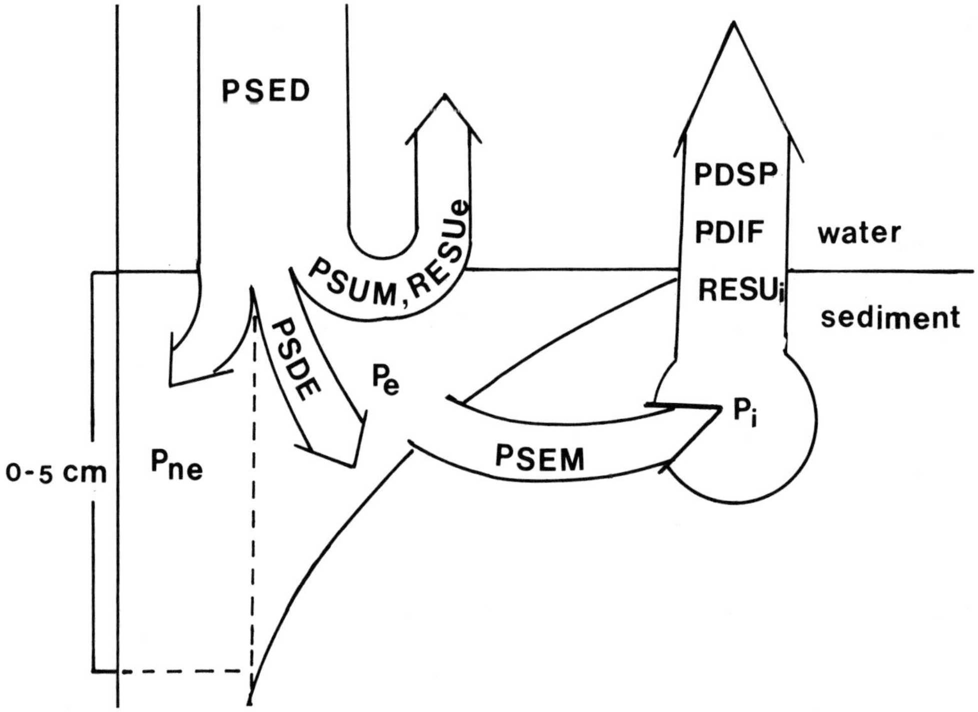


Fig. 3. Conceptual model of the pools and processes involved in the sediment-water exchange of phosphorus in L. Esrom. The symbols used:

Pools:

$P_e$ = the exchangeable, sedimentary phosphorus

$P_{ne}$ = the non-exchangeable, sedimentary phosphorus

$P_i$ = the interstitial and sorbed phosphate

Rates:

PSDE = the exchangeable fraction of the sedimentating phosphorus

PSUM = the exchangeable phosphorus mineralized on the sediment surface and released without transport limitation to the bottom water

PSEM = the exchangeable phosphorus mineralized within the sediment

$RESU_e$ = the exchangeable phosphorus removed from the sediment by resuspension

PDSP = the desorption of interstitial and sorbed phosphate

$RESU_i$ = the interstitial and sorbed phosphate removed by resuspension

PDIF = the interstitial and sorbed phosphate removed by diffusion to the overlying water

The development of the submodel is based on the conceptual model shown in Fig. 3. It is a refining of models previously described by Jørgensen et al. (1975) and Kamp-Nielsen (1976). The sediment model is consisting of the state variables: $P_e$ - the exchangeable phosphorus in the sediment - and $P_i$ - the interstitial and sorbed phosphate. Based on the principle of mass conservation the model is expressed by the differential equations:

(1) $\frac{dP_e}{dt}$ = $PSDE-PSUM-PSEM-RESU_e$

and

(2) $\frac{dP_i}{dt}$ = $PSEM-PDSP-RESU_i$ (aerobic sediment surface)

(3) $\frac{dP_i}{dt}$ = $PSEM-PDSP-RESU_i-PDIF$ (anaerobic sediment surface)

The symbols are defined in the caption to Fig. 3.

Other equations:

(4) PSDE = PSED $(\frac{F}{F+1})$, where PSED is the total sedimentation of phosphorus in g P x $m^{-2}$ x $d^{-1}$ and F is the ratio exchangeable/non-exchangeable phosphorus in the total sedimentating phosphorus shown in Fig. 2.

(5) PSUM = $0.56\ e^{0.203\ T} \frac{PSED}{PSED + 1} 10^{-3}$ where PSUM is the rate of mineralized surface phosphorus in g P x $m^{-2}$ x $d^{-1}$ and T = the bottom temperature in °C shown on Fig. 1. The equation is a development of the zero order equation: PSUM = $0.56\ e^{0.203\ T}$ experimentally determined by Kamp-Nielsen (1975 a). This use of a hyperbolic function gives a more realistic description of the process at low sedimentation rates, where the amount of sedimentated organic matter must limit the surface mineralization.

(6) PSEM = $k\ P_e\ \theta^{T-20}$, is the first order reaction previously used to express the mineralization of exchangeable phosphorus in g P x $m^{-2}$ x $d^{-1}$ (Jørgensen et al., 1975). k and θ are constants.

(7) $RESU_e$ = RESP$(\frac{F}{F+1})$, where RESP is the total resuspended sedimentary phosphorus in g P x $m^{-2}$ x $d^{-1}$.

(8) PDSP = $-(0.6 \ln P_s + 2.27)\ 10^{-3}$, where $P_s$= the phosphate concentration in the bottom water in g P x $m^{-3}$, is the equation experimentally derived by Kamp-Nielsen (1974, 1975 a) describing the time dependent sorption equilibrium in g P x $m^{-2}$ x $d^{-1}$.

(9) $RESU_i$ = RESP $\frac{P_i}{P_e}$ $(\frac{F}{F+1})$

The interstitial and sorbed phosphate removed by resuspension in g P x $m^{-2}$ x $d^{-1}$.

(10) PDIF = $(1.21(\frac{P_i-4.0x10^{-3}}{15.7x10^{-3}} - P_s)-1.7)$ $10^{-3}$

The interstitial and sorbed phosphate removed by diffusion in g P x $m^{-2}$ x $d^{-1}$. The equation has been experimentally derived by Kamp-Nielsen (1974, 1975 a). The concentration gradient across the sediment-water interface is converted from $(P_i^! - P_s)$, where $P_i^!$ is the concentration of interstitial and sorbed phosphate in the upper cm in g P x $m^{-3}$, to:

$P_i^!-P_s=(\frac{P_i-4.0x10^{-3}}{15.7x10^{-3}} - P_s)$

according to Fig. 4.

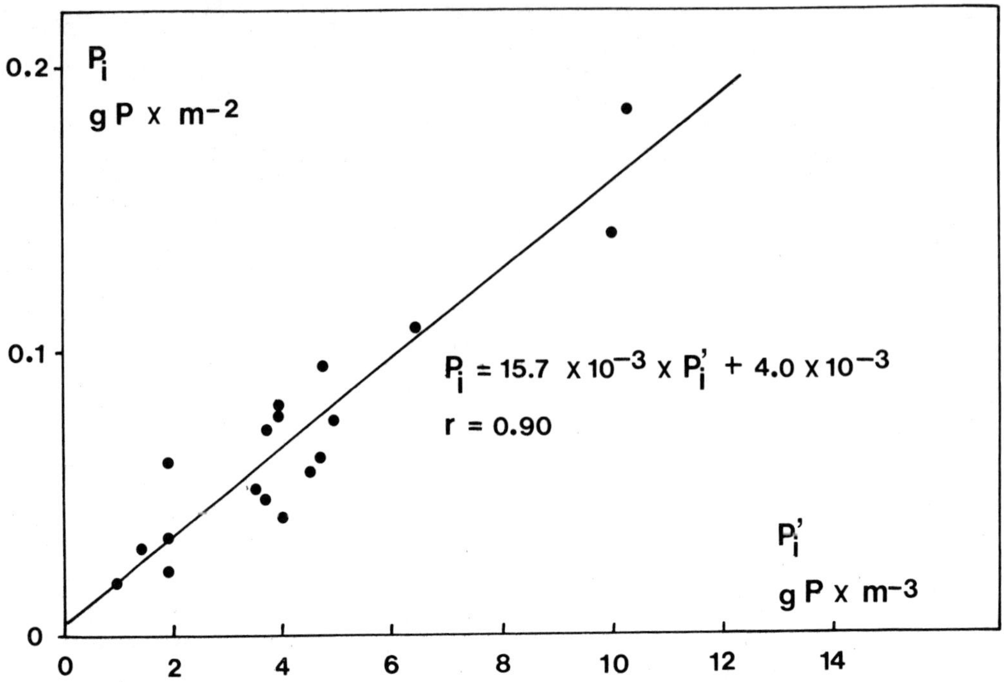


Fig. 4. The relation between the concentration of interstitial phosphate in the upper centimeter of the sediment - $P_i$ - and the total interstitial phosphate per $m^2$ - $P_i$.

(1) can now be written as:

$$(11)\quad \frac{dP_e}{dt} = PSED\left(\frac{F}{F+1}\right) - 0.56\, e^{0.203T}\left(\frac{PSED}{PSED+1}\right) - k\, P_e\, \theta^{T-20} - RESP\left(\frac{F}{F+1}\right)$$

(2) can be written as:

$$(12)\quad \frac{dP_i}{dt} = k\, P_e\, \theta^{T-20} + (0.6 \ln P_s + 2.27)10^{-3} - RESP\, \frac{P_i}{P_e}\left(\frac{F}{F+1}\right),$$

which is valid when the sediment surface is aerobic.

(3) can be written as:

$$(13)\quad \frac{dP_i}{dt} = k\, P_e\, \theta^{T-20} + (0.6 \ln P_s + 2.27)10^{-3} - RESP\, \frac{P_i}{P_e}\left(\frac{F}{F+1}\right) - \left(1.21\left(\frac{P_i - 4.0\times10^{-3}}{15.7\times10^{-3}} - P_s\right) - 1.7\right)10^{-3}$$

which is valid when the sediment surface is anaerobic. This situation is assumed to occur under the following conditions:

a) the oxygen concentration is $\leq$ 0.5 g $O_2$ x $m^{-3}$

b) PSED $\geq 13 \times 10^{-3}$

c) T $\geq$ 12

Case b) and c) is assumed to introduce anaerobic sediment surface due to increased oxygen consumption by the sediment in both cases cf. Jørgensen et al. (1975) and Kamp-Nielsen (1975 a).

Using the table functions PSED, RESP, T and F as driving variables, $\theta$ = 1.047 (used by Chen (1974) as an average for decomposition of organic matter) and a calibrated value of k = 5 x $10^{-4}$ the seasonal variations in $P_i$ and $P_e$ are simulated and the results are shown together with measured values on Fig. 5 and Fig. 6, respectively.

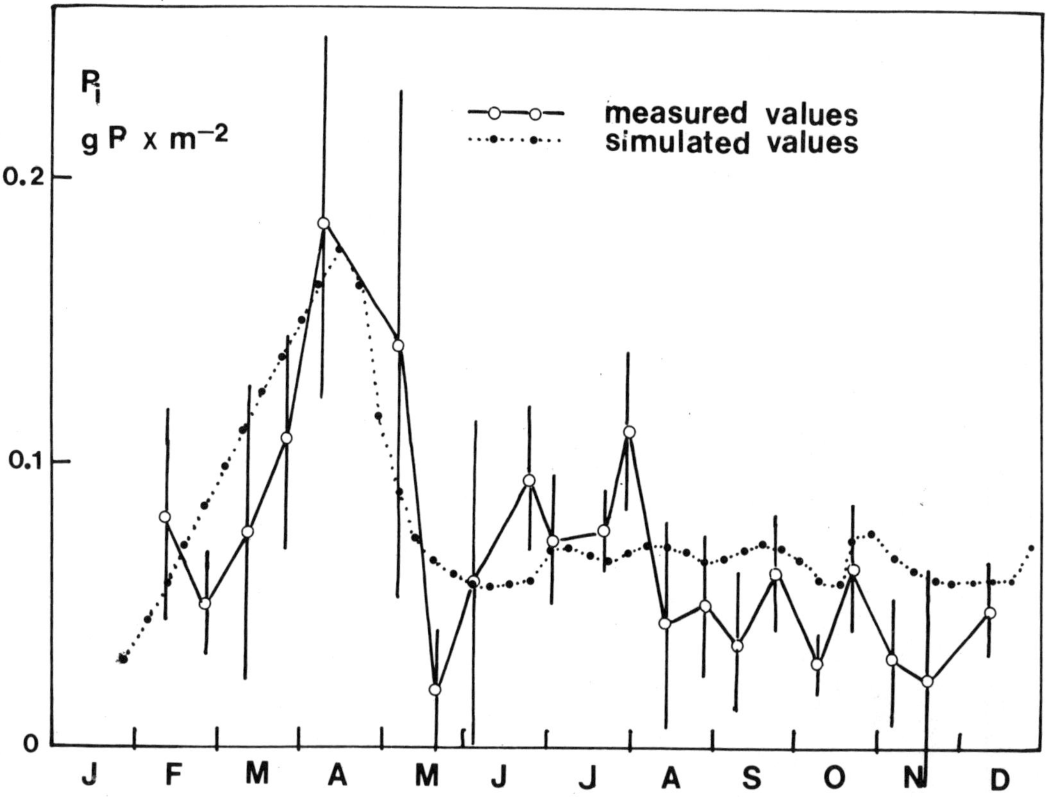


Fig. 5. The simulated and measured seasonal variation in interstitial phosphate - $P_i$. The measured values are shown with standard deviations.

## Discussion

The use of a sediment submodel consisting of two state variables - exchangeable phosphorus and interstitial phosphate - has by visual inspection given a reasonable simulation of the temporal variation in sedimentary phosphorus fractions. Considering $P_e$, 75% of the simulated values are lying within the S.D. of the measured values and with respect to $P_i$, 71% of the simulated values are within the S.D. of the measured values. It must be brought in mind that these results have been achieved with a model based on experimentally derived parameter values only except for $\theta$, which has been taken from the literature and k, which has been calibrated.

Comparing Fig. 5 and 6 it is seen that the maxima and minima in $P_i$ and $P_e$ are appearing simultaneously. Model verifications revealed that a poor response in simulated $P_e$ to changes in $P_i$ was due to the small flux through $P_e$ relatively to the size of $P_e$, whereas the more distinct fluctuations in simulated $P_i$-values were due to a greater flux through $P_i$ compared to the size of this pool. A more pronounced effect of $P_i$ on $P_e$ could be achieved by splitting up $P_e$ and introducing a new state variable in the model - the sedimentating phosphorus mineralized on the sediment surface. The proposed improvement is supported by the observation that the greatest seasonal amplitude in $P_e$ is measured in sediment depth 0-1 cm. Another possible improvement is the application of a more realistic description of the sorption equilibrium e.g. the application of a Freundlich expression and an irreversible transformation of adsorbed phosphate to apatite. A more causal approach to the modelling of sediment-water exchange might be provided by simulation of the various profiles instead of unlocalized pools, as done by Vanderborght & Billen (1975) on nitrogen

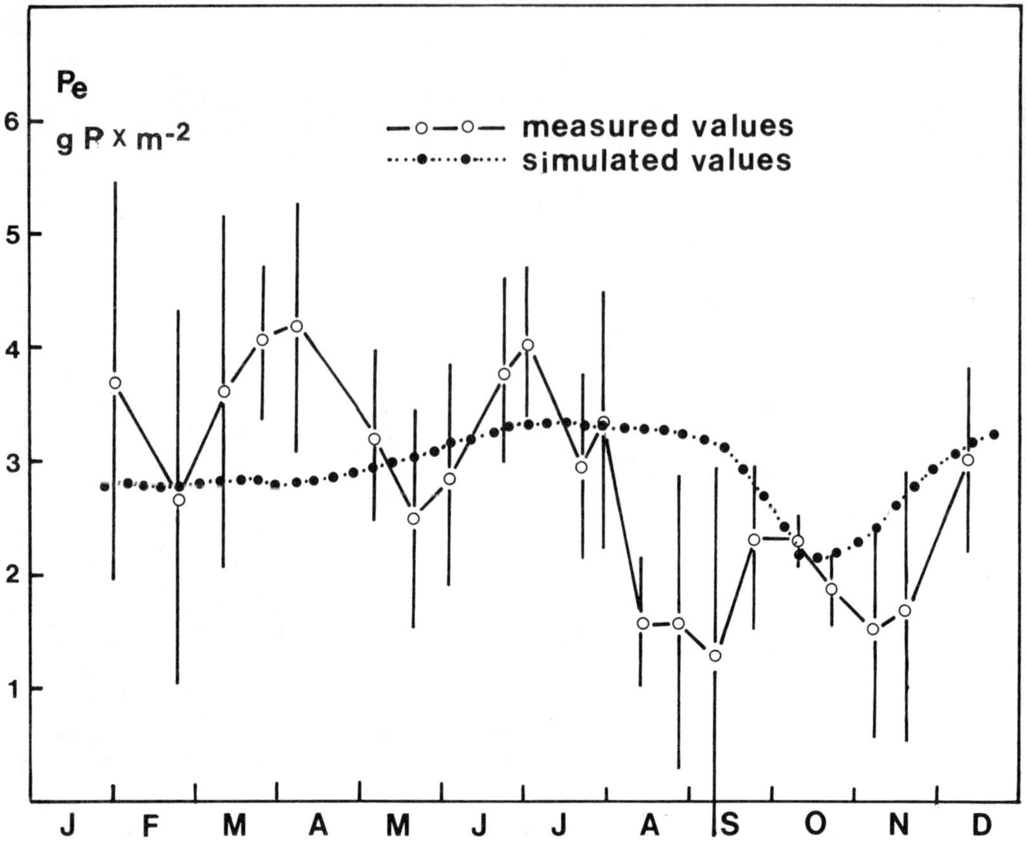


Fig. 6. The simulated and measured seasonal variation in exchangeable, sedimentary phosphorus - $P_e$. The measured values are shown with standard deviations.

metabolism in sediments. However, a sediment submodel in a simplified edition of the model described in this paper has given a satisfactory simulation of the transients in the long term behaviour of L. Lyngby with great amplitudes in external loading of phosphorus (Pedersen & Pedersen, 1976, Jørgensen, 1977).

Using the model the importance of the sediment in the phosphorus cycling can be estimated. The gross primary production of 260 g C $m^{-2}$ $yr^{-1}$ requires some 6.3 g phosphorus $m^{-2}$ $yr^{-1}$ using a C:P ratio of 106:1. The net external loading amounts to about 0.5 g phosphorus $m^{-2}$ $yr^{-1}$ (Jónasson et al., 1974), and the total annual release from the sediment is calculated to 2.0 g phosphorus $m^{-2}$ $yr^{-1}$ from the model. This means that 8% of the required phosphorus is supplied from external loading, 32% through mineralization in the sediment and 60% recycled in the water body via grazing and microbial decomposition. Based on oxygen consumption by the sediment Hargrave (1972) calculated that 31% of the gross primary production in L. Esrom was decomposed in the sediment.


Acknowledgements

This study has been supported by the Danish State Research Foundation. M.Sc. Henning Mejer, Sv. E. Jørgensen and Mogens Friis have kindly assisted with programming of the model.

## References


Andersen, J.M., 1976. An ignition method for determination of total phosphorus in lake sediments. Wat. Res. 10:329-331.

Bosselmann, S., 1974. The crustacean plankton of Lake Esrom. Arch. Hydrobiol. 74:18-31.

Chen, C.W., 1974. In: Water quality for river-reservoir systems. Rep. Hydrologic engineering Center, Corps of engineers, U.S. Army. Davis. Calif. 210 p.

Hargrave, B.T., 1972. A comparison of sediment oxygen uptake, hypolimnetic oxygen deficit and primary production in Lake Esrom, Denmark. Verh. Internat. Verein. Limnol. 18: 134-139.

Jónasson, P.M., 1972. Ecology and production of the profundal benthos in relation to phytoplankton in Lake Esrom. Oikos, Suppl. 14:1-148.

- -, E. Lastein & A. Rebsdorf, 1974. Production, insolation and nutrient budget of eutrophic Lake Esrom. Oikos 25:255-277.

Jørgensen, S.E., 1976. An eutrophication model for a lake. Ecol. Model. 2,2. (In press).

- -, 1977. Examination of a lake model. Ecol. Model. (In press).

- -, L. Kamp-Nielsen & O.S. Jacobsen, 1975. A submodel for anaerobic mud-water exchange of phosphate. Ecol. Model. 1:133-146.

Kamp-Nielsen, L., 1974. Mud-water exchange of phosphate and other ions in undisturbed sediment cores and factors affecting the exchange rates. Arch. Hydrobiol. 73:218-237.

- -, 1975 a. A kinetic approach to the aerobic, sediment-water exchange of phosphorus in Lake Esrom. Ecol. Model. 1:153-160.

- -, 1975 b. Seasonal variation in sediment-water exchange of nutrient ions in Lake Esrom. Verh. Internat. Verein. Limnol. 19:1057-1065.

- -, 1976. Model considerations on the sediment-water exchange of phosphorus. Naturvårdsverkets PM series. 694:161-166.

Lastein, E., 1976. Recent sedimentation and resuspension of organic matter in eutrophic Lake Esrom, Denmark. Oikos 27:44-49.

Murphy, J. & J.P. Riley, 1962. A modified single solution method for determination of phosphate in natural waters. Anal. Chem. Acta. 27: 21-26.

Pedersen, T. & M.H. Pedersen, 1976. Lake Models. A report from Dept. sanitar. Eng., Acad. Eng. 99 p.

Vanderborght, J.P. & G. Billen, 1975. Vertical distribution of nitrate concentration in interstitial water of marine sediments with nitrification and denitrification. Limnol. Oceanogr. 20:953-961.

Williams, J.D.H., J.K. Syers, D.E. Armstrong & R.F. Harris, 1971: Characterization of inorganic phosphate in noncalcareous lake sediments. Soil. Sci. Soc. Amer. Proc. 35:556-561.



Key words: sediments
sedimentation
exchange
phosphorus
modelling

# SEDIMENTS AS A SOURCE OF PHOSPHATE FOR ALGAL GROWTH

H.L. Golterman

Limnological Institute, Nieuwersluis, The Netherlands

Summary

The shallow lakes (mean depth 3.5 m) of The Netherlands receive phosphate loadings of more than 6 g $m^{-2}y^{-1}$ of $PO_4$-P. Because of biological and chemical binding about two third of the incoming phosphates accumulates in the bottom sediments. It is feared that these phosphates would become a major source of phosphate for algal growth, if in future the loading is to be restricted. In algal cultures the sediments proved to be an excellent source of phosphate, if sediments and algal cells are mixed without a physical barrier between the two. Since not all the sediment phosphate appeared to be available, we tried to separate chemically the available from the not available compounds. So far good results have been obtained employing an extraction with 0.01 M of NaNTA (pH $\simeq$ 7). The phosphate thus extracted was similar in quantities to that by algae such as Scenedesmus sp.
In an artificial eutrophic enclosure (40 m diameter) algal blooms occurred after flushing of the pond with relatively phosphate poor ground water. During these blooms the NTA extractable phosphate decreased by about fifty percent.

## Introduction

Most of the Dutch lakes are situated in the combined delta of the Rhine and Meuse. They receive large amounts of nutrients from these rivers. The Rhine carries about 70.000 tonnes of $PO_4$-P per year, of which about 80% is transported into the North Sea. Of the remaining 14.000 tonnes about 50-60% accumulates in the Dutch lakes. A second source is the sewage of surrounding population (400 persons/$km^2$) and industry. The average annual loadings of the lakes is at least 6 g $m^{-2}y^{-1}$ of $PO_4$-P. (Kolenbrander 1974). The mean depth of most of these lakes is between 2 and 3.5 m, so they do not stratify. Either water circulates between the euphotic zone and sediment layers constantly, or the euphotic zone extends to the sediments. Algal densities up to 100 mg $m^{-3}$ of chlorophyll a are regularly recorded in several lakes, and densities of over 300 mg $m^{-3}$ have been recorded, e.g. the Border Lakes between the newly reclamed polders in the IJsselmeer and the old land. (Fig. 1)

In many of these eu- or even hypertrophic lakes phosphate is no longer the factor limiting the density of algal populations; instead they represent cases where nitrogen or irradiance control algal growth. These lakes are typical illustrations of Lund's (1970) argument, that the more eutrophic lakes have a tendency to move away from phosphate limitation. Nitrogen limitation may become more and more important due to the fact, that in eutrophic shallow lakes the sediments cause nitrogen to be lost from the system by denitrification. This means that an excess of nitrogen in a certain period will be removed from the system, while an excess of phosphate will be adsorbed in the sediments and will return in the phosphate cycle in periods of relative phosphate depletion.

Public opinion is slowly beginning to recognise that phosphate replacement in laundry detergents and phosphate removal from sewage purification, are essential steps without which subsequent algal growth causes the production of as much organic matter as was destroyed in the sewage works. Although economic arguments against phosphate removal still caution us against a too great optimism, the question must now be raised at what rate will a lake recover, in case phosphate input rates are returned to their pre-industrial level. The mean loading might decrease to 1 g $m^{-2}y^{-1}$, if in the countries upstream the same measures are taken.

In earlier work we have already shown (Golterman et al 1969) that sediments are a perfect source of phosphate for algae (see fig. 2 and 3). The work reported here is an extension of that

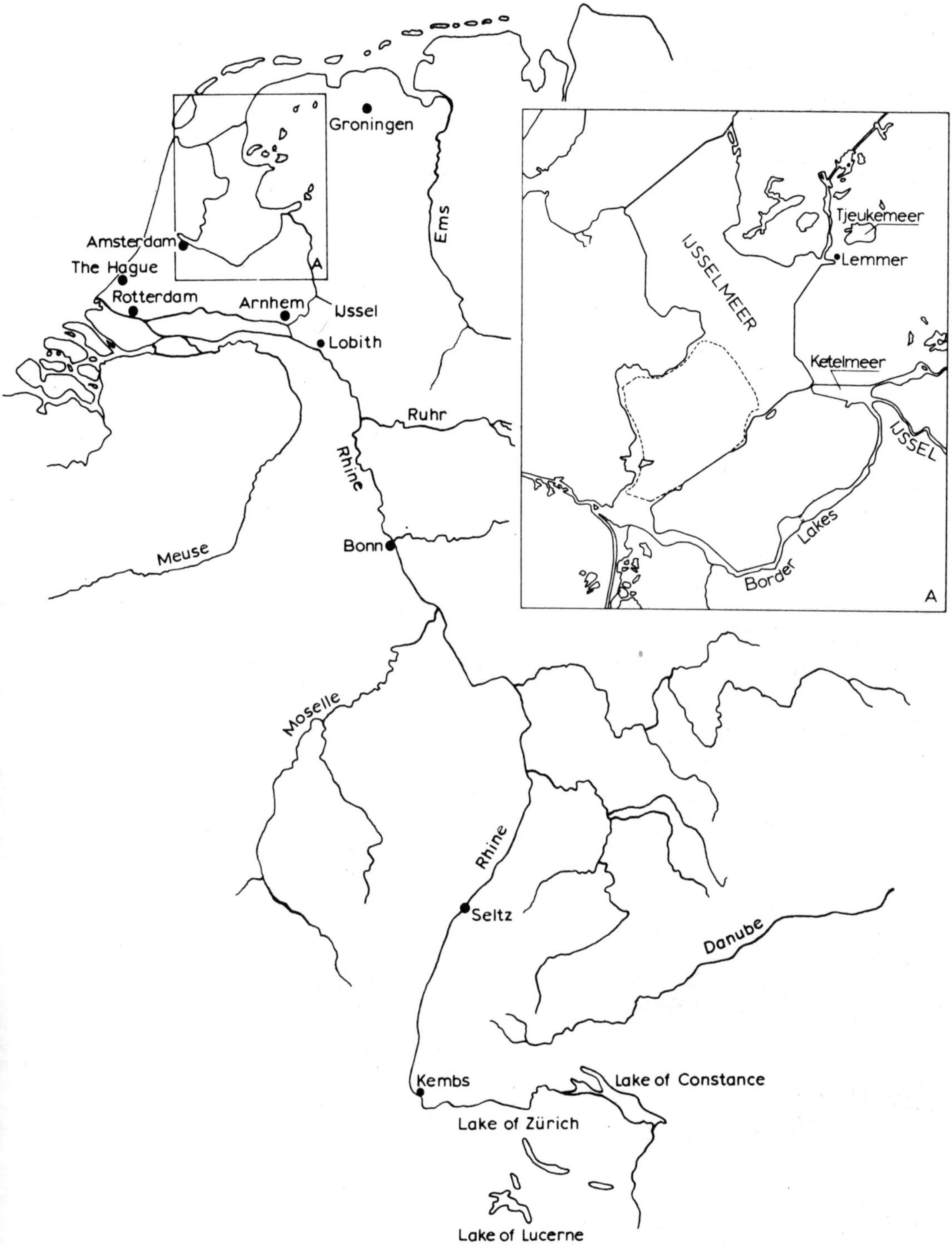


Fig. 1 Course of the Rhine.
Insert: IJsselmeer with border lakes.

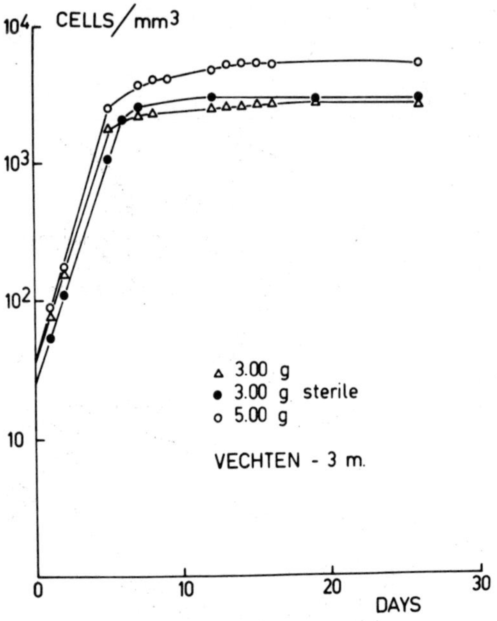


Fig. 2. Growth of Scenedesmus cells on a culture solution with sediments from Lake Vechten. Sediments from 3 m depth. One sample was sterilised with UV radiation. (From Golterman et al, 1969.)

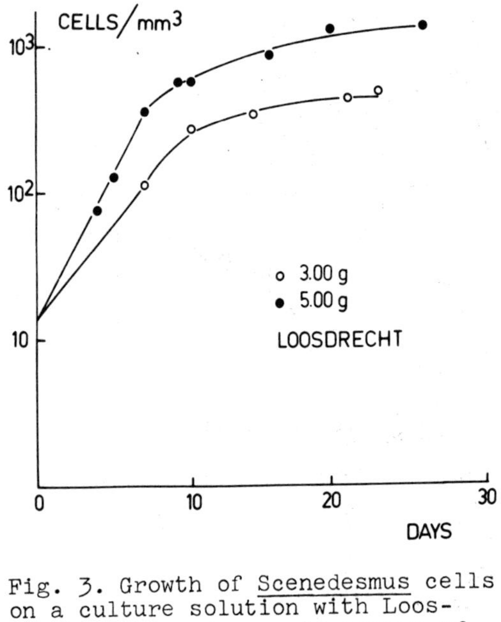


Fig. 3. Growth of Scenedesmus cells on a culture solution with Loosdrecht lake mud as only source of phosphate. (From Golterman et al, 1969.)

work with algal cultures, and shows that not all sedimentary phosphate is available. It describes a chemical fractionation of the different compounds, and the extent to which these sources can be used in a natural situation.

Bio-assays with sediments as phosphate source.

Sediments of several Dutch lakes which differ in chemical or mineral composition (clay, peat or gyttja) were separately added to Rodhe's culture solutions. Scenedesmus was inoculated and there was a good growth in all cases, although there was always some unused phosphate. Fitzgerald (1970) found that there was no algal growth if the sediment was separated from the algae by dialysis tubes. We confirmed this result with the same sediments that caused good growth when not separated from the algae mixed. Even when inorganic phosphate was added to the sediments in the dialysis tubes no growth occurred, although there was a good growth with dialysis tubes containing inorganic phosphate alone. We assume that low concentrations of phosphate in the tube were caused by the adsorption of phosphate onto the sediments which decreased the diffusion rate thus reducing rate of algal growth. This does not mean, that "physical" contact between algal cells and sediment particles is necessary; good growth can also be obtained by mixing the sediments with a tenfold excess of agar - agar solution (40°C) and slicing the block (after cooling and solidifying) in slices of 0.5 cm. The growth rate in some, but not all cases was reduced. This technique was used in these experiments, in which we wanted to estimate the different chemical fractions in the sediments after algal growth, which required separation of algae and sediments.

Although all sediments gave good growth, not all the phosphate present could be utilised by the Scenedesmus. This fact suggests, that certain compounds are, and others are not available for the growth of this algae. Because chemical fractionation is much quicker and easier for estimating the portion of available phosphate present in a given sediment than a bio-assay, we tried to establish a chemical fractionation, that would measure the same amounts, and presumably therefore the same compounds, as can be utilised by the algae.

The first fractionation scheme we

used, that of Chang and Jackson (Jackson, 1958) gave variable results on replicate samples. The mean values were different from those obtained in the bio-assays. The method of Williams et al (1970) was more reproducible but again gave mean values different from those of the bio-assays. In both these methods strongly alkaline and acid solutions are used. We suspected, that milder extractants should be used, and aimed to extract the phosphate salts of calcium and iron, if possible by chelation of these cations, without disturbing other structures, such as clay bound and organically bound phosphate. We found that with a 0.01 M solution of NTA (nitrilotriacetic acid) considerable quantities of iron, calcium and phosphate were released to the solution. Because not all extractable phosphate appeared in the first extraction, we extract normally 0.5-1.0 g of dry sediment with 3 portions of 200 ml, at pH=7.
From both experiments in table I we can see, that the amount of phosphate extracted with 0.01 M NTA was equal within the expected experimental error (2-5%) to the amount of phosphate used by Scenedesmus. In the second experiment of this table the sediments were extracted once more, while the agar-agar blocks were separated from the culture and re-inoculated again. Again in both cases equal amounts of phosphate were extracted, 4-5% of the originally extracted amounts.

Table I. 2 g sediments of Lake Vechten (wet weight = 0.58 g dry weight) was solidified with agar - agar and sliced (0.5 cm). The slices were a) extracted with 200 ml of 0.01 M NTA at pH=7; b) suspended in Fernbachflasks and inoculated with Scenedesmus. The amount of phosphate utilised by Scenedesmus could be estimated, because 1 mg $PO_4$-P is equivalent to $10^{10}$ cells. Results expressed per g dry weight.

| | NTA-extractable ($\mu$g $PO_4$-P) | | Scenedesmus cells ($x\ 10^9$) | |
|---|---|---|---|---|
| 1) | 412 | 307 | | |
| 2) | 203 | 265 | 6.7 | 6.5 |
| 3) | 57 | 83 | is equivalent to | |
| | 672 | 655 | 670 $\mu$g | 650 $\mu$g |

In a second series, a fourth NTA extraction was carried out, while the blocks were recovered from the culture, washed and used again.

| | NTA-extractable ($\mu$g $PO_4$-P) | Scenedesmus cells ($x\ 10^9$) |
|---|---|---|
| 1) | 260 | |
| 2) | 271 | 6.15 |
| 3) | 96 | is equivalent to |
| | 627 | 615 $\mu$g |
| | re-extraction | re-inoculation |
| | 29 | 26 $\mu$g |

Because these sediments might have a complex mixture of several phosphate binding mechanisms, of which adsorption on clay might be one, we studied phosphate behaviour after adsorption on clay, both with chemical and biological extractants. For this purpose we used a clay collected from the Kainji River (Uganda), which has never been polluted. At the place where the sediment was collected the river is swept free from clay in the wet season, so the sediment must have been in position for less than a year. From the experiment given in table II we can see, that a small fraction (ca. 10%) of the phosphate present can be extracted by NTA and a slightly larger proportion could be used by Scenedesmus. By suspending this clay in a phosphate solution the clay can be enriched with freshly adsorbed phosphate. This amount together with the already present extractable amount, could be extracted for ca. 100% while a slightly lower recovery was made with Scenedesmus cells. (See table II.)

An extraction with NTA does not separate between iron bound- and calcium bound phosphate (in this paper further respectively called "Fe-$PO_4$" and "Ca-$PO_4$") a separation, which for biological reasons might be desirable. Therefore we tried other chelators, with larger stability coefficients for calcium and iron. EDTA and DTPA (diethylene-triamino-pentacetate) were used, but did not give as good separation between the calcium- and iron bound phosphates as might be expected from their stability coefficients. Another disadvantage with DTPA with normal lake sediments was, that it extracted more phosphate than could Scenedesmus. Adsorbed phosphate on the Kainji river clay appeared to be 95% extracted by DTPA, 90-100% by NTA and only 75% by EDTA.

A promising method for the separation of iron- and calcium bound phosphate is to use a solution of 0.01 M Ca-NTA (NTA solution with stoicheiometric amounts of $CaCO_3$). This extractant makes only the "Fe-$PO_4$" soluble, "Ca-$PO_4$" is then extracted with

Table II. Clay from the Kainju River (Uganda) was extracted with solutions of NTA, EDTA and DTPA (0.01 M, pH=7). The clay was furthermore enriched with phosphate by adsorption. Both samples were used for chemical extractions and Scenedesmus cultures. From the natural clay 650 µg of $PO_4$ - P (± 25; duplicates) could be extracted by fusion with $NaKCO_3$.

Results in µg $PO_4$-P per g dry weight of clay

| | NTA extractable | Scenedesmus available |
|---|---|---|
| natural clay | 72 | 90 |
| "enriched" clay (+ 425 µg) | 479 | 493 |
| | DTPA extractable | |
| "enriched" clay (+ 834 µg) | 734 = 88% | 800 = 96% |

Mean recoveries of freshly adsorbed phosphate with:
DTPA: 88 - 96%
NTA : 90 - 100%
EDTA: 75%

Na-NTA. No "final" extraction scheme can yet be suggested; it has been found that the pH adjustment is critical for different sediments (usually a pH of 6 or 7 is necessary), while the Ca-NTA solution must have an excess of $Ca^{2+}$, because if calcium is added stoicheiometrically all the calcium is chelated, so that some "$Ca-PO_4$" may go into solution.

Besides these direct measurements of the quantities that can be used by the algae much can be discovered from the algal growth rate (Golterman, Bakels and Möglin, 1969). In batch cultures we found the relation between growth rate and phosphate concentration to follow a saturationcurve of the type:

$$\beta = \beta_{max} \frac{PO_4}{C_1 + PO_4}$$

in which:
$\beta$ = growth rate
$\beta_{max}$ = maximum value for $\beta$ in these experimental conditions
$C_1$ = half saturation constant.

Comparing the growth rate obtained with almost insoluble compounds such as $FePO_4$ and $Ca_5(PO_4)_3(OH)$ we could define an "apparent" phosphate concentration, i.e. the (lower) phosphate concentration (as $KH_2PO_4$) that would give the same growth rate as found with the insoluble compounds. With sediments growth rates were obtained, from which "apparent" phosphate concentrations could be calculated. These were in agreement with the phosphate concentrations in the lake water.

Equilibrium growth rates were not obtained in all cases however. It can be seen from the results in figure 2, that the same growth was obtained when using 3 or 5 g of sediment, strongly suggesting that an equilibrium situation occurred. With peaty mud from Lake Loosdrecht (figure 3) 5 g of mud gave a faster growth rate than 3 g. Originally we thought, that not sufficient sediment was used. Recent work with cultures with different amounts of $KH_2PO_4$ showed, that with very low phosphate concentrations, the growth rate and the saturation constant $C_1$, are strongly pH-dependent. (See fig. 4) Because the different sediment samples used contained quite different amounts of $CaCO_3$, we think, that in our older work variation of pH may account for some of the inconsistencies.

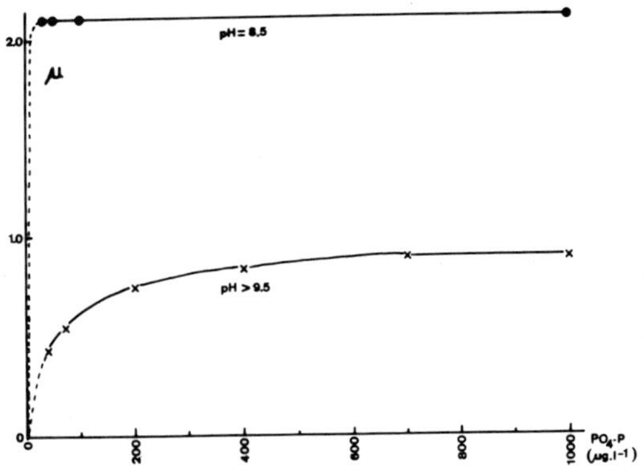


Fig. 4. Relation between growth rate of Scenedesmus ($day^{-1}$) and phosphate concentration at different pH values (pH = 8,5 this study; pH > 9.5 from Golterman et al, 1969)

Helfrich and Kevern (1973) described a transport of labelled freshly adsorbed phosphate from kaolinite particles to Pandorina morum. Because they worked with relative thick cell suspensions, no growth could take place and only 30% of the label was transported. It seems likely, that with growing cultures a larger amount could have been utilised by the algae, as in their experiments the labelled phosphate distributed itself over the different phosphate fractions (clay, dissolved, algal) and such an equilibrium might be shifted if the algae were able to growth. Porcella et al (1970) showed that algae could use phosphate from sediments in a laboratory reactor. They did not find a significant relation between biomass production and the phosphorus concentration in grams "available" phosphorus per gram sediment. This negative result may be caused by the relative large amounts of sediments they used (so that no depletion occurred) and because they measured "availability" by the Jackson method for extraction (see my comments on page ).

Experiments in lake enclosures.

The above experiments show a capacity for phosphate uptake from sediments by algae: they do not show the extent to which this process will occur in nature. It seems likely, that utilisation of sediment - phosphate is less important in lakes with a high phosphate loading than in lakes with a low, or decreasing phosphate loading.

To examine this problem an experimental pond (enclosure) was made in the Veluwe Border Lake. A plastic sheet was fixed around a circular floating device (diameter about 40 m, depth 1.5 m), while the lower part was dug into a circular trench of about 20 cm deep, which was filled in again with the sediments, fixing the sheet in the lake bottom. The whole structure was supported by wooden piles pushed into the lake. Although the lake is only 1.5 m deep, the sheet has a height of about 10 m enabling the outside waterlevel to move 0.5 m upwards or downwards, without flushing water into or from the enclosure. The enclosure can be flushed through with "polder water", relatively poor in phosphate.

Kouwe and Golterman (1976) reported that the enclosure was flushed through six times in 1975 (once with 1700 $m^3$ and five times with 3500 $m^3$) and that on every occasion sufficient phosphate was available to cause algal development again up to chlorophyll a concentrations about the same as in the lake itself, provided that nitrate was added as nitrogen source. The development time was about a week or slightly shorter, except in October when growth was less rapid.
The experimental data are summarised in table III; a detailed account is given by Kouwe and Golterman (1976). During the experiment the Ca-NTA extractable iron bound - phosphate decreased from 190($\pm$10) $\mu$g to 82 $\mu$g per gram dry sediment. This happened in the period 18-8 to 1-9, in which period the largest increase in Part-P was found. After 1-9 the concentration of "Fe-$PO_4$" remained constant and in the period 25-9 to 7-10, when the concentration of Part-P was increasing rapidly again, the amount of "Ca-$PO_4$" decreased from 163 $\mu$g to 114 ($\pm$ 10) $\mu$g per gram dry sediment. The decrease of the "Fe-$PO_4$" concentration took place mainly in the 2 to 4 cm layer, where 74% disappeared. The average for the 0-6 cm layer was 62%. The decrease of the "Ca-$PO_4$" was 37%, with a maximum in the 4 to 6 cm layer (51%).

The sum of "Fe-$PO_4$" and "Ca-$PO_4$" decreased by about 40%; it can be estimated that 1000 to 1600 mg $m^{-2}$ of $PO_4$-P has been removed from the sediments, while the increase in Tot-P in the water was 1200 mg $m^{-2}$ during this period. Although this is a very small percentage of the amount accumulated in the lake (20 years loading of 2 to 4 g $m^{-2}y^{-1}$), it is nevertheless a considerable fraction of the amount considered to be available.

The experiment was therefore repeated in 1976, but unfortunately during the winter the enclosure had to be removed. Although not all experimental data are yet available, it is clear that the sediments had accumulated phosphate again during the winter, so that we started with about the same high concentration as in 1975. Again during 9 flushings the bottom delivered sufficient phosphate to enable repeated massive algal growth and depletion was not reached again.

Besides the large enclosure two smaller ones (diameter 6 m) were built. The first one was not flushed through

Table III. Increase in concentrations of chlorophyll a, $PO_4$-P in the periods after flushing the enclosure in 1975.

| period | chlorophyll $\mu g\ l^{-1}$ | % | $PO_4$-P $\mu g\ l^{-1}$ | % | Part-P $\mu g\ l^{-1}$ | % | days after N supply |
|---|---|---|---|---|---|---|---|
| 18/7-18/8 | 168 → 206 | 23 | 451 | 278 | 3 | 1 | - |
| 22/8- 1/9 | 123 → 193 | 57 | -125 | -89 | 231 | 513 | 5 |
| 3/9-16/9 | 63 → 213 | 238 | - 34 | -77 | 11 | 4 | 7 |
| 18/9-23/9 | 68 → 96 | 41 | 0 | 0 | 20 | 23 | 5 |
| 25/9- 7/10 | 42 → 139 | 231 | 0 | 0 | 95 | 142 | 12 |
| 9/10-22/10 | 54 → 180 | 233 | ca.23 | - | 145 | 100 | 13 |

and remained green all through summer, suggesting that no improvement will occur if the phosphate rich water is simply diverted, leaving the sediment store untouched. The second enclosure was flushed through and was not supplied with nitrate. This pond remained clear for about three months, suggesting that the development of nitrogen fixing blue green algae takes some months. After the three months the pond became suddenly green, with a bloom of (non nitrogen fixing) Oscillatoria. Although this strongly suggests, that the nitrogen fixers could not compete with the blue greens already present, it is also possible that the unusual warm period with very little wind has caused a temporal stratification in the small pond. However it is clear that the role of the bottom in the phosphate metabolism cannot be understood without a better insight into the relationships between nitrogen supply and blue green algae.

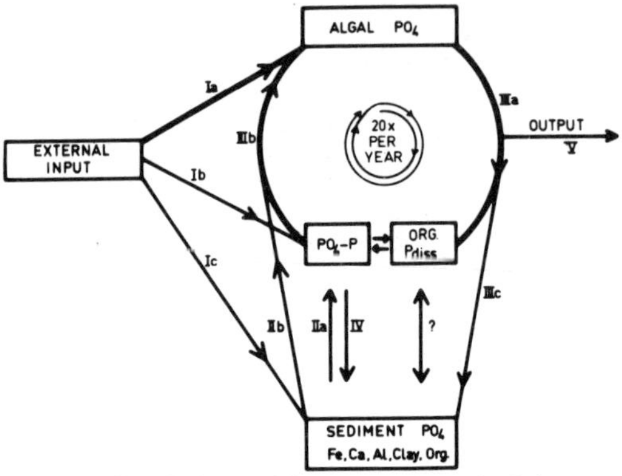


Fig. 5. Schematic representation of the phosphate cycle. (From Golterman, 1975.) The indicated turn-over time may vary between 10 and 40 times a year.

## Discussion

The number of publications stating that sediments act as a sink for phosphate is probably about as large as the number strongly denying this effect. It is clear that the role of sediments can only be understood from evaluating their influence on the whole phosphate metabolism of the lake. This can be represented as in figure 5. From this figure it can be seen, that reaction IIa will be of little or no importance as long as reactions Ia or Ib provide sufficient phosphate to meet algal demands. It is also clear, that the output reaction (V) will be inconsiderable if the phosphate binding reactions (Ic and IV) are strong. Output of phosphate can be important if the algae themselves are washed away. This factor in lake restoration is often overlooked. It is also clear that in establishing phosphate loadings a clear distinction must be made between output by algae, and that as dissolved inorganic phosphate particularly in winter. The recycling of the phosphate between the algal and dissolved fractions may also control the sedimentation, process IIIc It should be remembered however, that not all the organic phosphate entering the sediments in organic form will remain as such. It seems likely that these compounds will be converted into "Fe-$PO_4$" or "Ca-$PO_4$", which compounds may also undergo mutual interconversions. It is not impossible that in laboratory experiments or bio-assays such as reported in the section 2 of this paper an artificial shift occurs between these fractions It is possible - perhaps likely -, that in a lake one of the compounds controls the phosphate metabolism, while in the laboratory experiments, due to changed concentrations of iron and calcium, the other compound controls the phosphate exchange processes.

In this work the availability of sediment phosphate was studied by algal bio-assays, using Scenedesmus as a test organism. All sediments yielded a good algal growth, although not all the phosphate appeared to be available. It seems that the phosphate bound into the clay lattice (the natural clay phosphate) is not available for algae, although it is available for agricultural crops. (Golterman, 1973.) Phosphate freshly adsorbed onto clay is available to algal cultures. It was found that with NTA, an equal amount of phosphate could be extracted as the amount that can be used by algae. Iron bound phosphate is entirely available, while the availability of apatite depends on the crystal size. Whether this potential availability can indeed be used in the real lake situation depends of course strongly on the hydrodynamics of the system.

Experiments in artificial ponds (enclosure type) supported these results; in these ponds it is essential to obtain the same hydrodynamics as in a lake under study.


Acknowledgement

This account is based on the experiments and reports of several students, i.e. Mr. W. van Riemsdijk, Mr. E.J. Middelbeek, Miss B.L.J. van Leeuwen, Mr. F. Kouwe, Mr. W. Bruggeman, Mr. B. Rigter, Mrs. M. van Engel-'t Hof. Practical help and theoretical stimulus was given by Mr. A. Wisselo. The work has been financially supported by the Nederlandse Vereniging van Zeepfabrikanten, since 1976. Discussions with Ir N.T. de Oude are greatly appreciated. The enclosure was build by the "Afdeling Waterloopkunde van de Dienst der Zuiderzeewerken", which department also took care of the flushing operations. The author wishes to thank Dr R.S. Clymo for his critical help with the manuscript.


References


Fitzgerald, G.P., 1970. Aerobic Lake Muds for the Removal of phosphorus from Lake Waters. Limnol. Ocean., 15: 550-555

Golterman, H.L., Bakels, C.C. & Jakobs-Möglin, J., 1969. Availability of mud phosphates for the growth of algae. Verh. Int. Ver. Theor. Angew. Limnol., 17: 467-479

Golterman, H.L., 1973. Natural phosphate sources in relation to phosphate budgets: a contribution to the understanding of eutrophication. Water Res.: 7: 3-17.

Helfrich, L.A. & Kevern, N.R., 1973. Availability of Phosphorus-32, Adsorbed on Clay Particles, to a Green Alga. Michigan Academician: 71-80

Jackson, M.L., 1958. Soil chemical analysis, Prentice Hall, Englewood Cliffs, N.Y. 498 p.

Kolenbrander, G.J., 1974. Een schatting van de fosfaataccumulatie in Nederland in 1970. Rapport 10-74 van Inst. voor Bodemvruchtbaarheid.

Kouwe, F.A. & Golterman, H.L., 1976. Rol van bodemfosfaten in het eutrofiëringsproces. $H_2O$, 9: 84-86

Lund, J.W.G., 1970. Primary production. Water Treat. Exam., 19: 332-358

Porcella, D.B., Kumagai, J.S. & Middlebrooks, E.J., 1970. Biological effects on sediment-water nutrient interchange. J. of the San. Eng. Div.: 911-926



Key words: phosphate
algal growth
enclosures
sediments
lake restoration

SIGNIFICANCE OF OXIC VS ANOXIC CONDITIONS FOR LAKE MENDOTA SEDIMENT PHOSPHORUS RELEASE

G. F. Lee

Center for Environmental Studies, University of Texas-Dallas, Richardson, TX. 75080, USA[1]

W. C. Sonzogni

Great Lakes Basin Commission, Ann Arbor, Michigan, USA

R.D. Spear

U.S. Environmental Protection Agency, Edison, New Jersey, USA

Summary

Laboratory leaching tests conducted on Lake Mendota sediments have shown that under oxic and anoxic conditions, phosphorus is released from the sediments to the water. The initial rate of release under anoxic conditions is much more rapid than under oxic conditons; however, over extended periods of time, oxic release approximately equaled that obtained under anoxic conditions. It is believed that oxic release of phosphorus is primarily related to mineralization reactions and that this mode of release plays a much more important role in lake self-fertilization than generally thought.

Another approach used to estimate the relative significance of oxic vs anoxic release is one in which an overall mass balance of the total phosphorus within the lake is assessed at intervals of every few weeks over the annual cycle. From this, a total mass of phosphorus present in the water body or parts thereof can be assessed. From estimates of the amounts of phosphorus derived from external sources, it is possible to obtain an estimate of the net flux of phosphorus to and from the sediments. This approach has been utilized in a three year study of Lake Mendota located near Madison, Wisconsin. This study has shown that there can be substantial phosphorus released from the sediments under oxic conditions which can play a major role in stimulating late spring-early summer blue green algal blooms. Substantial amounts of phosphorus are released from the sediments during the summer under anoxic conditions. Little of this phosphorus reaches the euphotic zone during the summer recreational period except during periods of thermocline migration caused by high winds. The results of both the laboratory and field studies indicate that ferric iron probably plays a relatively minor role in controlling phosphorus release from Lake Mendota sediments.

Introduction

Increasing emphasis is being placed today on controlling excessive fertilization of lakes and impoundments. One of the primary concerns in any control program is that of the role of lake sediments in recycling nutrients. The basic issue is one of whether the sediments of the lake are sinks for phosphorus or do they act as buffers maintaining a certain phosphorus level in the overlying waters.

In order to gain some insight into the potential significance of lake sediments in controlling overlying water phosphorus concentrations, a series of studies has been conducted on Lake Mendota near Madison, Wisconsin. Two experimental approaches have been used. The first involved taking sediment samples into the laboratory and incubating them for a period of time during which phosphorus measurements were made. The other approach involved looking at overall phosphorus dynamics in the lake where the total mass of phosphorus present in the lake was

1. Also with EnviroQual Consultants & Laboratories, Plano, Texas USA

determined in relation to the annual phosphorus input and removal by the outlet. From the net phosphorus input and the total content of the lake, it is possible to discern the flux of phosphorus to and from the sediments. This paper summarizes the results of a series of studies designed to elucidate the overall role of Lake Mendota sediments in controlling the overlying phosphorus concentrations.

## Literature review

The classical limnological model for phosphorus dynamics in lakes and impoundments is one of involving the iron cycle where, under oxidizing conditions, ferric iron combines with orthophosphate and is transported to the sediments. Under reducing conditions, ferric iron is reduced to ferrous with liberation of phosphate to the overlying waters. In addition, there is phosphate transport from the water column to the sediments through the sedimentation of planktonic organisms, detrital minerals and precipitates. This model is based primarily on the classical work of Einsele (1936, 1937, 1938a, 1938b, 1941) and Mortimer (1942), conducted in the 1930's and 1940's. A number of other studies conducted since that time (using radioisotopes), principally those of Hayes (1963), have shown that there is very rapid exchange of phosphate among various phosphorus pools in lakes.

Lee (1970) presented a review of the various reactions governing the exchange of materials between water and sediments. He pointed to the importance of mixing processes within sediments, the overlying waters and between the sediments and water in controlling the release of phosphorus and other compounds from natural water sediments. This conclusion was based to a considerable extent on the fact that most lake sediment interstitial water phosphorus concentrations are considerably higher than the overlying water indicating that the chemical and biological processes associated with the release of phosphorus to the interstitial water were proceeding at a greater rate than the mixing of the interstitial water with the overlying water.

Lee (1970) also reported that the classical picture of well defined laminae or varves representing annual deposition layers in lake sediments may be a rare occurrence rather than the general situation. It appears that most lake sediments are mixed from a few cm to tens of cm in depth each year as a result of physical, chemical and biological processes. Based on these observations, it was felt that the classical work of Mortimer (1942) on Lake Windermere sediments may not present an accurate representation of the overall phosphorus release from lake sediments because of the experimental approach used which did not provide for an accurate simulation of the mixing processes that occur within the sediments and between the sediments in the overlying waters. It was therefore decided to conduct a series of studies on the release of phosphorus from Lake Mendota sediments under situations where mixing would not be the rate controlling step. A complete description of these studies and a comprehensive review of the literature on the exchange of phosphorus in water and sediments is presented in the Ph. D dissertation of Spear (1970). Only an overall summary of the approach and results are presented here.

## Laboratory studies

### Experimental procedure

The experimental procedure for these studies consisted of taking samples of sediments from several locations in Lake Mendota using an Ekman sediment grab sampler. The sediments were placed in a reaction flask into which could be passed nitrogen gas, carbon dioxide, various mixtures of these two gases or air. The anoxic experiments involving $N_2$ or $N_2$-$CO_2$ mixtures were purified to remove traces of a residual oxygen. The system was continuously stirred with magnetic stirrer and teflon coated stirring bar. Distilled water was used as the leaching solution for the majority of the tests; however, after contact with the sediments, the water quickly dissolved some of the calcium carbonate present in the sediments. Therefore, it is best characterized as calcium-magnesium carbonate leaching solution. The pH of the leaching solution was adjusted by the partial pressure of $CO_2$ in the gas which passed over the surface of the water-sediment slurry.

Two lakes were studied in this investigation--Mendota, a hard water eutrophic lake located in south central Wisconsin and Trout Lake, a moderately soft water, mesotrophic lake located in north central Wisconsin. Part of the Lake Mendota data will be presented in this paper. This lake has a surface area of 38 square kilometers and a maximum depth of 24 meters. Thermal stratification becomes well established by late May-early June and persists through September. The hypolimnion becomes anoxic usually about mid-July.

For the anoxic studies it was found that the filtration of the samples of the slurry had to be conducted under anoxic conditions in order to prevent loss of phosphorus in the ferric hydroxide precipitate which rapidly formed upon exposure to air. This was done by passing purified nitrogen gas over the top of a covered funnel. Soluble orthophosphate was measured on all samples as a function of time using 0.45 micron pore size membrane filters to distinguish between soluble and particulate phosphorus. The phosphomolybdate method with stannous chloride as a reducing agent was used for phosphorus measurements in accord with Standard Methods, 12th Ed.(APHA, et al., 1965). Additional details on the experimental procedures are available in the dissertation by Spear (1970).

Results

Sediments were taken from three different locations within Lake Mendota--from the deep hole area (Station 1), from the slope of the deep hole (Station 2), and from University Bay (Station 3). University Bay is separated from the deep hole area by a sill. The sediments from all three areas were taken from depths which are normally below the thermocline during the summer stratification period. These sediments had 82 to 86 percent water, 31 to 36 percent carbonates, 12 to 15 percent organic matter and 51 to 52 percent clastics (defined as the difference between the organic content of the sediments and the carbonate content).

Figures 1, 2 and 3 show the rates of soluble orthophosphate release from these three locations under oxic and anoxic conditions at a temperature of 24°C. Several things are evident from these figures. First, as expected, anoxic release took place much more rapidly, reaching a plateau within about 200 hours after initiation of the leaching experiment. Oxic release, however, while initially slower, did over the 1200 to 1400 hour incubation period release soluble orthophosphate which appeared to plateau at approximately one half the concentrations reached in the anoxic tests. It is somewhat surprising to see the relatively large amount of oxic release that took place in these tests since Lake Mendota sediments from these locations typically contain from 10,000 to 20,000 ppm iron (Lee, 1962). It is evident that the ferric iron formed upon aeration of these sediments was not controlling phosphorus release. A number of other similar paired experiments were run by Spear (1970) utilizing sediments from these locations. Examination of the effect of temperature on the rate of phosphorus release under oxic and anoxic conditions showed that the rate of release was markedly affected by temperature under both conditions.

Under anoxic conditions, temperature affected not only the rate of release but also the apparent plateau, with approximately half as much phosphorus being released at 11°C as 24°C. Under oxic conditions the effect of temperature was substantially greater with an eight fold reduction in the rate of release during the first 800 hours. It does appear, however, that over extended periods of time the lower temperature runs could reach the same plateau as the 24°C run.

In order to investigate the potential impact of pH and addition of air on the release of phosphorus from Lake Mendota sediments under anoxic conditions, an experiment was conducted in which the sediments were kept anoxic for approximately 480 hours by bubbling purified nitrogen gas through the solution. At this time the gas used to maintain anoxic conditions was changed to a mixture of $N_2$-$CO_2$, which resulted in the pH decreasing from about 8 to approximately 7. Examination of Figure 4 shows that the typical anoxic release pattern was

Figure 1. Phosphorus released from Lake Mendota sediments under oxic and anoxic conditions - Station 1

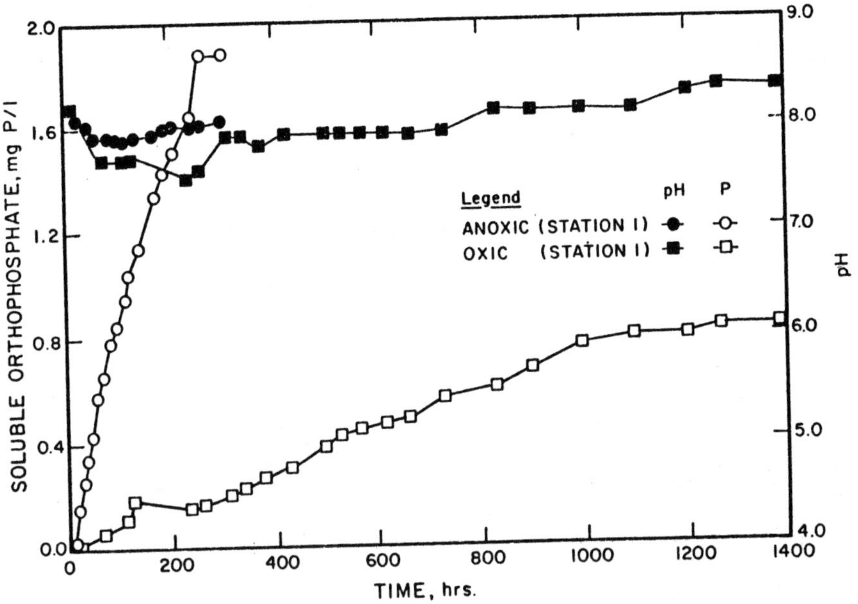


Figure 2. Phosphorus released from Lake Mendota sediments under oxic and anoxic conditions - Station 2

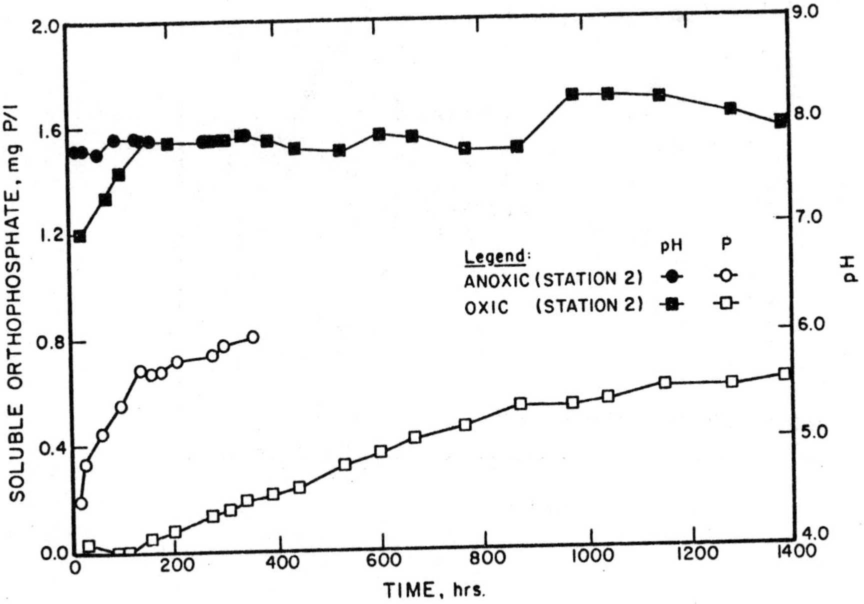

Figure 3. Phosphorus released from Lake Mendota sediments under oxic and anoxic conditions - Station 3

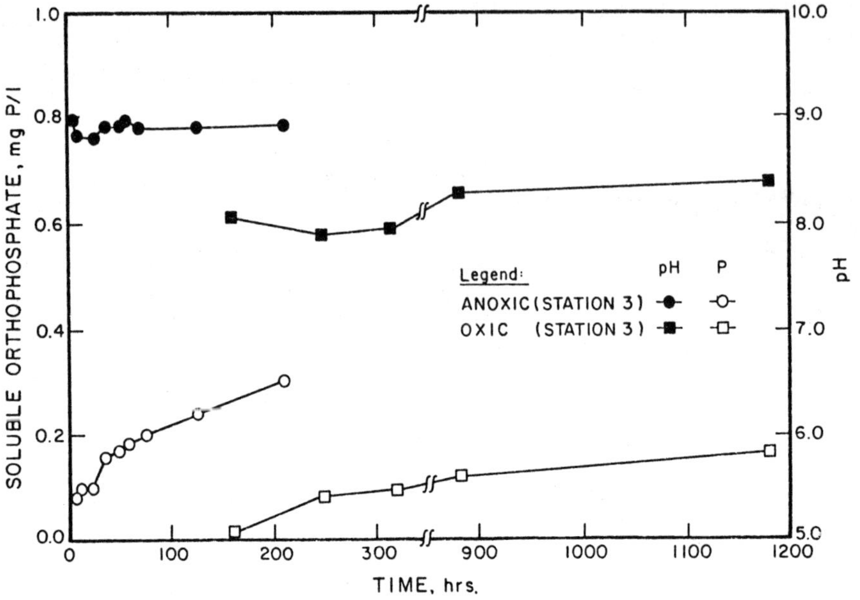


Figure 4. Effect of selected parameters on phosphorus release from Lake Mendota sediments, anoxic-oxic conditions - Station 1

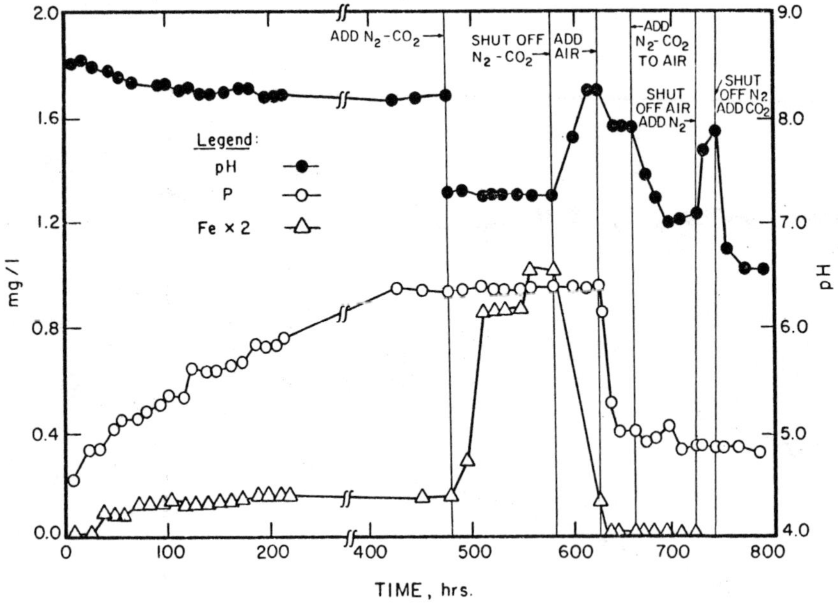

obtained during the first 480 hours. Changing the pH from 8 to 7 had no effect on the orthophosphate content of the solution. It did, however, markedly affect the amount of iron present in solution. At about 620 hours air was introduced into the test system. This resulted in a marked decrease in the orthophosphate and iron. This system was again made anoxic and the pH again adjusted to 7. This time, as in the past, there was no effect on the release of phosphorus.

This experiment demonstrates the classical limnological model for iron and phosphate behavior where under anoxic conditions phosphorus is released with the amount of release independent of pH in the neutral to slightly alkaline region. The introduction of air into the system caused the precipitation of ferric iron with the concomitant removal of the soluble orthophosphate. Based on these experiments, it is clear that the classical limnological model for iron control of phosphate holds for short periods of time. However, with extended periods of time, i.e., hundreds of hours, phosphorus is released under oxic conditons. This release appears to be related primarily to mineralization reactions of organic phosphorus. Measurements of the calcium content of the oxic test showed that it increased throughout the incubation period. This is believed to be due to a solubilization of calcium carbonate present in the sediments. Mineralization reactions result in the production of $CO_2$ which would interact with the solid calcium carbonate present in the sediments to form calcium bicarbonate. As noted in Figures 1 through 3, the pH during the course of the various tests did not change significantly. The sediments contained approximately 30 percent calcium carbonate and therefore represent a well buffered pH system.

Discussion

The implications of these results to the lake are significant. They indicate that potentially large amounts of phosphorus release could be occurring under oxic conditions. This is somewhat contrary to what is generally believed. If one examines the basis of the general position that anoxic release is of major significance, the conclusion is drawn that the reason for this position is primarily based on the fact that the thermocline acts as a trap for anoxic release. On the other hand, under oxic conditions, the lake is normally well mixed with the result that any phosphorus released from the sediments due to mineralization processes is circulated throughout the water column and therefore not readily detected because of the much greater dilution and uptake by algae.

Interestingly, the rates of phosphorus release under oxic conditions, while slower than anoxic conditions, are still sufficiently rapid to be of significance in internal cycling of nutrients. Also the rates of release under both oxic and anoxic conditions from these sediments are much more rapid than would be expected under quiescent conditions such as those frequently used by various investigators (a sample of sediment placed in a carboy and the phosphorus released to the overlying waters measured as a function of time). The lake probably behaves somewhere between the quiescent conditions and the completely slurried conditions used in these studies. During periods of storms shallow water sediments often are readily suspended and thereby provide the opportunity for oxic release of phosphorus associated with the particulate material as well as from the interstitial waters.

Since it is impossible to simulate the hydrodynamic conditions that exist in lakes in the laboratory with a known degree of reliability, it was felt that further laboratory studies on the potential significance of the overlying waters would be of limited value. Therefore, the whole lake mass balance approach was used, the results of which follow.

Field studies-mass balance

It was felt that the best approach to take for assessing the actual role of lake sediments in controlling phosphorus in the overlying waters is the mass balance approach where the total phosphorus present in the water column can be examined as a function of time. By correcting the total water column phosphorus for the net phosphorus input, it should be possible to determine the overall net fluxes of phosphorus to and from the sediments. This was the approach used by Sonzogni (1974). At periodic intervals over a three year period measurements were made of the total phosphorus content of Lake Mendota waters. Also, soluble orthophosphate was measured as well as various nitrogen species. This paper will present a summary of the phosphorus data. The complete data are available in Sonzogni (1974).

Coincident with measurement of the total phosphorus content of the lake, estimates were made of the phosphorus input to the lake. Table 1 presents a summary of these estimates. According to these estimates Lake Mendota receives approximately 60,000 kg/yr of total phosphorus of which slightly over half is in the dissolved reactive form. The principal source of phosphorus for this lake is rural runoff, with a substantial part of this runoff derived from dairy manure spread on frozen land. The amount of phosphorus contributed to the lake each year via this source is highly variable, dependent primarily on the climatological characteristics of the spring snow melt period. If the spring rains and snow melt occur at a time when the subsoil is frozen, then most of this phosphorus is transported over the surface of the soil to nearby tributaries which eventually reach Lake Mendota. However, under conditions where the subsoils are not frozen at the time of snow melt and spring rains, a substantial part of the phosphorus will go into the soil and become fixed on soil particles. In order to account for the yearly variability of phosphorus loading, a range of possible values is given in Table 1 for total phosphorus loading.

Table 1 is based on prediversion wastewater figures. In 1971 the wastewaters from several small towns located in the Lake Mendota watershed were diverted around the lake. Following diversion, wastewater discharges of total P decreased from 24 percent of the total (Table 1) to approximately 2 percent. The rural runoff post-diversion figures increased to 63 percent of the total P input to the lake.

Sonzogni (1974) has estimated that approximately 15,000 kg of total P leaves Lake Mendota each year via its outlet. Of this, approximately 11,000 kg is in a dissolved reactive (DRP) form. Therefore, the net phosphorus input to Lake Mendota each year is on the order of 46,000 kg/yr total P (TP).

Procedure

Water samples were taken in the region of the deep hole at 1 m intervals during the spring, summer and fall and at 2 m intervals during the winter. Measurements were made of Secchi depth, dissolved oxygen and temperature. The lake was sampled approximately weekly, except during the winter, when sampling was conducted at bi-weekly intervals. The sampling program was initiated in the summer of 1970 and continued until July, 1973. Dissolved reactive phosphorus was measured by filtration through 0.45 μ pore size membrane filters followed by the Murphy and Riley (1962) procedure. Total phosphorus was initially digested with persulfate in accord with the procedures presented in Standard Methods, 13th Ed. (APHA *et al*., 1971).

When a thermocline was present, the depth of the thermocline was determined by examining the temperature and chemical profile. In subsequent computations of hypolimnion volume, the metalimnion was considered to be part of the hypolimnion. During periods of ice cover the lake was arbitrarily divided into two layers consisting of the top 17 m and bottom 5 m.

Figures 5 through 8 present the results of the total and dissolved reactive phosphorus content of Lake Mendota from August, 1970 through November, 1973. In addition, where appropriate, the two forms of phosphorus present in the hypolimnion and epilimnion are also presented. Table 2 presents the average content per month for the study period for each of the two forms of phosphorus

Figure 5. Dissolved reactive phosphorus and total phosphorus content of Lake Mendota , 1970

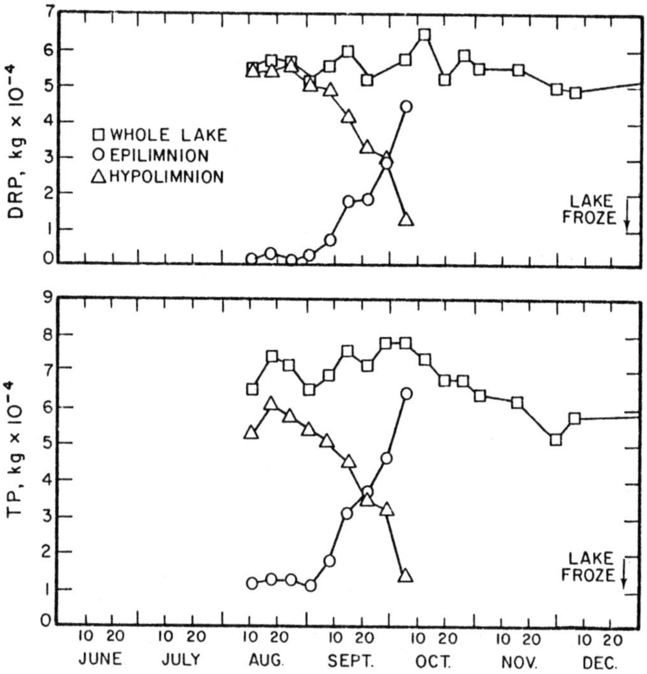


Figure 6. Dissolved reactive phosphorus and total phosphorus content of Lake Mendota, 1971

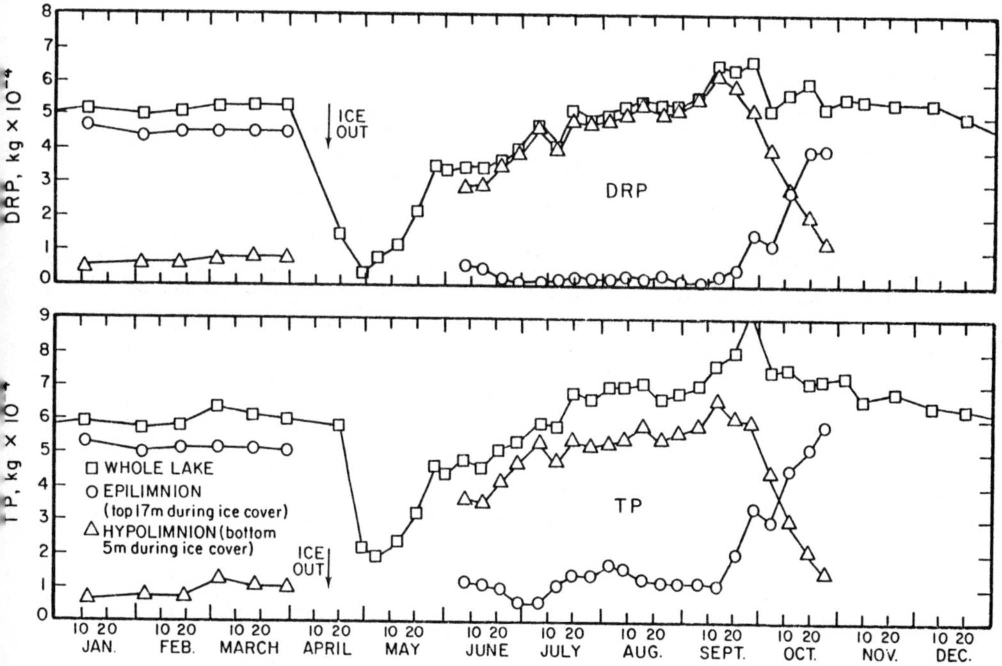

Figure 7. Dissolved reactive phosphorus and total phosphorus content of Lake Mendota, 1972

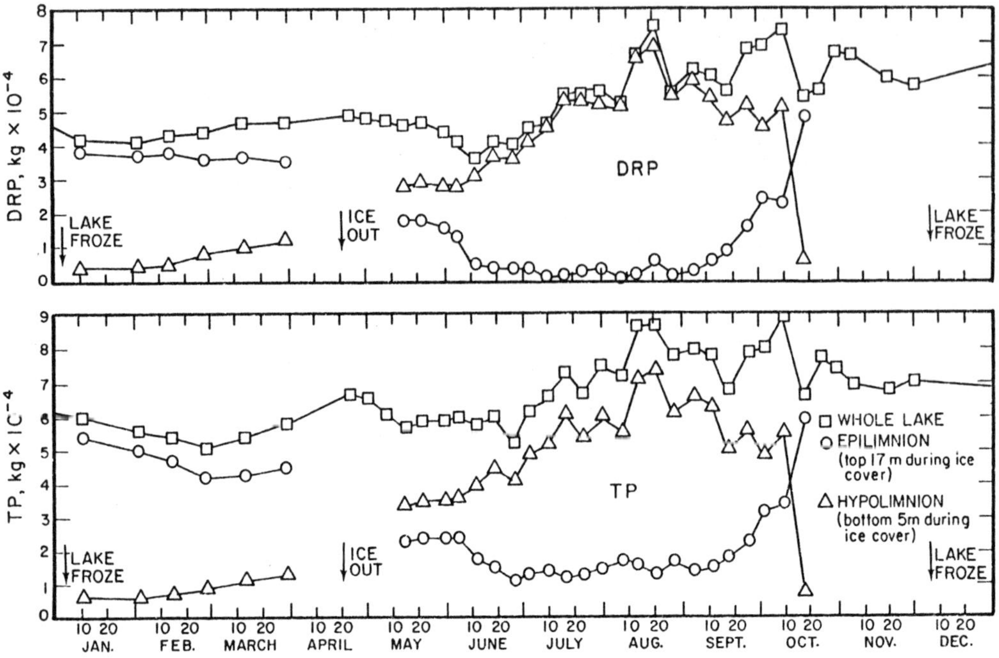


Figure 8. Dissolved reactive phosphorus and total phosphorus content of Lake Mendota, 1973

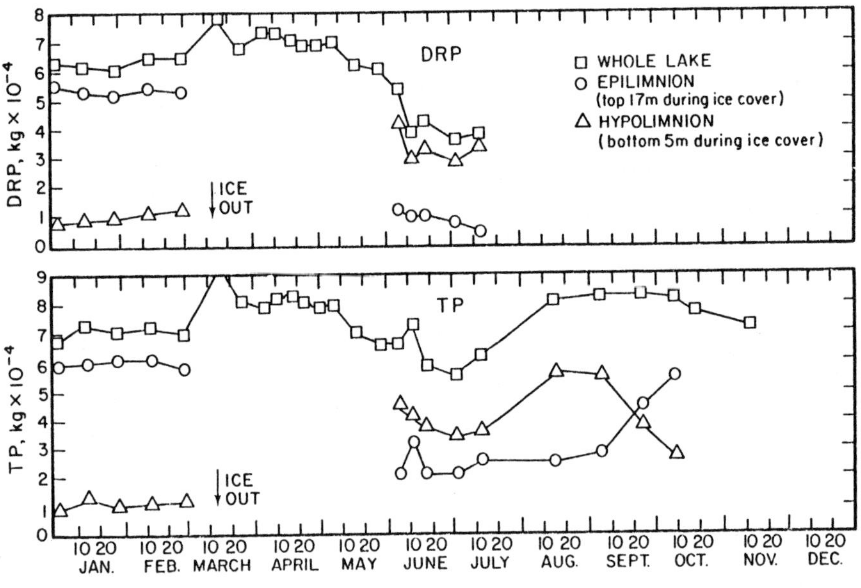

Table 1. Summary of pre-diversion phosphorus sources for Lake Mendota

| Source | DRP (kg/yr) | TP (kg/yr) | %Estimated Contribution TP |
|---|---|---|---|
| Wastewater discharge | 9,500 | 15,000 | 24 |
| Urban | 4,000 | 7,000 | 11.5 |
| Rural | 15,000 | 30,000 (4,500-46,500) | 48.5 |
| PPt | 1,000 | 1,000 | 1.5 |
| Dry fallout | 500 | 3,000 | 5 |
| Groundwater seepage | 300 | 300 | 0.5 |
| Base flow | 5,500 | 5,500 | 9 |
| Woodland runoff | 0 | 0 | 0 |
| Marsh drainage | 0 | 0 | 0 |
| Total | 35,800 | 61,800 | |

After Sonzogni (1974).

Table 2. Mean monthly, winter and annual phosphorus contents for Lake Mendota

| Mean Content | 1970-1971 kg x $10^{-4}$ | | 1971-1972 kg x $10^{-4}$ | | 1972-1973 kg x $10^{-4}$ | |
|---|---|---|---|---|---|---|
| | DRP | TP | DRP | TP | DRP | TP |
| Aug | 5.7 | 7.0 | 5.3 | 6.9 | 6.2 | 8.1 |
| Sept | 5.5 | 7.2 | 6.3 | 8.0 | 6.2 | 7.6 |
| Oct | 5.8 | 7.2 | 5.6 | 7.3 | 6.4 | 7.7 |
| Nov | 5.3 | 5.9 | 5.5 | 6.9 | 6.1 | 6.9 |
| Dec | 4.9 | 5.8 | 5.2 | 6.4 | 6.2 | 6.9 |
| Jan | 5.2 | 5.9 | 4.2 | 6.0 | 6.2 | 7.1 |
| Feb | 5.0 | 5.8 | 4.3 | 5.4 | 6.5 | 7.1 |
| March | 5.3 | 6.1 | 4.7 | 5.6 | 7.2 | 8.4 |
| April | 1.7 | 4.6 | 4.9 | 6.4 | 7.1 | 8.1 |
| May | 1.9 | 3.0 | 4.6 | 6.0 | 6.4 | 7.2 |
| June | 3.6 | 4.8 | 4.2 | 5.8 | 4.5 | 6.6 |
| July | 4.8 | 6.3 | 5.1 | 6.9 | 3.9 | 6.2 |
| Winter[a] | 5.1 | 5.8 | 4.6 | 5.9 | 6.3 | 7.0 |
| Annual[b] | 4.6 | 5.8 | 5.0 | 6.5 | 6.1 | 7.3 |

[a]Average for Dec, Jan and Feb

[b]Average for Aug through July

After Sonzogni (1974)

measured. Examination of these figures and this table shows that the total phosphorus in August 1970 and 1971 was approximately 7 x $10^4$ kg. Examination of the three years of record shows that approximately 80 percent of the phosphorus present in the lake at any time is in the dissolved reactive form.

The greatest year-to-year variability of the phosphorus at any time of the year occurred shortly after ice-out. In the spring of 1971 a massive diatom bloom occurred in Lake Mendota which caused the total phosphorus content of the lake to decrease from slightly over 5 x $10^4$ kg to about 1.7 x $10^4$ kg in association with the settling of this bloom. Following the bloom, Figure 6 shows a more or less linear increase in phosphorus through September. In the period mid-April through early July the bottom waters of the lake were oxic. This year was a particularly dry year in which there was below normal precipitation in the spring and early summer. Lake Mendota had the best water quality during that period that it has had in many years. This appears to be the result of the diatom bloom tying up the phosphorus in the cells and transporting it to the sediments. It is evident from Figure 6 that there was significant oxic phosphorus release arising from mineralization reactions of the phosphorus present in the sediments in the form of algal cells.

Examination of the spring 1972 and 1973 shows that the 1971 pattern was not repeated. In 1972 ice-out was accompanied by an increase in the phosphorus content, probably due to transport from the lake's watershed. In 1973 the total phosphorus content of the lake was the greatest of any of the four spring seasons monitored. This spring was one of the wettest springs in many years and the high phosphorus represents transport from the watershed.

During periods of summer stratification, usually beginning in late August, there is an inverse relationship between the total mass of phosphorus in the hypolimnion compared to the epilimnion. This is the result of thermocline migration, where during this time rapid downward movement of the thermocline results in the transport of phosphorus from the hypolimnion to the epilimnion. This downward movement was accompanied by a decreased hypolimnetic volume accounting for the decrease in the total mass of total and soluble orthophosphate present in the hypolimnion during this time.

The year-to-year variability in the phosphorus content of the lake as shown in Table 2 is strongly influenced by algal population dynamics and input from the watershed. The primary factor controlling the phosphorus input to Lake Mendota is the precipitation and snow melt with the associated runoff to tributaries of this lake. It is important to emphasize that during this period Lake Mendota received in excess of 98 percent of the total phosphorus input from diffuse sources such as urban stormwater drainage, rural runoff, atmospheric precipitation and dustfall.

Interestingly, the annual average total phosphorus content of Lake Mendota, 4 to 6 x $10^4$ kg, is approximately equal to the normal annual input. This results in phosphorus residence time for Lake Mendota of about one year (Sonzogni et al., 1976).

During the summer stratification period the primary controlling factor for algal blooms has been found by Stauffer and Lee (1973) to be the passage of cold fronts, in which high intensity winds tend to sharpen and lower the thermocline. Each time the weather frontal system passed, there was generally a period of high intensity wind which sharpened the thermocline and moved it downward, 0.5 to 1 m. This migration downward results in the transfer of sufficient amounts of available phosphorus from the hypolimnion to the epilimnion to cause a phytoplankton bloom about a week after passage of the front.

Some of the variability noted in Figures 5 through 8 during the summer reflects these algal blooms. Following the bloom, Secchi depth increases and total phosphorus content decreases due to the uptake of phosphorus by the algae and their subsequent settling to the bottom of the lake.

Table 3 presents an estimate of the net increase of the total phosphorus content of Lake Mendota during summer stratification for 1971, 72 and 73. In 1971 the average daily total phosphorus increase was about 187 kg/day; for 1973 it was approximately half this amount (92 kg/day). These differences probably reflect the fact that 1971 was a somewhat peculiar year in terms of typical phosphorus input from the watershed and the massive diatom bloom that occurred at ice-out. The bloom probably removed the phosphorus entering the lake in the spring providing the opportunity for an overall greater increase in phosphorus during summer stratification than was possible in the other years. Also, the overturn, which usually occurs in October, did not result in a significant change in the total phosphorus content of the lake. This is significant since classical models of phosphorus dynamics indicate that phosphorus tends to be removed

Table 3. Net increase in the TP content of Lake Mendota during summer stratification

| Year | Duration of stratification (days) | TP increase (kg) | Average daily TP increase (kg/day) |
|---|---|---|---|
| 1971 | 155 | 29,000 | 187 |
| 1972 | 146 | 16,000 | 110 |
| 1973 | 120 | 11,000 | 92 |

After Sonzogni (1974).

during fall overturn in association with the oxidation and precipitation of iron.

## Discussion

The mass balance approach for Lake Mendota and the laboratory results of Spear (1970) both indicate that under certain conditions, mineralization processes occurring under oxic conditons can result in significant release of phosphorus from lake sediments. As discussed earlier, in both the laboratory and in the lake, the ferric iron present in the surface sediments does not appear to be controlling the release of soluble orthophosphate. It further appears that at fall overturn the mixing of the ferrous iron-rich hypolimnetic waters in Lake Mendota does not have a major influence on the total phosphorus present in the water column.

The phosphate scavenging power of hydrous ferric oxide appears to be dependent on a number of factors. It has been found (Lee, 1975) that the formation of the ferric hydroxide floc in the presence of the phosphate was much more effective than a free form floc. This phenomenon likely accounts for the fact that the ferric hydroxide floc is formed relatively rapidly upon aeration of the sediments and therefore the phosphorus that is released under periods of days to weeks later comes in contact only with an aged floc. The same kind of phenomenon would be occurring in the surface sediments where the overlying waters are oxic.

## Conclusions

It must be concluded from these studies that at least for Lake Mendota sediments and quite possibly for many other lake sediments, the classical limnological model of ferric iron controlling phosphorus release from sediments and phosphorus content of the overlying waters is inappropriate to explain major phosphorus fluxes within the water body. It is further evident that while the sediments of lakes represent potentially significant sources of phosphorus over limited periods of the year, on an annual cycle the sediments of most lakes are sinks for phosphorus. Therefore, significant reductions of the input phosphorus from external sources will likely result in significant improvement in water quality arising from a decrease in the frequency and severity of noxious algal blooms in those water bodies where phosphorus is the key limiting nutrient or can be made so by input reduction.

## Acknowledgment

The support for the studies which serve as a basis for this paper was derived from several grants from the US Environmental Protection Agency. In addition, support was given these studies by the University of Wisconsin Department of Civil and Environmental Engineering where the studies were carried out. Further support was given by the Center for Environmental Studies at the University of Texas at Dallas and EnviroQual Consultants & Laboratories of Plano, Texas. We wish to acknowledge the assistance of Gerald Jayne and M. Torrey for their help in connection with the studies which serve as a basis for this paper.

## References

American Public Health Association, American Water Works Association, Water Pollution Control Federation, 1965. Standard methods for the examination of water and wastewater. New York.

American Public Health Association, American Water Works Association, Water Pollution Control Federation, 1971. Standard methods for the examination of water and wastewater, 13th ed. Washington.

Einsele, W., 1936. Uber die Beziehungen des Eisenkreislaufs zum Phosphatkreislauf im Eutrophen See. Arch. Hydrobiol. 29:664-686.

Einsele, W., 1937. Physikalisch-chemische Betrachtung Einiger Problems des Limnischen Mangan- und Eisenkreislaufs. Verh. Int. Ver. Limnol. 5:69-84.

Einsele, W., 1938a. Uber Chemische und Kolloidchemische Vorgange in Eisen-Phosphat-Systemen unter Limnochemischen und Limnogeologischen Geschichtspunkten. Arch. Hydrobiol. 33:361-387.

Einsele, W. & Vetter, H., 1938b. Untersuchungen uber die Entwicklung der Physikalischen und Chemischen Verhaltnisse im Jahreszyklus in Ainem Massig Eutrophen See (Scheinsee bei Langenargen). Int. Rev. Hydrobio. 36: 285-324.

Einsele, W., 1941. Die Umsetzung von Zugefuhrtem Anorganischer Phosphat im Eutropen See und Ihre Ruckwirkungen auf Seinen Gesamthaushalt. Z. Fisch. 39:407-488.

Hayes, F.R., 1963. The role of bacteria in the mineralization of phosphorus in lakes. In: C.H. Oppenheimer (Ed.): Symposium on marine microbiology. C.C. Thomas, Springfield, pp. 654-663.

Lee, G.F., 1962. Studies on the iron, manganese, sulfate and silica balances and distributions for Lake Mendota, Madison, Wisconsin. Trans. Wisc. Academy of Sciences, Arts and Letters. LI: 141-155.

Lee, G.F., 1970. Factors affecting the transfer of materials between water and sediments. Univ. of Wisconsin, Eutrophication Information Program, Literature Review No. 1. Madison, Wisconsin.

Lee, G.F., 1975. Role of hydrous metal oxides in the transport of heavy metals in the environment. Proc. of Symposium of Transport of Heavy Metals in the Environment. In: Progress in Water Technology. 17:137-147.

Mortimer, C.H.,1941-42. The exchange of dissolved substances between mud and water in lakes. J. Ecol. 29:280-329; 30:147-201.

Murphy, J. & Riley, J.P., 1962. A modified single solution method for the determination of phosphate in natural waters. Anal. Chim. Acta. 27:31-36.

Sonzogni, W.C., 1974. Effect of nutrient input reduction on the eutrophication of the Madison lakes. PhD dissertation, Univ. of Wisconsin-Madison.

Sonzogni, W.C., Uttormark, C. & Lee, G.F., 1976. A phosphorus residence time model: theory and applications. Water Research. 10:429-435.

Spear, R.D., 1970. The release of phosphorus from lake sediments. PhD dissertation, Univ. of Wisconsin-Madison.

Stauffer, R.E. & Lee, G.F., 1973. The role of thermocline migration in regulating algal blooms. In: E.J. Middlebrooks, D.H. Falkenborg & T.E. Maloney (Eds.): Modeling the eutrophication process. Utah Water Research Lab., Utah State University, Logan.

Key words: Lake Mendota
Eutrophication
phosphate
sediments
self fertilization

# PHOSPHORUS FLUX ACROSS THE SEDIMENT-WATER INTERFACE

P. A. Neame

Institute of Ecology, University of California, Davis, California, USA[1]

Summary

A comparison is made between the magnitudes of the possible fluxes of phosphorus across the sediment-water interface in Castle Lake, California. Dissolved phosphorus cycling across the interface, as illustrated by $^{32}P$ tracer experiments, formed the major part of the total phosphorus flux. This flux amounted to approximately 650 $\mu$g P $m^{-2}$ $day^{-1}$, and appeared to be largely mediated by the stirring activities of macrofauna in the oxidized zone. Input of phosphorus to the sediment in settling detritus averaged 380 $\mu$g P $m^{-2}$ $day^{-1}$ over the July-October measurement period, with maximum input occurring in September. Losses of phosphorus in the forms of insect emergence and fish feeding averaged 33 and 23 $\mu$g P $m^{-2}$ $day^{-1}$, respectively.

## Introduction

The importance of phosphorus as a limiting nutrient in many aquatic systems has long been recognized. The need for management of such systems necessitates greater understanding of the details of the cycling of this nutrient. The participation of sediments in nutrient cycles has received limited attention, and their importance in any particular system has been a matter of debate. Some authors (e. g. Schindler et al., 1973) have reported minimal phosphorus fluxes out of lake sediments, while others (Stumm and Leckie, 1971) have stressed the participation of sediments as a buffer for phosphate concentrations in the overlying water.

Interchange of phosphorus among the many pools of phosphorus in the sediments and overlying water is controlled by a variety of physical, chemical and biological factors. This discussion stresses the importance of water movements in the transport of nutrients across the sediment-water interface.

The research reported here was aimed at measuring phosphorus movement across this interface. The study was carried out in Castle Lake, California, a mesotrophic glacial cirque lake located at an elevation of 1706 m in the Klamath Mountains of northern California. Previous studies of nutrient dynamics in this system have focused primarily on the water column, with little attention being given to any possible recycling from the sediment (Richey, 1974).

Phosphorus can move across the interface in a variety of ways; in this case the fluxes of interest are the detritus input to the sediment, dissolved phosphorus cycling through the surficial sediment layers, and phosphorus movement out of the sediment by means of fish feeding and emergence of aquatic insects. The estimates of each of these fluxes will be discussed in turn, and then evaluated in terms of the total flux into and out of the oxidized sediment layer.

## Methods

The sediments for this study were collected using a stratified plexiglas box sampler (Milbrink, 1968) or plexiglas core tubes collected by SCUBA. Dry weight of sediment was determined after drying at 60°C to constant weight. Organic matter was determined by loss on ignition at 550°C for 1 hr. Soluble reactive phosphate was measured using a molybdenum blue technique (Strickland & Parsons, 1972). Particulate phosphorus in macrofauna, sediment and sedimenting detritus was determined by digestion in 70% perchloric acid (Chapman & Pratt, 1961). Sediment interstitial water was sampled using a Reeburgh (1967) filter press.

Emerging insects were collected in two traps constructed according to a design recommended by Mundie (1971), and suspended 1 m above the sediment. The samples were collected frequently, sorted into taxonomic groups, dried and weighed. Sedimenting detritus was collected in cylindrical containers (collecting area 15.5 $cm^2$) suspended just above the sediment.

Dissolved phosphorus release from the sediment was estimated by tracer kinetics analysis. A set of three undisturbed sediment cores was collected and oxygen profiles of each were determined using a micro-Winkler technique (Fox & Wingfield, 1938). Inorganic $^{32}P$ was then injected into the sediment at depths below the interface, so that the tracer was in the oxidized zone in cores number 1 and 2, and in the anoxic region in core 3. In each case the tracer was spread in a uniform layer across the core at the injection depth. At intervals,

1. Present address: Marine Ecology Lab, Bedford Institute of Oceanography, Dartmouth, Nova Scotia, Canada

Figure 1. Compartment model for phosphorus fractions in sediment and overlying water.

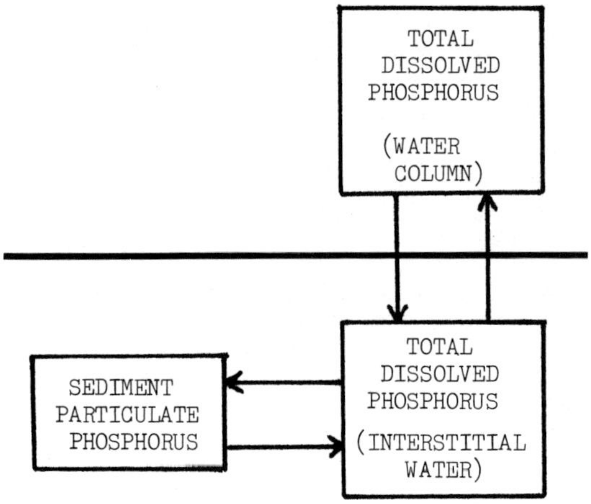


samples were taken from the overlying water, filtered through acid-washed, 0.45 μ membrane filters, and their tracer content determined by scintillation counting. In order to estimate rates of phosphorus movement through the sediment surface, a simple 3-compartment system was postulated for modeling purposes (Figure 1). Calculation of the transfer coefficients between these three compartments was carried out using the finite difference technique described by Watt & Hayes (1963). The transfer coefficients were calculated on the basis of the initial observed data points for the appearance of the tracer in the overlying water. Uptake of tracer by the particulate phase in the sediments was determined in a separate experiment, with the decrease in tracer in the interstitial water being measured by scintillation counting. All $^{32}P$ counts were corrected for decay and counting efficiency.

## Results and discussion

The major part of the sampling for this study was carried out in the middle of the northern shallow area of Castle Lake. This is a large flat area of visually homogenous sediment, lying at a depth of about 4 m, which makes it an optimum location for sampling by SCUBA. The sediment surface is cohesive but easily penetrated by the sampling equipment. The top 2 to 3 cm of sediment appears green-brown in colour, with aggregations of detritus of up to 5 mm in diameter, and many chironomid tubes. Deeper sediment is a fine silt-like grey material with little or no structure. The surface 2 to 3 cm layer is well oxygenated, and the bulk of the macrofauna population occupies this zone. Data on changes in oxygen uptake rates in poisoned cores and in cores to which additional macrofauna had been added indicated that the macrofauna actively pump water through this surface oxidized layer (Neame, 1975).

The dry weight of the sediment ranges from 16 mg $cc^{-1}$ in the top 3 cm to 40 mg $cc^{-1}$ below 10 cm. Organic matter was about 60% at the surface, decreasing to 30% below 8 cm. Particulate phosphorus appeared to be constant with depth, averaging 0.88 μg P $mg^{-1}$ dry sediment (Table 1). Total soluble phosphorus concentration in the interstitial

Table 1. Interstitial water total soluble phosphorus and total particulate phosphorus profiles in Castle Lake sediment.

| Sediment Layer (cm) | Interstitial Water Total Soluble P | | Total Particulate Phosphorus | |
|---|---|---|---|---|
| | μg $l^{-1}$ | μg $cc^{-1}$ x $10^{-2}$ | μg $mg^{-1}$ | μg $cc^{-1}$ |
| 0-1 | 14.1 | 1.23 | 0.86 | 13.2 |
| 1-2 | 19.1 | 1.28 | 0.84 | 14.2 |
| 2-3 | 21.0 | 1.11 | 0.92 | 15.0 |
| 3-4 | 24.9 | 1.27 | 0.86 | 17.2 |
| 4-5 | 15.9 | 0.76 | 0.84 | 19.0 |
| 5-6 | 22.8 | 1.12 | 0.88 | 22.7 |
| 6-7 | 19.9 | 0.98 | 0.93 | 27.0 |
| 7-8 | 53.0 | 2.54 | 0.93 | 29.8 |

Figure 2. Total phosphorus content of sedimenting detritus.

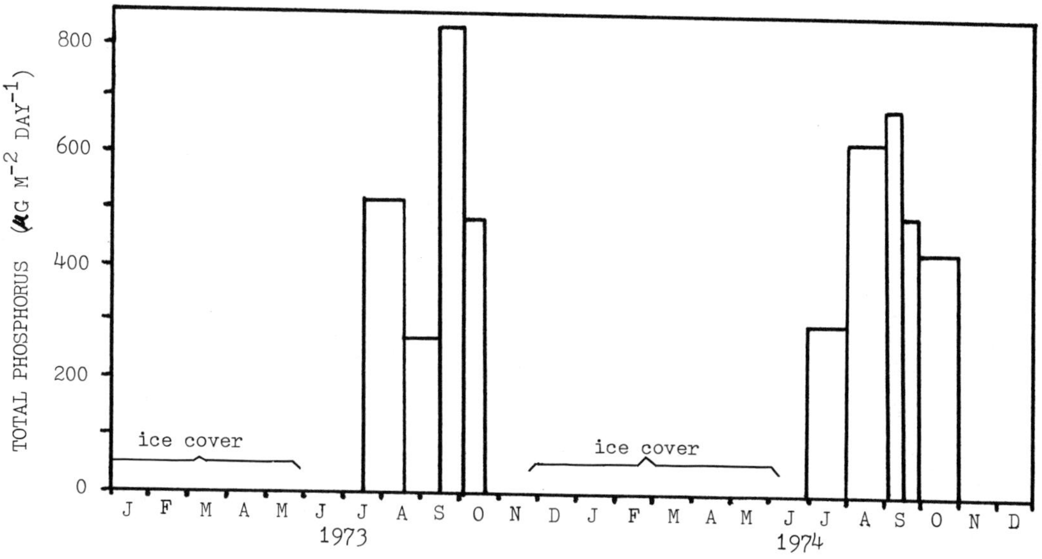


Figure 3. Total phosphorus content of emerging aquatic insects.

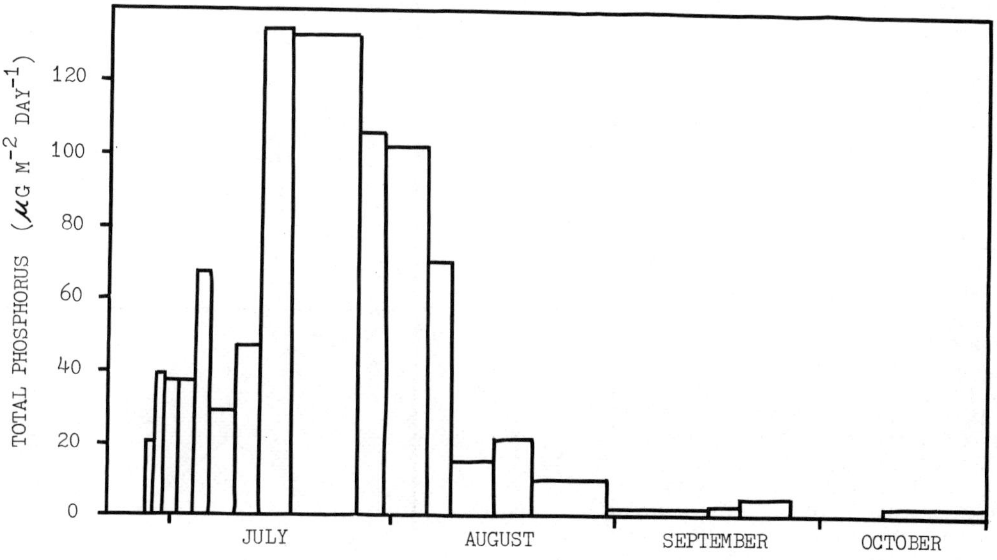


water increased somewhat with depth down to 4 cm, and then became more erratic. Soluble phosphorus in the overlying water was more than an order of magnitude less than these values.

Detritus input

The data on the phosphorus content of sedimenting detritus is presented in figure 2 for the two years studied. The variability among individual traps was very high, being about 67% overall. It did not appear in these samples that resuspended sediment from the bottom formed an important part of the collected material. The detritus input appears to peak in September, when falling temperatures and decreasing light may be contributing to decreases in both zooplankton and phytoplankton populations. The average rate of phosphorus input to the sediments over the July-October measurement period was 380 $\mu$g P $m^{-2}$ $day^{-1}$.

Insect emergence

Loss of nutrients from lake sediments as a result of the emergence of aquatic insects can be an important pathway of material flow in the lake ecosystem, although much of this may eventually return in the form of detritus. Estimates of the rate of loss of macrofauna phosphorus from the sediment were made using data on the numbers and biomass of insects collected in two emergence traps (Figure 3). Only Ephemeroptera and Chironomidae were collected in these traps. The Odonata and presumably the Trichoptera move to the edge of the lake to emerge. The values for emergence therefore underestimate somewhat the actual loss through this pathway. Average phosphorus flux over the ice-free portion of the year was 33 $\mu$g P $m^{-2}$ $day^{-1}$.

Fish feeding

It was not possible to obtain direct measurements of the magnitude of phosphorus removal from the sediments due to the feeding activities of the trout populations in the lake. However, making a number of assumptions about feeding locations and assimilation rates, it can be roughly estimated that not more than 23 $\mu$g P $m^{-2}$ $day^{-1}$ is removed from the sediment by this mechanism. This would be an average value over the course of the year, and may be higher in winter, when both the brook and rainbow trout are feeding on the benthic insect populations.

Soluble phosphorus flux

It had been initially planned to use $^{32}P$ as a tracer for the measurement of molecular diffusion of phosphorus out of the sediment. However, as previously noted, a parallel study of oxygen movement in these sediments made it apparent that water was circulating rapidly through the sediment. Soluble phosphorus was being pumped through the surficial sediment, and was not just diffusing out. It was therefore necessary to calculate transfer coefficients for the phosphorus movement across the interface, in order to obtain valid estimates of phosphorus movement in both directions.

The time course of observed tracer release to the overlying water is presented in figure 4 for the three cores studied. The predicted tracer release, based on transfer coefficients calculated using the initial data points, is also given in each case. The calculated profiles roughly approximate the observed data. Those for core 1 are very close, while for core 2 there is a larger decrease in observed counts before the plateau is reached than was predicted. Both curves for core 3 show similar trends, with the observed counts increasing more

Figure 4. Observed (●——●) and calculated (··········) profiles of tracer release in three undisturbed sediment cores.

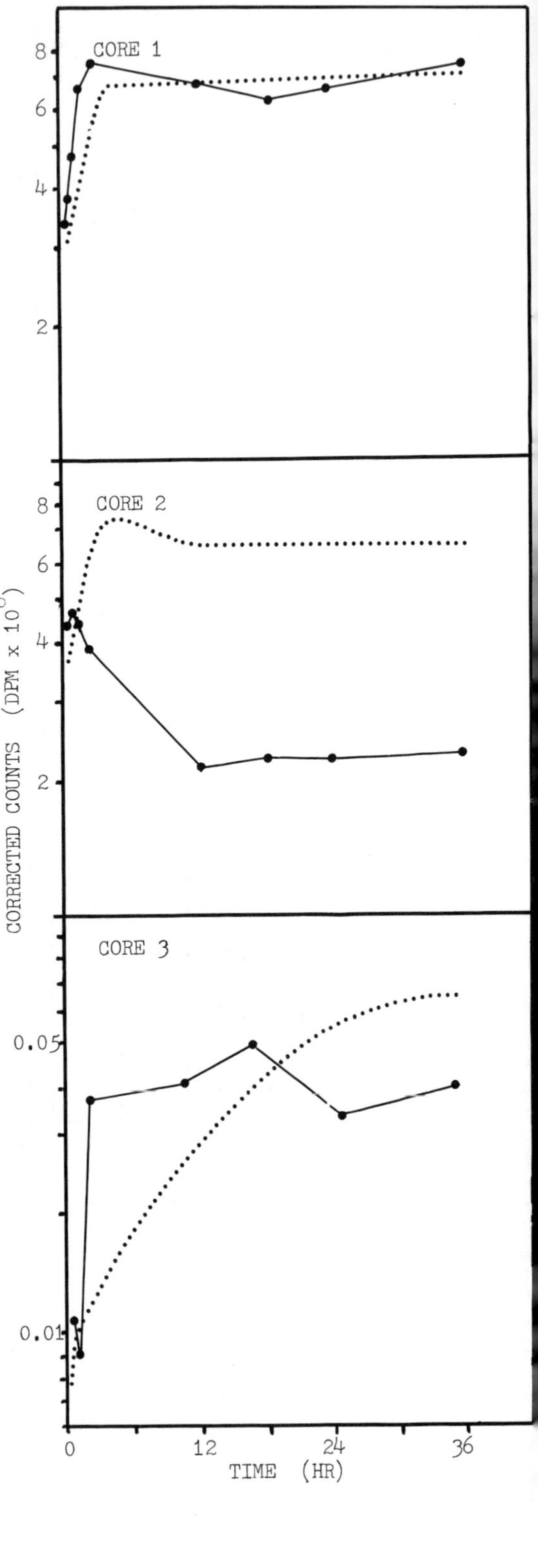

rapidly and then leveling off sooner than the calculations predict. In this core, in which the tracer was injected into the anoxic sediment layer, the observed and predicted tracer release was two orders of magnitude less than for the previous two experiments. In all three cases the orders of magnitude and the general trends of the two sets of results are similar. The observed movement of tracer into the overlying water is thus consistent with a model which incorporates a dissolved phosphorus flux both into and out of the sediment. No further attempt was made to "fit" a more complicated model to this data.

The goal of these experiments was to estimate the dissolved phosphorus flux across the sediment surface. Using the calculated transfer coefficient values for each sediment layer, and the measured interstitial and overlying water dissolved phosphorus pools, the dissolved phosphorus flux through the top 3 cm of oxidized sediment can be estimated to be in the range of 650 $\mu$g P $m^{-2}$ day$^{-1}$, for the time of year when these experiments were carried out (early fall). This flux will undoubtedly change over the course of the year, along with changes in macrofauna biomass, temperature and dissolved phosphorus concentrations in the water.

udget

The estimates of the phosphorus fluxes are ummarized in table 2 in the form of a budget or phosphorus movement across the interface. he magnitudes of each of these estimates ill change throughout the year. The annual atterns in sedimenting detritus and insect mergence have been indicated. The circulaion of water through the sediment and its onsequent dissolved phosphorus cycling ppear to be a function of macrofauna biomass, hich in turn varies over an annual cycle, nd also from year to year.

The magnitudes of the dissolved phosphorus luxes into and out of the sediment are pproximately equal. On a theoretical basis, t could be expected that soluble phosphorus utput would exceed input if the sediments re acting as a buffer to maintain water olumn concentrations at a given level. The igures for sedimenting detritus average 6% er day of the total water column phosphorus, o the water column would be rapidly depleted f phosphorus were not such a recycling echanism in effect.

In this instance the use of $^{32}$P as a tracer id not clearly indicate the magnitude of net hosphorus movement either in or out of the ediment. What it does show is that there is large movement of dissolved phosphorus hrough the surface oxidized sediment layers, argely as a result of bioturbation or acrofauna stirring activities, although hysical mechanisms of turbulence may also lay some role (Neame, 1975). This situation provides the potential for rapid equilibration of dissolved phosphorus compounds between interstitial and overlying water. The results also indicate that sedimentation and soluble phosphorus flux are the major components of benthic phosphorus movement in this system.

Table 2. Summary of measured and calculated phosphorus fluxes across the sediment-water interface.

| | | $\mu$g P $m^{-2}$ day$^{-1}$ |
|---|---|---|
| Input | Sedimenting detritus (average) | 380 |
| | Dissolved phosphorus circulation | 626 |
| | Total | 1006 |
| Output | Insect emergence (average) | 33 |
| | Dissolved phosphorus circulation | 681 |
| | Fish feeding (estimate) | 23 |
| | Total | 737 |


## Acknowledgements

I wish to thank Dr. C.R. Goldman, Division of Environmental Studies, University of California, Davis, for his encouragement and support of this research. This study was supported by National Science Foundation grants GB 3571 and BMS 72-02246 A01 to Dr. Goldman.


## References


Chapman, H.D., & P.F. Pratt, 1961. Methods of Analysis for Soils, Plants and Waters. University of California, Div. of Agricultural Sciences.

Fox, H.M., & C.A. Wingfield, 1938. A portable apparatus for the determination of oxygen dissolved in a small volume of water. J. Exp. Biol. 15:437-445.

Milbrink, G., 1968. A microstratification sampler for mud and water. Oikos 19:105-110.

Mundie, J.H., 1971. Insect Emergence Traps. In: A Manual on Methods for the Assessment of Secondary Productivity in Fresh Waters. W.T. Edmondson & G.G. Winberg (Ed.). IBP Handbook No. 17. Blackwell. p. 80-93.

Neame, P.A., 1975. Benthic Oxygen and Phosphorus Dynamics in Castle Lake, California. Unpublished Ph.D. thesis. University of California, Davis.

Reeburgh, W.S., 1967. An improved inter-

stitial water sampler. Limnol. Oceanogr. 12:163-165.
Richey, J.E., 1974. Phosphorus Dynamics in Castle Lake, California. Unpublished Ph.D. thesis. University of California, Davis.
Schindler, D.W., V.E. Frost, & R.V. Schmidt, 1973. Production of epilithiphyton in two lakes of the Experimental Lakes Area, Northwestern Ontario. J. Fish. Res. Board Can. 30:1511-1524.
Strickland, J.D.H., & T.R. Parsons, 1972. A Practical Handbook of Seawater Analysis. Fish. Res. Board Can. Bull. 167 (2nd ed.).
Stumm, W., & J.O. Leckie, 1971. Phosphate exchange with sediments: Its role in the productivity of surface waters. Proc. 5th Int. Water Pollution Res. Conf. Vol. 2: III-26.
Watt, W.D., & F.R. Hayes, 1963. Tracer study of the phosphorus cycle in sea water. Limnol. Oceanogr. 8:276-285.

Key-words: phosphorus, flux, kinetics, sedimentation, bioturbation.

# THE ROLE OF IRON IN THE EXCHANGE OF PHOSPHATE BETWEEN WATER AND SEDIMENTS

L. Lijklema

Twente University of Technology, Enschede, The Netherlands

## Summary

The solubility of iron and retention of phosphates by sediments rich in iron are complex functions of redox potential and pH. Reducing conditions in sediments have been shown to cause increased solubility of both iron and phosphate in sediments. At high redox potential, however, iron remains insoluble while phosphate can be released in large quantities from iron-hydroxyphosphate complexes due to competition of hydroxyl ions with phosphate ions. The adsorption isotherm of the iron(III)-phosphate system is controlled by the phosphate-iron ratio, the age of the complex formed and pH. Adsorption and desorption processes due to variations in pH also exhibit hysteresis.

These factors suggest that the phosphate binding capacity of the oxygenated sediment layer may vary due to fluctuations in the pH of the surrounding water. Major factors controlling the pH in the overlying water are the rate of photosynthesis, alkalinity, water depth and carbondioxide exchange rates at the surface. A reasonable prediction of this pH can be made. The extraction of phosphates from iron-rich sediments due to an increased pH in the overlying water thereafter depends upon the intensity of water-sediment contact (hydrodynamic conditions and compactness of the sediment), the buffering capacity of the sediment, and interstitial water and the rates of several (bio)-chemical reactions.

Another situation in which iron plays a major role in controlling the phosphate exchange occurs when anoxic bottom layers develop during stagnant periods in summer. Upon mixing of aerobic and anaerobic layers due to increased turbulence, fresh iron(III) hydroxide will be formed due to oxidation of dissolved iron(II) salts. The pH resulting from the mixing of these layers will control the quantity of phosphate being adsorbed on the freshly formed iron-hydroxide. This quantity can be lower than the original quantity bound in the sediments, resulting in a net release of phosphates.

## Introduction

It has been shown that phosphate and iron are strongly associated in many soils and sediments. The release of phosphate and iron(II) under reducing conditions suggests their interrelation. Although the usual extraction procedures are afflicted with severe shortcomings, the information obtained through chemical analysis of lake sediments often indicates fixation of phosphate by iron.

The sediments of Lake Brielle and the water reservoir "De Braakman", both in the South-West of the Netherlands, also exhibit a strong iron-phosphate relationship. The investigation of these sediments has initiated the research presented herein. In particular the data on the Braakman reservoir are significant since there are no inputs into the reservoir with the exception of a relatively short period during winter when the reservoir is filled with water. Hence all the variations in the phosphate concentration of the overlying water are due to interaction with the sediments. These phosphate concentrations tend to exhibit a yearly pattern with an increase in phosphate concentration in late summer. The trend over the years is towards higher peak concentrations occurring earlier in the season. This paper will discuss mechanisms that may explain these observations.

## Iron-phosphate chemistry

The solubility of iron-phosphate compounds is controlled by redox potential and pH. Under reducing conditions the iron concentration in interstitial waters will depend upon the relative concentrations of anions present and pH. Thus the phosphate, carbonate, sulfide or even the hydroxyl ion concentration will set an upper limit to the iron(II) concentration. In anaerobic sediments of Lake Brielle a phosphate concentration of several p.p.m. is not unusual and although the net flux of phosphate is towards the sediments, release of phosphate will occur occasionally.

For the transport-mechanism of phosphate in the water column, however, the iron(III)-phosphate interaction is the most important factor. Figure 1 shows the adsorption isotherm of phosphate on freshly precipitated iron(III)-hydroxide as a function of pH.

Due to the initial neutralisation of positive charges on the iron(III)-hydroxide particles the adsorption of the first phosphate ions is very strong and results in immeasurably low equilibrium concentrations. The increasing positive charge of iron(III)-hydroxide with decreasing pH is reflected in the inflection of the curve at higher adsorption when the pH is lower. Adsorption past the inflection point is mainly due to inter-

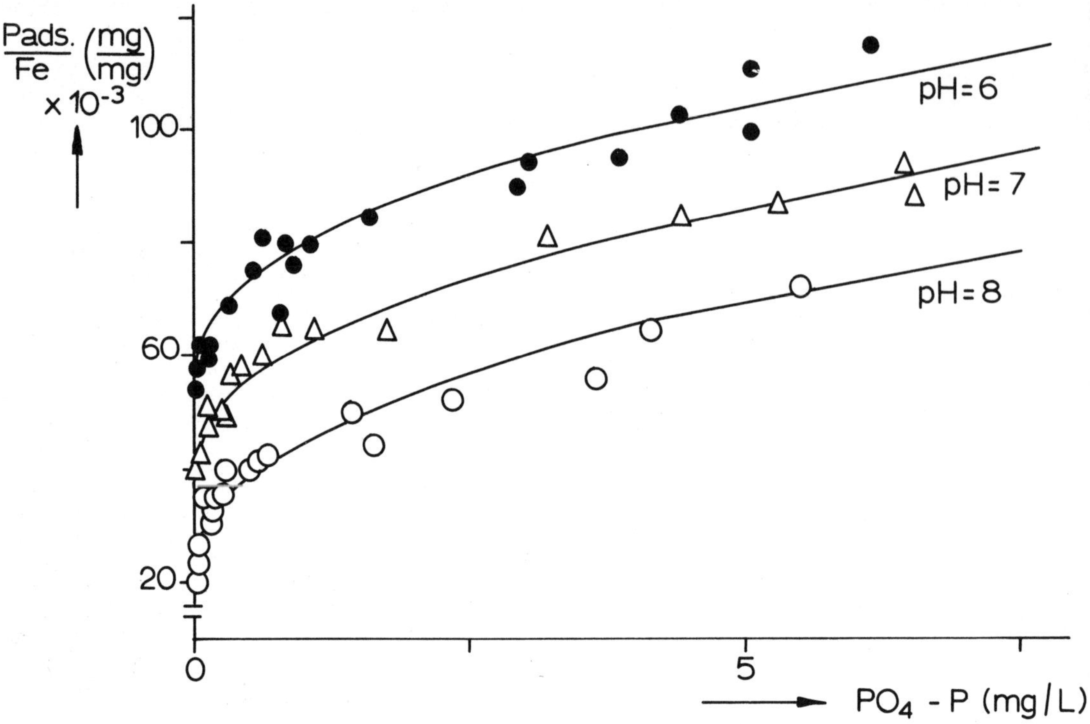


Figure 1. Adsorption isotherm of phosphate on freshly precipitated iron(III)-hydroxide.

action of neutral hydroxyl groups on the iron particles and results in a further increase of adsorption, which is approximately proportional to the square root of the phosphate concentration. This can be explained by the assumption of binuclear adsorption complexes, for which support from infra-red studies is also available (Atkinson et al., 1973): thus:

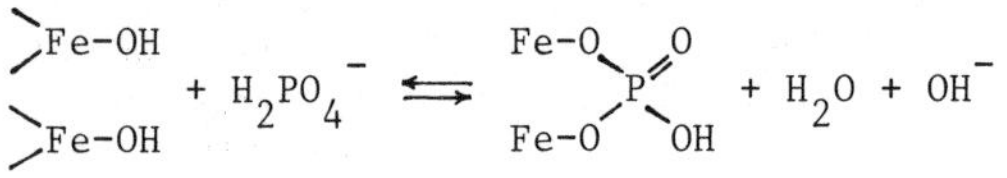

Variation of pH after equilibration will result in desorption and adsorption of phosphate accompanied by some hysteresis, but essentially the process is reversible. Ageing of the iron-hydroxide prior to adsorption reduces the adsorption capacity.

The shape of the isotherms shows that desorption of phosphate due to increasing pH is strongly dependent on the P : Fe ratio. At a low P : Fe ratio, an increase in pH will result in a smaller desorption of phosphate than at higher ratios with other conditions being equal. This observation is important when considering the role of iron in the vertical transport of phosphate.

Vertical transport in aerobic system

In a well mixed water body both the water column and the top few millimeters of the sediment will be generally aerobic. Due to photosynthesis, the pH of the water will tend to increase to a value depending on the initial pH, the alkalinity, water depth and the quantity of carbondioxide assimilated. Figure 2 presents calculated pH values for a system with an alkalinity of 2 mval per litre and a carbon assimilation of 7 grams of C per $m^2$. Exchange of carbondioxide with the atmosphere has been neglected.

Shallow waters in particular can exhibit a a substantial change in pH. Although vertical mixing is reduced near the sediments and reduction processes and carbondioxide production in the sediments tend to oppose a rising pH, the top layer of the sediments may eventually contact water of a higher pH; this in turn may be the cause of extraction of phosphate from iron-bound phosphato-hydroxy complexes accumulated in the top layer of sediments. Such a release will be controlled by the change in pH, the initial P : Fe ratio in the sediments and the total amount of iron-phosphates extracted relative to the water column. Figure 3 gives an impression of the relative importance of the pH change, the P : Fe ratio and the quantity of iron-phosphate extracted. These calculations are based upon the adsorption isotherms of Figure 1, a water depth of 2.5 m and an initial pH of 7.

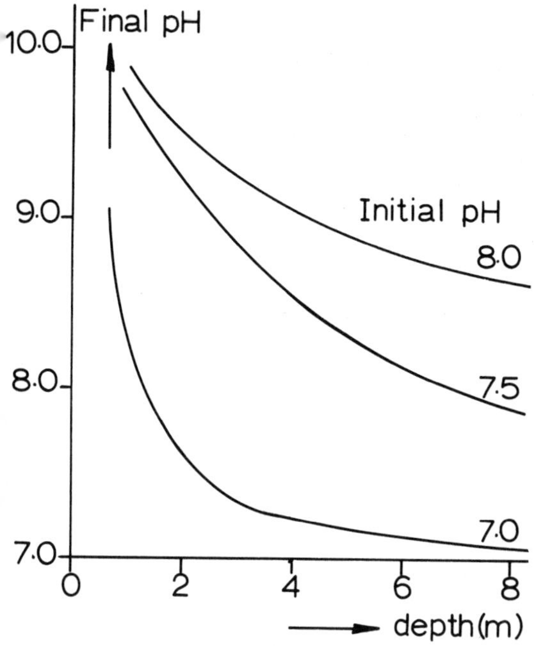


Figure 2. pH change due to photosynthesis as a function of mixing depth.

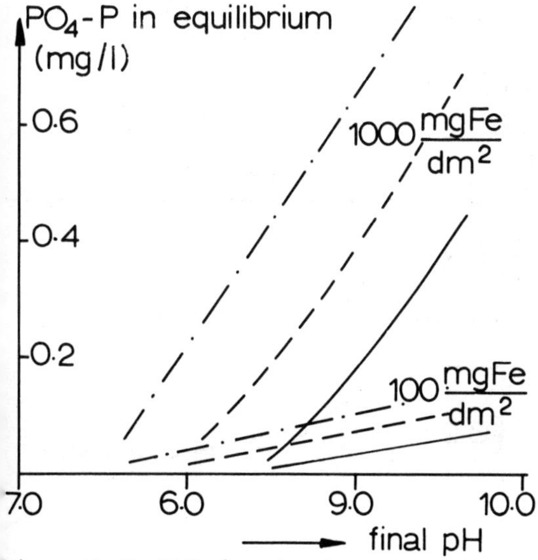


Figure 3. Equilibrium phosphate concentrations after extraction of iron(III)-phosphate-hydroxy complexes at different pH. Initial P : Fe ratio in complex: -·-·- = 0.04; ----- = 0.03 and ——— = 0.02.

The amount of iron-phosphate equilibrated with the overlying water dominates the equilibrium P-concentration. This amount depends upon the degree of mixing and resuspension of sediments caused by storms in the summer half year, which may induce P-release from iron-rich aerobic sediments. Evidence for such a relationship could not be found by correlating the weekly measured P-concentration in Lake Brielle and wind data (correlations calculated by Environmental Section, Deltadienst, Rijkswaterstaat). The data, however, were limited to moderate and weak-wind episodes. Moreover, Lake Brielle is rather deep with an average depth of about 5 m. In contrast, the shallower Lake IJsselmeer (1-2 m depth) showed a definite correlation between P-total and wind at certain locations (ZZW, RIZA, RIJP, 1976); this may be due to resuspension of particulate P as well. Nevertheless, the existence of resuspension as such is indicated. Resuspension will be more important for phosphate release in shallow lakes, due to the higher degree of erosion, the higher average rise in pH due to photosynthesis (Figure 2) and the higher proportion of sediment extracted per volume of water (Figure 3).

Future research will focuss on direct measurement of the effect of storms on the phosphate exchange between sediment and water.

## Vertical transport in partly anaerobic system

When anoxic bottom layers develop during stagnant periods in summer, reduction of iron(III)-hydroxy-phosphate complexes at the interface water-sediment can result in hypolimnetic waters rich in iron(II) and orthophosphate ions. The concentrations of both ions will depend upon their relative abundancy in the sediments and the interstitial water. Such observations have been made in Lake Brielle as well as in the Braakman reservoir.

Although during stagnant periods, exchange between epilimnion and hypolimnion is limited, there will still be fluxes of iron(II), o-phosphate and oxygen. These are related to the phenomena of oxidation and reduction, sedimentation and phosphate adsorption and desorption. In addition, algae contribute to the vertical movement of phosphate. Figure 4 outlines these processes.

Upon increased mixing due to reduced stability or induced by wind, the mixing process will be intensified. Once more, the P : Fe ratio and the pH are very important in controlling the phosphate budget in such a situation. To study the complex interactions a simulation model has been developed. The variables oxygen, iron(II), iron(III), o-phosphate, phosphate adsorbed to iron, the carbonate system (for the description of pH), algae and BOD have been included for the three vertical zones presented in Figure 4. Processes described are oxidation and reduction of iron, algal growth, respiration and decay, adsorption and desorption of phosphate on iron(III), sedimentation, atmospheric exchange of oxygen and carbondioxide and vertical turbulent transport across interfaces. Forcing functions were light for algal growth and wind for vertical mixing.

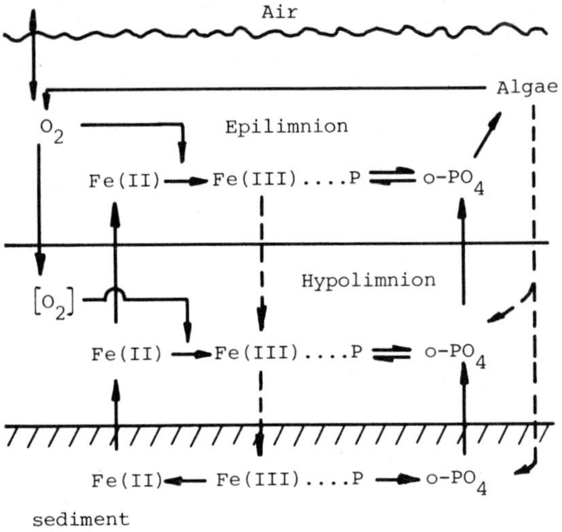


Figure 4. Iron and phosphate cycle in stagnant water.

Correct parameters for such a model are difficult to establish, but simulations clearly demonstrate at least qualitatively some features of the iron-phosphate interaction considered here.

Figure 5 indicates the effect of increased turbulence during a short period on the distribution of the iron species.

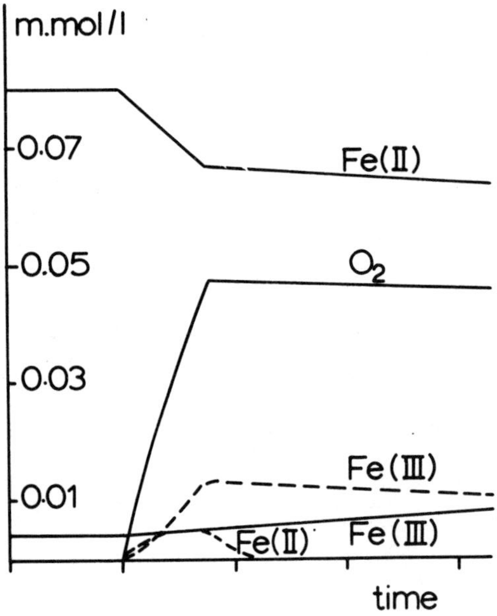


Figure 5. The effect of mixing upon iron distribution. —— hypolimnion, ---- epilimnion.

The iron(II) dispersing into the epilimnion is rapidly oxidized and results in an increased Fe(III) concentration that eventually will decrease by sedimentation through the hypolimnion towards the sediments. Fe(II) in the hypolimnion decreases first due to mixing and thereafter by a relatively slow oxidation. The oxidation rate is lower in the hypolimnion due to the lower pH.

The simulation presented in Figure 5 has also been run with different phosphate levels using each time the same iron concentrations in order to evaluate the effect of the P : Fe ratio on the phosphate transformations.

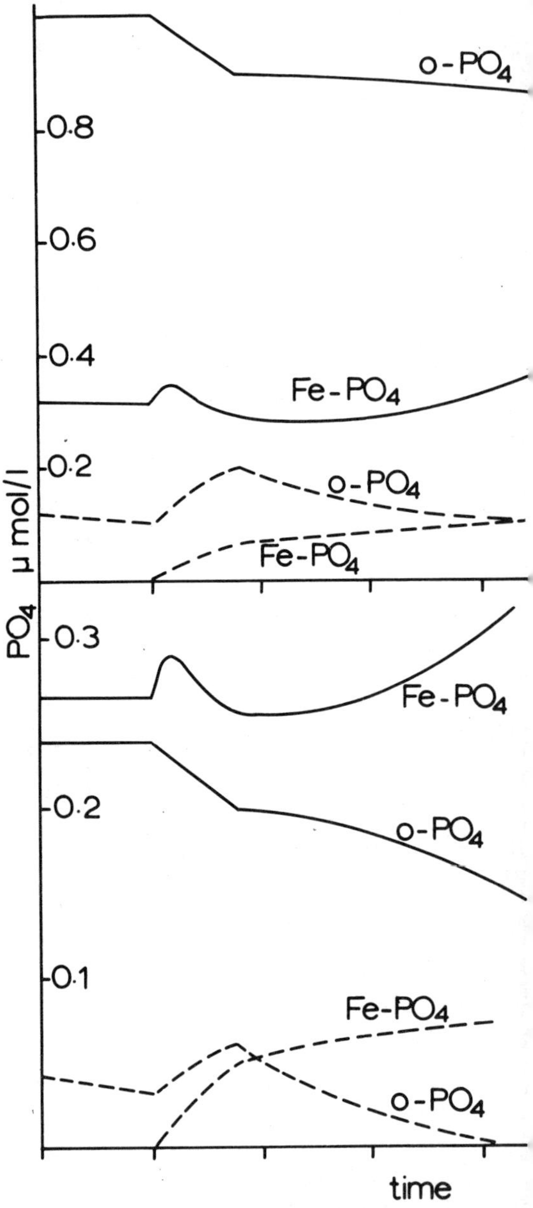


Figures 6, 7. The effect of mixing upon the phosphate distribution. —— hypolimnion, ---- epilimnion.

Figures 6 and 7 show results with initial phosphate-P concentrations of about 30 μg/ℓ and 10 μg/ℓ in the hypolimnion, respectively. In both situations the shift from iron(II) to iron(III) results in an increased fraction of the phosphate to become adsorbed onto the iron(III). After sedimentation, however, the resulting phosphate removal is much more effective at a phosphate concentration of 10 μg/ℓ than at 30 μg/ℓ. This difference in behaviour is mainly due to the very effective adsorption of the first phosphate ions as shown in the adsorption isotherms (Figure 1). The initial sharp increase in adsorbed phosphate in the hypolimnion is due to a lower pH, while the subsequent decrease is due to mixing. The later increase in sorbed phosphate is caused by iron(II) oxidation and to a lesser extent by sedimentation from the epilimnion. Uncertainty about the relative rates of these processes can alter the shape of such curves; nevertheless, the net effect will be a reduction of soluble phosphate to an extent depending upon pH, mixing and P : Fe ratio.

An explanation of the trend towards increasing phosphate releases in subsequent years as observed in Braakman Reservoir (see Introduction) can now be hypothesized. Originally the water-tight bottom of the newly constructed reservoir was covered with a layer of sand containing some iron. Filling with polder water with a relatively high P : Fe ratio during winter caused a gradual increase in this ratio in the sediments, since a large fraction of the incoming phosphate and iron precipitate. The stagnant anaerobic layers that develop normally during summer near the bottom will also have a trend towards higher P : Fe ratio, resulting in a higher final o-phosphate concentration upon mixing as shown in the simulation. The overall increase in eutrophication will tend to increase organic loading and thus enhance the reducing capacity of the sediments. This will lead to anoxic bottom conditions occuring earlier in the season, thus shifting phosphate regeneration from the sediments from September, as observed in the past, to August, as noted in recent years.

## Acknowledgement

I am grateful to Drs. W. Bassie for his cooperation and the supply of information on the Braakman reservoir and to my colleagues of the Deltadienst and RIZA for their interest and the correlation studies performed.

## References

Atkinson, R.J, R.L. Parfitt & R.St.C. Smart, 1974. Infra-red study of phosphate adsorption on goethite.
J. Chem. Soc. Farad. Trans. 1:1472-1479.

ZZW, RIZA, RIJP, 1976. Bijlagen VI.3.1.1 en VI.4.1.1 van de Nota: Waterstaatkundige werken en waterkwaliteit IJsselmeergebied.

## Key Words

Phosphate, Iron, Sediments, Vertical transport, pH.

CONSIDERATION OF PHOSPHORUS RELEASE FROM SEDIMENTS IN A LAKE MODEL[1]

F.A. DiGiano

University of Massachusetts, Amherst, U.S.A.

P.D. Snow

Union College, Schenectady, New York, U.S.A.

Summary

A model was developed to describe both the release and sedimentation rates of phosphorus in a eutrophic lake having a phosphorus-laden sediment. The release rate constant was considered as a transfer coefficient relating the flux of phosphorus across the sediment/water interface to the concentration gradient between the interstitial and overlying waters. The sedimentation rate constant described the assumed first-order dependency on phosphorus in the overlying water. A unique aspect of this research was the in-situ evaluation of release and sedimentation rate constants in Lake Warner (Hadley, Massachusetts). Here, emplacement of a tank enabled isolation of an overlying water-sediment volume. Incorporation of an anion-exchange column removed phosphorus from the overlying water, thus causing release of phosphorus from the sediment. This then allowed for calculation of rate constants in a complete-mix, batch reactor.

It was demonstrated that because of phosphorus storage in the sediment, reducing the phosphorus input to the lake would continue to produce substantial phosphorus release from the sediments for ten years or more. The model can also be used to predict the non-steady state approach to a new equilibrium phosphorus concentration with and without chemical suppression of sediment phosphorus.

## Introduction

The importance of lake sediments as a phosphorus (P) source rather than sink remains open to question. This issue becomes critical in deciding upon a control strategy for bringing about a reduction in P concentration. In examining the impact of wastewater diversion on the rather shallow, Lower Madison, Wisconsin lakes, Sonzogni & Lee (1974) concluded that the observed decrease in lake P was fairly well described by a simple dilution model. Based on their data, phosphorus residence time was less than the hydraulic residence time thus indicating that the sediments act as a sink rather than as a source. In contrast, Emery et al. (1973) reported that although P loading to Lake Sammamish had been reduced by 40 percent, there was still no significant recovery after four years; however, the sedimentation model developed by Vollenweider (1969) predicted recovery in 3.3 years. Studies on Lake Shagawa, Minnesota revealed similar results (Larsen et al., 1974). Here, installation of tertiary waste treatment greatly reduced the P loading, yet P leaving the lake was greater than that entering. Clearly, the role of the sediments as a phosphorus source rather than sink was implied.

## Proposed Model

Working independently from Lorenzen (1973) but at about the same time, we developed a model to account for both release and sedimentation of P in shallow eutrophic lakes (Snow & DiGiano, 1976). Our proposed model differs from that developed by Lorenzen in two important respects. First, the release rate of P across the sediment/water interface is described by

$$R_r = \frac{\varepsilon A}{V_L} K_1 \ (P_i - P_L) \qquad (1)$$

where the concentration gradient between the interstitial water, $P_i$, and the overlying water, $P_L$, provides the driving force for mass transfer. This mechanism is in agreement with that proposed by Kamp-Nielsen (1974), Chen et al. (1973) and Wildung & Schmidt (1973). Secondly, the proposed model includes a functional relationship between the solid phase and adjacent interstitial phase phosphorus, i.e. $P_s$ and $P_i$, respectively. This interplay allows for modeling of the change in release rate as the supply of P in the sediment shifts.

While not explicitly indicated in the Lorenzen model, a mass balance on the sediment phase (see Eq. 6) requires that a sediment

1. This paper is based on the Doctoral Dissertation of P.D. Snow (1976)

volume, $V_S$, be selected. More specifically, this volume is based on the exchange depth from which phosphorus can be released. Values ranging from 0.5 (Fillos & Swanson, 1975) to 30 cm (Wentz & Lee, 1969) have been cited. However, most values range between 5 and 10 cm. In addition, the exchangeable fraction of the solid phase P must be estimated. Again, a wide range of values have been reported with a low of 12 percent (Wildung & Schmidt, 1973) and a high of 50 percent (Wentz & Lee, 1969).

The major objective of this research was to evaluate the release and sedimentation rate constants, $K_1$ and $K_2$, respectively by in-situ lake measurements. This was considered important because although Lorenzen had proposed a release model, no actual rate data was presented. Moreover, the use of in-situ methods was selected in order to maintain the physical, chemical and biological integrity of the sediment/water system. Field investigations were essential in determining the relationship between $P_i$ and $P_s$. Anticipating the significance of the sediment release mechanism, chemical suppression studies were also undertaken. Lake Warner, located near the University of Massachusetts campus, provided a convenient field research site.

## Lake Warner

Lake Warner is a man-made river-lake having an average detention time of only 10 days, a surface area of 27.5 ha (68 acres) and an average depth of 1.6 m (5.2 ft). The major contribution of flow is from the Mill River which drains 8314 ha (32.1 sq mi). As the Town of Amherst grew over the past 10-15 years, overloading of the primary treatment facility caused waste-water to be bypassed into the Mill River. During the summer months, algal blooms occur frequently in Lake Warner. As a result, dissolved oxygen is far in excess of saturation at the surface but near zero at the bottom. Total P varies between 60 and 105 μg/ℓ with the higher values found near the bottom; ortho-P ranged from 30 to 50 μg/ℓ near the bottom.

Interstitial total P, measured by centrifuging sediment samples, ranged from 300 to over 600 μg/ℓ with the ortho-P/total P ratio remaining at about 0.8. The experimentally determined relationship of interstitial $P_i$ and exchangeable sediment $P_s$ for use in the proposed model is given in Figure 1. The exchangeable fraction was estimated as 25 percent based on earlier isotopic exchange studies on Lake Warner sediments (Ku & Feng, 1975). Measurements of Eh in the interstitial water ranged form -200 to -350 mv, with lower values being associated with higher levels of $P_i$.

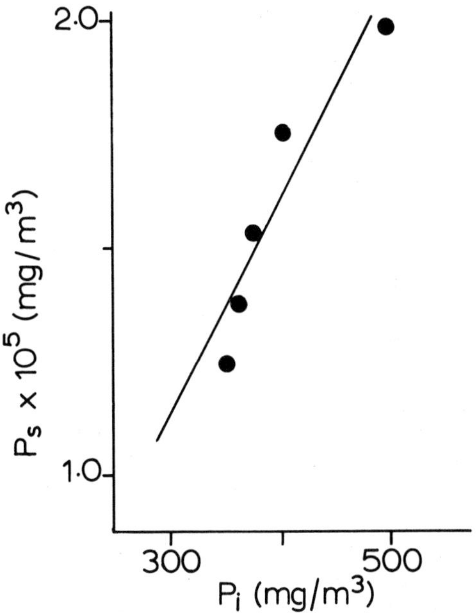


Figure 1. Adsorption equilibrium between sediment solid and interstitial phase P.

Ku & Feng (1975) also examined the equilibrium relationship between solid and liquid phase P at high and low oxidation-reduction potentials (ORP) for Lake Warner sediment. Low ORP was established by a biological system (glucose substrate addition) under ideal laboratory conditions. Decreasing ORP at a given solid phase loading greatly increases the equilibrium liquid phase concentration. Thus, extrapolating this result to field conditions, release of interstitial P should be greatly enhanced in a reduced ORP environment.

## Evaluation of Rate Constants

A water tight, fiberglas tank, enclosing 3.4 sq m (36 sq ft) of sediment surface area, was embedded into Lake Warner. As shown in Figure 2, the lake water within the tank was then pumped through an activated carbon bed and an anionic exchanger (quaternernary ammonium groups on a styrenedivinylbenzene polymer) for removal of P before return to the tank. Over a period of several days, the level of total P was thus reduced from about 60 to 30 μg/ℓ. The ion exchanger system was then dicontinued and the tank water mixed by a pump. Measurements of total P with time were used to obtain the rate constants $K_1$ and $K_2$.

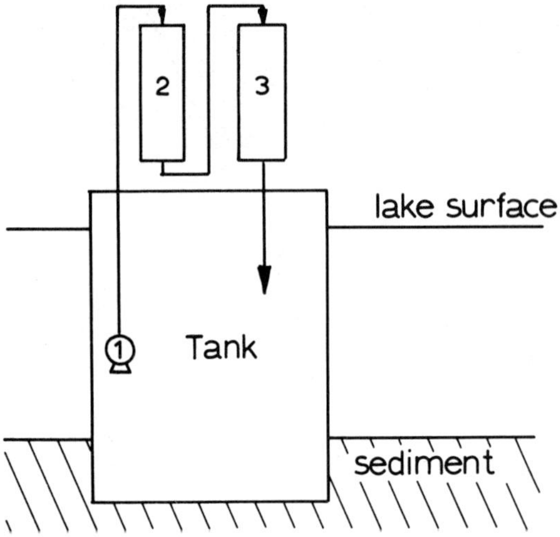


Figure 2. Schematic of device used for in-situ measurements of P release: 1 = pump, 2 = granular activated carbon, 3 = anion exchange resin.

The isolated lake volume, $V_B$, created by the tank can be modeled as a completely mixed reactor:

$$V_B \frac{dP_L}{dt} = \varepsilon A\ K_1\ (P_i - P_L) - K_2\ P_L\ V_B \qquad (2)$$

Removing phosphorus from the tank should cause release of more phosphorus from the sediment. From the initial rate, the value of $K_1$ can be estimated as

$$K_1 = \frac{\left.\frac{dP_L}{dt}\right|_{\text{Initial Rate}}}{\frac{\varepsilon A}{V_B}\ (P_i - P_L)} \qquad (3)$$

Here, it is reasonable to assume that $P_i$ will remain relatively constant.

As a result of simultaneous sedimentation, $P_L$ should reach a steady-state value given by:

$$P_L\Big|_{SS} = \frac{\frac{\varepsilon A}{V_B} K_1 P_i}{\frac{\varepsilon A}{V_B} K_1 + K_2} \qquad (4)$$

from which $K_2$ can then be approximated.

These estimates of $K_1$, $K_2$ and $P_i$ were then used in the model to calculate $P_L$ with time. After comparison of the calculated and experimental values, each of the three constants $K_1$, $K_2$ and $P_i$ was then incremented slightly to generate a new set of values for $P_L$. This computational scheme was repeated until the sum of the squared differences between the calculated and experimental values was minimized. The experimental and predicted increase in P after discontinuation of the ion exchanger is shown in Figure 3 along with the best fit values of $K_1$, $K_2$ and $P_i$.

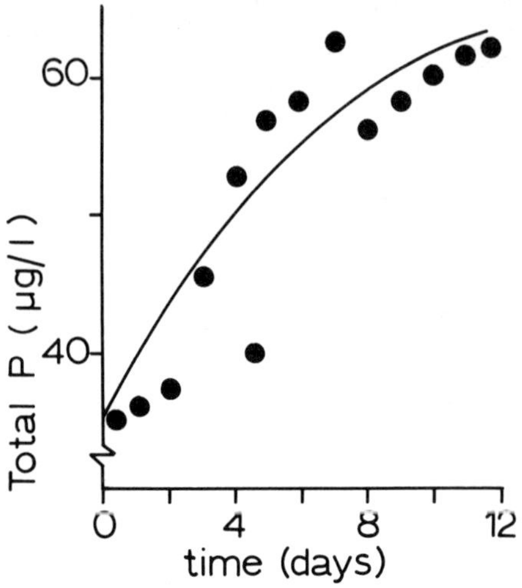


Figure 3. Release of Total P from sediment in test tank after discontinuation of ion exchange. Predicted results shown by solid line: $K_1$ = 0.091 m/day, $K_2$ = 0.176 day$^{-1}$, $P_i$ = 430 µg/ℓ.

Although these results were obtained in December, similar values were found in August and again in September. The release rate calculated on the basis of a $P_L$ of 90 µg/ℓ and $P_i$ of 425 µg/ℓ is 26 mg/sq m/day. This compares quite closely to values reported by Fillos & Swanson (1975) as also obtained in Lake Warner.

## Suppression of Phosphorus Release

Six, watertight tubes, each enclosing an area of 0.073 sq m (0.785 sq ft), were embedded into Lake Warner. Ion exchange was not used for P removal. Instead, differing dosages of alum and/or kaolinite (finely ground) were added to each tube in order to suppress the P level in the overlying water. Total P in the control tube (no chemical addition) remained high (63 µg/ℓ) throughout the 40 day observation period. In sharp contrast, treatment with 4 gm of alum and 16 gm of kaolinite depressed total P to less than 30 µg/ℓ. Total P in the surrounding lake water remained higher than in the control tube because of the constant input of P to the lake.

Addition of suppressants can be considered to decrease the effective interstitial P from $P_i$ to $P_i^*$ resulting in a lower release rate according to Equation 1. Although it may also be reasonable to assume a decrease in $K_1$ due to a chemical change in the interface, this was not considered. The effective interstitial $P_i^*$ can be calculated from the steady state value of $P_L$ using Equation 4.

Table 1 provides these values for each suppressant dosage combination. The most effective dosage combination gives a $P_i^*$ of 172 μg/ℓ. Projecting these results, 55,000 Kg/(Km)$^2$ of alum and 220,000 Kg/(Km)$^2$ of kaolinite would be required to most effectively lower $P_i$.

Table 1. Results of Suppression Study.

| Dosage in Grams Alum | Kaolinite | $P_i^*$ μg/ℓ |
|---|---|---|
| 4 | 0 | 318 |
| 1 | 0 | 305 |
| 1 | 8 | 298 |
| 0.25 | 2 | 225 |
| 4 | 16 | 172 |
| 0 | 0 | 410 |

## Application of the Model

The objectives of applying the model to Lake Warner were two-fold: first, to predict the transient response of lake P due to the recent history (past 10-15 years) of increased P loading caused by urban development and bypassing of primary effluent and second, to predict lake recovery by either reducing the P loading by elimination of the wastewater bypass or by suppressing the sediment P by chemical addition. The validity of the model can only be judged by comparing the present observed level of P with that predicted from the historic profile of P loading.

Prior to urbanization and the bypass of wastwater, influent P was assumed as 40μg/ℓ. Based upon the derived rate constants, $K_1$ and $K_2$, the corresponding steady state value of $P_i$ is estimated from Equation 4 to be 185 μg/ℓ. Based on estimates of the annual. bypassed wastwater volume and the Mill River discharge rate, average input P was assumed to increase to 90 μg/ℓ. However, the lack of records for the Amherst Treatment Plant and the Mill River make this only a crude estimate. Moreover, the actual field situation would not correspond to a step increase in influent P of 40 to 90 μg/ℓ. Instead, the amount of bypassed wastewater no doubt increased yearly as the treatment plant became more overloaded. Also affecting model predictions is the assumption that phosphorus stored within the first 10 cm of sediment can be released. Despite these modeling deficiencies, prediction of the transient lake response over a long period (several years) should still provide a valuable guideline for interpreting the present state of lake.

Figure 4 shows the model to predict a rather slow response of lake P to the step increase in loading. That is, despite the short detention time of 10 days, $P_L$ only increases from 40 to 65 μg/ℓ in three years. A simple hydraulic residence time model would predict this value in less than one month. Finally, after almost 14 years, a new steady state condition is reached. Most notably, interstitial P has increased from 185 to 407 μg/ℓ. Clearly, during this period, the sediment acted as vast phosphorus sink. Due to the adsorptive capacity of the sediment, interstitial P only gradually increased. Recent measurements in Lake Warner indicate a $P_L$ of 80-90 μg/ℓ and a $P_i$ of close to 400 μg/ℓ. The time frame predicted by the model for these values to be reached is on the order of 10 years. This corresponds quite well to the actual period of rapid growth in Amherst which should also be a reasonable indicator of increased volumes of bypassed wastewater. Thus, the similarity of the predicted and actual results lends support to the validity of the modeling approach.

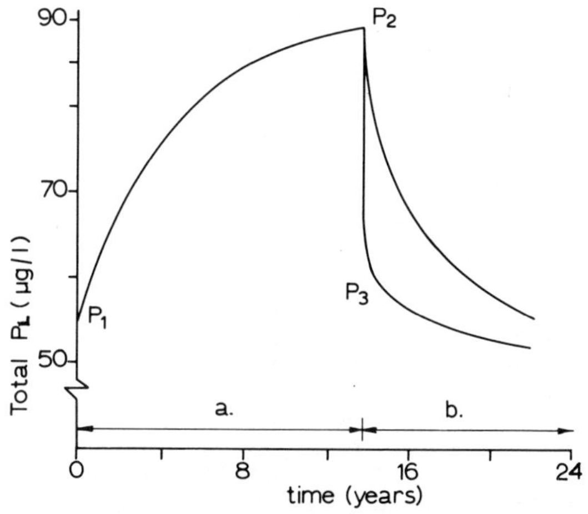


Figure 4. Application of phosphorus exchange model to Lake Warner with a step increase in P loading for "a" years and a step decrease for "b" years: $P_1$ = initial $P_i$ (185 μg/ℓ); $P_2$ = $P_i$ after "a" years (407 μg/ℓ); $P_3$ = $P_i^*$ by addition of chemical suppressants after "a" years (300 μg/ℓ).

To simulate a decreased P loading to the lake with elimination of the discharge of bypassed wastewater to the Mill River, the value of influent P was reduced from 90 to 50 μg/ℓ. This step response was initiated with a corresponding interstitial $P_i$ of 407 μg/ℓ. As indicated in Figure 4, the model predicts a rapid initial decrease in P from 90 to 75 μg/ℓ. However, many more years are required to reduce $P_L$ to the new steady state value of 50 μg/ℓ. This slow recovery is due to the vast storage of P in the sediment. The driving force for release is caused by the correspondingly large interstitial P concentration and the continually decreasing concentration of overlying lake water P (see Equation 1).

As earlier developed, the effect of adding chemical suppressants has been considered in the model as a mechamism for reducing the effective interstitial P. This is illustrated in Figure 4 by the predicted response with the initial value of $P_i^*$ set at 300 μg/ℓ rather than 407 μg/ℓ. Because the release rate has now been decreased, rapid recovery is predicted and the new steady state of lake P is reached in about one year.

## Factors Affecting Model Results

The relative importance of the release rate, sedimentation rate, hydraulic residence time and exchange depth can be understood from inspection of the transient response model. Mass balances for the overlying water and sediment in a completely mixed reactor, respectively are given by:

$$V_L \frac{dP_L}{dt} = Q\,(P_0-P_L)+K_1\ \varepsilon A(P_i-P_L)-K_2P_LV_L \quad (5)$$

$$V_S \frac{dP_S}{dt} = K_2P_LV_L - \varepsilon AK_1\,(P_i-P_L) \quad (6)$$

The rate of change of $P_L$ is directly influenced by the rate of change of $P_i$. As seen from Equation 6, this latter rate is determined by the relative magnitude of the release and sedimentation rates, the adsorption equilibrium relationship (Figure 1) and the assumed sediment exchange depth, $D_e$ (since $V_S = A\,D_e$). In model computations, Equations 5 and 6 must be solved simultaneously.

Over a small time step, the value of $P_i$ can be considered constant such that integration of Equation 5 yields:

$$P_L = \left[P_{L_0} - \frac{C_1}{C_2}\right] e^{-C_2t} + \frac{C_1}{C_2} \quad (7)$$

where:

$$P_{L_0} = P_L \quad \text{at} \quad t = 0$$

$$C_1 = \frac{P_0}{\overline{T}} + \frac{\varepsilon\, K_1\, P_i}{D}$$

$$C_2 = \frac{1}{\overline{T}} + \frac{\varepsilon}{D} K_1 + K_2$$

After a relatively short time (on the order of days), the exponential term in Equation 7 becomes relatively unimportant. The approach to a new steady state value of $P_L$ continues due to a gradual shift in the pseudo-steady state constant, $C_1$. This occurs as a result of either an increase or decrease in $P_i$ caused by a corresponding step loading increase or decrease.

The importance of $\overline{T}$ and the exchange depth, $D_e$, on the rate of change of $C_1$ can be seen for a given lake depth, D. An increase in $\overline{T}$ will proportionally increase the stored sediment phosphorus. Thus, the response to a step decrease in P loading will be retarded because $P_i$, and subsequently $C_1/C_2$, will remain large for a longer time. Similarly, an increase in the assumed exchange depth will produce the same effect.

The value of $C_2$ depends upon the hydraulic residence time ($\overline{T}$), release rate ($K_1$), lake depth (D) and sedimentation rate ($K_2$). However, $C_2$ is relatively insensitive to $\overline{T}$. For example, although Lake Warner has a $\overline{T}$ of only 10 days, the order of decreasing magnitude of the factors comprising $C_2$ is sedimentation (0.176); residence time (0.1); and release (0.048). For a deeper lake, the constant $C_2$ also becomes insensitive to the release rate from the sediment.

## Conclusions

Using experimentally determined release and sedimentation rate constants and appropriate equilibrium sediment-interstitial P relationships, the proposed model predicts the long term approach to a new steady-state in a lake. When P loading is increased, the resultant increase in sediment P provides a substantial reservoir to be released into the interstitial water and subsequently into the lake water upon future decrease of input loading. To illustrate its application, the proposed model predicts a period of over 10 years for lake P to be reduced in Lake Warner from 90 μg/ℓ to 50 μg/ℓ given a decrease in input from 90 to 50 μg/ℓ.

Generalization of these findings to all shallow eutrophic lakes must be carefully considered. Release and sedimentation rate constants as well as equilibrium relationships between the sediment and interstitial water phosphorus must be estimated in each case. Necessarily, the oxidation state of the sediment will greatly affect the release rate constant and the equilibrium relationship assumed.

## Notation List

| | |
|---|---|
| A | sediment surface area ($L^2$) |
| D | Depth of lake (L) |
| $D_e$ | Sediment exchange depth (L) assumed as 10 cm |
| $K_1$ | P sediment release rate constant (L/T) |
| $K_2$ | P sedimentation rate constant ($T^{-1}$) |
| $P_i$, $P_i^*$ | interstitial P ($M/L^3$) |
| $P_L$ | lake P ($M/L^3$) |

| | |
|---|---|
| $P_L\|_{SS}$ | steady-state lake P ($M/L^3$) |
| $P_0$ | input P to lake ($M/L^3$) |
| $P_s$ | exchangeable P in sediment ($M/L^3$) assumed as 25% of total |
| $Q_0$ | flowrate through lake ($L^3/T$) |
| $R_r$ | rate of release of P from sediment ($M/L^3T$) |
| $\overline{T}$ | detention time (T) |
| t | response time (T) |
| $V_L$ | volume of lake ($L^3$) |
| $V_B$ | volume of tank ($L^3$) |
| $V_S$ | volume of sediment ($L^3$) |
| $\varepsilon$ | porosity (assumed as 0.84) |

References


1. Chen, Y.S., J.N. Butler & W. Stumm, 1973. Kinetic Study of Phosphate Reaction with Aluminum Oxide and Kaolinite. Environ. Sci. and Technol. 7:4:327-332.

2. Emery, R.M., C.E. Moon & E.B. Welch, 1973. Recovery of a Mesotrophic Lake After Nutirent Diversion. Jour. Water Poll. Control Fed. 45:5: 913-925.

3. Fillos, J. & W.R. Swanson, 1975. The Release Rate of Nutrients from River and Lake Sediments. Jour. Water Poll. Control Fed. 47:5: 1032-1042.

4. Kamp-Nielsen, L., 1974. Mud-Water Exchange of Phosphate and Other Ions in Undisturbed Sediment Cores and Factors Affecting the Exchange Rates. Arch. Hydrobiology. 73:218-237.

5. Ku, W.C. & T.H. Feng, 1975. Equilibrium Adsorption of Inorganic Phosphate by Lake Sediments. Report No. Env. E. 45-75-1. Dept. of Civil Eng., University of Massachusetts/ Amherst.

6. Larsen, D.P., K.W. Malueg, D.W. Schults & R.M. Brice. Response of Eutrophic Shagawa Lake, Minnesota, U.S.A. Eutrophication and Lake Restoration Branch, Corvallis, Oregon, (unpublished manuscript).

7. Lorenzen, M.W., 1973. Predicting the Effects of Nutrient Diversion on Lake Recovery. In: Modeling the Eutrophic Process, Proc. of Workshop at Utah State. NTIS No. PB227065, pp. 205-210.

8. Snow, P.D., 1976. Mathematical Modeling of Phosphorus Exchange Between Sediments and Overlying Water in Shallow Eutrophic Lakes. Ph.D. Dissertation, University of Massachusetts/Amherst.

9. Snow, P.D. & F.A. DiGiano, 1976. Mathematical modeling of phosphorus exchange between sediments and overlying water in shallow eutrophic lakes. Report no. Env. E. 54-76-3. Dept. of Civil Eng., University of Massachusetts/Amherst.

10. Sonzogni, W.C. & G.F. Lee, 1974. Diversion of Wastewaters from Madison Lakes. Jour. of Env. Eng. Div., Amer. Soc. Civil Eng. 100:153-170.

11. Vollenweider, R.A., 1969. Possibilities and Limits of Elementary Models Concerning the Budget of Substances in Lakes. Archives of Hydrobiology. 66:1-39.

12. Wentz, D.A. & G.F. Lee, 1969. Sedimentary Phosphorus in Lake Cores-Observations on Depositional Pattern in Lake Mendota. Environ. Sci. Technol. 3:754-759.

13. Wildung, R.E. & R.L. Schmidt, 1973. Phosphorus Release from Lake Sediments. Report EPA-R3-73-024, Office of Res. and Monitoring, U.S.Environ. Protect. Agency, p. 185.


Acknowledgements

This research was supported by a grant from the Massachusetts Division of Water Pollution Control and conducted while Phillip D. Snow was a Ph.D. candidate in Environmental Engineering at the University of Massachusetts/Amherst.

Key Words

Modeling, Phosphorus, Sediments, Release Rate, Lake Recovery.

# EXPERIMENTAL INVESTIGATION ON THE RELEASE OF PHOSPHORUS IN RELATION TO IRON IN FRESHWATER/MUD SYSTEM

M.W. Banoub

Landesanstalt für Umweltschutz, 75 Karlsruhe, Hebelstr. 2, BRD

## Summary

The results of the following experiments show in some detail the movement of P and Fe as well as other ions between water and mud as the water changes between oxic and anoxic conditions. This is generally in agreement with previous findings of Einsele, Mortimer, Ohle and Gessner. The use of mud release experiments to indicate degrees of water pollution is discussed together with the use of an enhanced method for the liberation of the mud store by glucose addition. The use of mud as indicator of water pollution is envisaged to reflect the average condition rather than the transient state shown by direct water analysis. The study of some bacterial physiological groups affecting different processes occuring during anoxia in the water and in the mud itself shows the active role of bacteria. Their highest numbers are found at the same level in the mud as the highest concentrations of P and Fe in the interstitial water. The study of the Gnadensee muds at varying distances from its shore, showed a gradual change in its composition and its releasing pattern. This is attributed to a gradual mobilization of P and Fe in opposite directions which may be caused by the various plant communities of the reed, littoral and pelagial zones.

## Introduction

The present state of knowledge on P movement between lake-water and mud, has so far been related to changes in Fe, Mn, Ca, Mg, $CO_3$, $SiO_4$, $SO_4$, OH and H-ions. So far no gaseous form of P is claimed to occur in lakes, thus it is only between water and sediments that P seem to be moving. Iron has been found a major ion in the precipitation of soluble phosphates to the sediments in the form of $FePO_4$ and $Fe(OH)_3$ n.$PO_4$ under aerobic conditions (Einsele, 1936). The process is reversible in absence of oxygen. The role of $Fe(OH)_3$ as adsorbant for $PO_4$ ion further was emphasized by Ohle (1935-38), its presence as well as $FePO_4$ in natural waters was confirmed by Cooper (1948). In extreme anaerobic conditions where reduction affects the $SO_4$-S system, Einsele & Hasler (1948) found that FeS is precipitated leaving $PO_4$ ion free in water. Gessner (1939) proposed another method by which P sedimentation may be affected as coprecipitant on $CaCO_3$, which is formed during periods of active photosynthesis. The presence of $CaCO_3$ in above-equilibrium concentrations was also observed in hard-water lakes Excess $CaCO_3$ in water was found in colloidal form (Ohle, 1952). At the onset of anaerobic conditions iron as well as manganese are found as bicarbonates in feebly alkaline lakes (Hutchinson, 1957). In hard water lakes Ca and Mg bicarbonates are found in addition during hypolimnetic oxygen deficit. The origins of these bicarbonates are traced to $CaCO_3$, $CaSiO_3$, $MgSiO_3$ or in ironorganic complexes by base exchange when metabolic $CO_2$ acts on sediments (Hutchinson, 1957). Mortimer (1941-42) found that $PO_4$ is released from sediments in presence and absence of oxygen, though the release was more pronounced in the latter condition. He related the release and sedimentation of P and Fe, to the oxygen flux into the mud, and its effect on the formation or disruption of the oxidized microzone. He defined the oxidized-zone in mud as the depth where the redox potential reaches at least 0.2 V. Mortimer further suggested that upon reduction of the oxidized microzone anions and bases which were held as complexes with Ferri-ligno-proteins are released together with ferrous ions. The appearance of light-absorbing solutes in water above mud under reducing conditions have been attributed to an increase in total iron (Hutchinson, 1957), or to $Fe(OH)_3$ temporarily formed from $Fe^{++}$ and traces of $O_2$ still found in water (Mortimer, 1941-42).

Changes of iron concentration in water are attributed to release from sediments where the $Fe^{++}$ ion is released at a Redox E = 0.25 V.
A large part of $Fe^{++}$ may exist in complex form. According to Hutchinson (1957), 4 phases of stagnation above mud may be characterized.

1. High $O_2$ - low $PO_4$ - $Fe^{++}$ absent - $H_2S$ absent

2. Low $O_2$ - low $PO_4$ - $Fe^{++}$ absent - $H_2S$ absent

3. Very low $O_2$ - high $PO_4$ - $Fe^{++}$ present - $H_2S$ absent

4. No $O_2$ - high $PO_4$ - $Fe^{++}$ low - $H_2S$ high

The explanation of the release and sedimentation of $PO_4$ based on iron alone seems to be inconsistent. Total iron, $PO_4$ and ferrous ions sometimes appear to be released independently or succesively, for instance an experiment on Esthaite water by Mortimer (Quoted by Hutchinson 1957) and also some experiments done in this work. Hasler & Einsele (1948) found that manganese would be reduced before iron, and could wholly displace iron from the cycle. The significance of this is that when oxidation occurs again in presence of $PO_4$ there is no precipitation of $Mn_3 (PO_4)_4$ as in the case of Fe. This can only happen if appreciable amounts of Mn are found in the sediments to inhibit the iron-reduction and reoxidation in aerobic conditions. They have also suggested the addition of $MnO_2$ as a way of precipitating iron as oxide leaving $PO_4$ free in oligotrophic lakes. In a comprehensive review on sediments of Lake Constance Schmitz (1971) refered to a relation between adsorbed $PO_4$ and that in solution (interstitial water) which is described by Freundlich's relation. The adsorption of $PO_4$ on clay is accompanied by a slow decomposition with the liberation of $SiO_4$, a process called "Phosphatolysis" by Engelhard and Smolinski (1957). Experiments quoted by Schmitz on the exchange of P between sediment and water depend on the building up of chemical and adsorption equilibria.

## Experimental

Model testing of the release of P in relation to Fe

From the above mentioned models it is clear that anaerobia, presence of iron, $CO_2$ as well as other ions may affect directly or indirectly the release of P. To test their effect separate and in combination, prolonged anaerobia, aerobia, addition of iron and glucose (as a source of $CO_2$), were investigated.
In previous experiments it was found that the release of P from Gnadensee muds was mostly inorganic, its rate was enhanced by temperature, low oxygen and by the addition of leachates from dead reeds (Banoub, 1975). The release of P from mud occured at times without accompanying release of iron in any form. In the following experiment plexiglas cylinders (diameter 11 cm, height 30 cm) were filled to a depth of 10 cm with surface mud from Gnadensee (Lake Constance) containing 70-80 % carbonates (mostly calcium) and 6-7 % organic matter. One litre of lake water was added on top of the mud in each cylinder. The experimental setup (Banoub, 1975) includes a water tank with temperature kept thermostatically at 15°C, a sliding cover to keep the cylinders in the dark, and a nitrogen and air sources to adjust the system for either aerobic or anaerobic conditions.

Anaerobic condition was achieved by displacing air by nitrogen gas above the water surface and sealing the cylinders. Aerobic condition was reached by keeping the cylinders open under a draught of air to ensure efficient oxygen exchange. Glucose was substituted for reed leachates as a source of organic matter and to exclude any complexing compounds that would be present in leachates. Five operations were conducted, fig 1.

1 - Establishing an equilibrium under anaerobic conditions during which Ca (50mg Ca/l) and Fe (10mg Fe/l) were added. This operation did not cause any change in P concentration.
2 - Glucose was then added at the rate of 20mg C/l under the same anaerobic conditions. This caused release of both P and $Fe^{++}$. The latter slowly sedimented afterwards leaving P in water.
3 - The system was turned aerobic with further addition of glucose (20mg C/l). This also released P and Fe. The latter was mostly in the $Fe^{+++}$ form which quickly sedimented carrying back all the newly released P.

4-$FeCl_3$ were added under the same aerobic conditions (5mg Fe/l), co-precipitating almost all the P in the water.

5- The system was turned anaerobic again and glucose was added at the above concentration, which lead to the release of P without any appearance of Fe.

The P/Fe ratio by atoms of the released ions in operations (2) and (3) are 1.6 and 0.8 respectively, the former of which is undoubtedly too high to account for any iron phosphate dissolution alone, while adsorption of P on $Fe(OH)_3$ must be excluded due to the prolonged anaerobic conditions beforehand. The carbonates are the most probable source and they are known to adsorb P.

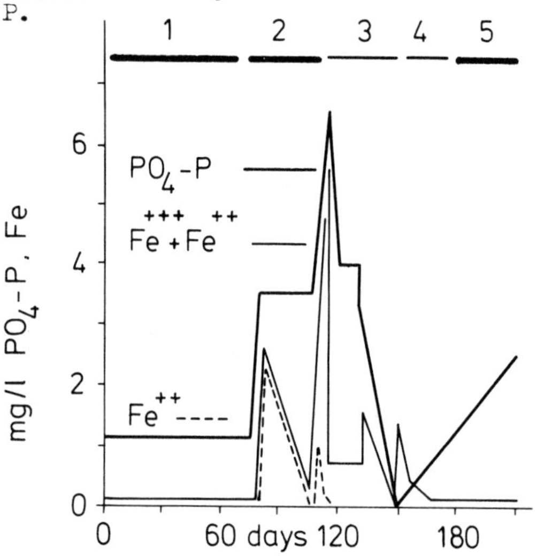


Fig 1
Experimental release and precipitation of P and Fe under various oxygen conditions:

1-Anaerobic period.
2-Anaerobic period, addition of glucose.
3-Aerobic period, addition of glucose.
4-Aerobic period, addition of Fe.
5-Anaerobic period, addition of glucose.

The experiment proves that the mechanisms of P release mentioned above are to a great extent true and that they explain the mobility of P and Fe in calcarious sediments. The principle of the oxidized micro-zone was not quantitatively established here, though its appearance in aerobic conditions was clearly seen. During anaerobic operations the mud swelled substantially, gas bubbles formed and might have helped in the release of materials from the mud. Th occasional release of P alone without accompanying release of Fe is thought to be dependent on the presence or absence of a well developed oxidized micro-zone. The decomposition of organic matter will affect the dissolution of mud carbonates (especiall Ca in this case) and thus releasing its adsorbed P with it. It seems tha iron could only be measured along wit P after intense anaerobia leading to the disruption of the oxidized micro-zone, which could allow ferrous ions for longer periods in solution before they are reprecipitated as sulphides.

Comparison of different muds

The study was further carried out on lake sediments of varying compositions. For the following experiment, four mud samples taken along a transe across a reed zone in Gnadensee (lake Constance), and a fifth mud sample from a newly dredged gravel-pit. The locations and composition of these muds are as follows :

1) 20m inside the reeds (Phragmites) org. matter 9.4%, $CaCO_3$ 73%.
2) Edge of the reeds with the lake, org. matter 3.9%, $CaCO_3$ 78%.
3) 50m inside the lake and away from the reeds, org. matter 2.6%, $CaCO_3$ 68
4) 500m inside the lake and away fro the reeds, org. matter 1.8%, $CaCO_3$ 68
5) Middle of a gravel-pit 14m deep, 220x170m area and 2 years old at Langenbrücken, org. matter 2%, $CaCO_3$ 28%.

Three operations were conducted on these muds according to the following scheme:

I- A period of aeration of 6 days.
II- A primary anaerobic period 13 da
III- Prolonged anaerobic period of 60 days.

During the aerobic condition, the pH tends to increase, conductivity remains almost the same, $PO_4$ decrease while Ca and Mg did not change appreciably. After the first anaerobic period the pH decreased, conductivity Fe, $PO_4$, $NH_4$ and $NO_2$ increased. The changes were more pronounced by the end of the second anaerobic period. The difference in the magnitude of the release by the 4 Gnadensee muds suggests that iron is released more from deeper lake mud than that from reed muds. The opposite is found wit regard to $PO_4$ release, while Ca and Mg did not show a clear trend. In each case $NH_4$ increased under anaerobic conditions. Nitrite was produced to a very marked degree by the reed mud (1). The difference in the release of Fe and $PO_4$ between the various muds of the Gnadensee seems to be correlated with the concentrations of both elements in their respective dry muds. Higher concentrations of P in dry mud are found nearer to the reeds and at the top of the deeper sediments. On the other hand,

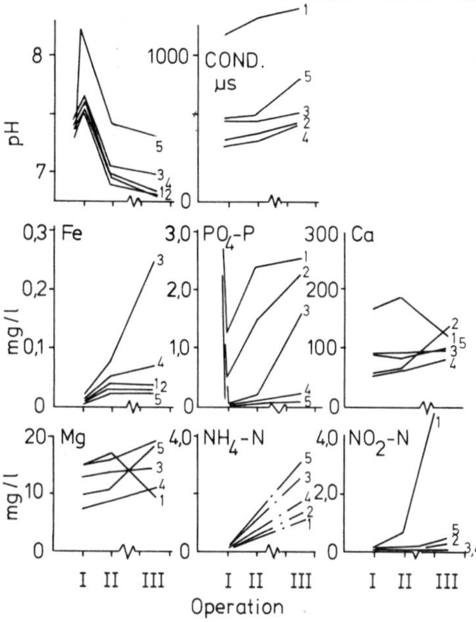


Fig 2

Release of Fe, P, Ca, Mg, $NH_4$, $NO_2$ from various muds: (1) 20m inside reeds, (2) Edge of reeds, (3) 50m outside the reeds, (4) 500m towards lake middle, and (5) middls of a gravel-pit in Langenbrücken.
(I) end of the aerobic period, (II) end of first anaerobic period and (III) end of the second anaerobic period (see text)

appreciably less iron is present in reed muds than in deeper sediments, fig 3. The mud from the gravel-pit released low concentrations of P and Fe even after prolonged anoxia and its apparently high ammonium-release may have originated from the high nitrate concentration of its water (20 mg $NO_3$-N/l). It appears from the above that the release of P and Fe from different muds depends on the concentration of both elements and their matrix in the solid phase. The difference between the Gnadensee muds is most probably due to the plant communties occupying each mud site. The muds from (1) and (2) are in the reed zone and are affected by the reed-rhyzomes as well as by their decay products. The mud in (3) is in a submerged hydrophyte zone, and the mud (4) is in the open water (Pelagial). The plant communities mentioned seem to affect the transport of Fe towards the deeper lake sediments and P towards shallower sediments. A similar P transport was also reported by Hutchinson (1957).

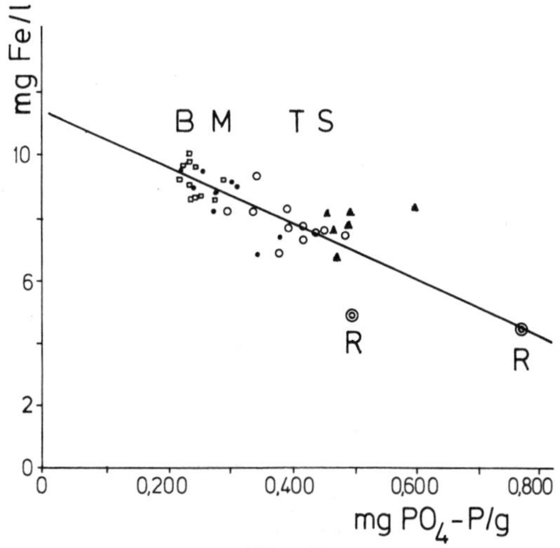


Fig 3

Variation of Fe with respect to P in the sediments of Gnadensee, B= 30 cm deep in lake sediments, M= 20 cm deep in lake sediment, T= 10 cm deep in lake sediment, S= upper centimetre of lake sediment and R= upper 10 cm of reed mud.

Bacterial and chemical changes in mud and water.

A study of the role of bacteria on the processes accompanying the exchange of materials between mud and water is here conducted. A surface mud sample (1.8% org. matter, 68% $CaCO_3$) was homogenized, divided into cylinders for aerobic and anaerobic study of the release of P, Fe, Mn, Ca, $NH_4$, $NO_2$, $NO_3$ and for the growth of some bacterial groups both in water and in mud . (Ritter, unpublished). The results, fig 4, show that in the anaerobic mud system there was no appreciable change in either Ca or Mg in water. The pH and conductance increased in the first 15 days of incubation then decreased afterwards. Release of ammonium ion from mud was very impressive, reduction of $NO_3$ continued till completion and nitrite showed an increase at the start. Iron and Mn (total Fe and total Mn) followed each other at a very low concentration. Phosphate increased to about 1 mg P/l after 64 days. In the aerobic mud system, the pH decreased at first then increased again. Calcium showed appreciable variations which seem to change inversely with conductance. This may suggest the formation of a colloidal form of calcium.

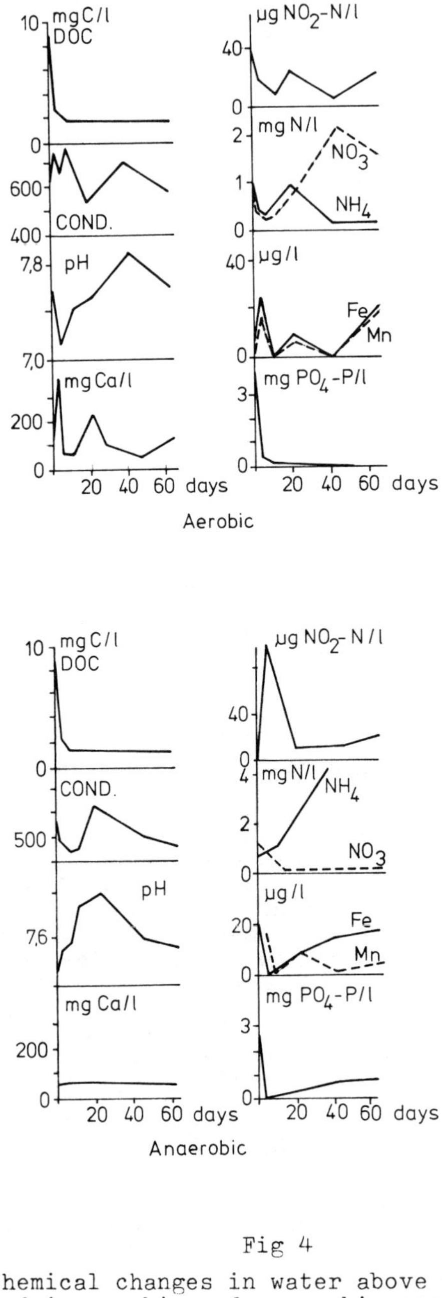


Fig 4

Chemical changes in water above the mud in aerobic and anaerobic conditions.

Phosphate decreased continuously to almost zero, while iron and manganese were found in water at very low concentrations which followed the revers trend of the pH changes. Ammonium wa released first, then oxidized to nitr The nitrate was however lost from the system at the end probably as N-gas b denitrification.

In this experiment it was intended to follow the variation of some bacte ial physiological groups in the water above the mud as well as in the inter stitial water before and after the termination of the experiment. In general, an increase in bacterial numbers occured upon the first contac between water and mud followed by a gradual decrease towards the end of the experiment. The following bacter ial groups: heterotrophs, nitrate - reducers, ammonifiers and phosphatase positive showed an increase in the first 10 days after which they decrea ed again. The nitrate-reducers and sulphate-reducers increased to a constant number in the first 10 days and remained till the end of the experiment. There were no wide differences between the two mud systems. The distribution of bacterial groups in the 10 cm of mud is shown in fig 5, where highest numbers of bacteria wer found few millimetres below the mud-surface. Higher bacterial numbers were found in the system that was kep anaerobic. Phosphate-releasing bacte ia were most abundant at the mud surface and almost absent below 2 cm depth in mud. Interstitial water, fig 6, shows that at 1 cm below the mud surface, higher $NO_2$, $PO_4$ and Fe concentrations were found, while ammonium increased with depth in mud. With such high concentrations of ions and bacteria only few millimetres fro the water, a physical barrier must in effect be controlling the release of these ions by diffusion.

## Mud releases as a criterion for water quality

The chemical approach for the deter mination of water quality of a lake, river or effluent, is to perform some chemical analyses of its water, quali is then based on one or several chemical indicators of pollution. The accuracy of this method, especially in case of effluents depends on the frequency of sampling so that the decision would be a fair representation of the average condition. The number of samplings needed depends on the degree of variability of the

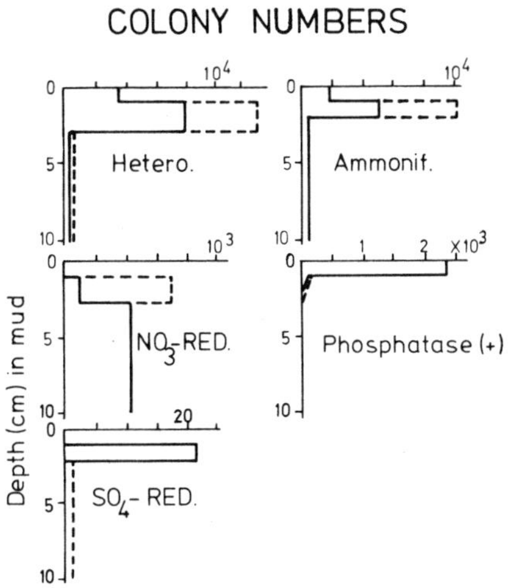


Fig 5

Vertical distribution of bacteria in mud after termination of the release experiment; colony numbers of heterotrophs, ammonifiers, nitrate-reducers, sulphate-reducers and phosphatase(+), aerobic (—), anaerobic (--) groups. (after Dr. R. Ritter)

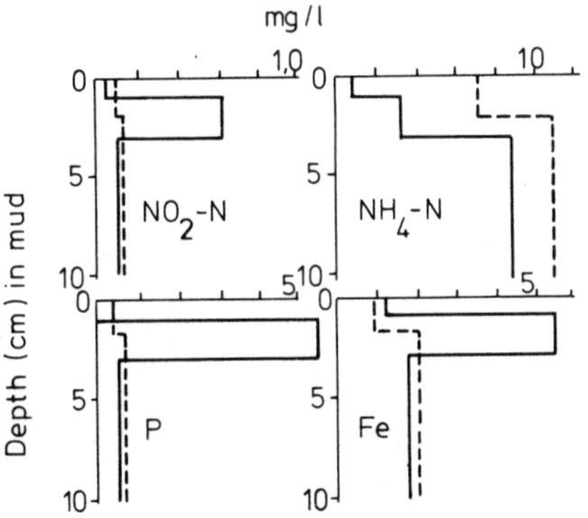


Fig 6

Vertical distribution of $NO_2$, $NH_4$, $PO_4$-P and reactive iron in mg/l of the interstitial water of the mud experiments after 60 days.
— aerobic, --- anaerobic experiments.

ffluent water. The expences of arrying out such samplings and their ubsequent analyses increase enormousy when the investigation covers a ide area. As the sediment of each ake, river or stream is nothing but the accumulation of material which was once carried by the water itself, it is thus a record of the past and present of the water frequently occupying that site. The experiments on the release of materials from mud indicate the future as well as the past, and may define the source of pollution (organic, metallic or toxic) if that would be required.

A simplified version of the above experiments uses the inner plexiglas tubes of gravity corers as sampling and reaction vessels at the same time. The sampled mud together with the overlying water constitute the mud/water system after being closed with rubber bungs from both ends. At first, the water is aerated to saturation, and $O_2$ uptake by the mud determined. The water is then aerated to saturation again, glucose is added at the rate of 20mgC/l of the overlying water, then the tube is sealed for 2 weeks after which the water was siphoned and analysed. The results of some of these experiments are shown, fig 7, in comparison with direct water analyses of the effluent water. Some differences in water

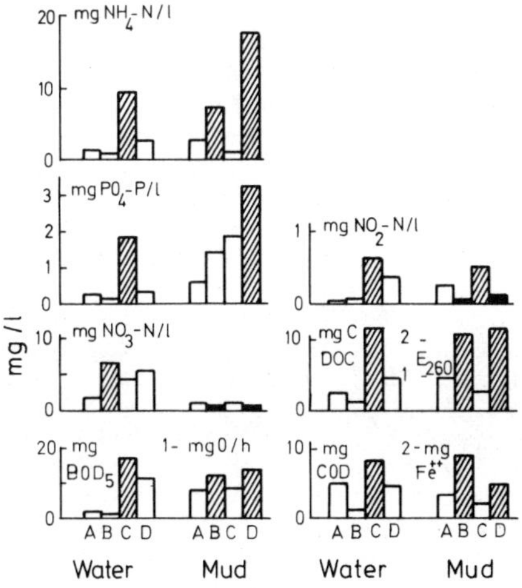


Fig 7

Comparison between direct water analyses and mud-release analysis for 4 effluents: A, B, C and D. While water analysis shows that sample C is more polluted than the others, mud releases from samples B and D indicate that their muds are highly polluted.

quality rating of the sampling sites by both methods are found as has been expected. The muds from samples B and D released higher $NH_4$, $PO_4$ and Fe, and consumed more oxygen than the others. All these changes classify them in a higher pollution level than the others. Direct water analysis on the other hand indicated that sample C contained higher $PO_4$, $NH_4$, DOC and demanded more COD and BOD, however, after aeration all these contaminants dropped to a low level and the mud showed upon anoxia lower releases than the others. In other words, although sites B and D contained relatively cleaner water at the time of sampling, yet their muds showed more pollutants reserve than the rest.


Acknowledgement

I am greatly indebted to Dr. R. Ritter for lending me a free hand to use his unpublished bacterial data.


References


Banoub,M.W.,1975. Experimental studies on material transactions between mud and water of the Gnadensee (Bodensee). Verh.Internat.Verein Limnol. 19:1263-1271.

Cooper,L.H.N.,1948. Some chemical considerations on the distribution of iron in sea water. J.Mar.biol. Ass.U.K. 27:314-321.

v.Engelhardt,W.&A.v.Smolinski,1957. Die Reaktion von Polyphosphaten mit Tonmineralien. Kolloid.Z.151:47-52.

Einsele,W.,1936. Über die Beziehungen des Eisenkreislaufes zum Phosphatkreislauf im eutrophen See. Arch. Hydrobiol.19:664-686.

Gessner,F.,1939. Die Phosphatarmut der Gewässer und ihre Beziehung zum Kalkgehalt.Internat.Rev.ges.Hydrobiol.38:202-211.

Hasler,A.D.& W.G.Einsele,1948.Fertilization for increasing productivity of natural inland waters.Trans.13th. N.Amer.Wild L.Conf:527-554.

Hutchinson,G.E.,1957. A Treatise on Limnology,Vol.I,John Wiley & Sons Inc.,New York,pp 1015.

Mortimer,C.H.,1941 & 1942. The exchange of dissolved substances between mud and water in lakes. J. ecol.29:280-329. 30:147-201.

Ohle,W.,1935. Organische Kolloidgele in ihrer Wirkung auf den Stoffhaushalt der Gewässer. Naturwissenschaften.23:480-484.

- 1938. Die Bedeutung der Austauschvorgänge zwischen Schlamm und Wasser für den Stoffkreislauf der Gewässer. Vom Wasser.13:87-97.

Ohle,W.,1952. Die hypolimnische Kohlendioxid-Akkumulation als produktionsbiologischer Indikator. Arch. Hydrobiol.46:153-285.

Schmitz,W.,1971. Bodensee-Sedimente. Bericht Nr.9, Internat.G.S.K. für den Bodensee,pp 38.



Key words: Phosphorus
Iron
Mobility
Sediment
Pollution

# PRELIMINARY STUDY OF THE SILICA CONTENT AND DIATOM ABUNDANCE IN KAINJI LAKE, NIGERIA

H. A. Adeniji

Kainji Lake Research Institute, New Bussa, Kwara State, Nigeria

## Summary

This paper reports on the study of the silica content and the diatom abundance near the bottom sediment in Kainji Lake. The observed annual variation in the silica content was from 2.6 to 11.0mg./lit. The peak was observed around March/April and the lowest was observed in July. High silica content which was observed in the deoxygenated hypolimnion may indicate anaerobic reduction of silica compounds in the lake's sediment. The seasonal abundance of the diatoms varied from 100 to 10225cells/ml. The highest abundance was observed in June and the lowest occured in October. The seasonal variations in the diatom abundance was observed to be inversely related to silica concentration. This indicates the possible release of silica from the lake's sediment and the use of the silica so released, by diatoms in the formation of their frustules.

## Introduction

### Scope

This study investigates the variations in the silica content near the bottom sediment in the Kainji Lake and its effects on the diatom abundance at this level in the lake.

### Problem

Diatoms have been observed in significant abundance in the river Niger before impoundment and also in the Kainji lake since its formation in 1968. It is also known that diatoms form one of the major pelagic primary producers in this lake. Since it is believed that the siliceous cell walls of dead diatoms contribute to the silica load in the bottom sediments of this lake, it became necessary to study how the silica in this sediment is recycled into the lake water above the sediment and its resultant effect on the diatom abundance.

This present study is a preliminary effort which was made to investigate this problem. The study which covers a period of one year, April 1971 to March 1972, was made at a sampling station on the lake as shown on Fig. 1. Many authors such as Imevbore (1970), El Zarka (1973) and Henderson (1973) have described the Kainji Lake in detail. A general description of the lake was also given by Adeniji (1975). Data were collected from the sampling station at intervals of about six weeks. These include data on the vertical distribution of

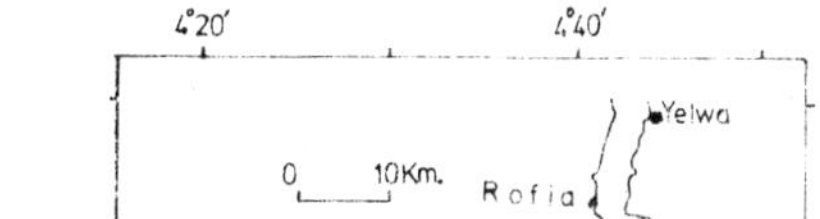


Fig. 1. Map of Lake Kainji

temperature and dissolved oxygen concentration. The silica concentration and diatom abundance were also studied at a distance of about one metre from the top of the bottom sediment. Most of the methods of collection and sample analysis that were used in this study have been described by Adeniji (1975) but the silica content was analysed by using the Heteropoly blue method of the Hach technique.

### Previous works

The study of silica in the Kainji Lake area started before the lake was formed. White et al. 1965) studied the chemistry of the river Niger and its tributaries in the region now covered by the lake. They recorded silica concentrations ranging from 0.3 to 1.5mM/lit. Imevbore (1970) also studied the chemistry of the river Niger in the Kainji reservoir area. He indicated that the silica content in the Kainji reservoir area ranged from 1 to 200mg./lit.

The study of the diatoms in the Kainji area was also not neglected. Eaton (1965) who did a quantitative analysis of the algae in the river Niger before impoundment showed that the phytoplankton in the Kainji Lake area was dominated by blue-green algae and diatoms of the genera Melosira and Nitzschia. This dominance by diatoms was also reported in the Kainji Lake especially during the rainy season as shown by Adeniji (1973). Though diatoms have been observed in significant abundance in the Kainji Lake since its formation, it was interesting to note that no study had yet been done on the relationship between silica and diatom abundance in the lake. This therefore prompted the preliminary study of the variations in the silica content near the bottom sediment of this lake and its effect on the diatom abundance at that level. The results obtained from the study are in this paper.

## Results and Discussion

Fig. 2 shows the annual variation in the silica content of the Kainji Lake water near the bottom sediment. The silica concentration was at its peak, 9.7 to 11.0mg./lit., during April/March, but it started to drop thereafter reaching its lowest concentration of 2.6mg./lit in July. From August it started to rise but with a slight drop in December before it reached the peak again in March.

The high silica content which was observed in March and April occured at a time when this station was thermally stratified as shown on Fig.3. This was also the time that the point of sampling near the bottom sediment was anoxic. These observations may therefore suggest the possibility of the release of silica from silica-containing compounds, especially from dead diatom cell walls in the sediment, into the overlying water body during thermal stratification and in the absence of oxygen. It is interesting to note that Mortimer (1941-42) found in his experimental tanks set up over mud from Lake Windermere, that the silica content rose from a few to 20 or 30mg./lit. under anaerobic conditions and that this rise was much greater in the anaerobic tanks.

The rise in the silica content during September and October may either be due to the release of silica from the sediments, or to its introduction into the lake from

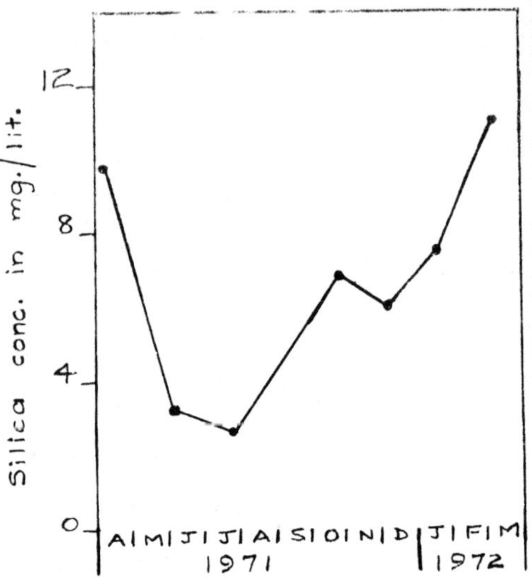


Fig. 2. Silica content at station S near the bottom sediment.

the inflowing river Niger and its tributaries. Fig. 4 shows that by September

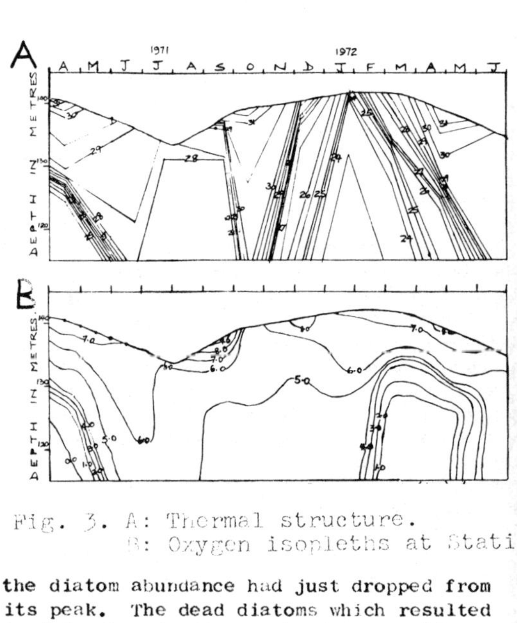


Fig. 3. A: Thermal structure.
B: Oxygen isopleths at Stati

the diatom abundance had just dropped from its peak. The dead diatoms which resulted in this drop in diatom abundance would sink and remain in the sediment where the silica in their cell walls will be released into the water above the sediment thereby increasing its silica content. Adeniji (1975) observed that there was an influx of

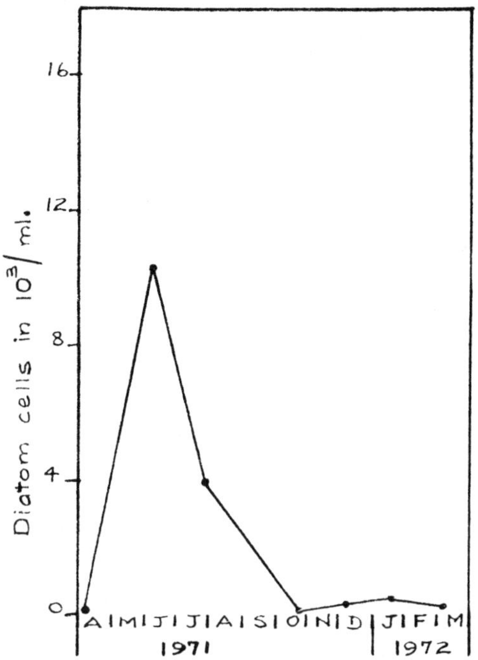


Fig. 4. Diatom abundance at station S near the bottom sediment.

turbid water into the lake between August and October. In fact this has always been a regular feature in this lake since it was formed. Imevbore (1970) further showed that the fine particulate clay which caused this turbidity contributes to the ionic composition of the lake water when this clay is worked upon by bacteria. He observed that the silica ions were one of the ions released in this process. He also noted that the second flood that enters the lake between December and February has a higher salinity due largely to bicarbonate ions but contains a lower silica content. There is therefore an indication that the sediment and the turbid inflow contribute to the rise in silica content in the lake at that time.

The drop in the silica content in June and July may either be due to the high abundance of diatoms at that time as shown on Fig.4, or to the introduction of oxygen at this depth when stratification has been destroyed Fig. 3. The use of silica in the formation of cell walls by these abundant diatoms could have resulted in a lowering of the silica content at that depth. In fact, Yoshimura (1930) showed that the observed annual variations in the silica content present in Takasuka - numa lake were due largely to the development of diatom populations. The introduction of oxygen into the deoxygenated hypolimnion which contained high silica content has been shown by Mortimer (1941 - 42) to reduce the silica content at that level. He also observed a slight drop in the silica content of his experimental tanks when oxygen was accidentally introduced into the anaerobic tank through a crack in the parafin seal of the tank. He attributed this reduction in the silica content to the precipitation of a ferric silicate complex.

The slight drop in the silica content which was obsered in December as shown on Fig. 2. is believed to be due to the vertical mixing that occured in the lake then as shown on Fig. 3. A slight increase in diatom abundance was also observed at that time Fig. 4. This might have been stimulated by the high silica content in the preceding month Fig. 2., or by the vertical mixing of the lake water which introduced diatoms from the epilimnion into the sampling area near the bottom mud.

The low level of diatom abundance observed between October and March could be due to

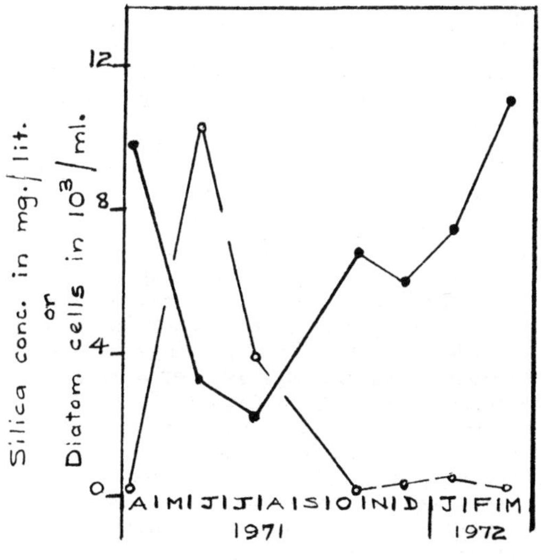


Fig. 5. Silica content •—•—• related to Diatom abundance o— —o at station S near the bottom sediment.

the very low water transparency during that period. Adeniji (1975) has shown that the lake water is generally more or less turbid throughout the year with maximum transparency not exceeding 2.5metres. He also showed that phytoplankton abundance in this lake dropped significantly at low water transparency and that it started to rise again in April when the water transparency improved.

There is no doubt that the diatoms in this lake make use of the silica available in the lake in the formation of their cell walls. Hutchinson (1957) indicated that the development of diatom blooms constitutes the most important mechanism by which silica is removed from lake waters. It is therefore interesting to observe that the annual variations in the silica content near the bottom sediments in the Kainji Lake is inversely related to the seasonal abundance of the diatoms in this part of the lake as shown on Fig. 5.

This preliminary study has not looked into the process whereby silica is released from the bottom sediments in this lake. It is however believed that financial assistance from the Kainji Lake Research Institute, who financed this preliminary work, will be available in the nearest future for research activities to be extended to this important aspect of the sediment.

## References

Adeniji, H.A. 1973. Preliminary investigation into the composition and the seasonal variation of the plankton in Kainji Lake, Nigeria. Geopys. Monogr. Ser. 17: 617 - 619.

Adeniji, H.A. 1975. Some aspects of the limnology and the fishery development of Kainji Lake, Nigeria. Arch. Hydrobiol. 75, 2: 253 - 262.

Eaton, J.W. 1965. Algal investigations. In: E. White (Ed.): The first scientific report of the Kainji biological research team. University of Liverpool. p. 8 - 16.

El Zarka, S.E. 1973. Kainji Lake, Nigeria. Geophys. Monogr. Ser. 17: 197 - 219.

Henderson, F. 1973. A limnological description of Kainji Lake, 1969 - 1971. F.A.O. Technical paper No. 10.

Hutchinson, G.E. 1957. A treatise on limnology. Chapman and Hall, London.

Imevbore, A.M.A. 1970. The chemistry of the river Niger in the Kainji reservoir area. Arch. Hydrobiol. 67, 3: 412 - 431.

Mortimer, C.H. 1941 - 42. The exchange of dissolved substances between mud and water in lakes. J. Ecol. 29: 280 - 329; 30: 147 - 201.

White, E., Eaton, J.W., Lutz, P., & Osinowo, A. 1965. Water chemistry. In: E. White (Ed.): The first scientific report of the Kainji biological research team. University of Liverpool. p. 3 - 7.

Yoshimura, S. 1930. Seasonal variation of silica in Takasuka-numa, Saitama. Jap. J. Geol. Geogr. 7: 101 - 113.

Key words: Bottom sediment
Diatom Abundance
Lake Kainji
Silica content
Silica recycling

# CONTRIBUTION ON THE ROLE OF SUSPENDED SOLIDS TO THE CHEMISTRY OF LAKE KAINJI

A.M.A. Imevbore and Funso Adeniyi

Department of Biology, University of Ife, Nigeria

## Summary

The suspended solids (mostly clay and silt) influence the chemistry of Lake Kainji water considerably. It was found that substantial amounts of $Sio_3$, $Po_4$ and $No_3$ were contributed by the solids to the water.

## Introduction

Previous studies have provided a body of data on the general limnology and water chemistry, (Henderson 1973, Imevbore, 1970a, 1975a and Adeniji 1975) plankton, (Adeniji 1973a and b, Karlman, Imevbore 1975b & c) primary production, (Karlman 1973, Imevbore 1975c) thermal stratification (Henderson, 1973a) and the mixing pattern of the waters of Lake Kainji (Zimmerman et al, 1976). Asseez (1970, 1970a) also measured the suspended solids in the lake during the first six months following the closure of the dam in 1968 and found values between 127-184 mg/1. He also described the material as mostly clay and silt.

These studies have shown that the seasonal occurrence of floods dominate the events in the lake. There are two floods annually. The major one called the White Flood coincides with the peak of the local rains between July and September while the subsidiary flood (Black Flood) enters the lake between November and February. The differences in their origin and colour patterns have been described (Nedeco 1959 and Imevbore 1970b).

As in every reservoir, it is expected that much of the river sediment is trapped within the reservoir. Based on the 1959 river flow, Nedeco and Balfour Beatty (1961) assessed the annual sediment load likely to enter Lake Kainji to be 4.6 x $10^6$ tons and with a volume of 6.5 x $10^6$ $m^3$. This represents only 0.04% of the maximum volume of the lake. At this rate, they calculated that it would take 2000 years for the lake to be silted up if all the material was trapped in the lake.

The significance of this material does not seem to lie in its quantity but its quality because Henderson and Imevbore have shown separately that the material brought into the lake screens off a very large proportion of the light reaching the surface of the lake. Secchi disk readings provided by the two workers showed that the transparency was generally less than 2m for most of the year and that the material responsible for this high turbidity of the water is the colloidal clay (Kaolinite) in suspension. The aim of this paper is to describe how the suspended solids affect the chemistry of Lake Kainji water. This is important in view of the findings of other workers about how clays entering a lake release adsorbed substances such as phosphates, iron and organic compounds in the water (Golterman 1973, 1973a). There is also evidence that suspended particles provide sites for microbial activity in lakes (Stotzky, 1967).

## Lake Kainji

Much has been written already on the environmental parameters of Lake Kainji (see Imevbore and Adegoke, 1975). The lake has an area of 1250$Km^2$, maximum volume 15$Km^3$, at 108m above sea level and with a maximum depth of 55m. The mean depth is 11m Fig. 1 shows the shape of the lake. The annual water-level fluctuation in 10m (Imevbore 1975a). This provides an unstable shore-line which migrates and could expose as much as 100,000 (ha) of the lake bed. Because of both the shallow mean depth and the large ratio of outflow

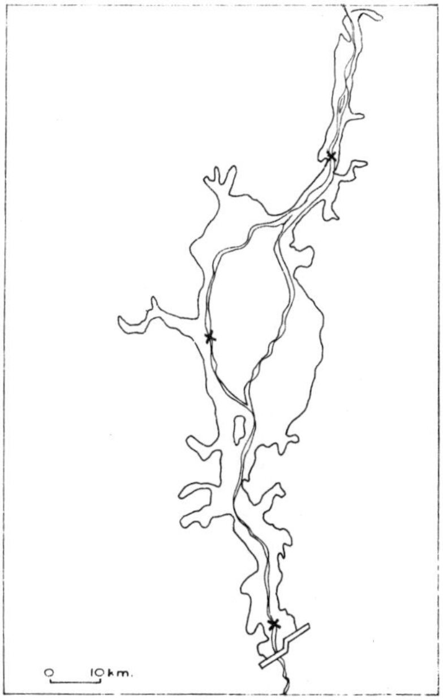


Fig. I. Map of The Kainji Reservoir Showing Sampling Station (X)

to inflow, the reservoir water tends to mix well (Zimmerman et al, 1976). The seasonal changes in the concentration of dissolved ions show an increase after the July-September floods, then a peak at low water when the last shrinks to about one third of its maximum volume. Regarding the chemistry of the water, the $HCo_3^-$ ion is the predominant ion. $Ca^+ + Mg^{++}$ are dominant over Na + K while $No_3$-N, $Cl^-$ and $So_4^{2-}$ occur in relatively smaller quantities. This condition is associated with the presence of sediments of the Nupe Sandstones (Upper Cretaceous) (Imevbore, 1970) Quantities of over 30mg/l for $Sio_3$- and up to 4mg/l for $Po_4$-P have been recorded while $No_3$-N rarely exceeds 110 µg/l. This relative paucity of $No_3$-N is ofcourse typical of most African freshwaters (Talling and Talling, 1965).

## Methods

Water samples were taken at different depths with a Ruttner type sampler from three major stations located in the northern (140Km from dam) central (60Km from Dam) and southern basins respectively. Previous studies by Henderson and Imevbore have confirmed that within limits, each station may be taken as representative of its basin.

The total suspended matter in 200ml of lake water was determined quantitatively by means of evaporation to a constant weight at a temperature of 105$^{o}$C. Then the dissolved substances were removed by filtering the same volume of water through a 0.45 µmm millipore membrane filter under pressure. The residue left after evaporation was dried to a constant weight at 105$^{o}$C. The difference between the weight of the total suspended water and that of the dissolved substances gave the weight of the suspended solids. This is the same as the non-filterable residue in the APHA Handbook (Standard Methods, 1965).

Water temperature was measured using a Grants thermistor probe; transparency with a Secchi disc, light penetration with Selenium photocells; Chemical factors $K^+$, $Na^+$, $Ca^{2+}$, $HCo_3$-, $Sio_2$-Si, $Po_4$-P, $No_3$-N, $No_2$-N, $O_2$, Colour, BOD Organic matter, were all analysed following APHA Standard methods and IBP Handbook No 8 (Golterman and Clymo, 1969) Particulate $No_3$-N and particulate organic matter were measured by the difference between the total in unfiltered water and the concentration in the filtered water. Electrical conductivity ($K_{20}$) was measured using a portable meter. Chlorophyll a, b and c were determined according to Strickland and Parsons (1965) as follows:

$$Chla = 15.6D_{665} - 2.0D_{645} - 0.8D_{630}$$

$$Chlb = 25.4D_{645} - 4.4D_{665} - 10.3D_{630}$$

$$Chlc = 109D_{630} - 12.5D_{665} - 28.7D_{645}$$

Margalef pigment ratio was calculated thus $\frac{D_{430}}{D_{665}}$

Weather data as well as inflow and outflow figures were obtained from the Dam authority officials. The chemical results are part of the survey of the chemical composition of the lake waters and will be published in full later.

## Results

### (a) The available suspended solids

The results obtained show that the muddy and turbid waters of the White Flood carry suspended solids in the range of (58-200mg/l with a mean of over 156 mg/l). In terms of the vertical distribution of the material, it was found that the surface and middle layers of the lake carry more or less uniform material which gets higher at the bottom. Asseez (1970) attributed this to the existence of a strong bottom current in the lake.

The waters of the Black Flood carry less material in suspension (46-96 with mean of 74mg/l) but the relationship between depth and material carried is the same with the White Flood. Between March and June, the water is fairly clear carrying solids in the range of (12-37 mg/l)

The data on the occurrence of suspended solids in the three basis of the lake agree very much with the view of Asseez that the flow across the lake is a strong one. Generally it can be concluded that the suspended solids are contributed both by the River Niger and the tributary streams and that a good proportion of the material does not infact settle within the lake.

Fig. 2 plots the amount of suspended solids available in the surface water (Central basin) against the transparency and water level. The relationship between suspended solids and transparency is a clear one and is discussed below.

### (b) Transparency of the water

Fig. 2 shows that transparency of the water increases during the floods and decreases between March and June when waterlevel is low. The relationship between transparency and suspended solids is described in Fig. 3. The relationship is a linear one with the equation

$Y = -0.007 + 1.25x$ where $y$ = transparency (m) and $X$ = suspended solids in mg/l

The suspended solids also affected the colour of the lake. The colour suspended solids relationship is described by the equation

$$y = 0.11x + 0.72$$

Fig. 2 Waterlevel transparency and suspended solid in Kainji Lake

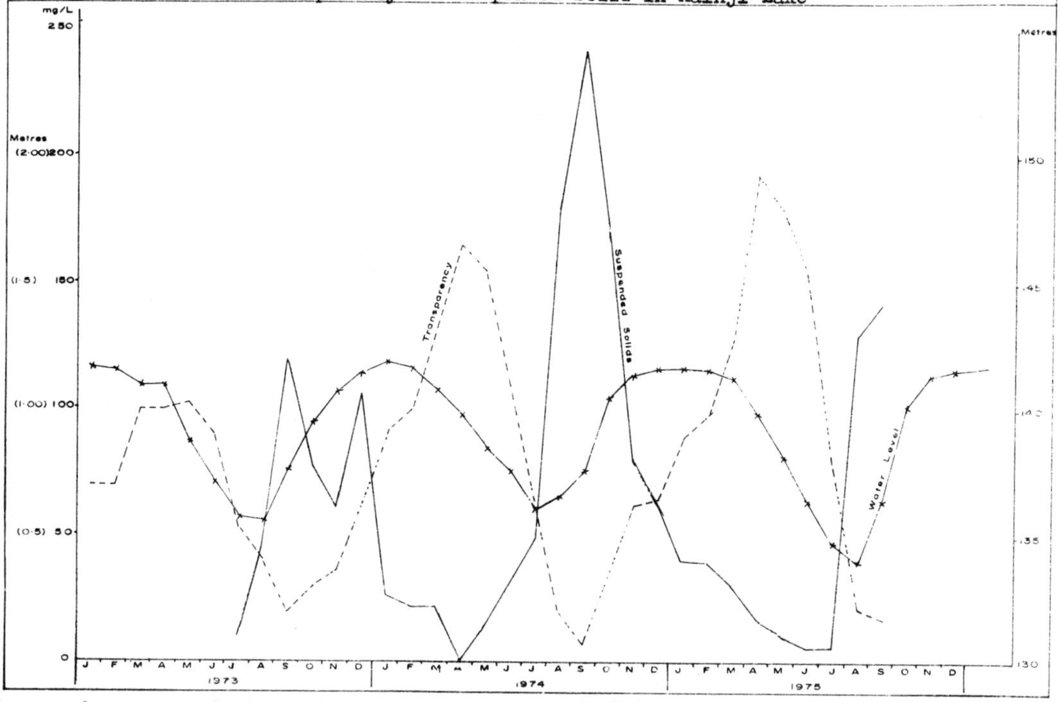


where y = colour in pt units and X is suspended solids.

Fig. 4. also shows that the suspended solids ensure that nearly all the light which penetrates into the water is absorbed between the surface and 2m during the floods. By contrast, the light penetrates as deep as 5m during clear water. However, most of the year over 80% of the light reaching the surface is absorbed within the top 1m. Under these conditions the effect of turbidity and rapid absorption of light upon the photosynthetic processes are bound to be considerable.

Fig. 3 Relationship between transparency and suspended solids

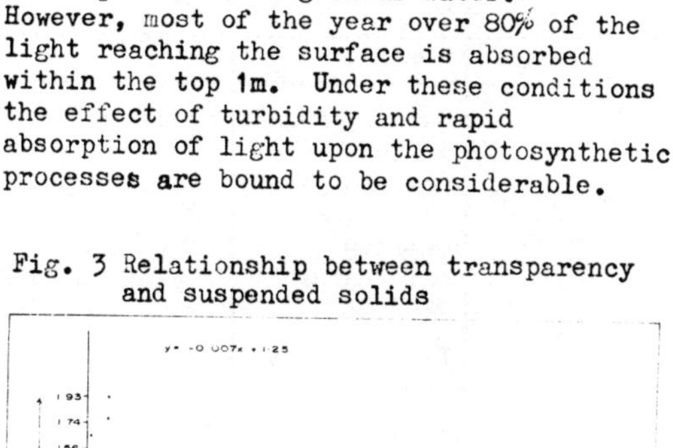


## (c) Water turbulence

According to Perkins (1974) the height to which a particle of sediment is carried into suspension is dependent upon the turbulence in the water. Some evidence about the turbulence in the lake has been obtained from the annual pattern of thermal stratification takin place in the lake (Fig. 5). A stable stratification was recorded between February and June while the water was generally cooler and mixed from July to January. The depth/time distribution for suspended solids during the same period is presented in Fig. 6. The degree to which this pattern is broadly similar to that of the temperature is due to the parallel response of both to the turbulence in the water. At the same time the turbulence itself must further delay the settlement of the material which is already in suspension. In any case the presence of fine suspended material leads to a decrease in the settling rates of larger particles (Perkins, 1974)

## (d) Interaction between suspended solids and water chemistry

The concentrations of the major dissolved compounds are generally low. Estimates of $Po_4$-P gave values between 0-4.28 mg/l while $Sio_3$ values are as high as 40 mg/l. The high $Sio_3$ value is apparently due to erosion. The values for dissolved iron and $No_3$-N are small (not more than 0.07 mg/l for $Fe^{2+}$)

Some information on the contribution of the suspenced matter to the water chemistry can be obtained from Table 1 which shows a high input from the particulate material. In view of its biological significance, the case of $No_3$-N is particularly interesting.

Fig. 7. Regression lines between suspended solids and various factors

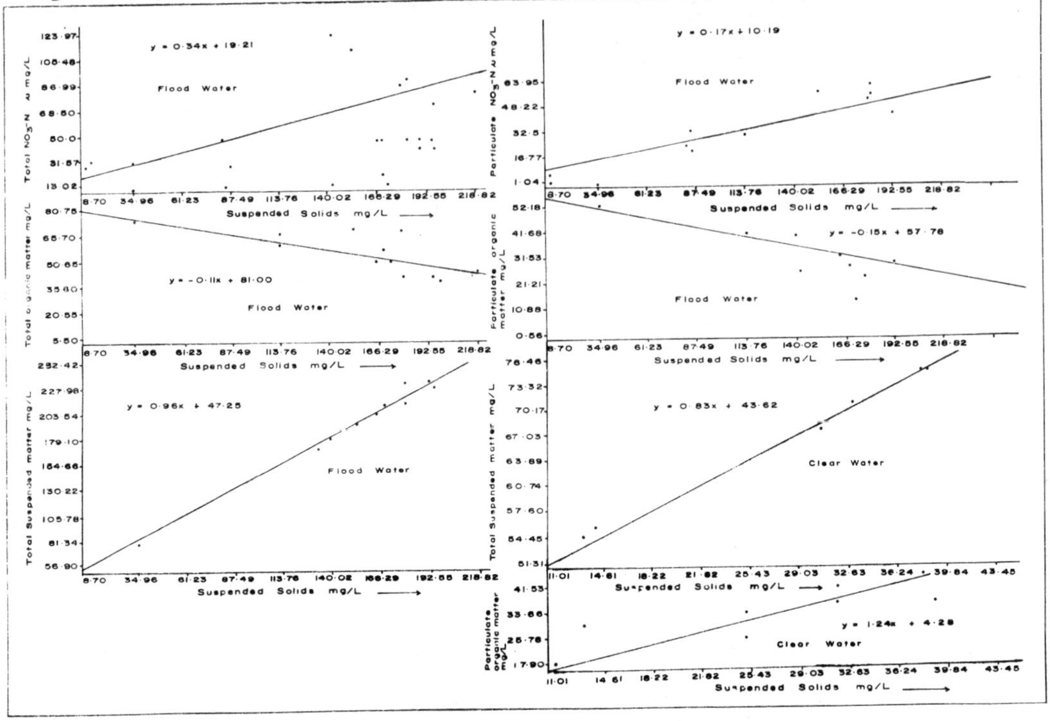


Table 2 summarises the results of correlation factors (r) calculated for various ions with the suspended solids present in surface water. Similar results for the near bottom water have been presented in Table 3.

Table 2. Correlation between suspended solids (mg/l) and other features in Lake Kainji (surface water).

| Correlation | +ve | -ve |
|---|---|---|
| Significant ($r > 0.3$) | Colour, $Fe^{2+}$<br>$No_3$-N, $No_2$-N<br>Assimilation Number<br>Organic matter (dissolved)<br>Organic matter (particulate)<br>Outflow discharge | Transparency<br>$Na^+$, $HCo_3^-$<br>pH, $O^{++}{}_2$,<br>$K^+$, $Mg^{2+}$<br>Productivity |
| Not Significant ($r < 0.3$) | Rainfall, BOD<br>$Po_4$-P, $Co_2$,<br>Chlorophyll a, b,<br>$Ca^{2+}$<br>Margalef's ratio | $Cl^-$, $Sio_3$,<br>Total Org. matter<br>Chlorophyll a<br>Water Temp.<br>Conductivity |

Table 3. Correlation between suspended solids (mg/l) and other features in Lake Kainji (bottom water).

| Correlation | +ve | -ve |
|---|---|---|
| Significant ($r > 0.3$) | Colour, $No_3$-N<br>Inflow discharge<br>Water temp.<br>Outflow discharge<br>$Cl^-$, $O_2^{++}$<br>Productivity | Transparency<br>$Sio_3$, $HCo_3^-$<br>BOD, $Co_2$<br>Org. Matter (particulate) |
| Not Significant ($r < 0.3$) | pH<br>Chlorophyll a,b,c | Water level<br>Org. matter (total)<br>$Ca^{++}$, $Fe^{2+}$<br>$Po_4$-P, $No_2$-N,<br>$K^+$, $Na^+$ |

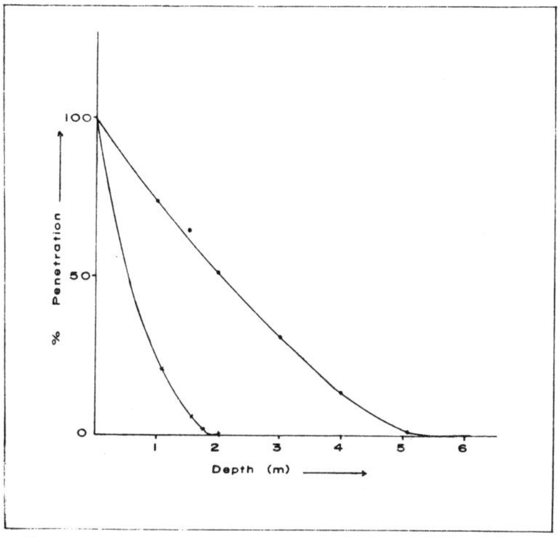


Fig. 4 The percentage penetration of light into the lake water (a- Flood water b-clear water).

## (e) Interaction between external solids and water in relation to internal cycling

In addition to the contribution from the river, there is recycling of nutrients within the lake itself as a result of mineralisation by bacteria and the suspension of detritus. In order to obtain some idea of the relative importance of the suspended solids from these two sources, separate correlation factors have been calculated for the period of floods (July-February) when the lake is mixed and the river brings in external material as distinct from the period of clear water (March-June) when outflow exceeds inflow and period of the lake is strongly thermally stratified. Fig. 7 describes the interesting distinctions which exist in the relationship between suspended solids and

Fig. 5 The depth/time profiles for water temperature (central basin)

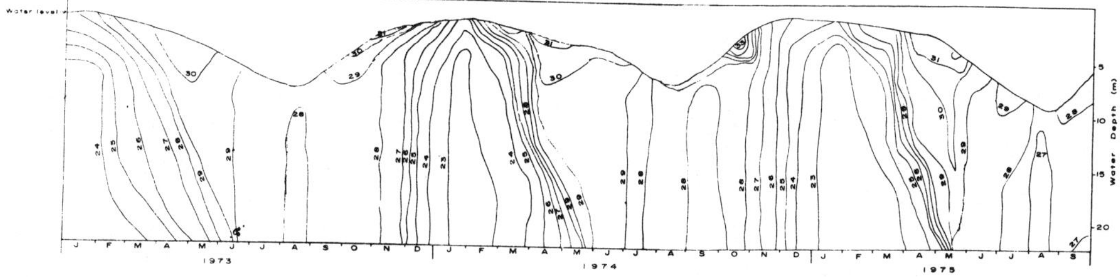


Fig. 6 The depth/time profiles for suspended solids (central basin)

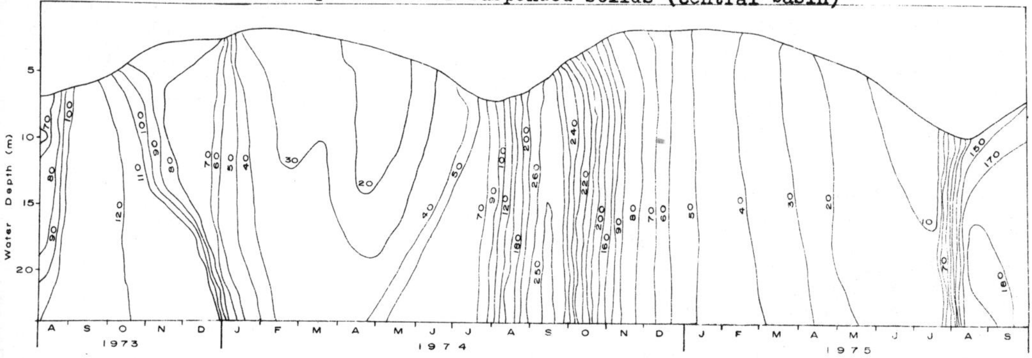

various other factors during the two periods. The different relationships found confirm that the interaction which takes between the suspended solids and the water during the two periods are different.

## Discussion

The properties of the environment which are of particular importance to the suspended solids in the lake are the annual floods and the seasonal pattern of thermal stratification which is itself dependent upon the factors of wind, air temperature and the water regime of the lake. Mclachlan (1974) has pointed out that the lake-level changes of reservoirs cause instability of the shore-line and promote the interaction between the aquatic and terrestial ecosystems. This is particularly the case at Kainji where the annual fluctuation in water level is 10m. This allows materials to be transferred from the seasonally exposed land into the lake in addition to the river sediment carried from upstream soil drainage.

Nedeco _et al_ (1961) have shown that 80-90% of the river material is silt and colloidel clay which remain in suspension for long periods. As clay greatly reduces the erosion capacity of water it means that the sediment load would have been much greater if the river was carrying other types of soils.

As has been shown, the light penetration into the lake is greatly hampered most of the year by the clay materials suspended in the lake. It is clear therefore under the conditions existing, the implications for light sensitive aquatic species will be considerable at Kainji. For this reason, the results of light extinction coefficients will be described in greater detail later.

Anyone who works on the lake is very quickly aware of the seasonal changes in the colour of the lake water. Using data from LANSAT-1 (Abiodun 1976) has described the colour patterns during the White Flood of 1972. His results demonstrate an intricate pattern dominated by the influence of suspended solids. It would be interesting to obtain comparative data for the Black Flood since the colour differences between the two floods has earned them different names.

In terms of the relative importance of physical and chemical weathering, the small ratio of dissolved to total load suggests that physical weathering is predominant (see table 3) This must be the reason for the generally low electrical conductivity of the water ($K_{20}$ 60 $\mu$mho/cm). In this respect, the Niger basin is different from the Amazon basin where Gibbs (1967) found that the high ratio of dissolved to total load suggested a relief dominated by chemical weathering.

The physical factors produced a high degree of turbulence lasting about seven months of the year. Under such conditions the interaction between water and suspended solids will be intimate. One of the controlling factors in this interaction is the density of the solids in suspension. Although we did not measure the size of the particles, we recall Henderson's observation that even in the laboratory less than 50% of the material settled after long periods. In the lake, long settling time of the maternal must be further enhanced by the two floods which occur annually. However, when the floods are over the material tends to settle making the water clearer. This coincides with the period when the water is thermally stratified. Henderson (1973) has suggested that appreciable decomposition of the clays is likely to take place under the reducing conditions prevailing then. Evidence from Golterman's (1973a) work suggests that the clay may thus be contributing ions like $Sio_3^{2-}$ and $Po_4^{-}$-P to the water in this way. It will therefore be interesting to study the proportions of these ions which occur in the particulate form in the lake water as Lake Kainji may be different in this respect from Lakes Rudolf and George which are known to have no particulate $Po_4$-P. Talling and Talling(1965)

In the case of the $No_3$-N available, the proportion of the total $No_3$-N which is in the particulate form is generally high. For example as much as 50% of the $No_3$-N present in the water during the floods is in the particulate form. This suggests a high contribution from the sediments present into the lake. The proportion of this material contributed by the external material could not be separated from that contributed by the resuspended detritus because the resuspended material is brought up at the same time as the floods bring in sediment from the river. Therefore a distinction between the interaction of the water with external and internal sources of suspended solids has not been attempted. Nevertheless it seems clear that the concentration of this nutrient from the suspended solids. This must be reason why the lake can sustain high levels of algal biomass (up to 180 mg/m$^3$ Chla). The results indicate that the suspended solids exerts a profound effect on the biological production within the lake.

Table 1. Showing the total and particulate concentrations for suspended matter, organic matter and $No_3$-N (particulates in brackets)

| Months | Suspended Matter mg/l | Organic Matter mg/l | $No_3$-N µg/l |
|---|---|---|---|
| August | 238.76 (203.17) | 43.47 | 84.53 (63.93) |
| September | 290.05 (209.82) | 62.64 (13.81) | 104.08 (32.90) |
| October | 226.1 (186.9) | | 46.7 (25.7) |
| November | 123.6 (83.9) | 126.1 (73.9) | 29.3 (18.5) |
| January '75 | 81.4 (45.9) | 127.4 (80.5) | 36.9 (24.4) |
| February | 78.71 (40.3) | 50.0 (43.70) | 33.8 (9.01) |
| March | 70.3 (32.5) | 52.8 (34.2) | 24.1 (11.0) |
| May | 54.2 (12.4) | 35.8 (26.2) | 41.8 (21.4) |
| July | 57.3 (9.3) | 30.4 (37.5) | 35.6 (3.8) |
| August | 200.8 (158.5) | 54.9 (27.4) | 134.0 (62.2) |
| September | 211.1 (166.2) | 53.9 (25.6) | 44.4 (26.6) |

Acknowledgements

The authors would like to express their sincere gratitude to the United Kingdom Overseas Development for grants to support the Kainji Research Project of the University of Ife.

References

Abiodun, A.A. (1876), Satellite survey of particulate distribution patterns in Lake Kainji. Remote Sens Environ 5 109-123.

Adeniji, H.A. (1973), Preliminary investigation into the composition and seasonal variation of the plankton in Lake Kainji Nigeria. Gephys Monogr. Ser. 17: 617-619.

Adeniji, H.A. (1975) Some aspects of the limnology and Fishery Development of Kainji Lake, Nigeria. Arch. Hydrobiol. 2 253-262.

APHA (1965) Standard Methods for the examination of Water Sewage, and Industrial Wastes, 12th edn. American Public Health Association, New York.

Asseez, L.O. (1970) A preliminary investigation of suspended sediments in Lake Kainji Nigeria. Bull int. Ass. Scient. Hydrol. 15 61-70.

Asseez, L.O. (1970a) Effects of organic acids on the flocculation of suspended sediments in tropical man-made lakes. J. Hydrol. 11 253-257.

Gibbs, R.J. (1967a) The geochemistry of the Amazon river system; Part I. The factors that control the salinity and the composition and concentration of the suspended solids. Bull. Geol. Soc. Am. 78 1203-32.

Golterman, H.L. (1973a) Deposition of river silts in the Rhine and Meuse Delta. Freshwat. Biol. 3, 267-281

Golterman, H.L. (1973b) Natural phosphate sources in relation to phosphate Budgets: Eutrophication. Water Research. 7, 3-17

Golterman, H.L. (1975) Chemistry. In River Ecology 725pp ed. B.A. Whitton p.39-80. Blackwell Scientific publication Oxford.

Golterman, H.L. and Clymo, R.S. (1969) (Eds.) IBP Handbook 8: Methods for chemical Analysis of Freshwaters Blackwell Scientific Publications, Oxford.

Halstead, L.B. (1971) "The shoreline of Lake Kainji, Nigeria: a preliminary survey", J. Min Geol.6: 3-22.

Henderson, F. (1973) A limnological Description of Kainji Lake 1969-71. FAO Technical Report F1/DR/NIR.66/524.

Henderson, F. (1973a) Stratification and circulation in Lake Kainji. Geophys Monogr Ser. 17 489 - 494.

Imevbore, A.M.A. (1970a) Some general features of the Kainji Reservoir Basin In Kainji Lake Studies I. ed. S.A. Visser 17-25. Ibadan.

Imevbore, A.M.A. (1970b) The chemistry of the River Niger in the Kainji Reservoir Area. Arch Hydrobiol. 67, 412-431.

Imevbore, A.M.A. (1975a) The chemistry of Kainji Lake Waters. August 1968-September 1969. In Lake Kainji Lacustrine Studies 1. (eds.) A.M.A. Imevbore and O.S.Adegoke Ife University Press.

Imevbore, A.M.A. (1975b) Phosphate absorption as an index of phytoplankton activity the use of Lake Kainji. In Lake Kainji Lacustrine Studies 1 (Eds.) A.M.A. Imevbore and O.S. Adegoke. Ife University Press.

Imevbore, A.M.A. (1975c) Preliminary measurements of photosynthetic production in Lake. In Lake Kainji Lacustrine Studies 1 (eds.) A.M.A. Imevbore and O.S. Adegoke. p. Ife University Press.

Karlman, S.G. (1973) Pelagic primary production in Kainji Lake: F1: SF/NIR 24. FAO Technical Report No. 3 1-59 UNDP/FAO

McLachlan, A.J. (1974) Development of some Lake Ecosystems in tropical Africa with special reference to the invertebrates. Biol. Rev. 49 365-397.

Nedeco (1959) River studies and Recommendations on Improvement of Niger and Benue. Federal Government of Nigeria. Amsterdam Holland publishing company.

Nedeco and Balfour Beatty (1961) Niger Dams Project Report 2, Hydrology and Reservoir Operation.

Perkins, E.J. (1974) The Biology of Estuaries and Coastal Waters. Academic press. London.

Stotzky, G. (1967): Clay minerals and microbial ecology. Trans. N.Y. Acad. Sci. 30 11-21.

Strickland, J.D.H. and Parsons, T.R. (1965) A mannual of sea water analyses. Bull. Fish. Res. Bd. Canada 125 - 203p.

Talling, J.F. and Talling, I.B. (1965) The chemical composition of African lake waters Int. Rev. ges. Hydrobiol. Hydrogr. 50 (3), 421-463.

Zimmermann, U. Baumann, U., Imevbore, A.M.A. Henderson and F., Adeniji, H.A. (1976) Study of the mixing pattern of Lake Kainji using stable isotopes. Catena 3, 63-76.

Key words: Lake Kainji
suspended solids
phosphate
silicate
nitrate

# SEDIMENT RELEASE OF PHOSPHORUS IN LOUGH NEAGH, NORTHERN IRELAND

R. J. Stevens & C. E. Gibson (presented by B. H. Rippey[1])

Freshwater Biological Investigation Unit, Antrim, Northern Ireland[2]

## Summary

Lough Neagh, a shallow eutrophic lake, was monitored weekly during 1973-1976. An accurate input-output phosphorus budget was calculated for the Lough during 1973-1974 from concentration and flow data for the six major inflowing rivers and one outflowing river. Rapid increases in the phosphorus content of the water column in the Lough occurred each summer. As the phosphorus loading due to river inputs was very small during these summer periods the increases in phosphorus concentration of the water column were due to sediment release. Sediment release rates were estimated to be 14-38 mg $m^{-2}$ $day^{-1}$ of total phosphorus, similar to values reported for anaerobic cores from other lakes. Phosphorus release occurred when the water overlying the sediment was oxygen depleted but not anoxic. If anoxic water existed above the sediment it must have occupied a layer, possibly only a few centimetres deep, which could not have been located accurately by an oxygen probe.

## Introducticn

Lough Neagh is a shallow freshwater lake situated 54° 35' N and 6° 25' W. It has a surface area of 383 $km^2$, a mean depth of 8.9 m, a capacity of 3.45 x $10^9$ $m^3$ and a water turn-over time of approximately 1.3 y. The Lough has a catchment area of 4,500 $km^2$. Six major rivers drain the catchment area with only one river draining the Lough to the Atlantic Ocean.

Lough Neagh is among the most eutrophic of the world's major lakes (Millway, 1970; Wood & Gibson, 1973). Mean values for some biological and chemical parameters are shown in Table 1. Phosphorus (P) appears to be the nutrient limiting algal growth (Wood & Gibson, 1973). Biomass production is therefore dependent on the rate of supply of orthophosphate from river inputs, recycling and sediment release (Stumm & Leckie, 1971; Golterman, 1973). River inputs to and output from Lough Neagh have been measured by Smith (1976). Using this information and data obtained from the weekly monitoring of the Lough it is hoped to quantify changes in the Lough P content caused by sediment release.

Table 1. Limnological parameters for Lough Neagh.

| Parameter | Value[1] |
|---|---|
| pH | 8.2 |
| Total alkalinity | 110 mg $CaCO_3$ $l^{-1}$ |
| Conductivity | 290 µS $cm^{-1}$ |
| Chlorophyll a | 25 µg $l^{-1}$ |
| Sodium | 17.8 mg $l^{-1}$ |
| Potassium | 2.4 mg $l^{-1}$ |
| Calcium | 30.5 mg $l^{-1}$ |
| Magnesium | 6.4 mg $l^{-1}$ |
| Chloride | 22.4 mg $l^{-1}$ |
| Sulphate | 16.9 mg $SO_4$ $l^{-1}$ |
| Soluble silica | 2.7 mg $SiO_2$ $l^{-1}$ |
| Total phosphorus | 133 µg P $l^{-1}$ |
| Soluble reactive phosphorus | 39 µg P $l^{-1}$ |
| Nitrate-nitrogen | 0.47 mg N $l^{-1}$ |
| Total organic nitrogen | 0.75 mg N $l^{-1}$ |
| Total phosphorus loading | 1.4 g $m^{-2}$ $yr^{-1}$ |
| Productivity | 550 g C $m^{-2}$ $yr^{-1}$ |

1 Mean values of measurements during 1973-75

## Experimental

(a) Lough Neagh monitoring, 1973-1976. The Lough was sampled weekly at north, middle and south locations. Temperature and dissolved oxygen were determined in situ. Composite 0-10 m tube samples were taken for laboratory analysis. Soluble reactive P was determined on a 0.45 µ membrane-filtered sample by an automated Murphy & Riley (1962) procedure. Total P and total soluble P were determined by persulphate digestions of unfiltered and filtered samples respectively (American Public Health Association, 1971). The difference between total P and total soluble P was called particulate P. The difference between total soluble P and soluble reactive P was called soluble organic P. Chlorophyll a was determined by methanol extraction using the simplified procedure of Talling & Driver

1. Present address: New University of Ulster Limnology Laboratory, Magherafelt, N. Ireland
2. Present address: Department of Agriculture, Muckamore, Antrim, N. Ireland

(1963). Total iron and total manganese were determined by perchloric acid digestion of unfiltered samples followed by neutralisation, chelation with diethyldithiocarbamic acid and methyl isobutyl ketone extraction (Nix & Goodwin, 1970).

(b) Lough Neagh P budget. An accurate input-output P budget was possible for Lough Neagh (Stewart, 1976; Smith, 1976). Six major rivers drain 88% of the catchment area, while only one river drains the Lough to the Atlantic Ocean. During 1973 and 1974 these major rivers were sampled every fifteen days. Continuous records of flow rate were available for each river. Using measured con- concentrations and flow rate at the time of sampling, linear regression equations of log load on log flow were computed for each river for soluble reactive P and total P. Loadings could then be predicted for every day of the year from daily flow data. Weekly input and output for the Lough were calculated from these predicted daily loadings for specific periods each year.

Results and Discussion

(a) Phosphorus content of the water column. Monthly mean observations of soluble reactive P and total P are shown in Fig. 1 for 1973-1976.

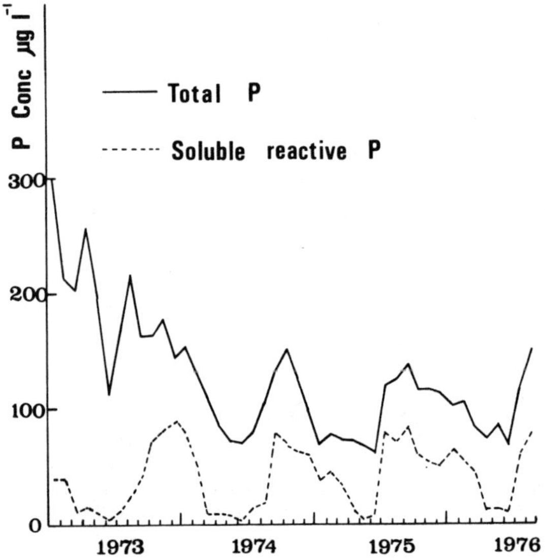


Fig. 1. Monthly mean concentrations of soluble reactive phosphorus and total phosphorus in Lough Neagh, 1973-1976.

Within the period June to September each year there was a large increase in both soluble reactive P and total P concentrations. In Fig. 2a the weekly mean water column contents of soluble reactive P and total P are plotted for two-month periods in 1973 and 1974. Increases in soluble reactive P and hence total P content during 10th-24th September 1973 and 21st August-4th September 1974 were of the order of hundreds of tonnes.

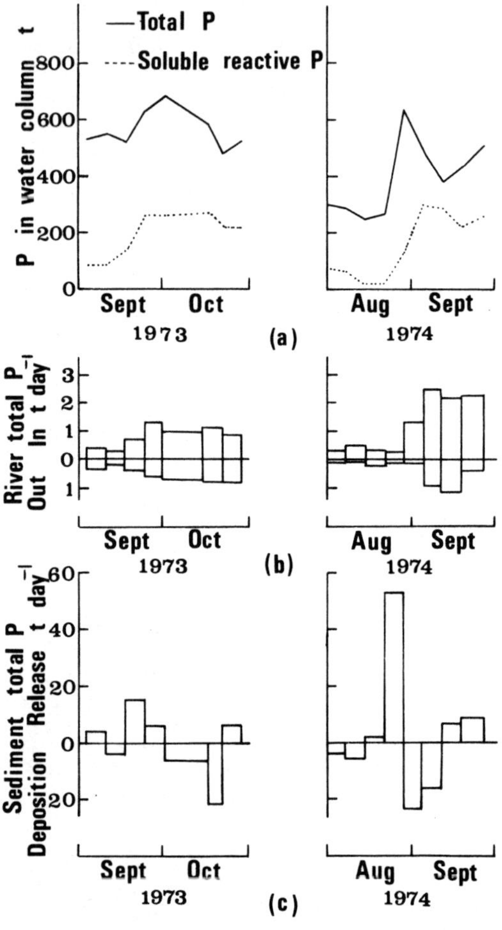


Fig. 2. Phosphorus fluxes during periods of sediment release in Lough Neagh, 1973 and 1974.

(a) Weekly changes in content of total phosphorus and soluble reactive phosphorus in the water column
(b) River inputs and output of total phosphorus
(c) Resultant contribution from or to the sediment.

The total P river inputs and output for the Lough during these two-month periods in 1973 and 1974 are shown in Fig. 2b. River inputs exceeded output so there was an overall accumulation of P in the Lough. The net river inputs minus output during 10th-24th September and 21st August-4th September 1974 were only 3 and 10 tonnes respectively. To account for the large increase in P content of the water column there must have been release of P from the sediment.

The sediment contribution was calculated using the equation: net gain from the sediment during the time $t_1$ to $t_2$ = P content of the water column at time $t_2$ - P content of the water column at time $t_1$ - (river inputs - output during the time $t_1$ to $t_2$). Results for 1973 and 1974 are shown in Fig. 2c. During P release the influence of river inputs minus output to the overall increase in P content of the water column was small. The total P input to Lough Neagh was 528 and 519 tonnes $P.yr^{-1}$ in 1973 and 1974 respectively (Smith, 1976). Phosphorus release from sediments therefore constitutes a significant internal loading.

(b) Phosphorus release rate from sediment. The P release rate during specific periods each year was calculated as shown in Table 2. The area of active sediment was assumed to be the surface area of the Lough. River inputs and output for 1975 and 1976 have not yet been calculated. Release rates for 1975 and 1976 are therefore slightly overestimated. During a release period the increase in soluble reactive P content of the water column was always greater than the increase in total P. Some conversion of soluble organic P or particulate P to soluble reactive P may have occurred concomitantly with sediment release, therefore the release rates for soluble reactive P are too high. Alternatively particulate P may have sedimented out of the water column and therefore the release rates for total P are too low.

The release rates recorded in Lough Neagh are in excess of rates estimated for diffusional transport from an aerobic undisturbed sediment (0.27 mg P $m^{-2}$ $day^{-1}$, Stumm & Leckie, 1970). The rates are of the same order as those reported for anoxic conditions in the Western basin of Lake Erie (13 mg P $m^{-2}$ $day^{-1}$, Burns 1976), and for anaerobic undisturbed sediment cores from Lake Furesø (17 mg P $m^{-2}$ $day^{-1}$, Kamp-Nielsen, 1974).

Rippey (1976) recorded release rates of about 10 mg P $m^{-2}$ $day^{-1}$ in oxygenated undisturbed Lough Neagh sediment cores only when the pH of the overlying water was increased from the original value of 7.5 to 8.5-9.0. As a pH shift was not measured during nutrient release, phosphorus release rates in Lough Neagh therefore suggest that the mechanism of release is associated with oxygen deficiency in the water column.

(c) Aeration status during phosphorus release. Soluble reactive P, total P and dissolved oxygen concentrations and temperature are shown in Fig. 3 for a depth profile taken on 2nd July 1976. Phosphorus was being released from the sediment while the dissolved oxygen concentration in the water just above the sediment was apparently 50% saturation. Continuous recording of dissolved oxygen at a fixed point 0.6 m above the sediment surface showed that the minimum oxygen concen-

Table 2. Phosphorus release rates from Lough Neagh sediments

| Period | P increase in water column | | Rivers | | Sediment release | Release rate |
|---|---|---|---|---|---|---|
| | | | Inputs | Output | | mg P $m^{-2}$ $day^{-1}$ |
| | | t | t | t | t | |
| 1973 | TP: | 79 | 7.0 | 3.8 | 76 | 14 |
| 10-24 Sept | SRP: | 166 | 4.4 | 0.4 | 162 | 30 |
| 1974 | TP: | 214 | 11.7 | 2.1 | 204 | 38 |
| 21 Aug-4 Sept | SRP: | 266 | 6.0 | 0.4 | 260 | 48 |
| 1975 | TP: | 224 | | - | 224 | 28 |
| ?-23 July | SRP: | 348 | - | - | 348 | 43 |
| 1976 | TP: | 176 | - | - | 176 | 19 |
| ?1 June-15 July | SRP: | 183 | - | - | 183 | 20 |

TP = Total phosphorus; SRP = soluble reactive phosphorus

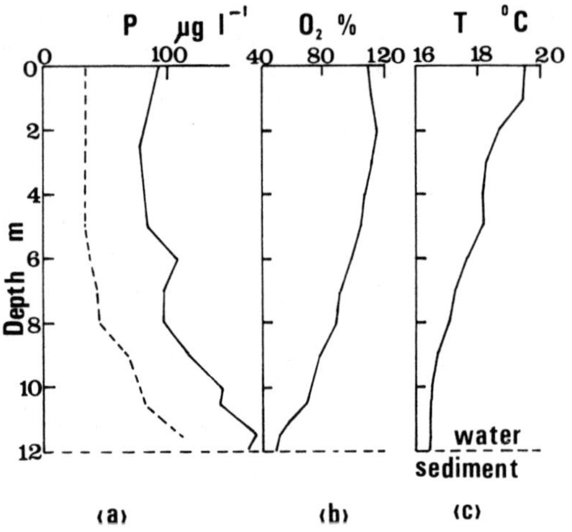


Fig. 3. Concentrations of (a) soluble reactive phosphorus and total phosphorus, (b) dissolved oxygen and (c) temperature in a depth profile of Lough Neagh, 2nd July 1976.

tration during the two weeks preceding 2nd July 1976 was 50%. Transfer of nutrients from sediment to water column would only be expected when the oxygen concentration at the sediment/water interface falls below 20% saturation (Mortimer, 1971). The lowest oxygen concentration recorded by routine weekly sampling of Lough Neagh during 1973-1975 was 30%. When measuring dissolved oxygen in a depth profile with an oxygen probe, it is experimentally difficult to locate the sediment/water interface. A shallow layer of anoxic water, possibly only a few centimetres in depth, may have existed immediately above the sediment and may not be detectable by the measurement technique. The reduction of the oxidised surface sediment layer would allow upward diffusion of nutrient rich anoxic interstitial water into the water column (Oláh, 1975). Complete mixing within the water column may have been hindered by the temperature-density gradient at the 8 m depth (Fig. 3c).

(d) Forms of sediment phosphorus. Simultaneous release of phosphorus, iron and manganese was recorded in depth profiles taken in 1975. The concentrations of total P, total iron and total manganese immediately above the sediment surface were 240, 600 and 340 $\mu g.l^{-1}$ respectively. Rippey (1976) has shown that P released from Lough Neagh sediment cores had been in association with ferric oxides rather than with calcium carbonate. Analyses of Lake Erie sediments showed that P regenerated under anoxic conditions was associated with the non-apatite inorganic P fraction (Nriagu & Dell, 1974; Williams, Murphy & Mayer, 1976).

## Conclusions

Release from sediment dramatically increased the P concentration in Lough Neagh in four successive years. The release occurred when the water immediately above the sediment was oxygen depleted but not necessarily anoxic.

## References

American Public Health Association , 1971. Standard methods for the examination of water and wastewater. 13th ed. pp. 489-495 American Public Health Association, Washington.

Burns, N.M., 1976. Nutrient budgets for Lake Erie, 1970. J. Fish. Res. Board Can. 33:520-536.

Golterman, H.L., 1973. Vertical movement of phosphate in freshwater. In: Environmental Phosphorus Handbook: Eds. E.J. Griffith, A. Beeton, J.M. Spence & D. T. Mitchell. Wiley, New York, pp. 509-538.

Kamp-Nielsen, L., 1974. Mud-water exchange of phosphate and other ions in undisturbed sediment cores and factors affecting the exchange rates. Arch. Hydrobiol. 73: 218-237.

Millway, C.P., 1970. Eutrophication in large lakes and impoundments. O.E.C.D., Paris, 560 p.

Mortimer, C.H., 1971. Chemical exchanges between sediments and water in the Great Lake - speculations on probable regulatory mechanisms. Limnol. Oceanogr. 16:387-404

Murphy, J. & Riley, J.P., 1962. A modified single solution method for the determinati of phosphate in natural waters. Analytic clim. Acta 27:31-36.

Nix, J. & Goodwin, T., 1970. The simultane us extraction of iron, manganese, copper, cobalt, nichel, chromium, lead and zinc from natural water for determination by atomic spectroscopy. Atomic Absorption Newsletter 9:119-122.

Nriagu, J.O. & Dell, C.I., 1974. Diageneti formation of iron phosphates in recent lak sediments. American Mineralogist 59: 934-946.

Oláh, J., 1975. Metalimnion function in shallow lakes. Symp. Biol. Hung. 15:149-15

Rippey, B., 1976. The behaviour of phosphorus and silicon in undisturbed cores of Lough Neagh sediments. This vol. Page 348.
Smith, R.V., 1976. Domestic and agricultural contributions to the inputs of phosphorus and nitrogen to Lough Neagh. Water Res. In Press.
Stewart, D.A., 1975. Mathematical modelling of the ecosystem of Lough Neagh. Ph.D. thesis, The Queen's University of Belfast.
Stumm, W. & Leckie, J.O., 1971. Phosphate exchange with sediments; its role in the productivity of surface waters. Advances in Water Pollution Research 2:III 26/1-26/16.
Talling, J.F. & Driver, D., 1963. Some problems in the estimation of chlorophyll a in phytoplankton. Proceedings, Conference on Primary Productivity Measurement, Marine and Freshwater, Hawaii, 1961. U.S. Atomic Energy Comm., Div. Tech. Inf., TID - 7633, 142-146.
Williams, J.D.H., Murphy, J.P. & Mayer, T., 1976. Rates of accumulation of phosphorus forms in Lake Erie sediments. J. Fish. Res. Board Can. 33:430-439.
Wood, R.B. & Gibson, C.E., 1973. Eutrophication and Lough Neagh. Water Res. 7:173-187.

Key words: sediments
nutrient release
oxygen depletion
phosphate
Lough Neagh

# THE BEHAVIOUR OF PHOSPHORUS AND SILICON IN UNDISTURBED CORES OF LOUGH NEAGH SEDIMENTS

B. Rippey

The New University of Ulster Limnology Laboratory, Magherafelt, Co. Derry, N. I.

## Summary

The sediments of L. Neagh have been established to be the seat of important processes in the cycling of the algal nutrients, phosphorus and silicon. Sediment cores and suspensions were used to investigate the sediment-water interaction of these elements and to indicate the sediment components and mechanisms responsible for the observed behaviour,in fully aerated systems.

Mass transfers of silicon from the sediment were measured which were similar to those reported for other lake sediments. pH, agitation of sediment surface and aqueous concentration were found to influence the release rate,but temperature was the most important. The source of the flux was indicated to be amorphous silica (diatom frustules) dissolution. The dissolution reaction was complicated by back reaction of the dissolved silicon with the clay minerals present in the sediment.

Phosphorus exchange was strongly influenced by solution concentration, sediment agitation and greatest of all pH. The basic ferric phosphate 'type' of salt was established to be the source of the released material and a ligand exchange mechanism indicated.

## Introduction

The behaviour of phosphorus and silicon in a lake system is greatly influenced by biological processes. In algal cells,these elements have a high chemical potential relative to the aqueous state, so that when the cells die transformations, chemical and biological, take place through which the elements find a new lower energy state, more in equilibrium with the environmental conditions. It is becoming increasingly clear that in many lakes the sediments are an important site for these processes.

The dominant role of sediments in the cycling of phosphorus in Lough Neagh, Northern Ireland, has been shown by field observation (Stevens & Gibson 1976); the same is true for silicon in L. Neagh (Stevens & Gibson pers. comm.). Andersen (1975), Bengtsson (1975) and Holden & Caines (1974) have also established from field data the importance of sediments in phosphorus, silicon and nitrogen cycles.

A knowledge of the chemical constitution of the various sediment phosphorus and silicon components and the mechanisms of their environmental interaction, must then form part of the understanding of lake phosphorus and silicon cycles. Sediment cores and suspensions were used to describe the characteristics of the sediment-water interaction and to indicate the components and mechanisms responsible for the observed behaviour.

Attention was focussed upon fully oxidized systems because L. Neagh is nearly always well oxygenated, and almost never drops below 50% oxygen saturation.

## Experimental

The limnological characteristics of L. Neagh have been described in Gibson et al. (1971), Wood & Gibson (1973) and Stevens & Gibson (1976). The composition of surface sediment collected from a site at the average depth of L. Neagh, 8.9 m, is shown in Table 1.

Table 1. Composition of surface sediment in L. Neagh

| | |
|---|---|
| Water content | 85% of wet weigh[t] |
| Loss on ignition | 20% of dry weigh[t] |
| Fe | 6% (d.w.) |
| Ca | 0.20% (d.w.) |
| P | 0.15% (d.w.) |

The sediment, a gyjtta, has a thin brown oxidized microzone which the redox profile, measured according to Hargrave (1972), shows to be approximately 5 mm. deep; the redox potential falls away steeply to the reduced zone which possesses a sulphide smell and is black to grey-black in colour.

Sediment cores

The exchange of Si and P between sediment and water was investigated using sediment cores. Undisturbed cores were taken with a Gilson corer from a site at the average depth of the lake and held in large water baths at temperatures controllable to ± 0.1°C. pH was regulated by sodium borate - boric acid buffer systems (Golterman & Clymo 1971), and oxygen maintained at saturation by slowly bubbling air through the core supernatant, which also ensured that the water was homogeneous. Samples of the supernatant were taken for analysis by the methods in Golterman & Clymo (1971) for soluble reactive phosphorus, total phosphorus (perchloric acid digestion) and soluble reactive silicon.

Two variants of this standard procedure were used to show up the effects of sediment disturbance and chemical concentration on the exchange rate, as measured by the normal method described above. Sediment agitation was crudely reproduced by stirring the supernatant twice a day in order to bring the surface sediment particles into suspension. Chemical concentrations were kept low by continual dilution of the supernatant, by removing half every three to five days, and replacing it with an equal volume of buffer-distilled water mixture. However most of the results pertain to the standard procedure described first.

Sediment suspensions

Sediment suspensions were used to remove diffusion effects from the exchange process, in order to isolate the reactions taking place at the solid-solution interface. No change in soluble silicon concentration was observed when the suspensions were aerated, so fresh mud was used directly. Suspensions were made up in 50 : 50 buffer: distilled water mixtures and shaken by hand a few times a day during the experiment. Samples were taken for soluble silicon analysis. As the surface sediment grab samples used in the phosphorus experiments contained oxidized and reduced sediment, precautions were taken to ensure chemical homogeneity of the solid phase. A stock sediment was prepared in $10^{-2}$M $KNO_3$ and vigorously aerated for two to three days to produce thoroughly oxidized material, which could then be diluted for individual experiments:

(a) The pH variation of phosphorus solubility of oxidized sediment was found by adjusting the pH in eight samples to cover the range pH 3-11, and equilibrating for twenty-four hours on an orbital shaker before sampling for soluble reactive phosphate.

(b) When alkali, KOH, was added to oxidized suspensions of L. Neagh sediments the soluble reactive phosphate concentration rose rapidly; the kinetics of this reaction were investigated in suspensions maintained homogeneous by stirring. The stock suspension was diluted and equilibrated for one hour before the reaction was commenced by addition of KOH. Samples were taken for soluble reactive phosphate analysis at appropriate times.

## Results and Discussion

It has been shown that borate interacts with natural organic matter causing aggregation Ghassei & Christman 1968). The possibility that such an interaction might modify the release of Si and P was eliminated, as comparison with cores controlled by NaOH addition each day revealed no differences.

Silicon, cores

Fig. 1. shows the change of Si concentration with time for five cores held under identical conditions. The linear increase of concentration with time allows the Si release rate between sampling dates to be calculated and these may be averaged to obtain a figure for the whole experiment; this was 68.3 mg Si/$m^2$.d with a coefficient

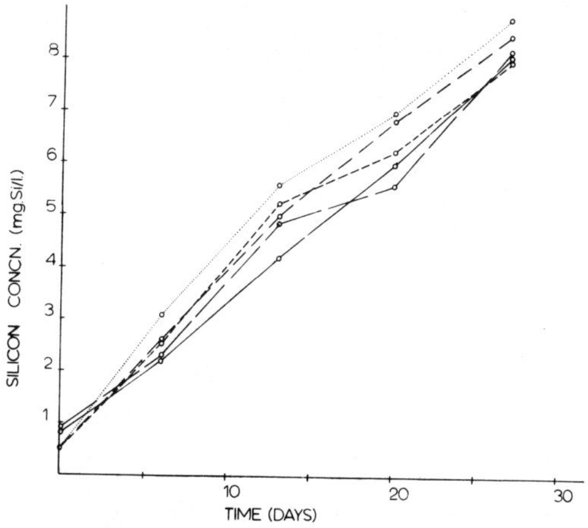


FIG.1 CONCENTRATION OF SILICON IN CORES

TEMP. = 10°C PH = 8.9 $O_2$ = 10[illegible]%

of variation of 7%. The precision of the experimental technique was high enough to reveal the major influences upon the release rate; pH, temperature, sediment agitation and concentration were investigated.

The results from seven cores covering the pH interval 7.90 to 9.05 are shown in Table 2.

Table 2. The relation between Si release rate and pH; T = 20°C, $O_2$ = 100%

| pH | 7.90 | 8.20 | 8.45 | 8.60 | 8.80 | 8.85 | 9.05 |
|---|---|---|---|---|---|---|---|
| mgSi/m$^2$.d | 94 | 104 | 102 | 109 | 107 | 105 | 114 |

This corresponds to an increase of 21% across the pH interval; the pH range of L. Neagh is 7.8 to 9.1. Six cores were used to evaluate the effect of sediment agitation and maintenance of low concentration upon release, by treating two in the standard way, two with mechanical agitation of the sediment surface, and two with dilution every three to five days. The results are shown in Table 3.

Table 3. Six cores, T = 15°C, pH =8.5, $O_2$ = 100% showing release rate, mgSi/m$^2$.d

| Standard procedure | | Agitation | | Dilution | |
|---|---|---|---|---|---|
| 77 | 80 | 94 | 96 | 115 | 103 |
| Av. = 79 | | Av. = 95 | | Av. = 109 | |

Disturbance of the sediment surface caused a

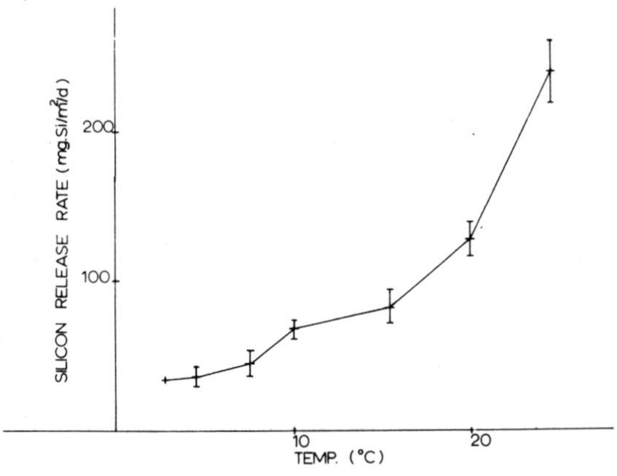


FIG.2 VARIATION OF SILICON RELEASE RATE WITH TEMP.
PH = 8-9 $O_2$ = 100% 95% CONFIDENCE LIMITS

21% increase in release rate and maintenance of low concentration 39%. Temperature was found to be the most important variable in the cores; Fig. 2 shows the variation of release rate with temperature to be an increase of 260% from 3 to 20°C.

The magnitude and temperature dependence of these release rates agree reasonably closely with those of Andersen (1974) and Kamp-Nielsen (1974). The sediment components responsible for Si release were investigated using sediment suspensions. The characteristics of the dissolution reaction from natural sediment was compared with the known behaviour of the silicon bearing components of the sediment, that is, quartz, primary minerals (feldspars), secondary minerals (clays) and amorphous silica (diatom frustules).

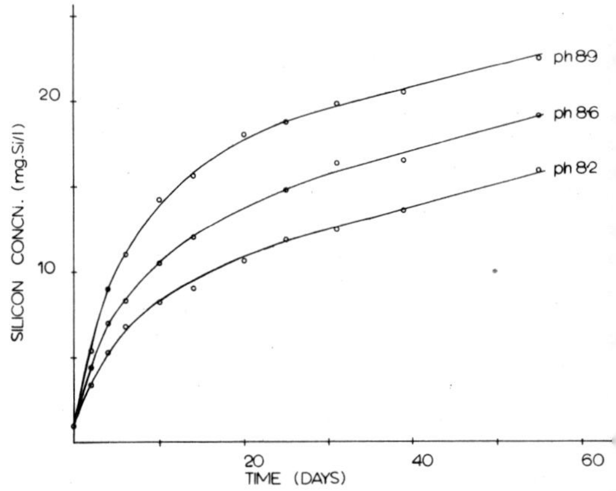


FIG.3 CONCENTRATION OF SILICON IN SUSPENSIONS

EFFECT OF VARYING PH

Fig. 3 illustrates the most significant features of the results. These are an increase of initial solution rate and extent of solution with pH over range pH 6 to 9. This information can be used to indicate the active sediment components. Quartz can be dismissed as its solubility is only 5.04 mgSi/l at 25°C (Lier et al., 1960, Morey et al., 1962), and so can the primary minerals, for Wollast (1967) has shown that the initial solution rate of K-feldspar decreases rapidly with increasing pH over the range four to ten. Clay minerals cannot be primarily responsible for the Si dissolution as Siever & Woodford (1973) have shown that for kaolinite, illite and montmorillonite, the extent of solution, given by the area under the concentration-time curve, decreases with increasing pH from four to ten. Only amorphous silica shows an increasing solution rate with pH (Greenberg, 1957, O'Connor & Greenberg, 1958).

However, Greenberg (1957) and O'Connor & Greenberg (1958) found that their results obeyed first order kinetics, as did Hurd (1972) for diatom frustules. The results displayed here do not obey first order kinetics or any other simple rate law, second order or parabolic; a complex rate expression is followed. Wollast (1967) found similarly shaped curves for the solution of K-feldspar in water and interpreted them quantitatively in terms of two opposing reactions; surface breakdown of the feldspar grains releasing silicon and cations and leaving essentially $Al_2O_3$, followed by the back reaction of the silicon with the $Al_2O_3$ to form an amorphous hydrated aluminosilicate. Siever & Woodford (1973) also interpreted their data on clay mineral dissolution in the same manner, that is, breakdown of the parent material followed by back reaction of the silica, with or without cations, to form a new amorphous clay mineral.

The results presented here can be interpreted in the same terms. Dissolution of the amorphous silica fraction of the sediment takes place, increasing the concentration of dissolved silicon, until a back reaction with one of the clays present, probably kaolinite, takes place, which slows up the rate of increase of dissolved silicon.

Phosphorus, cores

Results from five identical cores are shown in Fig. 4. Although the concentration increase with time is only approximately linear and the variation between the cores is larger compared to silicon, the same method of calculating the release rate was used; this gave 10.4 $mgP/m^2.d$ with a coefficient of variation of 19%. Releases were calculated in terms of total phosphorus, although soluble phosphorus gave almost identical values.

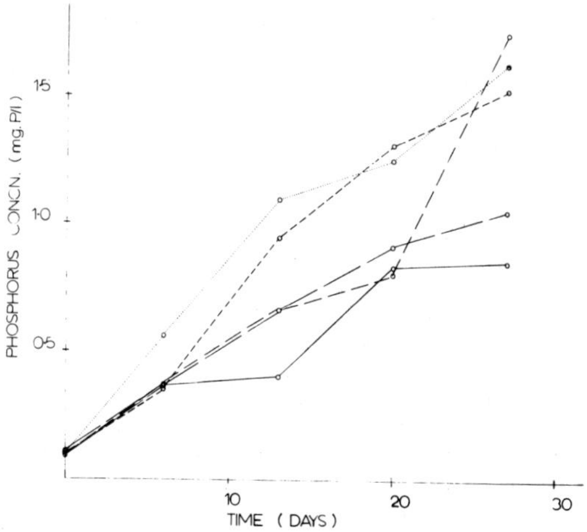


FIG. 4 CONCENTRATION OF PHOSPHORUS IN CORES

TEMP = 10°C PH = 8·9 $O_2$ = 100%

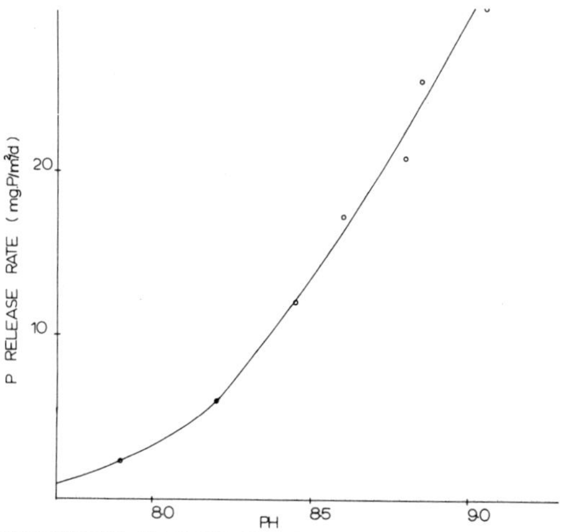


FIG. 5 VARIATION OF PHOSPHORUS RELEASE RATE WITH PH AT 20°C.

Above pH 8 the P release rate increases, Fig. 5, and the rate is also sensitive to stirring and supernatant concentration, Table 4.

Table 4. Six cores, T = 15°C, pH = 8.5, $O_2$ = 100% showing P release rate $mgP/m^2.d$.

| Standard procedure | | Agitation | | Dilution | |
|---|---|---|---|---|---|
| 10.8 | 8.6 | 12.5 | 15.0 | 21.9 | 15.4 |
| Av. = 9.7 | | Av. = 13.8 | | Av. = 18.7 | |

Agitation of the sediment surface caused a 42% increase in rate, and maintenance of low concentration 82%. These factors and the strong dependence of release rate on pH, together with variations in the sediment, contribute to the large variability between cores.

Both increases and decreases of P release rate with pH above 8, have been obtained; Andersen (1975) and Kamp-Nielsen (1974, respectively. The increase of P release rate with pH found in the present study, suggests that iron or clay bound P rather than Ca-P controls P solubility in L. Neagh sediments.

Sediment suspensions were used to isolate more fully the important phosphorus component. The relationship between P solubility of oxidized mud and pH is a solubility minimum between pH 5 and 8 increasing above and below this, Fig. 6.

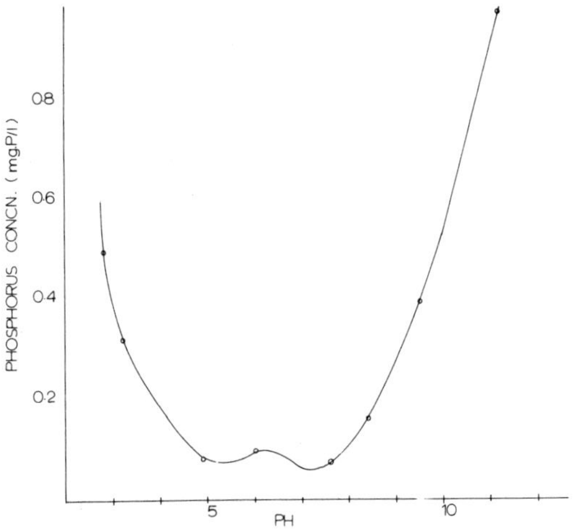


FIG. 6 VARIATION OF PHOSPHORUS SOLUBILITY WITH PH

This data can be compared to the known behaviour of pure phosphate components, that is, clays (Muljadi, et al., 1966) and the Ca and Fe systems (Stumm & Morgan, 1970). It is clear that the sediment exhibits the same pH behaviour as the Fe-P system; the basic ferric phosphate or alternatively ferrosoferric hydroxyphosphate (Nriagu & Dell, 1974) undergoes acid and basic, hydrolysis at the lower and higher pHs respectively.

The kinetics of the P solubilization reaction in the sediment, induced by pH increase, are shown in Fig. 7. Two analyses of the data are presented, one, according to first order kinetics, log ($C_f$-C) against time, and the other, second order, 1/($C_f$-C) against time ($C_f$ = final equilibrium concentration). First order kinetics can be seen to apply according to:

$$-\log(C_f-C) = 0.03\ \text{Time} + 0.77,\quad r = 0.997\ (P = 0.05).$$

The first order rate constant is calculated to be $1.2 \times 10^{-3} sec^{-1}$. This can be compared to 0.02 to $0.46 \times 10^{-3} sec^{-1}$ obtained by Atkinson et al., (1972) for the isotopic exchange of orthophosphate on goethite. They proposed a coordinative interaction between orthophosphate and Fe (III) and a ligand exchange mechanism.

Other evidence for a coordination model is found in the fields of infrared spectroscop[y] of iron oxides (Parfitt et al. 1975) and potentiometric acid-base titrations of oxide suspensions (Yates & Healy, 1975). Yates & Healy (1975) showed the existence of a OH/$PO_4$ ligand exchange on goethite.

$$\text{Fe-OH} + PO_4 = \text{Fe-}PO_4 + \text{OH} \quad --- \quad (1)$$

This evidence and the data presented for L. Neagh sediments allow one to postulate the ligand exchange of OH for $PO_4$ in the basic iron phosphate compound as the mechanism of P release due to pH increase; that is, the reverse of (1).

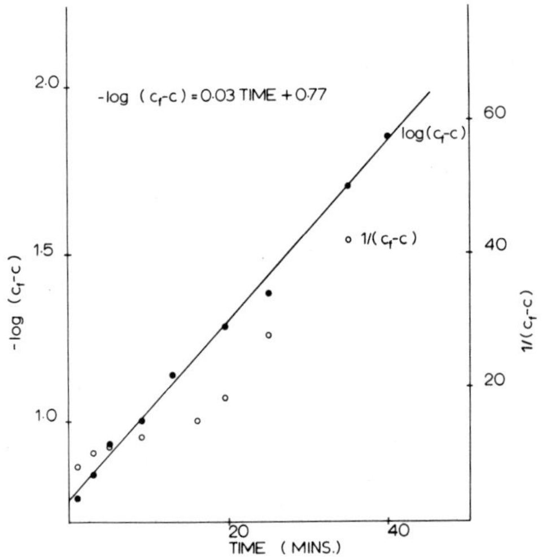


FIG. 7 PHOSPHORUS KINETICS 1ST AND 2ND ORDER PLOTS

## References


Andersen, Jens Moller, 1974. Nitrogen and phosphorus budgets and the role of sediments in six shallow Danish lakes. Arch. Hydrobiol. 74:528-550.

Andersen, Jens Moller, 1975. Influence of pH on the release of phosphorus from lake sediments. Arch. Hydrobiol. 76:411-419.

Atkinson, R. J., Posner, A.M. & Quirk, J.P., 1972. Kinetics of isotopic exchange of phosphate at the α-FeOOH aqueous solution interface. J. inorg. nucl. Chem. 34:2201-2211.

Bengtsson, L., 1975. Phosphorus release from a highly eutrophic lake sediment. Verh. Internat. Verein. Limnol. 19:1107-1116.

Ghassemi, M. & Christman, R. F., 1968. Properties of yellow organic acids of natural waters. Limnol. Oceanogr. 13:583-597.

Gibson, C.E., Wood, R. B., Dickson E.L., & Jewson, D.H., 1971. The succession of phytoplankton in Lough Neagh 1968-70. Mitt. Internat. Verein. Limnol. 19: 146-160.

Golterman, H. L. & Clymo, R.S., 1971. Methods for the chemical analysis of fresh waters. Blackwell Scientific Publications.

Greenberg, S.A., 1957. The depolymerization of silica in sodium hydroxide solutions. J. Phys. Chem. 41:960-965.

Hargrave, B.T., 1972. Oxidation-reduction potentials, oxygen concentration and oxygen uptake of profundal sediments in a eutrophic lake. Oikos 23:167-177.

Holden, A.V. & Caines, L.A., 1974. Nutrient chemistry of Loch Leven, Kinross. Proc. Royal Soc. of Edin. (B) 74:101-121.

Hurd, D.C., 1972. Factors affecting solution rate of biogenic opal in sea-water. Earth Planet. Sci. Lett. 15:411-417.

Kamp-Nielsen, L., 1974. Mud-water exchange of phosphate and other ions in undisturbed sediment cores and factors affecting the exchange rate. Arch. Hydrobiol. 73:218-237.

Lier, J.A., van de Bruyn, P.L. & Overbeek, J.T.G., 1960. The solubility of quartz. J. Phys. Chem. 64:1675-1682.

Morey, G.W., Fournier, R.O. & Rave, J.J., 1962. The solubility of quartz in water in the temperature interval from 25-300°C. Geochim. et Cosmochim. Acta 26:1029-1043.

Muljadi, D., Posner, A.M. & Quirk, 1966. The mechanism of phosphate adsorption by kaolinite, gibbsite and pseudoboehmite. I. The isotherms and the effects of pH on absorption. J. Soil Sci. 17:212-229.

Nriagu, J.O. & Dell, C.I., 1974. Diagenetic formation of iron phosphates in recent lake sediments. Amer. Mineral. 59: 934-946.

O'Connor, R.L. & Greenberg, S.A., 1958. The kinetics for the solution of silica in aqueous solutions. J. Phys. Chem. 62:1195-1198.

Parfitt, R.L., Atkinson, R.J. & Smart, S. St. C., 1975. The mechanism of phosphate fixation by iron oxides. Soil Sci. Soc. Amer. Proc. 39:837-841.

Siever, R. & Woodford, N., 1973. Sorption of silica by clay minerals. Geochim. et Cosmochim. Acta 37:1851-1880.

Stevens, R.J. & Gibson, C.E., 1976. Sediment release of phosphorus in Lough Neagh, Northern Ireland. This volume, page 343.

Stumm, W. & J. J. Morgan, 1970. Aquatic Chemistry. Wiley-Interscience. 521-525.

Wollast, R., 1967. Kinetics of the alteration of K-feldspar in buffered solutions at low temperatures. Geochim. et Cosmochim. Acta. 31:635-648.

Wood, R.B. & Gibson, C.E., 1973. Eutrophication and Lough Neagh. Water Res. 7:173-187.

Yates, D. E. & Healy, 1975. Mechanism of anion adsorption at the ferric and chromic oxide/water interfaces. J. Colloid. Interface Sci. 52:222-228

Key words: sediments
phosphate
silicon
exchange
Lough Neagh

# BENTHAL NITRIFICATION IN RIVERS

J. H. N. Garland

Water Research Centre, Stevenage, Herts, U.K.

## Introduction

About 4.34 mg of oxygen are required to oxidise 1 mg of ammoniacal nitrogen ($NH_3$-N) to the nitrate anion by bacterial action (Montgomery & Borne 1966; Wezernak & Gannon 1967). At temperatures above 15°C and a pH value of 8.0, and at normal atmospheric temperatures and pH 8.5, this weight of ammoniacal nitrogen will give rise to more than 0.025 mg of molecular $NH_3$-N per litre, the concentration which may be tolerated by freshwater fish for a long period without apparent harm (E.I.F.A.C. 1970). It is also more than twice the concentration recommended as a standard of quality for sources of water to be treated for potable supply (W.H.O. 1971). These statements indicate that $NH_3$-N is a pollutant having a potentially large demand on the oxygen resources of streams and affecting their suitability for fishery purposes and as sources of raw water for domestic use.

## Nitrification in the River Trent

An input-output analysis of $NH_3$-N and oxidized-N loads calculated from water quality and flow data collected in the Trent basin over the period 1964-1971 indicated that inorganic nitrogen compounds were conserved but that, in an average total mass flow of 60 tonnes/d, 9.5 t/d of $NH_3$-N were transformed in the river into oxidized-N (Garland & Hart 1973). A detailed annual mean daily oxygen balance for the river in 1970 indicated an average loss of 6.9 t/d of $NH_3$-N at the expense of 30 t/d of oxygen and an average loss of 10 t/d of oxygen in satisfying the carbonaceous oxygen demand derived from discharges of effluent. Thus nitrification accounted for 75 per cent of the total oxygen demand exerted in the self-purification process (Garland 1973).

Following these observations a series of 'time of travel' surveys was conducted at approximately monthly intervals over a 24 km reach in the mid-Trent to study the dynamic behaviour of $NH_3$-N in the river. It was found that the kinetics of the nitrification reactions could be well represented by the first-order consecutive-reaction scheme

$$NH_3 \xrightarrow{K_N} NO_2' \xrightarrow{K_Y} NO_3'$$

characterized by rate coefficients $K_N$ and $K_Y$ which were determined by fitting the integrated forms of the rate equations

$$\frac{-d\,[NH_3\text{-N}]}{dt} = K_N[NH_3\text{-N}]$$

and

$$\frac{d\,[NO_2'\text{-N}]}{dt} = \frac{-d\,[NH_3\text{-N}]}{dt} - K_Y[NO_2'\text{-N}]$$

to the observed data. Values of $K_N$ were found to vary between 0.44 and 2.99 $d^{-1}$ and were significantly related to the temperature of the river water, T(°C), the average velocity of flow, V(m/s), and the concentration of *Nitrosomonas*, X(thousand/ml), at the head of the reach by the equation

$$K_N = 0.0833T + \frac{2.1234}{V} + 0.0646X \quad (r^2 = 0.97)$$

over the temperature range of 6.1-22.9°C, a velocity range of 0.44-1.2 m/s, and bacterial-number range of 0.2-16.0 thousand/ml. Values of the rate coefficient, $K_Y$, for the oxidation of $NO_2'$-N to $NO_3'$-N were consistently larger than values found for $K_N$, indicating that the oxidation of $NH_3$-N to $NO_2'$-N is the rate-controlling step in the formation of $NO_3'$-N. Values of $K_Y$ varied between 2.0 and 7.1 $d^{-1}$ and were not found to be significantly related to any other variable but could be fitted by a log-normal distribution having a geometric mean of 3.7 $d^{-1}$ and a standard deviation of $\log_{10} 1.378$.

## The benthal nature of the nitrification process

The observation that the first-order rate coefficient, $K_N$, for the oxidation of $NH_3$-N to $NO_2'$-N depended upon the inverse of velocity of flow suggested that the extent of the oxidation of $NH_3$-N depended upon the period of contact of a body of water with the bed of the stream. The role of the benthos was confirmed in qualitative experiments which showed that nitrification in the samples of water withdrawn from the river during the surveys did not occur over periods comparable to the retention period of water in the study reach but could be observed taking place rapidly after introducing small quantities of sediment collected from the bed of the stream. Attempts were therefore made to examine the distribution of nitrifying bacteria both in suspension and attached to deposited sediment. The two approaches adopted were (a) a direct determination by serial-dilution culture and MPN estimates and (b) a calculation of the mass flow of Nitrosomonas indirectly from the results of BOD determinations routinely made on river water in the presence and absence of the first-stage nitrification inhibitor, n-allythiourea. It was concluded on the basis of a thorough enumeration of populations of Nitrosomonas and Nitrobacter in samples of river water and sediments that approximately 80 per cent of the nitrifying bacteria in the river are to be found in the top 1 cm of river bed material, and that the numbers in suspension would account for about 5 per cent of the observed losses of $NH_3$-N in the winter and about 20 per cent in the summer (Curtis et al. 1975).

The application of bacterial kinetics to the results of BOD measurements employs the equation

$$X_o = \frac{Y(S_o - S_5)}{\left[\exp \int_o^t u.dt\right] - 1}$$

derived from the Monod growth relationships,

$$\frac{dX}{dt} = \mu X \text{ and } - Y.\frac{dS}{dt} = \frac{dX}{dt} \text{ when}$$

$$\mu = \mu_{max}.\frac{S}{S + K_s},$$

where $X_o$ is the initial concentration of Nitrosomonas in the $BOD_5$ bottle, $S_o$ and $S_5$ are the concentrations of $NH_3$-N at the beginning and end of the test, $\mu$ is the specific growth-rate coefficient, and Y is the yield coefficient for the conversion of substrate to bacterial mass. Knowing the dilution factors employed in the BOD determination it is possible to relate the results of the calculation to concentrations of Nitrosomonas and $NH_3$-N present in the river water when the sample was collected.

The amount of $NH_3$-N consumed in five days was calculated from the difference between the carbonaceous and total $BOD_5$ of the sample, assuming that complete oxidation of $NH_3$-N to $NO_3'$-N occurred. The integral

$\int_o^t \mu.dt$ was evaluated approximately as

$$5\frac{(\mu_o + \mu_s)}{2},$$

where $\mu_o$ and $\mu_s$ are the initial and final values of the growth-rate coefficient at concentrations of $NH_3$-N of $S_o$ and $S_5$, respectively. Values of the maximum growth-rate coefficient ($\mu_{max}$) and Michaelis constant ($K_s$) were taken from the paper by Knowles et al. (1965), and a value of 0.09 for Y was derived from the work of Wezernak & Gannon (1967).

The results of this exercise indicated a mean daily input from the tributaries of 53.8 kg of Nitrosomonas (range 35-72 kg) over the period 1968-1972 and a mean daily output from the R. Trent of 127.2 kg (92-150 kg), yielding a nett production of 73.4 kg (36-99 kg) per day in the main river. The average loss of $NH_3$-N over the same period was 8.1 t/d (5.7-9.9 t/d) which should have produced 725 kg of Nitrosomonas (513-891 kg) at a yield coefficient of 0.09. Thus, on average, about 10 per cent of the $NH_3$-N was being oxidized by bacteria in suspension.

The ability of nitrifying bacteria to become attached and to grow on particulate surfaces is well known in percolating filters treating sewage, where high rates of nitrification can be achieved in the short periods of contact of any given body of waste water with the media. The key to successful nitrification in the activated-sludge process also lies in retaining solids in the aeration tank for a sufficient period to allow colonization and growth of these bacteria (Downing et al. 1964). It is not surprising therefore that the nitrifiers should behave similarly in rivers, becoming part of the attached flora and exploiting a niche in which they are bathed in a constant flux of nutrients, substrate, and oxygen. In addition, the surface of the bed of a stream is an obvious site at which to exploit the slow release of $NH_3$-N from decaying, organically rich, benthic deposits in well-aerated streams.

## Modelling the nitrification process in rivers

It is evident that for the purposes of water-quality management an empirical model of the nitrification process, based on first-order consecutive reactions and a predictive relationship between the dominant rate coefficient and measurable characteristics of the river, would probably suffice to predict changes in the $NH_3$-N content of river waters.

Apart from detecting changes in the concentration of $NH_3$-N in a downstream direction (other than those caused by dilution by clean tributaries) a simple qualitative indication that nitrification is occurring would be given by the appearance of consistent differences in the results of the inhibited and total $BOD_5$ tests. The need to model the nitrification process clearly depends upon the state of the river. The process is unlikely to be of much significance in clean streams having concentrations of less than about 0.5 mg/l $NH_3$-N. In addition, the phenomenon is likely to be of transitory occurrence in the history of a river since it would be inhibited by lack of oxygen in grossly polluted rivers and limited by lack of substrate in clean streams. In an era where the quality of rivers is improving as a result of remedial actions being taken at sewage-treatment works the development of significant nitrification in streams may therefore be regarded as a welcome sign, since the process cannot of itself cause anoxic conditions to occur and the natural removal of $NH_3$-N is of benefit to downstream users.

Acknowledgement

Published by permission of the Director, Water Research Centre

References

Curtis, E. J. C., K. Durrant & M. M. I. Harman, 1975. Nitrification in rivers in the Trent basin. Wat. Res. 9:255-268.

Downing, A. L., H. A. Painter & G. Knowles, 1964. Nitrification in the activated-sludge process. J. Proc. Inst. Sew. Purif. 130-158.

European Inland Fisheries Advisory Commission (E.I.F.A.C.) Working Party on Water Quality Criteria for European Freshwater Fish, 1970. Report on ammonia and inland fisheries. Technical Paper No. 11 (Wat. Res., 1973, 7:1011-1022).

Garland, J. H. N., 1973. Nitrification in rivers: studies in the Trent basin. In: A. James (Ed.): Symposium on the use of Mathematical Models in Water Pollution Control. Vol. 2, pp. 15-54. Department of Civil Engineering, Division of Public Health Engineering, Newcastle upon Tyne.

Garland, J. H. N. & I. C. Hart, 1972. The Trent Research Programme. Vol. 4. Effects of pollution on river quality. Water Resources Board, Reading.

Knowles, G., A. L. Downing & M. J. Barrett, 1965. Determination of kinetic constants for nitrifying bacteria in mixed culture, with the aid of an electronic computer. J. gen. Microbiol. 38:263-278.

Montgomery, H. A. C. & B. J. Borne, 1966. The inhibition of nitrification in the BOD test. J. Proc. Inst. Sew. Purif. 357-368.

World Health Organization, 1971. International Standards for Drinking Water. 3rd edition. W.H.O. Geneva.

Wezernak, C. T. & J. J. Gannon, 1967. Oxygen-nitrogen relationships in autotrophic nitrification. Appl. Microbio 15:1211-1215.

Keywords

benthos : nitrification : rivers : kinetics ammonia : Nitrosomonas : Nitrobacter : model water-quality

# IMPORTANCE OF THE DENITRIFICATION PROCESS FOR THE RATE OF DEGRADATION OF ORGANIC MATTER IN LAKE SEDIMENTS

Jens Møller Andersen

Freshwater Biological Laboratory, DK 3400 Hillerød, Denmark

## Summary

Rates of denitrification and oxygen uptake was measured at 8°C in 15 sediment cores from two Danish lakes, Esrom sø and Bryrup Langsø with 3 concentrations of nitrate in the overlying water. The rate of sediment oxygen uptake was not influenced by the nitrate concentration in the water (up to 10 mg N/l). Mean values of oxygen uptake for the two lakes were 560 and 720 mg $O_2/m^2d$, respectively. In Bryrup Langsø rates of oxygen uptake (mg $O_2/m^2d$) were linearly correlated to wet weight ($g/m^2$) of Chironomus plumosus, which dominated the profundal fauna (oxygen uptake = 480 + 5.7x(wet weight); r = 0.83). For each concentration of nitrate in the water the rate of denitrification was also linearly correlated with wet weight of Chironominae and with sediment oxygen uptake. The denitrification process contributed significantly to the degradation of organic matter in the sediment if concentration of nitrate in the water was above 2 mg N/l. At nitrate concentrations of 6-10 mg N/l, oxygen uptake of the sediment was less than the oxygen content of the nitrate denitrified. Therefore, if nitrate is present in considerable concentrations in the water, rates of degradation of organic matter in sediments measured by the oxygen uptake method are likely to be underestimated.

## Introduction

By the process of denitrification bacteria use nitrate as their terminal electron acceptor, and nitrate is reduced primarily to elemental nitrogen ($N_2$). It is well established that denitrification can be a very important sink for plant available nitrogen in lakes.

Rates of denitrification in lakes have been calculated from nitrogen mass balances by Vollenweider (1968), Brezonik & Lee (1968), Andersen (1974), and Larsen (1975). Rates of denitrification up to 50 g $NO_3^-$-N/$m^2$ year or about half of the total nitrogen loading have been measured. Denitrification can take place either in the anoxic water (Goering & Dugdale 1966, Brezonik & Lee 1968, and Koike et al. 1972) or in anaerobic sediments (Keeney et al. 1971, Chen et al. 1972, Keeney 1973, and Andersen 1976).

The process of denitrification can of course affect the nitrate concentration in lake waters, but not only nitrate is used by the denitrification process. Usually organic matter is oxidized and if the final product of the denitrification process is $N_2$, then for each $NO_3^-$ ion denitrified 2.5 oxygen atoms have been used to oxidize the organic substrate. This means that a rate of denitrification at a sediment surface of say 100 mg $N/m^2d$ will not only be very important for the nitrogen budget of a lake, but it also involves a degradation of organic matter corresponding to an oxygen uptake of 286 mg $O_2/m^2d$. This must be compared with oxygen uptake rates of about 500-1000 mg $O_2/m^2d$ often found for eutrophic lake sediments. Therefore, if the process of denitrification does not affect the rate of sediment oxygen uptake, the degradation of organic matter in lake sediments can be significantly enchanced by the denitrification process.

If, on the other hand, the rate of degradation of organic matter in the sediment is unaffected by the denitrification, the oxygen uptake of the sediment will be restrained by the denitrification process. Therefore, this investigation attempts to consider the effect, if any, which denitrification in lake sediments has on sediment oxygen uptake.

## Materials and methods

Profundal sediments from two lakes were sampled by taking 15 undisturbed sediment cores with perspex tubes (diam. 5.2 cm) with a Kajak bottom sampler (Kajak et al. 1965). Sediment was usually about 25 cm deep, and the height of the overlying water was

24.0 cm. Three quarters of the overlying water was removed every second day. In 5 of the cores the water removed was replaced by nitrate free lake water, in 10 it was replaced by lake water enriched with $KNO_3$ (3 mg N/l in 5 cores and 10 mg N/l in 5 cores).

The enriched and decanted water was analyzed for ammonia (Solorzano 1968) and nitrate (Strickland and Parsons 1968) and pH was measured. Simultaneously, the rates of sediment oxygen uptake of all cores were measured by the method described by Hargrave (1972). Homogeneity of the water column was maintained by a vertically rotating magnetic stirrer (length 20 mm, diam. 6 mm, stirring rate 375 rpm) placed 5 cm below a rubber stopper, which prevented exchange of oxygen with the atmosphere. Temperature during all experiments was 8 ± 2°C. Oxygen in the overlying water usually decreased from 10 mg/l to about 4 mg/l during the incubation period of about 48 hours. Samples were withdrawn from each core 5 times. Thus, cores were incubated for a total of 10 days.

Denitrification rates were calculated from the decrease in nitrate concentration in the water. Processes other than denitrification, however, also affect dissolved nitrate. Firstly, nitrate is produced by nitrification. The rate of nitrification was considered to be independent of nitrate concentration and increases in nitrate in cores where initial nitrate concentration was zero, were used to correct for this source of additional nitrate.

A second process which could alter nitrate concentration is the reduction of nitrate to ammonia. However, there was no significant difference between the changes in ammonia in cores with and without nitrate, and thus the reduction of nitrate to ammonia was assumed to be unimportant in this study.

Assimilative uptake of nitrate by microorganisms, a third pathway for nitrate in sediments and water, was not considered in the measurements. The importance of this process has been discussed (Andersen 1976). It will result in overestimations of the rates of denitrification, but the overestimation is not considered to surpass 10-20%.

Relations between sediment oxygen uptake and numbers of Chironominae in sediment cores were investigated by sieving (0.2 mm). The animals were counted, and their wet and dry weight measured.

## Results and discussion

Some data for the lakes considered in this study are given in Table 1. Table 2 summarizes the results of the oxygen uptake - denitrification experiments. There is no effect of nitrate in the overlying water on sediment oxygen uptake up to the highest nitrate conc. used (10 mg N/l). In both lakes denitrification increased the rate of breakdown of organic matter (expressed as the sum of oxygen uptake and denitrification in mg O/m² d) significantly at the 95% confidence level at nitrate concentrations larger than 2 mg N/l. At 0-10 mg N/l in the overlying water the uptake of oxygen from the nitrate exceeded the uptake of molecular oxygen from the water.

Table 1. Some data for Bryrup Langsø and Esrom sø. Sediment data are average numbers of the uppermost five cm

| | | Esrom sø | Bryrup Langsø |
|---|---|---|---|
| | Area (ha) | 1730 | 32 |
| | Depth average (m) | 12.3 | 5 |
| | max. | 22 | 9 |
| | Alkalinity (meq/l) | 2.5 | 1-1.5 |
| | Approx. gross primary production (g C/m²y) | 250 | 400 |
| | Range of nitrate conc. in lake water (mg N/l) | 0-0.5 | 0.2-3 |
| sediment | dry weight (% of wet weight) | 5.7 | 5.7 |
| sediment | loss on ignition (mg/g DW) | 350 | 380 |
| sediment | Kjeldahl nitrogen (mg N/g DW) | 20 | 23 |

Table 2. Average rates of oxygen uptake and denitrification of undisturbed sediment samples from Bryrup Langsø and Esrom sø at different concentrations of nitrate in the overlying water. To compare with the $O_2$-uptake the amount of oxygen in the nitrate denitrified is calculated. Standard errors of the means of 5 cores are given

| | conc. of nitrate | rate of denitrification | oxygen in nitrate denitrified | rate of $O_2$-uptake | sum of denitrification and $O_2$-uptake |
|---|---|---|---|---|---|
| | mg N/l | mg N/$m^2$ d | mg O/$m^2$ d | mg $O_2$/$m^2$ d | mg O/$m^2$ d |
| Bryrup Langsø | 0 | 0 | 0 | 707± 25 | 707± 25 |
| | 3.0-2.0 | 105± 13 | 298± 38 | 662± 74 | 960±112 |
| | 9.9-5.8 | 419± 51 | 1200±150 | 790± 66 | 1990±210 |
| Esrom sø | 0 | 0 | 0 | 540± 40 | 540± 40 |
| | 3.0-2.5 | 73± 16 | 208± 42 | 590± 70 | 798±112 |
| | 9.5-7.0 | 195± 38 | 560±108 | 550± 40 | 1110±138 |

About 80% of the wet weight of the profundal fauna in Bryrup Langsø consisted of Chironomus plumosus. In an attempt to explain the variation between cores in sediment oxygen uptake and denitrification the average oxygen uptake for each core is shown as a function of wet weight of C. plumosus (g/$m^2$) in Fig. 1. A reasonably good linear correlation is obtained. However, the increase in oxygen uptake with increasing density of C. plumosus can hardly be ascribed to the respiration of the animals only. At 17°C Walshe-Maetz (1954) found a rate of respiration of C. plumosus of about 200 µl $O_2$/h per gram wet weight corresponding to about 7 mg $O_2$/gd. If the rate of respiration of C. plumosus in this work where temperature was 8±2°C is about half of that found by Walshe-Maetz the rate of respiration of C. plumosus becomes about 200 mg $O_2$/$m^2$d at a density of 70 g wet weight per $m^2$. The increase predicted by the regression equation is 400 mg $O_2$/$m^2$d. The oxygen consumption not accounted for is probably due to the pumping of water by C. plumosus into the sediment, where the oxygen is used by the microorganisms.

Further evidence for this indirect effect of the Chironominae is shown in Fig. 2, where the rate of denitrification is plotted as a function of wet weight of C. plumosus for two different concentrations of nitrate in the overlying water. C. plumosus obviously enchances the denitrification, but not because it is performing the denitrification process itself, but probably by the pumping of water rich in nitrate from the overlying water into the sediment, where it is used by denitrifying bacteria.

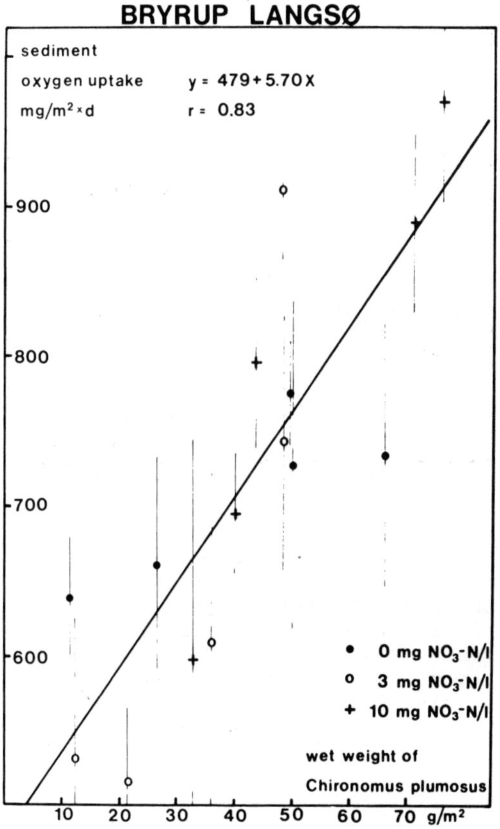


Fig. 1. Bryrup Langsø. Rate of oxygen uptake of undisturbed sediment cores as a function of density of Chironomus plumosus in the cores. Vertical bars indicate standard errors of the mean of 5 determinations on each core. Temperature 8±2°C. Sampling date 16.2.1976

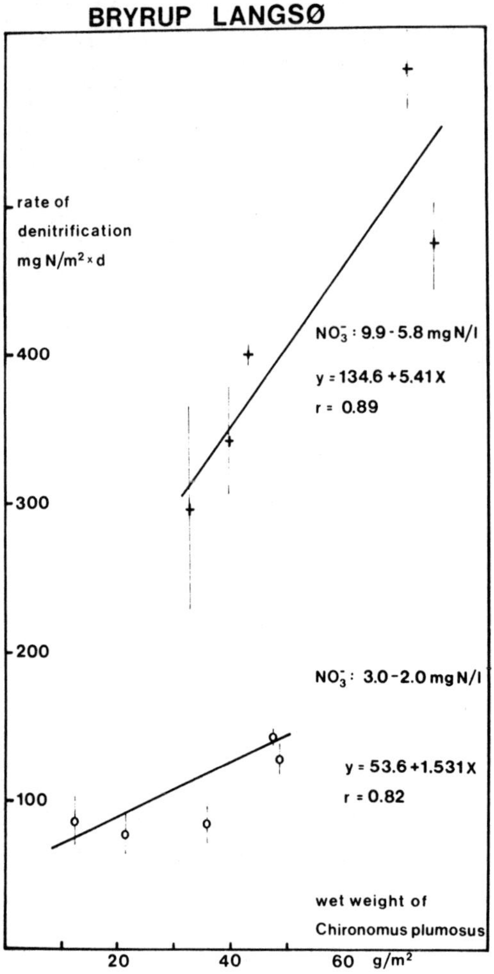


Fig. 2. Bryrup Langsø. Rate of denitrification of undisturbed sediment cores as a function of density of Chironomus plumosus at two levels of nitrate concentration in the overlying water. Vertical bars indicate standard errors of the mean of 5 determinations on each core. Temperature 8±2°C. Sampling date 16.2.1976

Rates of denitrification as a function of sediment oxygen uptake are shown in Fig. 3 and 4 for Bryrup Langsø and Esrom sø. For each concentration of nitrate in the overlying water, there is a clear, positive correlation between rate of denitrification and oxygen uptake. In part this positive correlation is caused by the pumping of oxygen and nitrate in the water into the sediment by the Chironominae. However, even without benthic animals a positive correlation must be expected, as a high oxygen demand of a sediment generally results in a more reduced sediment, which favours the process of denitrification.

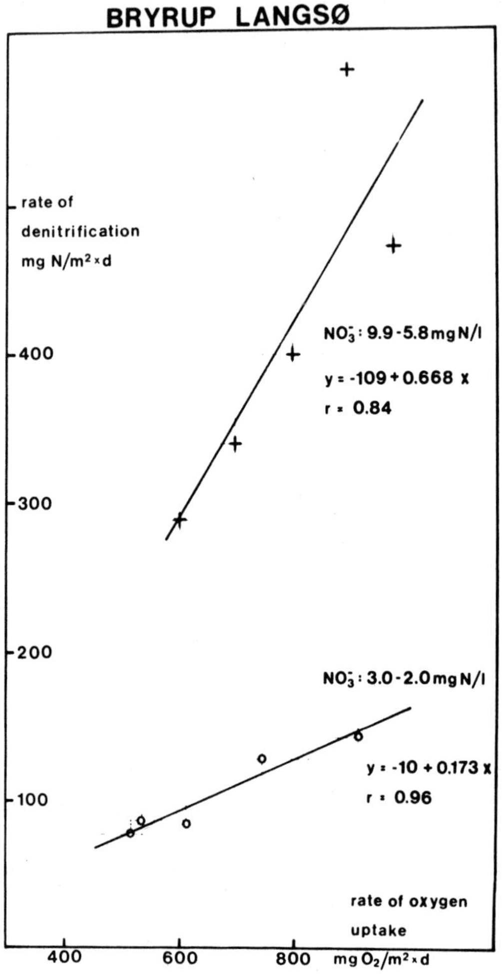


Fig. 3. Bryrup Langsø. Rate of denitrification of undisturbed sediment cores as a function of sediment oxygen uptake at two levels of nitrate concentration in the overlying water. Vertical and horizontal bars indicate the standard errors of the mean of 5 determinations on each core. Temperature 8±2°C. Sampling date 16.2.1976

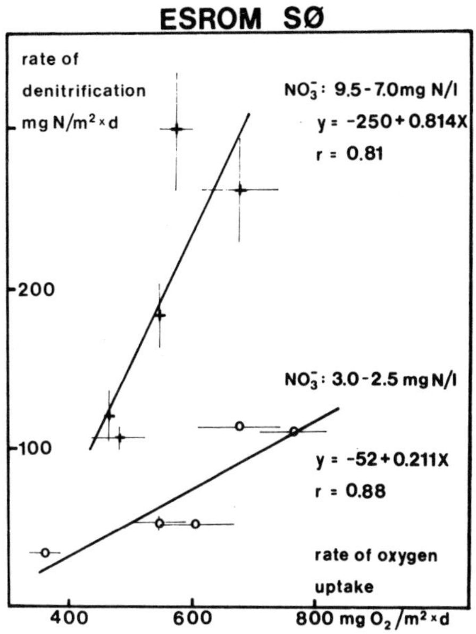


Fig. 4. Esrom sø. Rate of denitrification of undisturbed sediment cores as a function of sediment oxygen uptake at two levels of nitrate concentration in the overlying water. Vertical and horizontal bars indicate the standard errors of the mean of 5 determinations on each core. Temperature 8±2°C. Sampling date 23.1.1976

## Conclusions

The main results of this work are:

1) The rate of oxygen uptake of lake sediments is unaffected by a simultaneously occurring denitrification in the lake sediment.

2) The process of denitrification contributes significantly to the breakdown of organic matter at nitrate concentrations above 2 mg N/l in the overlying water, and at 10 mg N/l denitrification is more important than oxygen uptake.

3) For each concentration of nitrate in the overlying water there is a positive linear correlation between rate of denitrification and rate of oxygen uptake of the sediment.

4) Not only rate of sediment oxygen uptake but also rate of denitrification is positively correlated to the density of Chironomus plumosus in the sediment. Apart from their own oxygen uptake they enchance the oxygen and nitrate uptake of the microorganisms in the sediment by their pumping of water into the sediment.

## References

Andersen, J.M., 1974. Nitrogen and phosphorus budgets and the role of sediments in six shallow Danish lakes. Arch. Hydrobiol. 74(4):528-550.

- -, 1976. Rates of denitrification of undisturbed sediment from six lakes as a function of nitrate concentration, oxygen and temperature. Arch. Hydrobiol. (In press).

Brezonik, P.L. & G.F. Lee, 1968. Denitrification as a nitrogen sink in Lake Mendota. Environ. Sci. Technol. 2:120-155.

Chen, R.L., D.R. Keeney, D.A. Graetz & A.J. Holding, 1972. Denitrification and nitrate reduction in lake sediments. J. Environ. Qual. 1:158-162.

Goering, J.J. & V.A. Dugdale, 1966. Estimates of the rate of denitrification in a subarctic lake. Limnol. Oceanogr. 11:113-117.

Hargrave, B.T., 1972. A comparison of sediment oxygen uptake, hypolimnetic oxygen deficit and primary production in Lake Esrom, Denmark. Verh. Internat. Verein. Limnol. 18: 134-139.

Kajak, Z., K. Kacprzak & R. Polkowski, 1965. Tube bottom sampler for taking samples of micro and macro benthos, and for sampling of undisturbed structures of mud samples for experimental purposes. Ekol. Pol. (B):11:159-165.

Keeney, D.R., 1973. The nitrogen cycle in sediment-water systems. J. Environ. Qual. 2:15-29.

- -, R.L. Chen & D.A. Graetz, 1971. Importance of denitrification and nitrate reduction in sediments to the nitrogen budgets of lakes. Nature 233(3):66-67.

Koike, I., E. Wada, T. Tsuji & A. Hattori, 1972. Studies on denitrification in a brackish lake. Arch. Hydrobiol. 69:508-520.

Larsen, V., 1975. Nitrogen transformation in lakes. Paper presented at IAWPR-conference on nitrogen as a water pollutant, Copenhagen, Denmark 18-20th Aug. 1975.

Solorzano, L., 1969. Determination of ammonia in natural waters by the phenolhypochlorite method. Limnol. Oceanogr. 14:799-810.

Strickland, J.D.H. & T.R. Parsons, 1968. A practical handbook of sea-water analysis. Fish. Res. Bd. Can. No. 167, Ottawa.
Vollenweider, R.A., 1968. Scientific fundamentals of the eutrophication of lakes and flowing waters with particular reference to nitrogen and phosphorus as factors in eutrophication. OECD report DAS/CSI/68. 27.
Walshe-Maetz, B.M., 1954. Le métabolism de Chironomus plumosus dans des conditions naturelles. Physiologia comp. Oecol. 3:135-154.

Key words: denitrification
sediment
respiration
benthos
degradation

# DENITRIFICATION IN SEDIMENT-WATER SYSTEMS OF VARIOUS TYPES OF LAKES

T. Tirén

Royal College of Forestry, Department of Forest Soils, Uppsala, Sweden[1]

## Summary

Denitrification rate in a demarcated sediment-water system was measured in situ using a bell jar. The water above the sediment was gently stirred with a magnetic bar. $^{15}$N-labelled $NO_3$-N was added at start of the experiment and the fate of labelled nitrogen was followed.

$NO_3$-N and $O_2$ consumption, $NH_4$-N release and the rate of formation of $N_2$ were measured. $N_2O$ was rarely observed in small quantities.

In two eutrophic lakes, Lake Norrviken and Lake Ramsjön denitrification rate was about 60 and 80 mg/$m^2$·day, respectively. In the unpolluted, eutrophic Lake Erken and the oligotrophic, mesohumic Lake Botjärn measured rates were about 25 and 5 mg $N_2/m^2$·day, respectively. These values however, must be considered as experimental data received from field experiments, indicating what may happen in a relatively undisturbed sediment-water system, but they say little about what actually happens in the lake as a whole.

Nutrient rich sediments normally have a very high denitrifying capacity, but the extent of these processes in situ is not clear. On basis of experimental data and field observations some of the regulating factors are discussed.

Key-words: Denitrification, $^{15}$N, in situ measurements, nitrate distribution, lake sediment.

## Introduction

Even though the existens of the microbiological denitrification process is well documented from marin as well as brackish and limnic environments investigations in limnic systems have become common only in recent years. In a literature review on denitrification in natural waters Brezonik (1975) pointed out that surprisingly little is known about the quantitative importance of denitrification. Estimations of the significance of denitrification in balancing the nitrogen budget indicate that the process may well be of great importance in the annual nitrogen budget for lakes (Table 1).

Because of the crudeness of the budgeting technique in situ measurements of denitrification with labelled nitrogen are well justified to get more reliable data of the significance of the process.

Table 1. Estimated losses of nitrogen due to denitrification.

| Lake | % of annual loading | References |
|---|---|---|
| Mendota USA | 11 | Brezonik & Lee, 1968 |
| Mälaren Sweden | 57 | Ahl, 1973 |
| Norrviken - " - | 41-61 | Ahlgren, 1973 |
| Six shallow Danish lakes | 0-54 | Andersen, 1974 |
| Hald Denmark | 23 | Larsen, 1975 |
| Glenstrup | 61 | - " - |
| Ramsjön Sweden | 60 | Ryding & Forsberg, 1976 (report at this symposium) |

The investigation has been carried out at the Department of Plant Ecology and Forest Soils, Royal College of Forestry, Uppsala in close cooperation with the Institute of Limnology, University of Uppsala.

I will here give a brief description of the methods used and also discuss results and observations from varius types of lakes. A detailed description of the in situ experiments and mass spectrometric technique is given elsewhere (Tirén et al., 1976).

## Material and Methods

### Description of the lakes

Norrviken is a strongly eutrophicated lake in central Sweden. During 1969 pollution by domestic and industrial waste water ceased and recovery now is proceeding (Ahlgren, 1973 and in a report at this symposium). The deeper part of the lake is normally termal stratified both in summer and winter.

Ramsjön is a small, shallow and likewise strongly eutrophicated lake situated about

1. Present address: Institute of Limnology, Box 557, S-751 22 UPPSALA 1, Sweden.

40 km W of Uppsala. It was formely very heavily polluted by domestic sewage, which is now chemically treated. The lake is not stratified and during the winter total oxygen demand and fish killing sometimes occures.

Erken is a moderately eutrophic lake situated about 50 km E of Uppsala and is practically unaffected by pollutants. Due to relatively strong wind influence, the summer stratification is often broken up and the circulation periods are often extended. Physical and chemical data from the lake from 1930 to the present has been reviewed by Boström & Pettersson (1973).

Botjärn is a small, oligotrophic lake situated about 200 km NW of Uppsala in the nothern conifer region.

Some data from the four lakes are collected in Table 2.

Field technique

In order to enclose a sediment-water system, a Plexiglass cylinder developed and described by Edberg & von Hofsten (1973), was used (Fig. 1). The apparatus was lowered slowly to the lake bed and the lid was closed atomatically when the cylinder section was placed on the bottom. Hereby, 10-12 l of water was enclosed in contact with a comparatively undisturbed bottom sediment. By using a magnetic stirrer the enclosed water could be suitably mixed and through two PVC tubes additives could be introduced and samples extracted.

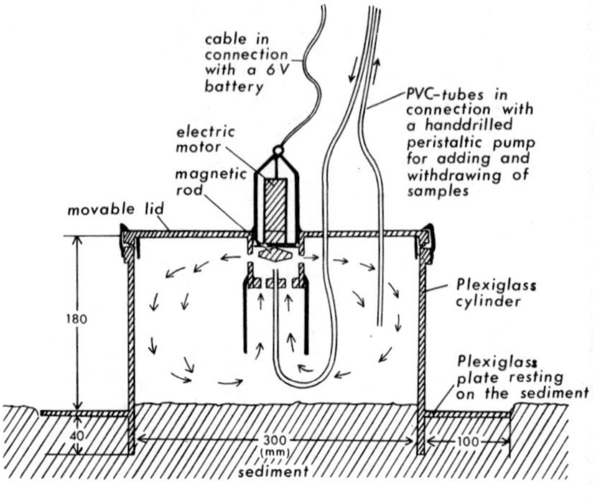


Fig. 1. The cylinder resting on the bottom sediment, with closed lid. (After Tirén et al., 1976).

About 3 mg/l (Norrviken, Ramsjön), 1 mg/l (Erken) or 0.5 mg/l (Botjärn) of $^{15}N$-labelled nitrate nitrogen (33.4 atomic % of $^{15}N$) were pumped down into the cylinder. In order to obtain an estimate of the enclosed water volume and to check the dilution caused by extraction of samples and the admixture of interstitial and leaking water, each cylinder was supplied eith 4 µCi $^{36}Cl$ in the form of NaCl solution.

Samples were taken as a rule once every

Table 2. Data for the lakes where in situ experiments have been performed.

| | Norrviken[1] | Ramsjön[2] | Erken[3] | Botjärn[4] |
|---|---|---|---|---|
| Area (ha) | 267 | 40 | 2490 | 9.7 |
| Depth, mean (m) | 5.4 | 1.8 | 9 | 3.3 |
| Depth, max. (m) | 12.5 | 3.5 | 21 | 14.4 |
| Transparency min. values (m) | 0.3 | 0.2 | 2.1 | |
| pH | 7.5-9.3 | ~8 | 8.1-8.5 | 5.5-5.7 |
| Conductivity (mS/m, 20°C) | 37 | 19 | 25 | 2.3-2.7 |
| Tot-N (mg/l) | 2.5-5.5 | 1.0 | 0.7-0.8 | 0.3-0.5 |
| Tot-P (mg/l) | 0.17-0.30 | 0.5 | 0.01-0.09 | 0.01-0.02 |

1. Ahlgren, (1969) (unpublished, stenciled report)
2. Forsberg & Ryding, (1973)
3. Widell, (1970) and Segerfeldt, (1967)
4. Ramberg et al., (1973)

24 hours. In the Norrviken experiment stirring took place about 1 hour per day, while in the other experiments the stirrer was in continous operation.

The conditions just above the sediment surface are important for evaluating the role of sediment for the lake's nitrogen metabolism. Working from ice, it was possible to get profiles down to the last few centimeters above the sediment surface. The samples were extracted using PVC-tubes and a handdrilled peristaltic pump.

Physical and chemical analyses

The composition of the dissolved gases and collected marsh gas, as well as the isotopic composition of the nitrogen gas was analysed by Jan Thorin, who used a LKB 9000 mass spectrometer equipped with a double collector system for measuring isotope ratios. The sampling technique, the calculations concerning the isotopic composition in the nitrogen compounds and the chemical analyses are described in Tirén et al., (1976).

## Results and discussion

Volume of water in the cylinders

Figure 2 examplifies from the various experiments the volume of water involved in the "mixing" as calculated with simplyfying assumptions (Tiren et al., 1976) from the $^{36}Cl$ values. Most likely a combination of mass transport and molecular diffusion have been responsible for the increased volume. In the Norrviken experiment, however, the molecular diffussion was possibly the main process (linear increase of "mixed" volume). Diffrences with respect to structure of the sediment, bioturbation and stirring time may have been of importance as regards the exchange of interstitial water.

Oxygen and nitrogen transformations

Figure 3 illustrates what happens in the enclosed sediment-water system after the addition of labelled nitrate. In eutrophic lakes the molecular oxygen is consumed relatively rapidly. The consumption of nitrate is accompanied by a corresponding increase of labelled nitrogen gas. A small temporary accumulation of nitrite was also observed. In all the experiments there was a net ammonium release.

It was possible from the values obtained to make calculations of the oxygen and nitrate consumption, the net ammonium release and the rate of denitrification emanating from the labelled nitrate.

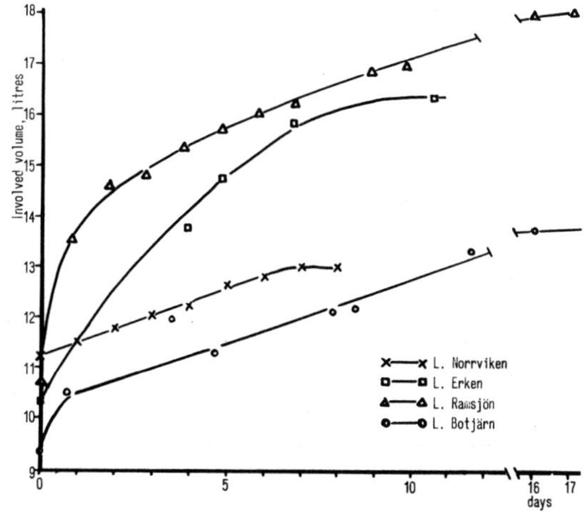


Fig. 2. The volume of water involved in "mixing". (Partly after Tirén et al., 1976).

The production of labelled nitrogen gas corresponds closely to the changes in nitrate and nitrite content. A slight decrease in the proportion of labelled nitrogen gas in the latter part of the experiments may be due to slow adjustment to equilibrium with any marsh gas present in the containers, or to leakage. No one of these possibilities seems likely because at end of the end of the experiments marsh gas was present only in the Norrviken and Ramsjön experiments and leakage between the lid and the cylinder was prevented by a greased rubber packing.

However, the changed proportion also can be due to the production of nitrogen gas from sources of unlabelled nitrogen. Johnston et al. (1974) and Ripl (pers. comm.) have obtained results indicating denitrification under conditions where no molecular oxygen or nitrite or nitrate is present. They proposed the presence of some form of organic bound oxygen in the sediment.

From the experiments it is clear that the rate of denitrification may be considerable at the temperature interval of 3-6°C. In reports on denitrification in soils and in sewage treatment experiments one often finds high temperature optimum and very low activity at temperatures lower than 5°C (Stanford et al., 1975). However, Goering and Dugdale (1966) reported high denitrification activity in a subarctic lake at about 2°C. The small and slow temperature changes in lake water obviously provide better possibilities for a kryophilic microflora to exist.

From the experiments it is also evident that conditions for denitrification exist in the sediment, in spite of oxygen con-

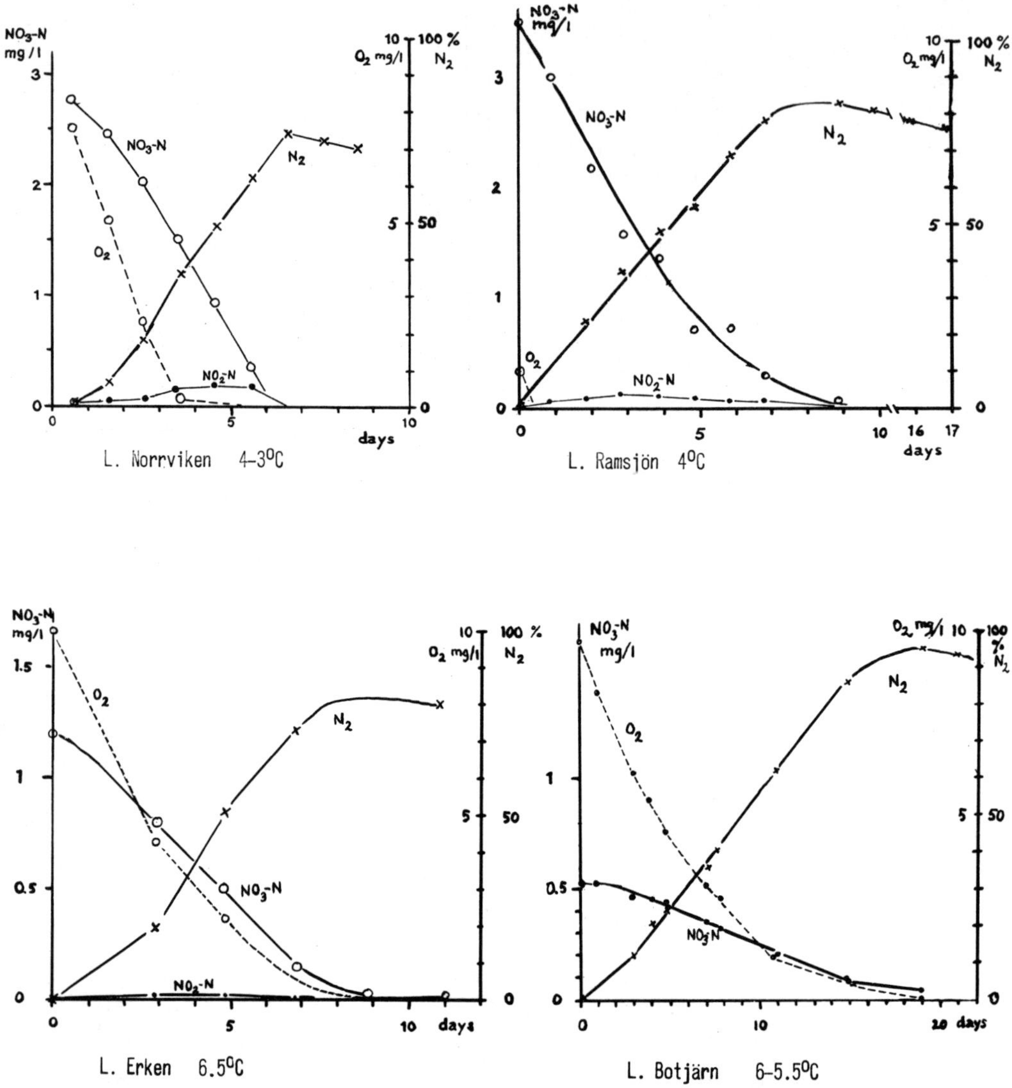


Fig. 3. Changes in oxygen, nitrate, nitrite and labelled nitrogen gas. The $N_2$ figures given are percentages of total labelled nitrate nitrogen at start of the experiments.

tents of up to 6-8 mg/l in the free water near the bottom. This is important because the presence of oxygen may stimulate the nitrification and the newly formed nitrate can be easily transported by diffusion or currents to microenvironments where conditions are favourable to denitrification (Andersen, 1974). Thus in shallow, eutrophic lakes there is often a considerable denitrification (Andersen, 1974 and Ryding & Forsberg, pers. comm.).

The rate of formation of nitrogen gas corresponds to about 20, 30, 8 and 1-2 $g/m^2$·year for the lakes Norrviken, Ramsjön, Erken and Botjärn, repectively. These values, of course, are not identical with the true denitrification on an annual basis, but they do indicate the potential capacity for denitrification in the sediment in lakes of various nutrient status. In annual budget calculations for the years 1961-62, 1970, 1971 and 1972 Ahlgren (1973) gives the following values for net denitrification in Lake Norrviken: 30, 7, 3, 3 g $N/m^2$·year, respectively. During 1974 there was likely no net denitrification at

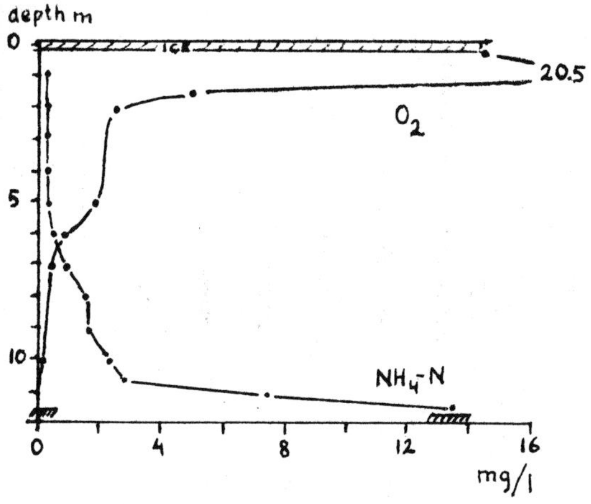


Fig. 4. L. Norrviken, mars 1976. Oxygen and ammonium distribution at a station in the south part of the lake. Samples are taken down to a depth less than 5 cm above the bottom.

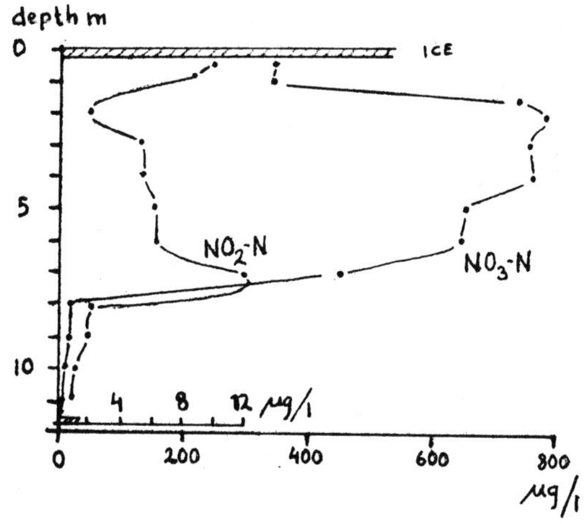


Fig. 5. L. Norrviken, mars 1976. Nitrite and nitrate distribution (the same time and place as in figure 4).

all and, futhermore, there was no significant nitrogen fixation either (Ahlgren, pers. comm.). For Ramsjön Ryding & Forsberg calculated the value 7,8 g N/m$^2$·year (presented at this symposium). Widell (1970) estimated the nitrogen loading to 850 mg/l for the year 1965 for Lake Erken. Even if the value is uncertain it indicates a very low nitrogen loading and in the annual nitrogen budget nitrogen fixation is important (Granhall & Lundgren, 1971 and Lännergren, 1972). The experimental denitrification value obtained in the present investigation thus seems very high and it is not clear to what extent it reflects the true denitrification. However, it is evident that it is a complicated task to estimate the nitrogen budget for Lake Erken. For Lake Botjärn there are no budget calculations finished yet.

From the experiments performed it is evident that the denitrifying capacity, of even an undisturbed lake sediment, is much higher than the calculation of net denitrification may indicate. The supply of nitrate to the sediment must be one of the most important regulatory factors.

Hereto we have made only a few measurements in the water above the sediment and we can not discuss any results yet. Normally, the rates seem to be some orders of magnitude slower in the water without contact with the sediment.

ome observations from icecovered lakes

he vertical and horisontal nitrate distributions were studied in late winter. The strong gradient of oxygen and ammonium in the south part of Lake Norrviken is evident from figure 4. Just above the bottom the ammonium values are of the same order of magnitude as in the interstitial water. A series of profiles taken across the south basin indicated the existence of density currents near the bottom. Thus, the high concentrations just above the bottom are a result of processes carried out over a larger bottom area. However, the existence of such strong gradients indicates that there are relatively small watermovements under the ice at this site. Under such conditions, the bottom influence on denitrification in the overlying water are probably strongly reduced.

The supersaturation of oxygen just under the ice (Fig. 4.) was due to primary production of green algae.

At about 7 m depth there was a strong gradient in the nitrate content and at the same depth there was a small, but distinct, top of nitrite (Fig. 5.), possibly indicating high denitrification activity in the water column at that depth.

In Lake Ramsjön there was a strong horisontal gradient in the nitrate content with the highest values near the outlet point of the sewage plant and the lowest values near the outlet of the lake (Fig.6). This can be a combined effect of dilution and denitrification. The existence of a denitrifying process seems resonable from the vertical profiles (Fig. 7) and the fact that the oxygen content was very low near the bottom. Primary production caused excess of oxygen in a thin layer just under the ice.

In Lake Igelsjön, situated about 30 km NNW of Uppsala, (meandepth of about 1 m, maximum depth of 1.8 m, area 2,6 ha, watershed 6,6 $km^2$ covered almost exclusively by conifers; data from Andersson et al., 1975), there was an inflow of nitrate-rich water and, after passage of the about 300 m long lake, there was very low nitrate content in the outflowing water (Fig. 8 ). This situation was observed both at the start and the end of march, 1976. The lake was ice covered and the primary production had not started yet. There was no oxygen near the bottom and only 1.5 mg/l in a thin layer just under the ice, here due to inflowing oxygen-rich water. There was a strong vertical gradient in the nitrate profiles (Fig. 9), which could be a combined effect of high denitrifying capacity in the bottom sediment and the stratfiation of inflowing water.

The examples mentioned demonstrate that denitrification under specific conditions can significantly diminish the nitrate content in lakes. For a discussion about the relationship between nitrate assimilation and dissimilation, see Tirén et al., (1976).

Finally, from the discussion it is clear that the diffrence between what here is called in situ measurements and pure laboratory experiments is one of degree rather than one founded on a principle. The results obtained in the in situ experiments can, therfore, only with great caution be transferred to what really happens in the lake.

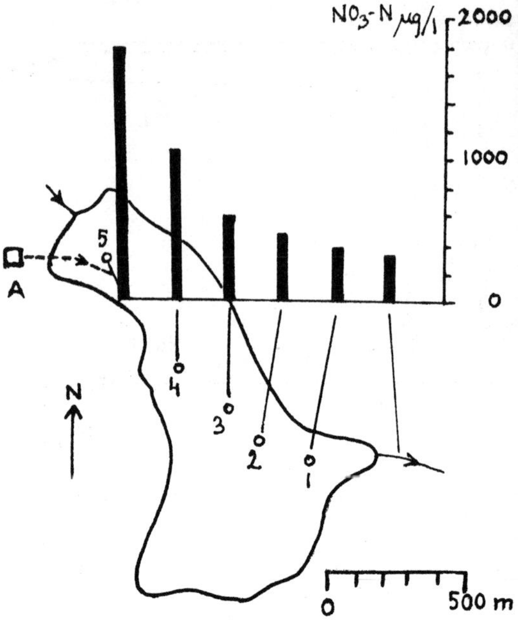


Fig. 6. L. Ramsjön, mars 1976. Horisontal nitrate distribution in the ice covered lake. Values are means in the layer 0.5 - 1.5 m depth. A = sewage treatment plant with outlet near station 5.

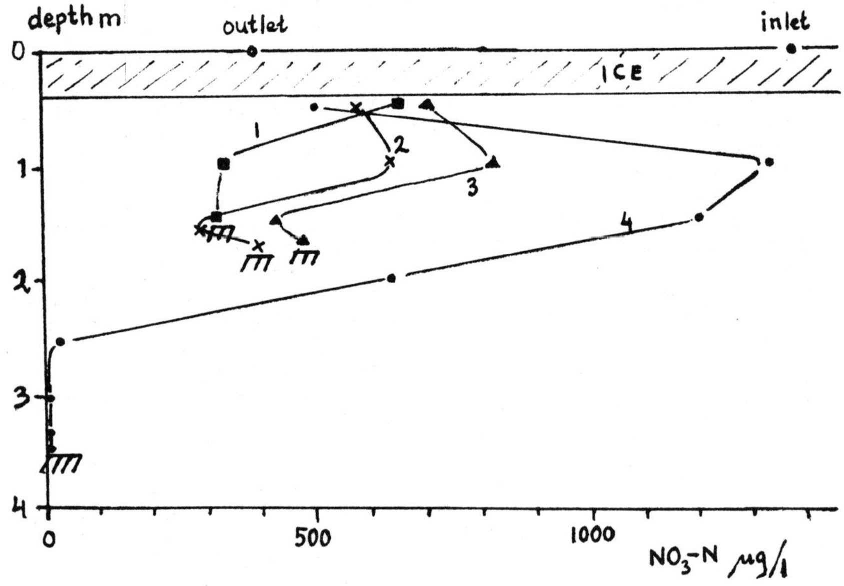


Fig. 7. L. Ramsjön, mars 1976. Vertical distribution of nitrate at station 1-4. Samples are taken to a depth less than 5 cm above the bottom. Concentrations in natural inlet and outlet are indicated.

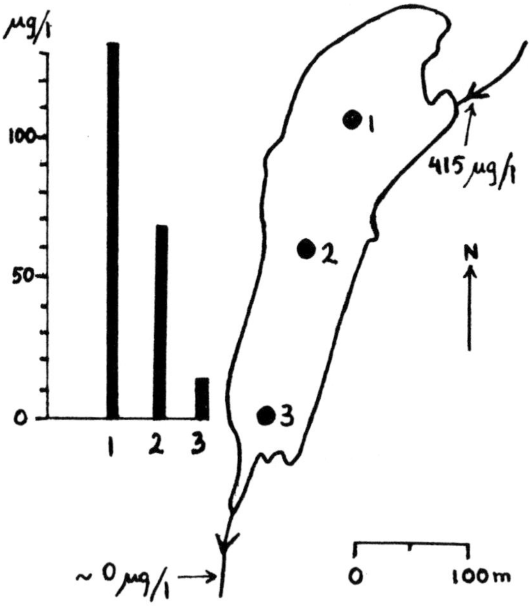


Fig. 8. L. Igelsjön, mars 1976. Horisontal nitrate distribution in the ice covered lake. Values are means in the layer 0.5 - 1.5 m (prctically the whole water column). Concentrations in inlet and outlet are marked.

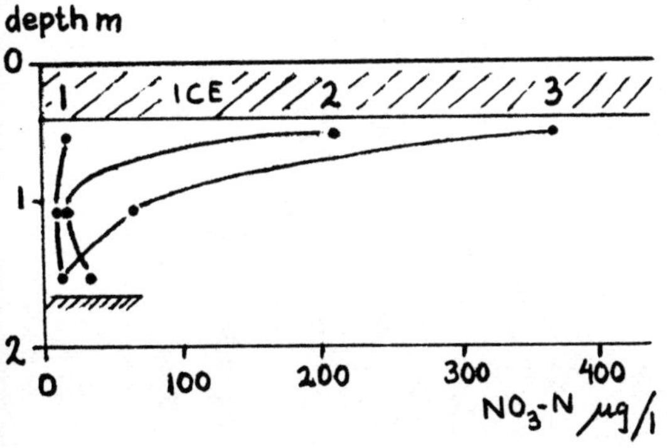


Fig. 9. L. Igelsjön, mars 1976. Vertical distribution of nitrate at station 1-3. Samples are taken to a depth less than 5 cm above the bottom.


References

Ahl, Th., 1973. Mälarens belastning och vattenkvalitet. Scripta Limnologica Upsaliensia, Coll. 9, No 332. (Summary in English).

Ahlgren, I., 1973. Limnologiska studier av sjön Norrviken. III. Avlastningens effekter. Scripta Limnologica Upsaliensia, Coll. 9, No 333.

Andersen, J. M., 1974. Nitrogen and phosphorous budgets and the role of sediments in six shallow Danish lakes. Arch. Hydrobiol. 74:528-550.

Andersson et al., 1975. Igelsjön Nuvarande status och förslag till åtgärder. Stenciled report, Institute of Limnology, Uppsala.

Boström, B. & K. Pettersson, 1973. Fysikaliska och kemiska data från Erken sedan 1930-talet. Scripta Limnologica Upsaliensia, Coll. 9, No 344.

Brezonik, P. L. & G. F. Lee, 1968. Denitrification as a nitrogen sink in Lake Mendota. Environ. Sci. Technol. 2:120-125.

Brezonik, P. L., 1975. Denitrification in natural waters. Conference on Nitrogen as a Water Pollutant. Proceedings. Volume 2. IAWPR specilized conference, Copenhagen.

Edberg, N. & B. von Hofsten, 1973. Oxygen uptakes of bottom sediments studied in situ and in the laboratory. Water Research 7:1285-1294.

Forsberg, C. & S.-O. Ryding, 1973. Naturvårdsverkets RR-undersökning 2. Ytvattenkvaliteten i RR-undersökningens sjöar, augusti-oktober, 1972. Vatten 29:464-468.

Goering, J. J. & V. A. Dugdale, 1966. Estimates of the rates of denitrification in a subarctic lake. Limnol. Oceanogr. 11: 113-117.

Granhall, U. & A. Lundgren, 1971. Nitrogen fixation in Lake Erken. Limnol. Ocenogr. 16:711-719.

Johnston, D. W., A. J. Holding & J. E. McCluskie, 1974. Preliminary comparative studies on denitrification and methane production in Loch Leven, Kinross, and other freshwater lakes. Proc. R.S.E. (B) 74(8):123-133.

Larsen, V., 1975. Nitrogen transformation in lakes. Conference on Nitrogen as a Water Pollutant. Proceedings. Volume 2. IAWPR specilized conference, Copenhagen.

Lännergren, Ch., 1972. Kvävefixering och primärproduktion i Erkens pelagial. Stenciled report, Institute of Limnology, Uppsala (Abstract in English).

Ramberg et al., 1973. I. Klotenprojektets undersökningsområde. Scripta Limnologica Upsaliensia, Coll. 9, No 320.



Key words: denitrification
$^{15}N$
in situ measurements
nitrate distribution
lake sediment

Theme 4 - Water quality, health and management aspects of sediments

# ROLE OF SEDIMENTS IN THE PROCESS OF RECOVERY OF A EUTROPHICATED LAKE

I. Ahlgren

Institute of Limnology, Uppsala, Sweden

## Summary

The highly eutrophicated Lake Norrviken was completely relieved of all wastewater effluents in 1969, having served as a recipient of industrial and domestic waste waters for nearly a hundred years. Total phosphorus concentration at autumn overturn in 1969 was about 470 $\mu g\ l^{-1}$. In 1975 it had decreased to about 150 $\mu g\ l^{-1}$. It is demonstrated that this decrease on average follows a simple dilution curve and that the rates of phosphorus release from and accumulation in the sediments have been about equal and thus tend to counterbalance. Only in one out of six years after the diversion of sewage effluents has a net release of phosphorus from the sediments taken place; in the four last years there have been net deposits calculated on whole-year bases. There are, however, large seasonal variations; during the summer there is always a net release maintaining the algal bloom. It is shown that this net release is continuously decreasing, from a rate of 9.2 mg P $m^{-2}d^{-1}$ in 1971 to 1.6 mg P $m^{-2}d^{-1}$ in 1976. A recovery of the lake is thus taking place not only by dilution but also by a decreasing internal loading of nutrients from the sediments.

## Introduction

It is well known that nutrients stored in the sediments of eutrophicated lakes may severely delay their response to decreased external nutrient loading. Cases have been reported where no recovery has taken place during a period of ten years after that the loading was decreased (Björk, 1972). Due to the "internal" loading caused by leakage of nutrients from such sediments, eutrophication is often said to be irreversible. In many cases this may be true but need not necessarily mean that no recovery at all is possible. This paper will examine some aspects of the recovery process in Lake Norrviken in central Sweden after the complete diversion in 1969 of all sewage effluents from the lake.

The lake has a very long eutrophication history, the first signs of cultural eutrophication being reported in the early years of this century (Sondén, 1914). Prediversion conditions have been described by Rodhe (1948), I. Ahlgren (1967) and G. Ahlgren (1970). In 1969 a research project was started in order to study the extent to which the lake could recover from its polyeutrophic condition. Since 1970 the project is being financed by the National Swedish Environment Protection Board.

Nutrient release from lake sediments has been studied in laboratory experiments by e.g. Tessenow (1972), Austin & Lee (1973), Kamp-Nielsen (1974, 1975), Jørgensen et al. (1975), Jacobsen & Jørgensen (1975). It is, however, difficult to base conclusions about the quantitative role of sediment release under natural conditions only on laboratory experiments.

Recently Vollenweider (1976) stressed that the best way of evaluating the role of sediments in the nutrient cycles of lakes, would be to make detailed studies of input, output and variations of the contents of the water column over the year.

In the present study both the experimental and the input-output methods have been used, this paper being devoted to an examination of the latter method. This type of treat-

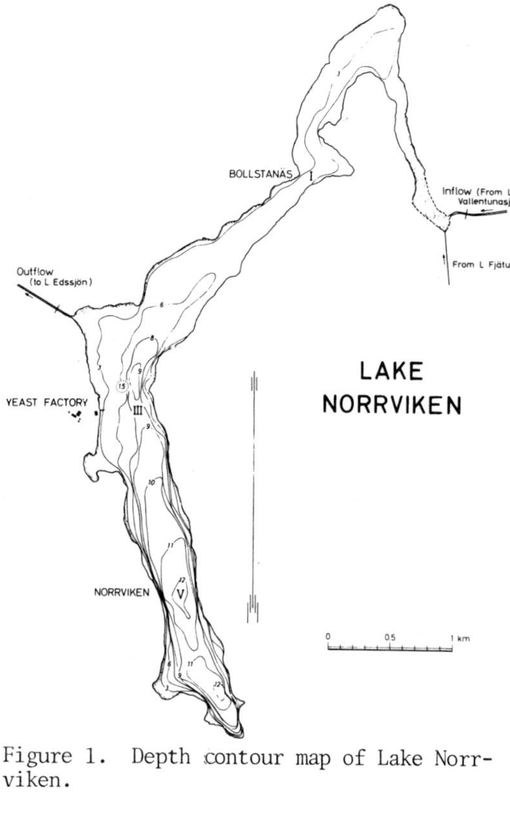


Figure 1. Depth contour map of Lake Norrviken.

ment of data has been used recently by Bengtsson (1975) for the lake S. Bergundasjön, and also by several other authors of this volume.

In the most simple case the concentration of an element will decrease according to a dilution equation after a decrease of the input (Rainey, 1967). This dilution process will be examined with regard to potassium and phosphorus concentrations in Lake Norrviken.

## The lake and its drainage area

Lake Norrviken is situated 20 km north of Stockholm, Sweden. Morphometric and hydrologic data are given in Table 1.

Table 1. Morphometric and hydrologic data of Lake Norrviken.

| | | |
|---|---|---|
| Surface area | 2.67 | $km^2$ |
| Volume | 14.3 | $10^6 m^3$ |
| Max depth | 12 | m |
| Mean depth | 5.4 | m |
| Drainage area | 91 | $km^2$ |
| Detention time (average) | 10 | months |

The bedrock of the drainage area is mainly granites and gneisses. The overlying soil, however, has a high content of calcium carbonate transported to the area during the last glaciation. The water therefore is rich in calcium and bicarbonate (alkalinity 2 meq $l^{-1}$).

The main basin of the lake lies in a north-south direction and is sheltered by a steep shore from the SW winds which prevail in summer. A stable thermal stratification is therefore always established. The thermocline is usually developed at a depth of about 6 m by the end of May and is successively eroded down to about 9 m by the end of August. Complete circulation takes place around mid September. The bottom area below 6 m is about 37% and below 9 m about 22% of the total bottom area. The corresponding hypolimnetic volumes are 15% and 6% of the total lake volume, respectively.

The hypolimnetic water rapidly becomes anoxic. Before the diversion of sewage effluents the total water volume became anoxic during the ice-covered period (December 1-April 20) but this has not occurred during the last three winters.

## Phosphorus budget

The phosphorus budget for Norrviken before and after diversion of sewage effluents is summarized in Table 2. The external loading was about 4 g P $m^{-2}year^{-1}$ in 1961/62 (Ahlgren 1967) and probably even higher immediately before the diversion. After the diversion it has decreased to about 0.5 g P $m^{-2}year^{-1}$. In 1961/62 the lake retained about 50% of the loading. Since the average concentration in the lake did not change during the year the total retention took place in the sediments. This is normal for a lake in balance with its loading, its sediment acts as a nutrient sink. After the diversion the lake retention has been negative or near zero. If the changes in phosphorus content of the water are subtracted from the lake retention, it is found that the sediment retention was negative in 1971 but positive in all other years. With the exception of 1971 this means that the sediment has not acted as a phosphorus source over longer periods. As will be shown later, however, during periods with thermal stratification and anaerobic conditions in the hypolimnion, sediment release may be an important nutrient source.

## Dilution rate

The rate of change in the concentration of a conservative element after a change of the input rate to a continuously stirred volume may be described by the following equation:

$$\frac{dC}{dt} = \frac{q \cdot C_i}{V} - \frac{q \cdot C}{V} \quad (1)$$

or in integrated form:

$$C_t = C_i + (C_0 - C_i) \cdot e^{-\frac{q \cdot t}{V}} \quad (2)$$

$C_i$ = concentration in influent
$C$ = concentration in stirred volume
$C_0$ at time zero
$C_t$ at time t

Table 2. Summary of phosphorus budget of Lake Norrviken before and after diversion of sewage effluents (g $m^{-2}year^{-1}$).

| | | 1961-1962 | 1970 | 1971 | 1972 | 1973 | 1974 | 1975 |
|---|---|---|---|---|---|---|---|---|
| Total P loading | (a) | 4.05 | 2.09 | 0.42 | 0.45 | 0.46 | 0.75 | 0.48 |
| Loss through outflow | (b) | 2.06 | 2.36 | 0.90 | 1.01 | 0.43 | 0.94 | 0.40 |
| Lake (water+sediment) retention | (a-b) | 1.99 | -0.27 | -0.48 | -0.56 | 0.03 | -0.19 | 0.08 |
| Change in P content of water | (c) | 0.00 | -0.27 | -0.13 | -0.61 | -0.13 | -0.32 | 0.03 |
| Sediment retention | (a-b-c) | 1.99 | 0.00 | -0.35 | 0.05 | 0.16 | 0.13 | 0.05 |
| Sediment retention, % of loading | | 50 | 0 | -83 | 12 | 35 | 17 | 10 |

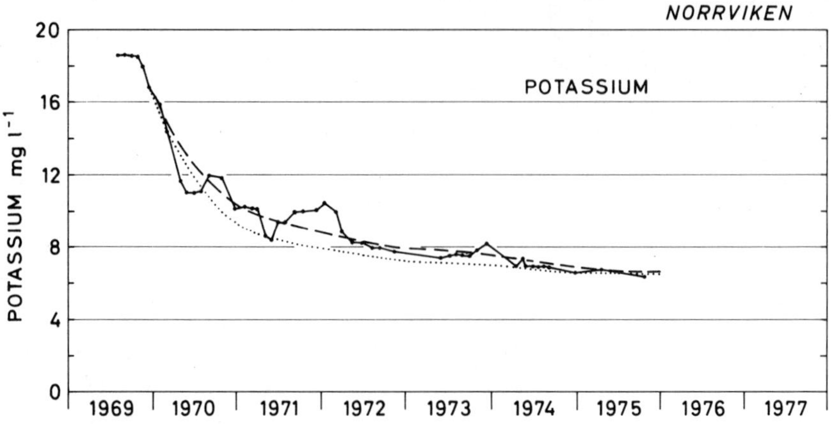


Figure 2. Mean potassium concentration in Lake Norrviken after diversion of sewage effluents rich in potassium. Dotted line shows concentrations calculated by equation (2) and dashed line values calculated according to equation (3) with $t_{rel} = 1.3$.

q = water flow in inflow and outflow
V = volume

It may also be used to describe concentration changes in lakes provided they are well mixed (Rainey, 1967).

Few elements in natural waters are strictly conservative since most of them take part in biological and chemical transformations. In Norrviken potassium had been increased far above the normal concentration by the effluents of a yeast factory and this element may be used as a "tracer" to estimate the effective residence time of the water in the lake. In 1969 its concentration was about 19 mg $l^{-1}$. Compared with this concentration the biological uptake and exchange processes are small and may be neglected. Figure 2 shows the variations of the mean potassium concentration after the diversion. The dotted line describes the calculated decrease assuming an average concentration of 6.5 mg $l^{-1}$ in the inflowing water. The calculations are based on the mean water flow for each year, i.e. no seasonal variations are considered and the equation predicts a dilution rate which is slightly faster than that observed. If the effective retention time for the water is assumed to be 1.3 times longer than the theoretical ($t_{rel} = 1.3$, found by trial and error) then a calculation according to the modified equation:

$$C_t = C_i + (C_0 - C_i) \cdot e^{-\frac{q \cdot t}{V \cdot t_{rel}}} \qquad (3)$$

results in a better approximation of the course of concentration decrease.

Phosphorus is certainly not a conservative element but is involved in rapid exchange processes both with the biotic and abiotic environment. Therefore it is not expected

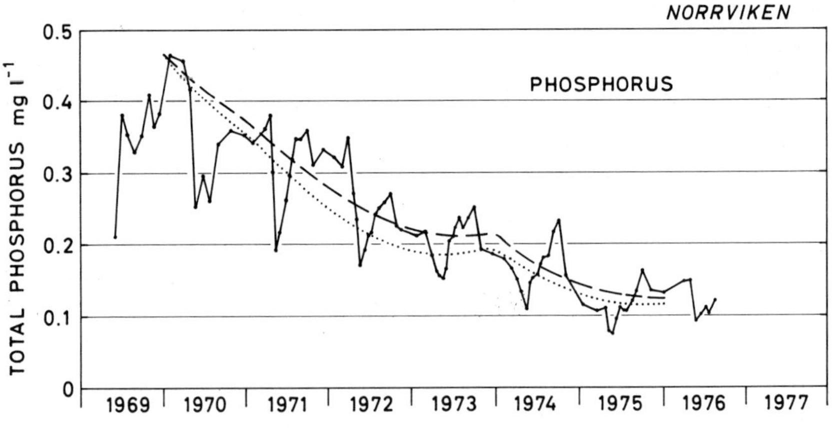


Figure 3. Mean phosphorus concentrations in Lake Norrviken after diversion of sewage effluents in 1969. Dotted and dashed curves are concentrations calculated analogously to those in Figure 2.

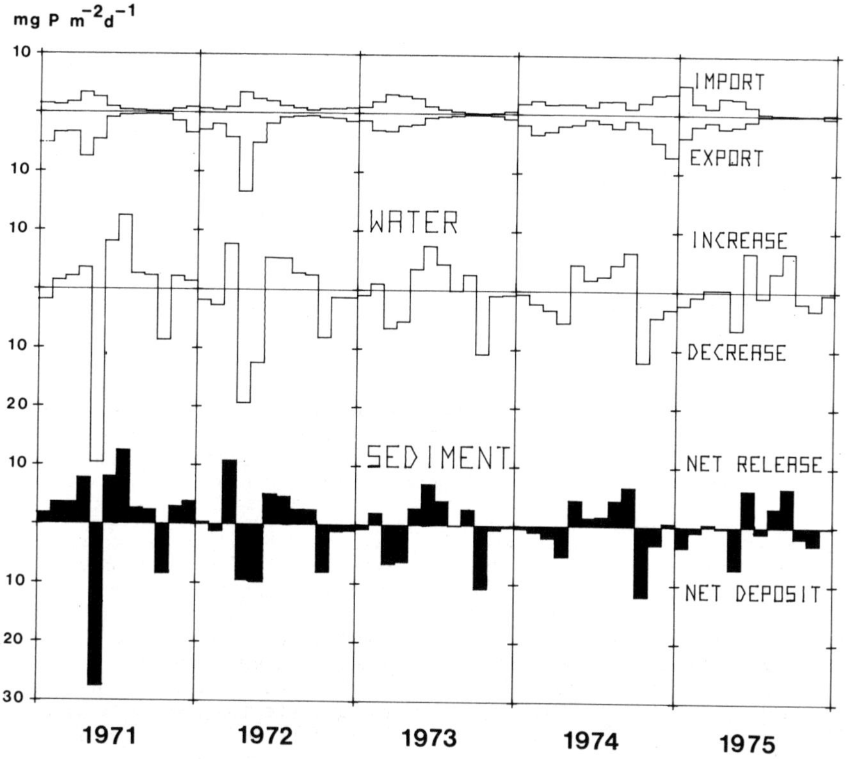


Figure 4. Detailed phosphorus budget for Lake Norrviken. The values for sediment net release and net deposit are calculated as the difference between import and export+change of content in the water.

*a priori* that the phosphorus concentration would decrease according to a simple dilution expression. However, if dilution curves are calculated in the same way as for potassium, using measured annual averages of water flow and phosphorus concentrations in the influents, we get the curves depicted in Figure 3. The broken line, calculated with $t_{rel}$ for water = 1.3, follows the measured mean concentrations quite well. Large seasonal variations occur, caused partly by the flow regime but mainly by sedimentation and release from the sediments. A minimum is always found in May after the diatom maximum and rapid sedimentation and a subsequent increase with a maximum in October caused by release from the sediments. The agreement between calculated and observed concentrations is possible only because sedimentation and release from the sediments tend to counterbalance. Furthermore, the maximum flows (and thus maximum transports to and from the lake) occur in the autumn and early spring when the phosphorus concentration in the lake is most uniform.

## The role of the sediments in the phosphorus budget

A more detailed calculation of the phosphorus balance of the lake is presented in Figure 4, giving monthly averages expressed as mg P $m^{-2}d^{-1}$. Since the data from 1970, the first year after the diversion of sewage, are incomplete they have not been included.

In 1971 there was a considerable net release of phosphorus both in winter and summer and net deposits occurred only in May, after the diatom maximum, and in October during the breakdown of the *Oscillatoria agardhii* population. In the last two years there was a net release only during the summer months and the amplitudes of the rates also decreased. An average of the net fluxes regardless of their signs (absolute values) may serve as an index of the phosphorus mobility and gives the series 7.1, 4.8, 3.7, 3.6, and 2.8, respectively, for the years 1971 through 1975.

During the summer stratification period the mean total phosphorus concentration in the lake increases in a more or less regular way, as shown in Figure 3. This is further illustrated in Figure 5 where values are found for May-August of 1971-1976. The straight lines, found by regression analysis, show the rates of concentration increase. Their slopes continuously decrease from 1971 to 1976 and are illustrated in Figure 6 expressed as mg P $m^{-2}d^{-1}$.

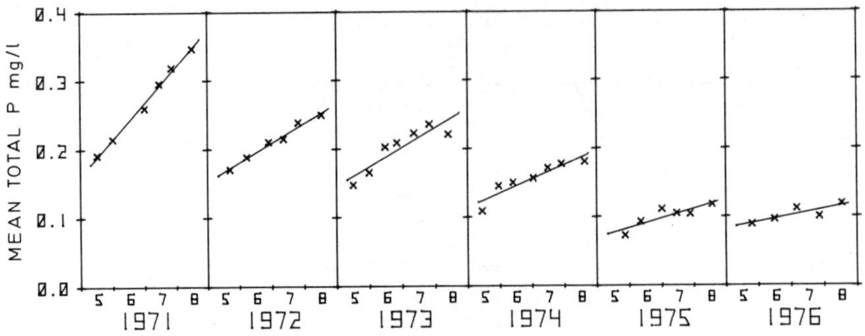


Figure 5. Mean total P concentrations during May-August (stratification period). Straight lines are calculated by regression analysis. Their slopes give the rate of net release of phosphorus from the sediments and are illustrated in Figure 6.

In summer, the import and export values are small and tend to counterbalance. The net increase of phosphorus in the water is thus the difference between gross release (+resuspension) from the sediments and gross sedimentation (+absorption?). A decrease of these net release rates does not necessarily mean that the gross release rates have decreased. It could equally well be caused by an increase in the rate of sedimentation.

Since the summer of 1973 sedimenting matter has been regularly collected in a cylinder placed at a depth of 9 m (2 m above the bottom). During the summer stratification period this is below the thermocline. (Results of these measurements are to be published by B. Ulén.) Figure 7 shows the rates of accumulation of particulate phosphorus during the summer of 1973-1975. The mean rates during May-August are 18.7, 30.8 and 33.9 mg P $m^{-2}d^{-1}$, respectively, i.e. an increase that might explain the decreasing net release rates.

However, there is also a tendency towards decreased hypolimnetic concentrations of both dissolved orthophosphate and total phosphorus during the summer stagnation periods, as is shown in Figure 8 for orthophosphate. These values support the alternativ of a decreased release from the sediments rather than an increased sedimentation. Laboratory experiments for the study of nutrient release from sediment cores from Lake Norrviken have given rather inconclusive results but show a tendency towards decreased rates of phosphorus release when the values from 1970 are compared with those from 1975 and 1976 (Ulén in prep.).

Therefore the measured rates of phosphorus

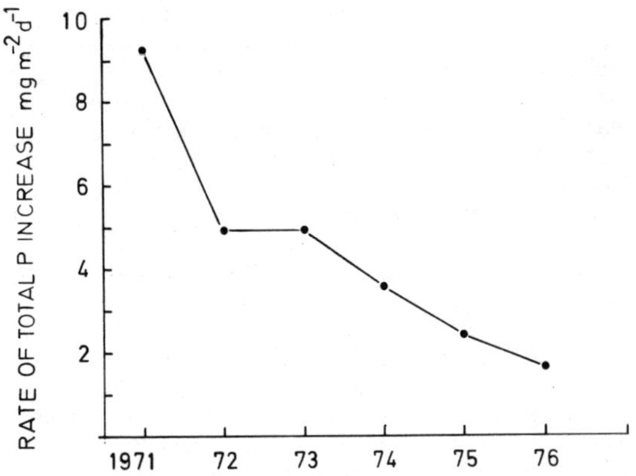


Figure 6. Rates of net release of phosphorus from the sediments.

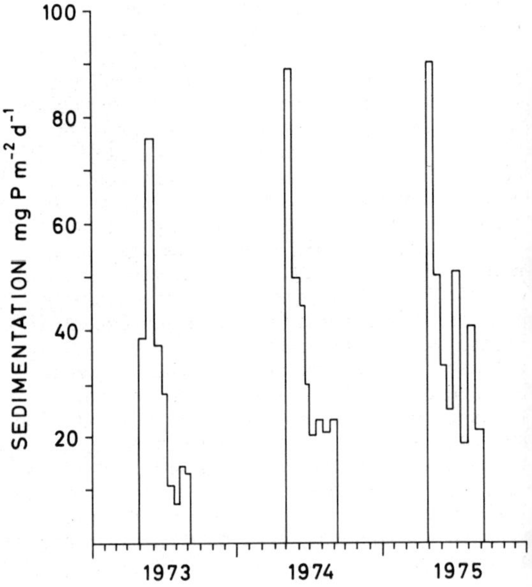


Figure 7. Rates of accumulation of particulate P in sediment traps placed below the thermocline in Lake Norrviken.

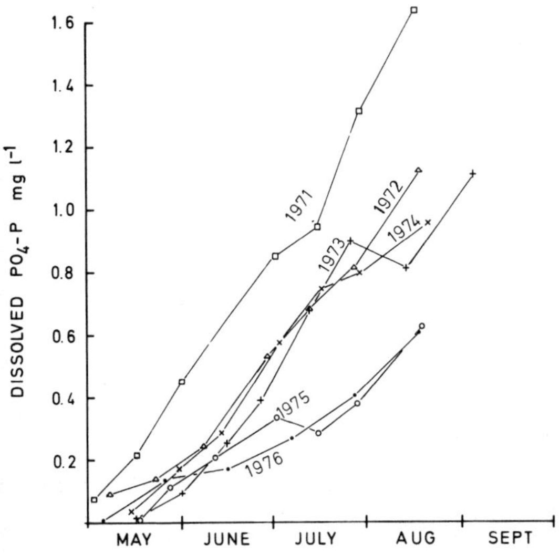


Figure 8. Concentrations of dissolved orthophosphate at a depth of 11 m (0.5 m above bottom) during the period with a stable thermal stratification in 1971 to 1976.

accumulation in the sediment trap must be considered unreliable and probably highly dependent on a varying resuspension of particulate matter, and the conclusion must be that the decreasing net release rates reflect decreasing gross release rates.

A recovery of the lake is thus taking place, not only by dilution of nutrients in the lake water but also by way of a decreasing internal loading of nutrients from the sediments. There is at present no reason to believe that this process has reached an equilibrium but it seems most probable that the recovery will continue.

The phosphorus concentration in the lake is still high enough to cause severe blooms of bluegreen algae but their intensity and duration have decreased. A more detailed description of the recovery of the lake is being prepared for publication elsewhere.

## References

Ahlgren, I., 1967. Limnological studies of Lake Norrviken, a eutrophicated Swedish lake. I. Water chemistry and nutrient budget. Schweiz. Z. Hydrol. 29:53-90.

Ahlgren, G., 1970. Limnological studies of Lake Norrviken, a eutrophicated Swedish lake. II. Phytoplankton and its production. Ibid., 32:353-396.

Austin, E.R. & G.F. Lee, 1973. Nitrogen release from lake sediments. J. Wat. Pollut. Control Fed., 45:870-879.

Bengtsson, L., 1975. Phosphorus release from a highly eutrophic lake sediment. Verh. Internat. Verein. Limnol. 19:1107-1116.

Björk, S., 1972. Swedish lake restoration program gets results. Ambio 1:153-165.

Jacobsen, O.S. & S.E. Jørgensen, 1975. A submodel for nitrogen release from sediments. Ecol. Modelling 1:147-151.

Jørgensen, S.E., L. Kamp-Nielsen & S.O. Jacobsen, 1975. A submodel for anaerobic mud-water exchange of phosphate. Ibid., 1:133-146.

Kamp-Nielsen, L., 1974. Mud-water exchange of phosphate and other ions in undisturbed sediment cores and factors affecting the exchange rates. Arch. Hydrobiol. 73:218-237.

Kamp-Nielsen, L., 1975. A kinetic approach to the aerobic sediment-water exchange of phosphorus in Lake Esrom. Ecol. Modelling 1:153-160.

Rainey, H.R., 1967. Natural displacement of pollution from the Great Lakes. Science 155:1242.

Rodhe, W., 1948. Sjön Norrvikens vattenbeskaffenhet år 1946-47 och vattenblomningens bekämpande med kopparsulfat sommaren 1947. Vattenhygien 2:38-61.

Sondén, K., 1914. Anteckningar rörande svenska vattendrag med hänsyn till beskaffenheten af vattnet i desamma. Stockholm 1914.

Tessenow, U., 1972. Lösungs-, Diffusions- und Sorptionsprozesse in der Oberschicht von Seesedimenten. I. Ein Langzeitexperiment unter aeroben und anaeroben Bedingungen im Fliessgleichgewicht. Arch. Hydrobiol. Suppl. 38, 4, 353-398.

Vollenweider, R.A., 1976. Rotsee, a source, not a sink for phosphorus? A comment to and a plea for nutrient balance studies. Schweiz. Z. Hydrol. 38:29-34.

Key words: eutrophication
recovery
sediments
phosphorus
dilution rate

CHEMICAL EVOLUTION OF MUDS IN A EUTROPHIC LAKE

E. Davaud

Department of geology, University of Geneva, Switzerland

Summary

The geochemical study of two piston cores taken in lac de Morat, a eutrophic Swiss lake, shows that chemical associations are different in recent and older sediments. These differences are due to natural processes such as diagenesis, but also to anthropogenic influences such as eutrophication and excess loadings of heavy metals.

Introduction

The lac de Morat is a small eutrophic lake, located in the western part of the Swiss plateau in a depression caused by glacial erosion. Its depth does not exceed 45 meters and its surface is about 25 square kilometers.

The sediments consist essentially of silt-size precipitated calcite (offen accounting for more than 50 % of the sediments), of sand-and silt-sized quartz and of clay minerals. The organic matter content of the surficial sediments is about 3 % by weight. This rather high concentration can be related to the eutrophic state of the lake.

The two cores studied were sampled every 5 centimeters in the topmost meter of sediment, and every 25 centimeters from one to seven meter below the interface.

The samples were wet sieved at 60 $\mu$ and the silt and clay fraction were analysed by X-ray fluorescence and atomic absorption for 23 elements. The determinations include major elements expressed as their oxides, such as $Al_2O_3$, CaO, $Fe_2O_3$, $SiO_2$, MgO, organic carbon, and trace elements : Ni, Mn, Cr, Co, Y, Ba, Sr, V, Zn, Cu.

Special attention was given to heavy metals known to be possible pollutants such as Hg, Sn, Pb.

Discussion

In order to check whether the chemical associations between all these elements change with time, we have split the samples into two stratigraphic subpopulations. This subdivision is based on palynologic and sedimentologic considerations.

The first subpopulation includes all the samples from the upper sixty centimeters of the cores, representing the two last centuries (group A).

The second group (B) includes all the samples collected between seventy centimeters and seven meters depth. The lower parts of the two cores are from the subboreal period, as proved by palynologic data (Jan du Chene ,1974 pers.comm.).

Scatter diagrams of some of the more interesting chemical elements show evidence for changes in the nature of the statistical associations between the two subpopulations.

The first diagram (fig. 1) shows the relationships between organic carbon and zinc. This metal is well correlated in the upper part of the cores (group A, light dots) and poorly in the lower part (group B, black dots).

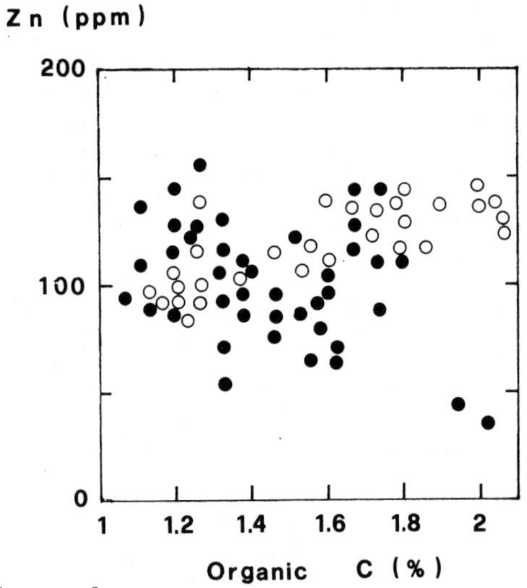


Figure 1.

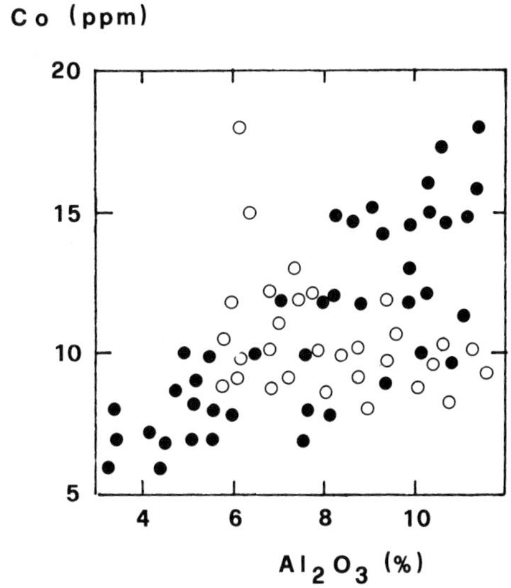


Figure 2.

By contrast, the figure 2 shows that cobalt is more closely associated with alumina in old sediments (group B black dots) than in recent ones (group A, light dots).

In order to clarify these differences we have computed two correlation coefficient matrices, one for each subpopulation, and checked statistically whether there are significant differences between the way chemical elements are associated in recent and old sediments.

This computation gives the following results :

1. Good correlations are observed in recent sediments between the main components of clay minerals (Al,Fe, Mg,K,Si) and some trace elements such as nickel,vanadium,yttrium,chromium and boron. These correlations increase strongly in older sediments.

2. Copper, zinc and phosphorus are linked with organic carbon in recent sediments but are associated with clays in older sediments.

3. Pollutants such as mercury,lead, cobalt and tin are statistically associated with clay minerals in older sediments (group B) but show no significant relationships with these components in the upper part of the cores (group A).

4. High correlations exist between calcium, strontium and organic carbon in recent sediments. The correlations tend to decrease strongly in older sediments.

In order to demonstrate the differences between the two correlation coefficient matrices in a more synoptic manner, we have tried to express them graphically (fig. 3 and 4), using a multivariate analysis algorithm (Davaud,1976).

In this type of diagram the distance between two variables is a function of their degree of association. The first diagram (fig. 3) portrays statistical associations in recent sediments and shows a bipolar configuration.

The left pole includes major elements (Al,K,Fe,Si,Mg) and some trace elements (Cr,B,V,Y,Ni). It can be interpreted as the clay pole. The right pole includes carbonates and organic matter.

On the same diagram (fig. 3) certain pollutants lie between clay and organic matter poles. This may mean that they are linked with both these petrographic components.

The high correlation between calcium and organic carbon does not have any chemical significance and can be explained by the fact that both -precipitated carbonates and organic matter- are dependent on organic productivity.

In contrast to figure 3 ,the figure 4, a summary for the older sediments, shows :

1. A greater clustering of certain elements around the clay mineral pole.

2. The displacement of pollutants towards the neighbourhood of the clay mineral pole.

3. The displacement of copper, zinc and phosphorus from the vicinity of organic matter towards the clay pole.

4. No good relationship between the carbonates and organic carbon.

All these chemical evolutions can be related both to diagenesis of organic matter under reducing conditions and to the increasing amount of heavy metals in the lake during the last century. The first factor is a natural

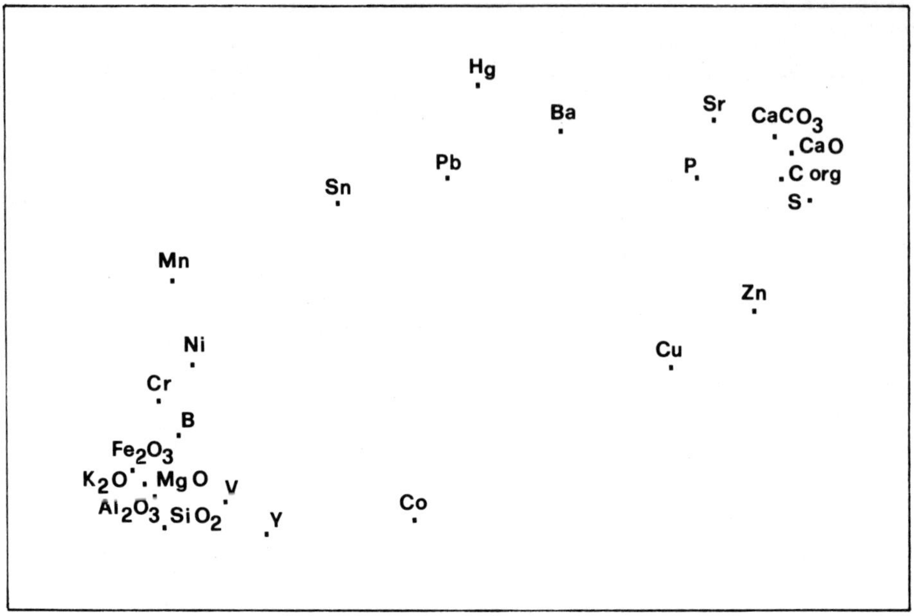


Figure 3. Geochemical associations in recent sediments (group A) The distance between two elements is a function of their degree of statistical linear dependence. Two well-correlated elements will be located close to each other. The horizontal and vertical axis have no meaning.

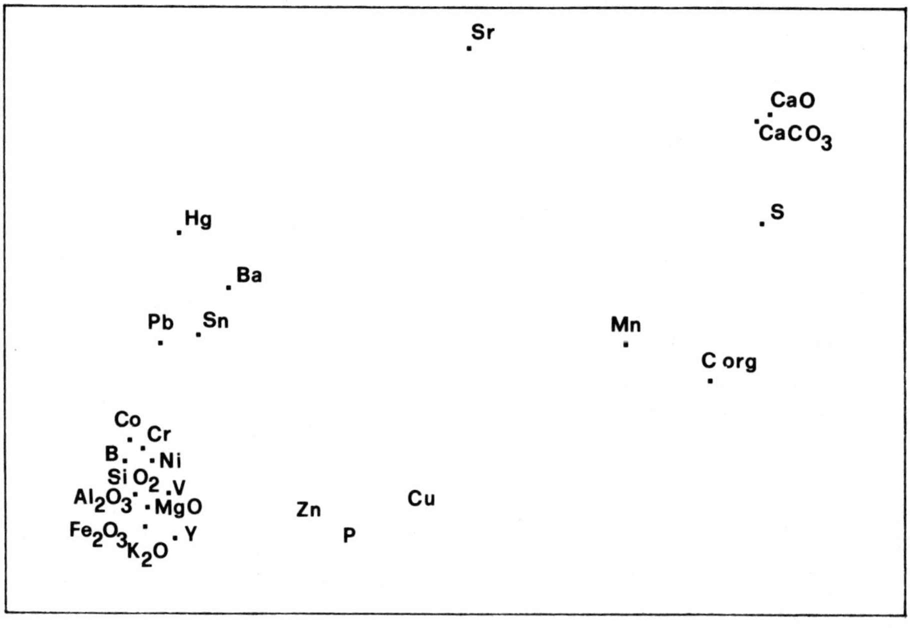


Figure 4. Geochemical associations in older sediments (group B). Same scale as figure 3.

and permanent phenomenon, while the second is recent and due to human activity.

Following this interpretation we can now explain the main statistical differences we have observed between older and recent sediments.

In the top part of the cores the organic matter is altered by bacteria and releases sulfur, phosphorus and probably trace elements such as copper and zinc. It is presumed that part of these elements return to the overlying water and the rest are sorbed by clay particles.

The weaker statistical relationship between organic carbon and calcium, shown by the older sediments, is probably due to organic matter being partly destroyed after sedimentation, while the precipitated calcite is more stable.

Heavy metals such as mercury and lead are present at very low concentrations in uncontaminated sediments and are sorbed by clays (Förstner, Müller, 1974). Perhaps when these elements are introduced in excess in a natural environment, the clay particles become saturated and the excess must react with other petrographic components such as organic matter (Thomas,1972.1973).

This possibility of two different chemical linkages explains the very low correlation coefficients we have observed between heavy metals and clays or organic matter in recent sediments.

## Conclusions

The chemical differences between recent sediments and older ones are quite significant in lac de Morat. Some of these changes are due to human activity, but others can be related to the natural evolution of muds.

In most cases these two phenomena occur together, in such a way that discriminating between natural and anthropogenic changes becomes difficult.

It seems therefore very important to mesure and understand the effects of diagenesis when trying to mesure the impact of human activity on lake sediments.

## References


Davaud,E., 1976. Contribution à l'étude géochimique et sédimentologique de dépots lacustres récents. Thèse 1745 Université de Genève.

Förstner,U.,Müller,G., 1974. Schwermetalle in Flüssen und Seen als Ausdruck der Umweltverschmutzung. Springer Verlag, Berlin.

Thomas,R.L., 1972. The distribution of mercury in the sediments of Lake Ontario. Can. journ. of Earth sci. 9, p.636-651.

Thomas,R.L., 1973. The distribution of mercury in the sediments of Lake Huron. Can. journ. of Earth sci., 10/2, p.194-204.



Key words: diagenesis
organic matter
heavy metals
statistical analysis
lac de Morat

PRESENCE OF BLUE-GREEN ALGAE IN SEDIMENTS OF LAKE BRIELLE

F.I. Kappers

National Institute for Water Supply, Voorburg, The Netherlands

Summary

Analysis of plankton has shown that certain species of water-bloom-forming blue-green algae (Cyanophyta) were not present in the plankton all the year round. Aphanizomenon flos-aquae, Microcystis aeruginosa, Oscillatoria agardhii and O. tenuis are present throughout the year in the sediment. Anabaena flos-aquae does not have a benthic phase.

## Introduction

Since 1973 the National Institute for Water Supply has undertaken research into the factors leading to the nuisance growth of blue-green algae. The research is being undertaken in Lake Brielle, where extensive water blooms form annually. Analysis of the plankton has shown that certain species of blue-green algae were not present in the plankton all the year round. The question therefore arose as to whether these algae spent part of the year in the sediment.
There is little information on the matter in the literature. Chernousova (1968) states that blue-green algae are found in the sediment of the Dnjeper storage reservoirs in the winter. This sediment is an important source of blue-green algae, which causes water bloom in the summer.

If large amounts of blue-green algae do in fact live through the winter in the sediment, it could be necessary to remove a layer of this material in order to prevent water blooms.

## Features of the lake

The damming of the former tidal River Brielse Maas in 1950 created a freshwater lake with a volume of 23 x $10^6$ cubic metres, an average depth of five metres and a maximum depth of fifteen metres. The lake consists of two parts, a narrow part 8 kilometres long (the supply channel) and a part of greater width 11 kilometres long (fig. 1). Water remains in the supply channel for 5-10 days. Little mixing occurs. In the wide part the residence time varies between 50-200 days. There is more mixing in this part. At the inlet near Spijkenisse phosphate-rich water enters from the River Old Meuse. Considerable sedimentation occurs in the supply channel, but the improvement in the water quality resulting from this is partly offset by discharges of polluted water into the broad part of the lake. Surplus water can be removed through the outlet at the western-end . Klomp (1973) calculated that in the second half of 1971 the total gross phosphate loading amounted

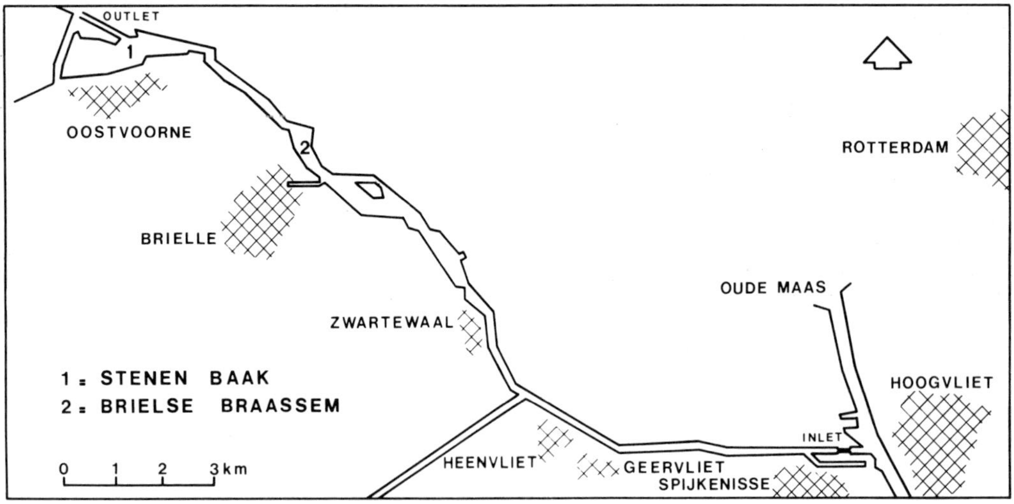


Fig. 1.: Sediment sampling locations in Lake Brielle

to 4.2 $gP/m^2$ per half-year. 40% originated from local discharges.

In the supply channel few algae are found in the plankton. The number of algae increases just downstream of Zwartewaal. In the western part water bloom occurs each year.

In 1972, 1973 and 1974, 2 $mgFe^{++}/l$ was added to the water entering from the Old Meuse in an attempt to lower the phosphate content and thus limit the production of algae. The effect of this measure on eutrophication was, however, only slight, since it was counteracted by the effect of the local discharge.

Three points were selected for sampling. The first one was situated in the supply channel, near the inlet. The second location, Brielse Braassem, is downstream of Brielle, the largest local discharge. The third point, the Stenen Baak, is the deepest point in the lake.

The choice of the first point was not a particularly appropriate one, due to the phosphate removal experiment. At this point a thick layer of sedimented iron at a depth of about two metres can be found.

Samples of undisturbed sediment were taken once a month at four points in a transect stretching from the deepest point to the southern bank in the Brielse Braassem. At these four points the water is 10, 6, 4 and 2 metres deep. Between the deepest point and the northern bank the sediment is mainly blue clay, while to the south peat, loam and sand are found.

A similar transect was sampled at the Stenen Baak. Samples were taken at depths of 14, 10, 6, 4 and metres. Between the deepest point and the northern bank the sediment is mainly coarse sand, while on the southern side loam, clay and sand are found.

A Jenkin mud sampler was used to collect undisturbed sediment samples. When the sampler was raised, the perspex tube containing the sample was frozen in immediately. For benthos analysis the frozen core was cut into one centimetre pieces and then defrosted (Kappers, 1976).

A single analysis was made of the sediment to estimate the nutrient contents. Both at Brielse Braassem and at Stenen Baak the greatest quantities of phosphorus were found at the greatest water depths, with a much lower content at shallower depths (Table 1). This was also true for the total of nitrogen.

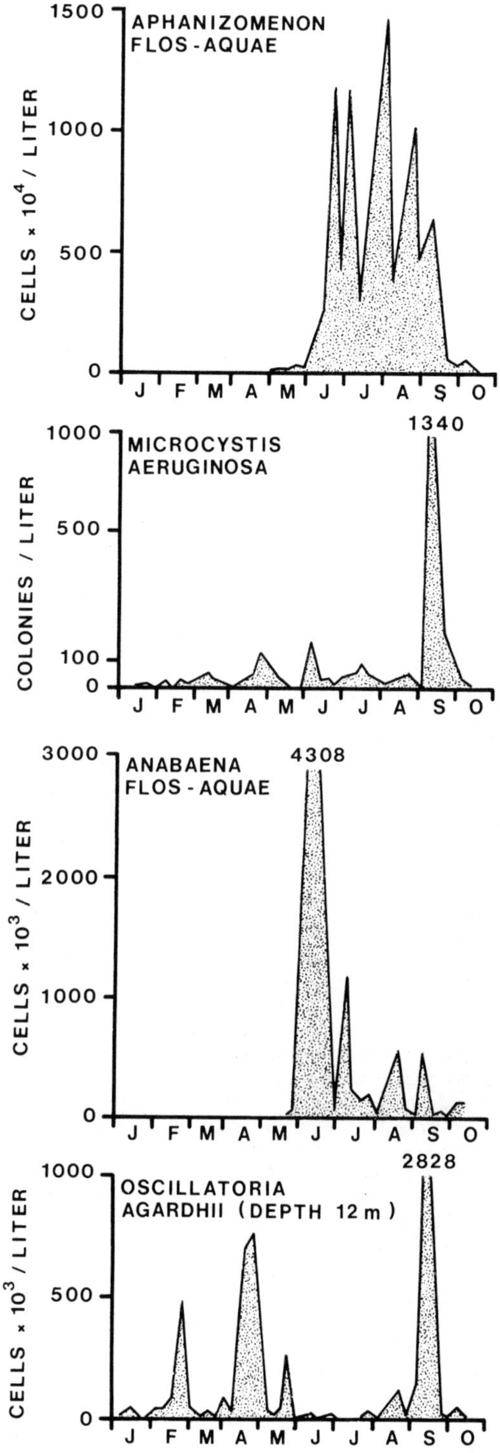


Fig. 2.: Blue-green algae in plankton (depth 1m); Stenen Baak 1974.

Table 1. Sediment analysis October 1974

| Sampling location | water depth (m) | pH | per 100 g dry weight $P_2O_5$mg | total N mg |
|---|---|---|---|---|
| Brielse | 10 | 7.1 | 106 | 30 |
| Braassem | 8 | 7.1 | 73 | 29 |
| | 6 | 7.2 | 57 | 19 |
| | 2 | 7.4 | 28 | 12 |
| Stenen | 14 | 7.2 | 74 | 40 |
| Baak | 10 | 7.6 | 27 | 4 |
| | 6 | 7.6 | 24 | 5 |
| | 2 | 7.7 | 17 | 3 |
| | 1 | 7.9 | 16 | 3 |

Results

The following algae were found in the sediment throughout the period of investigation: Aphanizomenon flos-aquae, Microcystis aeruginosa, Oscillatoria agardhii and Osc. tenuis. The last of these is not found in the plankton. Occasionally Anabaena flos-aquae was found. Aphanizomenon appeared in easily recognisable filaments. In August and September many spores of Aphanizomenon were found in the sediment. Microcystis formed colonies. A number of cells were dividing. Some colonies from the sample taken on May 9 were examined in greater detail; 26% of the cells were found to be dividing.

In the sediment near the inlet numbers were small; algae of other taxa were also rare at that point. The numbers were too small to be processed in the calculations and are not considered further. A check carried out in the sandy bottom towards the northern shore at Stenen Baak also showed only negligible numbers of blue-green algae. Throughout the period more blue-green algae were found at Stenen Baak than at Brielse Braassem.

Aphanizomenon is found in the sediment throughout the year (Fig. 3). The greatest numbers develop in the summer months. Aphanizomenon was almost entirely absent from the plankton in 1973/74 outside the summer period (Fig. 2). The largest numbers of colonies of Microcystis, both in the sediment and in the plankton, were found in September. Microcystis was present in small quantities throughout the year in both sediment and plankton. Oscillatoria agardhii was found in the plankton mainly at lower depths in 1973/74. It was always present in the sediment. Very little was found in the plankton in June and July 1973/74. Anabaena flos-aquae was not present in significant quantities in the sediment. It appeared in very large quan-

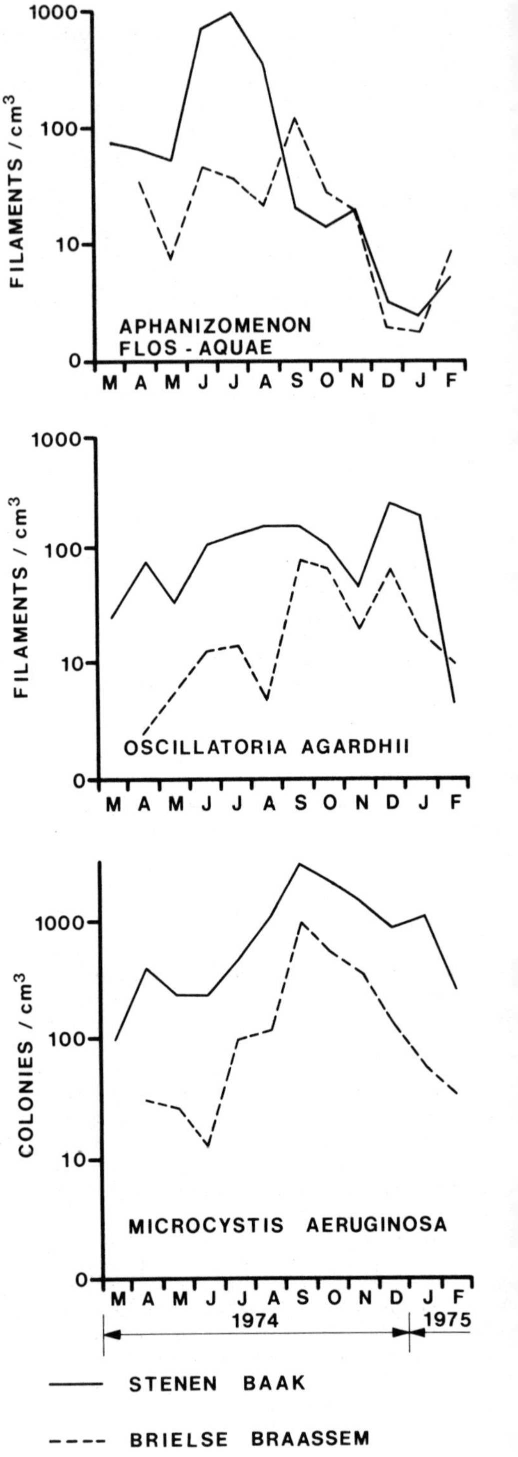


Fig. 3.: Blue-green algae in sediment. The points are the means of the eight levels ($b_1$, $b_2$, $b_3$, $b_4$ both at 2-5 m and at 10-14 m)

tities in the plankton in June, while in the rest of the summer only small numbers were found.

Averaged over the whole of 1974, 60% of Anabaena was found in the plankton in the upper water layer; 68% of Oscillatoria agardhii, on the other hand, was found near the bottom. Aphanizomenon and Microcystis were more evenly spread over the depth (Fig. 4).

The undisturbed sediment samples were examined at four levels: the water layer just above the bottom ($b_1$), the water-sediment interface ($b_2$), 1 cm below the interface ($b_3$), and 2 cm below the interface ($b_4$). Fig. 5 shows the average numbers of colonies/filaments at each level. Fewer blue-green algae were found at all layers in the sediment from Brielse Braassem than in that from Stenen Baak. Levels $b_1$, $b_2$ and $b_3$ did not differ significantly from one another in numbers found, though there was a gradual fall off from $b_1$ to $b_3$. Level $b_4$ contained markedly fewer algae. This picture did not change in the course of the year.

In order to make data easier to process, the transects were divided into two zones: shallow zone with more sand, where the water depths are 2 and 4 metres in the Brielse

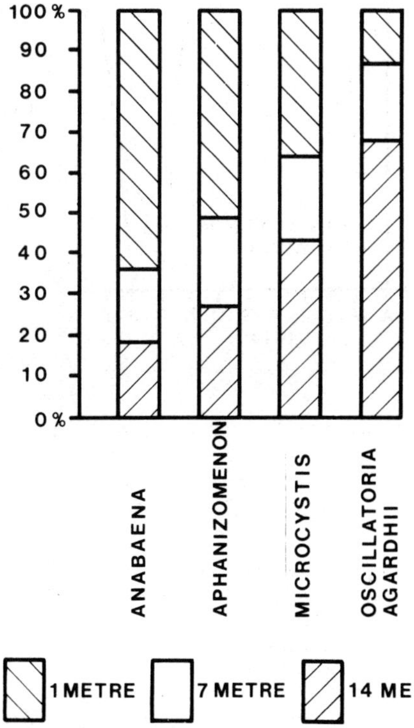


Fig. 4.: Blue-green algae in plankton vertical distribution. Stenen Baak 1974.

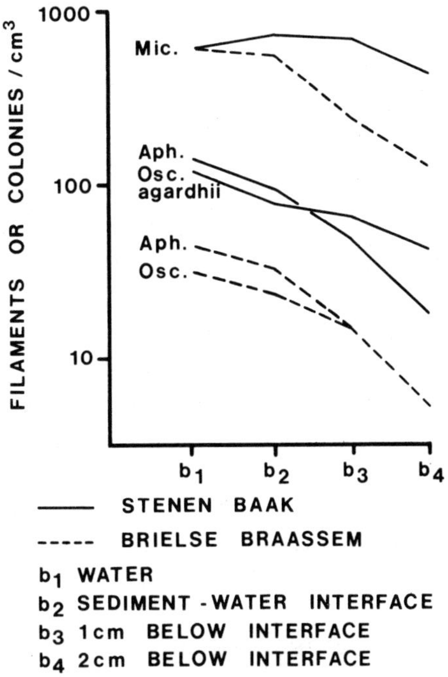


Fig. 5.: Vertical distribution of blue-green algae in sediment as the means of the two zones

Braassem and 2, 4 and 6 metres in the Stenen Baak; and a deeper zone, where the water depths are 6 and 10 metres in the Brielse Braassem and 10 and 14 metres in the Stenen Baak. A number of differences were found (Fig. 6). Aphanizomenon was found over the whole transect, but the most filaments were found in the shallow zone. In May and December the inverse was true: the largest numbers of filaments were found in the deeper part. In the case of Microcystis the differences between the shallow and the deeper zones were less pronounced, but here, too, there were usually more colonies in the shallow zone than in the deeper zone, the inverse being true in May and December. There was no significant difference for Oscillatoria agardhii and Osc. tenuis between the shallow and deeper zones.

There appears to be no significant difference between the four levels $b_1$ - $b_4$ of the undisturbed sediment sample linked with the position in the shallow or deeper zone.

## Conclusions

It is clear from the foregoing that research into the ecology of blue-green algae must involve investigation of the phase occurring in the sediment as well as the plankton phase.

Different patterns of behaviour were observed with the different species of blue-green algae. Anabaena flos-aquae does not appear to have a benthic phase of any importance. Aphanizomenon flos-aquae shows a benthic phase throughout the year and a plankton phase only in summer. Microcystis aeruginosa is present throughout the year in both the plankton and the benthos. Oscillatoria agardhii mainly lives near or in the sediment.

Because of the mild winter it was not possible to check the effect of temperature on this pattern.

It is advisable, therefore, that within the framework of the qualitative water management programme of Lake Brielle the bottom will be dredged from time to time not only to neutralize the effect of e.g. phosphates and heavy metals, but also because of the presence of blue-green algae.

Acknowledgement

I like to thank Mr. W. Vermeulen for carrying out the biological analyses.

## References

Chernousova, V.M., L.A. Sirenko & V.V. Arendarchuk, 1968. Localization and physiological state of mass species of blue-green algae during late-autumn and spring periods. In: Tsve Tenie Vody, Akademiya nauk Ukrainska SSR, Kiev.

Kappers, F.I., 1976. Blue-green algae in the sediment of the Lake Brielse Meer. Hydrobiol. Bull. (in press).

Klomp . R., 1973. Enige fysisch-chemische aspecten van de gedragingen van rivierwater in een sterk geëutrofieerd milieu. $H_2O$ 6 (4): 92-97.

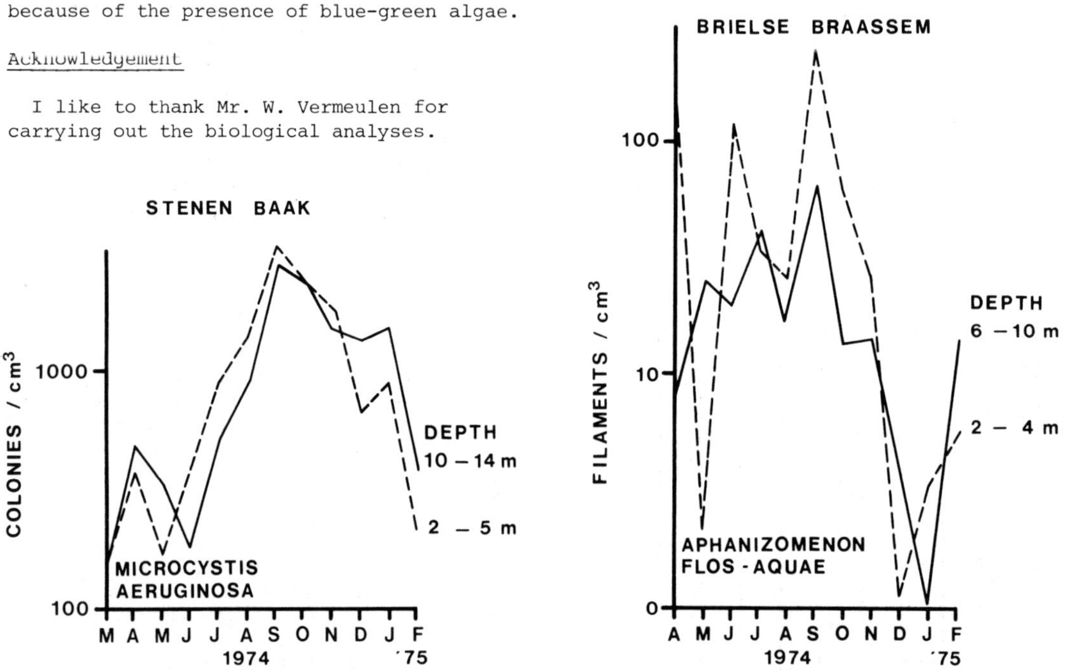


Fig. 6.: Horizontal distribution of blue green algae in sediment. The points are the means of the four levels.

Key words: blue green algae
eutrophication
plankton
benthos
reservoirs

THE INFLUENCE OF PHOSPHATE AND NITROGEN IN SEDIMENT ON RESTORATION OF LAKES

Sven Erik Jørgensen

The Royal Danish School of Pharmacy, Universitetsparken 2, DK-2100 Copenhagen Ø, Denmark

## Introduction

The title of this symposium is "Interaction Between Sediments and Water" - but why devote a symposium just to the interface sediment-water? The answer is of course that the interaction processes sediment-water is very important for the chemical composition and the ecology of the entire water phase, which I shall attempt to demonstrate by two studies, where eutrophication of lakes is considered.

## Results

During the last half decade several different lake-models have been developed with the goal to enable the prediction of ecological changes as a result of changes in the nutrient budget for a lake. See e.g. Chen (1970), Chen and Orlob (1972) (1975), Larsen et al. (1972), Dahl Madsen et al. (1974), Schofield and Krutchkoff (1974) and Scavia et al. (1976).

All these models have a simple constant or first order process description of the interaction processes ⇄ water phase. However, a more detailed description of these interaction processes is possible by means of a submodel based upon Kamp Nielsen (1974) (1975), Jørgensen et al. (1975) and Jacobsen et al. (1975). This submodel distinguishes between exchangeable and unexchangeable nutrients and between aerobic and anaerobic exchange of nutrients. Furthermore, the submodel considers a biologically active layer on the top of the sediment. The equations of the submodel are shown in Lars Kamp Nielsens paper on phosphorus, presented at this conference and in Jacobsen et al. (1975) for nitrogen.

However, it is obvious to question whether it is necessary for the total lake model to use a very complicated submodel for the mud - water exchange of nutrient, or whether a simpler submodel as previously used is sufficient.

To answer this question, two different versions of a lake model were tested - one with a first order reaction for the interaction processes and one with the more complicated description indicated above. Both versions were calibrated to fit best possible with measured data in Glumsø Lake. For further details about this lake and the lake model see Jørgensen et al. (1973) and Jørgensen (1976). Both versions were used to predict how the production and transparency of the lake would change if phosphorus was reduced in the waste water to 0,4 mg $l^{-1}$ and 0,1 mg $l^{-1}$ respectively. The lake is at this stage receiving mechanical-biological waste water with an average concentration of 5-6 mg $l^{-1}$. The volume of the lake is 420,000 $m^3$ and the flow rate of streams going to the lake 530,000 $m^3/y$.

Table 1 gives the results. As seen, there is a significant difference between the two predictions.

Table 1. Prediction by means of two model versions.

Case A: P in waste water reduced to 0.4 mg $l^{-1}$
Case B: P in waste water reduced to 0.1 mg $l^{-1}$

| | | 3rd year after changed loading A | 3rd year after changed loading B | 9th year after changed loading A | 9th year after changed loading B |
|---|---|---|---|---|---|
| Lake model with submodel | $gC/m^2 yr.$ | 600 | 500 | 500 | 320 |
| | transparency (cm) | 50 | 60 | 60 | 75 |
| Lake model with first order reaction mud-water interchange | $gC/m^2 yr.$ | 850 | 800 | 610 | 480 |
| | transparency (cm) | 30 | 35 | 50 | 60 |

It is a general problem, in the development of lake models, to find the right balance between a too complicated and a too simple description of the lake system. It was attempted to find this balance by simulating the following situation: by use of a lake model, it is desired to be able to predict the consequences of a changed phosphorus loading ($P_{in}$). The steady state values of soluble phosphorus in the lake ($P_s^o$) and the total phosphorus in the lake and sediment ($P_t^o$) as consequence of the loading, are calculated. Two kinds of what is called ecological buffer capacities have been considered:

$$\beta = \frac{P_t^o}{P_s^o} \quad \text{and}$$

$$\beta^1 = \frac{P_{in}}{P_s^o}$$

The question set up is: how are β and $\beta^1$ dependent upon the complexity of the total lake model and the submodel for interaction processes between mud and water. The complexity was measured by means of Shannon's index.

$$H = \sum_j a_j \frac{P_j}{P_t} \log_2 \frac{P_t}{a_j, P_j} \text{ (bits)}$$

$$\text{where } P_t = \sum_j a_j P_j$$

and the Exergy

$$E - RT \sum_j a_j \left(P_j \cdot \ln \frac{P_j}{P_j^{eq}} - (P_j \quad P_j^{eq})\right) \text{ KJ m}^{-3}$$

$P_j$ is the phosphorus concentration for component j in the total lake model and $a_j$ is a volume ratio, which transfers all phosphorus concentrations to the lakewater.

It was found that there is a good relation between the buffer capacity and the exergy, while Shannon's index seems to be a poor measurement for the ecological buffer capacity.

For further details about these examinations see Jørgensen et al. (1977).

The results of interest for the sediment are summarized in table 2. Four cases are studied:

1. A lake model with three trophic levels but without any interaction processes mud-water.
2. As 1) but considering a first order reaction between mud and water.
3. As 1) using the submodel, see above.
4. As 3) using an active sediment-layer only 1/10 of the actual measurement.
5. As 2) using an active layer only 1/10 of the actual measurement.

Table 2.

| Case | β | $\beta^1$ | E | H |
|---|---|---|---|---|
| 1 | 1.12 | 1.12 | 1.32 | 0.62 |
| 2 | 10.7 | 3.88 | 31.7 | 0.72 |
| 3 | 10.9 | 3.88 | 31.8 | 1.52 |
| 4 | 3.85 | 3.88 | 4.83 | 2.24 |
| 5 | 2.45 | 3.88 | 4.73 | 1.69 |

## Discussion

Table 1 shows that it is important when lake models are used as a tool to set up pollution impact statements, to apply the more detailed description of the interaction processes sediment-water.

The application of a first reaction is a simplification which will lead to a different prediction from the one found with the more complicated submodel.

The explanation of this difference is obvious.

The submodel applied distinguishes between exchangeable and unexchangeable phosphorus, and the prediction is based upon a lower P-loading. Since about 60% of the settled phosphorus is unexchangeable, the prediction will give lower P-concentrations in the water and consequently lower production and higher transparency by use of the complicated submodel.

The exergy is a measurement for the mass flow in the lake and therefore it is not surpricing that the exergy is better related to the ecological buffer capacity than the Shannon index; which indicates only the number of bits.

From the results in table 2 it is seen that the ecological buffer capacity increases with increased complexity, although the difference between case 2 and case 3 is small. Furthermore it is very important to include the sediment processes in the total model, as the buffer capacity changes considerably when the sediment is included in the total model, compare cases 1, 2 and 3. The significance of the sediment is also demonstrated when case 4 is compared with case 3 and case 5 with case 2.

When the active layer is reduced the buffer capacity is reduced significantly.

The buffer capacity measures the response of the lake to changed conditions (in this case changed P-loadings). It can therefore be concluded, on the basis of the results in table 2, that a lake model used to predict changes in a lake through changed conditions, must include a good description of the mud-water processes.

## Conclusion

Two versions of a lake model were used for prediction of the production and the transparency of the lake by changed P-loadings. It was found that the version containing a more complicated and better description of the mud-water interactions give a significantly lower production and better transparency than the version using a simple first order reaction for the interaction processes.

The response by changed conditions of P-loading in a lake was also "measured" by means of a new concept: ecological buffer capacity, and it was found that 1) it is of importance to consider the interaction processes mud-water, 2) the more complicated submodel (table 1) gives a higher buffer capacity.

The examinations were only carried out in context with phosphorus and one lake, which has a high nutrient level, but the results seem to justify the efforts used to give a more detailed description of the processes going on in the sediment water interface.

## References


Chen, C.W., 1970. Concepts and Utilities of Ecological Models. J. Sanit. Eng. Div. ASCE 96 No. A5.

Chen, C.W. & G.T. Orlob, 1972. Ecologic Simulation for Aquatic Environments. Off. Water Resourc. Res. C-2044. WRE-1-0500.

Chen, C.W. & G.T. Orlob, 1975. Ecologic Simulation for Aquatic Environments. In: B.C. Pattern (Editor), Systems Analysis and Simulation in Ecology, Vol. III. Academic Press, New York, N.Y., pp. 476-587.

Dahl-Madsen, K.I. & K. Strange Nielsen, 1974. Eutrophication Models for Ponds. VAND, 5: 24-31.

Jacobsen, O.S. & S.E. Jørgensen, 1975. A Submodel for Nitrogen Release from Sediments. Ecol. Modelling, 1: 147-151.

Jørgensen, S.E., 1976. A Eutriphication Model for a Lake. Ecol. Modelling 2: 147-165.

Jørgensen, S.E., O.S. Jacobsen & I. Høi, 1973. A Prognosis for a Lake. Vatten, 29: 382-404.

Jørgensen, S.E., L. Kamp-Nielsen & O.S. Jacobsen, 1975. A Submodel for Anaerobic Mud - water Exchange of Phosphate. Ecol. Modelling, 1: 133-146.

Jørgensen, S.E. & H. Mejer. Will be published in Ecol. Modelling 3.

Kamp-Nielsen, L., 1974. Mud-water Exchange of Phosphate and other Ions in Undisturbed Sediment Cores and Factors Affecting the Exchange Rate. Arch. Hydrobiol., 2: 218-237.

Kamp-Nielsen, L., 1975. A Kinetic Approach to the Aerobic Sediment-water Exchange of Phosphorus in Lake Esrom. Ecol. Modelling 1: 153-160.

Larsen, D.P., H.T. Mercier & K.W. Malveg, 1972. Modelling Algal Growth Dynamics in Shagawa Lake, Minnesota. In: E.J. Middelbrooks, D.H. Falkenberg & T.E. Maloney (Editors), Modelling the Eutrophication Process. Ann Arbor, Mich. ... _ ... Publishers, Ann Arbor.

Scavia, D. & R.A. Park, 1976. Documentation of Selected Constructs and Parameter Values in the Aquatic Model CLEANER. Ecol. Modelling, 2: ... _ ...

Schofield, W.R. & R.G. Krutchoff, 1974. Bulletin 60 Stochastic Model for a Dynamic Ecosystem. VPI - WRRC - Bull, 60, Virginia Water Resources Research Center.



Key words: Lake restoration
phosphate
nitrogen
loading
lake models

HEAVY METAL CONTENT OF LAKE AND RIVER SEDIMENTS IN SWITZERLAND

J.P. Vernet*, F. Rapin* & G. Scolari**

* Department of Geology, University of Geneva, Switzerland
**Department of marine geology of B.R.G.M., Orléans, France

Summary

Sediment samples from the main Swiss rivers and lakes were analyzed in 1973-1975 to evaluate their heavy metal content.

Strong Hg contamination was found in lac Léman and its main tributary, the Rhône river. Weaker contamination was found in lac de Joux and lac de Morat.

Apart from Hg, the other pollutants include chromium, zinc, cobalt, copper, tin and cadmium, with lead as an accessory.

This study has led to the localization of the main anthropogenic sources of heavy metals to Swiss lakes and rivers.

## Introduction

A study of heavy metal pollution in sediments was undertaken by the GEOLEM group on lac Léman (Switzerland) in 1970. The work was later extended to the whole Rhône river basin (Vernet et al., 1976) and to other Swiss lakes and rivers (Vernet & Thomas, 1972a and b; Scolari & Vernet, 1975). The geochemical cycle of elements, particularly Hg in the food chain, was also reviewed by Vernet et al. (1975) and Vernet & Thomas (1972c).

The present paper is intended as an overview of the heavy metal content and pollution of several lakes and rivers of Switzerland.

## Sampling and analytical methods

The sediment samples collected from rivers were sieved through a 80 micron nylon screen and over dried at 40°c to preserve organic matter and minimize mercury volatilization. Mercury was analyzed by atomic absorption and the other trace metals by optical spectrophotometry.

The minus 80 micron fraction (or, in some cases, minus 175 micron) was chosen because it includes the clay, silt and part of the very fine sand size classes; in other words, virtually all of the sediment absorbant fraction. As the silt fraction makes up a large part of the river sediments (between 18-80% for the Rhône river; Vernet, 1969) and the clay fraction only 0-5%, the minus 80 micron fraction is considered to be more representative than the minus 2 micron fraction, analyzed by Förstner & Müller (1974) in German rivers. Because of this difference in sediment sizing, results from these last authors cannot be directly compared with ours.

Altogether, more than 1000 samples were taken in the main rivers of Switzerland and in the Léman, Joux, Bienne and Morat lakes.

## Results

### Natural background

Based on samples taken in rivers upstream of potential contamination sources, and on lake sediment cores the tentative natural background values of heavy metals in our sediments are given in table 1.

Table 1. Natural background values of heavy metals in the -80 micron fraction of sediments of Swiss rivers and lakes (in ppm or g $tonne^{-1}$).

| | | | |
|---|---|---|---|
| Hg | < 0.05* | Sn | 5 |
| Cd | 0.3 | Ni | 50 |
| Cu | 25 | V | 50 |
| Zn | 75 | Y | 25 |
| Cr | 35 | Co | 10 |
| Pb | 50 | | |

*corresponds to the original detection limit of the method. More recent investigations indicate values close to 0.02 ppm.

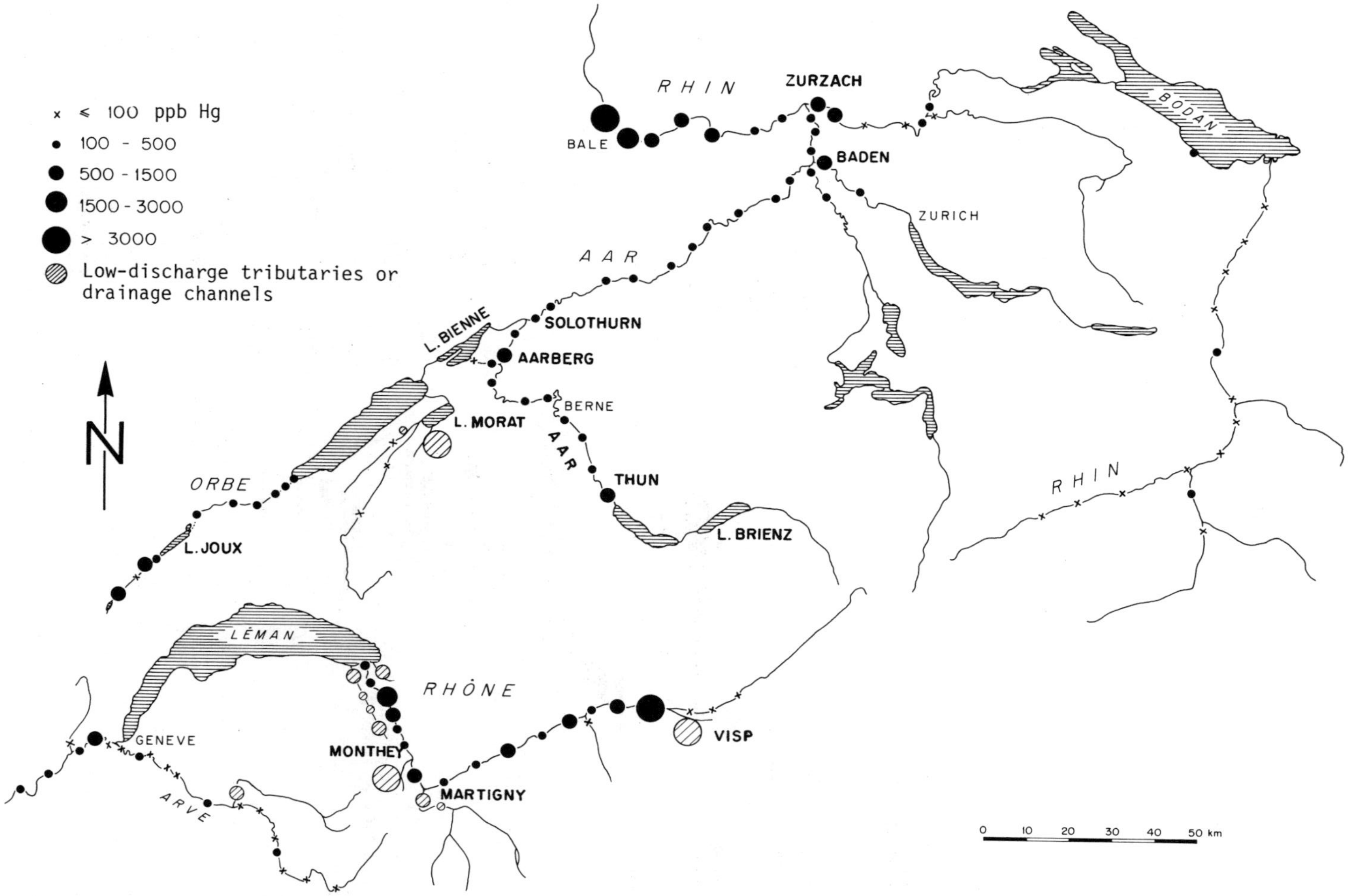


Fig. 1 - Mercury content of river sediments in Switserland in 1973

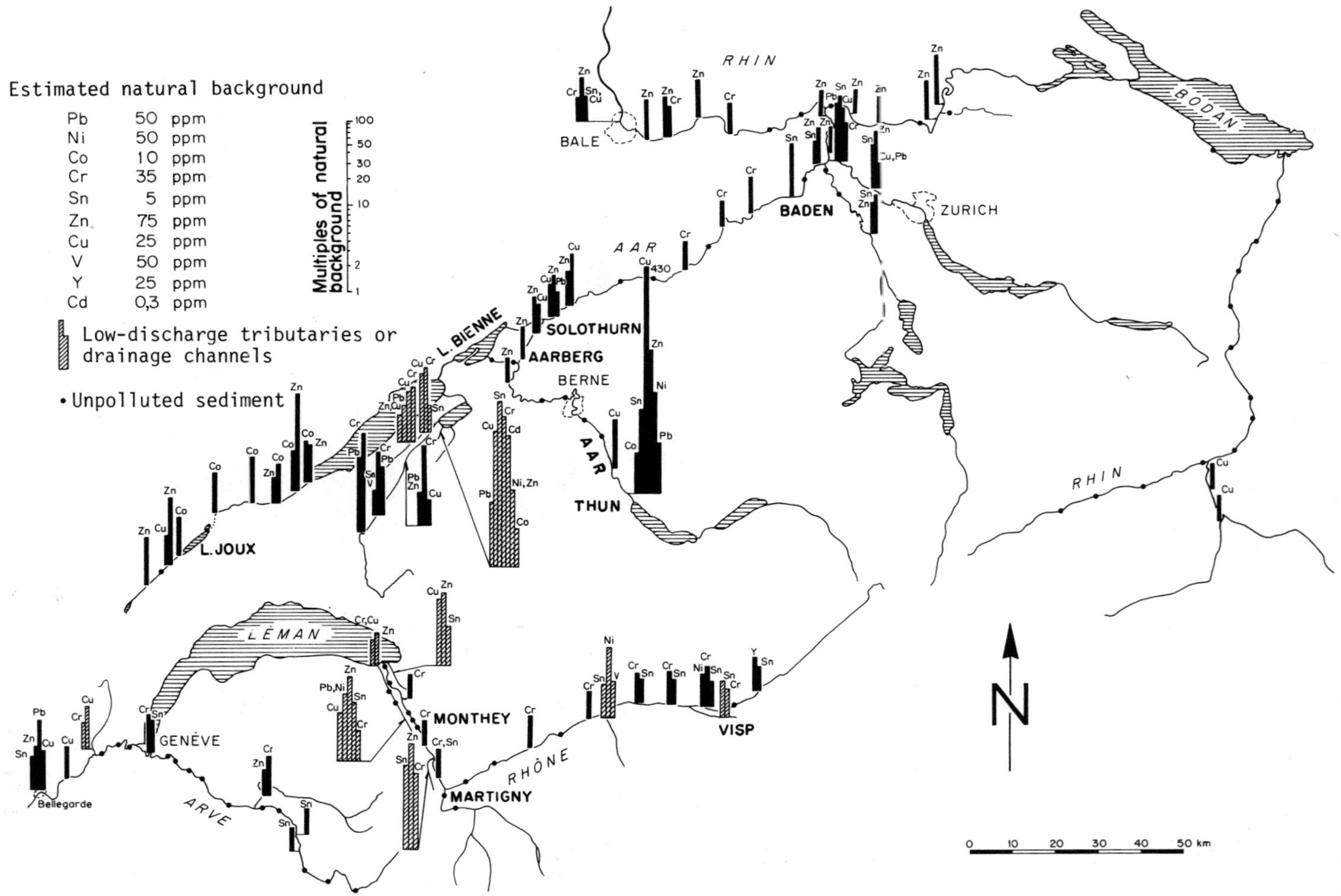


Fig. 2 - Heavy metal content of river sediments in Switserland in 1973

In the case of Cd, the background value is similar to that proposed by Turekian and Wedepohl (1961) and accepted by Förstner and Müller (1974) in their study of German rivers.

Owing to the provisional nature of the values shown in table 1, the contamination threshold has been set at twice the background values shown here.

Heavy metal contamination of river sediments

The results of sediment analyses carried out on the main rivers of Switzerland are shown in fig. 1 for Hg and fig. 2 for other trace metals. In order to estimate their relative contamination, sampling sites were subdivided into two categories based on water discharge: samples taken on the main rivers are represented by black dots or blocks, and grey is used for those taken from low-discharge drainage channels or streams.

The Hg contamination of Swiss rivers can be summarized as follows (fig. 1):

1) The Rhône river is not polluted in its upper reaches, the first mercury source being located in the industrial zone of Visp. There are several smaller anomalies between Visp and Monthey, and an important contamination has been located around this latter locality, coming from anthraquinone and chlor-alkali plants. Between 1973 and 1975, drastic steps were taken to reduce mercury losses and the most recent surveys show a significant decrease in the Hg content of the sediments (Vernet & Ribordy, 1976). Downstream from Geneva, there are small anomalies related to sewage treatment plant effluents.

2) Hg contamination of the Orbe river sediments is more or less constant all along its course. This is attributed to watch-making industries located upstream from lac de Joux (see also fig. 3).

3) Contamination of lac de Morat is caused by a small river draining an industrial area (see also fig. 4).

4) There is an almost constant contamination in the Aar river, with three possible Hg sources at Thun, Aarberg and Baden.

5) The Rhine sediments are not enriched in Hg upstream from the Bodan (Lake of Constance). Contamination starts at the junction with the Aar and increases towards Bâle, where one of the largest concentration of chemical industries in Europe is located.

The pollution status of Swiss rivers by other trace metals is as follows (fig. 2):

1) In the Rhône upstream from lac Léman, there are several anomalous areas. The dominance of Cr, associated with Sn in the upper reaches and Zn downstream, can be ascribed to several chromium-plating industries (Vernet & Ribordy, 1976).

Downstream from the lake, the industrial area of Bellegarde is characterised by Pb contamination.

2) The Orbe basin is characterized by the association Zn-Co.

3) The sediments of the tributaries of lac de Morat are contaminated by several heavy metals. The main association is Cr-Pb, with one occurrence of Cd reaching 30 times the background level.

4) The Aar upstream from Solothurn is characterized by Zn-Cu pollution, with a major source at Thun. Cr, Zn and Sn are strongly enriched at the junction with the Rhine.

5) On the Rhine, contamination starts downstream from the Bodan, with Zn clearly the dominant pollutant.

Mercury contamination of lake sediments

In a previous paper, Vernet & Thomas (1972a) have shown that the sediments of numerous Swiss lakes, varying in location and type, contain anomalous amounts of mercury. Three lakes only will be examined here as examples.

A contour map of the Hg content of sediments from lac de Joux (fig. 1 for location) is given in fig. 3. The rather high values on the southern shore of the lake are due to point sources and to the inflow of the Orbe river. According to Rapin (1976), most of these inputs can be traced to effluents from the watch-making industry.

Fig. 4 shows that the Hg inputs to lac de Morat come from three sources, the most important of which is the influx, at the SW tip, of the Broye river (Davaud, 1976).

In the case of lac Léman, the map of the Hg distribution in the sediments (fig. 5) clearly demonstrates

Fig. 3 - Mercury content of Lac de Joux sediments in 1975

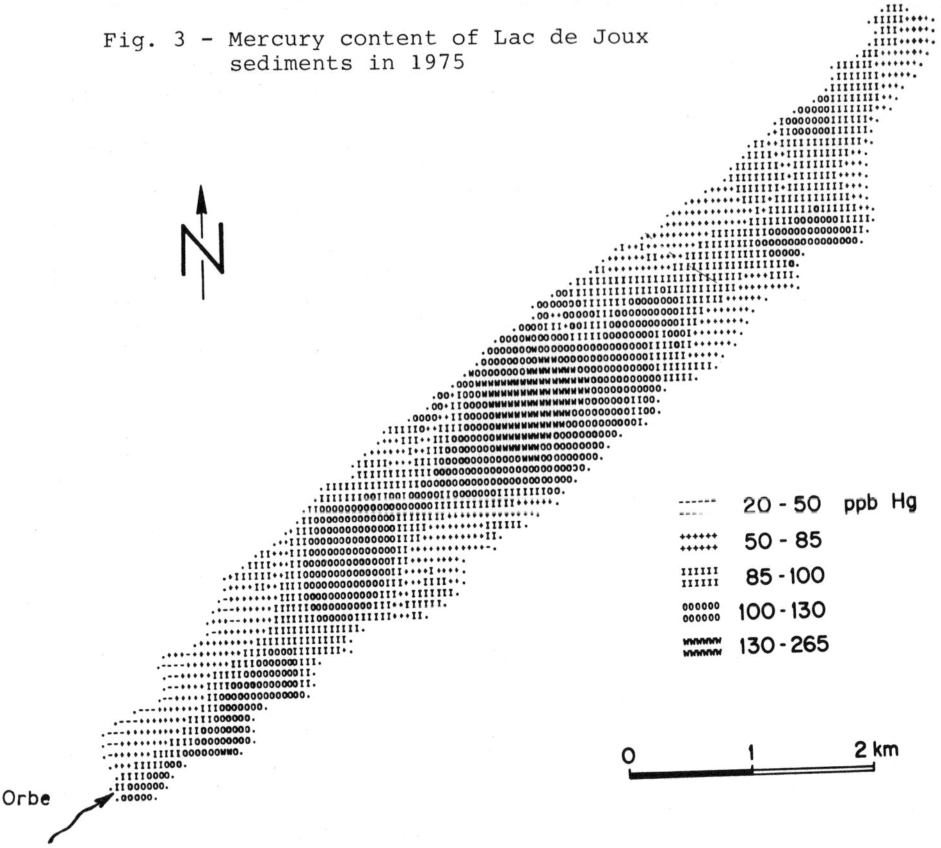


Fig. 4 - Mercury content of Lac de Morat sediments in 1975

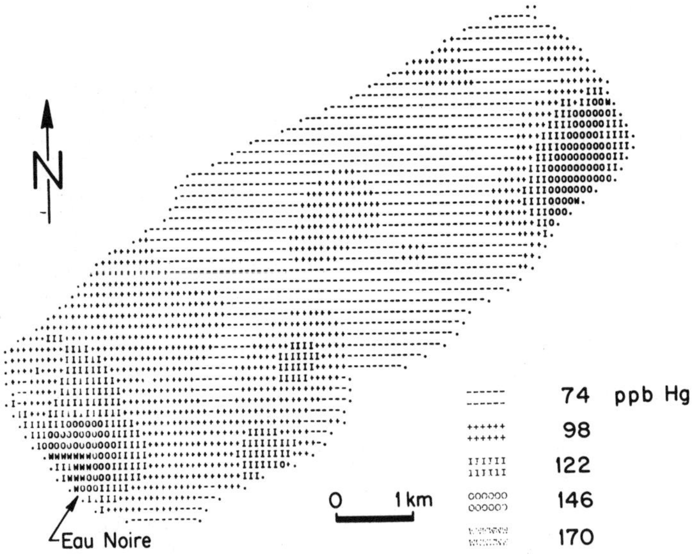

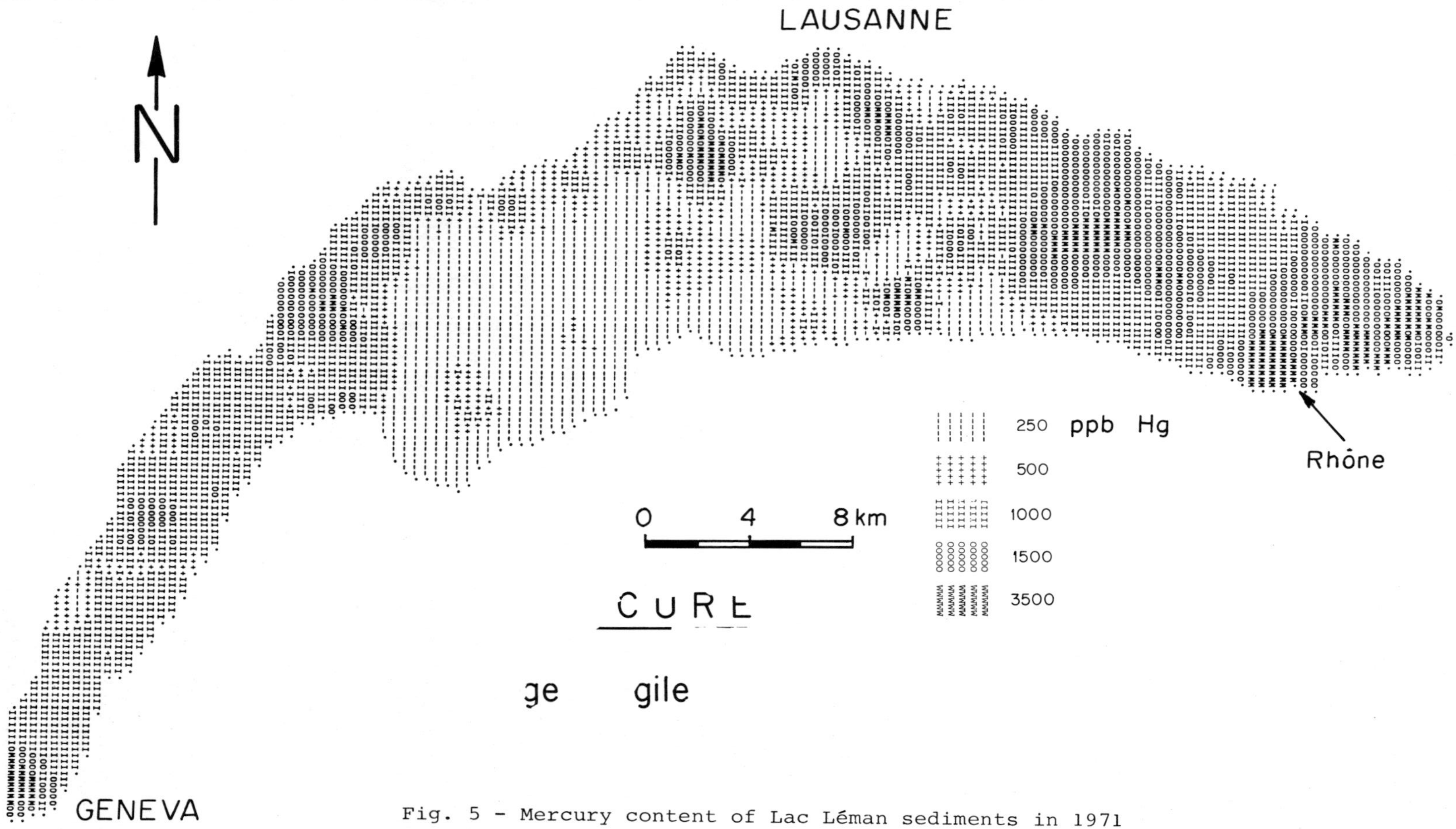


Fig. 5 - Mercury content of Lac Léman sediments in 1971

that the main input is from the alpine Rhône. The contaminated sediment is carried in suspension by the prevailing currents along the northern shore and eventually deposited in the narrow western part of the lake (Jaquet et al., 1975). In addition, several smaller sources have been located in municipalities along the northern shore.

Table 2. Average Hg content in the fine fraction of recent sediments of some Swiss lakes.

| Lake | Hg (ppb) | No of samples | Authors | Fraction analyzed (microns) |
|---|---|---|---|---|
| Léman | 651 | 392 | Vernet (unpubl.) | -175 |
| Morat | 79 | 65 | Davaud (1976) | - 80 |
| Joux | 93 | 46 | Rapin (1976) | - 80 |
| Bienne | 341 | 132 | Vernet (unpubl.) | -175 |
| Background (estimated) | 50 | -- | This paper; Ribordy, 1974. | - 80 |

A summary of the mercury content of some Swiss lakes is given in table 2 above. It is obvious that, even considering a background value of 50 ppb which is probably too high, all these lakes show mercury contamination. Values of about 70-100 ppb seem fairly representative of basins subject to small industrial or domestic inputs, whereas averages around 500 ppb indicate more serious contamination, originating from industry and large urban communities.

## Conclusions

Monitoring of mercury and other heavy metals in the sediments of Swiss rivers and lakes has led to the definition of background values, as well as to the delimitation of geographic areas characterized by anomalous concentrations.

The most severe mercury contamination occurs in the Rhône and Rhine basins. In both instances, the cause can be traced back to effluents from chemical industry.

Apart from mercury, the metals chromium, zinc, cobalt, copper, tin and, less frequently, lead, are in excess in numerous areas. These concentrations can be related to industrial concentrations (Alpine Rhône, lac de Joux, Thun, NE Switzerland).

## Acknowledgments

We wish to gratefully acknowledge the financial support given by the "Fonds national suisse de la recherche scientifique" and the "Bureau de recherches géologiques et minières, Service géologique national" of France

## References

Davaud, E., 1976. Contribution à l'étu de géochimique et sédimentologique de dépôts lacustres récents (Lac de Morat, Suisse). PhD thesis, University of Geneva, 129 p.

Förstner, U. & G. Müller, 1974. Schwermetalle in Flüssen und seen Springer Verlag. Berlin.

Rapin, F., 1976. Le lac de Joux, son origine et son environnement quaternaire. Unpublished M Sc thesis, University of Geneva, 123 p.

Ribordy, E., 1974. Etude de trois sondages dans le Petit Lac (lac Léman). Unbuplished M Sc thesis, University of Geneva, 63 p.

Scolari, G. & J.P. Vernet, 1975. Premiers résultats de la pollution par le mercure et autres métaux lourds dans les sédiments du bassin du Rhône et du Léman. Bulletin ARPEA 71: 21-57, Neuchâtel, Switzerland.

Turekian K.K. & K.H. Wedepohl, 1961. Distribution of the elements in some major units of the earth's crust. Bull. Geol. Soc. Am. 72: 175-192.

Vernet, J.P., 1969. Etude pétrographique des matières en suspension dans le Rhône et ses affluents. Bull. Soc. Vaud. Sc. Nat. 330 (70).

Vernet, J.P. & R.L. Thomas, 1972a. Levels of mercury in the sediments of some Swiss lakes, inclusing lake Geneva and the Rhône river. Eclog. Geol. Helv. 65(2): 293-306.

Vernet, J.P. & R.L. Thomas, 1972b. The occurrence and distribution of mercury in the sediments of the Petit Lac (Western lake Geneva). Eclog. Geol. Helv. 65(2): 307-316.

Vernet, J.P. & R.L. Thomas, 1972c.
Le mercure dans l'environnement et le rôle de la géologie sédimentaire. Bull. BRGM, Sect. III, 3: 43-61.
Vernet, J.P., A. Chappuis, P.Y. Favarger & E. Davaud, 1975. Teneur en mercure des poissons du Léman, campagne 1975. Unpublished rept., GEOLEM, Geneva, 32 p.
Vernet, J.P. & E. Ribordy, 1975. Teneurs en métaux lourds des sédiments du Rhône valaisan et de ses affluents. La Murithienne, 92:9-20, Sion, Switz.
Vernet, J.P., G. Scolari & F. Rapin, 1976. Teneurs en métaux lourds des sédiments de rivières suisses, du Rhône français et de ses principaux affluents. To be published in Bull. BRGM.

Key words: pollution
heavy metals
sediments
Swiss waterways

PHYSICO-CHEMICAL FORMS OF TRACE METALS AND THE SEDIMENT-WATER INTERFACE

Barry T. Hart and Simon H. Davies

Water Studies Centre, Caulfield Institute of Technology, Melbourne, Australia

Abstract

An analytical scheme useful in the investigation of the physico-chemical forms of trace metals in natural waters (including interstitial waters) is reported. The scheme involves a combination of membrane filtration (0.4 μm = 400 nm), ultrafiltration (~ 2 nm pore size) and ion exchange (using Chelex-100) separation.

A broad characterisation of the trace metals into the following forms is possible:

. labile, including free ionic species and labile inorganic and organic complexes.
. bound, including non-labile inorganic and organic complexes.
. colloidal, including trace metals associated with, or existing as, colloidal species defined to be < 400 nm but > 2 nm in size.

A preliminary investigation to determine the forms of Fe, Cd, Cu, Pb and Zn in a sediment interstitial water sample and a sample of the overlying water, both taken from an urban creek in Melbourne, Australia, has been undertaken, and the results are reported and discussed.

## Introduction

A knowledge of the transfer of trace metals across the sediment-water interface is essential to an understanding of trace metal cycles in aquatic ecosystems.

The physico-chemical system involved in such cycles can be represented by Figure 1, which shows three reservoirs - the water, sediments and suspended material, and the interactions between them. In this paper, we will be concerned only with the sediment-water interactions.

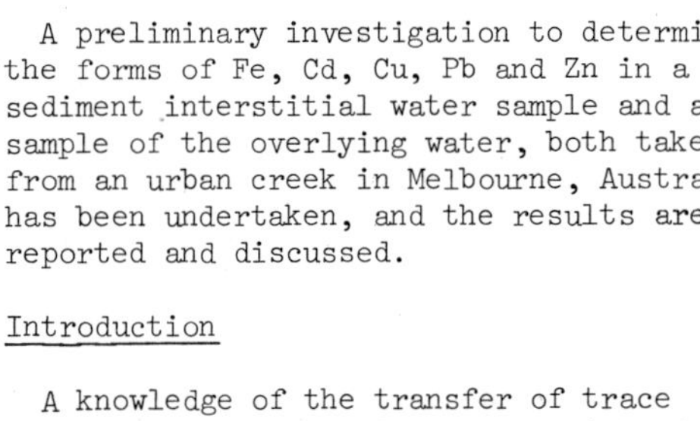


**Figure 1: Trace Metal Cycle in an Aquatic System (excluding interations involving the biotic components)**

We believe little progress can be made in understanding these real system interactions until methods are developed for defining (at least in an operational sense) the way in which the trace metal exists in each of the interacting phases.

This paper discusses some preliminary work aimed at identifying the physico-chemical forms of Fe, Cd, Cu, Pb and Zn in interstitial waters and the associated water column.

## Experimental

### Sampling methods

The samples were taken from Mordialloc Creek, a small, rather polluted creek in Melbourne, Victoria. At the time of sampling (13.00 hours, 16/VIII/76), the water depth was approximately 15 cm.

Ten core samples were taken with a perspex corer from an area measuring 1 m x 0.2 m. The top 5 cm of each core was kept and the ten samples pooled. The sediment was black in colour and, when treated with a small amount of dilute HCℓ, gave off $H_2S$. Unfortunately, Eh measurements were not taken; however, at the same sampling site one week earlier, the sediments showed Eh values near 200 mV. The bulk water sample was obtained by filling an acid washed polythene bottle held approximately 5 cm from the bottom. A separate sample was taken for pH, chloride and total organic carbon (TOC) determinations.

Sediment and water samples were returned immediately to the laboratory. The interstitial water was removed from the sediment by centrifugation for 20 min at 3000 rpm (1000 g). This water was then filtered through a 0.4 μm Nucleopore membrane filter to remove any particulate material. Approximately 60 g water was obtained by this process from 300 g of wet sediment. The filtration was completed approximately 3½ hours after the samples were taken.

### Separation and analysis

The separation scheme used for the bulk and interstitial waters is shown in Figure 2. Both samples were filtered through acid washed 0.4 μm Nucleopore filters before use.

All samples were analysed for Fe and Zn, using direct aspiration flame atomic absorption spectrometry (AA-flame). Background correction was employed for each analysis. Anodic stripping voltammetry (ASV) (PAR 174A

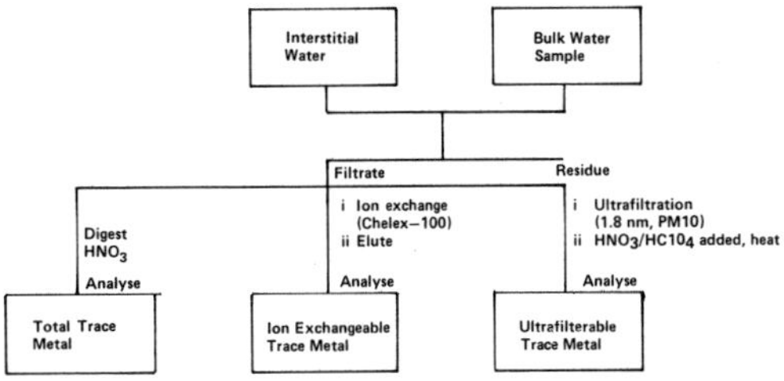


**Figure 2: Scheme for Determination of Physico-chemical forms of Trace Metals in Waters**

with hanging mercury drop electrode) was used in the analysis of Cd, Cu and Pb.

Details of the separation techniques used are as follow :

Total filterable - A portion of the filtrate was acidified with $HNO_3$ to pH 1 to 2. For AA-flame analyses, samples were used directly. For the ASV analyses of the bulk water sample, a 20 ml portion of the sample was placed in a teflon beaker and boiled to almost dryness, then made up to approximately 20 ml and analysed. It was found necessary to digest the interstitial water samples for ASV analysis with a fuming $HNO_3/HC\ell O_4$ mixture. The experimental conditions have been described previously (Hart & Davies, 1976).

Ultrafilterable - A 65 ml capacity Amicon® ultrafiltration cell with a PM10 filter was used. According to the manufacturers, this filter has an approximate pore diameter of 1.8 nm.

Before use, the cell was rinsed with sample (2 x 5 ml portions of the sample for the interstitial water, and 2 x 65 ml portions of the sample for the bulk water). The first 5 ml of filtrate was discarded, and the next 10 ml collected and acidified to pH 1 with $HNO_3$. The filtration took 3 hours for the interstitial water and 2 hours for the bulk water. For ASV analyses, the samples were digested with a fuming $HNO_3/HC\ell O_4$ mixture and analysed as outlined above. For AA-flame analyses, samples were used directly.

Ion exchangeable - The ion exchangeable fraction was obtained by allowing the water sample to remain in contact with Chelex-100 (Na form) resin for 24 hours in an acid washed polythene bottle. After this time, the resin was filtered off (0.4 µm Nucleopore), washed with doubly distilled-deionised (Milli-Q system) water and eluted with 1 M $HNO_3$.

The eluate was then analysed by either AA-flame or ASV.

We have previously found this "batch" ion exchange technique gives close to quantitative recovery for Fe, Cd, Cu, Pb and Zn (Hart & Davies, 1976a).

## Results and discussion

The results obtained from analysis of the interstitial and bulk water samples are recorded in Tables 1 and 2.

Table 1. Trace Metal Concentrations in Three Fractions Separated from Bulk & Interstitial Water Samples from Mordialloc Creek (16/VIII/76).

| Fraction | Concentration (µg/ℓ)* | | | | |
|---|---|---|---|---|---|
| | Zn | Cd | Cu | Pb | Fe |
| Bulk Water | | | | | |
| Total filterable ($M_{tf}$) | 90 | 1.3 | 10 | 7.1 | 60 |
| Ultra filterable ($M_{uf}$) | 90 | 1.4 | 9.3 | 7.5 | < 30 |
| Ion exchangeable ($M_{ie}$) | 35 | 0.9 | 9.0 | 3.0 | < 30 |
| Interstitial Water | | | | | |
| Total filterable ($M_{tf}$) | 400 | 14.7 | 65 | 59 | 8800 |
| Ultra filterable ($M_{uf}$) | 155 | 2.4 | 57 | 19 | 230 |
| Ion exchangeable ($M_{ie}$) | 95 | 2.8 | 28 | 11 | < 30 |

* Detection limits: Zn 0.5, Cd 0.02, Cu 0.2, Pb 0.1, Fe 30.

Table 2. General Parameters for Bulk and Interstitial Waters from Mordialloc Creek (16/VIII/76).

| Parameter | Bulk Water | Interstitial Water |
|---|---|---|
| pH | 7.66 ± 0.02 | 6.95 ± 0.02 |
| Cℓ (mg/ℓ) | 300 | 7120 |
| TOC (mg/ℓ) | | |
| - filtered sample (< 400 nm) | 24.0 | 148.0 |
| - ultrafiltered sample (< 2 nm) | 3.1 | 25.5 |

On the basis of this one sample, we do not believe it prudent to seek detailed explanations of the results. However, there are several noticeable trends in the results which are worthy of comment. Before considering these, we will briefly discuss the possible forms of the trace metals associated with each fraction.

The total filterable metal fraction ($M_{tf}$) will contain trace metals existing in a free ionic form as inorganic or organic complexes, and in and associated with colloidal forms (Stumm & Bilinski, 1973).

The fraction designated 'ultrafilterable' ($M_{uf}$) will contain all trace metals existing as free ions or as complexes, provided they are smaller than ~ 2 nm in diameter. This fraction will exclude the greater proportion of the colloidal iron and manganese oxides and perhaps also metal-humic acid complexes. On the basis that fulvic acid has a molecular weight ~ 1000 (Perdue et al., 1976), we have assumed that such complexes would not be retained. Work is in progress to confirm this assumption. It is relevant that the system sampled in this study, Mordialloc Creek, receives effluent from a secondary sewage treatment plant; thus the organics present in our samples are likely to be of sewage origin. The greater proportion of the organic material discharged from this plant is not retained by an ultrafilter (Hart & Waite, unpublished data).

The metal retained by the Chelex resin is referred to as the 'ion exchangeable' metal ($M_{ie}$). This fraction will include the free ionic forms of the trace metal, and that trace metal associated with any labile complex whose stability constant is less than that resulting from the combination of the trace metal with the Chelex.

Useful information can also be obtained from the differences between $M_{tf}$ and $M_{ie}$, and $M_{tf}$ and $M_{uf}$. The difference ($M_{tf} - M_{ie}$) probably represents the trace metal fraction strongly bound to organic molecules such as fulvic acid, and possibly also to colloidal iron oxides. The fraction less than 400 nm but greater than 2 nm in size (i.e. $M_{tf} - M_{uf}$) will probably represent trace metals associated in one or more of the following ways :

i) in colloidal form, e.g. Fe as FeOOH.
ii) associated with humic acid type organic molecules.
iii) associated with colloidal sized clay particles or iron and manganese hydroxy -oxides. [N.B. The iron and manganese oxides may be present as discrete particles or as oxide films on clay colloids (Jenne, 1968; Lee, 1975)].

To assist the following discussion, the results are presented diagrammatically in Figures 3 to 7. These figures show the relative proportion of each trace metal associated with each of the broadly characterised physico-chemical forms; namely,

A - labile forms (determined by $M_{ie}/M_{tf}$)

B - strongly bound organic or inorganic forms (determined by $\frac{M_{tf} - M_{ie}}{M_{tf}}$)

C - trace metals existing as, or associated with, colloidal species (determined by $\frac{M_{tf} - M_{uf}}{M_{tf}}$).

The results given in Table 1 show that the trace metals are considerably more concentrated in the interstitial water than in the bulk water. Other workers have also noted the generally greater concentration of trace metals in interstitial waters compared with the overlying waters (Brooks et al., 1968; Presley et al, 1972; Elderfield and Hepworth, 1975).

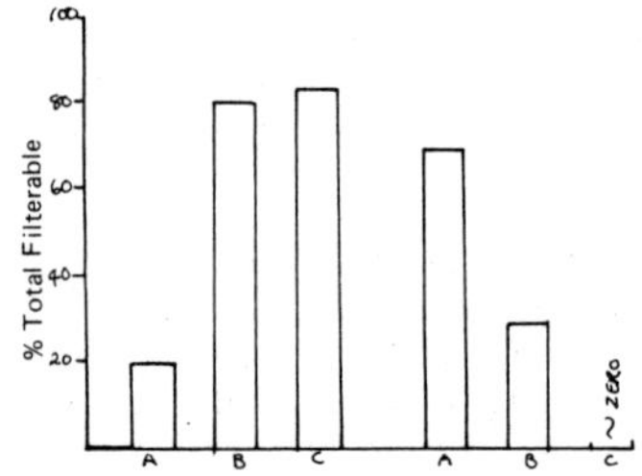

Figure 3: Proportion of Cd in labile bound and colloidal forms.

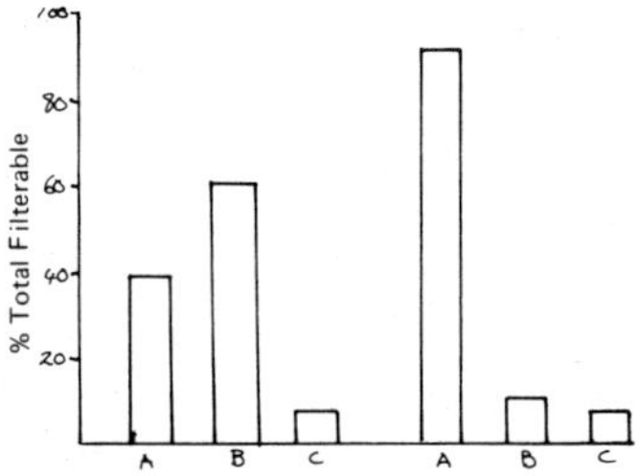

Figure 4: Proportion of Cu in labile bound and colloidal forms.

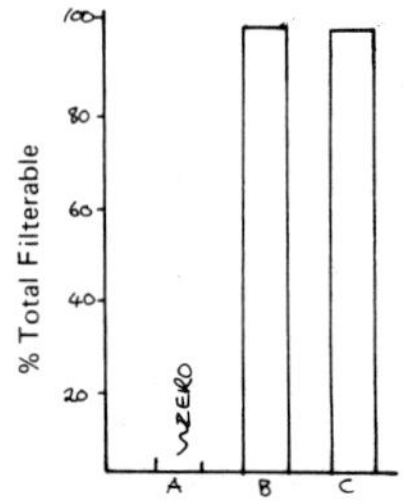

Figure 5: Proportion of Fe in labile, bound and colloidal forms in the interstitial water.

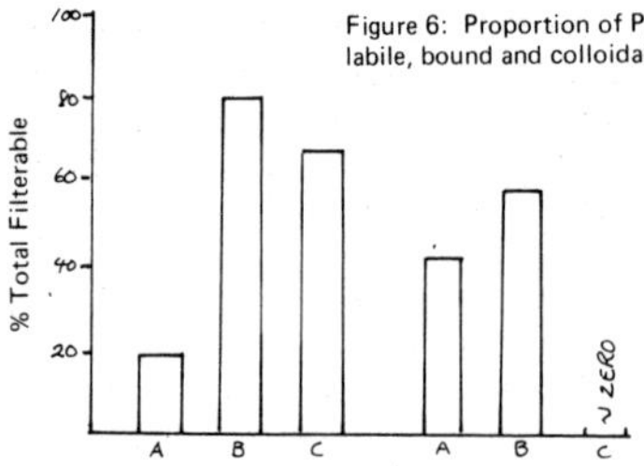

Figure 6: Proportion of Pb in labile, bound and colloidal forms.

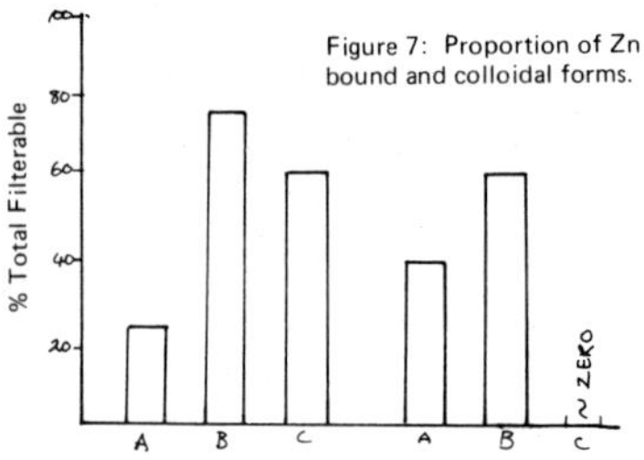

Figure 7: Proportion of Zn in labile, bound and colloidal forms.

Cd

In the bulk water sample, around 70% of the Cd was present in labile forms. In the interstitial water, a noticeably different distribution exists, with the major proportion (80%) of the Cd being associated with the colloidal fraction. This large increase in association with the colloidal phase may be explained in terms of sorption of Cd species by colloidal iron oxide, since the interstitial water contains a considerable amount of colloidal iron (~ 8600 μg Fe/ℓ).

Cu

The major proportion of the Cu in the bulk water sample (~ 90%) is present in a labile form. This is in marked contrast to the situation in the interstitial water sample, where it would appear the Cu is almost evenly distributed between labile species (~ 40%) and more strongly bound complexes (~ 45%).

Copper is known to form more stable complexes with natural organic materials than either Cd, Pb or Zn (Schnitzer & Khan, 1972; Gamble & Schnitzer, 1973; Rashid, 1974).

Although we lack a detailed knowledge of the organic materials present, and of their ability to form metal-organic complexes, it seems reasonable to postulate that the difference between the distribution of Cu and of the other trace metals might be explained in terms of the ability of Cu to form strong complexes with the natural organic materials present.

Fe

Iron was found to be considerably more concentrated in the interstitial water than in the bulk water (8800 μg/ℓ cf. 60 μg/ℓ).

Interestingly, there was very little iron (< 30 μg/ℓ) in an ionic or labile form in either the interstitial or bulk water. This result was not surprising in the bulk water sample, as, in well-oxygenated waters, one can calculate (assuming no organic complexes are present) that the concentration of soluble ferric iron species should be < 1 μg/ℓ (Hem, 1975). The lack of Eh data on our interstitial water sample makes difficult the estimation of the predominant forms of iron.

However, inspection of the appropriate Eh-pH diagram (Hem, 1975) shows that, provided the Eh is in excess of ~ 175 mV, $Fe(OH)_3$ should be the predominant species (pH of sample 6.99). In the week preceding these experiments, the Eh of Mordialloc Creek sediments was consistently > 200 mV. Thus it seems reasonable to assume the interstitial water sample under discussion was low in ion exchangeable Fe because conditions were sufficiently aerobic to prevent the production of the soluble ferrous species.

Over 97% of the Fe is present in the colloidal form in the interstitial water sample. Due to the high detection limit in the Fe analyses (30 μg/ℓ), we are unable to comment on the bulk water sample, except to indicate that at least 50% of the Fe is present in the colloidal form.

A review of the available literature suggests that the dominant colloidal forms of iron are likely to be :

colloidal hydrous iron oxides and/or Fe-organic complexes.

The relative significance of these two forms is still a matter of some dispute (Jenne, 1968; Plumb & Lee, 1973; Lee, 1975; Perdue et al, 1976).

Iron oxides, particularly in colloidal forms, are acknowledged to be extremely important in the transport of trace metals in aquatic systems (Jenne, 1968; Lee, 1975). It is therefore probable that a large proportion of the differences in trace metal distribution between the bulk water and the interstitial water can be accounted for by differences in the amount (and form) of the colloidal iron oxide fraction.

Pb, Zn

These two trace metals appear to behave in a similar manner, and are discussed together. Figures 6 and 7 show that, in the bulk water sample, approximately 40% of Zn and Pb are present in a labile form. This distribution is somewhat different in the interstitial water sample. Here the major proportion (~ 60%) is associated with the colloidal phase, 20 to 25% in labile forms and 10-15% in more tightly bound forms.

The difference in the distribution of forms of these two trace metals in the interstitial and bulk waters is probably due in large measure to the very much greater amount of colloidal iron oxide present in the interstitial water.

References

Brooks, R.R., B.J. Presley & I.R. Kaplan, 1968. Trace elements in interstitial waters of marine sediments. Geochim. et Cosmochim. Acta. 32:397-414.

Elderfield, H. & A. Hepworth, 1975. Diagenesis, metals and pollution in estuaries. Mar. Pollut. Bulletin. 6:85-87.

Gamble, D.S. & M. Schnitzer, 1973. The chemistry of fulvic acid and its reactions with metal ions. In: P.C. Singer (Ed.): Trace metals and metal-organic interactions in natural waters. Ann Arbor Science Publishers Inc., Michigan. p.265-302.

Hart, B.T. & S.H.R. Davies, 1976. A new dialysis-ion exchange technique for determining the forms of trace metals in water. Aust. J. Marine and Freshwater Research. (In press).

Hart, B.T. & S.H.R. Davies, 1976a. A batch method for determining ion exchangeable trace metal fractions in natural waters. (Paper in preparation).

Hem, J.D., 1975. Role of hydrous metal oxides in the transport of heavy metals in the environment (discussion of paper by G. F. Lee). In: P.A. Krenkel (Ed.): Heavy metals in the aquatic environment. Pergamon Press. p. 149-153.

Jenne, E.A., 1968. Controls on Mn, Fe, Co, Ni, Cu and Zn concentrations in soils and water : The significant role of hydrous Mn and Fe oxides. Advances in Chemistry Series, Vol.73: Trace inorganics in water. American Chemical Society, Washington, p. 337-387.

Lee, G.F., 1975. Role of hydrous metal oxides in the transport of heavy metals in the environment. In: P.A. Krenkel (Ed.): Heavy metals in the aquatic environment. Pergamon Press. p. 137-147.

Plumb, R.H. & G.F. Lee, 1973. A note on the iron-organic relationship in natural water. Water Res. 7:581-585.

Presley, B.J., Y. Kolodny, A. Nissenbaum & I.R. Kaplan, 1972. Early diagenesis in a reducing fjord, Saanich Inlet, British Columbia, II Trace element distribution in interstitial water and sediment. Geochim. et Cosmochim. Acta. 36:1073-1090.

Purdue, E.M., K.C. Beck & J.H. Reuter, 1976. Organic complexes of iron and aluminium in natural waters. Nature. 260:418-420.

Rashid, M.A., 1974. Humic compounds of a sedimentary environment : their chemical nature and geochemical significance. Geol. Survey Canada. Paper 74-30:123-132.

Schnitzer, M. & S.U. Khan, 1972. Humic substances in the environment. Marcel Dekker, New York.

Stumm, W. & H. Bilinski, 1973. Trace metals in natural waters : difficulties in interpretation arising from our ignorance of their speciation. In: S.H. Jenkins (Ed.): Advances in water pollution research - Proceedings of 6th International Conference on Water Pollution Research, Jerusalem, 1972. Pergamon Press, Oxford. p. 39-52.

Key words: trace metals
species
sediments
interstitial waters
fractionation

# UPTAKE AND RELEASE OF HEAVY METALS IN THE BOTTOM SILT OF RECIPTIENTS

P. Literáthy and F. László

Institute for Water Pollution Control, Research Centre for Water Resources Development, Budapest, Hungary.

Summary

Toxic heavy metals which may be present in the effluents in both solid and dissolved forms, are extremely liable to be incorporated in the suspended sediment. The suspended sediment is sorted in the process of settling, as a result of which heavy metals tend to accumulate in the bottom silt. Under changing flow conditions, e.g. during a flood, turbulence may increase enough to counteract settling, thus stirring up the bottom silt and carrying it into suspension again. Bottom silts enriched in toxic heavy metals present a permanent source of hazard to the recipient and thus the quality of surface waters polluted by such substances can only be charactrized by evaluating the analytical results on the water and the bottom silt in combination.

The first problems were encountered in the analytical results on the bottom silt, since the data on samples taken from practically the same point - expressed in the conventional mg per kg of dry substance - were found to scatter widely. This was traced back to sampling uncertainties, due primarily to the highly variable number of sand grains contained in the samples. Therefore the fraction of particles smaller than 1 mm soluble in hydro-chloric-nitric acid and decomposed on ignition was adopted as the basis to which the analytical results were related.

The extensive program of sampling and analysis performed on one of the recipients in Hungary has provided the basis for intensive studies on the variations of the heavy metal pollution. In the course thereof the phenomena of settling and stirring up have been explored over the reach downstream of the point of discharge. Special attention has been devoted to the scouring of pollutants in the bottom silt during floods and to their accumulation. Among the most toxic heavy metals, lead was found to be most stable in the bottom silt, followed by cadmium and mercury.

## Introduction

Without going into details of the hydrodynamical and methodological aspects, some questions have to be discussed for understanding the subsequent problems.

Solid particles of a water course may be present in the form of suspended, rolled or bottom sediment. The quantitative and qualitative distribution of the particles is a function of the settling velocity and the flow conditions. Settling velocity of the particles is essentially determined by Stokes law, according to which settling velocity is proportional to the square of particle diameter and to the density of the particle.

Metal pollutants of waters may occur practically in particles of any size, either alone or adsorbed to each other or to the natural particles.

Due to density differences, settling velocity of the particles of the same size is not identical. Flow conditions also play an important role, in addition to size and density of particles. Turbulence may be predominant over the settling process resulting in the stirring up of the settled bottom sediment.

In the range of very small particles the Brownian movement becomes dominant, with the result that particles smaller than a few tenths of $\mu$ are not able to settle out. Below 10 $\mu$ settling velocity is very slow, and may take some days or more.

Before analyzing the water samples the "dissolved" and suspended sediment fraction is separated by membrane-filter of 0.45 um pore-size. This means that colloids may pass as dissolved fraction, but the non-settling particles in the range from 0.45 $\mu$m up to size between 1.0 or 10 $\mu$m get to the suspended sediment.

No upper limits of particle size exists for bottom sediments. Particles with diameters of several mm or cm which are practically absent in the suspended matter can often be found in the bottom sediment. These are mostly of natural origin. Consequently their removal by sieving from the bottom sediment samples and investigation of the particles smaller than 1 mm only means an approach with a view to better comparison of the sediments.

The majority of the significant metal pollutants is present in the sediment, mainly in the range between 0.45 and some tens and hundreds µm. Since the quantity of sand particles - not causing pollution in itself - can be highly variable in the bottom sediment samples, a comparison of analytical results of sediment samples is difficult on the basis of the widespread characteristic mg/kg of dry substance. Further restriction of the size-range of particles may not be successful, as a consequence of the above mentioned facts. Only one way remains in the interest of better comparability: the result should be referred to a certain fraction of the samples.

Results and discussion

In the course of the analytical investigation of the sediment samples, the fraction characterized by the sum of the part soluble in a mixture of hydrochloric and nitric acid plus the ignition loss was found to be the best applicable reference basis, subsequently noted by RB.

The reference basis defined this way contains "accumulated" the manmade pollution: certain compounds of the studied metals as well as organic matter, in addition to the natural calcium, magnesium etc. compounds. The residue, on the other hand, is mainly of geochemical origin and was found to contain 80-95% silica. A special advantage of using this reference basis is that the disturbing effect of the sand particles can be eliminated. It is well indicated by table 1 containing the results expressed in both forms of two bottom sediment samples taken in the same section.

But it should be mentioned that in the case of unpolluted streams, where the composition of bottom sediment is controlled by geochemical conditions and natural processes this proportion

Table 1. Comparison of bottom sediments taken practically at the same sampling point. Characterization is based on different reference basis.

| Det. | mg/g Dry subst. | | mg/g RB | |
|---|---|---|---|---|
| | No 1 | No 2 | No 1 | No 2 |
| 25.11.74 | | | | |
| RB | 112 | 221 | | |
| Fe | 16.8 | 30.9 | 150 | 140 |
| Mn | 0.58 | 1.26 | 5.2 | 5.7 |
| Zn | 0.28 | 0.63 | 2.5 | 2.8 |
| Hg | 0.003 | 0.006 | 0.027 | 0.02 |
| Cd | 0.013 | 0.024 | 0.116 | 0.10 |
| Pb | 0.134 | 0.309 | 1.2 | 1.4 |
| 18.06.75 | | | | |
| RB | 99 | 236 | | |
| Fe | 14.5 | 36.1 | 147 | 153 |
| Mn | 0.56 | 1.6 | 5.6 | 6.8 |
| Zn | 0.28 | 0.83 | 2.8 | 3.5 |
| Hg | 0.0064 | 0.017 | 0.065 | 0.07 |
| Cd | 0.0088 | 0.021 | 0.089 | 0.08 |
| Pb | 0.089 | 0.286 | 0.9 | 1.2 |

Det. = Determinants
Dry subst. = Dry substabce
RB = Acid soluble part + Ignition los

is a fairly stable one, exemplified by the Syrian section of the Euphrate River, where the so called reference basis (RB) varied from 27 to 30% over a round 70 km long reach upstream of the about 80 km long reservoir, while from 36 to 37% over a distance of 100 to 150 km downstream thereof.

The picture obtained for one of the most highly polluted streams in Hungary is an entirely different one. Figure 1 shows the map of the studied area.

The ignition loss and the acid solubl part of the bottom sediment are seen to be highly variable, especially downstream of the sources of pollutio sometimes between 5-60%. In the case cadmium pollution the results related to the dry matter show no definite trend over the reach downstream of th discharge point, whereas the new meth of representation reveals the rapid precipitation and settling of cadmium, further the rather uniform, but substantially higher level of concentration over the downstream reach than upstream of the source. This can be see on Figure 2.

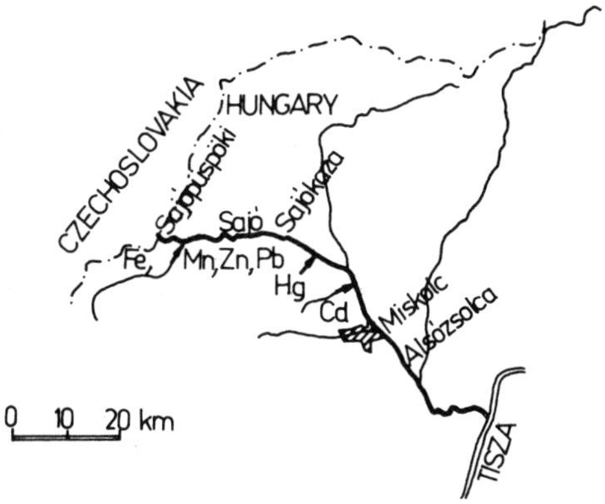


Fig. 1 Map of the studied area.

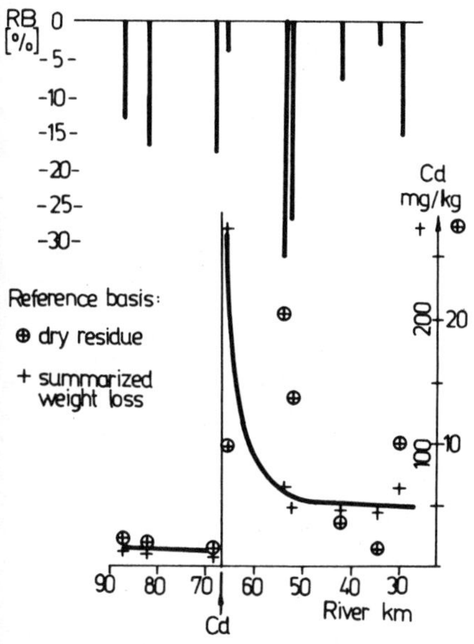


Fig. 2 Characterization of bottom sediments along a river section on different reference basis.

In one of the sampling programs the later samples were taken weekly in the regular sampling network. In every quarter samples were continuously taken during one week at Sajópüspöki and Alsózsolca as well as at the same time additionally at Kazincbarcika and at the mouth of the Sajó river. The bottom sediment samples were taken either separately or joined to the one-week-water sampling.

In cases of bottom sediment sampling samples were retrieved from the upper 1-2 cm layer of sediment. It should be noted that "core" sampling was also used to estimate the stratification of pollutants. It was found on this basis that changes occurring during 1-2 years or over a shorter time can be featured by the analysis of the upper 1-2 cm layer.

Heavy metals have been determined by atomic absorption spectrophotometry. Mercury was determined by cold vapour flameless technique, cadmium, lead and chromium by graphite furnace while the other metals were measured in airacetylene flame. The determinations were performed for the total metal content of the samples.

During the intensive data collection period on metal pollutants besides sampling sections on the Sajó the main effluent discharges were also subject to a parallel study. Results of the weekly measurements for the four descriptive sections are shown in Figure 3. The figure also indicates the average concentration and water discharge values in the important pollution sources for selected metals. In the investigations none of the sections of the Sajó had higher maximum chromium and copper content than the 10% of the limit values, so these two metals are not shown in the figure. Of the 6 metals in figure 3, iron, manganese, zinc are less dangerous, but lead, mercury and cadmium are extremely dangerous because of their toxic effect. Limit values were assigned to each metal. Thus it can be seen that despite the considerable background pollution average concentrations of iron and manganese are only slightly above the limits, while in the case of zinc even maximum values or pollution source average values do not reach it (5 mg/l). In the case of the other three metals, average lead concentration never reaches the toxic limit value, a fact that can be explained mostly by the phenomenon that of all the metals lead occurs in the suspended particles of the largest density, so these particles will quickly be incorporated in the sediment by fractionated settling, and are the most difficult to be stirred up again. The spreading of lead in the longitudinal section at increased turbulence and higher water discharges is also mostly in the form of rolled silt. The most serious problem is caused by mercury having an average border section concentration amounting to, and after receiving sewages from the main Hungarian effluent discharger reaching far beyond the limit value. It is especially remarkable, that among the three metals

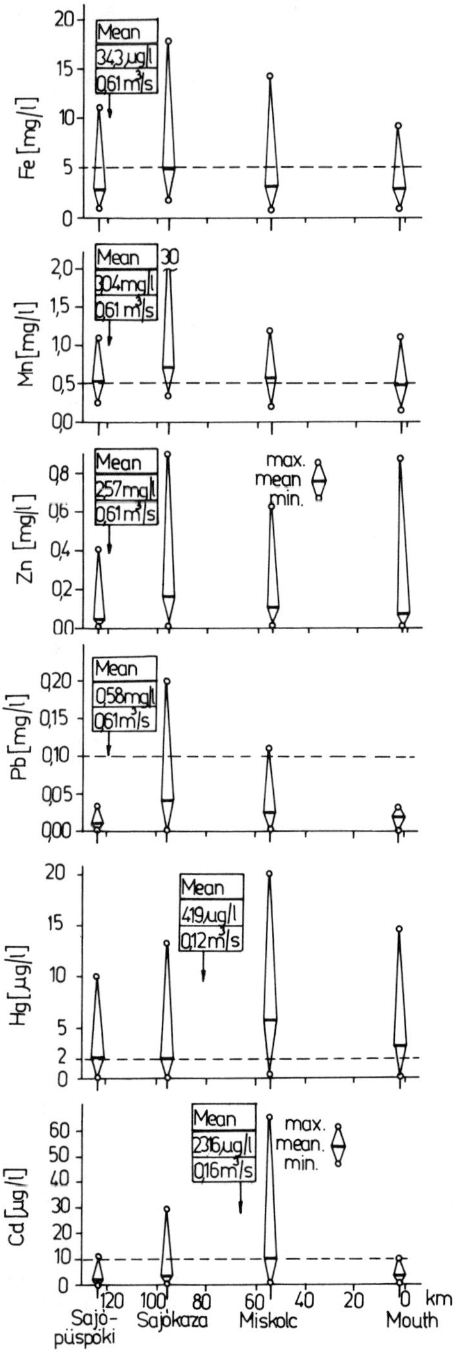


Fig. 3 Minimum, maximum and mean values of metal concentrations at typical sections, as well as the metal concentrations and discharge of the main pollution sources during 1974.

mercury has the least inclination to settle as is shown by the low rate of concentration reduction while getting to the mouth section. On the basis of model experiments settling rate of lead was found to be roughly two time as high as that of mercury. As to cad mium the situation is intermediate be tween mercury and lead.

Short term fluctuations of mercury an lead content were also studied during a week. During the investigation in October, 1974, there were higher wate: discharges and a flood period began just at the end of sampling. Values for mercury concentration indicate that sometimes shock type pollution can be observed at Sajópüspöki, but especially the rising of the minimum pollution over the toxic limit due to the flood at Alsózsolca between the 13th and the 14th is remarkable. (See Figure 4a.) In October, 1974 general- ly higher cadmium concentrations coul be observed than the toxic limit valu The 1975 March investigations yielded even higher maxima at Alsózsolca. (See Figure 4b.)

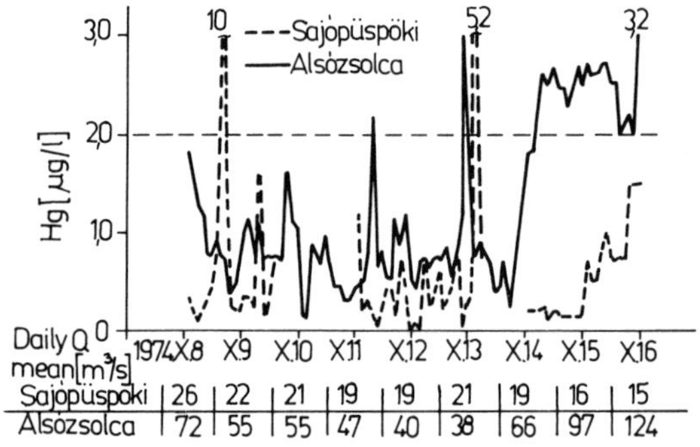


Fig. 4a

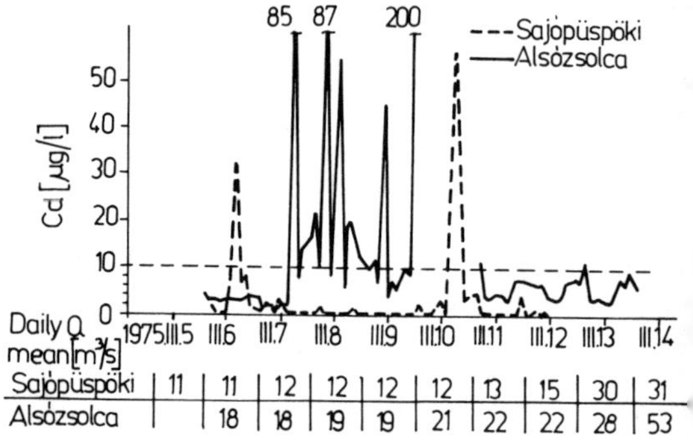


Fig. 4b

Fig. 4 Characterization of short perio fluctuation of total mercury and cad- mium pollution based on continuous sampling at two sampling points along the Sajó river.

It is even more remarkable that in the early morning hours on 6th and 10th

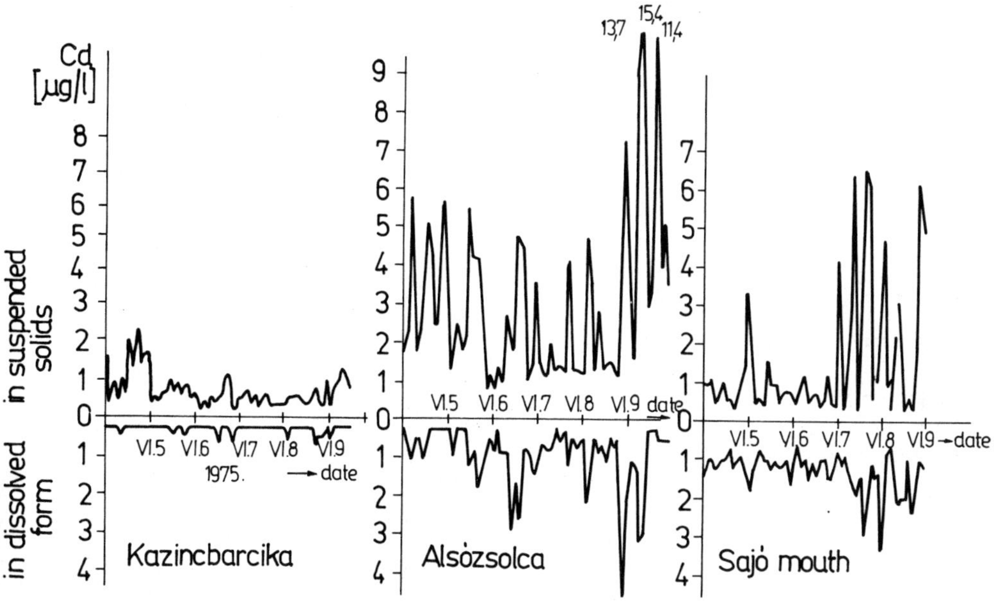


Fig. 5 Cadmium concentration in dissolved and solid phase upstream and downstream of the pollution source.

of March at Sajópüspöki concentrations much beyond the toxic limit were measured, while when referring to annual data in Figure 3 the maximum hardly reached 10 µg/l.

Variation of cadmium content in dissolved and suspended solids is shown in Figure 5. During a one-week investigation, cadmium pollution occurred in relatively lower concentrations.

Similarly to the above, pollution arriving to the border section in peaks is compensated upstream of Kazincbarcika, and the equilibrium dissolved form suitable for the ion environment of the Sajó is developed, while most of the suspended solids have settled down. There are considerable fluctuations in both the suspended and the dissolved cadmium content at Alsózsolca due to the Hungarian pollution and the larger part of cadmium is present in suspended solids. Down to the mouth section cadmium content in the suspended solids considerably declines, the dissolved form becomes more compensated and it can be well seen that there is a considerable increase in the equilibrium concentration of the dissolved form in the different ion environment in comparison with Kazincbarcika. This phenomenon is mostly caused by one of the tributaries of the Sajó river which has high streamflows.

In order to evaluate the results it must be taken into account that the harmful effects of parallely present toxic metals are additive, so the situation becomes particularly critical in the Sajó section between Kazincbarcika and the mouth of the above mentioned tributary.

The presence of pollutants in the bottom sediment can be regarded as the "memory"-unit of the water course, as it conserves those pollutants inclined to settling that may be impossible to "catch" by water sampling. At the same time this means that bottom sediment sampling has not to be done so often as water sampling.

Metal content in the bottom sediment of the longitudinal section of the Sajó can be described by Figure 6. Concentration and % composition were given in the reference basis (RB) described before. The figure also represents metal concentrations beyond which the mixing up of 1 g RB per liter is not probable in the Sajó, so metals in the bottom sediment below this value were further disregarded. Chromium and copper were dropped also in the case of bottom sediment in this way, though their quantities are increased in the bottom sediment of the Sajó by the first discharger. Referring to observations made at water investigations iron, manganese and zinc though occurring in considerable quan-

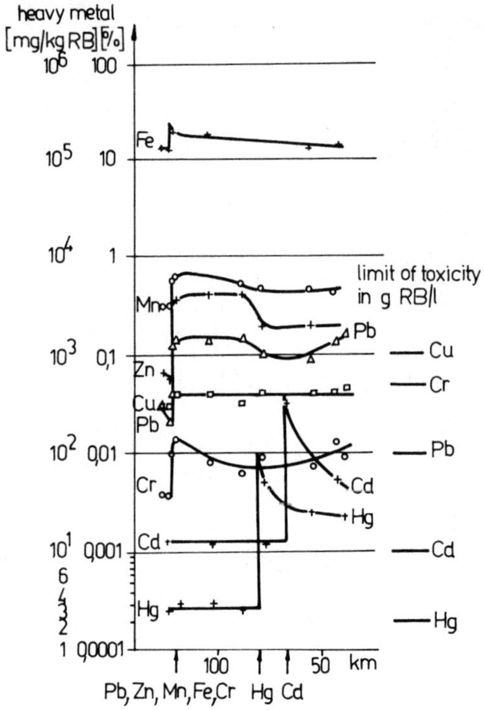


Fig. 6 Metal concentrations of bottom sediments along the Sajó river.

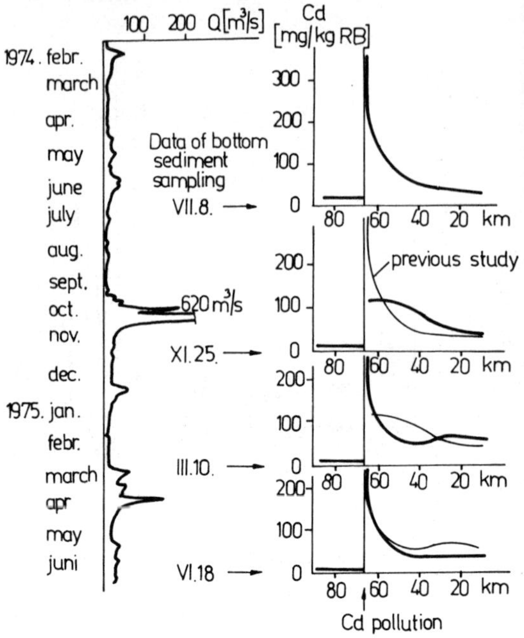


Fig. 7 Spreading of the polluted bottom sediment zone at various discharges.

tities will be disregarded in the discussion of results related to the bottom sediment further on as well. Lead, mercury and cadmium contents in the bottom sediment all along the Hungarian section of the Sajó are higher than the presented value. Compared to the background pollution upstream of the pollution source, the quantities of cadmium and mercury show an order of magnitude increase.

Owing to the variability of flow conditions in the recipient the settling out of heavy metals will also be variable and will take place over different distances below the particular sources of pollution.

Variations of cadmium content in the bottom sediment following the period of water discharge changes are shown in Figure 7. It can be well seen that accumulation near the polluting source formed before the flood of October-November, 1974, was washed out by the flood and "spread" along the downstream section. Besides the accumulation that was already characteristic in March, 1975, on this section before the final washing out to the Tisza, higher values were found, but by June the situation turned to be similar to that before the flood.

Studying the mercury pollution patter similar changes were found.

In the evaluation of the results of the investigations on metal pollutant mass fluxes and material balances for the two most dangerous metals, mercur and cadmium were prepared.

At both Sajópüspöki and Alsózsolca there are gauging stations to measure water discharge continuously. Regarding the fact that the polluting sources of these two metals are between these two points, the material balances could be calculated.

Our work included calculating mass fluxes for material balances on the basis of different data series. Varying of mass fluxes at selected water discharges is especially remarkable, as it proves accumulation and stirrin up processes discussed above.

Calculation of material balance for mercury pollution indicates that effluent discharge from a chemical factory forms the most important polluting source on the Hungarian section.

Using average concentrations of weekly measurements and time compensated sewage quantities average mercury load is found to be 53 mg/s. All the other mercury loads on the Sajó section between the two locations represent only 7 per cent of the load from the main pollution source, so in the further calculations these will be neglected.

Results were divided into two groups on the basis of streamflow. In 1974 the water discharge was lower than the mean value during months, and exceeded the mean streamflow during 5 months. Using the equation:

amount crossing the border + pollution source = reserved in the exit section + difference

entering and leaving mass quantities at lower water discharges than the mean are as follows

$$\begin{aligned} 587\ \text{kg} + 973.5\ \text{kg} &= 769.2\ \text{kg} + K \\ K &= 791.3\ \text{kg} \end{aligned}$$

It means that the quantity of mercury rejected in the system - accumulated in the bottom sediment - is 791.3 kg.

At water discharges higher than the average entering and leaving mass quantities during the 5 months are as follows

$$\begin{aligned} 336\ \text{kg} + 695\ \text{kg} &= 1568\ \text{kg} + K \\ K &= -\ 537\ \text{kg} \end{aligned}$$

i.e. mercury quantity leaving the system by stirring up of bottom sediment amounts to 537 kg.

To sum up mercury pollution, during 1974 923 kg of mercury arrived in the Sajó at Sajópüspöki, 1668 kg Hg entered the river from the Hungarian discharger and 2337 kg left the system at the exit cross section. This means that downstream of the sewage inlet 254 kg mercury, 15 per cent of the Hungarian mercury load accumulated in the bottom sediment.

Material balance for cadmium can be calculated in the same way as that for mercury. There exists only one major pollution source which discharges 421 mg/s cadmium on annual average. The proportion of the total amount of all the other cadmium loads is only 2.8 per cent that is why those sources were neglected. Material balance of cadmium pollution at water discharges lower that the mean flow:

$$\begin{aligned} 177\ \text{kg} + 7747\ \text{kg} &= 1544\ \text{kg} + K \\ K &= 6380\ \text{kg} \end{aligned}$$

thus 6380 kg cadmium is retained in the system.

For discharges exceeding the mean flow (during 5 months):

$$\begin{aligned} 1568\ \text{kg} + 5529\ \text{kg} &= 3550\ \text{kg} + K \\ K &= 3551\ \text{kg} \end{aligned}$$

This means that cadmium accumulates in the bottom sediment even at higher stream discharges. Compared to pollution inlets, however, accumulation takes only 64 per cent of pollution, while during low waters 82 per cent is typical. The difference is still greater if we take into account that the background pollution crossing the border is by one order of magnitude larger at high waters than at low discharges.

To sum up the 1974 material balance of the cadmium pollution it can be stated that besides the 1746 kg of cadmium crossing the border section, the Sajó river is loaded by 13277 kg cadmium from the Hungarian discharger. 9933 kg cadmium accumulates in the bottom silt.

From the presented results it is concluded that in case of metal pollutants complex water quality investigations should be performed which include studying the dissolved and solid forms. Analysis of suspended and bottom sediments has an important role in evaluating material balances.

Key words: sediments
heavy metals
uptake
release
bottom silts

# THE EXCHANGE OF $^{65}Zn$ BETWEEN SUSPENDED SEDIMENT AND ALGAE

P.C. Keulder

Department of Botany, University of the Orange Free State, Bloemfontein, South Africa

## Summary

An account is given on the adsorption of $^{65}Zn$ on montmorillonite and illite as reference clays. The adsorption isotherms are determined under standard conditions, as well as the influence of various ions, which are present in algal growth media, on the adsorption isotherms.

It is followed up by experiments on the exchange of $^{65}Zn$ between the alga Scenedesmus obliquus (Turp.) Krüger and the clay types montmorillonite and illite. These experiments are done in batch cultures and the influence of hydrogen, calcium, magnesium, sodium, potassium, cupri, ferrous and phosphate ions are determined on the exchange of zinc.

## Introduction

Suspended colloidal material, mainly as clay minerals, could play an important role in the nutrition of planktonic algae by exchange processes in the surface layers of water bodies.

Clay minerals in the Orange River System, where such a suspension occurs almost trhoughout the year, were identified as montmorillonite and illite, both oh the three layer type. Organic particulate matter was present in a neglible small quantity.

In this paper the adsorption of one element, zinc, is studied under controlled conditions and the exchange of the zinc ions between the clay particles and a unicellular alga is followed.

Zinc was studied, since it is an useful indicator of contact with galvanized water pipes in urbanised areas, and a very convenient radio-isotope ($^{65}Zn$), a gamma emitter with a halflife of 245 days, is available.

## Materials and Methods

The green unicellular alga, Scenedesmus obliquus (Turp.) Krüger was used as the test organism. It was chosen since it could stand rather severe conditions in the chemical environment without drastic effects. The reference clays, montmorillonite (No 20) and illite (No 36) were purchased from Wards Science Establishment and it was saturated with potassium ions.

Zinc uptake experiments were done in batch cultures (2 litre) of Scenedesmus obliquus in a nutrient medium described by Soeder et al (1967). Control treatments were without clay and $^{65}Zn$ (carrier free) was added at an activity of 50 µCi/ℓ. A sample taken during the experimental growth phase was centrifuged at 3500 x g for 15 minutes, washed twice with water, resuspended and the radio-activity counted with an autogamma-spectrometer.

The same procedure was followed with the clay treatments of the uptake experiments, except for the addition of 0,8 mg/ℓ clay to the cultures. The algae and clay were fractionated, before the radioactive counting, applying the density gradient technique (Lammers, 1962). The gradient composition was 80% sucrose and 30% Ficoll and samples were centrifuged at 4500 x g for 25 minutes.

In the adsorption experiments clay was suspended in Soeder's medium and agitated for 48 hours in the presence of 50 µCi/ℓ $^{65}Zn$. The adsorbed zinc was regarded as the total of the fixed zinc (fraction built into the lattice structure of the clay mineral) and the exchangeable zinc. The latter was measured as the radio-activity of the precipitate after the centrifugation of a sample at 3500 x g for 15 minutes.

In all the experimental treatments the pH of the nutrient medium was kept constant at a value of 7,5. Samples were aerated to ensure that anoxic conditions could not develope.

## Results and Discussion

### Effect of pH on zinc adsorption and uptake

A competition between the adsorption of hydrogen and zinc ions was indicated by McLean & Owen (1969) where they have measured an increase in cation exchange capasity of clay minerals with an increase in pH. The same phenomenom was observed with the reference clays illite and montmorillonite (figure 1). At a low hydrogen ion concentration a preference to zinc adsorption at montmorillonite above illite was observed, possibly due to the relative high cation exchange capasity of montmorillonite (Grim, 1953).

The uptake of zinc by Scenedesmus obliquus at a pH lower than 7,0 was dominated by hydrogen ions (figure 2). Similar results have been obtained by Rains et al (1964), Gutknecht (1961) and Rathore et al. (1970) with different

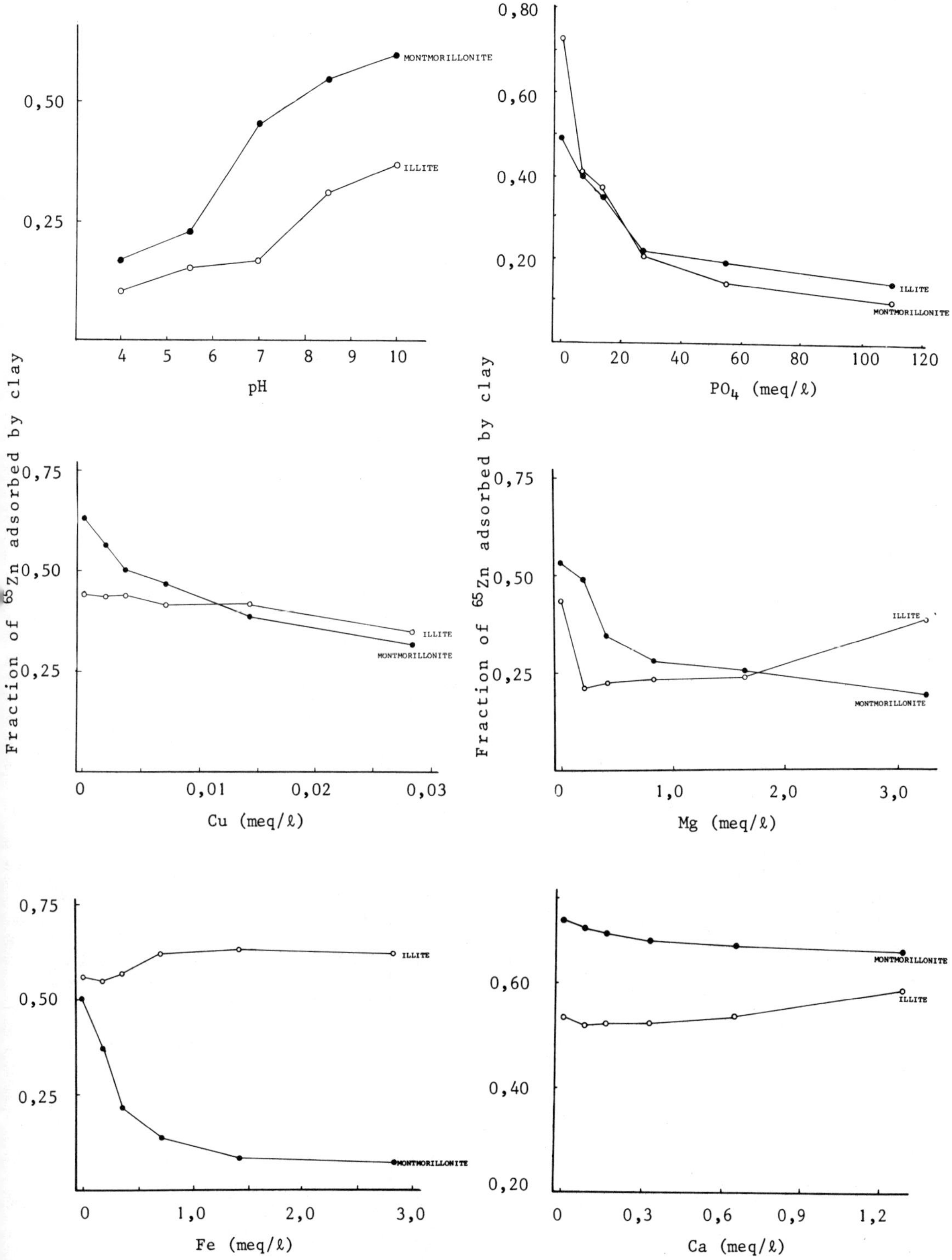


Figure 1 The influence of hydrogen, phosphate, cupri, magnesium, ferrous and calcium ions on the adsorption of zinc on montmorillonite and illite.

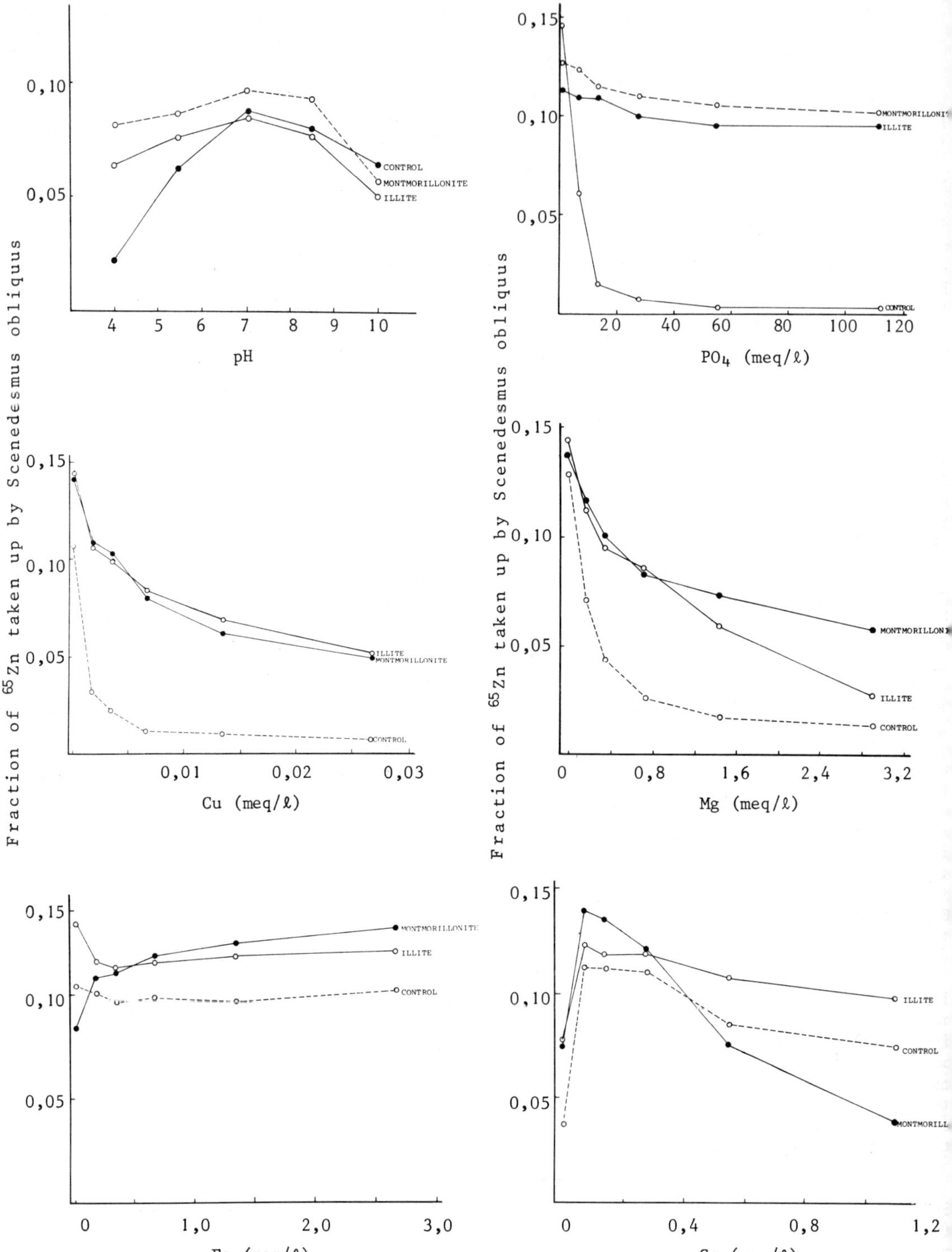


Figure 2 The influence of hydrogen, phosphate, cupri, magnesium, ferrous and calcium ions on the uptake of zinc by Scenedesmus obliquus in the presence of the clay minerals illite and montmorillonite. Control treatment without clay.

species of seed plants. At a pH above 7,0 the selectivity decreased due to the insolubility of zinc.

The uptake of zinc was stimulated with both clay treatments at a low pH. This stimulation in zinc uptake was due to the selectivity of the clay minerals for hydrogen during the adsorption process. Consequently more zinc was made available which could be taken up by the algae. No significant differences in zinc uptake between the control and clay treatments were measured at a high pH.

Effect of phosphate on zinc adsorption and uptake

The adsorption of zinc was considerably reduced in the presence of very low concentrations of phosphate ions at the montmorillonite treatment, but the effect was less spectacular with illite. This difference could be caused by the restricted interlayer volume of illite, in contrast to montmorillonite with its interlayer swelling character which allows for the occupation of phosphate ions.

The uptake of zinc ions by seed plants is also a process which is dominated by excessive quantities of phosphate (Rogers and Wu, 1948; Brown and Tiffen, 1962). In figure 2 an intensive competition was measured in the uptake of zinc and phosphate ions by Scenedesmus obliquus. The presence of the clay minerals, however, suppressed the competition effect. This suppression was due to the high affinity of the clay minerals for phosphate ions, with the result that phosphate ions were fixed to the clay and was not available to interfere with the uptake process. The difference in the adsorption character of montmorillonite and illite was not reflected in their effect on the uptake of the zinc by the algae.

Effect of copper and magnesium on zinc adsorption and uptake

Illite and montmorillonite have shown a lesser selectivity in regard to copper and magnesium ions than phosphate ions (figure 1) in respect to their adsorption. Copper and magnesium ions had only a small reducing effect on the adsorption of zinc. At a high concentration of magnesium, however, a synergestic effect was measured in regard to zinc adsorption.

The uptake of zinc was considerably reduced in the presence of copper and magnesium ions (figure 2). The clay minerals reduced this competition effect in the presence of copper. At high concentrations of magnesium ions in the illite treatment, however, this effect has shown no significant difference with the control treatment. The synergestic effect of a high activity of magnesium ions lead to a high adsorbed fraction of zinc ions, which caused only a slight stimulation of zinc uptake by the algae. The adsorbed zinc in the presence of a high magnesium activity was thus only partially available for the algae.

Effect of iron on zinc adsorption and uptake

Montmorillonite and illite differed considerably in respect to the effect of ferrous ions on the adsorption of zinc (figure 1). Montmorillonite has shown the typical competition effect as measured above, but no competition was indicated in respect to illite in the presence of ferrous ions. The ferrous ions could not replace the zinc ions adsorbed by the illite particles.

The basic uptake of zinc is not effected by ferrous ions (Malavolta et al, 1956) and similar results were obtained with Scenedesmus obliquus (figure 2). Although there was a remarkable difference in the adsorption character of illite and montmorillonite, these clay minerals had both the same stimulating effect on zinc uptake in the presence of ferrous ions. This stimulation was not as effective as measured in previous experiments.

Effect of calcium on zinc adsorption and uptake

Although montmorillonite was more effective than illite in its capasity to adsorb zinc ions, calcium concentration had no effect on the zinc-calcium ratio (figure 1). No competition occurred between calcium and zinc ions for adsorption sites on the clay minerals. It seemed that also in these experiments, as well as those described above, montmorillonite had a greater capasity to accomodate zinc ions than illite. This phenomenom is reflected in the relative high cation exchange capasity of montmorillonite.

The uptake of zinc was stimulated by the presence of calcium ions (figure 2). In the absence of calcium in the nutrient medium, the uptake of zinc was reduced considerably due to the decrease in ion retention of the cell membranes (Foote and Hansen, 1964).

Low concentrations of calcium has a stimulating effect on zinc uptake by Scenedesmus obliquus, due to its function of retaining of the zinc in the cells. At high calcium concentrations, however, a definite competition with zinc was measured.

The uptake of zinc was not altered in the presence of the clay minerals. This was possibly due to the fact that calcium had no effect on the availability of adsorbed zinc ions in the adsorption experiment.

## Conclusion

From the above results it is clear that

suspended minerals does certainly plays a role in the mechanism of mineral nutrition of planktonic algae. In the case of zinc, the presence of the clay particles made the ions more available in the most of the experiments, acting as a nutrient source. There was, however, treatments, although limited in number, where the clay acted as a nutrient sink.

The question whether the suspended clay particles acts as a nutrient source or sink, could be answered that the process is coupled to the specific ion ratio and the composition of the clay mineral species. Intensive investigations on ion types which are important in relation to eutrophication should be continued on the basis of these zinc experiments.

## References


Brown, J.C. and L.O. Tiffen, 1962. Zinc deficiency and iron chlorosis dependent on the plant species and nutrient-element balance in Tulare clay. Agron. J. 54: 356-358.

Foote, B.D. and J.B. Hansen, 1964. Ion uptake by soybean root tissues depleted of calcium by ethylenediaminetetra-acetic acid. Pl. Physiol. 39: 450-460.

Grim, R.E. Clay Mineralogy. McGraw-Hill, New York

Gutknecht, J., 1961. Mechanism of radioactive zinc uptake by Ulva lactuca. Limn. Oceanogr. 6: 426-421.

Lammers, W.T., 1962. Density gradient separation of plankton and clay from river water. Limn. Oceanogr. 7: 224-229.

Malavolta, E., J.P. Arzolla and H.P. Haag, 1956. Preliminary notes on the adsorption of radiozinc by young coffee plants grown in nutrient solution. Phyton 6: 1-6.

McLean, E.A. and E.J. Owen, 1969. Effects of pH on the contributions of organic matter and clay to soil exchange capasities. Soil Sc. Soc. Am., Proc. 33: 855-858.

Rains, D.W., W.E. Schmid and E. Epstein, 1964. Absorption of cations by roots. Effect of hydrogen ions and essential role of calcium. Pl. Physiol. 39: 274-278.

Rathore, V.S., S.H. Wittwer, W.H. Jyung, Y.P.S. Bajaj and M.W. Adams, 1970. Mechanism of zinc uptake in bean tissues. Physiol. Plant. 23: 908-919.

Rogers, L.H. and C. Wu, 1948. Zinc uptake by oats as influenced by application of lime and phosphate. J. Am. Soc. Agron. 40: 563-566.

Soeder, C.J., G. Schulze, and M. Thiele, 1967. Einflusz verschiedener Kulturbedingungen auf der Wachstum in Synchronkulturen von Chlorella fusca (Sh. et Kr.). Arch. Hydrobiol. (Suppl.) 33: 127-171.



Key words: zinc
adsorption
suspended sediment
algae
clay minerals

# RELEASE OF OXYGEN-CONSUMING AND TOXIC SUBSTANCES FROM ANAEROBIC SEDIMENTS BY WHIRLING-UP AND AERATION

D. Müller & U. Schleichert

Federal Institute of Hydrology, Koblenz, Germany

Summary

By means of a plain example is illustrated, that the whirling-up of anaerobic sediments in consequence of hydraulic phenomena releases large quantities of oxygen-depleting substances; under unfavorable conditions, the oxygen equilibrium of the river thus affected may be disturbed.

Own investigations are discussed, showing that

- in the case of sediments with an extremely high heavy metal content, the transition from the aerobic to the anaerobic state may lead to a toxic inhibition of the activity of water bacteria;

- in the simultaneous presence of sparingly degradable organic compounds and high heavy metal contents, the increase in bacterial activity associated with the adaptation of water bacteria to sparingly degradable compounds, and the inhibition of bacterial activity which depends on the heavy metal content, may become superimposed on each other, with the remaining bacterial activity being determined by the predominating process.

While the main problem connected with the whirling-up of oxygen-consuming substances is quantification, the processes associated with the release of toxic substances during the transition to the aerobic state are so complex that reliable predictions cannot be made at this time.

## Introduction

Sedimentation is one of the most important self-purification phenomena in flowing water. The metabolism of streams is essentially influenced by the following basic processes (Eppler & Heiß, 1975):

"(1) Reduction of the BOD in the overlying water body by sedimentation and adsorption of organic substances.

(2) Increase of the BOD in the water body by the whirling-up of settled sediments as a result of hydraulic phenomena.

(3) Increase of the BOD by diffusion of partially degraded organic substances into the overlying water.

(4) Oxydation of gaseous end products of anaerobic degradation which diffuse from lower strata into the aerobic layer.

(5) Oxygen transfer into the upper layer of the bottom sludge to meet the needs of aerobic degradation and respiration of the fauna inhabiting the said layer."

The following paper deals primarily with point (2) of this list, i.e., with the increase of the BOD in the water body by the whirling-up of settled sediments as a result of hydraulic phenomena. It will be shown that the hydraulic whirling-up of settled sediments can have still another effect on the metabolism of the stream; this effect is listed unter point (6) below:

(6) Release of toxic substances by whirling-up and aeration.

Our investigation was prompted by the special problems encountered by the Water Ways Administration of the Federal Republic of Germany wherever streams with heavy sludge deposits are used as navigable waterways or where sedimentation is influenced by regulation measures.

## Influence of river sludge whirled up by floods on the oxygen balance

It is evident that this influence does exist. Scientists concerned with mathematical water quality models have considerable difficulties with the quantification of this influence, since it depends to an exceptional extent on a multiplicity of hydrological and biological parameters. To discuss details of the mathematical treatment of this problem would go beyond the thematical framework of the Symposium. A plain example of this sediment influence on water quality will be of greater interest in the present context.

On 9 June, 1971, an extensive fish kill occurred in the Middle and Lower Rhine in consequence of river pollution by putrescible organic substances and the resultant complete oxygen depletion (Kleinsteuber, 1971). Fortunately, an oxygen depletion of this extent has, so far, been a solitary event in the Rhine river. As, however, the causal links leading to its occurrence are of exemplary importance, it may be worthwhile to treat this unique event in greater detail.

Fig. 1 shows water level, suspended sediment content and oxygen content of the Rhine during the period concerned. The unbroken line in the upper graph represents the water level which, originally, was between the longterm means MW (mean water level) and MNW (mean lowest water level). As a result of heavy rain in the catchment areas of several tributaries, the water level rose unusually fast within two days to values above MW. Flow doubled at the same time and reached a maximum of approximately 3000 m³/s.

A tremendous increase of the pollution load is typical of such a quick rise of the flow rate. In Figure 1, this pollution load is indicated in the form of suspended sediment values which are plotted as columns.

Simultaneously with the rise of the water level and the intensification of suspended sediment transport, the oxygen content dropped to 0 mg/l. For 24 hours, the Rhine remained free from oxygen. The oxygen-free zone then moved downstream and dissolved in the Lower Rhine.

The event confronted us primarily with two questions, namely: 1) Where may such masses of sludge come from? 2) Why does an oxygen depletion of this extent not occur more frequently with rising floods?

Fig. 1

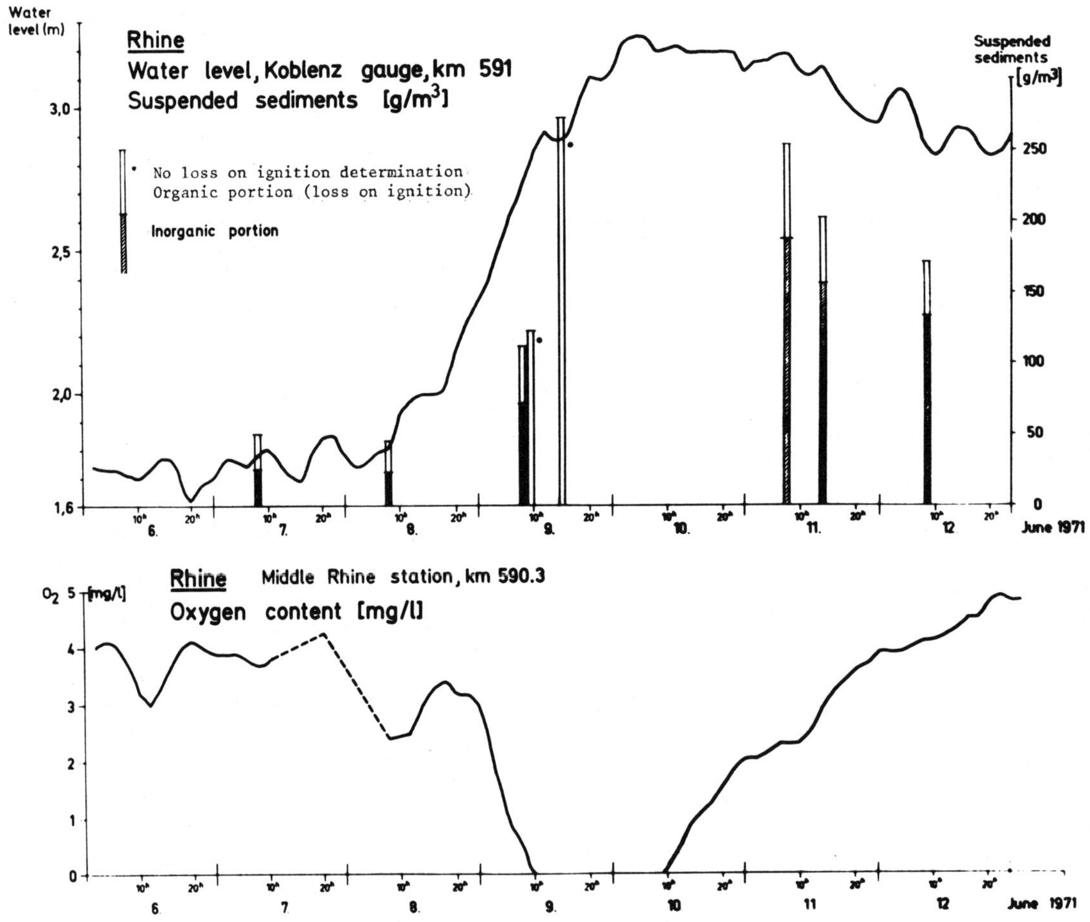

On the first question: The large tributaries of the Rhine are slow-flowing streams with relatively large quantities of sludge at the barrages along their course (Hellmann, 1972). Actionally the sludge quantities whirled up were rather small according to sludge probings. This, however, does not mean much in the context of our considerations. As according to Imhoff (1976) 1 litre of freshly whirled-up river sludge can consume within 24 hours an amount of oxygen equal to the quantity of dissolved oxygen in 100 l of water; sludge volumes that may seem small to the hydraulic engineer can already exert a catastrophic influence on river biology.

On the second question: The hydraulic whirling-up of river sediments normally does not have the catastrophic effects it produced in the example just described. The reasons are as follows:

(1) Normally, the flood waves in our rivers rise at a slower pace. Their erosive effect is, therefore, relatively small. As a result, the increased oxygen demand due to the whirling-up of sediments is usually compensated by increased flow, i.e., by dilution.

(2) The heavy rains which led to the rapid increase in flow occurred mainly in the catchment areas of those tributaries of the Rhine river which flow through agglomeration areas. In these agglomeration areas, waste water handling facilities, in particular storm water retention basins, had not been sufficiently developed at the time. If precipitation occurs evenly distributed over the entire catchment area of the Rhine, floods from rivers with heavy waste water and sediment loads are diluted by floods from less polluted rivers.

(3) In our country, floods in summer are rare. In the presence of winter temperature, the oxygen-consuming processes are much slower.

### Release of oxygen-consuming and toxic substances after aeration during a laboratory test

The example of the oxygen depletion in the Rhine river shows that hydraulic whirling-up of sediments may cause a release of large quantities of oxygen-consuming substances, if these sediments are rich in organic matter. As waste water in industrial agglomeration areas frequently has a high content of potentially toxic heavy metals besides large quantities of organic substance, it is of interest to know the influence which the sediments formed by the component substances of such water have on the metabolism of streams when they are whirled-up.

The metal contents of the sediments that are of interest here and which originated from the mouth of a tributary of the Rhine river, are shown in Figure 2. The figure also shows the heavy metal contents of the tributary in comparison with comparable sediments from the Rhine. In addition, the geochemical background is indicated. It can be seen that the heavy metal load (lead, zinc, copper and nickel) of the sediments from the tributary is higher by a factor of 5-10 than that of the Rhine river sediments.

Fig. 2

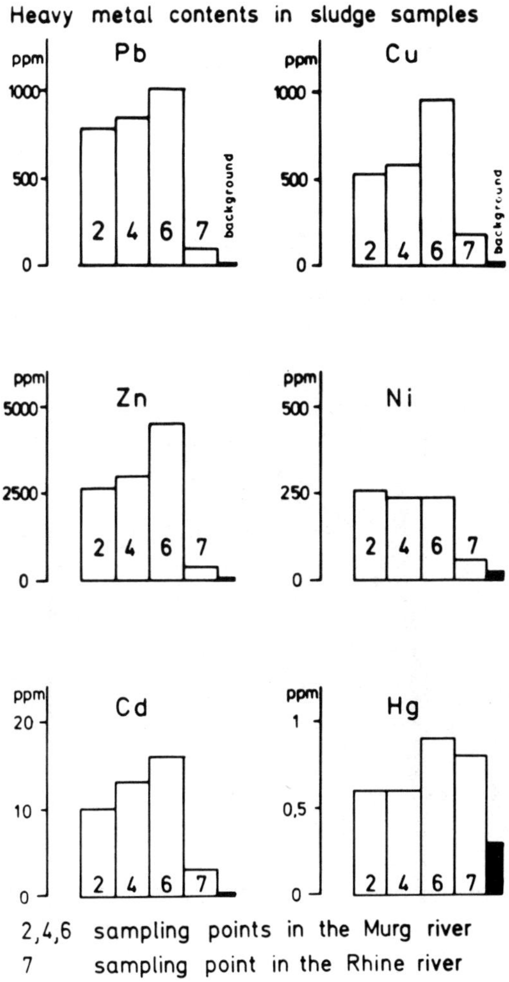


The effects such heavily polluted sediments have after whirling-up on the most important reaction of biological self-purification, namely bacterial oxygen consumption, were investigated at the laboratory.

The biological test used was the assimilation and depletion test (DEV, 1971) in which, inter alia, the influence of the sediments on the oxygen consumption associated with bacterial peptone degradation is determined. The test employs a dilution series which allows extrapolation of the test results toward the actual in situ dilution.

Figure 3 shows the effect of sediments on the test system. Sediment concentrations in g of dry weight per litre of dilution water are plotted on the abscissa. On the ordinate, the increase in bacterial oxygen consumption is recorded upward, and the inhibition downward. The results shown in Figure 3 are obtained through a test in which river sludge, anaerobic at the beginning, was aerated for a few hours or days.

Fig. 3

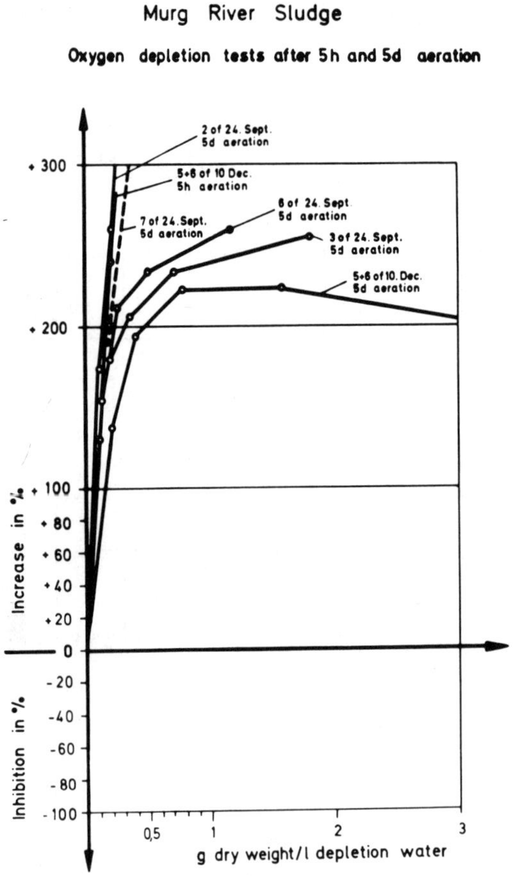


If concentrations are low, all sediments so aerated effect a considerable intensification of bacterial oxygen consumption. At these concentrations, sediments with severe heavy metal pollution do not differ from the less heavily polluted Rhine river sediments (broken line). With high sediment concentrations, however, the course of the depletion curve for these sediments from the tributary which were aerated for five days indicates an inhibiting effect. This effect is revealed by the fact that the depletion curve of these sediments does not rise proportionally to the quantity of the substances involved. If the tests are continued and the whirled-up sediments are aerated for additional time periods, the effects shown in Figure 4 take place. (Aeration periods of several weeks are of interest in cases in which the sediments remain aerobic after renewed deposition, or where the time of travel and presence in inland waters are relatively long.)

Fig. 4

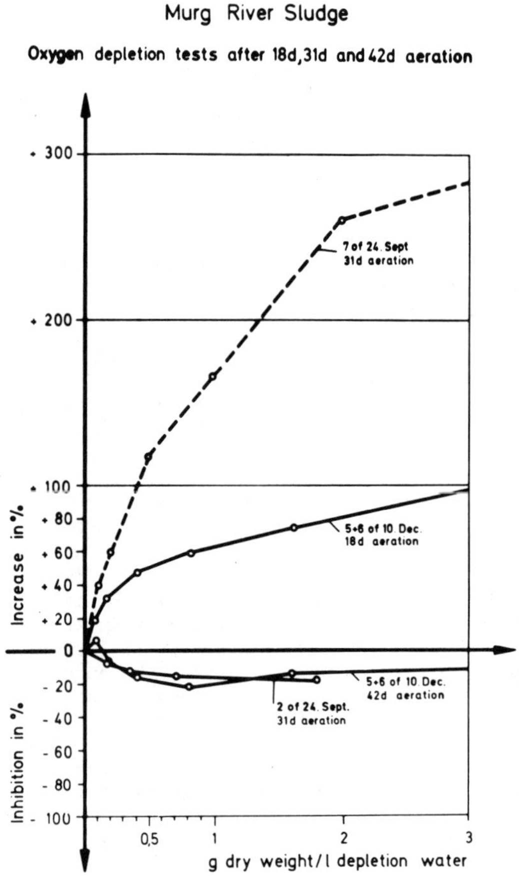

In the case of sediments from the Rhine river, the promotion effect decreases with increasing aeration time, as expected, because biological degradation of organic substances takes place during such periods. With sediments from the tributary, the same effect is noticed at the beginning; a promotion effect is still apparent after 18 days. However, if aeration continues, the effect is reversed into an inhibiting effect.

This toxicological behaviour of these river sediments can be explained as follows: the sludge of this river consists of a mixture of oxygen-consuming and toxic substances. The portion comprising the oxygen-consuming substances causes an increase in bacterial oxygen consumption, and the toxic portion an inhibition of bacterial activity. At the beginning of the test, which is equivalent to the whirling-up of the sludge in consequence of hydraulic phenomena, and with the resulting transition to the aerobic state, the consumption-promoting effect of degradable substances prevails at first. In the course of the aeration or with aerobic conditions in the river, the content in organic substances decreases quickly. As a result, potentially toxic heavy metals which are linked to the organic material in one way or the other, may be released.

This means that, besides consumption-promoting substances, the hydraulic whirling-up of river sediments may also release toxic substances which, due to their chemical properties as sparingly soluble compounds, are present in anaerobic river sediments.

For the detailed study of the changes in heavy metal solubility and of the effects on micro-organisms associated with the transition to the aerobic state, a second laboratory test was made, using sediments from another tributary of the Rhine. In contrast to the tributary from which the sediments for the aforementioned first test were taken, the second tributary is severely polluted with heavy metals and organic substances from waste waters discharged by the chemical industry.

Aeration of the originally anaerobic sediments was effected in the same way as in the first test. Additionally to the depletion test, biochemical oxygen demand, loss on ignition, heavy metal contents, and a number of other chemical parameters were determined. Figure 5 shows the values of three parameters, namely: loss on ignition as a measure of the organic substance, BOD, which characterizes the degradable portion of the organic substance, and the depletion curve which describes the effect on bacteria.

Fig. 5

**Changes induced by aeration of Main river sludge**

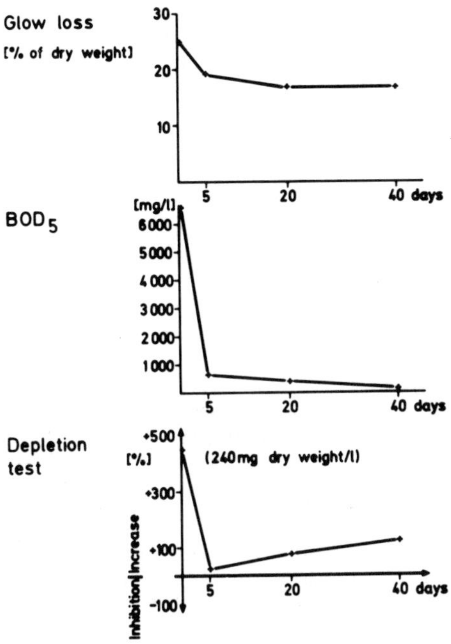


Loss on ignition and $BOD_5$ were as expected. Already five days after the beginning of the test, a large portion of the organic substance had been degraded. Further degradation proceeded only slowly. The effect on bacteria was similar during the first five days: the strong promotion effect decreased quickly. From the 5th to the 40th day, however, the characteristic depletion curve took an unexpected course. Instead of continuing to decline as in the case of the sediments of the first example, it began to rise again and continued to rise right up to the end of the test.

Considering the high content of organic substances from the chemical industry which the sediments used in this test had, the result can be interpreted as follows: after a decrease of the depletion-promotion effect during the first days of the test, the bacteria population of the aerated sediment adapted to the sparingly degradable compounds and degraded them partially. This means that part of the compounds that were sparingly degradable at the beginning, were in a degradable state during the depletion test.

The importance bacterial adaptation has in this context must be clarified by specific investigations. In the present study, primary interest concerns heavy metal behaviour which played a decisive role in the first test.

Figure 6 shows the result of the heavy metal analyses. For these analyses, the aerated sludge-and-water mixture was centrifuged after 1, 5, 20 and 40 days, and the supernatant water filtered through a membrane with pore sitz 0.45 µm. The sludge residue was dried and determined by way of fluorescent X-ray analysis. The membrane filtrate was analyzed for Hg, Cr, Cd, Cu and Pb by flameless atomic absorption. Zinc was determined by atomic absorption spectroscopy and calcium compexometrically. The glow loss was determined by one-hour glowing at 550°C.

After five days of aeration, the loss on ignition diminished from about 25% to 19% in accordance with the heavy reduction in BOD. The simultaneous increase in the quantity of the dissolved heavy metals zinc, lead, copper, and chrome may, therefore, be interpreted as indicating the release of organically bound heavy metals. Mercury and cadmium are not affected in this way. After twenty days of aeration, a reduction in dissolved heavy metals, again with the exception of cadmium and mercury, was observed. After additional twenty days of aeration, the heavy metal contents, with the exception of mercury and chrome, increased again, in some cases heavily. For example, the zinc content increased to 4660 µg/l.

The explanation of this unexpected behaviour may be found in the calcium content of the solution. In the course of aeration, calcium changes from the solid phase into the dissolved state. The chemical process involved consists of the transformation of calcium, present in the form of carbonate, into the more easily soluble calcium hydrogen carbonate by carbon dioxide which is released by the biological degradation of organic substances. In correspondence with the increase of the Ca contents in the dissolved phase, the Ca content decreases in the solid phase. The different development of the dissolved heavy metal contents is explained by the fact that the decisive process in the first phase is the release of organically bound metals. In the second phase between the 5th and 20th day of aeration, the process is practically completed. Now, precipitation of the released heavy metals in the form of carbonates prevails, and the contents diminish accordingly. During the third aeration phase between the 20th and the 40th day, heavy metals which form easily soluble hydrogen carbonates dissolve again to a larger extent, parallel to calcium, if the $CO_2$ supply persists. Chrome and calcium do not follow this pattern.

Calculation of heavy metal ppm after 40 days dissolved from the solid phase reveals the following gradation for zinc, cadmium, lead and copper which corresponds to the solubility of their carbonates:

| | Zn | Cd | Pb | Cu |
|---|---|---|---|---|
| µg/ppm | 1.9 | 0.4 | 0.2 | 0.1 |
| $-\log L_{MeCO_3}$ | 10.8 | 11.3 | 13.1 | 9.6 |
| $-\log L_{Me(OH)_2CO_3}$ | | | | 33.8 |

Before discussing the possible connexion between the results of the heavy metal analyses and the results of the toxicity tests some remarks on the heavy metal contents in the sediment, which were determined by way of fluorescent X-ray analysis during the aeration period, would seem to be indicated.

With three exceptions, the heavy metal contents of the dissolved phase, increased by re-dissolution, do not appear in the reduced contents of the sediment residue. This is due to the high solid content of 30 g/l and to the nature of fluorescent X-ray analysis which includes an error of ± 5 %. For example, in the case of copper with 400 ppm in the solid phase, 120 µg/l would have to dissolve before the reduction in the said solid phase can be registered with certainty.

The connexion between heavy metal content and the result of the depletion test, which is obvious on the first and the fifth, and even on the twentieth day of the test should be explained. During the first days of thee test, the degradation of sparingly degradable compounds and the associated release of heavy metals result in a pronounced reduction of the depletion-promoting effect. The precipitation of heavy metal carbonates during the second phase reduces the depletion-inhibiting influence.

If the heavy metal content should be the only decisive factor, a reduction of the promoting effect, and perhaps even an inhibiting effect, could be expected during the third phase which is characterized by increased dissolution of heavy metals. However,

Fig. 6

Changes in heavy metal contents induced by aeration of Main river sludge

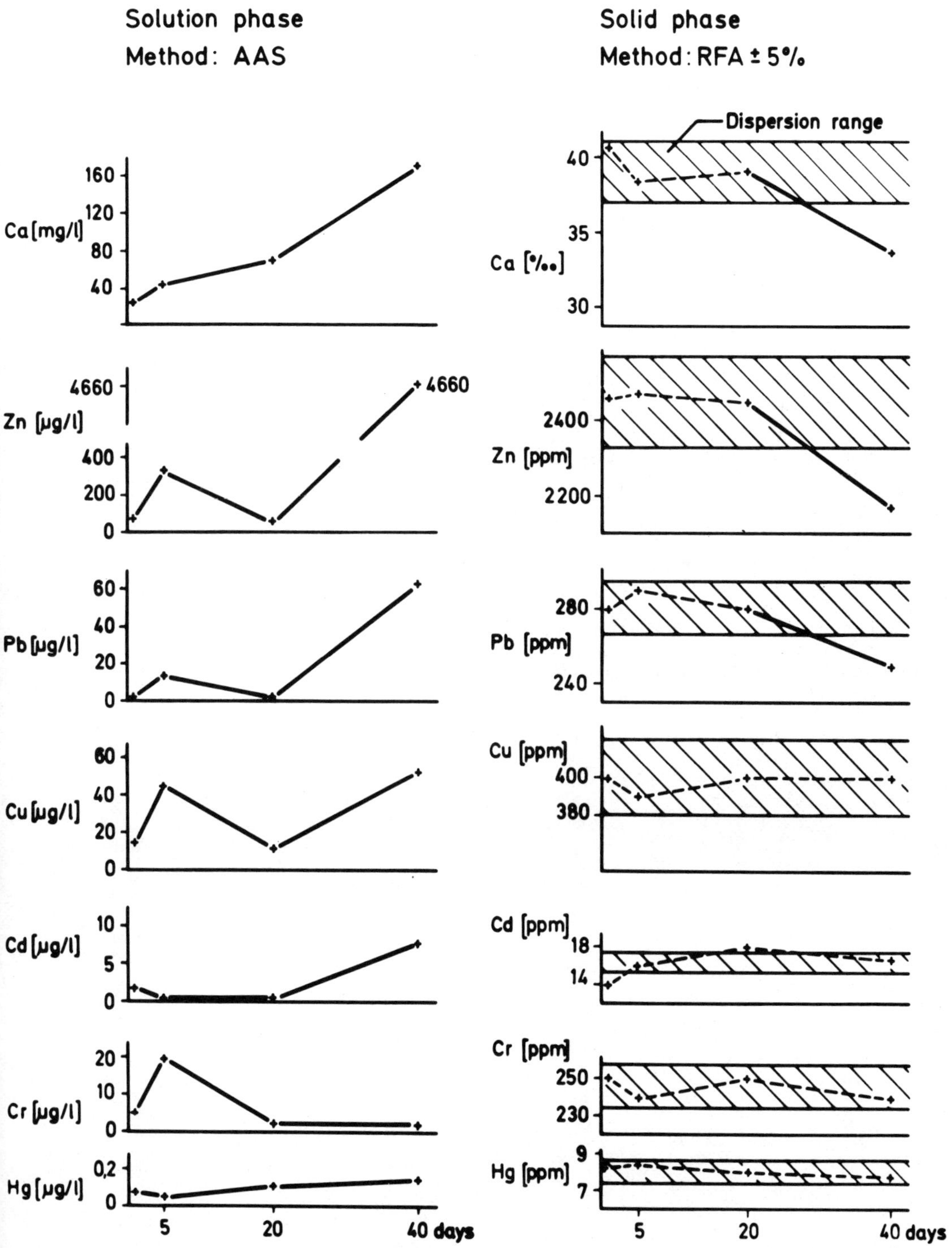

the depletion test reveals that the promoting effect increases during this time period. Therefore, the effect of the changing heavy metal contents on the bacterial degradation which takes place after the transition of the sediment into the aerobic state can, as yet, not be assessed unambiguously. Further investigations are required to clarify the interaction between sparingly degradable organic substances, heavy metals and easily degradable matter.

## References


DEV, 1971. Deutsche Einheitsverfahren zur Wasseruntersuchung. Verlag Chemie, Weinheim.

Eppler, B. & H.-J. Heiß, 1975. Einfluß des Bodenschlamms auf den Sauerstoffhaushalt von Gewässern. - Anhang zum Forschungsbericht SFB 80/E/70. Ausbreitungs- und Transportvorgänge in Strömungen. Universität Karlsruhe.

Hellmann, H., 1972. Herkunft der Sinkstoffablagerungen in Gewässern. 1. Mitteilung: Chemische Untersuchungen der Schwermetalle. DGM 16, H. 1, S. 14-18.
2. Mitteilung: Überlegungen und Ergebnisse aus der Sicht der Abwassertechnik. DGM 16 (1972), H. 5, S. 137-141.

Imhoff, K. & K., 1976. Taschenbuch der Stadtentwässerung.Oldenbourg Verlag, München.

Kleinsteuber, H., 1971. Dokumentation zum Fischsterben im Rhein vom 8.-10. Juni 1971. - Bundesanstalt für Gewässerkunde, Koblenz.



Key words: anaerobic sediment
reaeration
remobilisation
heavy metals
toxicity for bacteria
oxygen consumption

# THE IMPACT OF TACONITE TAILINGS ON DISSOLVED SOLIDS CONCENTRATIONS IN LAKE SUPERIOR

Russel H. Plumb, Jr., U.S. Army Corps of Engineers Waterways Experiment Station, MS-USA

G. Fred Lee, Center for Environmental Studies, University of Texas at Dallas, Richardson, TX, USA

## Summary

An iron ore mining beneficiation plant discharges approximately 60,000 metric tons per day of taconite tailings (finely divided crushed rock) to Lake Superior. These tailings are composed primarily of quartz and an amphibole cummingtonite, a few percent iron minerals and other trace minerals. Several years leaching tests have shown that taconite tailings solubility ranges from 0.03 to 1.2 percent, with the latter figure representing the maximum possible release under laboratory conditions simulating the Lake Superior environment. Under these conditions the alkalinity in the lake might increase up to 2 mg/l after 50 years of discharge. The total dissolved solids content of the lake is in the order of 57 mg/l. The maximum expected increase in dissolved solids after 50 years of operation would be in the order of 3 mg/l. These computations assume that all dissolved solids added to the lake are conservative, i.e., remain in solution and are only removed by outflow from the lake with no incorporation into the sediments.

A field study demonstrated that the concentrations of various major dissolved solids in the interstitial waters where taconite tailings have been deposited previously were elevated above the concentrations in areas where there had been no taconite tailings deposited. However, measurements immediately above the sediment-water interface showed that the rate of transport from the sediments to the overlying waters was sufficiently slow so that there was no detectable change in the concentrations of dissolved solids in the areas where the sediments were covered with taconite tailings compared to non-tailings area. It is concluded that past and future discharges of taconite tailings would have no significant impact on dissolved solids concentrations in Lake Superior.

## Introduction

Reserve Mining Company of Silver Bay, Minnesota initiated the operation of an iron ore beneficiation plant on the north shore of Lake Superior in 1955. Since then, more than 250 million tons of taconite tailings have been discharged directly to Lake Superior. Approximately one-half of the tailings settled out immediately to form a delta with a surface area of more than 1.3 sq km. The remainder of the tailings were carried to the bottom of a 305 m deep trench and created a benthic deposit covering more than 1,000 sq km (see Figure 1) (Plumb, 1973).

There is concern that the discharge of taconite tailings by Reserve Mining Co. could cause a measurable increase in the dissolved solids concentration in Lake Superior (U.S. Dept. of Interior, 1969a,b). As a result, several laboratory studies have been conducted to investigate the leaching of chemical constituents from these tailings (Plumb, 1973; Plumb and Lee, 1975; Klaysmat, 1970; Lemire, 1970 a,b,c; Glass, 1970). These studies generally show that only a limited release of chemical constituents occurs in Lake Superior water. However, the laboratory tests, while designed to simulate lake conditions, do not duplicate them. Therefore, to assess the applicability of these laboratory studies, several mathematical approaches have been used to estimate the increase in dissolved solids concentrations that would result from the dissolution of taconite tailings and assess the impact of the discharge on Lake Superior. This paper presents a summary of these approaches and a discussion of their results.

## Characteristics of taconite tailings

Taconite tailings are primarily composed of quartz, hornblende, cummingtonite, and a wide variety of

Figure I

Location of Reserve's Tailings Deposit in Lake Superior

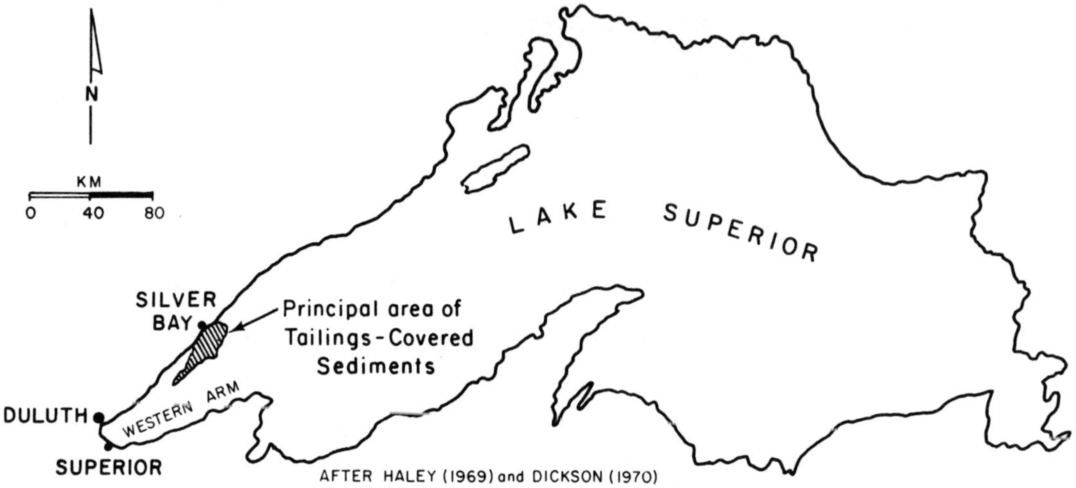


other iron magnesium silicates such as those listed in Table 1.

Table 1. Listing of minerals reported in eastern Mesabi taconite ores*

| Mineral | Formula |
|---|---|
| Quartz-Chert | $SiO_2$ |
| Stilpnomelane | $2(Fe,Mg)O,(Fe,Al)_2O_3$ $5SiO_2 \cdot 3H_2O$ |
| Minnesotaite | Complex Fe silicate structure similar to talc |
| Chlorite | Hydrous silicates of Al, Fe and Mg |
| Pyroxene group | Ferromagnesian silicates |
| Mica group | Ferromagnesian or potassium silicates |
| Feldspar group | Aluminum silicates |
| Actinolite | $Ca_2(Mg,Fe)_5Si_8O_2$ $(OH)_2$ |
| Calcite | $CaCO_3$ |
| Hedenbergite | Ferromagnesian silicate |
| Hornblende | Complex Fe,Mg silicate of Ca,Mg,Na,Ti and Al |
| Cummingtonite | $(Mg,Fe)_7SiO_3$ |
| Ferrohypersthene | Ferromagnesian silicate |
| Fayalite | Ferromagnesian silicate |

*After Gunderson and Schwartz (1962).

Table 2 presents typical chemical analyses of the Reserve Mining Co. tailings which have been discharged to Lake Superior. The beneficiation process involves wet grinding of the ore in ball and rod mills. The magnetic iron is separated from the quartz, iron magnesium silicates and other minerals by permanent magnets. Some hydraulic fractionation is also used involving hydroseparators. No chemicals are added to the ore as part of the beneficiation.

## Review of approaches

### Dissolution across the plant

An approach to estimate the effect of taconite tailings disposal on water quality in Lake Superior is based on the volume of discharge and the concentration changes that occur in the plant process water. This approach utilizes the period of record regarding the amount of dissolution of taconite tailings that has occurred during the processing of ore by Reserve Mining Co. The validity of this approach is that it represents a direct measurement of the dissolved solids increase in the process water that is being discharged to the lake.

Two variations of this method were tried by Gay (1970). In the first case it was assumed that the concentration of any constituent in the discharge would not be affected by any change in the lake. In the second variation Gay (1970) assumed that

Table 2. Typical chemical analysis of Reserve Mining Co. tailings*

| Element | Percentage** |
|---|---|
| Si | 33.59 |
| Fe | 14.93 |
| Al | 0.53 |
| Ca | 1.67 |
| Mg | 1.96 |
| Mn | 0.37 |
| Ti | 0.030 |
| P | 0.026 |
| Na | 0.040 |
| K | 0.040 |
| S | 0.030 |
| Pb | 0.007 |
| Zn | 0.004 |
| Ni | 0.002 |
| Cu | 0.004 |
| Co | 0.002 |
| C | 0.11 |
| H | 0.10 |
| O | 46.54 |
| Mo | <0.001 |
| V | <0.001 |

*After Haley (1969).
**Percentages are based on dry wt. of tailings.

the concentration difference between the plant influent and plant effluent would be constant. Equations for each condition were formulated and solved based on the present changes that occur in the plant process water and the assumption that the plant would operate for periods of 50 and 100 years (In 1955 Reserve Mining Co. had a life expectancy of 35 to 40 years). The predictions based on this approach are presented in Table 3.

The equations developed by Gay (1970) were also used by Klaysmat (1970) to estimate the influence of the tailings discharge on the increase of dissolved solids in the lake. Klaysmat (1970) predicted the changes that would occur in Lake Superior for iron, chloride, sulfate, silica and several species of nitrogen based on concentrations in the Reserve Mining Co. effluent. Again the calculated effect of the tailings discharge was only seen in the second or third decimal place and it was concluded that the continued discharge of taconite tailings would not have a significant effect on the dissolved solids concentrations in Lake Superior even after 100 years of continuous plant operation.

The magnitude of the changes predicted by Gay (1970) and Klaysmat (1970) are such that they would not be detectable in the lake in 100 years. In fact, the calculated changes in Table 3 are generally less than the detection limits of most present day procedures. However, these changes should be considered conservative since they only consider the dissolution of tailings during the time necessary to process the crude ore (approximately 30 minutes). If the tailings continue to dissolve after disposal, then the actual changes may be greater than the predicted changes in Table 3.

Laboratory leaching tests

Glass (1970) presents still another estimate of the dissolution of tailings. This study was conducted with

Table 3. Predicted effect of Reserve Mining Co. discharge on the dissolved solids in Lake Superior water*

| Component | Present intake concentration | Predicted lake value in 50 years of plant operation | Predicted lake value in 100 years of plant operation |
|---|---|---|---|
| | Equation 1 (assumes concentrations in plant effluent remain at present level) | | |
| Dissolved solids (mg/l) | 57.50 | 57.520 | 57.540 |
| Calcium (mg/l) | 13.70 | 13.704 | 13.706 |
| Magnesium (mg/l) | 2.80 | 2.801 | 2.803 |
| Iron (μg/l) | 5.00 | 5.022 | 5.039 |
| Manganese (μg/l) | 1.00 | 1.136 | 1.241 |
| | Equation 2 (assumes incremental increase in plant process water remains constant) | | |
| Dissolved solids (mg/l) | 57.50 | 57.520 | 57.660 |
| Calcium (mg/l) | 13.70 | 13.756 | 13.780 |
| Magnesium (mg/l) | 2.80 | 2.808 | 2.815 |
| Iron (μg/l) | 5.00 | 5.023 | 5.039 |
| Manganese (μg/l) | 1.00 | 1.113 | 1.292 |

*After Gay (1970).

fresh tailings obtained from Reserve Mining Co. that were suspended in Lake Superior water in plastic bottles at 20°C for 332 days. After nearly eleven months, silica, potassium, calcium, magnesium, sodium, suspended solids and total dissolved solids concentrations were determined on the sample. The observed release of soluble material ranged from 11 mg magnesium/kg tailings to 1110 mg total dissolved solids/kg tailings (Table 4).

Effect of calcium chloride. A slightly different approach was to determine whether the calcium chloride used by Reserve Mining Co. to prevent the ore from freezing during winter shipment to the plant site would have a measurable effect on the chloride content of the lake. Assuming a calcium chloride usage of 330 tons/month (U.S. Bureau of Mines, 1968), six months a year for a 16 year period, approximately $1.8 \times 10^{13}$ mg chloride would have been added to the lake. If the effect of the chloride was confined to five percent of the total lake volume, the calculated chloride increase would only be 0.03 mg/l. Such a small incremental change is within the experimental error of the chloride procedure and would probably not be seen. In fact, chloride increases have not been detected in the vicinity of Silver Bay, Minnesota, the site of the Reserve Mining Co. beneficiation plant (Haley, 1969).

Table 4. Predicted effect of tailings discharge to Lake Superior based on EPA report on the dissolution of taconite tailings*

| Species | Increase in mg/kg total tailings | Total mg released, total trailings 14 years | Changes that should have occurred in lake as a result of mg released - µg/l | | | | |
|---|---|---|---|---|---|---|---|
| | | | 100% | 20% | 15% | 10% | 5% |
| $SiO_2$ | 331 | $6.9 \times 10^{13}$ | 5.7 | 28.5 | 38.0 | 57.0 | 114.0 |
| Na | 37 | $8.0 \times 10^{12}$ | 0.6 | 3.0 | 4.0 | 6.0 | 12.0 |
| K | 70 | $1.4 \times 10^{13}$ | 1.2 | 6.0 | 8.0 | 12.0 | 24.0 |
| Ca | 282 | $5.9 \times 10^{13}$ | 4.9 | 24.5 | 32.6 | 49.0 | 98.0 |
| Mg | 11 | $2.1 \times 10^{12}$ | 0.2 | 1.0 | 1.3 | 2.0 | 4.0 |
| S.S. | 61 | $12.0 \times 10^{12}$ | 1.0 | 5.0 | 6.6 | 10.0 | 20.0 |
| T.D.S. | 1110 | $2.3 \times 10^{14}$ | 20.0 | 100.0 | 133.0 | 200.0 | 400.0 |

Assumptions:

1. All tailings respond as study suggests.
2. Tailings production is $210 \times 10^6$ tons since 1955.
3. The volume of Lake Superior is $1.25 \times 10^{16}$ l.
4. There is no flow out of Lake Superior.
5. There is instantaneous mixing in the lake for the total lake (100%) volume.
6. For the 20%, 15%, 10% and 5% lake volume calculations, the effect of the tailings is restricted to that portion of the lake volume.

*After Glass (1970).

The total amount of dissolved material that would have been released to the lake was calculated (Table 4) based on the assumptions that (1) all tailings respond as Glass' (1970) results suggest, (2) the production of tailings since 1955 has been 210 million tons and (3) all released material behaved conservatively and remained in the water column. The predicted concentration changes that should have occurred if the effect of the discharge was restricted to a finite fraction of the lake were calculated using the estimated mass release and a value of $1.25 \times 10^{16}$ liters for the volume of Lake Superior. Using this approach, the largest predicted increase was 0.4 mg/l total dissolved solids (Table 4) when the effect of the discharge was hypothetically restricted to five percent of the lake volume.

Estimated changes in the dissolved solids content of Lake Superior water were also calculated based on leaching data from a three year study by Plumb (1973) and discussed by Plumb and Lee (1975). The observed release of material from taconite tailings (expressed as mg material/g dry -325 mesh tailings) and the expected total release of dissolved solids based on a tailings production of 210 million tons since 1955 are presented in Table 3. The calculations of predicted concentration changes that should have occurred in Lake Superior water were based on a lake volume of $1.25 \times 10^{16}$ liters and the assumption that the effect of the discharge would be confined to the stated lake volume percentage.

Predicted changes for total dissolved solids, silica and alkalinity in five percent of the lake volume were 2.9 mg/l, 1.2 mg/l and 2.6 mg/l, respectively (Table 5). The predicted changes for total dissolved solids are less than the errors generally associated with the analytical procedures for this parameter but the 1 to 2 mg/l increases estimated for silica and alkalinity should not be detectable with standard procedures if the changes are real. However, available chemical data from Lake Superior indicates that these predicted changes have not occurred (Haley, 1969; Wagner & Wahlstrom, 1973).

taconite tailings will release proportionately less material than finer sized tailings. In fact, the solubility of coarse tailings is approximately one-half that of fine tailings (0.5 to 0.6 percent vs 0.9 to 1.2 percent) under the conditions of the leaching study (Plumb, 1973). Finally, total tailings had a solubility of 0.1 percent compared to 0.9 to 1.2 percent solubility for -325 mesh tailings (Plumb, 1973).

Another simplification in the Table 5 calculations is the assumption that all tailings are mixed in the lake. Actually, 50 percent of the tailings settle immediately to form a delta at the point of discharge (Haley, 1969). This means that the calculated total release is high and the predicted concentrations will, in turn, be high. Moreover, numerous studies have shown that substantial mixing occurs between the western arm of Lake Superior and the main body of the lake. Therefore, the assumptions that the volume of Lake Superior influenced by the Reserve Mining Co. discharge is limited to some small fraction, particularly five percent, is conservative.

Table 5. Predicted changes in Lake Superior water based on leaching studies by Plumb and Lee (1975)

| Species | Laboratory leaching results mg/g | Total mg released 17 years operation | Changes that should have occurred in Lake Superior | | | | |
|---|---|---|---|---|---|---|---|
| | | | 100% mg/l | 20% mg/l | 10% mg/l | 5% mg/l | (Percent volume) |
| T.D.S. | 9.0 | $1.8 \times 10^{15}$ | 0.14 | 0.73 | 1.40 | 2.90 | |
| $SiO_2$ | 4.0 | $8.0 \times 10^{14}$ | 0.06 | 0.30 | 0.60 | 1.20 | |
| Alkalinity | 8.0 | $1.6 \times 10^{15}$ | 0.13 | 0.65 | 1.30 | 2.60 | |
| $Mg^{++}$ | 0.29 | $6.0 \times 10^{13}$ | 0.005 | 0.03 | 0.05 | 0.10 | |
| $Ca^{++}$ | 1.6 | $3.3 \times 10^{14}$ | 0.02 | 0.10 | 0.23 | 0.45 | |
| $Na^{+}$ | 0.3 | $6.3 \times 10^{13}$ | 0.005 | 0.025 | 0.05 | 0.10 | |

The concentrations listed are the incremental increases that should have occurred in Lake Superior based on laboratory studies with tailings.

There are several reasons for the discrepancy between predicted and actual changes in Lake Superior. First, the predicted changes in Table 5 are based on laboratory studies in which tailings suspensions were constantly agitated. Such a condition would not occur in the lake. The quiescent conditions that would be expected in the lake would result in a 20 to 25 percent smaller release of dissolved solids from taconite tailings (Plumb, 1973). Second, the calculated predictions assumed that all tailings behave as -325 mesh tailings. This assumption is in error since larger sized

Additional factors that may tend to exaggerate predicted concentration changes in the lake are the various uncertainties regarding the pretreatment of tailings in the laboratory tests and the extent to which buried tailings will contribute dissolved solids to the lake. It is reasonable to expect that once taconite tailings have been covered by additional tailings or other sediments that the amount of dissolved solids that would be released would be diminished. Therefore, the results in Table 5 are conservatively high by a factor of 5 to 20.

Characteristics of water at sediment-water interface

A study was conducted by Glass (1973) and Glass and Poldoski (1975) to assess the impact of taconite tailings on the exchange of dissolved constituents across the sediment-water interface in Lake Superior. Fifteen sets of water and core samples were collected from a 220 sq km area within which the sediments have been influenced by the discharge of taconite tailings and a similar set of samples were collected in a reference area of the same size outside the zone of discharge. Water samples were collected at several depths between 3 cm and 30 m above the sediments and interstitial water samples were obtained from cores at depths of 1.3 cm, 3.8 cm and 6.3 cm below the sediment-water interface.

Higher concentrations of most elements (calcium, copper, manganese, magnesium, potassium and silica) were observed in the interstitial water in both sampling areas when compared to the overlying lake water concentrations. Because higher interstitial concentrations were measured in the tailings deposit compared to natural lake sediments, it was concluded that dissolved solids were diffusing out of the tailings deposit and affecting water quality (Glass, 1973; Glass & Poldoski, 1975). This conclusion was the result of evaluating individual water column profiles and assuming that mass transport proportional to the concentration gradient was occurring as a result of diffusion. However, when Glass' (1973) original data is averaged as a function of distance from the sediment-water interface and the tailings area data is compared to the reference area data, the data suggests that the quality of Lake Superior water is not being significantly affected by the tailings deposit (Table 6).

Glass (1973) expressed the concern that mixing in the lake would dilute the release of dissolved constituents from the taconite tailings deposit in Lake Superior. Therefore, cores were collected from the disposal region and the reference area and chemical migration into the water column was determined after a laboratory incubation period of approximately 70 days. This data was also averaged as a function of distance from the sediment-water interface and is presented in Table 7. Although several parameters suggest a decreasing concentration with increasing distance from the interface, concentrations over the tailings cores are essentially the same as concentrations over Lake Superior sediments for all parameters except manganese. Greater manganese concentrations existed in the upper water column (11 to 57 cm) over taconite tailings but greater concentrations existed in the lower water column (3.8 cm) over lake sediments. This data also demonstrates that diffusion of dissolved constituents from the tailings deposit is not having a significant effect on Lake Superior water quality even though tailings pore water may have higher concentrations than Lake Superior sediment pore water.

Comparison of tailings derived solutes to natural derived solutes

A slightly different approach to assessing the impact of taconite tailings on the bulk chemical content of Lake Superior water is to compare the contributions derived from taconite tailings with the contributions derived from natural weathering products. Collier (1968) has estimated that 96 percent of the total suspended solids entering Lake Superior from all Minnesota sources is derived from taconite tailings. However, on an annual basis, the tailings are a less important source of dissolved solids than are the natural additions from runoff, weathering and precipitation. Upchurch (1972) estimates that the annual natural addition of dissolved solids to Lake Superior is approximately $4x10^{12}$g/year. He gives values of 1 to $8x10^{11}$g/year for major ions such as chloride, sulfate, calcium, sodium and potassium (Table 8).

Additional estimations of the annual contribution of dissolved solids to Lake Superior from taconite tailings are given in Table 8. These are based on studies by Glass (1970), in-plant changes measured at Reserve Mining Co. (Klaysmat, 1970; Lemire, 1970a,b,c) and the leaching experiments of Plumb (1973). In-plant changes measured at Reserve

Table 6. Lake Superior water concentrations near the sediment-water interface-field data*

| Distance from interface cm | Ca mg/l | Cu µg/l | Mn µg/l | Mg mg/l | K mg/l | $SiO_2$ mg/l |
|---|---|---|---|---|---|---|
| | | | Tailings disposal area | | | |
| 3050 | 12.5 | 1.2 | 1.0 | 2.8 | 0.38 | 2.6 |
| 226 | 13.1 | --- | 0.7 | --- | 0.40 | 2.7 |
| 100 | 12.9 | 1.3 | 2.2 | 2.7 | 0.39 | 2.8 |
| 57 | --- | --- | --- | --- | --- | --- |
| 30 | 12.6 | 1.5 | 1.3 | 2.7 | 0.40 | 3.2 |
| 27 | 12.2 | 2.0 | 0.2 | 2.8 | 0.44 | 2.5 |
| 3.8 | 11.4 | 1.6 | 0.1 | 2.6 | 0.36 | 6.9 |
| -1.3 | 11.6 | 12.1 | 42 | 2.5 | 1.15 | 27.7 |
| -3.8 | 11.4 | 13.5 | 466 | 2.4 | 1.53 | 32.2 |
| -6.3 | 12.8 | 10.9 | 468 | 2.7 | 2.01 | 29.8 |
| | | | Reference area | | | |
| 3050 | 12.6 | 1.3 | 0.2 | 2.8 | 0.34 | 2.6 |
| 226 | 12.9 | --- | 0.2 | 2.7 | --- | 2.8 |
| 100 | 13.2 | --- | 0.3 | --- | 0.40 | 2.7 |
| 57 | 12.7 | 1.4 | 0.6 | 2.8 | 0.32 | 2.7 |
| 30 | 12.1 | 1.3 | 1.4 | 2.7 | 0.42 | 3.6 |
| 27 | 12.5 | 1.3 | 0.3 | 2.7 | 0.34 | 2.9 |
| 3.8 | 12.1 | 2.2 | 0.1 | 2.6 | 0.35 | 6.4 |
| -1.3 | 7.6 | 6.6 | 15 | 1.5 | 0.94 | 23.4 |
| -3.8 | 8.2 | 5.6 | 14 | 1.7 | 1.33 | 22.3 |
| -6.3 | 9.4 | 5.3 | 25 | 2.0 | 1.71 | 19.1 |

---indicates no data available.
A - sign in the distance column signifies depth below the sediment-water interface.
*After Glass (1973).

Table 7. Lake Superior water concentrations near the sediment-water interface-laboratory data*

| Distance from interface cm | Ca mg/l | Cu µg/l | Mn µg/l | Mg mg/l | K mg/l | $SiO_2$ mg/l |
|---|---|---|---|---|---|---|
| | | | Tailings disposal area | | | |
| 57 | 11.5 | 1.2 | 4.0 | 3.4 | 0.43 | 3.4 |
| 19 | 12.2 | 1.2 | 36 | 3.6 | 0.46 | 4.4 |
| 11 | 13.5 | 1.3 | 184 | 3.6 | 0.52 | 6.0 |
| 3.8 | 14.6 | 1.5 | 331 | 4.0 | 0.54 | 7.8 |
| -1.3 | 20.4 | 8.8 | 1502 | 5.0 | 1.56 | 32.4 |
| -3.8 | 20.8 | 7.4 | 2976 | 5.0 | 1.88 | 40.0 |
| -6.3 | 20.1 | 6.6 | 2101 | 4.7 | 1.93 | 40.6 |
| -8.9 | 15.4 | 4.9 | 1417 | 3.7 | 1.76 | 38.4 |
| | | | Reference area | | | |
| 57 | 10.2 | 1.1 | 0.4 | 2.8 | 0.41 | 3.4 |
| 19 | 12.9 | 1.1 | 21 | 2.8 | 0.42 | 4.2 |
| 11 | 12.7 | 1.4 | 135 | 3.1 | 0.46 | 5.4 |
| 3.8 | 14.8 | 1.4 | 488 | 3.6 | 0.51 | 7.8 |
| -1.3 | 24.2 | 5.0 | 1299 | 5.0 | 1.65 | 25.5 |
| -3.8 | 27.2 | 2.5 | 477 | 5.7 | 1.92 | 33.5 |
| -6.3 | 26.1 | 3.3 | 353 | 5.5 | 2.05 | 36.4 |
| -8.9 | 26.2 | 3.9 | 347 | 5.7 | 2.36 | 35.5 |

A - sign in the distance column signifies depth below the sediment-water interface.
*After Glass (1973).

Table 8. Comparison of dissolved solids entering Lake Superior from natural sources and from taconite tailings.

| | Natural additions to Lake Superior | | | | Additions to L. Superior based on taconite tailings studies: Reserve in-plant changes (Lemire, 1970b,c) | | | | | |
|---|---|---|---|---|---|---|---|---|---|---|
| | Upchurch (1971) | Upchurch (1972) | Glass (1970) | | Winter | | Summer | | Plumb & Lee (1975) | |
| | g/yr | g/yr | g/yr | Percent* | g/yr | Percent* | g/yr | Percent* | g/yr | Percent* |
| Dissolved solids | $4.7 \times 10^{12}$ | $3.9 \times 10^{12}$ | $1.7 \times 10^{10}$ | 0.4 | $5.4 \times 10^{9}$ | 0.1 | $1.5 \times 10^{9}$ | 0.03 | $2.0 \times 10^{11}$ | 4.6 |
| Chloride | $1.0 \times 10^{11}$ | $2.0 \times 10^{11}$ | - | - | $1.9 \times 10^{9}$ | 1.2 | $3.0 \times 10^{7}$ | 0.02 | - | - |
| Phosphate | $4.0 \times 10^{9}$ | - | - | - | - | - | - | - | - | - |
| Nitrate | $5.0 \times 10^{10}$ | - | - | - | $1.3 \times 10^{7}$ | 0.02 | $1.7 \times 10^{7}$ | 0.03 | - | - |
| Calcium | $7.3 \times 10^{11}$ | $8.0 \times 10^{11}$ | $4.2 \times 10^{9}$ | 0.55 | $6.8 \times 10^{8}$ | 0.09 | - | - | $2.7 \times 10^{10}$ | 3.5 |
| Silica ($SiO_2$) | $2.8 \times 10^{11}$ | - | $5.0 \times 10^{9}$ | 1.8 | $4.2 \times 10^{8}$ | 0.15 | $9.4 \times 10^{8}$ | 0.33 | $1.0 \times 10^{10}$ | 3.6 |
| Sulfate | - | $2.3 \times 10^{11}$ | - | - | $3.0 \times 10^{7}$ | 0.01 | $3.2 \times 10^{9}$ | 0.13 | - | - |
| Sodium + Potassium | - | $2.0 \times 10^{11}$ | $1.6 \times 10^{9}$ | 0.80 | $3.9 \times 10^{8}$ | 0.19 | $6.5 \times 10^{8}$ | 0.32 | $6.0 \times 10^{9}$ | 3.0 |

*Tailings additions expressed as a percentage of estimated annual additions to Lake Superior.

Mining Co. would only contribute a few hundredths to a few tenths of a percent of each species derived from the other sources listed in Table 8. The only exception was the 1.2 percent value for chloride although this value would drop to 0.4 to 0.5 percent if the calculations were adjusted to compensate for the use of calcium chloride during the winter to prevent freezing of the ore. However, as mentioned earlier, the estimates based on in-plant changes are probably conservative since they are the result of a relatively short contact period between taconite tailings and the process water.

The estimates based on Glass(1970) data are slightly higher. These data suggest that taconite tailings would contribute a quantity of dissolved solids equal to 0.4 percent of the annual natural load (Table 8). Tailings would add 0.55 percent of the calcium, 1.8 percent of the silica and 0.8 percent of the sodium plus potassium.

The leaching study data of Plumb (1973) yielded even higher results because they are based on the assumptions that all tailings are dispersed in the lake, all tailings are less than -325 mesh (less than 44 μ in diameter), and all tailings are continually mixed in the lake. The calculated values from this study were 4.6 percent of the dissolved solids, 2.5 percent of the calcium, 3.6 percent of the silica and 3.0 percent of the sodium plus potassium. These estimates are considered high by a factor of 5 to 20 because of the initial assumptions. Therefore, the upper limit of dissolved solids contributions from taconite tailings is on the order of one percent or less of the annual additions from natural sources.

The amount of dissolved solids entering Lake Superior as a consequence of the discharge of taconite tailings was also compared to natural additions by Wagner *et al.* (1973). The estimated natural loading was 4,871,000 metric tons per year (including 979,000 metric tons per year due to rainfall and 504,000 metric tons per year from all Minnesota tributaries) which is in good agreement with the value of 4,300,000 metric tons per year calculated by Upchurch (1971). By comparison, the Reserve Mining Co. effluent contributes 9,500 metric tons per year. Thus, the Reserve contribution is less than 0.2 percent of the total annual loading and less than one percent of the annual rainfall addition. Since the effluent contribution was such a small percentage of the annual loading and an even smaller percentage of the dissolved solids content of Lake Superior (less than 1x10$^{-3}$ percent), Wagner et al. (1973) concluded that the discharge of taconite tailings would not have a significant impact on the concentration of soluble constituents in Lake Superior.

## Discussion

The large quantity of taconite tailings discharged to Lake Superior over the past 20 years has created a man-made delta on the north shore of the lake and a benthic deposit covering several hundred sq km of lake sediments. The concern over the potential effects of the discharge on water quality in the lake has provided a sufficient data base to assess the impact of the tailings deposit on concentrations of soluble constituents. This data base includes analysis of the Reserve Mining Co. effluent, numerous laboratory leaching studies including one that lasted for three years and several years of field studies.

Depending on the length of the study, the particle size of the tailings and other experimental conditions, laboratory leaching studies indicate that taconite tailings are between 0.03 to 1.2 percent soluble (Plumb, 1973). The addition of this quantity of dissolved solids to the lake is less than one percent of the estimated natural runoff to the lake. In addition, if it is assumed that any soluble constituent released from taconite tailings has been mixed with more than five percent of the water in Lake Superior, it is predicted that the Reserve Mining Co. discharge would not have a measurable effect on soluble concentrations in the lake.

The fact that predicted changes in concentrations should not be measurable is supported by four years of field observations presented by Wagner and Wahlstrom (1973). Concentrations for eighteen parameters were grouped as being in the zone of discharge (Silver Bay area) or in the remainder of the western arm of Lake Superior and compared to determine differences (Table 9). A t-test comparison at the 0.05 testing level (95 percent confidence limits) demonstrated that 13 parameters were not statistically different. Sulfate, chloride, silica and total dissolved solids were lower in the zone of discharge and alkalinity was higher in the zone of discharge compared to the western arm of Lake Superior. When values near the Duluth-Superior harbor area were excluded from the western arm of Lake Superior group, there is no significant difference ($p=0.05$) between the Silver Bay discharge zone and the remainder of the western arm of Lake Superior for any of the parameters in Table 9.

Table 9 also presents data for many of the trace elements present in taconite tailings which are of potential concern in affecting water quality such as the heavy metals. Leaching studies by Glass (1973), Plumb (1973) and the in-lake measurements shown in Table 9 all support the conclusion that the heavy metals such as copper, zinc, cadmium, nickel and the algal nutrient phosphorus are not released from taconite tailings to Lake Superior water. As noted above, there is a release of manganese from tailings as measured by laboratory leaching tests. However, increased concentrations of manganese are not detected in the water as a result of Reserve Mining Company's operation. It appears that the Mn released from tailings is oxidized in Lake Superior water to $MnO_2$ which precipitates and settles to the bottom of the lake.

Lee (1975) has proposed that the release of phosphorus, copper, zinc and other trace heavy metals, except manganese, does not occur because of a hydrous oxide coating of ferric iron that develops on the surface of the taconite tailings particles. This hydrous oxide coating acts as a scavenger for many heavy metals and phosphate, which would tend to dissolve as part of the dissolution of the tailings iron silicate matrix. Support for this conclusion is derived from the fact that taconite tailings which have been dried, exposed to low pH or high organic content, show release of certain heavy metals. However, this release does not occur in Lake Superior water under the conditions that prevail in this lake.

## Conclusions

A review of the results of laboratory leaching tests designed to evaluate dissolved solids release from taconite tailings has shown that tailings solubility ranges from

Table 9. A comparison of Lake Superior water composition in the Silver Bay area and the western arm of the lake*

| Constituent | Silver Bay area (a) Number of Samples | Silver Bay area (a) Average Concentration | Western arm-L. Superior Number of Samples | Western arm-L. Superior Average Concentration | Differences |
|---|---|---|---|---|---|
| Ca mg/l | 168 | 12.89 | 378 | 12.85 | n.s. |
| Mg mg/l | 168 | 2.78 | 378 | 2.80 | n.s. |
| Na mg/l | 168 | 1.14 | 378 | 1.16 | n.s. |
| K mg/l | 168 | 0.489 | 378 | 0.498 | n.s. |
| Alk mg/l | 152 | 41.91 | 339 | 41.67 | a>b |
| $So_4$ mg/l | 168 | 2.74 | 378 | 2.85 | b>a |
| Cl mg/l | 167 | 1.37 | 358 | 1.47 | b>a |
| Si mg/l | 156 | 2.48 | 351 | 2.59 | b>a |
| Total-P μg/l | 84 | 4.5 | 185 | 4.2 | n.s. |
| Total Fe μg/l | 95 | 34.4 | 216 | 33.9 | n.s. |
| Cu μg/l | 167 | 1.6 | 399 | 1.5 | n.s. |
| Zn μg/l | 144 | 1.9 | 344 | 2.1 | n.s. |
| Ni μg/l | 154 | 3.0 | 370 | 3.7 | n.s. |
| Cd μg/l | 128 | 0.2 | 313 | 0.3 | n.s. |
| $NO_3$-N mg/l | 168 | 0.283 | 378 | 0.278 | n.s. |
| pH | 96 | 7.89 | 216 | 7.87 | n.s. |
| Turbidity mg/l | 96 | 0.82 | 216 | 1.00 | n.s. |
| T.D.S. mg/l | 96 | 54.47 | 216 | 54.99 | b>a |

n.s. - difference is not significant at p=0.05 testing level (95 percent confidence level).

* - after Wagner and Wahlstrom (1973).

0.03 to 1.2 percent and is dependent upon solid-liquid ratio, agitation and particle size. It is estimated that no measurable change should occur in the dissolved solids concentrations of Lake Superior if normal dilution of the discharge occurs. This projection has been verified with several years of field data demonstrating that concentrations in the discharge zone are not significantly different from concentrations in the open waters of Lake Superior. This suggests that the release of dissolved solids from taconite tailings is minimal compared to the dilution of the discharge that is occurring and the volume of the lake.

The solubility of tailings has caused a measurable increase in the interstitial water concentrations of the sediments in the discharge zone. However, an evaluation of laboratory and field data indicates that concentration dependent diffusion processes will not produce dissolved solids concentration changes in the lake.

Studies have shown that no increase in any dissolved chemical parameter has occurred in the western arm of Lake Superior during the 17 years Reserve Mining Co. has been operating. It therefore appears that the dissolution of taconite tailings will not have an adverse effect on dissolved solids and water quality in Lake Superior.

## Acknowledgments

This investigation was supported by the U.S. Environmental Protection Agency in the form of a training grant, the Water Chemistry Program of the University of Wisconsin at Madison where the study was conducted and the Reserve Mining Co. of Silver Bay, Minnesota. Additional support for the study was provided by the Center for Environmental Studies, University of Texas at Dallas and EnviroQual Consultants and Laboratories, Plano, Texas.

## References

Collier, C.R., 1968. Preliminary report on streamflow conditions and sedimentation in the vicinity of Silver Bay, Minnesota. U.S. Geological Survey Water Resources Div. St. Paul, Minn.

Dickson, F. 1970. Tailings inventory. Reserve Mining Co. memo. to

Ken Haley, June 28, 1970.

Gay, J.C., 1970. Prediction by means of a mathematical model of the effect of tailings on Lake Superior water chemistry. Reserve Mining Co., Silver Bay, Minnesota. In: Reserve Mining Co. report to Minneosta Pollution Control Agency Minneapolis, Minneosta.

Glass, G.E., 1970. The dissolution of taconite tailings in Lake Superior. National Water Quality Laboratory, U.S. Environmental Protection Agency, Duluth, Minn.

Glass, G.E., 1973. A study of western Lake Superior: surface sediments, interstitial water and exchange of dissolved components across the water-sediment interface. In: Studies regarding the effect of the Reserve Mining Co. discharge in Lake Superior. U.S. Environmental Protection Agency, Office of Enforcement and General Council, Washington, D.C., pp. 1028-1121.

Glass, G.E. & Poldoski, J.E., 1975. Interstitial water components and exchange across the sediment interface of western Lake Superior. Verh. Internat. Verein. Limnol. 19:405-420.

Gunderson, J.N. & Schwartz, G.M., 1962. The geology of the metamorphosed biwabik iron-formation, eastern Mesabi district, Minn. University of Minnesota Press, Minneapolis.

Haley, K.M., 1969. Statement of K.M. Haley, Vice President and Manager, Research and Development Div., Reserve Mining Co., Silver Bay, Minn. Presentation at the conf.on the matter of pollution of Lake Superior and its tributary basin, Duluth, Minn.

Klaysmat, A.W., 1970. Chemical effects of Reserve's discharge on Lake Superior. Reserve Mining Co. Silver Bay, Minn., Report to Minnesota Pollution Control Agency, Minneapolis, Minn.

Lee, G.F., 1975. Role of hydrous metal oxides in the transport of heavy metals in the environment. Proceedings of symposium of transport of heavy metals in the environment. In: Progress in Water Technology, 17:137-147.

Lemire, R.S., 1970a. Addendum to my memo on plant tailings water analysis dated March 10, 1969, revised March 13, 1969. Reserve Mining Co., Silver Bay, Minn. In: Reserve Mining Co. report to Minnesota Pollution Control Agency, Minneapolis, Minn.

Lemire, R.S., 1970b. Element changes in lake waters between the plant intake and its return to the lake. Reserve Mining Co., Silver Bay, Minn. In: Reserve Mining Co. report to Minnesota Pollution Control Agency, Minneapolis, Minn.

Lemire, R.S., 1970c. Plant tailings water analysis. Reserve Mining Co., Silver Bay, Minn. In: Reserve Mining Co. report to Minnesota Pollution Control Agency, Minneapolis, Minn.

Plumb, R.H., Jr., 1973. A study of the potential effects of the discharge of taconite tailings on water quality in Lake Superior. Ph.D. thesis, University of Wisconsin, Madison.

Plumb, R.H.,Jr. & Lee, G.F., 1975. Impact of taconite tailings on water quality in Lake Superior. I. Dissolved solids leaching tests. Submitted for publication.

U.S. Bureau of Mines, 1968. Report of tailings disposal at Reserve Mining Co. Plant, Silver Bay, Minn. Twin Cities Office of Mineral Resources, U.S. Bureau of Mines, Minneapolis, Minn.

U.S. Department of the Interior, 1969a. An appraisal of water pollution in the Lake Superior basin. U.S. Department of the Interior, Federal Water Pollution Control Administration, Great Lakes Region.

U.S. Department of the Interior, 1969b. Proceedings of the conference on the matter of pollution in Lake Superior and its tributary basin, Minnesota-Wisconsin-Michigan. U.S. Department of the Interior, Federal Water Pollution Control Administration, Washington, D.C.

Upchurch, S., 1971. Chemical characteristics of the Great Lakes.

Great Lakes Research Center, U.S. Lake Survey, Detroit, Mich.

Upchurch, S., 1972. Natural weathering and chemical loads in the Great Lakes. Proc. 15th conf. Great Lakes research.

Wagner, D.M. & Wahlstrom, P.T., 1973. Permanent station water chemistry data analyses 1969-1972. Project No. 23-1108, Reserve Mining Co., Silver Bay, Minn.

Wagner, D.M., Anderson, D.W., & Koschak, D.F., 1973. Comparison of dissolved solids discharged to Lake Superior by 16 Minnesota rivers and Reserve Mining Co. Reserve Mining Co. Project No. 23-1110-1, Reserve Mining Co., Silver Bay, Minn.

Key words: Great Lakes
dissolved solids
mine tailings

# SOME INFLUENCE OF DREDGING IN THE GREAT LAKES

P.G. Sly

Canada Centre for Inland Waters, Burlington, Ontario, Canada

## Summary

Dredging in the Great Lakes (a large freshwater system) is undertaken mostly to maintain navigation, although sand and gravel are exploited to a limited extent. Until recently, most dredging activities used open-lake disposal although the use of confined disposal in Lake Erie and the Detroit River remain notable exceptions. No major program has yet been undertaken solely for the removal or treatment of contaminated lake sediments. Following establishment of the Great Lakes Water Quality Agreement in 1972, open-lake disposal activities have been much reduced.

Since 1972, also, a number of unpublished studies have been made to assess the influence of dredging practices on water quality and the following comments refer principally to studies in the Canadian portion of the Great Lakes which, in most cases, have been undertaken in conjunction with regular dredging operations.

In studies before, during, and after maintenance dredging at Port Stanley (Lake Erie Figure I) and Bronte Harbour (Lake Ontario), it was shown that total and reactive phosphorus levels increased rapidly in the receiving waters both at the removal site and at the open-lake dumping site; similar increases in other nutrient elements and heavy metals were also observed. However, as a result of particle settling and dilution, elevated concentrations decreased rapidly and background conditions in the overlying waters were generally reestablished within a few hours. Because of influence of wave activity in Lake Erie the dumped materials were rapidly redistributed and no evidence was obtained to indicate a long term influence on water chemistry.

At Thunder Bay in Lake Superior, however, recent evidence suggests that harbour materials disposed of in deep water, below wave base, may continue to influence overlying waters for extended periods.

At Mitchell Bay, in Lake St. Clair containment of dredged materials in an artificial island was examined; particularly to assess the significance of sediment/water exchange processes. Tests were begun after dumped materials had settled. Despite seasonal variations in surface materials, pore water concentrations below about 1.5 m remained sensibly constant, concentrations of As, Cd, Cu, Pb and Zn were similar and concentration profiles of Fe, Mn, and P showed strong dependence on pH and redox potential; Hg in pore water remained independent of sediment values.

Dredging effects and ship turbulence are undoubtedly significant at the local level, in the Great Lakes; however, the scale of events remains small in comparison to sediment resuspension resulting from wind-wave action.

## Introduction

Figure I shows the distribution of major dredging activities in the Great Lakes with quantities averaged over a 10 year period. Dredging is concentrated in the lower lakes where shoreline materials remain easily erodable and where shallow water depths subject both sand and muds to the major effects of sediment transport.

In order to condense what is now known from studies on dredging and dumping in the Canadian portion of the Great Lakes the text which follows has been based on the conclusions of a more detailed manuscript by the author (to be published in 1977). After each statement of these conclusions a brief discussion of background material or comment is presented.

## Conclusions and supporting information

Dredging plant in the Great Lakes comprises mostly clamshell or suction dredges, with the occasional use of hopper dredges. Open-lake disposal is usually made from bottom dumping barges (scows) and not by means of a pumped slurry. Confined disposal, including land disposal, is usually made by slurry pumping (though for experimental purposes, at Mitchell Bay, some confined disposal was made using a clamshell dredge).

1) As an immediate result of sediment disturbance pore waters may be released, thereby adding soluble nutrient elements to the water mass.

   ———— During dredging and disposal activities the physical structure of bottom sediments is altered by shearing and turbulent mixing; in this, slurry pumping represents an extreme situation. Concentrated pore water solutions become mixed with the proximal water mass thereby locally and temporarily increasing the concentration of some dissolved species. Field data is strongly circumstantial.

2) During dredging/dumping activities ferrous iron becomes rapidly oxidized to ferric

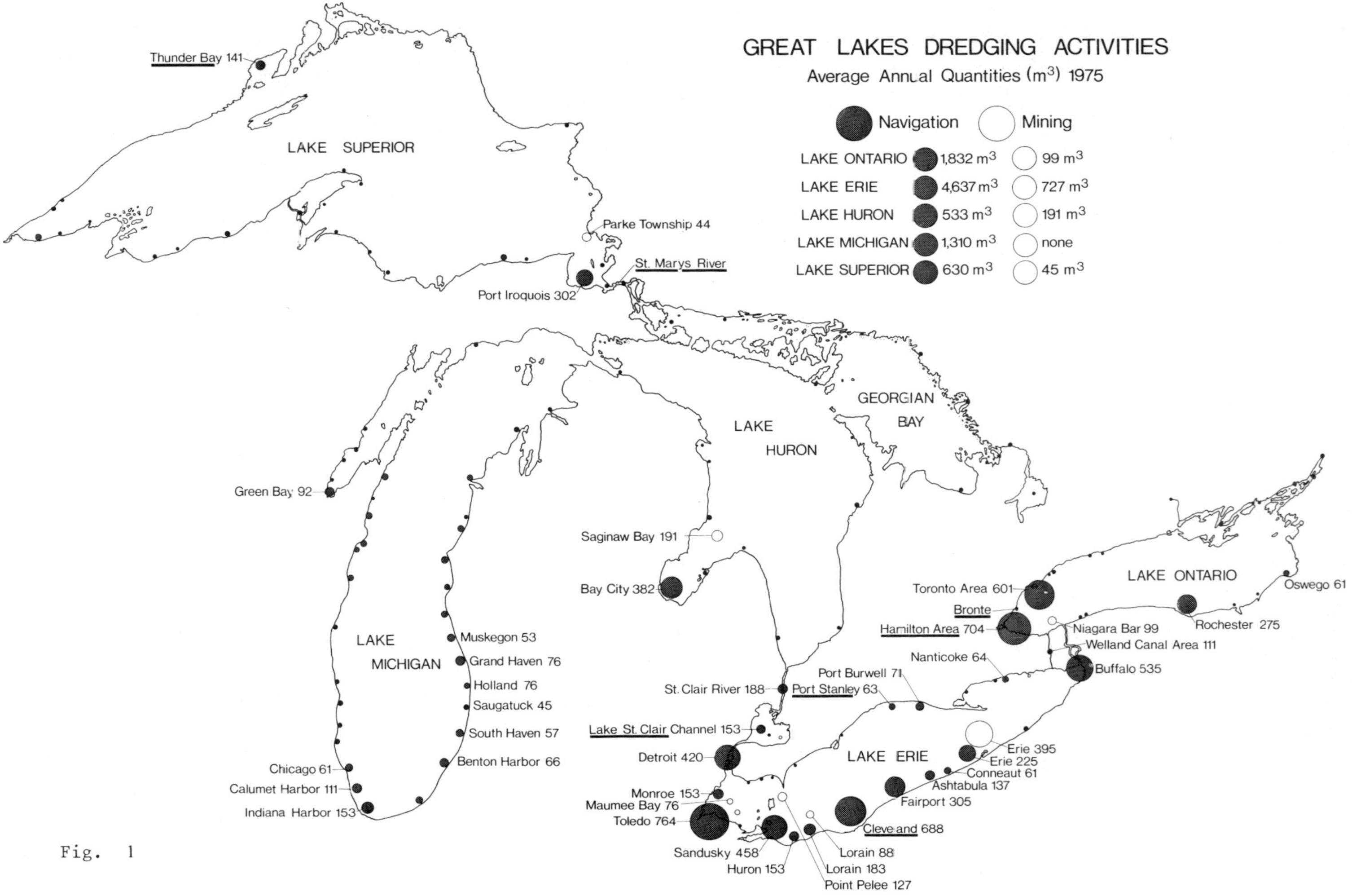
GREAT LAKES DREDGING ACTIVITIES
Average Annual Quantities (m3) 1975
Navigation
Mining
LAKE ONTARIO 1,832 m3 99 m3
LAKE ERIE 4,637 m3 727 m3
LAKE HURON 533 m3 191 m3
LAKE MICHIGAN 1,310 m3 none
LAKE SUPERIOR 630 m3 45 m3
LAKE SUPERIOR
Thunder Bay 141
Parke Township 44
St. Marys River
Port Iroquois 302
LAKE MICHIGAN
Green Bay 92
Muskegon 53
Grand Haven 76
Holland 76
Saugatuck 45
South Haven 57
Benton Harbor 66
Chicago 61
Calumet Harbor 111
Indiana Harbor 153
LAKE HURON
GEORGIAN BAY
Saginaw Bay 191
Bay City 382
St. Clair River 188
Port Stanley 63
Port Burwell 71
Lake St. Clair Channel 153
Detroit 420
Monroe 153
Maumee Bay 76
Toledo 764
Sandusky 458
Huron 153
Lorain 88
Lorain 183
Point Pelee 127
LAKE ERIE
Cleveland 688
Fairport 305
Ashtabula 137
Conneaut 61
Erie 225
Erie 395
Nanticoke 64
Buffalo 535
Welland Canal Area 111
Niagara Bar 99
Hamilton Area 704
Bronte
Toronto Area 601
LAKE ONTARIO
Rochester 275
Oswego 61

Fig. 1

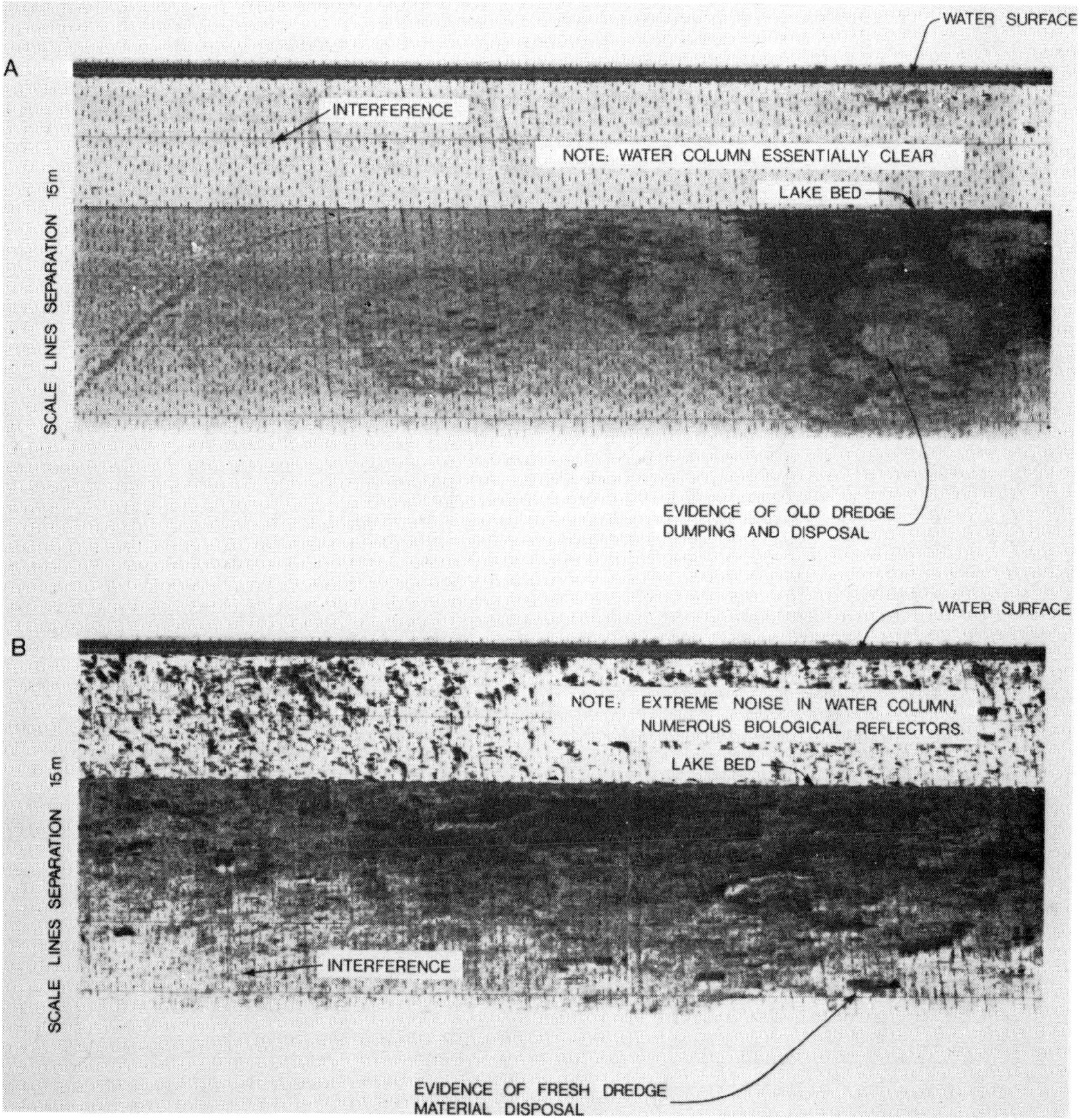


Fig. 2

A COMPARISON BETWEEN ADJACENT AREAS AT A DREDGED MATERIAL DUMP-SITE, THUNDER BAY, LAKE SUPERIOR, MAY 1975.

COMPARATIVE STARBOARD HAND SIDE-SCAN SONAR RECORDS OF SIMILAR BOTTOM SEDIMENT AREAS AT A DUMP SITE.

A LAKE BED IN AREA WHERE DUMPING HAS BEEN UNDERTAKEN PERIODICALLY, BUT NOT DURING 1974.

B LAKE BED IN AREA WHERE DUMPING HAS BEEN UNDERTAKEN WITHIN THE LAST YEAR.

form and through sorption reactions forms iron phosphates.

———— Burns and Nriagu (1976), have reported, in the central basin of Lake Erie, that Fe occurs strongly attached to clay particles rather than as a precipitated oxide; further, where anoxic conditions persist for some time, that large amounts of Fe are potentially releasable from clay materials as a result of resuspension. If similar conditions pertain to the dumped materials (Port Stanley) it may be appreciated that amorphous Fe - inorganic P complexes can rapidly form (Williams et al, 1971; Williams and Mayer 1972), as reflected in the high Fe/P concentrations observed by Chemex (1974 and 1975), Fe/clay relations may differ at other sites.

3) Release/precipitation reactions for P, Fe and Mn appear to conform to the well established pH/redox potential controls, with differential dissolution of Mn and Fe oxyhydroxides resulting in higher levels of Mn, relative to Fe, in the water.

———— Data from the Chemex studies, (1974) at Port Stanley and Bronte Harbour indicated that changes in concentration of Mn in waters at the dredging sites corresponded with pH/redox potential but, in part, remained independent of Fe/P response. The sustained high level of Mn in the water column and pore water after dredging appears to be related to differential dissolution of Mn and Fe oxyhydroxides particularly under conditions of partial $O_2$ depletion (Delfino and Lee, 1968; Burns and Nriagu, 1976). Concentrations of Fe, P and Mn in sediments show a marked increase at the surface of dredged areas, Chemex, (1974) and core data is interpreted as evidence that early settlement of new material has had the opportunity to scavenge elevated concentrations of Fe, P and Mn available in the partly entrained waters of the harbour dredging sites (see also Dean, 1970).

4) Nutrient release, undifferentiated, may persist for some time after open-lake disposal.

———— Following observations of open-lake dumping adjacent to Port Stanley it was found that elevated concentrations of P (both total and reactive P) persisted only for a few hours (Chemex, 1975) and because of the rapid dispersion of dumped material in the nearshore zone, no long term release could be determined. Evidence from a disposal site in Thunder Bay, Lake Superior, however, may indicate longer term effects. (The following text has been modified after the 1975 report by the International Working Group on the Abatement and Control of Pollution from Dredging Activities.)

"In May 1975 an acoustic survey was made at a lake disposal site in Thunder Bay, Ontario, The survey consisted of traverse lines which intersected at the disposal site marker buoy. In Figure 2 a side-scan sonar record illustrates the lake bed and overlying water in areas adjacent to the marker buoy.

Records A and B cover adjacent parts of an open-lake location. Wave action is minimal (at the time of recording) and circulation is of low intensity. The dumping has occurred in about 30 m of water and reworking of dumped materials as a result of wave turbulence may be considered negligible.

In Record A where dumping has occurred, but not during the last 12-18 months, the water column overlying the lake bed sediment is essentially clear (acoustically). The lake bed is generally flat but with a slightly irregular relief. The presence of old dumped materials can be seen in the darker, patchy nature of some of the bottom sediments.

In Record B the water contains numerous acoustic reflectors, typical of dense fish populations; and, additionally, the darker patches (and lighter shadow areas) imply more recently dumped materials on the lake bed. Dumping at this location has occurred within the last 12 months.

It may be implied that some form of food-source from the dumped materials has, directly or indirectly, attracted a large population of fish. In lake bed areas, where dumped materials are somewhat older, contributions from such sources are not evident. Hence a sharp distinction may be seen between the adjacent areas A and B. Under more usual conditions in which the water column would be affected by horizontal circulation, or under the influence of thermal stratification, it would be unlikely to observe such relationships. This example highlights the effect of dumped materials, acting as a food source for some time after emplacement; the source being associated with nutrients and organics which locally affect productivity and attract fish. (No samples were taken.)

From Lake Erie, Sweeney et al (1975), reported on a 1973 study of an open-lake disposal site in 18 m of water off Cleveland (Figure 1) which had been used for the dumping of dredged materials between 1925 and 1968. Samples from adjacent areas provided controls. Evidence of previously dumped materials was clearly shown by the persistence of elevated levels of total phosphorus, iron, mercury, cadmium, and oils and greases, and by relatively high BOD and COD values. In addition, population analysis, showed the presence of a particularly limited fauna of oligochaetes in areas of dumped materials.

These observations from Lake Erie add further support to those from Thunder Bay,

Lake Superior, and strongly suggest that some dumped materials disposed of in depths beyond wave influence may provide a persistent source of available materials for an extended period of time.

5) N availability appears to be high, especially from sites receiving high N loadings and locally increased BOD may be caused by the breakdown of additional organic materials.
——————— Studies by Chemex (1975) show that the concentration of N (several forms) increased significantly in waters at the disposal sites as a result of sediment dumping.
Burns (1976), in a study of Lake Erie nutrient budgets has shown that soluble inorganic nitrogen -SIN ($NH_3$+$NO^-_3$+$NO^-_2$) appears to be of limited availability in central Lake Erie epilimnion waters during the summer growth period, and that phytoplankton uptake is much dependent upon the SIN reserve available at the start of lake stratification. However, it is recognized that depleted SIN may not, in itself, limit phytoplankton concentrations although it could cause an increase in the presence of blue-green species, because of their ability to fix atmospheric nitrogen.
An increase in the concentration of nitrogen in epilimnion waters, such as at Port Stanley, should, therefore, give no major cause for concern. Organic carbon in dumped materials may also stimulate increased bacterial activity resulting in oxygen depletion, or local anoxia, in the hypolimnion at or near the sediment/water interface.

6) Although heavy metals in soluble form are available in pore waters their early release generally presents little concern since, mostly, concentrations are already within potable water standards and, in any case, the scavenging effects of oxides and hydroxides may rapidly co-precipitate P and other elements in the water mass soon after disposal. However, the occurrence of heavy metals in waters discharged from settling basins receiving pumped slurry, may give rise to concern. Additionally heavy metals may become available as a result of such physio-chemical effects as leaching and dehydration.
——————— Analysis of pore water concentrations in sediments from Port Stanley and Bronte Harbour (for such elements as As, Cd, Cu, Fe, Hg, Mn, Pb and Zn) showed that most values were below that required by drinking water standards (iron was slightly in excess). Studies on the quality of water returned to the lake after settlement of materials from a pumped slurry at a confined disposal site in Lake St. Clair (Coastal Engineering Laboratory, Queen's University, 1974, and Termarex/Barringer 1974) showed that, as a result of carefully regulated flow, up to 90% of the mercury contained in the slurry could be retained at sites receiving hydraulic discharge. Returning waters had, for example, mercury concentrations at levels close to those required for drinking water standards. To maintain this quality, however, receiving basins could not become filled to more than 2/3 their capacity without the use of secondary stilling ponds and flow velocities had to be kept below 20 cm/sec to inhibit resuspension of silt materials. The concentration of mercury in the returning waters was undoubtedly affected by the presence of fine particulates since analyses were made on total samples.
Studies at the same location (and upon sediments from Hamilton Harbour) in conjunction with leaching and elutriate experiments have given further indications of heavy metal availability under more severe conditions (Mudroch and Cheam, 1974). As an example, maximum elemental concentrations in pore waters at the confined disposal, island site, were: P 2.0 ppm; N 90 ppm; Fe 100 ppm; Mn 5 ppm; Hg 0.5 ppb; and Org-C 20 ppm. Based upon these case studies therefore it may be suggested that the release of materials subjected to leaching (associated with atmospheric exposure) represents the most severe conditions under which longer term release of contaminants may occur.
Often, however, long term release of contaminants is associated with conditions in which sediments are disposed of into open-lake sites below wave base, and where they remain largely unaffected by further physical redistribution. In such areas both physiochemical and biochemical processes control the long term distribution and concentration of heavy metals.

7) The release of compounds such as oils, grease and phenols, depends upon both their original concentration in the sediments, and upon the amount of sediment/water mixing prior to disposal. In most cases such contaminants fall rapidly to the lake bed in association with the particulates; in areas of frequent dumping, however, some tainting may occur. Information regarding the fate of other persistent organic contaminants such as PCBs, pesticides and herbicides is not yet available.
——————— The lack of suitable data in respect to the occurrence and availability of persistent organic contaminants is of particular concern and requires investigation. As with heavy metals, attention must be given not only to quantities present but also to biological availability and synergistic relationships.

8) Biochemical conversions of certain heavy metals (e.g. Hg, and possibly As, Pb and others), through the process of methylation provide a mechanism for the long term release of toxic species which may be

incorporated into the biota.

Saitoh and Cheam(1975), showed that quantities of methyl mercury in sediments from Lake St. Clair, Port Stanley, the Niagara River mouth and Hamilton Harbour varied from 0.4 to 18.5 ppb and were independent of total mercury concentrations (0.1-3.9 ppm). The upper limit of methyl mercury, expressed as a percentage of total mercury, approached 1% for Lake St. Clair sediments and these findings appear to conform to the conclusions of Cooley and McCarty (1976), and others, that rates of methylation are primarily related to the size of microbial populations and their activity. Jernelov (1969) suggested that methylation rates (Hg) may range between 1% and 10% per annum, a value which seems generally applicable to conversion of culturally enriched mercury in Great Lakes sediments (Vernet and Thomas 1972, and Thomas, personal communication). Methylation is also recognized as a potential release mechanism for arsenic and, more recently, Wong et al (1975) have shown that lead from Hamilton Harbour, Lake St. Clair, and other areas can be methylated at rates up to 6% per week (under laboratory conditions); although none of the bacterial isolates were able to produce $Me_4$ Pb from inorganic lead. The same authors have shown the isolates could also methylate selenium.

Wolery and Walters (1974), considered the problem of long term mercury release(by methylation) from Lake Erie sediments and concluded that there was a critical sediment depth below which mercury (and other metal) concentrations became largely unavailable. It was argued that sediments were subject to bioturbation by tubificids and that this reworking continuously presented mercury to microbial activity close to the sediment surface; critical depth appeared to be about 3-4 cms below the sediment interface (Jernalov, 1970, suggested that reworking by Anodonta may extend the critical depth to 9-10 cms). Assuming that mercury enrichment may decrease as a result of regulatory controls, methylation may continue for a period of 10-100 years under steady-state conditions. Under open-lake conditions, however, the effects of sediment accumulation and transport modify this. In the western basin of Lake Erie, for example, where accumulation rates exceed 0.4 cm/year, effective burial of a contaminated layer could be accomplished in about 10 years, if uninfluenced by resuspension. It is recognized, however, that wind/wave turbulence may resuspend sediments at shallow depth (Thomas et al, 1974, Thomas & Jaquet, 1976 and Thomas, Lam and Jaquet, 1976), and that such influences will tend to increase the effective critical depth. In addition, it is important to note that concentrations of mercury in the algae Cladophora showed a marked increase following storm activity in Lake Erie (Burkett, 1973), and that this reflected the significance of brief increases in the mercury content of the water mass due to periodic physical disturbance of the sediment surface.

Despite implications regarding longer term availability of heavy metals in the Great Lakes, and their potential incorporation in the food chain, published research remains limited. McKinnon et al.(1975), suggested that uptake of pollutants (methyl mercury) by yellow perch from the Ottawa River was dependent upon ration and respiration necessary to support metabolism and growth (being largely sensitive to body weight); and was largely independent of concentration in food and water. Huckabee and Goldstein (1975) in studies on large mouth bass showed that tissue retention of methyl forms (Hg) was longer than for inorganic forms derived from treated food supplies; they also implied, since methyl mercury compounds were almost never detected in water, that food chain accumulation mechanisms must predominate.

Delisle et al (1975), reported on aquaria experiments with goldfish and catfish and showed that fish exposed to grossly contaminated sediments showed a substantial increase in tissue levels of Zn,Pb,Cu and Cd; maximum concentrations were attained within two months and showed little variation for the remaining three months of the experiments.

Brown and Chow(1975) reported on concentrations of heavy metals in fish taken from Lake Huron (representing an environment essentially free of contaminant loadings) and from Toronto Harbour (an area subject to relatively high urban and industrial loadings). Concentrations of Cd,Cu,Pb,Zn, and Hg in the muscle tissue of Toronto Harbour fish were higher than those from Lake Huron by factors of 2,4,9,8 and 4, respectively. Maximum concentrations, however, remained within health standards.

On the basis of existing evidence, it seems that whilst elevated levels of many heavy metals may sustain long term release mechanisms which allow for their incorporation in the food chain, in most cases resulting concentrations do not pose health hazards in edible species. However, where methylation takes place bioaccumulation through direct assimilation and via the food chain may show rapid and sustained response resulting in tissue concentrations in excess of standards for human consumption.

9) Heavy metal availability, both in soluble form and as a source in biochemical conversion is strongly influenced by

speciation. Evidence from leaching and elutriate tests suggests that a fairly wide range of solubilities exist but data is non-specific.

——— No attempt was made to detail speciation of heavy metal forms in the dredging and disposal studies, cited.

(10) On a quantitative basis the contribution of particulates to lake systems by dredging is minor when compared to other loadings and lake-wide resuspension by storms. However, if dumped materials are disposed of in shallow areas subject to wave and current activity so that they become widely dispersed, it seems certain that they would contribute a measurable percentage to the total nutrient (and toxic) load. Since "dilution should not be the solution to pollution", dumped materials should be placed in as great a depth of water as possible having a sufficient resistance to DO depletion, with the minimum of mixing and dispersal during the activities of dredging and dumping.

——— Lake bottom disturbance, induced by shipping, is only of importance in shallow channels and shallow harbour areas. In most channels flow velocities are relatively high so that the main effect of disturbance by shipping is to cause temporary resuspension and to amplify the rates of existing sediment transport. In harbours, where waters remain partially entrained, disturbance by shipping has little effect, with resuspended materials settling quickly after slight redistribution.

Estimates of the amount of sediment redistributed by bottom disturbance from shipping in the Great Lakes have not been made, although they must exceed that produced by dredging and disposal by at least an order of magnitude. On this basis and despite the fact that coarser particulates are generally associated with low concentrations of available nutrients, heavy metals and other toxic materials, it seems likely that the passage of shipping in the Great Lakes contributes at least as much, if not more, to the release and redistribution of contaminant materials than dredging/disposal activities. Materials redistributed under relatively high flow conditions in channels are typically transported beyond the channel and become widely dispersed and dilute; on the other hand, unless deposited within the zone of wave induced influence, dumped materials tend to remain physically immobile.

Contaminant release from continuing sediment disturbance by biological mechanisms and from disturbance by wind/wave action and storm seiches appears to be by far the most important source of internal loading in the Great Lakes system; no estimate is available of a release associated with biological processes.

Chemex (1973), on the basis of turbidity measurements estimated, under the conditions prevalent off Bronte Harbour in November, that at least 1000 metric tons of material were naturally in suspension in the nearshore zone, for every kilometre of shoreline, at any point in time; this loading was many times that derived from the total daily dumping at the Bronte site.

Whilst concentrations of available nutrients and heavy metals were certainly higher in the dredged materials than in the naturally derived materials, such figures help to place the significance of these particulate loadings in perspective.

A more appropriate estimate of the impact of storms in terms of internal loading and nutrient availability, however, may be found in the work of Lam and Jaquet (1976). It was shown that substantial quantities of phosphorus were regenerated and settled as a result of storm action, with particularly large internal loadings occurring in the central basin of Lake Erie.

In 1976, Kemp et al, estimated that dredged/dumped materials accounted for about 4% of the annual lake loadings of silt/clay sized materials, and since such materials are associated with most of the available phosphorus fraction, and on the basis of the work by Williams et al (1976) it may be estimated that dredged materials could contain something in the order of 300-800 metric tons of non-apatite inorganic phosphorus (NAIP). If such dredged/dumped materials were fully mixed and dispersed during the process of disposal their total annual contribution would approach the same order of magnitude as the phosphorus regeneration and settling estimated for the 20 day period between cruise and model predictions by Lam and Jaquet (1976).

As earlier stated, however, unless dredged/dumped materials are disposed of within wave influenced areas, they tend to remain physically immobile and, further, since phosphorus release during dumped disposal in shallow water represents only a part of that potentially available, because of fall within an entrained phase (Krishnappen, 1975), it appears that short term phosphorus release from dredged/dumped materials represents only a fraction of that induced by lake-wide regeneration. Because of the brief period of sediment/water mixing and its limitation, even at depth, by entrained fall, short term contaminant availability from dredged or dumped materials is limited to chemical species which are easily released or rapidly exchanged. Long term release from dumped materials is controlled by processes such as biological reworking, methylation, diffusion and dewatering(and de-gasing) and the influence of Eh/pH conditions at the sediment/water interface. Reloading of contaminants to the overlying water mass would seem least under conditions of limited water movement coupled with adequate dissolved oxygen to withstand BOD/COD sediment requirements.

11) Harbours and channel areas effectively trap part of the "cultural loadings" to the Great Lakes, thereby lessening their total impact. Although it may be argued that by disposal of this material in open water there is no significant increase in total loading, such a view should be qualified. Harbour and channel areas trap both materials swept into the near-shore from the lake and from inflowing drainage systems, and by this role they lessen the effect of total loadings to the lake. Open-lake dumping of materials dredged from these locations is a form of reloading to the system where water quality may be effected by long term chemical release, or by the activities of storm action or bioturbation on such dumping. The use of open-lake disposal, therefore, may be seen as a trade-off between limited environmental impact and the expense of more costly disposal alternatives, and on this basis environmental impact assessments must consider each situation on a case by case basis.

———— It should be borne in mind that without maintenance dredging the capacity to receive such loadings will rapidly diminish in harbour areas (etc.). After steady state is achieved these locations no longer serve as traps. Maintenance dredging and careful disposal is therefore potentially beneficial in terms of reducing lake loadings.

## Acknowledgements

Much of the data referred to in this publication has been derived from reports completed by Chemex Ltd. of Calgary, Alberta, under contract to the Canada Centre for Inland Waters, (CCIW), on behalf of Environment Canada, and Public Works Canada. Additionally the author has drawn on the unpublished reports by A. Mudroch, V. Cheam, A. Zeman, and H. Saitoh, and upon the unpublished data of R.L. Thomas and J.D.H.Williams. I would like to express my thanks to these and other staff of CCIW for permission to include such material, and to acknowledge my appreciation of their assistance and discussion during the preparation of this manuscript; likewise to G.F. Lee, Univ. Texas.

The members of the International Working Group on the Abatement and Control of Pollution from Dredging Activities are acknowledged for their review and approval, and the Director, CCIW, is gratefully acknowledged for permission to publish this contribution.

## References

Brown, J.R. and L.Y. Chow, 1975. The comparison of heavy metals in fish samples taken from Baie du Dore, Lake Huron, and Toronto Harbour, Lake Ontario. Internat. Conf. on Heavy Metals in the Environment, Toronto. Abstr. C256-257 (also, in press).

Burns, N.M., 1976. Nutrient budgets for Lake Erie (1970). Jour. Fish. Res. Board Canada. 33: 520-536.

__________ and J.O. Nriagu, 1976. Forms of iron and manganese in Lake Erie waters. Jour. Fish. Res. Board Canada 33: 463-470.

Burkett, R.D., 1973. The use of Cladophora to monitor mercury occurrence in western Lake Erie waters. Unpublished Ph.D. Thesis, Ohio State University, Columbus, Ohio.

Chemex Labs Ltd., 1973. Turbidity monitoring of a dredge plume. Unpublished Report by Chemex Labs Ltd., Calgary, Alberta. Project 13-01-02, for Canada Centre for Inland Waters, Burlington, Ontario 8 p.

______________, 1974. Lower Great Lakes dredging impact study. 1973. Unpublished contract report for CCIW (SR01. KW109-3-0554) 27 p.

______________, 1975. Port Stanley dredging disposal study 1974. Unpublished contract report for Public Works Canada, Ontario Region. 45 p.

Coastal Engineering Laboratory, Queen's University, (Kingston, Ont.) 1974. Containment of mercury polluted dredged spoil in dyked areas. Unpublished report prepared under contract for the Canadian Federal Department of Public Works, Ottawa. 41 p.

Cooley, R.V. and P.L. McCarty, 1976. Kinetics of microbiological aerobic decomposition of methyl mercury. In Environmental Geochemistry Vol 2. Ed. J.O. Nriagu. Ann Arbor Science. 451-472.

Dean, W.E., 1970. Fe-Mn oxidate crust in Oneida Lake, N.Y. Proc. 13th Conf. Great Lakes Res. Internat. Assoc. Great Lakes Res. 217-226.

Delfino, J.J. and G.F. Lee, 1968. Chemistry of manganese in Lake Mendota, Wisconsin. Environ. Sci. Technol. 2: 1055-1148.

Delisle, C.E., B. Hummel and K.G. Wheeland, 1975. Uptake of heavy metals from sediments by fish. Internat. Conf. on Heavy Metals in the Environment, Toronto. Abstr. C262-264 (also in press).

Huckabee, J.W. and R.A. Goldstein, 1975. Transport of mercurials in freshwater food chains.
Abstract International Conference on Heavy Metals in the Environment,

Toronto C76-78 (also in press).
International Working Group, 1975. Report on the abatement and control of pollution from dredging activities (Great Lakes). Public Works Canada, Ottawa. 227 p.
Jernalov, A., 1969. Conversion of mercury compound. In "Chemical Fallout". Ed. Miller and Borg. Publ. Charles C. Thomas Springfield, Illinois.
___________, 1970. Release of methylmercury from sediments with layers containing inorganic mercury at different depths. Limnol. Oceanogr. 15: 958-960.
Kemp, A.L.W., R.L. Thomas, C.I. Dell and J.-M. Jaquet, 1976. Cultural impact on the geochemistry of sediments in Lake Erie. Jour. Fish. Res. Board, Canada 33: 440-462.
Krishnappan, B.G., 1975. Dispersion of granular material dumped in deep water. Inland Waters Directorate Report, Scientific Series #55, Environment Canada, Ottawa. 114p.
Lam, D.C.L. and J.-M. Jaquet, 1976. Computations of physical transport and regeneration of phosphorus in Lake Erie. Fall 1970. Jour. Fish. Res. Board Canada. 33: 550-563.
McKinnon, A.E., R.J. Norstrom, A.S.W. de Freitas, 1975. Bioaccumulation of methyl mercury: A predictive model with parameters for Ottawa River Yellow Perch (Perca flavescens). Abstr. Internat. Conf. on Heavy Metals in the Environment Toronto. C23-34 (also, in press).
Mudroch, A. and V. Cheam, 1974. Chemical changes in pore water from dredge spoil disposed on the pilot island Mitchell Bay, Lake St. Clair, Ontario. CCIW Unpublished Report. 8 p.
Saitoh, H. and V. Cheam, 1975. A study of mercury release from lake sediment. Canada Centre for Inland Waters, Unpublished Report. 19 p.
Sly, P.G., 1977. A report on studies of the effects of dredging and disposal in the Great Lakes. Inland Waters Directorate, Scientific Publication Series, Ottawa (in preparation).
Sweeney, R., R. Foley, C. Merckel and R. Wyeth, 1975. Impacts of the deposition of dredged spoils on Lake Erie sediment quality and associated biota. Jour. Great Lakes Res. Internat. Assoc. Great Lakes Res. Vol 1 (1) 162-170.
Termarex Ltd. (and Barringer Research Ltd) Toronto, 1974. Report on further study to provide data for use in predicting the redistribution of liquid and solid contaminants that would result from the disposal of contaminated dredgings at artificial island site, Mitchell Bay, Lake St. Clair. Including: To monitor atmospheric mercury release at artificial island site, Mitchell Bay, Lake St. Clair. Report prepared for the Environmental Protection Service of the Canadian Federal Department of the Environment, Ottawa (DSS contracts OSS4-0094, OSS4-0220). 88 p.
Thomas, R.L., 1974. The distribution and transport of mercury in the sediments of the Laurentian Great Lakes system. Proc. Internat. Conf. on Transport of Persistent Chemicals in Aquatic Ecosystems. NRC Ottawa, Canada. I 1-16.
___________ and J.-M. Jaquet, 1976. Mercury in the surficial sediments of Lake Erie. Jour. Fish. Res. Board Canada. 33: 404-412.
Vernet, J.P. and R.L. Thomas, 1972. Le mercure dans l'environnement et le role de la geologie sedimentaire. Bull. B.R.G.M. (2), 111, 3: 43-61.
Williams, J.D.H., J.K. Syers, S.S. Shukla, R.F. Harris and D. Armstrong, 1971. Levels of inorganic and total phosphorus in lake sediments as related to other sediment parameters. Environ. Science and Technol. 5: 1113-1120.
_______________ and T. Mayer, 1972. Sediment diagenesis and regeneration of phosphorus. In, Nutrients and Natural Waters Ed. H.E. Allen and J.R. Kramer. Wiley Interscience, Toronto. 281-315.
_______________, J.-M. Jaquet and R.L. Thomas, 1976. Forms of phosphorus in the surficial sediments of Lake Erie. Jour. Fish. Res. Board Canada. 33: 413-429.
Wolery, T.J. and L.J. Walters, 1974. Pollutant mercury and sedimentation in the Western Basin of Lake Erie. Proc. 17th Conf. Great Lakes Res. 235-249.
Wong, P.T.S., Y.K. Chau, P.L. Luxon, and B. Silverberg, 1975. Methylation of lead and selenium in the environment. Abstr. Internat. Conf. on Heavy Metals in the Environment. Toronto C220 (also in press).

Key words: Great Lakes
dredging
disposal
sediment
exchange

SUMMARY OF STUDIES ON THE RELEASE OF CONTAMINANTS FROM DREDGED SEDIMENTS ON OPENWATER DISPOSAL

G. Fred Lee*

Center for Environmental Studies, University of Texas at Dallas, Richardson, Texas 75080 USA

## Introduction

Approximately 400 million cubic yards of sediments are dredged from US waterways each year for the purpose of maintenance of suitable channel and harbor depth for navigation. The cost of this dredging is on the order of $200 million per year with the cost varying from a few tens of cents per cubic meter to tens of dollars per cubic meter of sediment dredged. For many years the approach used to determine the site of dredged material disposal was one primarily based on economic considerations. However, in the late 1960's emphasis began to shift toward considering the potential significance of the chemical contaminants present in the sediments.

The U.S. Federal Water Quality Administration adopted bulk chemical criteria in which the total chemical content of the sediment or any particular constituent was used as a basis for determining the method of dredged material disposal that would be employed. Arbitrary numerical concentrations were to be used to determine whether the sediment was "polluted" or not. In some instances the bulk criteria for an element utilized by the Federal Water Quality Administration was less than normal crustal abundance. The adoption of these criteria caused a significant increase in the cost of dredged material disposal in many parts of the U.S. (Lee, 1975a).

A critical examination of the potential validity of bulk content criteria would lead one to conclude that this was not a valid basis for classifying sediments as polluted and thereby cause expenditure of additional funds for dredged material disposal based on this classification. In response to this situation, the U.S. Army Corps of Engineers initiated a $30 million, five-year research program which was designed to evaluate the actual environmental impact of dredged material disposal. This program is now in its fourth year. Information on the nature of this program can be obtained from the Environmental Effects Laboratory, USAE Waterways Experiment Station, Vicksburg, Mississippi, 39180, USA.

One of the areas of emphasis of this program was an evaluation of environmental impact of disposal of contaminated sediments in aquatic habitats. The author has been involved in development of a leaching test (elutriate test) which is designed to detect possible water quality problems in the dredged material disposal site water column. The elutriate test consists of mixing one volume of sediment with four volumes of dredging site water for 30 minutes followed by a one-hour settling period and filtration of the supernatant through 0.45 μ pore size filter. The concentrations of "soluble" constituents released during the elutriation procedure are compared to the ambient concentrations at the disposal site in order to determine whether sufficient release has taken place to have an adverse effect on water quality at the disposal site water column. A list of pertinent references to these studies is found in the References section at the end of this paper.

## Conclusions

1. U.S. waterway sediments taken from many industrial areas contain elevated concentrations of chemical contaminants such as heavy metals, chlorinated hydrocarbon pesticides, PCB's and aquatic plant nutrients such as nitrogen and phosphorus which could, if made available during dredged material disposal, have a significant adverse effect on receiving water quality.

2. The elutriate test as originally proposed must be modified to establish a well-defined oxidation-reduction regime during the period of the test. It is recommended that since one of the primary purposes of this test is to protect aquatic organisms in the disposal site water column, the mixing period be conducted under aerobic conditions using compressed air for agitation.

3. The elutriate tests conducted on sediments from a wide variety of waterways in the U.S. in general show the release of manganese and ammonia, the uptake of zinc and little or no change in copper, cadmium, nickel, mercury, iron, phosphate, chlorinated pesticides and PCB's. Bioassays using Palaemontes pugio (grass shrimp) for marine sediments and Daphnia magna for freshwater sediments showed that there is little or no acute toxicity of unfiltered elutriates.

4. A comparison of the concentrations of contaminants predicted to be released at the dredged material disposal sites showed that in general, the elutriate test is a valid procedure for estimating potential water quality problems at openwater disposal sites from hydraulically dredged sediments. In each case for compounds which could be of significance at the dredged material disposal site, the concentration in the tests exceeded the concentrations at the disposal site water column.

5. Mechanically dredged sediments dumped by a barge should utilize a modified elutriate test which avoids slurrying the sediments in the water column. This more properly represents the actual conditions at the disposal site for this type of dredging and disposal operation.

6. The likelihood of chronic exposure of water column organisms to excessive concentrations of contaminants at an openwater disposal site where disposal is accompanied by dumping is remote due to the rapid dispersion, settling and intermittent nature of dredging and disposal operations that take place in most openwater situations.

7. Alternate methods of disposal involving onland or disposal within diked areas with overflow of the supernatant water to the nearby water course may be more ecologically damaging than open water disposal as a result of the limited removal of potentially significant contaminants that occurs in most diked disposal operations. Also of concern is the fact that the discharge from diked disposal operations often enters natural waters where there is limited opportunity for mixing and dilution of the contaminants in the overflow waters.

8. While the elutriate test is not designed to evaluate potential problems of long-term toxicity to benthic or epibenthic organisms, the results of this test combined with the results of other investigators working within the U.S. Army Corps of Engineers Dredged Material Research Program as well as studies sponsored by the U.S. Environmental Protection Agency all show that chemical contaminants in dredged sediments have minimal adverse impact on water quality when openwater disposal methods are used.

## References


Lee, G.F., 1975a. Significance of chemical contaminants in dredged sediments on estuarine water quality. Report submitted to U.S. EPA to be published Proc. Estuarine Pollut. Workshop, Pensacola. Mimeo available from Center for Environmental Studies, University of Texas-Dallas.

Lee, G.F., 1975b. Potential environmental problems of dredging and dredged material disposal. Proc. 6th Pacific Coast Conf. on Dredging, 62-68.

Lee, G.F., 1976a. Dredged material research problems and progress. Environ. Sci. & Tech. 10:334-338.

Lee, G.F., 1976b. Environmental impact of dredging, dredged material disposal and dredged material research in the U.S. Center for Environmental Studies, University of Texas-Dallas, Occasional Paper No. 10.

Lee, G.F., 1976c. Evaluation of the impact of chemical contaminants in dredged sediments. To be published in Proc. of World Dredging Conf., World Dredging Asso. Mimeo available from Center for Environmental Studies, University of

Texas-Dallas.

Lee, G.F. & Plumb, R.H., 1974. Literature review on research study for the development of dredged material disposal criteria. U.S. Army Corps of Engineers Dredged Material Research Program. Vicksburg, Mississippi.

Lee, G.F., Lopez, J. & Mariani, G., 1975a. Leaching and bioassay studies on the significance of heavy metals in dredged sediments. Pres. at Inter. Conf. on Heavy Metals in the Environment. Toronto, Canada. (In press).

Lee, G.F., Piwoni, M., Lopez, J., Mariani, G., Richardson, J., Homer, D., & Saleh, F., 1975b. Research study for the development of dredged material disposal criteria. U.S. Army Corps of Engineers Dredged Material Research Program, Vicksburg, Mississippi.

Lee, G.F., Lopez, J.M. & Piwoni, M.D., 1976a. An evaluation of the factors influencing the results of the elutriate test for dredged material disposal criteria. Proc. ASCE Specialty Conf. on Dredging and Its Environmental Effects, Amer. Soc. Civil Engr., 253-288.

Lee, G.F., Lanza, G.R. & Mariani, G.M., 1976b. Significance of sediment associated contaminants in water quality evaluation. Center for Environmental Studies, University of Texas-Dallas, Occasional Paper No. 11.

*Also with EnviroQual Consultants and Laboratories, Plano, Texas, USA.

Key words: dredging
environmental impact
pollution

# OBSERVATIONS OF INTERACTIONS BETWEEN WATER AND SEDIMENT WITH A 30-kHz-SEDIMENT ECHOSOUNDER

H.E. Müller

Geographical Institute of the University, Kiel, Germany.

Summary

By means of a 30-kHz-sediment echosounder longitudinal profiles were taken of the Blunker See, situated in the young moraine area of Schleswig-Holstein, in monthly intervals during two years. They reveal that "acoustic transparency" and "acoustic non-transparency" of the uppermost sediment layer are annually occurring phenomena that are closely connected with dynamic processes of the budget of substances. If during the stagnating phase of the lake the hypolimnetic water becomes oxygen-free, due to anaerobic decomposition processes in the water and in the sediment, the acoustic transparency of the uppermost sediment layer decreases. With oxygen saturation at the time of turnover, the sediment after a short time-lag becomes acoustically transparent again. The change from anaerobic to aerobic phase in the water and the uppermost sediment layer indirectly causes a change in the ultrasonic properties of the sediment. Acoustic non-transparency of sediments is due, inter alia, to their content of gas bubbles which are generated by metabolic gases that form with the anaerobic decomposition of organic matter. With oxygen saturation of the whole water body depending on circulation, an aerobic decomposition phase occurs, and the intermediate metabolic products that exist in the sediment as gas bubbles from the anaerobic phase are oxidized. By the disappearance of gas bubbles the uppermost sediment layer becomes acoustically transparent again.

## Introduction

In connection with investigations on the budget of substances, bathymetric echosounder profiles of 20 lakes in Schleswig-Holstein (Germany) were taken during the years 1973 to 1975. The profiles necessary to construct exact bathymetric maps were recorded by means of a 30-kHz-sediment echograph and later also by a 200-kHz-echograph (Müller, 1974, 1976a; Kiefmann & Müller 1976). Taking bathymetric profiles for mapping purpose by an echosounder was performed for the first time on these lakes in Schleswig-Holstein. The complete length of all profiles recorded during these investigations adds up to about 600 km.

According to its special design the 30-kHz-sediment echosounder was used to obtain records of different subbottom layers, and to give additional informations on the basis of the unconsolidated holocene sediments. However while taking the profiles and interpreting the results, it became obvious that this could be obtained only to a limited extent. The ultrasonic properties of the sediments in the lakes under survey revealed variations relating to space as well as to time. These facts made it mostly impossible to measure by echosounder exactly the expansion and thickness of the unconsolidated holocene sediments. Within one profile recorded "acoustic transparent" as well as "acoustic non-transparent" sediments were observed. Similar observations of such acoustic non-transparent sediments have repeatedly been made (Müller, 1974, 1975; Schubel & Schiemer 1973). But *in situ* observations concerning the temporal variation of the ultrasonic properties of holocene sediments are so far unknown to the author.

One of the reasons for the acoustical non-transparency of holocene sediment is their content of gas bubbles (Grim et al 1970; Schubel & Schiemer 1973; Stiles et al 1969). The interfaces gas/water respectively gas/sediment act as total reflectors to ultrasonic waves. Therefore the strata beneath a sedimental surface interspersed with gas bubbles cannot be recorded by means of the echosounder. The generation of gas bubbles in the sediment of the lakes under survey is due to the microbial decomposition of the organic matter (Ohle 1968), and is anaerobic in all cases.

The hypothesis that the temporal change of the acoustic properties of the uppermost sediment layer is closely connected to the annual dynamic of the budget of substances in the water body and sediment of the lakes, was substantiated during the observations.

## Investigations

During regular intervals during a two years' period ultrasonic profiles of the sediment were recorded on the lake Blunker See which is situated in the young moraine area of south-western Schleswig-Holstein (Fig. 1). At the same time physical and chemical investigations of the water body were performed. Here and at a couple of other lakes being echo-surveyed sediment samples were taken from appropriate and interesting parts.

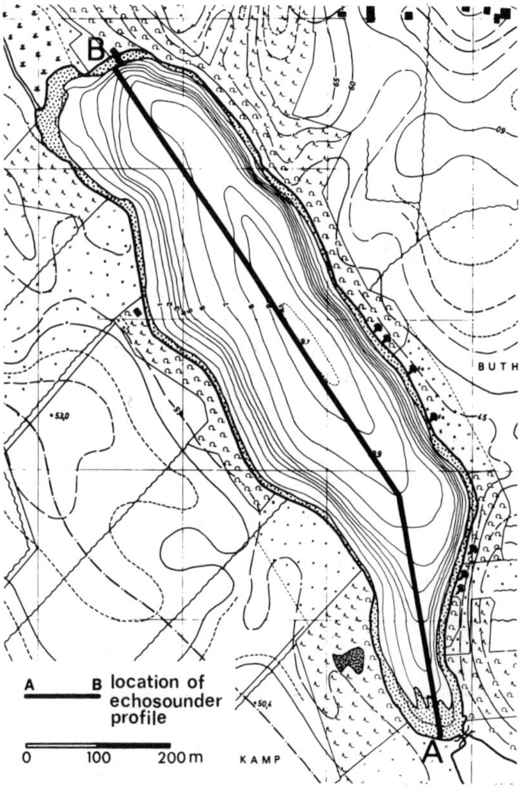


Fig. 1. Bathymetric map of the Blunker See and location of echosounder profile.

The echograms were always taken on exactly the same profile by means of the 30-kHz-sediment echosounder model LAZ 17 C with sediment-transceiver LVG 12 R 01 and Ni-transducer LSE 23, system ELAC, Kiel, Fed. Rep. of Germany. Results are displayed on a dry paper recorder. Using depth range 0 - 24 m there are 534 soundings p. min. with an acoustical pulse length of about 0.3 ms and a wave length of 4.8 cm at 1440 m/s sound velocity for fresh water. Thus the premises are given for obtaining a high depth resolution. The transducer mounted over the side of a small boat had a 17 x 26° beaming at the -3 dB point and a surface of 135 cm$^2$.

The physical and chemical observations include vertical profiles of temperature and oxygen saturation above the deepest part of the lake, in order to follow the phases of turn-over and stagnation, and to predict the conditions of aerobic and anaerobic decomposition of organic matter in water and sediment. Temperature as well as oxygen saturation were measured in situ electrometrically.

The sediment samples taken by a Lenz-bottom sampler were to give information about consistency and structure of the uppermost sediment layer. Size and dispersion of the gas bubbles in the sediment which are essential parameters to quantify their influence on the acoustic transparency or acoustic non-transparency, could be estimated only very roughly from the sediment taken by the Lenz-bottom sampler. Continuing these investigations a sampling procedure in which the sediment is frozen in situ, is being evaluated. The method conserves the structure of the extremely liquid sediment under hydrostatic pressure on the bottom of the lake. This procedure allows one to measure the size and dispersion of the enclosed gas bubbles.

## Results

The sediment echograms of the Blunker See, taken in regular monthly intervals, clearly show a rhythmical change of "acoustically transparent" and "acoustically non-transparent" phases during the year (Fig. 2 A-H, left. (The original recordings show much more details, but could not be reproduced in this paper)). A phase with layer recording in the sediment - i.e. acoustic transparency of the uppermost sediment layer - from winter to spring (Fig. 2 A-C, F and G) is followed by a phase of acoustic non-transparency of the uppermost sediment layer during summer and autumn (Fig. 2 D and E). This rhythmical change of the ultrasonic properties of the uppermost sediment layer is caused by circulation controlled dynamic processes of the budget of substances in the lake, especially the conditions of anaerobic and aerobic decomposition of organic

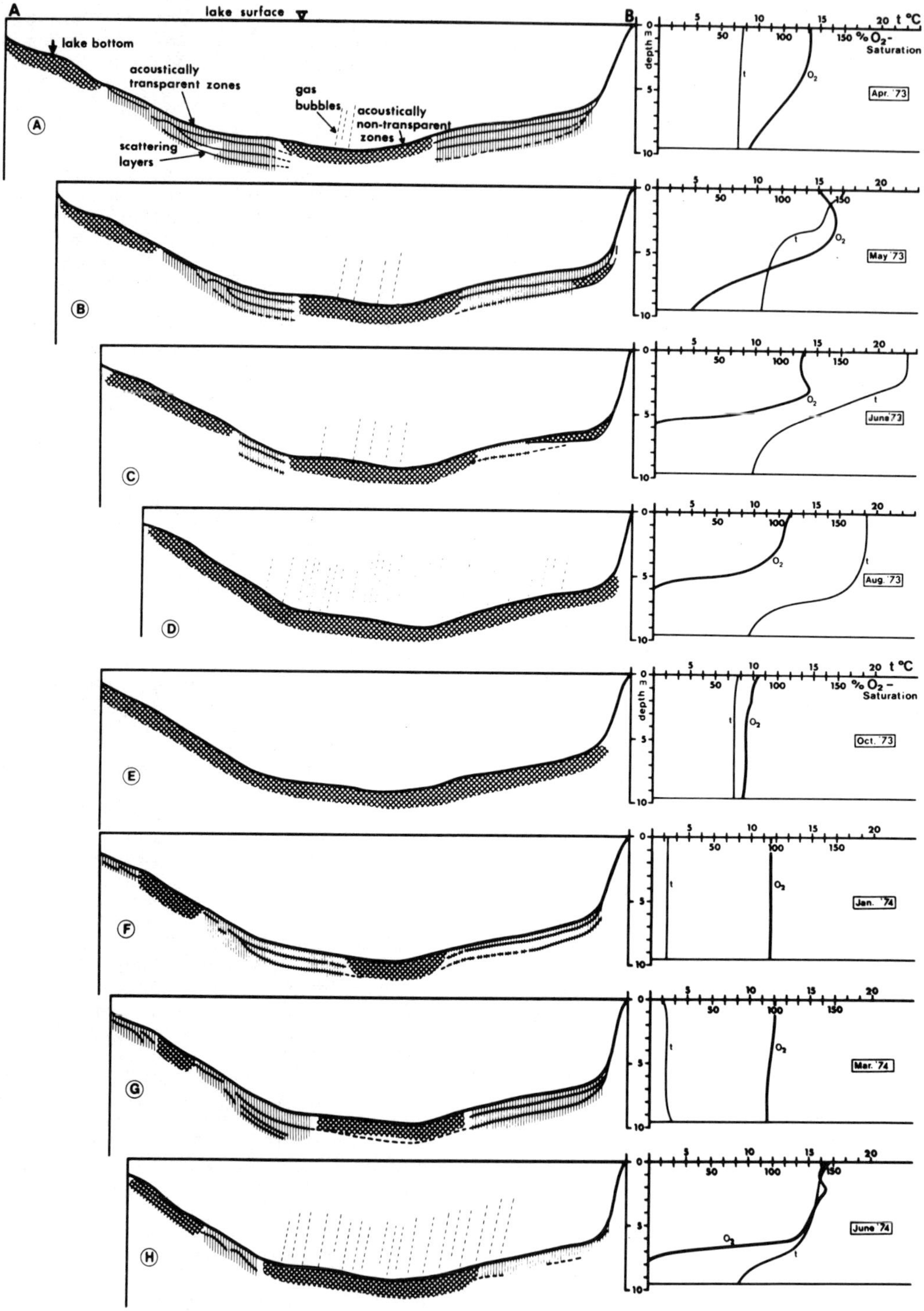


Fig. 2. Sediment echograms and vertical profiles of temperature and oxygen saturation from Blunker See (April 1973 - June 1974).

matter. In the anaerobic conditions gas bubbles are generated in the uppermost sediment layer, but they are not under aerobic conditions. The indicator for the occurrence of anaerobic decomposition processes in sediment and water body is the decrease of oxygen saturation caused by autolytic and microbial decomposition of organic matter which sinks from the epilimnion into the stagnating hypolimnion.

The yearly course of temperature and oxygen saturation (Fig. 2 A-H, right) gives the key to understand the change in the recordings of the sediment profiles during the annual rhythm. During turnover in spring (Fig. 2 A, F and G) the whole water body from surface to sediment is practically oxygen saturated, and the anaerobic decomposition processes in the uppermost sediment layer have given way to aerobic decomposition. Thus the generation of gas bubbles as a result of reducing processes, occurring under anoxic conditions and developing methane, hydrogen, nitrogen, hydrogen sulphide, is stopped. The reduced compounds of the metabolic gases are oxydized by the oxygen transported during the circulation at the overturn and - at least partially - removed from the uppermost sediment layer. Samples from such sediments taken by a Lenz-bottom sampler show in structure and consistence a sediment of a homogeneous macrostructure poor or void of gas bubbles. In this stage of the turnover in spring the uppermost sediment layer can be penetrated by the ultrasonic frequency in use, and deeper scattering layers can be made visible.

With the beginning of stratification (Fig. 2 B) a stagnating hypolimnion forms and oxygen consumption starts in the deep water (Fig. 2 B, $O_2$-curve). During this phase the organic matter is decomposed under anaerobic (reducing) conditions, bubbles of metabolic gases are formed in the uppermost sediment layer and prevent ultrasonic waves from penetrating this layer. It is not possible to get records from deeper layers (Fig. 2 C and D). Bottom samples taken during the summer stagnation phase show a somewhat sludge-pasty structure and a high content of gas bubbles, and due to the decrease of pressure (depth of water about 10 m) gas bubbles exploded spontaneously.

During the autumn turnover following the stagnation in summer, oxygen saturation in the whole water body increases again, and the metabolic gases forming gas bubbles in the uppermost sediment layer must be oxydized - as far as they do not escape - before it is possible to penetrate the sediment by means of the echosounder in use. Thus a certain timelag between the diagramm of oxygen saturation and the sediment echograms becomes understandable (Fig. 2 E).
Only in winter after the beginning of turn-over or in early spring it was possible to penetrate the uppermost sediment layer with the 30-kHz-echosounding device (Fig. 2 F and G). The sediment becomes acoustically non-transparent again due to anoxic conditions in the hypolimnetic water with anaerobic decomposition processes (Fig. 2 H).

To analyse the budget of substances in a lake connected with more or less advanced eutrophication, a knowledge of regions with sediments containing gas bubbles is indispensable.
That applies to lake basins with a simple morphological structure like the Blunker See considered here, but even more to lakes with more complicated structures, where due to the existence of compartments no general statements about budget of substances can be made.

Echosounder recordings of other lakes under survey reveal in different water depths areas of sedimentation with mud containing gas bubbles, close to other ones which can be identified from the echogram as poor or void of gas bubbles forming scattering layers beneath the surface. The importance of sediments containing gas bubbles in the budget of substances of a lake is well known, but so far it was not possible to explore the extent of zones with sediments containing gas bubbles. However now it can be done by means of the 30-kHz-echosounding device.

Sediments containing gas bubbles generally develop and last only if the conditions of decomposition of the organic matter in sediment a n d hypolimnion are anaerobic as revealed during the investigations on lake Blunker See. Recording acoustically transparent and acoustically non-transparent zones makes it possible to analyse the ability of lakes to mineralize or not the organic matter in an aerobic manner. A total aerobic mineralization of organic matter never is found in natural waters, a nearly complete mineralization however always requires aerobic conditions of decomposition. Highly eutrophicated water systems with a large load of organic matter, in Schleswig-Holstein always reveal sediment containing gas bubbles, at least during the summer if a stagnating hypolimnion exists. Lakes containing little organic matter mostly show sediment without gas bubbles and aerobic decomposition conditions in the hypolimnion.

Recordings of the lake bottom, taken by means of the 30-kHz-sediment echosounder help to solve various problems related to lake investigations. The profiles can be used to get statements related to the budget of substances in lakes particularly with regard to the interactions between sediment and water. It is possible to get information on the spatially and annually occurring changes of the sediment surface with its close connections to the dynamic processes of the budget of substances of the lake, thus opening the possibility of conclusions as to the load of a lake with sinking materials fit for decomposition (Müller 1976b).

## References


Grim, M.S., C.L. Drake & J.R. Heirtzler, 1970. Sub-bottom study of Long Island Sound. Bull. Geol. Soc. Amer. 81: 649-666.

Kiefmann, H.M. & H.E. Müller, 1976. Die Tiefenkarte des Grossen Plöner Sees als Beitrag zur Erforschung einer frühen Kulturlandschaft und der regionalen Morphogenese. Offa 32.

Müller, H.E., 1974. Profillotungen in Seen des schleswig-holsteinischen Jungmoränengebietes mit einem 30 kHz Sediment-Echolot. Wasser und Boden 16: 135-138.

Müller, H.E., 1975. Echographische Beobachtungen jahresrhythmischer Veränderungen im Sediment von Seen. Deutsche Hydrogr. Zeitschr. 28: 26-31.

Müller, H.E., 1976a. Zur Morphologie pleistozäner Seebecken im westlichen schleswig-holsteinischen Jungmoränengebiet. Zeitschr. f. Geomorphologie N.F. 20 (3): 350-360.

Müller, H.E., 1976b. Der Stoffhaushalt des Molfsees. Untersuchungen zum Problem der Gewässereutrophierung. Jahresbl. des Kommunalver. Schulensee-Rammsee-Molfsee für 1975. Kiel 1976. 32-43.

Ohle, W., 1968. Chemische und mikrobiologische Aspekte des biogenen Stoffhaushaltes der Binnengewässer. Mitt. Int. Verein. Limnol. 14: 122-133.

Schubel, J.R. & E.W. Schiemer, 1973. The cause of the acoustically impenetrable, or turbid, character of Cheasapeake Bay sediments. Mar. Geophys.. Res. 2: 61-71.

Stiles, N.T., L.R. Breslau & M.D. Beeston, 1969. The riverbed roughness and subbottom structure of the main shipping channel to Sai Gon, RVN (Nga Bay, Long Tau, Nha Be, and Sai Gon Rivers). Proc., 12th Conf. on the Naval Minefield, Naval Ordnance Laboratory Technical Report 69-69, 1: 327-368.



Keywords: Acoustical sediment observations
Budget of Oxygen
Echosounder sediment observations
Gas bubbles in sediments
Sub-bottom penetration

THE USE OF POLAROGRAPHY IN THE INVESTIGATION OF HYPOLIMNETIC WATERS AND THE SEDIMENT - WATER INTERFACE

W. Davison

Freshwater Biological Association, Ambleside, Cumbria, England

Summary

The suitability of polarography as a technique for investigating the ionic equilibria of hypolimnetic waters is discussed. The results to date show the possibility of simultaneously measuring dissolved oxygen and sulphide, ferrous and manganous ions. Manganous and ferrous ions, which often originate from sediments, appear to be particularly amenable to measurement, values for both concentrations being obtainable in less than 1 minute. Traditional methods for their determination have been shown to be in good agreement with the polarographic technique. Concentration - lake depth profiles for ferrous and manganous ions have been measured and their seasonal progression is shown. The possibilities of performing in situ measurements are discussed with a view to obtaining information about the water chemistry of the sediment - water interface.

Introduction

In order to perform a detailed study of the chemical exchange that occurs between water and sediment it is necessary to have some means of measuring chemical concentrations in a very narrow boundary region. Any well defined interface must necessarily have a negligible volume so the use of discrete sampling techniques is precluded because they would include water which does not reside in the boundary. An in situ measuring technique is therefore required. It should have the desirable properties of only sensing concentrations within a small well defined region and being able to perform measurements without disturbing the overall equilibrium of the system. The platinum electrode, as used for redox measurements, is capable of fulfilling these requirements (Hargrave, 1972) but suffers from the fact that quantitative interpretation of such measurements for natural waters is not possible (Stumm, 1967). Stumm & Morgan (1970) have stated that only a few elements - C, N, O, S, Fe, Mn - are predominant participants in aquatic redox processes, and so the determination of the concentration of their individual oxidation states is of primary importance. Recently polarography has been applied to measurements in the hypolimnion and its capability for measuring dissolved oxygen and sulphide, iron II and manganese II ions has been established (Davison, 1976; Davison, 1977; Heaney & Davison, 1977). This paper will (a) illustrate the features of the technique by describing the work which has already been performed on hypolimnetic water, (b) describe how the measurements may be applied to the study of solution equilibria and (c) examine the potential application of the technique to the study of the sediment water interface.

Experimental procedure

The water samples studied were collected from Esthwaite Water, one of the more productive of the English Lakes (Macan, 1970), which is 2.5 km long, has a maximum width of 0.6 km and a maximum depth of about 15 m. The lake is mixed in winter, but in summer there is thermal stratification and a deoxygenated hypolimnion develops. The alkalinity of the bottom waters is within the range 0.4 - 1.1 m eq $\ell^{-1}$ and the total ionic strength varies between 1.5 and 2.2 m mol $\ell^{-1}$.

Water samples from different depths were collected at a buoy over the deepest position in the lake using either a Friedinger or Ruttner sampler. Care was taken to prevent exposure of the sample to air by running the water to the base of darkened acid washed glass bottles so that at least three times the volume of the bottle was flushed through. Stoppers were replaced carefully to avoid trapping air bubbles. Measurements were made within 2-6 h of sampling. The samples were handled in a purpose-built perspex glove box in which the whole cell assembly was fitted. The cables to the cell were sealed into a rubber grommet. The polarographic capillary for the dropping mercury electrode passed through a special rubber stopper which was sufficiently flexible to allow successful transmission of drop timer vibrations. The argon-carbon dioxide mixture used for deoxygenation was also used to flush the glove box. Polarograms of the lake water were recorded immediately and then repeated after 5 minutes deoxygenation. Potentials were measured against a saturated calomel electrode. The pH was monitored while degassing, and kept constant to $\pm$ 0.1 units, at a value appropriate to the natural water which was in the range pH 6.5 to 7.0, by regulation of the proportions of the Ar-$CO_2$ gas mixture. The polarographic equipment and instrument settings used have been previously described (Davison, 1976). All operations were performed at 22 ± 2°C.

Polarographic measurements in the hypolimnion

Two polarographic techniques were used. The direct current (d.c.) sampled mode may be regarded as a more refined version of classical polarography, with an output that eliminates fluctuations due to drop growth. Differential pulse is a more advanced technique that considerably reduces the capacitive interference and gives a peaked output which approximates to the first derivative of the polarographic wave.

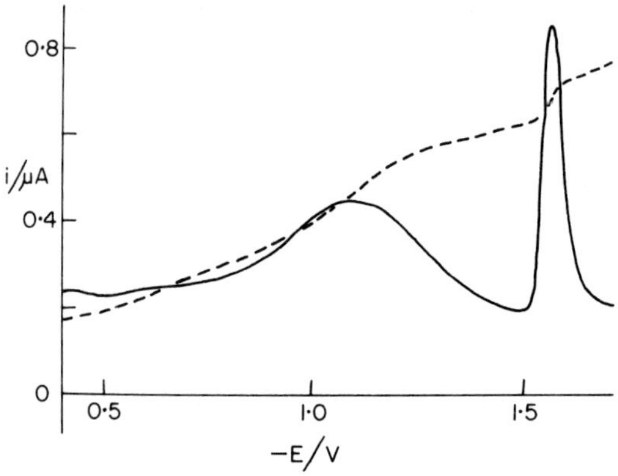


Figure 1. Polarograms (current, μA versus potential, V) of lake water with only manganese present. ----, d.c. signal; ——, differential pulse signal.

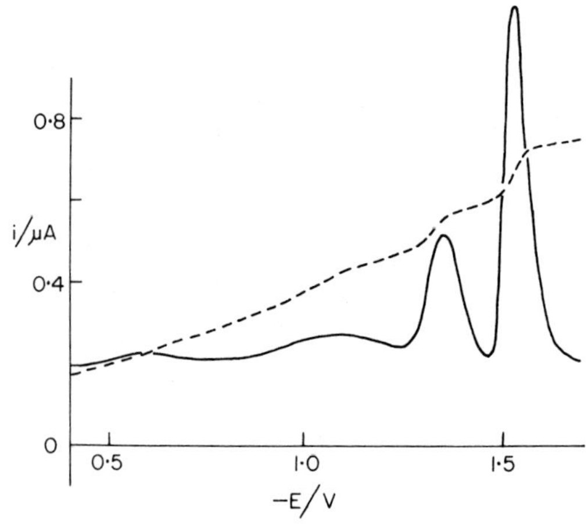


Figure 2. Polarograms (current, μA versus potential, V) of lake water with iron and manganese present. ----, d.c. signal; ——, differential pulse signal.

Figure 1 shows d.c. and differential pulse scans performed on the same sample of hypolimnetic water at an early stage in summer stratification, when manganese II was the dominant electroreducible species. The differential pulse signal is a well defined peak at approximately - 1.55 V which corresponds to the half wave potential of the d.c. wave. Figure 2 shows the equivalent scans when both ferrous and manganous ions are present. The ferrous peak occurs at approximately - 1.35 V and this separation of 0.2 V provides very good discriminations between the two elements. This is especially true for the differential pulse case, but the two d.c. waves are still clearly discernible. Later in the summer (Figure 3) the redox conditions progress to a further stage when soluble sulphides appear in solution giving two more signals at - 0.6 V and - 1.1 V. These additional peaks are well separated from the iron and manganese signals making the simultaneous measurement of the peaks due to all three elements quite possible.

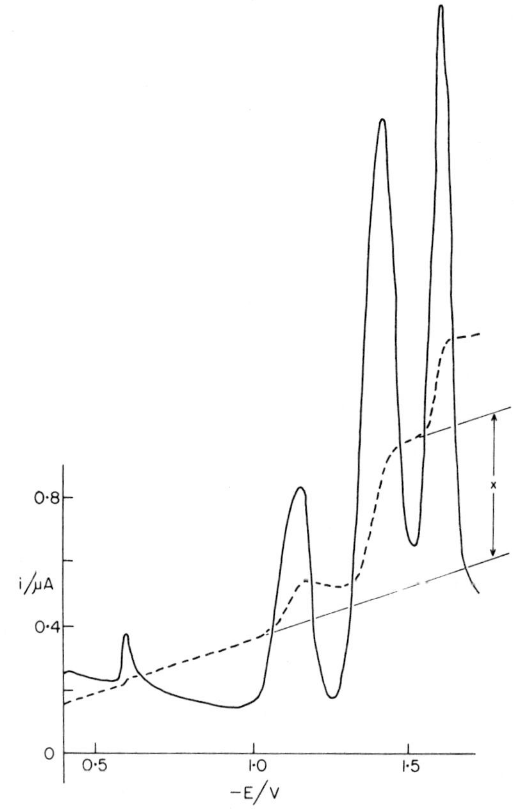


Figure 3. Polarograms (current, μA versus potential, V) of lake water with iron, manganese and sulphide present. ---- d.c. signal; ____, differential pulse signal.

Examination of the electrode reactions (Table 1) shows that the sulphide signal results from a different process to the others.

Table 1. Electrode reactions and half wave potentials.

| Electrode reactions | $E_{½}$/volts |
|---|---|
| $Hg + S^{2-} = HgS + 2e^-$ | - 0.6 |
| $O_2 + 2H_2O + 2e^- = H_2O_2 + 2OH^-$ | - 0.1 |
| $H_2O_2 + 2e^- = 2OH^-$ | - 1.1 |
| $Fe^{2+} + 2e^- = Fe$ | - 1.35 |
| $Mn^{2+} + 2e^- = Mn$ | - 1.55 |

The electrode reactions of oxygen, iron and manganese are all reduction processes, but the sulphide reaction has its electrons on the opposite side of the equation and is therefore an oxidation. Mercury and sulphide react at potentials more positive than - 0.6 V (Turner et al., 1975) to form mercuric sulphide. On scanning to more negative potentials this process becomes less favourable and eventually ceases with the result that a polarographic wave is observed at - 0.6 V. The two oxygen reactions show that oxygen is reduced successively, under neutral pH conditions, to hydrogen peroxide and eventually hydroxide. This results in two waves at - 0.1 V and -1.1 V.

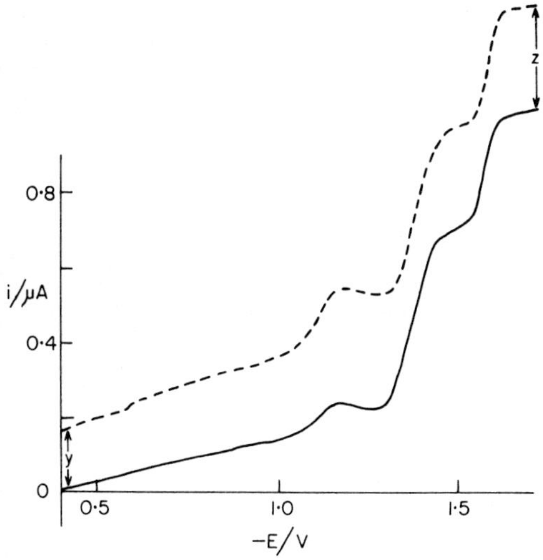


Figure 4. D.c. polarograms (current, μA versus potential, V) of lake water before, ----, and after, ____, deoxygenation.

Figure 4 shows the effect of oxygen on the d.c. signal. After recording the polarogram, oxygen was removed by degassing with an $Ar\text{-}CO_2$ gas mixture which did not alter the pH, and then another polarogram was measured. Since the scan starts at - 0.4 V it is already on the plateau of the first d.c. oxygen wave and so the difference in height, y, between the two lines at - 0.4 V reflects the concentration of dissolved oxygen which was present prior to deoxygenation. The second oxygen wave can be seen in the sulphide free d.c. signal of Figure 2 as a poorly formed wave prior to the iron signal, but cannot be observed in Figure 4 owing to the masking effect of the sulphide signal. However, the difference in height of the two scans at - 1.7 V, indicated by z, is twice that of y, showing that the oxygen reactions are obeying their theoretical equations given in Table 1. Figure 4 illustrates a further phenomenon. The deoxygenation procedure has removed sulphide as the volatile $H_2S$ gas and this is shown by the signals at - 0.6 V and - 1.1 V being much smaller for the continuous line, when compared with the broken one. This effect is more clearly illustrated by a comparison of the differential pulse measurements which are shown in Figure 5.

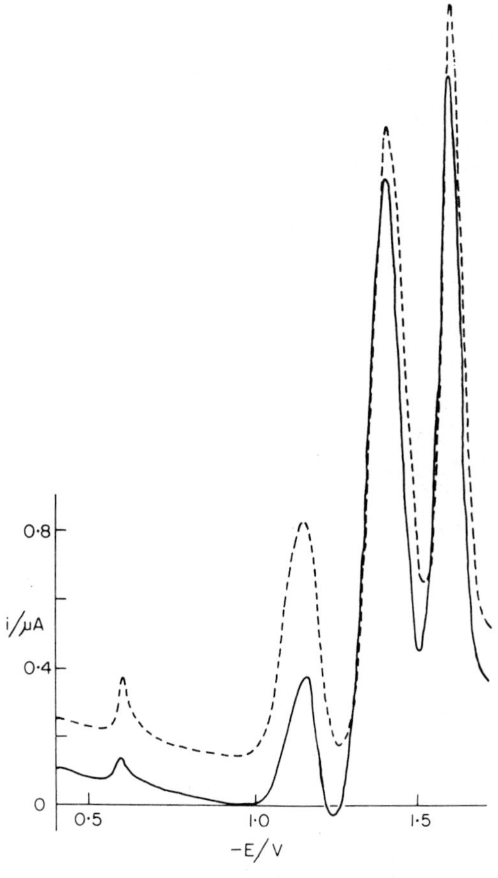


Figure 5. Differential pulse polarograms (current, μA versus potential, V) of lake water, before, ----, and after, ____, deoxygenation.

The peaks at - 0.6 V and - 1.1 V are markedly depressed after the solution has been deoxygenated. The effects due to oxygen are less pronounced for the differential pulse case than for the previously discussed d.c. case. The baseline is shifted, but in a non-quantitative fashion because it merely reflects the joint effect of the two very broad peaks which result from the elongated d.c. oxygen waves. The sensitivity of the technique may be roughly gauged from a knowledge of the approximate concentrations of Figures 4 and 5. They were : sulphide, $4.10^{-6}$ mol $\ell^{-1}$; ferrous iron, $10^{-4}$ mol $\ell^{-1}$; manganous ions $5.10^{-5}$ mol $\ell^{-1}$.

From this discussion of the polarography of hypolimnetic samples it is apparent that sulphide, iron II, and manganese II can all be measured simultaneously by a single differential pulse scan, which, with the instrumental settings used, takes 2 minutes 10 seconds to perform. Owing to the iron and manganese peaks being close together a single scan taking less than 1 minute could suffice for the measurement of these two elements. Oxygen can be measured in the same sample by a further d.c. sampled scan. The polarographic measurements have been compared quantitatively (Davison, 1977; Heaney & Davison, 1977) with other techniques and shown to be in good agreement. Depth profiles for iron II and manganese II have been measured by the technique and some examples are given in Figure 6.

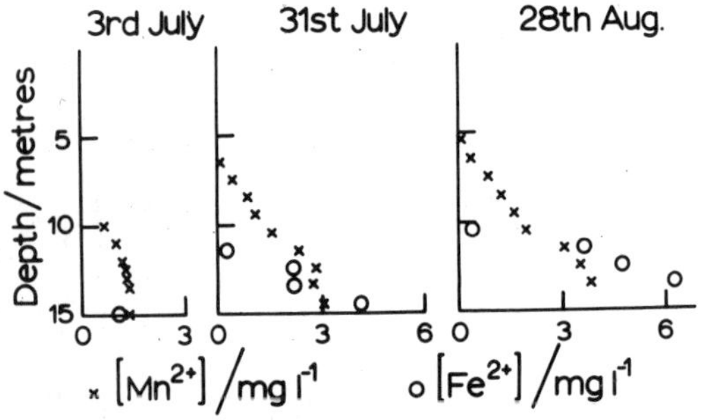


Figure 6. Concentration depth profiles for manganese II and iron II in Esthwaite Water.

These show the expected shape which has been well established for Esthwaite water by more conventional measurements (Mortimer, 1941-1942; Talling & Heaney, personal comm.).

## Applications to solution equilibria

The ability of polarography to perform useful analytical measurements of hypolimnetic waters has clearly been demonstrated. However, the special virtue of the technique, the fact that it measures only free equilibrium concentrations, extends its use beyond that of simple analysis. Since the measurement is actually performed at a small electrode probe, only the equilibrium conditions of the solution in the immediate vicinity are affected, the bulk of the solution remaining unchanged. Particular, well defined components in solutions are measured, namely the electroreducible species and those species that are in sufficiently rapid equilibrium, during the time scale of the measurements, that they cannot be distinguished from the electroreducible form. Particulate matter and any strongly bound organic complexes are therefore automatically excluded.

The measurement of soluble sulphide in the presence of ferrous iron is of particular interest because ferrous sulphide may be regarded as an insoluble salt; the negative log of its solubility product, $pK_s$, is 16.38 at 25°C (Sillen & Martell, 1964). Figure 3 shows simultaneous measurements of the concentration of ferrous and total sulphide ions in the same sample of hypolimnetic water. The measured values were $1.01 \cdot 10^{-4}$ mol $\ell^{-1}$ and $3.63 \cdot 10^{-6}$ mol $\ell^{-1}$ respectively, and the pH of the water was 6.85. Using $pK_1$ ($H_2S = HS^- + H^+$) = 7.00 and $pK_2$ ($HS^- = S^{2-} + H^+$) = 13.00 the measured solubility product may be calculated to be $10.8 \cdot 10^{-17}$ $mol^2$ $\ell^{-2}$. This value is in excess of the theoretical one of $4.2 \cdot 10^{-17}$ $mol^2$ $\ell^{-2}$. However, the difference is slight, and may be accounted for by the possible errors which exist in the calculation. These are the uncertainties in the literature values of $pK_s$, and of $pK_1$ and $pK_2$ for the sulphide equilibrium, and the temperature difference between the lake water and that appropriate to the stability constants. This calculation has therefore shown that the ferrous and sulphide content of the lake water are near the maximum possible for this pH, if the inorganic chemistry can be assumed to be simple. However, it may well be that the polarographic measurement of ferrous iron can include some complex species and if so, these should be taken into account in solubility product calculations.

Re-examination of Figure 3 reveals an anomaly. There are two peaks that have been attributed to sulphide. As already indicated the smaller one at - 0.6 V is the true sulphide peak while the one at - 1.1 V cannot be entirely attributed to sulphide species in solution. Unlike the - 0.6 V peak it is a cathodic signal, and it is not present in a laboratory prepared standard sulphide solution. It can, however, be reproduced in synthetic solutions which contain both ferrous and sulphide ions. It does not appear immediately the sulphide and ferrous components are mixed, but gradually increases in magnitude over a period of one hour. Figures 4 and 5 have already shown that it decreases as sulphide is removed from solution as $H_2S$. These facts strongly suggest the presence of an electroactive ferrous sulphide complex.

The d.c. signal (Figure 3) provides the clearest information. Preliminary measurements appear to indicate that the sum of the iron II signal, and the signal at - 1.1 V, that is x in Figure 3, remains constant as sulphide is removed. Furthermore the d.c. signal does not maintain a steady current plateau. This peaked effect of the d.c. wave and the contribution of iron II to the signal suggests that a ferrous sulphide species may be adsorbed prior to reduction (Delahay, 1965). No information on any relationship between the - 1.1 V signal and the - 0.6 V sulphide signal is so far available. It is clear, however, from these preliminary results that polarography has revealed a hitherto unknown complex, so demonstrating the usefulness of the technique for examination of solution equilibria.

## Applications to measurements at the sediment water interface

Since polarographic measurements take place at an electrode surface there is a possibility of using the technique to probe a sediment-water interface to obtain selective measurements of the chemical species present. In situ measurements of oxygen in seawater have already been made using the dropping mercury electrode (Foyn, 1967). The cell was rapid in its response and was able to resolve fine structure in the oxygen depth profiles. So the practical problems associated with using polarography in field conditions have already, to a large extent, been overcome. The discrete water body which is actually measured is dictated by the dimensions of the dropping mercury electrode. A free falling mercury drop usually has a diameter of less than 1 mm and the chemical equilibrium of the solution is not affected beyond the diffusion layer of this drop. Since the diffusion layer thickness is typically 1/20th that of the drop diameter (Lingane, 1958) the spacial region in which the polarographic measurements are made can be defined to 1 mm.

In conclusion it appears that the requirements that seemed necessary to make viable in situ measurements at an aquatic interface have been substantially met. Geometric discrimination to within 1 cubic mm should be sufficient for many studies and the polarographic technique can provide direct quantitative measurements of a single oxidation state for four of the six elements that were stated by Stumm & Morgan (1970) to be major participants in aquatic redox processes.

## References


Davison, W., 1976. Comparison of differential pulse and d.c. sampled polarography for the determination of ferrous and manganous ions in lake water. J. Electroanal. Chem. 72:229-237.

Davison, W., 1977. The polarographic measurement of $O_2$, $Fe^{2+}$, $Mn^{2+}$ and $S^{2-}$ in hypolimnetic water. Limnol. Oceanogr. in press.

Delahay, P., 1965. Double layer and electrode kinetics. John Wiley, New York.

Foyn, E., 1967. Density and oxygen recording in water. In: H.L. Golterman & R.S. Clymo (Ed.): Chemical environment in the aquatic habitat. N.V. Noord-Hollandsche Uitgevers Maatschappij, Amsterdam. p. 127-132.

Hargrave, B.T., 1972. Oxidation-reduction potentials, oxygen concentration and oxygen uptake of profundal sediments in a eutrophic lake. Oikos 23 : 166-177.

Heaney, S.I. & W. Davison, 1977. The determination of ferrous iron in natural waters with 2,2' bipyridyl. Limnol. Oceanogr. in press.

Heyrovsky, J. & P. Zuman, 1968. Practical polarography. Academic Press, London.

Lingane, J.J., 1958. Electroanalytical Chemistry. 2nd Ed., Interscience.

Macan, T.T., 1970. Biological Studies of the English Lakes. Longmans, London.

Mortimer, C.H., 1941-42. The exchange of dissolved substances between mud and water in lakes. J. Ecol. 29: 280-329, and 30: 147-201.

Sillen, L.G. & A.E. Martell, 1964. Stability constants of metal-ion complexes. Special Publications No. 17. The Chemical Society, London.

Stumm, W., 1967. Redox potential as an environmental parameter; conceptual significance and operational limitation. Advances in water pollution research. Vol. 1. Water Pollution Control Federation. Washington. D.C.

Stumm, W. & J.J. Morgan, 1970. Aquatic Chemistry. John Wiley, New York.

Turner, J.A., R.H. Abel & R.A. Osteryoung, 1975. Normal pulse polarographic analysis based on mercury anodisation : Sulphide and iodide. Anal. Chem. 47:1343-1347.



Key words: equilibrium
hypolimnion
polarography
sediment
water

STUDY OF TRANSPORT THROUGH THE SEDIMENT-WATER INTERFACE IN LAKES USING NATURAL RADON-222

D.M. Imboden & S. Emerson[1]

Swiss Federal Institute of Technology, Institute for Aquatic Sciences and Water Pollution Control, Dübendorf, Switzerland

## Summary

The flux of natural radon from the sediments in oceans and lakes can be used for tracing vertical mixing through the sediment-water interface and in the overlying water. A one-dimensional diffusion theory relates the standing crop of excess radon $\Sigma Rn^*$ to the activity of thorium-230 and radium-226 in the sediment. In the deep-sea, $\Sigma Rn^*$ which lies between 0.2 and 20 dpm $cm^{-2}$ depends mainly on the thorium activity. In contrast, due to the commonly much higher sedimentation rates, $\Sigma Rn^*$ in lakes is determined by the radium-activity.

Measured radon activities in Greifensee yield $\Sigma Rn^* = 0.2$ dpm $cm^{-2}$ in good agreement with the measured "escapable" radium activity of 0.6 dpm/g. The radium activity in sediments of other lakes suggest that the radon method is applicable to a larger class of lakes. $\Sigma Rn^*$ may also be used to determine boundary layers at the sediment-water interface of time varying thickness.

## Introduction

Interstitial water profiles of solutes near the sediment-water interface in lakes may exhibit large variations, including sign reversal in the concentration gradient. This may be the case for phosphate in highly eutrophic lakes (Emerson, 1976). As a result, the flux of solutes from the sediments into the overlying water often originates from the sediment surface or the top few millimeters of the sediment column. It is then difficult if not impossible to estimate this flux from chemical data on sediment cores. However, concentration profiles in the overlying lake water together with information on mixing processes make such a calculation feasible.

Radon-222, the daughter product of radium-226, is a radioactive noble gas with a half-life of 3.82 days (Table 1). Oceanographers have widely used the flux of excess radon (i. e. radon that is not in radioactive equilibrium with radium-226) from the sediments in order to study vertical transport in the bottom waters of the ocean (for example see Broecker, 1965; Broecker et al, 1967 and 1968; Chung and Craig, 1972). Thorium-230 and radium-226 are fairly reactive isotopes and occur in far greater quantities in sediments than in the water column, thus providing a source for emanation of radon from the sediments into the water.

Table 1. Half-life and decay constant of isotopes of the $^{238}U$ decay series.

| | half-life*) | $\lambda(yr^{-1})$ |
|---|---|---|
| U-238 | $4.51 \cdot 10^{9}$ yr | $1.54 \cdot 10^{-10}$ |
| Th-234 | 24.1 d | 10.5 |
| Pa-234 | 1.17 min | $3.1 \cdot 10^{5}$ |
| U-234 | $2.47 \cdot 10^{5}$ yr | $2.81 \cdot 10^{-6}$ |
| Th-230 | $7.52 \cdot 10^{4}$ yr | $9.2 \cdot 10^{-6}$ |
| Ra-226 | 1602 yr | $4.3 \cdot 10^{-4}$ |
| Rn-222 | 3.82 d | 66.2 |

*) Burton, 1975

Radon has only rarely been used as a tracer for mixing in lakes. For instance, Hesslein and Quay (1973) have artificially added radium to a small Canadian lake. Imboden and Emerson

1. Present address: Dep. of Oceanography, University of Washington, Seattle, Wash. 98195, U.S.A.

(1976) have used natural radon for the study of vertical transport of phosphate from the sediment-water interface into the euphotic zone in Greifensee, a highly eutrophic Swiss lake.

Due to the potential benefit of the radon method in lake research, it is important to ask the question in what systems the method is expected to be applicable. Or more specifically: Which parameters influence the radon flux from the sediments and in which lakes and in what depths can we expect to find enough excess radon to use it as a tracer for the study of mixing?

## Study of vertical transport using radon

Broecker (1965) has described the analytical method by which the radon-222 and radium-226 activity is measured. From the water sample which usually has a volume between 10 and 30 liters, all gases are extracted by cycling helium through it and trapping the gases at liquid nitrogen temperature. $CO_2$ and $H_2O$ is removed by pumping through ascarite. The radon is transfered to an alpha scintillation counter to determine its activity. After the first extraction, radium supported radon starts to grow in again. Ten or more days later, the stripping of radon is repeated in order to determine radium activity.

The precision of the radon data has been discussed by Chung and Craig (1972). For activities of excess radon of 1 dpm/liter or more, the error can be brought down to about 1 %. On the other hand, since radium activity in the ocean is between 0.2 and 0.3 dpm/liter, excess radon of 0.1 dpm/liter and less results in much larger errors.

In the deep-sea, the bottom can be assumed to be an infinitely large plane if measured with respect to the average migration distance of a radon isotope. Provided that vertical mixing in the near bottom ocean occurs by eddy diffusion, the transport equation for radon can be stated as one-dimensional equation.

$$(1) \quad \frac{\partial Rn}{\partial t} = D_E \frac{\partial^2 Rn}{\partial z^2} - \lambda_{Rn}(Rn - Ra)$$

where $D_E$ means the coefficient of vertical eddy diffusion assumed to be constant with depth, z is distance from sediment-water interface, $\lambda_{Rn}$ the decay constant of radon, Rn and Ra activity per unit column of radon-222 and radium-226. In the deep-sea, Ra is usually constant within the water layers of interest for this model. In addition, the flux of radon from the sediments, $F_{Rn}$, as well as the mixing regime ($D_E$) is commonly constant for time periods in the order of the half-life of radon. Therefore, the radon profile is expected to be at steady state, i.e. $\partial Rn/\partial t = 0$. The solution of eq. (1) for the activity of excess radon, $Rn^* = Rn - Ra$, is

$$(2) \quad Rn^* = Rn_0^* \exp\left(-z\sqrt{\lambda_{Rn}/D_E}\right)$$

The standing crop of excess radon

$$(3) \quad \Sigma Rn^* = \sqrt{D_E/\lambda_{Rn}}\, Rn_0^*$$

is related to the radon flux by the expression

$$(4) \quad \Sigma Rn^* = F_{Rn} / \lambda_{Rn}$$

Eq. (2) is used to calculate $D_E$ from vertical profiles of excess radon.

The application of the method in lakes introduces several complications. As shown by Imboden and Emerson(1976), bottom geometry cannot be neglected, which makes eq. (1) a two- or even three - dimensional equation (Lietzke and Lerman, 1975). Furthermore, fast changing mixing regimes and possibly varying $F_{Rn}$ may frequently violate the steady state assumption. However, if radon is used in combination with other tracers, it can still be a powerful tool for the study of lakes.

Standing crop of excess radon in the ocean are between 0.2 and 20 dpm $cm^{-2}$ (Broecker et al., 1968). Since in the near bottom ocean $D_E$ is usually larger than 1 $cm^2 sec^{-1}$, excess radon can be detected up to 100 or more meters above bottom, thus providing information on the average mixing regime in this layer. On the other hand, $D_E$ in lakes seems to be significantly smaller which compresses the profile of $Rn^*$ to a layer which is only a few meters in thickness. Due to the drastic difference between lakes and oceans in their boundary to volume ratio, the radon distribution in lakes may be strongly influenced by other sources such as by the usually high excess radon of surface and groundwater inlets. Under certain circumstances, this radon may serve as tracer for mixing studies in other parts of a lake, for instance in the thermocline.

In the next section, a general theory for calculating $F_{Rn}$ in oceans and lakes is derived. The last two sections deal with the applicability of the radon method to lakes and with some of the complications which may occur in those systems.

## Radon flux from the sediments

Consider the radioactive decay series

$$(5) \quad X_1 \rightarrow \cdots \rightarrow X_i \rightarrow \cdots \rightarrow X_n$$

where $X_i$ is the total activity of the element i in the sediment. $X_i$ is divided into three components:

$$(6) \quad X_i = L_i + A_i + W_i$$

where $L_i$ represents the detrital (lithogenous)phase, $A_i$ the authogenic phase and $W_i$ the pore water; all activities are conveniently expressed as dpm per unit volume of pore water.

The following assumptions are made:

1. There is no chemical exchange between the lithogenous phase and the pore water.
2. Following the decay of a nuclide i in the lithogenous phase the daughter product (i+1) has the probability $f_i$ to migrate into the pore water.
3. There is an immediate equilibrium between $A_i$ and $W_i$ expressed by a distribution factor $K_i$

$$(7) \quad A_i = K_i W_i$$

   where $K_i$ is assumed to be constant within the sediment column.
4. All activities in the sediment are in steady-state, i.e. at a fixed distance from the sediment-water interface the activities remain constant.
5. Porosity variations, sediment compaction, and advection within the pore water are neglected. $D_i$ is assumed to be constant.

Note that assumption 4 implies that the sedimentation rate s and all the activities in the fresh sediment are constant at least over a time period which is in the order of the longest half-life of the elements used for the present calculation (cf. below).

The time derivative of any function f at a fixed horizon T, denoted as $(\partial f/\partial t)_T$, and the derivative at a fixed distance from the sediment-water interface z, denoted as $(\partial f/\partial t)_z$, are related by the equation (e.g. Imboden, 1975, eq. 6 together with eq. 24)

$$(8) \quad \left(\frac{\partial f}{\partial t}\right)_T = \left(\frac{\partial f}{\partial t}\right)_z + s\frac{\partial f}{\partial z}$$

Assumption 4 means $(\partial f/\partial t)_z = 0$.

The lithogenous phase is not affected by diffusion and thus we simply have

$$(9) \quad \left(\frac{\partial L_i}{\partial t}\right)_T = (1-f_{i-1})\lambda_i L_{i-1} - \lambda_i L_i$$

where $\lambda_i$ is the decay constant of element i. Inserting (8) into (9) yields

$$(10) \quad 0 = \left(\frac{\partial L_i}{\partial t}\right)_z = (1-f_{i-1})\lambda_i L_{i-1} - \lambda_i L_i - s\frac{\partial L_i}{\partial z}$$

Due to eq. (7) another single equation is sufficient to describe both components, $W_i$ and $A_i$. It is convenient to start with the sum of the two activities since phase transitions between $A_i$ and $W_i$ cancel:

$$(11) \quad \left(\frac{\partial W_i}{\partial t}\right)_T + \left(\frac{\partial A_i}{\partial t}\right)_T = D_i\frac{\partial^2 W_i}{\partial z^2} + \lambda_i(W_{i-1} - W_i) + \lambda_i(A_{i-1} - A_i) + f_{i-1}\lambda_i L_{i-1}$$

where $D_i$ is the diffusion coefficient per unit bulk sediment of the element i. Eq. (7) and its derivative $(\partial A_i/\partial t)_T = K_i(\partial W_i/\partial t)_T$ together with (8) yields

$$(12) \quad 0 = \left(\frac{\partial W_i}{\partial t}\right)_z = \frac{D_i}{1+K_i}\frac{\partial^2 W_i}{\partial z^2} - s\frac{\partial W_i}{\partial z} + \lambda_i\left\{\frac{W_{i-1}(K_{i-1}+1) + L_{i-1}f_{i-1}}{K_i+1} - W_i\right\}$$

If the decay series (5) consists of n elements, we get a system of 2n coupled linear differential equations (eq. (10) and (12) for every i) which can be analytically solved. We are interested in calculating the flux of radon-222 through the sediment-water interface, $F_{Rn}$, which is given by the expression ($\phi$: porosity)

$$(13) \qquad F_{Rn} = \Phi D_{Rn} \left( \frac{\partial W_{Rn}}{\partial z} \right)_{z=0}$$

where z is downwards oriented and $F_{Rn}$ is positive for a flux from the sediments into the overlying water.

Due to the permanent growth of the sediment column, an isotope deposited at a constant rate may be prevented from changing significantly its original activity (by decay or diffusion) in the layer near the sediment surface from which radon can diffuse into the overlying water. Such an isotope has no influence on $F_{Rn}$. For the smallest oceanic sedimentation rate, $s = 10^{-4}$ cm yr$^{-1}$, the ratio $\lambda_i/s$ expressing the relative decay of the isotope i per 1 cm of sediment accumulation is 0.03 and 0.09 cm$^{-1}$ for $^{234}$U and $^{230}$Th, respectively.

Since thorium is removed much faster from the water than uranium, the activity ratio $^{234}$U/ $^{230}$Th in the uppermost layers of deep-sea sediments is much smaller than one (Ku; 1965, Duursma 1972; Burton 1975). In fact, for the calculation of $F_{Rn}$, all isotopes preceding $^{230}$Th can be disregarded. If $s > 10^{-2}$ cm yr$^{-1}$, which applies for most lakes, the size of $F_{Rn}$ is infact dominated by the activity of $^{226}$Ra in the sediment.

In order to calculate an expression for $F_{Rn}$ which is valid for the whole range of possible sedimentation rates in oceans and lakes, we have analytically solved the system of six differential equations originating from the reduced decay series

$$(14) \qquad ^{230}\text{Th} \rightarrow \; ^{226}\text{Ra} \rightarrow \; ^{222}\text{Rn} \rightarrow$$

Instead of listing the fairly long expressions for the activities as a function of depth we merely present the resulting expression for $F_{Rn}$, the radon flux out of the sediments. $F_{Rn}$ is a linear superposition of the boundary values ($L_i^o, A_i^o$) of the different isotope activities which are weighted with coefficients $a_i$:

$$(15) \qquad F_{Rn} = \Phi D_{Rn} \left\{ a_1 L^o_{Th} + a_2 A^o_{Th} + a_3 L^o_{Ra} + a_4 A^o_{Ra} + a_5 W^o_{Rn} \right\}$$

The coefficients are listed in Table

Fig. 1. The 5 contributions to $\Sigma Rn^*$ for a "standard sediment" containing 1 dpm/g of lithogenous and authigenic $^{230}$Th, respectively, and an equal activity of $^{226}$Ra (eq. 15).

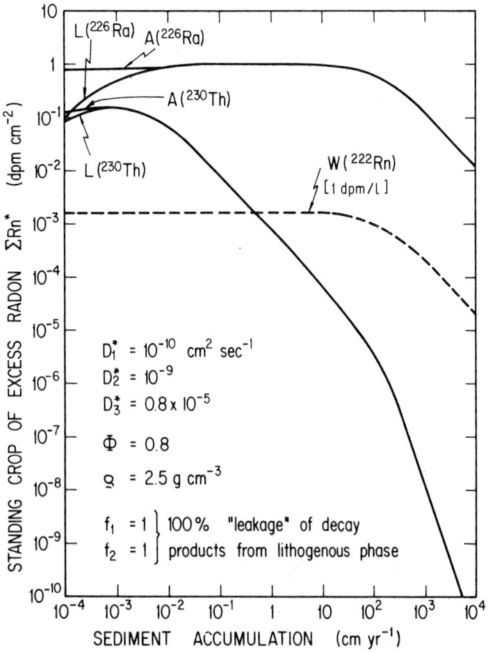


2; they depend on s, $f_i$, and $D_i^*$. Note that, as a consequence of assumption 1, there is no term containing $L^o_{Rn}$.

In order to give an idea of the relative size of the terms appearing in (15) we have calculated the five contributions to $\Sigma Rn^*$ for a hypothetic "standard sediment" in which each boundary activity of thorium and radium are taken as 1 dpm per dry sediment (Fig. 1). The corresponding values per pore water volume are thus

$$(16) \qquad L^o_{Th} = A^o_{Th} = L^o_{Ra} = A^o_{Ra} = \varrho (1-\Phi)/\Phi \times 1\,\text{dpm/g} = 0.625\,\text{dpm/cm}^3_p$$

where the densitiy of the dry sediment is chosen as $\varrho = 2.5$ g cm$^{-3}$, and $\Phi = 0.8$. The boundary value of radon in the water, $W^o_{Rn}$, is chosen as 1 dpm/l corresponding to typical values measured in oceans and lakes. $D_2^*$ is assumed as $10^{-9}$ cm$^2$ sec$^{-1}$ (Goldberg and Koide, 1963) and $D_1^*$ as

Table 2. Coefficients of eq. (15).

$1 = {}^{230}Th \qquad 2 = {}^{226}Ra \qquad 3 = {}^{222}Rn$

$$a_1 = \frac{\lambda_2 \lambda_3}{(1+K_3)} \left\{ (\beta_2-\beta_3) \frac{f_1}{X_{32}Y_{21}} + (\alpha_1-\beta_3) \frac{f_2(1-f_1)}{(\lambda_2-\lambda_1)Y_{31}} - (\alpha_1-\beta_3) \frac{f_1}{Y_{31}Y_{32}} - (\alpha_2-\beta_3) \frac{f_2(1-f_1)}{(\lambda_2-\lambda_1)Y_{32}} \right\}$$

$$a_2 = \frac{\lambda_2 \lambda_3}{X_{21}(1+K_3)} \left\{ \frac{(\beta_2-\beta_3)}{X_{32}} - \frac{(\beta_1-\beta_3)}{X_{31}} \right\}$$

$$a_3 = \frac{\lambda_3}{Y_{32}(1+K_3)} (\alpha_2-\beta_3) f_2$$

$$a_4 = \frac{\lambda_3}{X_{32}(1+K_3)} (\beta_2-\beta_3)$$

$$a_5 = \beta_3$$

The following definitions are used

$$X_{ij} = D_i^* \beta_j^2 + s\beta_j - \lambda_i$$

$$Y_{ij} = D_i^* \alpha_j^2 + \lambda_j - \lambda_i$$

$$\alpha_i = \lambda_i / s$$

$$\beta_i = \frac{s}{2D_i^*} \left[ -1 + \sqrt{1 + 4\frac{D_i^* \lambda_i}{s^2}} \right]$$

$$D_i^* = D_i / (1 + K_i)$$

$10^{-10}$ $cm^2 sec^{-1}$ due to the higher affinity of thorium to the adsorbed phase. For radon the value by Rona (1917) is extrapolated to 4°C, $D_3^* = D_3 = 0.8 \times 10^{-5}$ $cm^2 sec^{-1}$ $(K_3 = 0)$.

For the moment, the escape probabilities for lithogenous isotopes (thorium or radium) are put $f_1 = f_2 = 1$. With the exception of $s < 10^{-3}$cm $yr^{-1}$, both chemical components of each isotope have thus the same influence on $F_{Rn}$, i.e. $a_1 = a_2$ and $a_3 = a_4$. In fact, $F_{Rn}$ is not very sensitive on the choice of $f_1$ whereas $a_3$ is linearly related to $f_2$. We therefore can rewrite (15) in the form

$$(17) \quad \Sigma Rn^* = b_1 Th^\circ + b_2 Ra^\circ$$

where the term containing $W^\circ_{Rn}$ has been neglected. $Th^\circ$ and $Ra^\circ$ are total thorium and radium activity in the sediment per unit pore water volume.

In Fig. 2 are plotted the two terms of eq. (17) for $Th^\circ$ and $Ra^\circ$ each corresponding to 1 dpm/g = 0.625 dpm $cm^{-3}$ ($\Phi$ and $\varrho$ as before). The total escape probability for thorium and radium are assumed to be 50 % and 70 %, respectively. Ku (1965) has found a total radon leakage from Ra between 40 and 80 % in deep-sea sediments, most values lying around 70 %.

For s typical for the deep ocean, both thorium and radium have a similar influence on $F_{Rn}$ per activity unit present in the sediments. Broecker et al. (1968) have estimated Rn* under the assumption that the flux of $^{230}Th$ into the sediments corresponds to the $^{230}Th$ produced by the $^{234}U$ dissolved in the ocean uniformly distributed over the ocean floor. $Th^\circ$ would then be inversly related to s, $Th^\circ$[dpm $cm^{-3}$] $8 \times 10^{-3}/s$. Neglecting the term with $Ra^\circ$

$$(18) \quad \Sigma Rn^* \, [\text{dpm } cm^{-2}] \sim 9 \times 10^{-3}/s$$

Fig. 2. The two contributions to $\Sigma Rn^*$ for sediments containing 1 dpm/g of total $^{230}Th$ and $^{226}Ra$ activity, respectively (eq.17). The total escape probabilities are 50 % for $^{230}Th$ and 70 % for $^{226}Ra$.

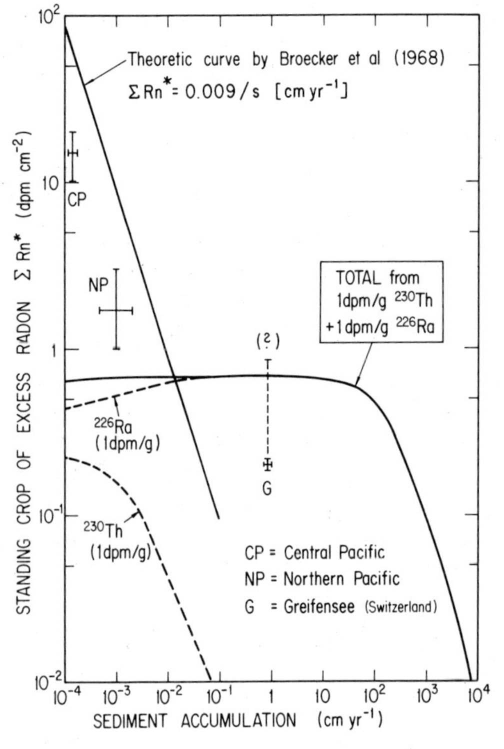

where s is in cm $yr^{-1}$. The corresponding curve is shown in Fig. 2 together with average measured values of $\Sigma Rn^*$ and s found in the Central and Northern Pacific. They are somewhat smaller than predicted by the theory. In fact, the $^{230}Th$-content of deep-sea sediments depends on s and ocean depth in a more complex way (Kuznetsov et al., 1968).

## Radon in lakes

If $s > 0.1$ cm $yr^{-1}$, which is valid for most lakes, the influence of thorium on $F_{Rn}$ can be neglected in comparison to that of radium (Fig. 2). Eq. (15) is therefore reduced to the two terms containing $a_3$ and $a_4$. With $K_{Rn} = 0$ we get

$$a_3 \sim f_2 \sqrt{\lambda_{Rn}/D_{Rn}}\ ,\ a_4 = \sqrt{\lambda_{Rn}/D_{Rn}}$$

and thus

$$(19)\quad F_{Rn} = \Phi \sqrt{D_{Rn}\ \lambda_{Rn}}\ Ra^e$$

$$= (1-\Phi)\, \varrho \sqrt{D_{Rn}\ \lambda_{Rn}}\ Ra'$$

where $Ra^e = (f_2\ L^o_{Ra} + A^o_{Ra}\ )$

means lithogenous and authigenic radium-226 activity in the sediment from which radon can leak into the pore water and Ra' the corresponding value expressed as dpm per unit dry weight of sediment.

The simple linear relationship between $F_{Rn}$ and Ra' provides a straight forward method to predict $F_{Rn}$ from analyzing sediment cores. Lake bottom topography may complicate the interpretation of radon profiles in the water column wherefore it may be an advantage to determine $F_{Rn}$ independently of the profile.

We have determined radon flux from sediments in Greifensee, the lake in which radon tracer methods have been applied as mentioned earlier. Two different methods were used to estimate the flux of radon from cores taken at different locations in the lake. In the first experiment, 10-15 cm of destilled water was carefully placed on top of the cores (cross section = 9 and 400 $cm^2$). After about 20 days the water was removed and analyzed for radon. The standing crop per unit area is given in Table 3 as $\Sigma 1$. The volume of water into which radon can diffuse is limited in the experiment whereas in the lake it is nearly infinite. Correction for this effect is, however, small as long as the measured activities are not much greater than 0.2 dpm $cm^{-2}$. In the second experiment, radon was directly extracted from about 30 g of wet sediment. The samples had been previously stripped of all radon and stored for about two weeks allowing the growth of a new standing crop. From these measurements the "leakable" radium activity in the sediment is determined as about 0.6 dpm/g. The corresponding standing crop are calculated ($\Sigma 2$ in Table 3). Similar or higher Ra-content in sediments of other lakes have been found (Vernet et al., 1975; Koide et al., 1976).

In the last column of Table 3 the calculated standing crops from radon profiles taken in the lake are given ($\Sigma 3$). Not only are they larger than $\Sigma 1$ and $\Sigma 2$ due to horizontal diffusion of radon from the sides (station A is located at the deepest point of the lake), but $\Sigma 3$ also shows significant temporal variations (cf. April 1975). One possible explanation, a boundary layer between sediments and water of varying thickness, is discussed in the last section.

## Radon flux through boundary layer

The one-dimensional diffusion model graphically represented in Fig.3 consists of three layers: sediment with porosity $\Phi$ (semi-infinite for $z \to -\infty$) a stagnant boundary layer of thickness d, and the water column (semi-infinite for $z \to \infty$). Radium activities are assumed to be constant within each layer corresponding to the implicit assumptions valid for eq. (16). In order to have a significant effect on Rn*, the boundary layer has to be poor in radium: $Ra_w$, $Ra_b \ll Ra_s$ (w = water, b = boundary layer, s = sediment). Vertical transport in the water is accomplished by eddy diffusion whereas in the other two layers it is by molecular diffusion : $D_{Rn,s}$ , $D_{Rn,b} \ll D_{Rn,w} = D_E$ .

The boundary layer, if there is any, is only a few centimeters in thickness. Thus, radon samples taken in the lake would all lie above z = d. The linear regression analysis of the logarithmic form of eq. (2) yields excess radon above the stagnant boundary layer, $Rn^*_d = Rn_d - Ra_w$,

Table 3. Comparison of different methods to determine $F_{Rn}$ from Greifensee sediments.

| | | $\Sigma 1$ (dpm cm$^{-2}$) | $\Sigma 2$ (dpm cm$^{-2}$) | Ra' (dpm g$^{-1}$) | $\Phi$ | $\Sigma 3$ (dpm cm$^{-2}$) |
|---|---|---|---|---|---|---|
| September 1974 | | | | | | |
| Station | 202 | 0.20 | 0.17 | 0.62 | 0.95 | |
| | 204 | 0.16 | 0.16 | 0.59 | 0.95 | |
| Okt.75/Jan.76 | | | | | | |
| Station | A | 0.11 | 0.13 | 0.64 | 0.965 | 0.19 |
| | B | - | 0.06 | 0.21 | 0.946 | 0.05 |
| April 75, Sta.A | | | | | | 0.95 |

$\Sigma 1 = \Sigma Rn^*$ from radon in destilled water placed on top of core
$\Sigma 2 = \Sigma Rn^*$ from Ra' using eq. (19) and $\varrho = 2.5$ g cm$^{-3}$, $D_3 = 0.8 \times 10^{-5}$ cm$^2$ sec$^{-1}$.
$\Sigma 3 = \Sigma Rn^*$ from radon profile measured in the lake

Fig. 3. One-dimensional diffusion model of $^{222}Rn$ for a boundary layer at the sediment-water interface.

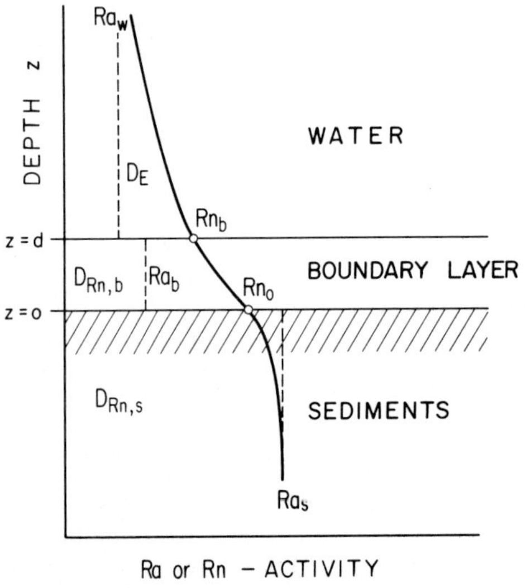


as boundary value. The corresponding standing crop of excess radon, $\Sigma Rn^*$, refers to the profile above z = d, only. We are thus specifically interested in the value of $Rn^*_d$ from the solution of the three-layer diffusion model.

A tedious but straight-forward evaluation of the three transport equations of the form (1) together with the made assumptions on the relative size of Ra and $D_{Rn}$ yields

$$F_{Rn} = Ra_s \, \Phi \sqrt{D_{Rn,s} \lambda_{Rn}} \; B \tag{20}$$

where $(\gamma = \sqrt{\lambda_{Rn}/D_{Rn,b}})$

$$B = \left[\cosh \gamma d + \Phi \sqrt{D_{Rn,s}/D_{Rn,b}} \sinh \gamma d\right]^{-1}$$

Eq. (20) differs from (19) by the reduction factor B which for $\gamma d \ll 1$ is equal to $\exp(-\gamma d)$.

If $\Phi = 0.95$,

$$D_{Rn,s} = D_{Rn,b} = 0.8 \times 10^{-5} \text{ cm}^2 \text{ sec}^{-1},$$

the reduction factor is B = 0.78 for d = 1 cm, B = 0.28 for d = 5 cm, and B = 0.08 for d = 10 cm, respectively. Relatively small boundary layer may therefore strongly influence $\Sigma Rn^*$ which could turn the radon measurements in lakes into a powerful tool for detecting varying boundary layers or layers of radium poor freshly deposited algae. On the other hand, if indeed the large $\Sigma Rn^*$ detected in April 1975 in Greifensee would correspond to the "normal" standing crop in contrast to the depressed values found later in the year, we would be left with an unexplained discrepancy between $\Sigma 3$ and $\Sigma 1$, $\Sigma 2$ (Table 3). Further experiments may possibly reveal other mechanisms to explain sudden variations of $\Sigma Rn^*$ in lakes.

## References


Broecker, W.S., 1965. An application of natural radon to problems in ocean circulation. In: T. Ichiye (Ed.): Symposium on diffusion in oceans and fresh waters. Lamont Geological Observatory, Palisades, N.Y. p. 116-145.

Broecker, W.S., Y.H. Li & J. Cromwell, 1967. Radium-226 and radon-222: Concentrations in Atlantic and Pacific Oceans. Science 158: 1307-1310.

Broecker, W.S., J. Cromwell & Y.H. Li, 1968. Rates of vertical eddy diffusion near the ocean floor based on measurements of the distribution of excess $^{222}$Rn. Earth Planet.Sci.Letters 5: 101-105.

Burton, J.D., 1975. Radioactive nuclides in the marine environment. In: J.P. Riley & G. Skirrow (Ed.): Chemical oceanography. Vol.2, 2nd edition, Academic Press, London. p. 91-191.

Chung, Y. & H. Craig, 1972. Excess-radon and temperature profiles from the eastern equatorial Pacific. Earth planet. Sci.Letters 14: 55-64.

Duursma, E.K., 1972. Geochemical aspects and applications of radionuclides in the sea. Oceanogr.Mar. Biol.Ann.Rev. 10:137-223.

Emerson, S., 1976. Early diagenesis in anaerobic lake sediments: chemical equilibria in interstitial waters. Geochim.Cosmochim. Acta 40: 925-934.

Goldberg, E.D. & M. Koide, 1963. Rates of sediment accumulation in the Indian Ocean. In: J. Geiss & E.D. Goldberg (Ed.): Earth Science and Meteoritics. North Holland, Amsterdam. p. 90-102.

Hesslein, R. & P. Quay, 1973. Vertical eddy diffusion studies in the thermocline of a small stratified lake. J. Fish.Res.Board Can. 30: 1491-1500.

Imboden, D.M., 1975. Interstitial transport of solutes in non-steady state accumulating and compacting sediments. Earth Planet. Sci. Letters 27: 221-228.

Imboden, D.M. & S. Emerson, 1976. Natural radon and phosphorus as a limnologic tracer: Horizontal and vertical eddy diffusion in Greifensee. Submitted to Limnol.Oceanogr.

Koide, M., K. Bruland & E.D. Goldberg, 1976. $^{226}$Ra chronology of a costal marine sediment. Earth Planet.Sci.Letters 31: 31-36.

Ku, T.L., 1965. An evaluation of the $^{234}$U/ $^{238}$U method as a tool for dating pelagic sediments. J.Geophys. Res. 70: 3457-3474.

Kuznetsov, Y.V., Z.N. Simonyak, A.P. Lisitsyn & M.S. Frenklikh, 1968. Thorium isotopes ( $^{230}$Th, $^{232}$Th) in the surface layer of the Indian ocean sediments. Geokhimiya 2: 218-227, (Transl.) Geochem.Int. 5: 169-177.

Lietzke, T.A. & A. Lerman, 1975. Effects of bottom relief in two-dimensional oceanic eddy diffusion models. Earth Planet. Sci.Letters 24: 337-344.

Rona, E., 1917. Diffusionsgrösse und Atomdurchmesser der Radiumemanation, Z. Physik.Chemie 92:213-218.

Vernet, J.P., E. Davaud, M. Cosandy & J. Berlie, 1975. Analyse de la radioactivité des sédiments des lacs Léman et de Morat. Eclogae geol. Helv. 68:87-96.



Key : Eddy diffusion
Boundary layer
Radon-222
Sediment-water interface
Vertical mixing

## ESTIMATING SUSPENDED SEDIMENT LOADS FROM MEASUREMENTS OF REFLECTED SOLAR RADIATION

Jerry C. Ritchie, Frank R. Schiebe and J. Roger McHenry

USDA Sedimentation Laboratory, Oxford, Mississippi 38655, USA

### Summary

One major problem faced by the limnologists, studying the dynamics of a body of freshwater, is to determine the amount of suspended sediment in the system. Even though suspended sediment affects the heat budget, the chemical equilibrium and the biological activity of freshwater lakes, there are no readily available methods for their synoptic determination. Present techniques for measuring suspended sediment load in lakes and reservoirs involve measuring turbidity or collecting samples for future gravimetric analysis for suspended sediments. These techniques are time consuming, expensive and do not give the synoptic view of the water body that is needed if the limnologist is to understand the dynamic processes involved in a water body. Three years of study at the US Sedimentation Laboratory, using a portable spectroradiometer, has shown that a quantitative relationship exists between surface suspended sediments and reflected solar radiation. This relationship is most significant in the wavelengths between 700 to 800 nm. These studies also have shown that the concentration of surface suspended sediment can be used to accurately estimate the total amount of suspended sediment in a vertical water column or the total amount of suspended sediment in a water body. Recent studies have shown that remote sensing techniques, from either airplanes or satellites, can be used to effectively delineate surface suspended sediment patterns and in many cases the concentrations of surface suspended sediments. These types of estimates are essential to fully understand, explain and model the physical, chemical, and biological cycles in lakes and reservoirs. This paper illustrates the application of remote sensing techniques in estimating the total suspended sediment load in four large reservoirs from measurements of reflected solar radiation.

### Introduction

Suspended sediments are a major problem in lakes and reservoirs. Suspended sediments affect the physical (Tyler, 1968; Robinson, 1971; Schiebe, et al., 1975), chemical (Keeny, 1973; Pionke and Chester, 1973; Romkens, et al., 1973), and biological (Ellis, 1936; Cardone and Kelley, 1961; Ritchie, 1972) properties of water bodies. To fully understand and interpret physical, chemical, and biological data, the limnologist needs an understanding of the amount and spatial distribution of suspended sediments in a water body. Present techniques involve the measurement of turbidity which does not give the quantitative information needed (National Oceanographic Instrument Center, 1974) or gravimetric determination of suspended sediments. These point sampling techniques are time consuming and expensive and neither gives the synoptic view of suspended sediment concentrations needed for proper data interpretation. This paper describes a technique to measure suspended sediments using reflected solar radiation. With this technique the limnologist can use aerial photographs or satellite images to obtain a synoptic view and to estimate the total amount of suspended sediment in a water body.

### Discussion

Three basic research projects were designed to evaluate the use of reflected solar radiation for estimating the suspended sediment loads in water bodies. These three studies were 1) an evaluation of the relationship between in situ measurements of reflected solar radiation and concentration of surface suspended sediments, 2) an evaluation of the relationship between concentrations of surface suspended sediments and total suspended sediment load in a water body, and 3) an evaluation of the use of aerial photography and/or satellite imagery to estimate the concentration of surface suspended sediments. This paper discusses projects 1 and 2 and indicates the type of results that are expected when project 3 is completed.

Project 1 was begun in 1973 with a study of four large (<200 $km^2$) flood control reservoirs in the northern part of Mississippi (US) to determine the relationship between solar radiation reflected from the water surface and the concentration of total solids in the surface waters.

Measurements of incident and reflected solar radiation were made from a small boat using a portable spectroradiometer mounted

50 cm above the surface of the water. Measurements were made on cloud free days at 25 nanometer (nm) intervals from 400 to 750 nm and at 50 nm intervals from 750 to 1550 nm. Samples were collected of the surface water at each spectroradiometer measurement station and the concentrations of total solids were determined gravimetrically after drying the samples at 105°C. A complete description of the methods and materials has been published (Ritchie, Schiebe, and McHenry, 1976).

These studies showed that the amount of solar radiation reflected from the surface water varied with both the wavelength and the concentration of suspended sediment in the surface water (Fig. 1). The amount of

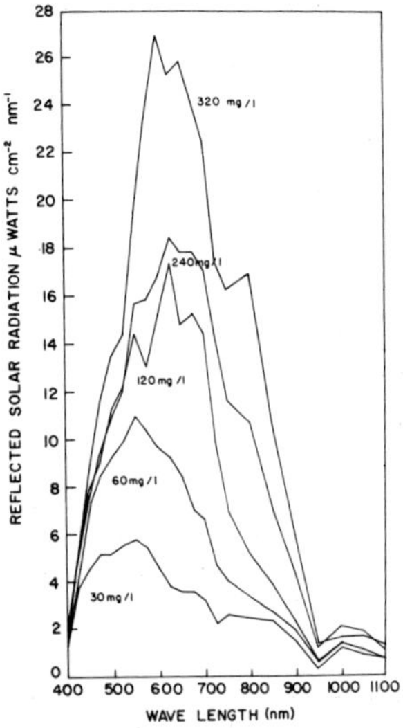


Figure 1. The relationship of reflected solar radiation with wavelength for different concentrations of suspended sediments in the surface water.

reflected solar radiation rapidly increased from 400 nm to between 500 to 600 nm for all concentrations of suspended sediment. Between 500 and 700 nm reflected solar radiation was related to the concentration of suspended sediments in the surface water. There was a shift from the maximum of reflected solar radiation in the green band for low concentration of suspended sediments to red as the concentrations of suspended sediments increased. Above 700 nm there was a decrease in the amount of reflected solar radiation with increasing wave length for all concentrations of suspended sediment. These data and others (Kaldo, 1972; Ritchie, et al., 1974) show that there is a relation between the amount of solar radiation reflected from surface water and the concentration of suspended sediments.

The relationship between reflected solar radiation and concentration of suspended sediment was then statistically analyzed, and it was found that a linear relationship best fit the data. It was also found that the correlation coefficient varied with wavelength with the highest correlation coefficient being for the wavelength between 700 and 800 nm (Ritchie, et al., 1974). Representative data (Fig. 2) show the type of fit observed. These plots showed that below 200 milligram per liter (mg/l) suspended sediments the linear relationship fit well, but that above 200 mg/l suspended sediment the deviation from the line of best fit was much greater. This probably represented a saturation of the amount of solar radiation that can be reflected by the system.

Further analysis showed the relationship between reflected solar radiation and concentration of suspended sediments was affected by zenith angle of the sun according to the Fresnel reflectivity relationship (List, 1971). These analyses (Ritchie, et al., 1975) showed that by accounting for reflectivity of water due to the sun angle, by the Fresnel reflectivity formula, the correlation coefficient for the relationship between reflected solar radiation and concentration of suspended sediments improved by 3 to 6 percent.

These studies of in situ measurements of the solar radiation reflected from the surface water have shown that reflected solar radiation from water surface can be used to estimate the quantity of suspended sediment in the surface water. Estimates are more accurate if zenith angle of the sun is used with the prediction equation.

Once it was determined that reflected solar radiation could be used to estimate the concentration of suspended sediment in the surface waters, the problem became one of determining if the concentration of surface suspended sediment could be used to estimate the total amount of suspended sediment in a water body. To answer this question, samples were collected at the surface and at 1 meter intervals in a vertical water column in each of the four reservoirs. The concentration of suspended sediments was determined gravimetrically for each sample.

Analysis of these data showed that the concentration of suspended sediment in the surface water could be used to estimate accurately the total suspended sediment load of a vertical column of water (Fig. 3) in these four reservoirs. Statistical analyses showed no significant statistical differences between reservoirs. The vertical distribution of suspended sediment in the water column was related to temperature so that there was normally an increase in suspended

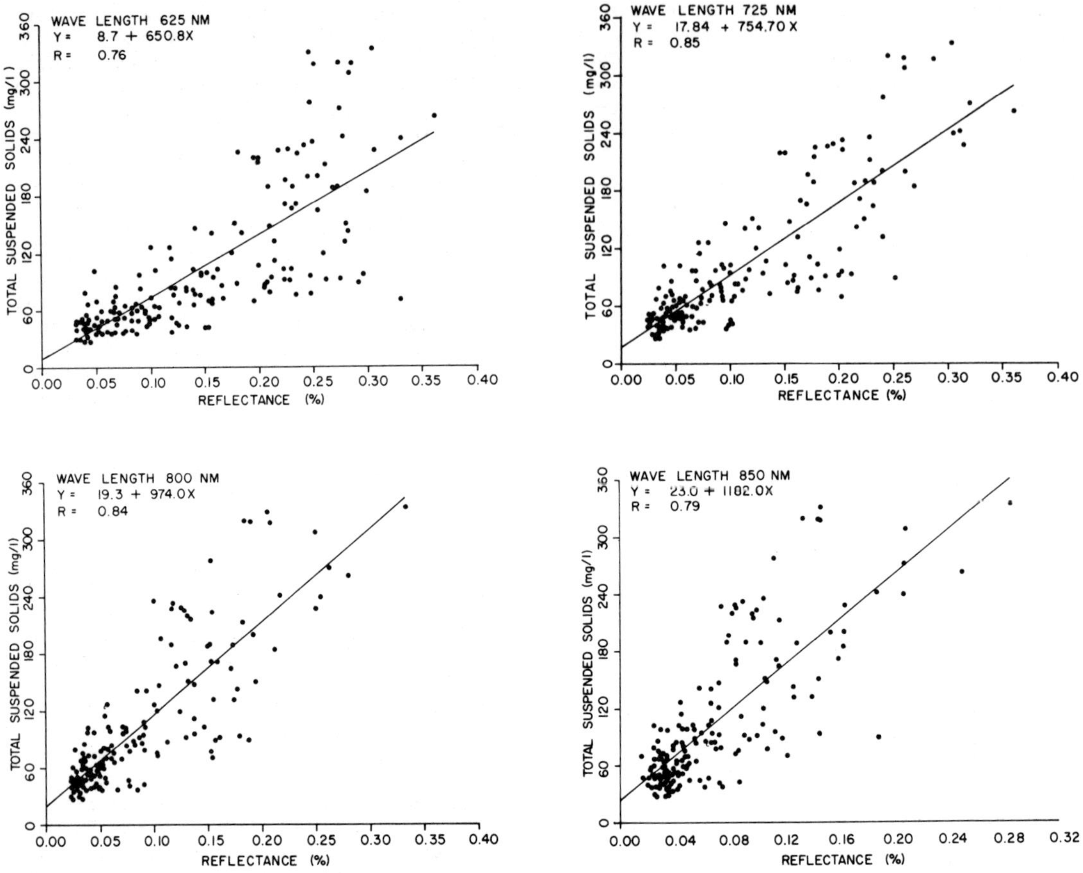


Figure 2. Relationship between reflected solar radiation and concentration of suspended sediment in the surface water at 500, 600, 700, and 800 nm.

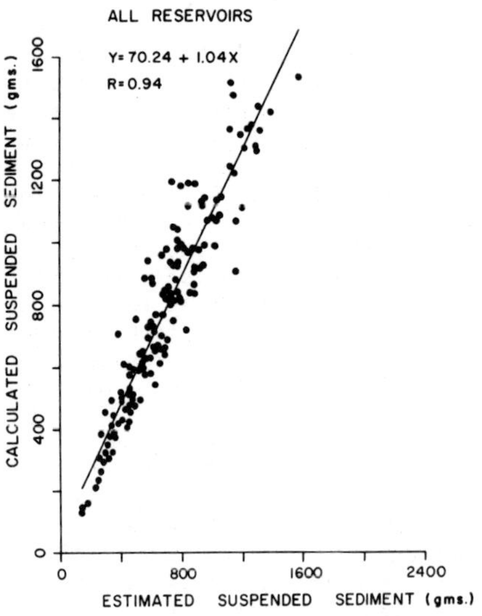


Figure 3. Relationship between calculated suspended sediment load and the estimated suspended sediment load for four reservoirs in north Mississippi.

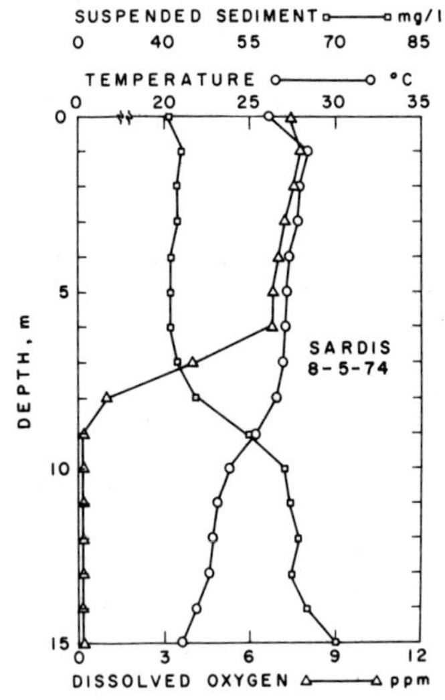


Figure 4. Vertical distribution of suspended sediment, temperature, and dissolved oxygen in Sardis reservoir August 5, 1974.

sediment below the thermocline in the water column (Fig. 4). However, the volume of water in the epilimnion was much greater than that in the hypolimnion so that the differing suspended sediment load below the thermocline did not significantly affect the total load estimate made by using the concentration of suspended sediment of the surface water.

Once it was shown that the concentration of surface suspended sediment could be used to estimate the total load of suspended sediment in a vertical water column, a study was made to determine if the total load of suspended sediment in a reservoir could be estimated (McHenry, et al., 1976). This study showed that the concentration of suspended sediments in the surface water can be used to estimate accurately the total load of suspended sediment in the reservoir (Fig. 5). The next step was to use solar radia-

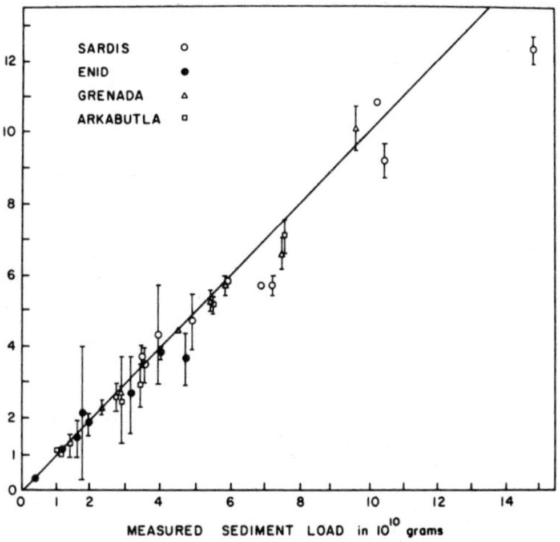


Figure 5. A comparison of the total suspended sediment load estimated from the suspended sediment in the surface water and the measured suspended sediment load for four north Mississippi reservoirs. The vertical bars are the standard deviations of the plotted means based on all site samples on a particular day.

tion reflected from the surface water to estimate total load of suspended sediment in the water column. One year's (August 1973 to August 1974) data were analyzed to determine the regression equation which best fitted the relationship between solar radiation reflected from the surface water and the concentration of suspended sediment in the surface water for the wavelengths between 700 and 800 nm. This band of wavelengths corresponds to the wavelength region of best fit and also to band 6 of the ERTS satellites. This regression equation and measurements of solar radiation reflected from the surface water made from August 1974 to August 1975 were used to estimate the concentration of suspended sediments in surface water for this time period. These estimates of surface suspended sediment concentration were used to calculate the total load of suspended sediments in the vertical water column. The calculated suspended sediment loads were compared with the measured suspended sediment loads (Table 1). The calculated suspended sediment load ranged from

Table 1. Comparison of the total suspended sediment load calculated from regression estimates of surface suspended sediment and total suspended sediment measured gravimetrically.

| Sample # | Suspended sediment load: Measured g | Suspended sediment load: Calculated g | Difference % |
|---|---|---|---|
| 271 | 924.5 | 442.5 | -52 |
| 272 | 984.0 | 372.0 | -62 |
| 273 | 1121.0 | 703.0 | -37 |
| 274 | 749.6 | 325.6 | -56 |
| 275 | 665.0 | 420.5 | -37 |
| 278 | 515.0 | 960.0 | +86 |
| 279 | 427.7 | 464.0 | + 9 |
| 285 | 664.0 | 286.0 | -57 |
| 286 | 1148.5 | 864.5 | -25 |
| 288 | 550.9 | 826.0 | +50 |
| 289 | 411.6 | 329.3 | -20 |
| 290 | 222.5 | 192.7 | +14 |
| 291 | 581.5 | 493.5 | +15 |
| 292 | 658.7 | 607.6 | + 8 |
| 293 | 425.9 | 453.6 | + 7 |
| 302 | 429.5 | 414.0 | + 4 |
| 303 | 329.4 | 324.8 | + 1 |
| 304 | 169.5 | 127.5 | +13 |
| 305 | 617.5 | 591.6 | - 2 |
| 306 | 688.2 | 573.5 | -17 |
| 330 | 819.5 | 837.5 | + 2 |
| 331 | 600.0 | 423.0 | +30 |
| 336 | 1443.0 | 1005.0 | +30 |
| 337 | 1300.5 | 910.0 | +30 |
| 338 | 1364.1 | 1124.8 | +18 |

overestimating the measured load by 86% to underestimating the measured load by 62%. The mean difference between the calculated and measured suspended sediment load was 2% with a standard deviation of ±36% for the 25 samples.

The results of our studies indicates that in situ measurements of solar radiation reflected from the water surface can be used to estimate the concentration of suspended sediments in the surface water and the total load of suspended sediment in a reservoir. Studies are now underway to determine if satellite imagery and aerial photographs can be used to determine the differences in solar radiation reflected from the water surface with different concentrations of suspended sediments. Since satellite images average over a large water area, we feel that suspended sediment loads estimated from these images will be comparable to our in situ estimates.

Other studies (Carlson and McCulloch, 1974; Klemas, et al., 1974; Kritikos and Younks, 1974) have shown that satellite and aerial images can be used to qualitatively measure surface patterns as suspended sediments. Other preliminary studies (Williamson, 1974; Johnson, 1975; Clark, et al., 1974; Pionke and Blanchard, 1975; Yarger, et al., 1974) have shown that quantitative estimates of the concentration of surface suspended sediments may be possible using satellite and aerial images. These studies indicate that it should be possible to use either satellite or aerial imagery to estimate the total suspended sediment load in a reservoir. Such data would be invaluable to the limnologist who is interpreting physical, chemical, and biological data.

## Acknowledgements

This paper is a contribution of the US Sedimentation Laboratory, Alabama-North Mississippi Area, Southern Region, Agricultural Research Service, U. S. Department of Agriculture. Jonathan May and Kathy Roush assisted in the field collection and laboratory analyses of the samples, Robert Wilson assisted in the computer analyses of the data, and Donna Bethany assisted in the preparation of the manuscript.

## References

Cardone, A.J. & D.W. Kelley, 1961. The influence of inorganic sediments on the aquatic life of streams. Calif. Fish and Game 47:189-228.

Carlson, P.R. & D.S. McCulloch, 1974. Aerial observations of suspended sediment plumes in San Francisco Bay and adjacent Pacific Ocean. J. Research, U. S. Geological Survey 2:519-526.

Clark, D.K., J.B. Zaitseff, L.N. Strees, & W.S. Glidden, 1974. Computer derived coastal water classifications via spectral signatures. Proc. 9th Int. Symp. on Remote Sensing of Environment 9:1213-1239.

Ellis, M.M., 1936. Erosion silt as a factor in aquatic environments. Ecology 17:29-42.

Johnson, R.W., 1975. Quantitative sediment mapping from remotely sensed multispectral data. p. 565-576. IN: F. Shahrokhi (ed), Remote Sensing of Earth Resources, Vol. IV, The University of Tennessee Space Institute, Tullahoma, Tennessee.

Kaldo, A.G., 1972. Study of suspended sediments in reservoirs by spectral luminance method. Hydrology: Selected Papers 2:101-110 (Translation of the State Hydrologic Institute (Trudy GGI) No. 191 (1972) 159-171.

Keeny, D.R., 1973. The nitrogen cycle in sediment water systems. J. Environ. Quality 2:15-29.

Klemas, V., M. Otley, W. Philpat, & R. Rogers, 1974. Correlation of coastal water turbidity and circulation with ERTS-1 and skylab imagery. Proc. 9th Int. Symp. on Remote Sensing of Environment 9:

Kritikos, H., L. Yorinks, & H. Smith, 1974. Suspended solids analyses using ERTS-A data. Remote Sensing of Environment 3:69-78.

List, R.J., 1971. Smithsonian meteorological tables. Smithsonian Institute Press, Washington, D. C. 527 p.

McHenry, J.R., J.C. Ritchie, & F.R. Schiebe, 1976. Estimating the suspended sediment load in reservoirs. Water Resources Bulletin (In Press).

National Oceanographic Instrument Center, 1974. Proceedings NOIC turbidity workshop. U. S. Department of Commerce, NOAA, Washington, D. C., 216 p.

Pionke, S.H. & B.J. Blanchard, 1975. The remote sensing of suspended sediment concentrations in small impoundments. Water, Air and Soil Pollution 4:19-32.

Pionke, H.B. & G. Chester, 1973. Pesticide-sediment-water interaction. J. Environ. Quality 2:29-45.

Ritchie, J.C, 1972. Sediment, fish and fish habitats. J. Soil and Water Conservation 27:124-125.

Ritchie, J.C., J.R. McHenry, F.R. Schiebe, & R.B. Wilson, 1974. The relationship of reflected solar radiation and the concentration of sediment in surface water of reservoirs, p. 57-72. IN: F. Shahrokhi (ed), Remote Sensing of Earth Resources, Vol. III, The University of Tennessee Space Institute, Tullahoma, Tennessee.

Ritchie, J.C., F.R. Schiebe, R.B. Wilson, & J. May, 1975. Sun angle, reflected solar radiation and suspended sediment in north Mississippi reservoirs, p. 543-554. IN: F. Shahrokhi (ed), Remote Sensing of Earth Resources, Vol. IV, The University of Tennessee Space Institute, Tullahoma, Tennessee.

Ritchie, J.C., F.R. Schiebe, & J.R. McHenry, 1976. Remote sensing of suspended sediments in surface water. Photogrammetric Engineering and Remote Sensing. J. Am. Soc. Photogrammetry (In Press).

Robinson, A.R., 1971. Sediment. J. Soil and Water Conservation 26:61-62.

Romkens, M.J.M., D.W. Nelson, & J.V. Mannering, 1973. Nitrogen and phosphorus composition of surface runoff as affected by tillage methods. J. Environ. Quality 2:292-295.

Schiebe, F.R., J.C. Ritchie, & J.R. McHenry, 1975. Influence of suspended sediment on the temperatures of surface waters of reservoirs. Verh. Internat. Verein. Limnol. 19:133-136.

Tyler, J.E., 1968. The secchi disc. Limnol. Oceanogr. 13:1-6.
Williamson, A.N., 1974. Interpretation of multispectral remote sensor data. J. Surveying and Mapping Division ASCE 100:143-153.
Yarger, H.L., J.R. McCauley, G.W. James, & L.M. Magnunson, 1974. Quantitative water quality with ERTS-1. 3rd Earth Resources Technology Satellite-1 Symposium, p. 1637-1651.

Key words: remote sensing
satellite
total suspended load
wave length

it fills the overflow, raises the valve by means of the float, and passes into the soil-pipe. Neilson, 1873; Gilchrist, 1874; Keyser, 1876; Moellmann, 1882, invented slight and unimportant improvements either in the form and position of the overflow or supply-valve, in connection with which each used a plain solid plunger.

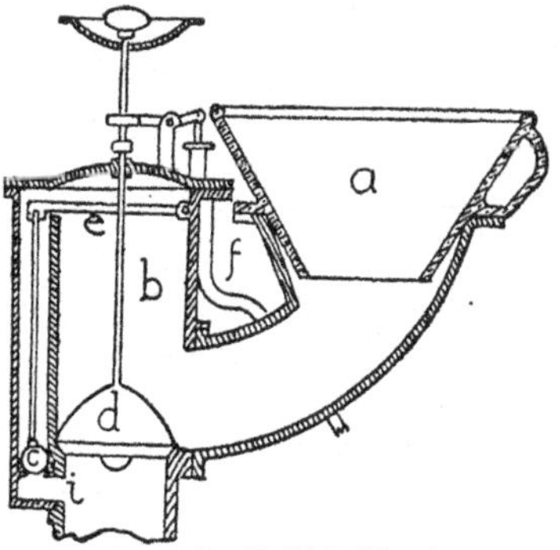

Fig. 159.—Budde's Closet.
*a*, Bowl. *b*, Plunger-chamber. *c*, Ball-valve in overflow. *d*, Plunger. *e*, Float for overflow. *f*, Supply-pipe. *h*, Supply-valve. *i*, Overflow.

E. S. Hutchinson's Closet has the plug in the same compartment that forms the bowl, in this respect being similar to closets that were in use more than one hundred years ago.

Stanton M. Howard has recently (1884) invented a closet in which the plunger moves in a horizontal position. The plug being conical, it fits tightly into the side outlet of the bowl. The plunger is operated by means of a rack, and ventilated. The pinion is weighted, and by this means the plunger is forced back against its seat.

## DEDUCTIONS.

It will have been noticed by those who have read the foregoing descriptions, that the mechanical faults and the sanitary objections to the plunger-closet are almost identical to those mentioned in connection with valve-closets. The utility of the plunger in a closet of this class depends upon its taking a water-tight seat. All the useful plunger-closets have a rubber flange or ring around the bottom of the plunger, and this form of plunger is generally effective, as long as the vulcanized rubber retains its elasticity; when it gets hard the plumber must be called in to put on a new one.

This form of closet is liable at any time to lose the water in the bowl and plunger-chamber, by foreign matter lodging between the

plunger and its seat; so it is never safe to have a closet of this kind connected to the soil-pipe without a trap; although the patentees usually claim it as one of the advantages of their closet, that it does not need a trap between it and the soil-pipe. The plunger takes and retains its seat by the action of gravity, and for this reason it is more simple, and a better mechanical device than the valve; requiring no levers, weights, or cranks to operate it, or keep it in position.

It is necessary for a plunger to be heavy, so it can take and keep its seat properly; this weight is often a decided inconvenience and objection where the closet is to be constantly used by delicate women and small children.

Taking a sanitary view of the subject, the greater number of the different patterns of plunger-closets should be considered inferior to the valve-closets, the plunger chamber being—as with a few exceptions it is—larger than the receiver of a valve-closet, and containing the plunger, float, lever, and frequently the supply-valve, upon the different surfaces of which filthy matter is liable to collect and decay.

Mr. E. S. Philbrick says of closets belonging to this class, "The construction of these (supply) valves is various, and their action is not always reliable, which has at times caused considerable annoyance. If these valves can be made more certain in their action, and the float-chamber made more accessible for frequent cleaning, the apparatus can be made a very satisfactory one; but it needs careful treatment and good care."

The supply-valves not being reliable, closets of this class must have overflows.

Where the plunger is hollow, the overflow is carried off through the plunger, but where the plunger is solid, a separate conduit has to be provided: and where this is trapped by a water-seal the water is liable to be drawn from the trap by siphon action whenever the closet is used. To prevent this siphon action, a vent must be coupled to the closet, somewhere between the seat of the plunger and the point where the closet is connected with the soil-pipe.

If the closet is conected with the soil-pipe without the interven-

tion of a siphon-trap, the breaking of the seal in the overflow would put the occupants of the house in direct communication with the soil-pipe, as the covers to the plunger-chamber are not intended to have their joints air-tight, except in one or two instances. In selecting a closet of this class, care should be taken to get one with a plunger-chamber at least enamelled, while a bowl and plunger-chamber in one piece of earthenware is the best. The mechanism should be simple, the plunger-chamber small, and with a top or cover that can be easily removed.

The closet should be supplied with water from a cistern or tank, and should be admitted to the bowl through a flushing-rim. There should be a branch-supply for washing the plunger-compartment and plunger, admitted through a ring or sprinkler.

A hollow plunger simplifies the question of the overflow. It is necessary for the plunger to have a vulcanized-rubber ring or flange around its bottom.

The closet must have a siphon-trap under it, with a coupling or hub at its crown, to which a pipe may be soldered, or into which a pipe may be caulked; the vent must not be less than two inches in diameter.

# SECTION V.

## HOPPER-CLOSETS.

UNDER this heading I have thought it best to group all closets that have a simple bowl and no mechanical device other than a water-seal between them and the soil-pipe. This branch of the subject may with advantage be treated under the two types of long and short hopper-closets.

The long hopper is connected with a trap which is placed beneath the floor, while the short hoppers have their trap, which is often formed in one piece with the bowl, placed above the floor.

*Long Hopper-Closets.*—This type of closet being apparently the simplest form, we would naturally expect to find it the earliest kind put into use, but although this may have been the case there is no evidence to that effect; and if it was so, it was considered until recent years too simple to put on record.

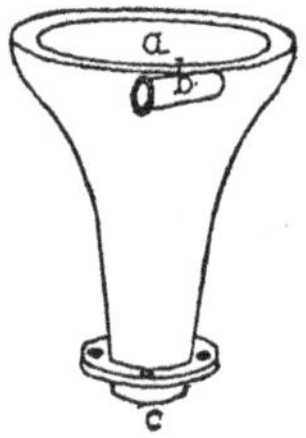

Fig. 160.
Philadelphia Hopper.

The opportunity for invention in this class is limited to the shape of the closet, the shape and position of its water-supply device and flushing-rim, or to the manner adopted for ventilating it.

Long hopper-closets are manufactured by all the firms making plumbers' supplies, and may be obtained of iron, plain, painted or enamelled, as well as in solid earthenware or porcelain.

The most common form is what is called the Philadelphia hopper. This has a bowl at the top, which suddenly changes to a diameter about the size of the soil-pipe. This closet is the cheapest and most objectionable form of its class. The sides are liable to become foul, the water being introduced in such a way that it can then flow around the bowl in a stream insufficient either to cleanse

oset.—A. G. Myers, of New York, manu-
; type and class. This hopper has Myers's
it zigzag flushing-rim, described in connec-
with the same firm's "All China Closet"
156). It is useless to enumerate the many
ent closets which are similar to the ones
d. All the prominent firms in England and
ica manufacture closets which vary from the
llustrated only in the mode of flushing, or
slight variation in their form. The long
r closet is the best form for exposed situa-
where it is necessary to place the trap some
ce below ground, according to the severity
climate. It is necessary in a closet of this
-cock placed at the same distance below
y means of a long key or rod.

" *Closet.*—The Maddock's "Wash-Out"
per class, and has a projecting lip formed
art of the rim, to better
al.

Fig. 164.
Maddock's Hopper.
*a*, Bowl.
*b*, Flushing-rim.
*d*, Outlet.
*e*, Projecting lip.

Rhoads long hopper
his country in the year
nded to be used without
or seat, in this manner
on of, or collections of
wood-work. A closet
is excellent for use in
the first case the wood-
communicate infectious
the wood-work would
usage.
ng hopper closets have vent-pipes at-
of the hopper just before it is connected
Among closets with the vent, may be
" closet, and I have been informed by



Mr. O. W. Norcross, that it was in use in the city from which it takes its name a number of years previous to the year 1876. In these closets the arrangement of the vent is for the same purpose, and in principle they seem to be identical. The receiver is a decided objection in the Smith closet, acting as a collector of filth: it is almost as objectionable as a receiver under a pan-closet.

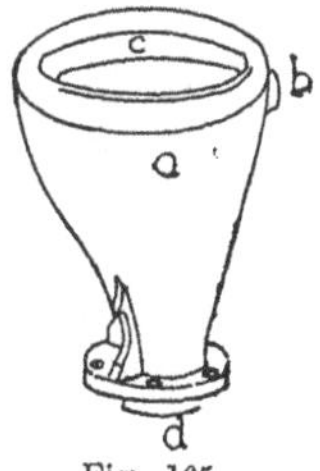


Fig. 165.
Rhoads's Hopper.
*a*, Bowl.
*b*, Supply.
*c*, Seat and flushing-rim.
*d*, Outlet.

The intention was to use these closets without a trap, in which case the vent near the soil-pipe, as in the Worcester closet, would be advantageous; but they are also intended to act as a vent for the room: for this purpose the Smith closet would have a slight advantage.

*The Phillips Closet.*—In 1882, a closet was invented which is a combination of the Worcester and Smith closets. The bowl is formed separately to set into a small hopper. The vent-pipe starts near the floor, runs along the side of the hopper, and has an opening into it near the bottom of the bowl. This

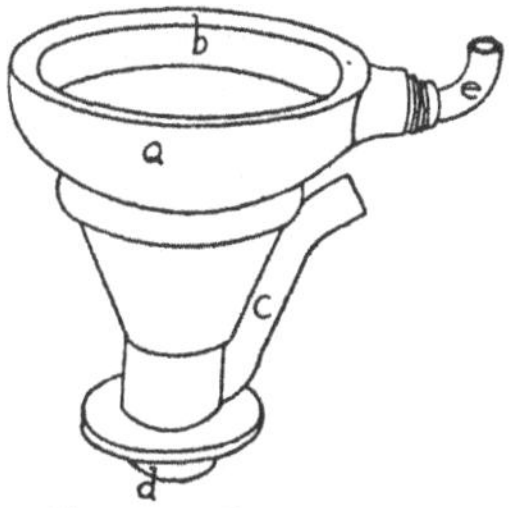


Fig. 166.—Perspective.

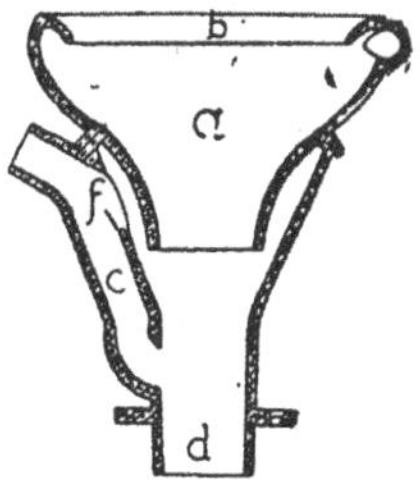


Fig. 167. Section.

Phillips Closet.
*a*, Bowl. *b*, Flushing-rim. *c*, Vent-pipe. *d*, Outlet. *e*, Supply-coupling. *f*, Upper opening into vent-pipe.

closet instead of having a metal hopper into which the bowl is set with the same objectionable putty-joint described in connection with the pan-closets, has the bowl, hopper and vent-pipe made in one piece of earthenware.

*Watson's Closet.*—Another closet of this kind, designed by Watson, is so arranged, by curving the outlet from the bowl, that the trap and its contents will be out of sight. This closet has a vent let into socket for a vent-pipe just before it joins the trap or soil-pipe. The bowls of this closet are made oval or square, and are much larger than this portion of the closet usually is. They can thus be used as slop-hoppers or urinals to better advantage than if the bowl were small.

There is no objection to using a vent-pipe in the position shown in the preceding closets; and it might be found advantageous in

Fig. 168.—Oval Bowl.

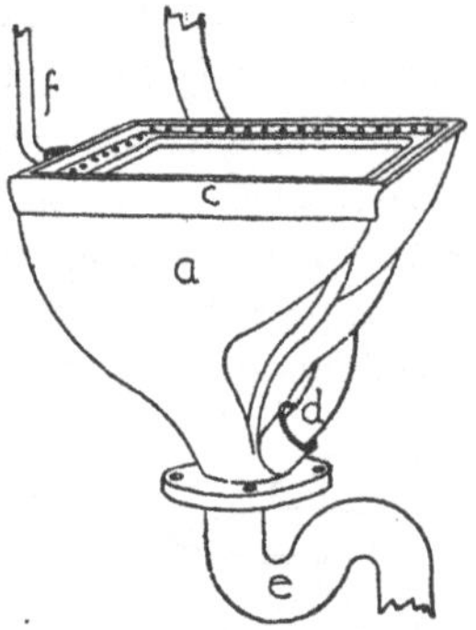


Fig. 169.—Square Bowl.

Watson's Closet.

*a*, Bowl. *b*, Vent. *c*, Flushing-rim. *e*, Trap. *f*, Supply.

taking off odors from the room, or odors created while some one was at stool. It must be remembered that a vent-pipe can never be a substitute for a proper siphon trap under the closet. Patentees sometimes claim that there is no further necessity for a trap, as the vent-pipe opening into a hot flue would carry off all noxious gases generated in what Mr. Eardley F. Bailey-Denton properly calls the inside sewerage system of the house.

The purpose which a vent-pipe of this kind is intended to fulfil—to carry off odors from the closet bowl or room—would be better served by an annular ring with slots in it, extending around the



the sides of the bowl, or to empty the trap properly. The tra
connected with the bowl just below the floor line.

*Foley's Closet.*—A good type of an improved form of long hop
closet was invented in 1883, by J. Foley. This closet as wel
other better forms of the class are straight or nearly so at the b
so fæcal matter will drop directly into the mouth of the trap.
water-supply is conducted through an annular branch, forme

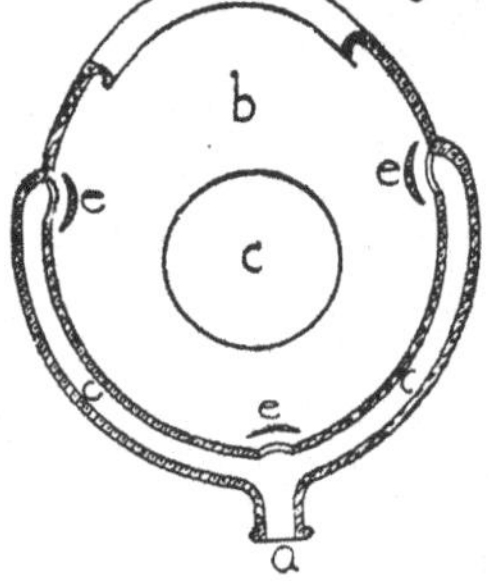


Fig. 161.

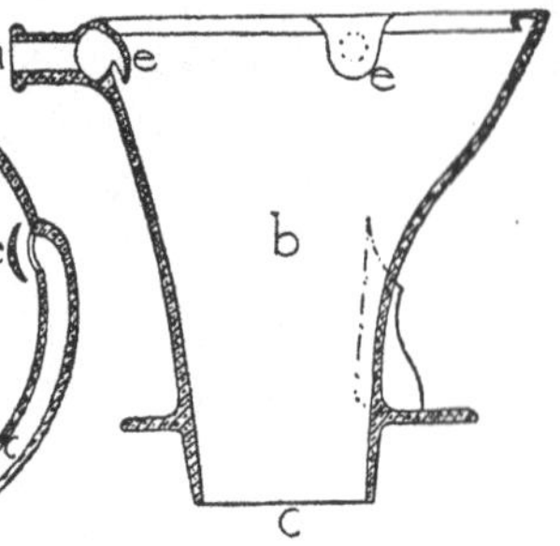


Fig. 162.

Foley's Closet.

*a*, Inlet for water. *b*, Bowl. *c*, Outlet. *e*, Fans.

one piece of earthenware with the bowl. This conductor ext
about half-way around the top of the bowl. The water is adm
to the bowl through three openings, one being at the back and
on each side. The water is distributed over the surface of
bowl by means of fans, instead of a flushing-rim, as is the cas
other closets of its form. I think a good flushing-rim washes
bowl more thoroughly than any arrangement of fans.

*Demarest Long Hopper.*—The long hopper-closet furnished u
this name by J. L. Mott & Co., is either round or oval in sh
and has one of Demarest's patent flushing-rims, the same th
used on the Hygeia closet (Fig. 139). This is in one piec
earthenware, and forms a very neat and effective closet when
plied, as it is intend to be, by a tank or cistern.

top of the bowl, and connected with some heated flue or pipe, and made either of metal separate from the closet, or of earthenware formed in one piece with the bowl.

*Fowler's Closet.*—Baldwin Latham describes a novel form of long hopper closet in the one designed by Fowler. The novelty in this closet consists in the trap being connected with the drain-pipe from the yard, sink, wash-basin, or other fixture.

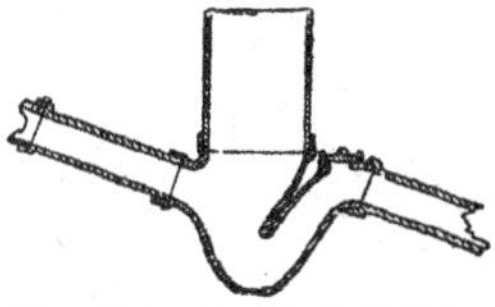

Fig. 170.—Fowler's Closet. *a*, Hopper. *b*, Trap. *c*, Drain. *d*, Inspection-cover.

In this manner only waste-water could be used in washing matter from the trap. Under some circumstances, in villages or country places a closet of this kind might be found useful. I consider the better long hopper-closets preferable to any other class or form, except the short hopper, and there are only two reasons why the short hopper is to be preferred. It has less distance between the bowl and trap, and thus less surface to become foul.

It appears to have been conclusively proved by the experiments of Messrs Edw. S. Philbrick and E. W. Bowditch made under the auspices of the National Board of Health, that the momentum of the water created by the distance between the top of the bowl, and the water in the trap was sufficient to force sufficient water out of the trap to break the seal. The vent-pipe (2") seems to have very little effect in preventing loss of seal by momentum, which must not be mistaken for siphon action. So it can be seen that a closet of this kind might lose its water-seal, and the party using it be none the wiser. The trap is such a distance from the closet seat, that it is impossible to see whether the trap has lost its seal or not.

## SHORT-HOPPER CLOSETS.

This type may be conveniently treated by grouping together those which have a bottom outlet in distinction from those which have a side outlet.

Mr. W. P. Gerhard has seen proper in his book,* to classify a number of the short-hopper closets, some of which have their outlet

* *House Drainage and Sanitary Plumbing.* Wm. Paul Gerhard, New York, 1882.

at the bottom and others at the side, under the name of "Wash-out Closets." I think it simplifies the matter to place them where they certainly belong, in the class of hopper-closets. Mr. Gerhard himself says: "A wash-out closet is in fact only a modified and improved form of hopper."

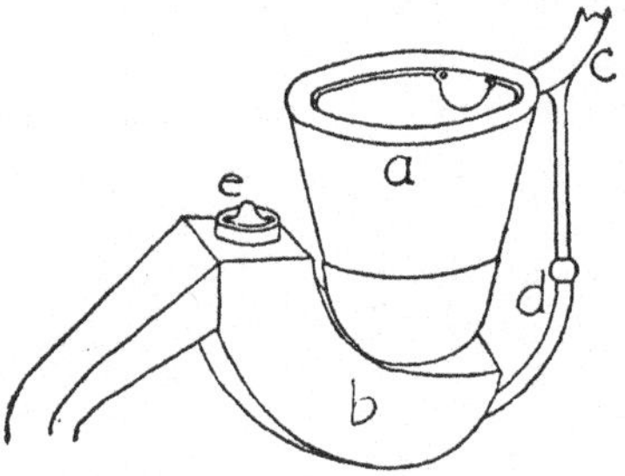


Fig. 171.—Thomas Smith's Closet.
*a*, Bowl. *b*, Trap. *c*, Supply-pipe. *d*, Branch to bottom of trap from supply-pipe. *e*, Inspection-hole.

*Thomas Smith's Closet.*—The earliest patent for a closet of this type was issued by Great Britain to Thomas Smith, in 1842. The patentee claims as a novelty the branch from the supply-pipe (the closet being supplied from a tank), which is connected with the bottom of the trap. This branch from the supply-pipe is intended to thoroughly wash out any matter which might be in, or would otherwise remain in the trap. In this closet the trap, instead of being circular in cross-section, is rectangular. It has an inspection-cover on the crown of the trap, and the water is spread over the bowl by means of a fan.

*French Short Hopper.*—F. Liger* tells us that this closet (Fig. 172) was designed in Paris, in 1872, for use in public places, being modelled according to English patterns. He states "that the gases from the sewer are cut off by a siphon-trap."

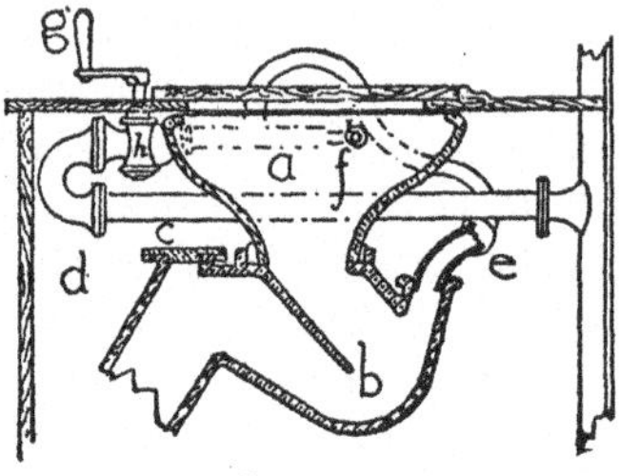


Fig. 172.
French Short-Hopper Closet.
*a*, Bowl. *b*, Trap. *c*, Inspection-hole. *d*, Supply-pipe. *e*, Branch from supply-pipe connected with bowl. *f*, Opening into bowl. *g*, Lever opening supply-valve.

In public places where this closet had been introduced, it was made to operate by opening the doors. The closet is arranged so it can

* *Fosses d'Aisance, Latrines, Urinoirs, et Vidanges.* F. Liger, Paris, 1875.

be supplied with water, either by means of a supply-valve or from a tank. The peculiarities of this closet consist in the branch from the main supply-pipe for flushing the trap, and in the partition of the trap being so placed as to form a screen, and hide, rather than get rid of, any disagreeable matter that might possibly remain.

*Hellyer's "Vortex" Closet.*—S. S. Hellyer, of London, received patents from Great Britain in 1878, and from the United States in 1880, for a short hopper closet which he has named the "Vortex." The novelty of this closet is based on the shape of the bowl and flushing-rim. The flushing-rim is so arranged as to throw a strong

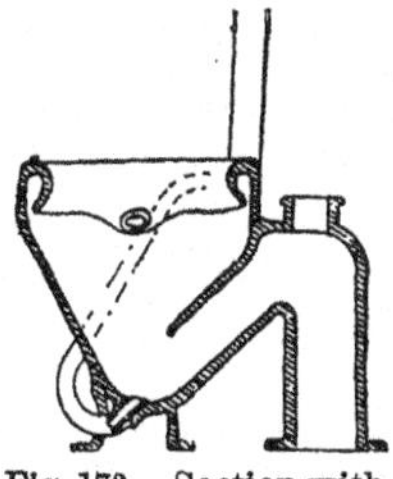

Fig. 173.—Section with Jet.

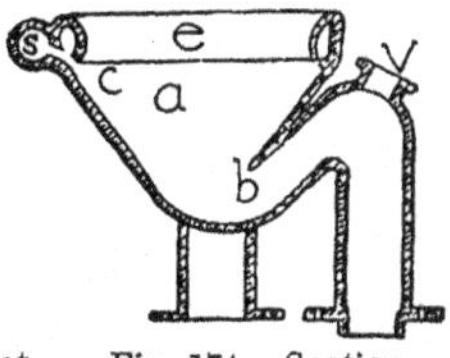


Fig. 174.—Section.

Fig. 175.—Perspective.

Hellyer's "Vortex" Closet.

*a*, Bowl. *b*, Trap. *c*, Opening for jet of water. *e*, Flushing-rim. *s*, Supply. *v*, Vent.

jet of water directly into the bottom of the bowl, this being also the bottom of the trap, by this means forcing whatever water or foreign matter there is through the trap into the soil-pipe. It will be seen by the illustration (Fig. 173) that it is intended in some cases to have a branch from the main supply-pipe feeding the bottom of the trap, as described in the Smith closet (Fig. 171). There are openings in the flushing-rim around the sides of the bowl, by which the surface of the bowl is washed. The trap is properly vented. This branch connecting with the bottom of the trap is unnecessary in the better forms of short-hopper closets, where they are flushed from a tank, through a properly-arranged flushing-rim.

*Wm. Smith's Closet.*—The "Siphon-Jet" closet, manufactured in San Francisco, and patented in 1876 and 1878 by Wm. Smith, also depends upon the branch from the supply-pipe to cleanse the trap and force the waste matter into the soil-pipe. This closet is fur-

nished in one piece of earthenware, or with an earthenware bowl and iron trap. It is connected with a tank, or may be connected directly with the supply-pipe, the flush being governed by a patent valve.

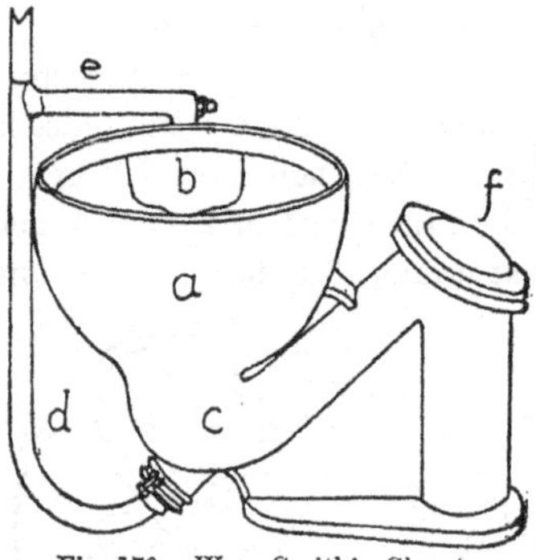


Fig. 176.—Wm. Smith's Closet.
*a*, Bowl. *b*. Fan. *c* Trap. *d* Branch supply-pipe. *e*, Supply-pipe. *f*, Inspection-hole.

Mr. J. Pickering Putnam, in 1884—within the last month—has patented a closet with a jet supply to the trap.

The jet is large. It forms a part of the tank, and the water is held in position by a vacuum formed in the top of the tank itself. It claims to be noiseless, as the water enters below the water line. I should think this large receptacle would become foul by the water from the closet. From the illustrations it appears complicated and to have many corners and crannies to collect dirt.

*English Closet.*—Baldwin Latham describes a short-hopper closet in which the flushing-rim is connected with the bottom of the trap. He says: "this form of closet with bye flushing-pipe was introduced to remedy the defect, which, as already referred to, attends the hopper-closet, that is the cleansing of the trap after the use of the closet." This evil mentioned by Latham has been remedied by properly arranged and devised water-supply and flushing-rims, which are now manufactured as a part of the better short-hopper closets.

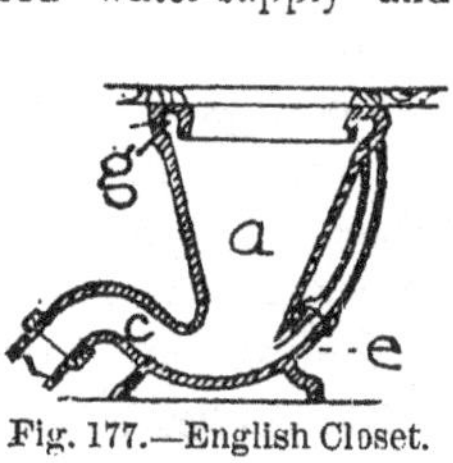


Fig. 177.—English Closet.
*a*, Bowl. *c*, Trap. *e* Branch-supply. *g*, Flushing-rim.

Closets manufactured as described, with a branch from the supply-pipe connecting with the bottom of the trap, would be peculiarly susceptible to the evil which it is claimed sometimes takes place when the supply-pipe is attached directly to the bowl of the closet, a valve only intervening; viz., the liability of foul air or filthy water being drawn from the closet

when a faucet is opened on some floor below, the supply-valve of the closet being open at the same time. For the reason stated care should be taken always to connect closets of this kind with a separate tank or cistern when—and they are simple and useful—they are used.

*Hellyer's "Short Artisan Hopper."*—This closet is composed of two pieces of earthenware, the bowl and trap, which are joined together by means of clamp-screws, putty or cement being put into the joint. The bowl of this closet is oval in shape, with its back edge nearly straight, so the bowl need not become foul by fæcal matter dropping on it. This closet has a flushing-rim which serves its purpose well, and the trap is properly vented at the crown. In

Fig. 178.

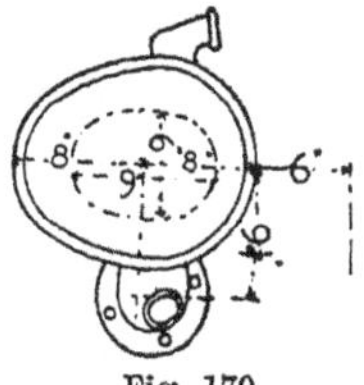

Fig. 179.

Hellyer's Short Artisan Hopper.

connection with his patent flushing-rim, Hellyer arranges an annular rubber band, in a groove made for the purpose on the top of the flushing-rim. The seat of the closet is intended to fit down on this band and thus prevent smells or cold air from coming into the room from beneath the seat. It is the custom in England to have the "tell-tale" pipe from the closet-safe run through the wall into the outer air. A draught through this pipe might become disagreeable to a person occupying the seat of the closet. This closet has an opening at the crown of the trap, to which a vent must be attached, and in use it is simple and effective.

Fig. 180.—Flushing-rim.
*i*. India-rubber ring.
*c*, Flushing-rim.

*Demarest's Short-Hopper Closet.*—The J. L. Mott Iron Works manufacture a short-hopper closet which in appearance and in its manner of working is almost identical with the Hellyer short hopper. The novelty consists in the manner of connecting the bowl with the trap, and the trap with the vent-pipe, and in the Demarest

patent flushing-rim (see Fig. 180). The top of the trap and the bottom of the bowl are both encircled by a projecting ring or flange, over and around which a metal collar fits. This collar when bolted makes a very strong joint. The space between the collar and the joints has to be filled with putty or cement. The coupling for the vent-pipe is a two-inch zinc-coated iron quarter-bend, which has a hub on the free end, into which the spigot of the vent-pipe can be caulked. If a lead vent-pipe is to be used, a brass coupling must be connected with the trap instead of the iron one described.

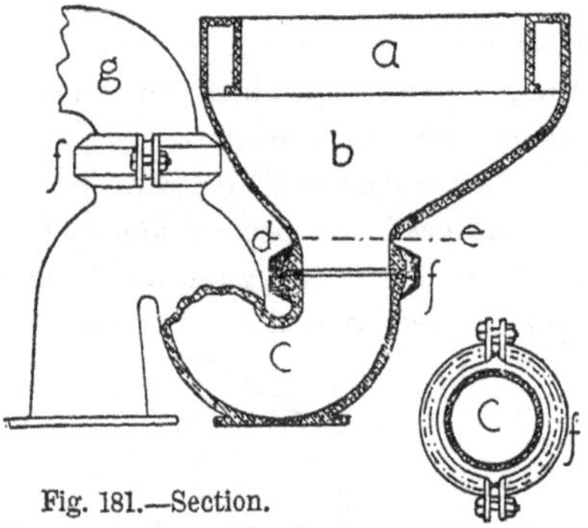


Fig. 181.—Section.
Demarest's Short-Hopper Closet. Fig. 182.
Section through line *d-e*.
*a* Flushing-rim. *b*. Bowl. *c*, Trap. *g*, Vent-pipe. *f*, Collar or clamp.

*Roe's Closet.*—Baldwin Latham, in his work on sewerage, illustrates two short-hopper closets. Mr. Latham states that in this closet exists the first instance of a vent-pipe being properly placed on the trap. The reader of the foregoing pages will probably

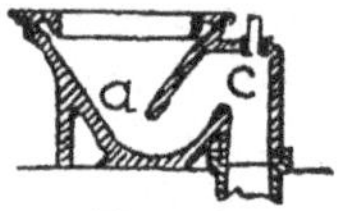

Fig. 183. Roe's Closet. Fig. 184. Straight-Back Hopper.
*a*, Bowl. *b*, Trap. *c*, Vent.

recollect that the vent to the trap was used or invented in connection with the Wilkins closet forty years ago. Figure 184 is shown as an example of this class, in which the back is made straight, so fæcal matter will drop directly into the trap, and not adhere to the sides of the bowl.

*Carmichael's "Wash-down" Closet.* Mr. W. P. Buchan, in 1879, invented a closet of this class, which he informs us is fur-

nished in one or two pieces of earthenware, the joint being made above the water-line.* This closet is supplied with a flushing-rim to wash the sides of the bowl, and also with a jet of water which strikes the water in the bowl so as to force it through the trap. The trap has an inspection-hole on top. This closet seems to have deservedly met with success in England, where there seems to be a tendency, which should be followed in this country, to adopt short-hopper closets in preference to all styles or patterns. Mr. Buchan calls the short-hopper closets which have an outlet at the bottom wash-down closets, while those which have an outlet from the side he calls wash-out closets.

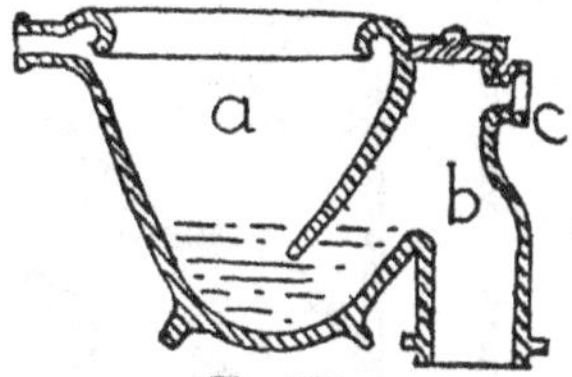

Fig. 185.
Buchan's "Carmichael" Closet.
*a*, Bowl. *b*, Trap. *c*, Vent.

*A. G. Myers's "Niagara" Closet No.* 2.—A short-hopper closet has been quite recently introduced into the market by A. G. Myers, of New York. This closet is made in one piece of earthenware, or it is furnished with a part of the trap separate from the bowl. It is flushed by carrying the water-supply through or against a fan. The bowl of this closet is waved or corrugated in circles parallel to the floor. The idea is that the water, coming in contact with the inequalities of the surface, will wash the bowl more thoroughly than would be the case if it were smooth. It appears to me that the under side of the wavy surface would not be washed at all, and if this is the case, more harm than good would be done by making a bowl in this manner.† The trap in this closet has a hub for a vent-pipe, which is placed, in accordance with the suggestion of Hellyer, on the side of the trap nearer the soil-pipe than is usual. This position is intended to prevent foul matter from being forced up into, and lodg-

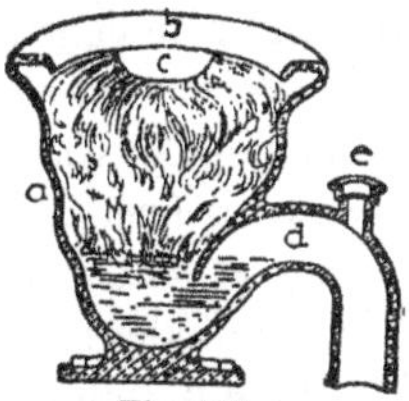

Fig. 186.
"Niagara" No. 2.
*a*, Corrugated bowl.
*b* Fender. *c*, Pan.
*d*, Trap. *e*, Vent.

* *Plumbing.* W. P. Buchan, London, 1883.

† Since the above was written, Mr. Myers has changed the form of this closet, and put it on the market in the form of Hellyer's short hopper (see Figs. 178–179).

ing in the vent-pipe. It is claimed that the vent-pipe is sometimes made useless in this manner.

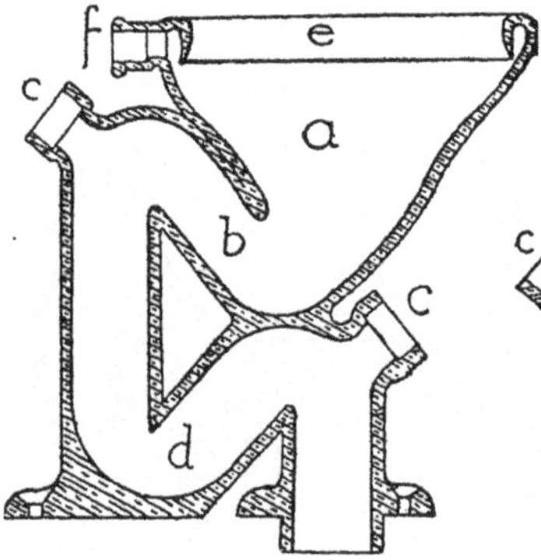


Fig. 187.—Single Trap.
*a* Bowl. *b*, Trap No. 1.
*d* Trap No. 2. *c*, Vents.
*e*, Flushing-rim. *f*, Water-supply inlet.

*Owens's Closet.*—Henry Owens, in 1878, invented a

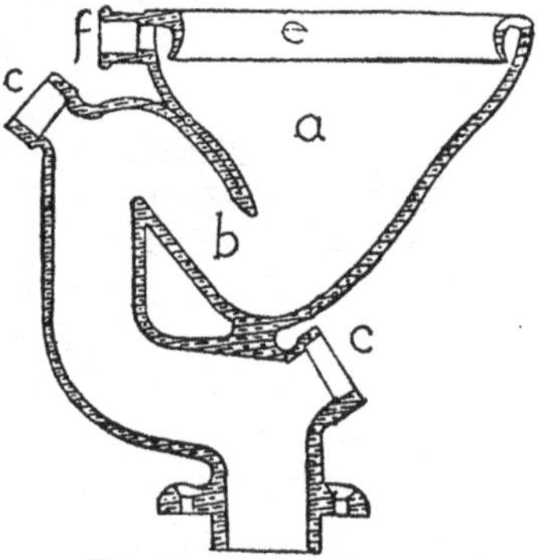


Fig. 188.—Double Trap.
Owens's Closet.

closet which is made in one piece of earthenware, and is designed so as to have either two traps or a trap and an offset between it and the soil-pipe. The traps, and also the offset have openings for vent-pipes. The closet with the double trap would have a confined air-space which could only be ventilated by an inlet as well as an outlet pipe for the air. I think as there is nothing gained by the double trap, the closet with the trap and offset is to be preferred.

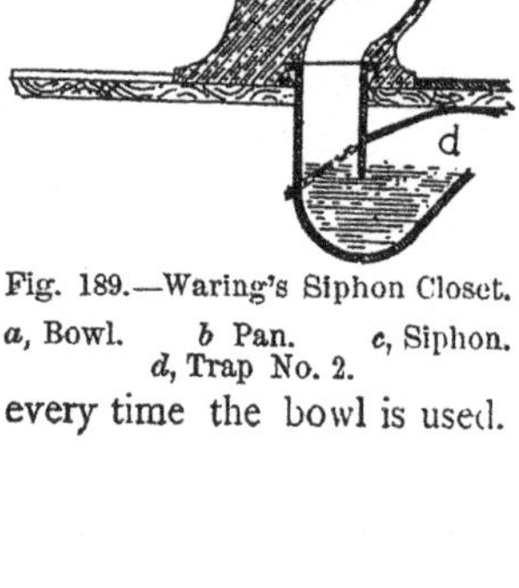


Fig. 189.—Waring's Siphon Closet.
*a*, Bowl. *b* Pan. *c*, Siphon.
*d*, Trap No. 2.

*Waring's Siphon Closet.*—This closet, which was designed by Mr. Geo. E. Waring, Jr., of Newport, R. I., has a very deep water-seal trap at the the bottom of the bowl. The closet is set over a weir, which is placed beneath the floor. The contents of the bowl are siphoned out every time the bowl is used. From its shape it is necessary that

this closet should be filled to within a short distance of the top of the bowl, before the necessary siphon action which empties the bowl can take place. This excessive amount, or, more properly speaking, the height to which the water rises in the bowl, is exceedingly disagreeable sometimes. The weir below the floor is in shape or form very much like the objectionable D-trap. The first trap by reason of its siphon action cannot have a vent-pipe in the usual position. The closet is gracefully formed in one piece of white earthenware.

*Boyle's "Tidal Wave" Closet.*—In the year 1882 a closet with the double trap was invented by J. E. Boyle. An air-pipe runs from the space between the two traps, up and through the tank. This air-pipe is governed by a valve which opens only when acted on by a pressure from below, so this vent-pipe would not interfere with forming of the partial vacuum, which would be necessary before the closet could be emptied by siphon action. When two-thirds of the service-box has run off, the vacuum is broken and the remaining third fills the bowl. The bowl and its double trap is manufactured in one piece of earthenware. The lower trap is properly ventilated. The closet must be supplied from one of Boyle's patent cisterns or tanks.

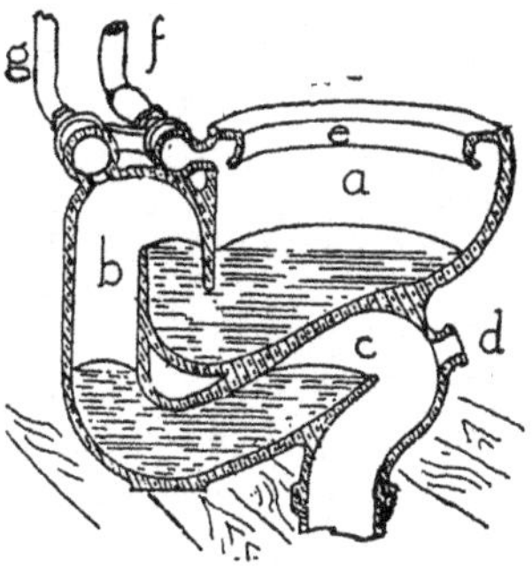

Fig. 190.—Boyle's Closet.
*a*, Bowl. *b*, Siphon. *c*, Trap. *d*, Vent. *e*, Flushing rim. *f*, Supply pipe. *g*, Air pipe.

William Bunting, of New York, received patents from the United States, in 1878, for a closet in which the siphon action is created by a piston which works in a chamber situated above the trap.

*A. G. Jennings's Side-Outlet Hopper.*—The first short hopper with a side outlet was invented in 1852, by A. G. Jennings, of London, in which he claims as novelties "the basin (bowl) and trap combined in one piece of earthenware, so formed that the basin is made to contain a quantity of water.....The escape may

be from the face, side, or front. The basin may be of any shape to hold water in the bottom." In 1876 an improvement was made on this closet, for which a patent was issued. He says: "It has been found to have the disadvantage that the water tends to run around and around, instead of passing out in a stream." To remedy this defect, a rib, which is shown in the illustra- by a dotted line, was introduced. This rib causes the water to be thrown in a mass toward the outlet, carrying whatever waste matter there was in the bowl with it through the trap. An opening for a vent-pipe, which is not shown in the illustration, is now manufactured in the usual position on the crown of the trap. The water is supplied through a fan instead of a flushing-rim, and it is intended to be connected with a tank.

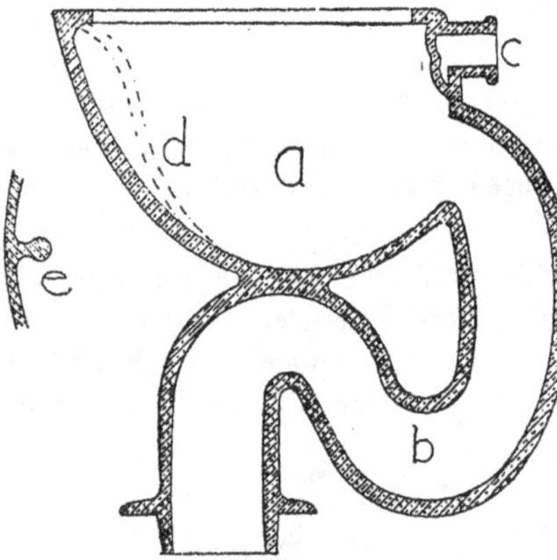


Fig. 191.—Jennings's Side-Outlet Hopper.
*a*, Bowl. *b*, Trap. *c*, Supply. *d*, Dotted lines showing where rib *e* was added in 1876.

*Buckland & Rees's Closet.*—The second closet of this class was invented in England in 1862, by Buckland & Rees. This closet has its fan so shaped that a portion or half the water entering the bowl is driven to each side, "instead of being driven with a spiral motion, as is usually the case, the greater part of the water being allowed to flow directly to the bottom of the basin, and thence through the aperture or discharge-pipe on the other side." This closet has its trap and bowl in separate pieces. The "National" side-outlet closet manufac-

Fig. 192.
Buckland & Rees's Hopper-Closet.
*a*, Bowl. *b*, Trap. *c*, Fan. *f*, Supply.

tured in this country is apparently a combination of the two closets which have just been described.

*Hellyer's Side-Outlet Hopper.*—I illustrate two forms of short hopper of the type under discussion, which were invented by S. S. Hellyer, of London. He acknowledges that he was unable to make closets of this kind act satisfactorily. The fæcal matter would strike against the sides of the outlet, and leave it in a foul condition. To obviate the tendency which the waste had to remain in the bottom of the bowl, the supply-pipe was divided into

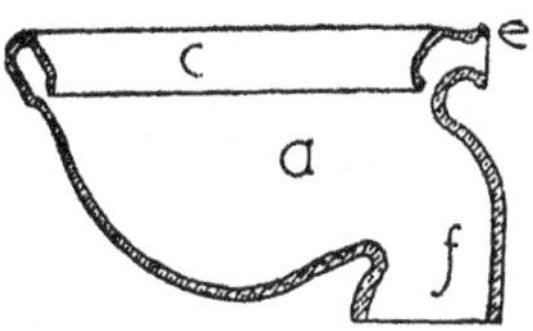


Fig. 193.

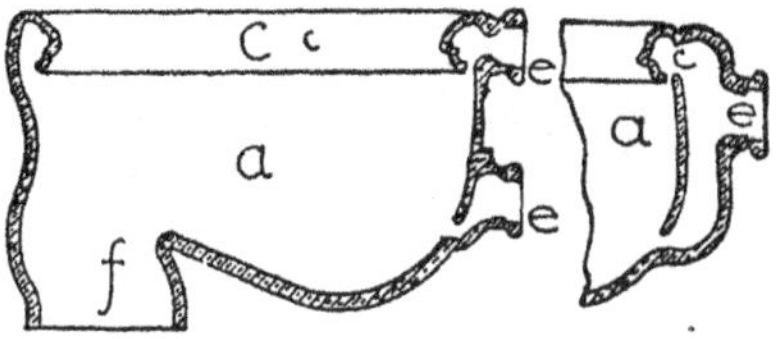


Fig. 194.

Hellyer's Side-Outlet Short-Hopper.

*a*, Bowl. *c*, Flushing-rim. *e*, Inlet for supply. *f*, Outlet.

two parts, one of which entered the bowl near the bottom, and in this manner forced the water and foreign matter through the outlet. Hellyer concluded that this particular form of hopper was not good for the purpose which it was intended to serve. The bowl is formed so it can be placed in an ordinary trap that is placed above the floor.

*Bostel's "Brighton" Closet.*—A closet of this class, which has recently been extensively used in this country and in Europe, was invented in 1877, by D. T. Bostel, of Brighton, England. This closet is made in one piece of earthenware, the bowl and trap being formed in a manner similar to the Jennings short hopper (Fig. 191), the difference being in the flushing-rim, inspection-hole, and in the vent-pipe. The supply-pipe is divided so as to enter the flushing-rim at two points. Flowing around the flushing-rim, the water enters the bowl through small holes, which are purposely made larger on the side of the bowl, opposite to, or furthest from the outlet. By this arrangement the greatest volume of water is thrown

where it is most needed, and the matter in the bowl is carried into the trap, if not through it. There is an opening with a cover, to be used for the purpose of inspecting the mouth of the trap, and form-

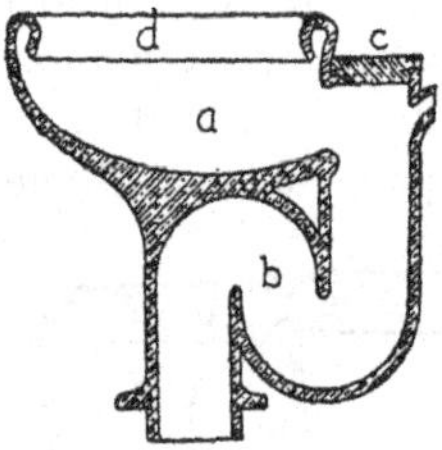


Fig. 195.

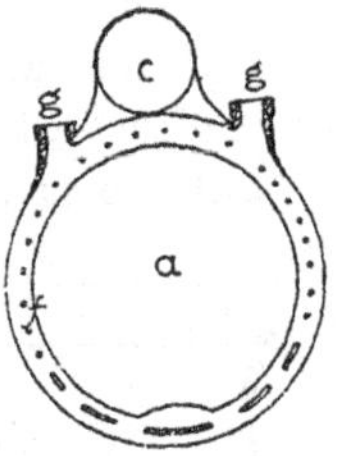


Fig. 196.

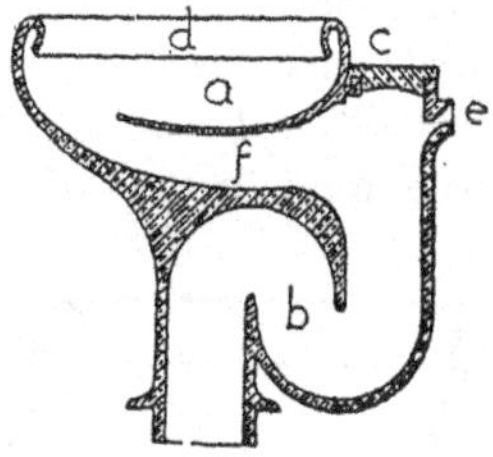


Fig. 197.

The "Brighton" Closet.

*a*, Bowl. *b*, Trap. *c*, Inspection-cover. *d*, Flushing-rim. *e*, Vent. *f*, Slots in flushing-rim. *g*, Double-supply.

ing a top to the side outlet. The opening for a vent-pipe is just below the inspection-cover. A vent-pipe in this position will only serve to carry off locally-generated gases, and must never be connected with the soil-pipe, as it would form a communicating duct between the inside sewerage system and the house. In this closet a vent-pipe to prevent siphonage, back pressure, and to carry off gases generated in the soil and other waste pipes, would have to be connected with the branch from the soil-pipe below the floor, as there is no opening for the purpose near the trap. There is sometimes a false bottom or screen manufactured in this closet. I can see no good purpose served by this screen, while it increases the space and surface between the bowl and trap, which is liable to be fouled by matter in its passage to the trap. This closet is arranged to receive its water from a tank.

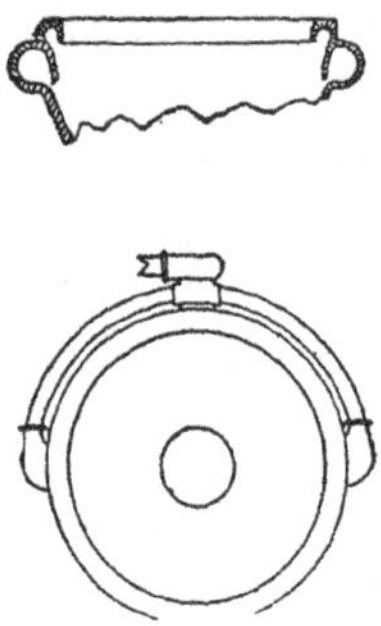

Fig. 198. Stidder's Branch Supply.

*Stidder's Closet.*—In 1878, one Stidder invented, and received patents from Great Britain, for a supply-pipe divided so the water

could enter the flushing-rim at three points. This flushing device was intended for use with a short-hopper closet.

*The Lambeth "Flush-out" Closet.*—A closet of this form and class, which was designed and is manufactured by Doulton & Co., of England, is made in two pieces of earthenware. This closet has an inspection-hole so that the compartment over the top of the trap can be examined without difficulty. There is also an opening in the crown of the trap, which may either be used as an inspection-hole or as a place into which a vent-pipe can be inserted. The flushing-rim is similar to the one described in connection with the Lambeth "Trapless" closet (Fig. 61). The openings in the flushing-rim, opposite the side-outlet, are so arranged that a large body of water is thrown directly into the bottom of the bowl, and in this manner the bowl and trap are cleaned by the scour of the water. The J. L. Mott Iron Works, of New York, have recently introduced a closet of this kind into the market, which has an inspection-opening over the side outlet, and a brass coupling in the usual position on the soil-pipe side of the trap, to which a vent-pipe must be soldered or caulked. These closets are furnished in one piece of earthenware, or with an earthenware bowl and an iron trap, and are called the "Inodora" and "Purita."

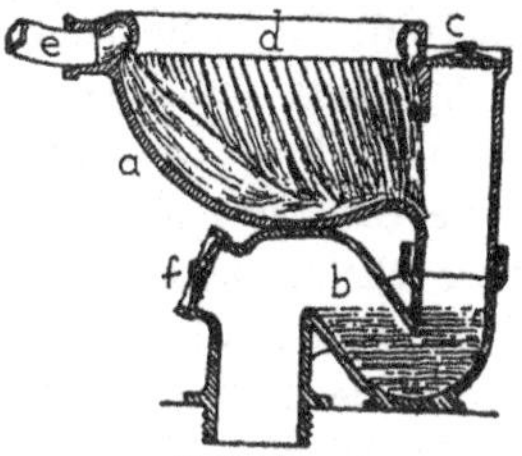


Fig. 199.
The Lambeth "Flush-out" Closet.

*a*, Bowl. *b*, Trap. *c*, Inspection-cover. *d*, Flushing-rim. *e*, Supply. *f*, Opening for vent.

DEDUCTIONS.

As stated before, *I consider the simple short-hopper closet, with a bottom outlet, the best form of water-closet in use at the present time*, when it is properly flushed from a tank, through a well-arranged flushing-rim, and has a trap properly ventilated. Although I have not found it necessary, a vent-pipe encircling the bowl and connected with a heated flue may sometimes be found advantageous.

In this form of closet we find one simple in its construction, effective in its action, immediately discharging the fæcal matter into

the soil-pipe, and at the same time cleanly and healthful, having no concealed chambers or compartments to become foul and generate unwholesome gases. In fact, it leaves little to be desired as an improvement on water-closets. Closets with bottom outlet and of the short-hopper pattern, with effective flushing-rims, are manufactured by the majority of dealers in plumbers' supplies in the United States and Great Britain, and the French have also made an effort to manufacture and introduce them into Paris. The short-hopper closets of the form under consideration are recommended by such prominent authorities on sanitary plumbing as S. S. Hellyer,* of England; W. P. Buchan,† of Scotland, and Prof. P. Putyo‡ in a work recently published in Belgium. By a comparison with other forms which are and have been in use, the cleanliness, simplicity and effectiveness of the short hopper can easily be appreciated by those who are least expert on the subject. With the valve-closet we have the bowl and its usually faulty connection with the receiver. The valve itself and its seat, nicely adjusted with all necessary cranks, spindles, journals, springs, levers, weights, etc., according to the pattern of closet, are all liable to get out of order, and all parts which are within the receiver and the sides of the receiver are liable to become foul. Portions of the valve are formed of perishable rubber, the overflow is more or less complicated, and the trap, when one is used, is necessarily placed below the floor, with few exceptions.

The plunger-closet has its plunger, plunger-compartment, and, in all except a very few cases, floats and supply-valves with overflows, all connected with the bowl, and between it and the trap.

In the classes just mentioned, all the described mechanism is used in place of the simple bowl and trap of earthenware. And with what gain? The speedy and effectual removal of waste matter is accomplished more quickly and thoroughly by the simple than by the complex closet. In the plunger and valve closet small particles of filthy matter and foul sediment collect in the compart-

---

* *The Plumber and Sanitary Houses.*

† *Plumbing.*

‡ *L'Hygiene dans la Construction des Habitations Privees,*

ments which are under the valve or around the plunger, and concealed from view. These particles decay and generate foul and in some instances poisonous gases.

The whole surface of the short hopper, except in the side-outlet form, is visible, and any particles that cling to the bowl are in plain view, and I think all will acknowledge that it is better to have such foulness where it can and will be cleaned off by a careful housekeeper, than to have it out of sight, where it will remain and undergo decay; fair without, but foul within. When these closets (the simple short hopper) have their seats in the proper position, the tank placed at a sufficient height, between seven and eight feet, one-and-a-quarter-inch supply-pipe, which must not be diminished by bends, I have never known them to fail to wash the bowl perfectly whenever flushed. The plunger and valve closets are supposed by many to form a mechanical seal between the bowl and the soil-pipe, but this is not the case with the closets of this class which are in the market, as the overflow only interposes a water-seal, except in the closets mentioned in preceding pages, which is the same seal as the trap of the short hopper. Although the experiments of Dr. Fergus, of Glasgow, and Doremus, of New York, have apparently proved that gases in a concentrated form will pass through a water-seal trap, nevertheless I think it can be said without exaggeration, that the majoity of sanitarians have accepted the experiments of Dr. Neil Carmichael, of Scotland, which have since been confirmed by Dr. Wernich, as conclusively proving that "water-traps are therefore, for the purpose for which they are employed, that is for the exclusion from the house of injurious substances contained in the soil-pipes, perfectly trustworthy;" in other words, organic germs will not pass through a water-seal trap. With the traps vented and the soil-pipe open, we need have no fear of gases passing through the trap, provided the water seal remains intact.

The advantage which a short hopper has over a long hopper has been mentioned in connection with the description of the latter type, as having less surface to become foul between the seat and the trap, and the possibility which exists of the trap under the long

hopper losing its seal by the impetus attained by the water. The advantage which this form has over the hopper with the side outlet is having the mouth of the trap always visible, and being without the concealed space formed by the side-outlet between the bowl and trap, which should always be avoided. It is needless to say that the pan-closet, as it is usually manufactured, with its container which becomes almost filled with filth, the faulty connection between bowl and container, which allow gases generated to enter the room, its imperfect flush and leaky supply-valves, is not to be compared with the short-hopper closet.

## SECTION VI.

### MISCELLANEOUS CLOSETS.

THE two closets of which the following is a description, do not belong to either of the classes which have been enumerated; at the same time they appear worthy of description.

The first is a closet of German pattern, described by Baldwin Latham in his work on sewerage.

*German Closet.*—This closet consists of a side-outlet hopper, with a siphon trap above the floor, in the bowl of which there is a tilting-basin that can be filled with water. The inside basin conceals the outlet from the bowl into the trap. When this basin is lifted, or tilted, its contents are discharged through the trap into the soil-pipe.

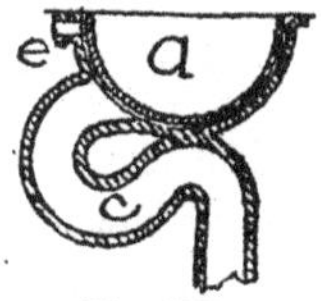


Fig. 200.
German Closet.
*a*, Tilting-basin.
*c*, Trap.
*e*, Supply.

This closet in its manner of operation is almost exactly like the tilting wash-basins patented and manufactured by J. G. Jennings, of London, and I should think the space between the two basins was liable to become very foul.

The second closet that belongs to the miscellaneous class, which I have selected for description, is one invented by John Tylor & Son, of London, in the year 1873.

This closet has a large receiver (in every way equal to, and as objectionable as the receiver under the ordinary pan-closet), into which the bowl projects.

The projecting outlet from the bowl turns at a right angle with itself after entering the receiver, and over its outlet is fitted a horn-like valve which revolves on a short arm and has its upper end open. This valve has a rubber band stretched over its larger end, where it fits against a metal seat that is clamped to the outlet of

the bowl. In this manner the water-closet bowl is kept filled with water to the height of the small end of the horn. The overflow, which, according to Mr. E. F. Bailey-Denton, is now supplied with a ball-valve, takes place through the horn.

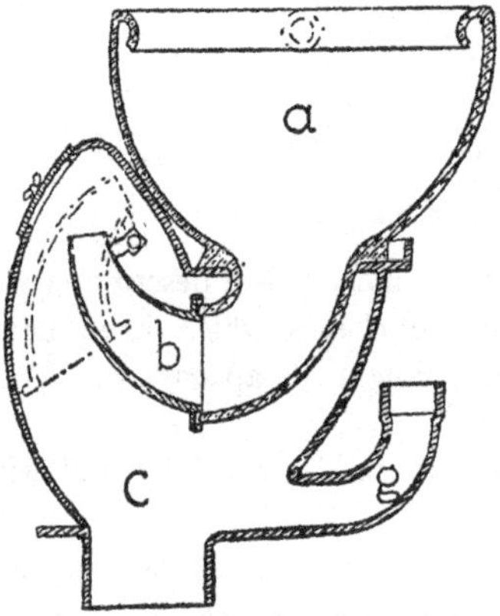

Fig. 201.—Tylor & Son's Closet. *a*, Bowl. *b*, Horn-valve. *c* Receiver. *g*, Bath-waste.

There is an inlet cast on the receiver of this closet into which bath or water-basin wastes may be caulked.

Neither of the above described closets probably will ever come into extended use, as they are inferior to most plunger and valve closets, and for this reason far inferior to the class of hopper-closets.

LATRINES.

The class of which the following is a description is in reality a combination of closets, rather than a single one, and is always intended for use in public places, such as depots, schools, hospitals, barracks, asylums, or any place where the usual closet will most probably be neglected or abused, and where a janitor can take charge of flushing and emptying them; the plug, valve, or tank, which forms the mechanism of the closet are under lock and key, and accessible only to the janitor in charge of the building.

*The Lambeth Improved Latrine.*—Doulton & Co., of England, manufacture a latrine that consists of a series of stone-ware bowls connected with each other and the main drain by suitable stone-ware pipes. The plunger is a simple hollow one that is formed by an iron tube. The plunger forms the overflow and has a perfect seat over the siphon-trap. The plunger is lifted by a handle that is kept under look and key by the janitor.

The bowls are filled from one cock. When it is necessary to

empty the latrine, the handle is raised and the outlet pipe (5" diameter) is opened, when the contents of the whole row of bowls and connecting-pipes will be discharged through the trap into the soil-pipe, in a few moments.

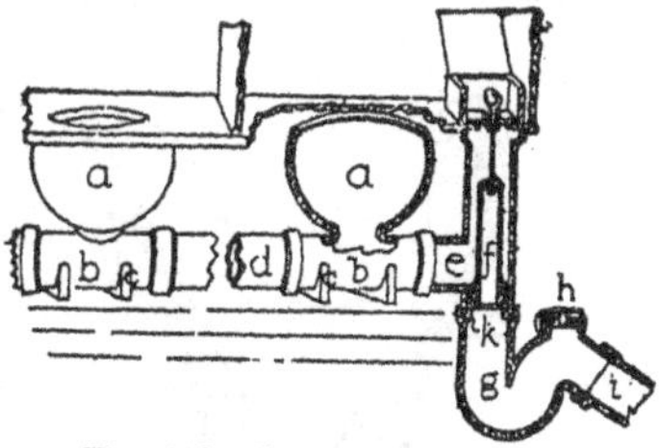

Fig. 202.—Lambeth Latrines.
*a*, Bowls. *b*, Connecting-pipes jointed to bowls. *d* Connecting-pipe. *e*, Plunger-chamber. *f*, Plunger. *g*, Trap. *h*, Inspection-cover.

The trap of this latrine is supplied with an inspection-hole, and it should also have, when used, an opening for a vent-pipe on the side of the trap next to the soil-pipe.

*Jennings's Patent Latrines.*—These latrines have been extensively used in public buildings in England. They are composed of bowls made on a special spherical pattern, connected with each other by means of short pipes, the flanges of which are bolted together. The bowls rest on short legs, so the whole system is raised above the floor. The bowl is divided in hemispherical pieces. This form and arrangement is convenient for shipping, as the upper half will set into the lower one. The bowls are furnished either in highly glazed vitrified stone-ware or enamelled iron. The plug is one of Jennings's hollow plungers opening over a simple siphon-trap. The plunger should be raised once a day, and oftener if ample water can be obtained. The water-cock as well as the plunger should be in charge of the janitor.

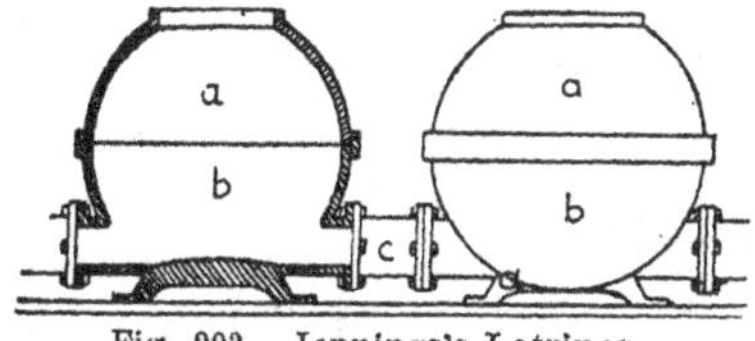

Fig. 203.—Jennings's Latrines.
*a*, Upper part of bowl. *b* Lower half. *c*, Connecting-pipe. *d*, Feet.

*Tylor's Latrine.*—In the year 1873, Tylor & Son, of London, received patents for a latrine of a peculiar form. The bowl is formed so as to set into and rest upon a receiver. This receiver

has on each side at the bottom a projecting spigot that could be cemented or caulked into the bell of a pipe. In this way a continuous line of latrines can be formed. The bowl, which is oblong,

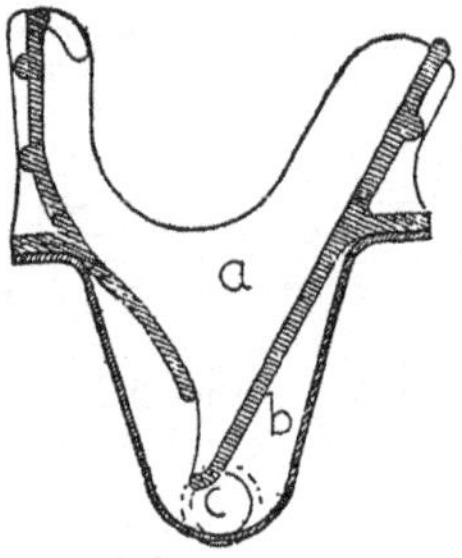


Fig. 204.

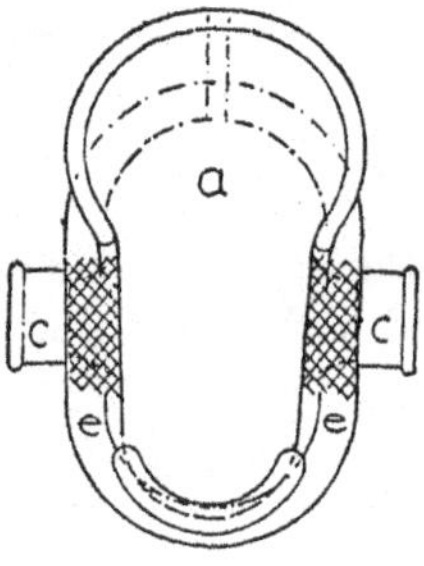


Fig. 205.

Tylor's Latrine.

*a*, Bowl. *b*, Receiver. *c*, Spigot ends.

has foot-rests on each side, and is intended to be used by parties crouching, no seat being provided. The patentee says they were invented for the foreign market. In public parks where water-closet seats would be liable to abuse, or where they might communicate disease, latrines of this or a pattern to accomplish the same end would be found useful.

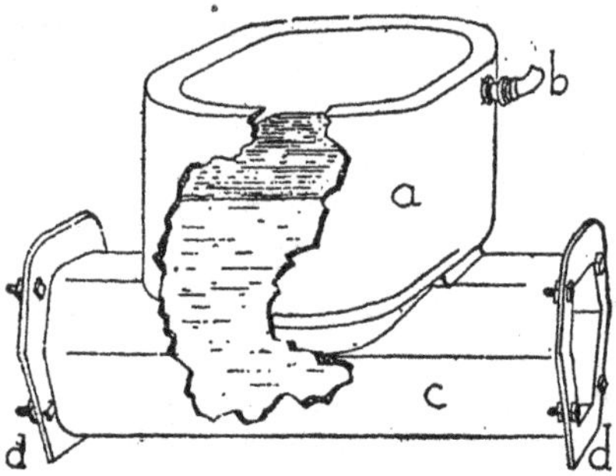


Fig. 206.—Mott's Latrine.

*a*, Bowl. *b*, Supply. *c*, Connecting-pipe.

*Mott's Latrines.*—The J. L. Mott Iron Works, of New York, manufacture latrines in this country, of plain, zinc coated, or enamelled iron. The bowl is formed in one piece with a section of the connecting pipe. The pipes are octagonal in section, and have flanges on the end by which they can be bolted together. The water is

held at the same level in each of the bowls by means of a hollow trapped plunger. This plunger as well as the compartment in which it is placed is of the same kind as the one described in connection with Demarest's closet (Fig. 138). The same firm manufactures a number of trough-closets. In closets of this kind there is no bowl; but a trough long enough to receive the number of seats desired. The outlets to the trough are closed by a hollow trapped plunger according to Demarest's patent. The plunger operates in a chamber separated from the trough, and it is lifted by a combination of levers, which are clearly shown in the illustration. (Fig. 207.)

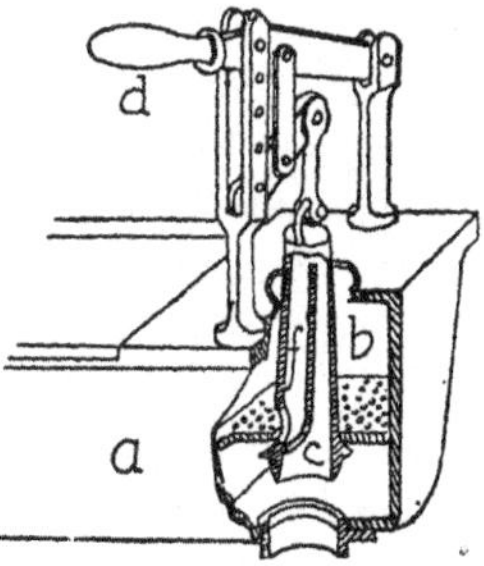

Fig. 207.—Trough-Closet. *a* Trough. *b*, Plunger-chamber. *c* Plunger. *d*, Lever-handle. *f*, Overflow.

*Parfitt's Latrines.*—A. G. Myers, of New York, manufactures a latrine formed in the shape of a long wooden box lined with lead, the cover to this box being divided into as many seats as it is necessary to have. This box is connected by a six-inch outlet and trap to the main soil-pipe. The novel feature in this latrine is the arrangement of a tilting-tank above the end farthest from the outlet. The tank is connected with, and is to be filled from wash-basin or

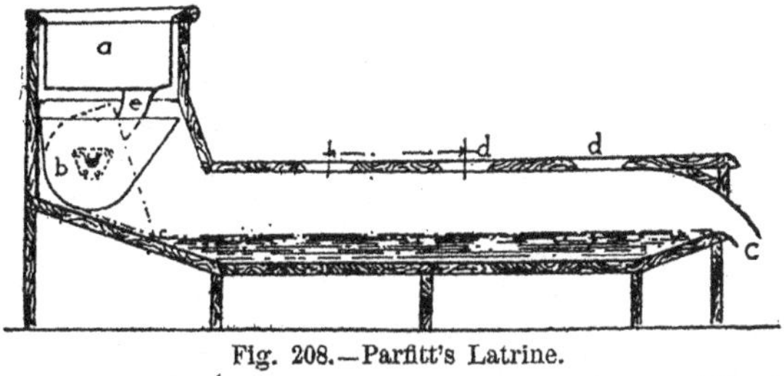

Fig. 208.—Parfitt's Latrine.
*a*, Sink or wash-bowl. *b*, Tilting-tank. *c*, Outlet. *d*, Seats. *f*, Lead lining.

slop-sink wastes. The tank being balanced on journals is so arranged that one end will be heavier than the other when it be-

comes filled with water. Tilting automatically it will empty its contents into the trough, and at the same time force the contents of the latrine through the trap.

In the place of latrines it is now a common, and I think the best practice, to use a row of long or short-hopper closets, either flushed by a tilting-tank, or a tank which is operated by connection with the seat or door of the water-closet room.

A tilting-tank can have its water-supply so arranged that the closet may be flushed at either short or long intervals. The time between the flushes should be regulated according to the abundance of the water-supply as well as according to the use the closet receives.

One or two arrangements for opening the supply-valve in the tank automatically will be described further on. Where rows of either long or short-hopper closets are used in place of a latrine, each bowl has the advantage of a separate and ventilated trap between it and the soil-pipe. While with the ordinary latrine we have each bowl with its connections to become foul, and these foul surfaces will be exposed to the room whenever the latrine is emptied. When foul matter is dropped into one bowl of a latrine, there it must remain, no matter how great or offensive the stench may be.

In hospitals it is often imperative to have the fæcal matter from patients discharged directly into the soil-pipe, in which case a detached closet, one of a row, would be found most serviceable.

I do not think any case could arise in which a row of hopper-closets would not be superior to a latrine unless it were in the case of a public park, where the closet would be open to the most common forms of public use and abuse.

# SECTION VII.

## SUPPLY-PIPE CONNECTIONS.

THE common manner a few years ago of joining the supply-pipe to the water-closet was by enlarging the mouth of the lead supply-pipe so it would fit against, or making it small enough to enter, the nozzle of the bowl; when the two were fitted, a piece of putty was put around it, and the whole covered with a piece of cloth held in place by a string. The joint is sometimes covered with a piece of sheet-rubber, the rubber being wrapped with wire instead of cord. Joints of this kind are rarely tight for any length of time, and they cause a great deal of trouble by their leakage. Years ago efforts were made to perfect the joint between the supply-pipe and the bowl; but the plumbers as a class do not appear to have taken advantage of the best methods.

As early as 1829, Tyler, of London, invented a nozzle which was to be bolted to the side of the bowl by means of a plate made in one piece with the nozzle, and shaped the same as a portion of the surface of the bowl. This plate had ridges, and the bowl corresponding grooves. When the plate was bolted to the bowl, and the space between the two was filled with red lead, or a cement of red and white lead, it made a good joint. The metal nozzle, being made of brass, either had threads to which a pipe could be screwed, or a lead pipe could be connected with it by means of a wiped solder-joint. Before making a solder-joint, the surface of the metal had to be made clean.

Fig. 209.
Tyler's Joint.
*a*, Brass nozzle and plate. *b*, Opening into bowl with raised rim. *c*, Bowl.

In 1848, Armstrong invented a mode of connecting the supply-pipe with the closet-bowl. A metal ferule, with screw-threads on

the outside edge, is imbedded in the nozzle of the bowl. Working on the thread is a clamp nut. A second brass ferule, with a flange around the larger end, is firmly screwed up against a rubber washer that rests on the first ferule. The lead supply-pipe can be connected by the usual wiped solder-joint to the brass nozzle.

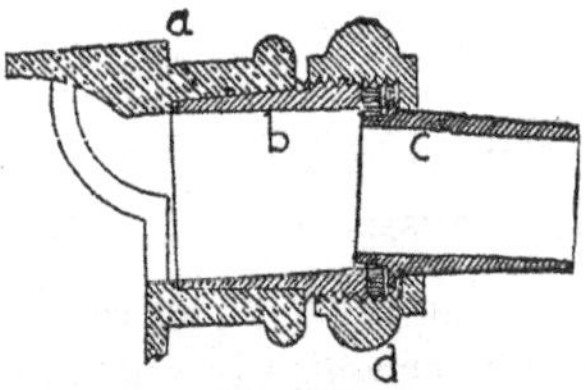


Fig. 210.
Armstrong's Connection.
*a*, Nozzle of bowl. *b*, First ferule. *c*, Second ferule. *d*, Clamp nut. *e*, Rubber washer.

A novel method of connecting the bowl with the water-supply pipe was invented in this country by W. S. Carr, of New York. A piece of brass tubing is bent or cast so as to form a right-angle with one end made smaller than the earthenware opening into which it is intended to fit. This metal tube has a flange or shoulder that fits against the outer edge of the earthenware nozzle, there being a rubber washer between the brass flange and the earthenware. Where the brass pipe bends, a hole is left through which a hook passes. This hook is made to fit around the earthenware where the orifice enters the bowl. Where the hook passes through the bend of the pipe there is a nut, by which the pipe can be screwed up tight against the earthenware nozzle. A lead pipe may be joined to this connecting pipe by a wiped solder-joint if it is brass, or it may be connected with iron or brass pipes by a screw-joint.

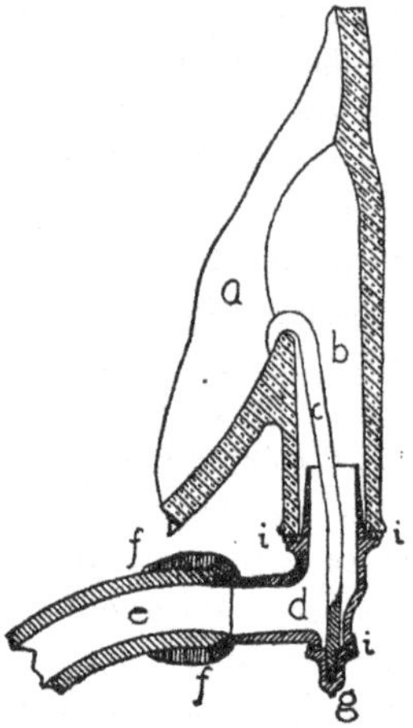


Fig. 211.—Carr's Connection.
*a*, Bowl. *b*, Earthenware nozzle. *c*, Hook. *d*, Brass connecting-pipe. *e*, Lead pipe. *f*, Solder-joint. *g*, Nut. *i*, Rubber washer.

I find patented in this country a manner of connecting the supply-pipe by putting it directly into the nozzle which projects from the bowl, when a piece of vulcanized-rubber tubing, which fits tightly around the pipe, is

stretched over the earthenware nozzle. This manner of connecting the supply-pipe might be employed as a temporary expedient.

J. L. Mott has patented an ingenious mode of connecting the supply-pipe with the closet bowl. A small metal plate is used, which is zinc-coated or brass, and has a conical hole in the centre of it. The inlet formed on the bowl has a conical projection of earthenware that corresponds to the hole in the metal plate. A piece of lead pipe is fitted over the conical projection, when the plate is put over the pipe and screwed fast by means of clamp-screws introduced through slots which are made in the earthenware for the purpose.

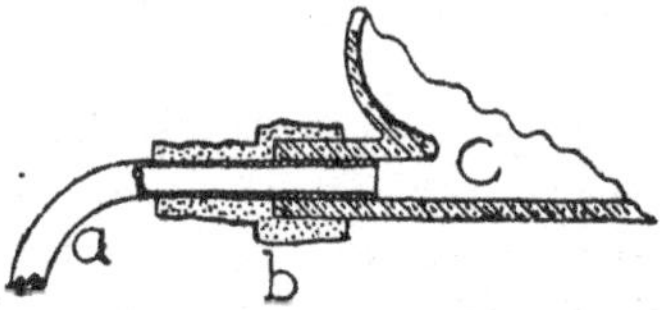


Fig. 212.
*a*, Lead pipe. *b*, Rubber-tubing. *c*, Bowl.

*Soil-Pipe Connections.*—With the most approved forms of water-closets where the trap is above the floor and forms a part of the closet instead of a part of the soil-pipe system, it becomes very important to have the joint between the water-closet and the branch from the soil-pipe made gas-tight; otherwise the trap and vent-pipes would be useless.

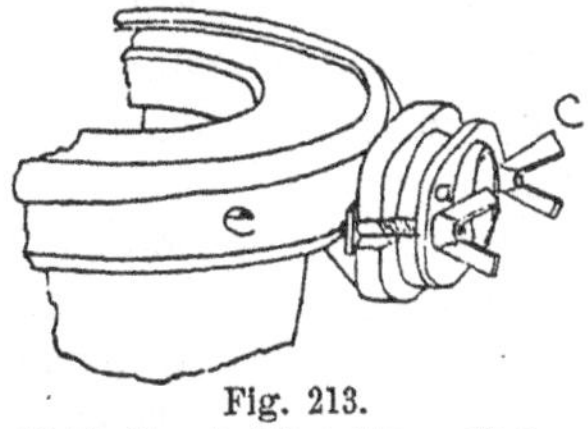


Fig. 213.
Mott's Supply-Pipe Connection.
*a*, Metal plate. *c*, Clamp-screws.
*e*, Bowl.

The common method of bringing the mouth of a trap, or a bell on the branch from the soil-pipe, and beating a part of the sheet-lead safe into it, then putting a little putty or cement on the floor, and screwing the closet down, is very defective. Where the sheet-lead is beaten into the bell it is a very imperfect joint, even when the space is filled with putty. The sheet-lead has no power to resist the packing of even

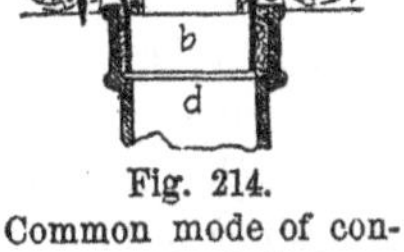


Fig. 214.
Common mode of connecting the Soil-pipe.
*a*, Outlet. *b*, Sheet-lead.
*c*, Putty. *d*, Soil-pipe.

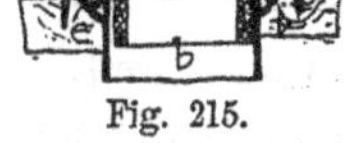


Fig. 215.
Hellyer's Connection,
*a*, Outgo of closet.
*b* Soil-pipe.
*c*, Cement.
*i*, India-rubber ring.

purty into the bell of the pipe. A joint of this kind might be opened, even when gas-tight in the first instance, by jarring on the floor, by any movement of the closet itself, or by the jarring of the soil-pipe.

Hellyer in his work describes a method of connecting a lead soil-pipe to the "Vortex" closet. A soft india-rubber ring is sent out with these closets, for fixing between the flange of the outgo and the "tafted" (bent over and beaten down) edge of the soil-pipe or lead safe. This india-rubber ring can be made to perfectly seal the jointing, and it will allow for considerable movement in the closet or the floor. The space between the flange of the closet which is screwed to the floor and the "tafted" edge of the soil-pipe or safe can be caulked with spun yarn and a cement composed of red and white lead.

Buchan in his work on plumbing illustrates his method of connecting the "Carmichael" closet to the soil-pipe. A lead branch is brought through a hole left in the floor for the purpose, and beaten out flat upon the floor by a dresser. The outlet of the closet, instead of having a flange, in the usual manner, has three grooves formed in the earthenware. A piece of lead is placed around the bottom of the closet outgo, and being forced into the grooves, and the depressions filled with solder, it would probably form a tight joint. The lead on the outlet of the closet and the tafted edge of the pipe are joined by solder.

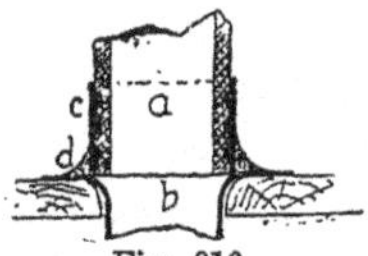

Fig. 216.
Buchan's Connection.
*a*, Outlet of closet.
*b*, Soil-pipe branch.
*c*, Lead around outlet.
*d* Solder-joint.

What is generally considered the best American practice is a modification of Hellyer's and the common methods. A Y-branch is brought from the soil-pipe to within a limited distance of the floor. Into the hub of the Y-branch a short section of brass pipe is caulked, by putting in a gaskin, then pouring in molten lead and driving it firmly in with a caulking tool. A short section of lead pipe is put in, just long enough to be carried through the hole made for the purpose in the floor, and beaten over so as to form a flange about two inches wide on the top of the floor, and at the same time reaching the brass pipe. When the lead pipe has been properly

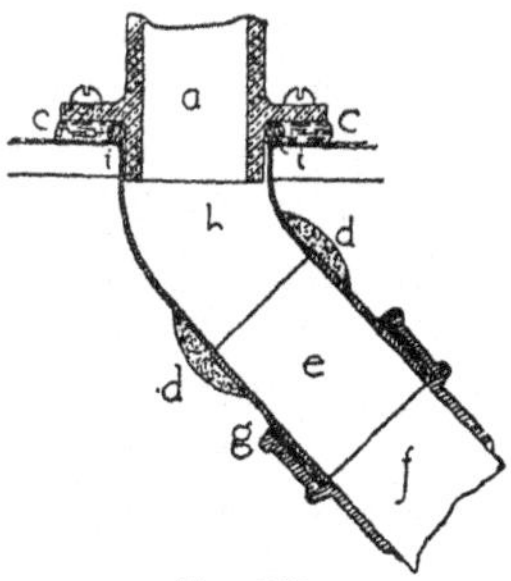


Fig. 217.
Improved Connection.
*a*, Outlet of closet. *b*, Lead pipe. *c*, Putty. *d*, Solder joint. *e*, Brass pipe. *f*, Iron pipe. *g*, Outlet-joint. *i*, India-rubber ring.

soiled and shaved, and the brass tube filed bright, they can be joined together by a wiped solder-joint. Where the pipe is turned over the floor, it is usually joined to the safe by a solder joint. There is a soft rubber ring stretched around the outgo of the closet just beneath the flange which is screwed to the floor. The space between the flange and the tafted edge of the lead pipe or the safe can be filled by a cement composed of white and red lead. This method of joining the closet to the soil-pipe is recommended by the most prominent sanitarians, but I think it objectionable in having the short lead pipe where it is exposed to the action of gases generated in the soil-pipe. There is every reason to believe that this acts on lead injuriously; in fact, reasons which are urged against lead soil-pipe may with equal force be urged against this short piece of lead; the whole inside sewerage system being only as efficient or as perfect as is its weakest point. I also object to having the closet depend for its stability on the floor, which is constantly undergoing expansion and contraction, according to the moisture and dryness of the air, and is subjected to vibrations from movements that may occur upon it. The above methods are all imperfect in their mode of connection. I think the methods described below are the best that have been introduced.

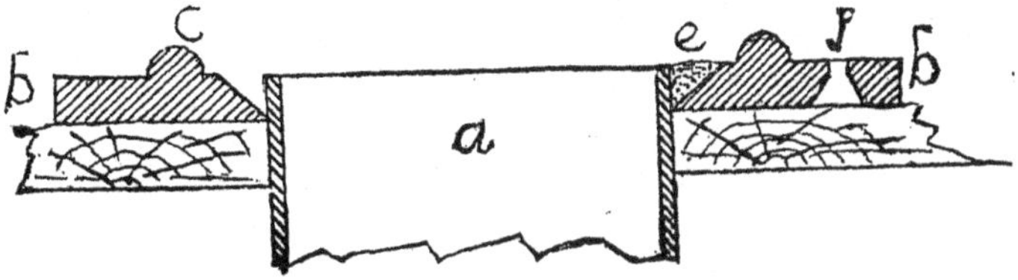


Fig. 218.

Fig. 218 represents the method adopted by the Meyer Sniffen Co. in connecting the Brighton closet with the soil pipe.

A brass plate that rests upon the seat is soldered to a short section of lead pipe, as shown in the cut. The flange on the outgo of the closet is bolted to this plate, being first bedded in red lead. There is a circular ridge on the brass plate that fits into a corresponding depression in the flange of the closet.

The Durham Drainage Company, of New York, have connected with the branch from the soil-pipe an iron fitting with a large plate or flange formed on it to receive the closet. When the outgo of the closet has a soft rubber ring put just beneath its flange, and it is screwed down tightly, a permanent gas-tight joint would be formed, the space between the two flanges which was not filled by the rubber ring being filled with cement. In section its appearance would be the same as Fig. 214, except an unyielding plate, which forms a part of the soil-pipe, would take the place of the yielding floor, and an iron pipe the place of the lead one. This method may be used where the trap, flange, and outlet are made in porcelain or earthenware.

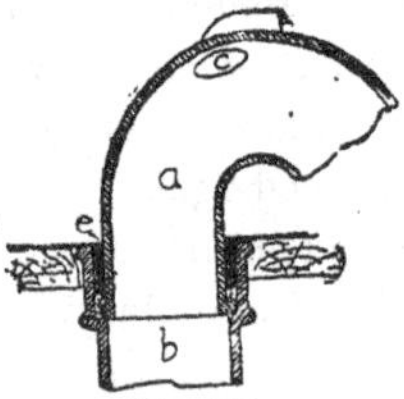

Fig. 219.
*a*, Trap. *b*, Branch from soil-pipe. *c*, Vent. *e*, Caulked joint.

The other method, which can only be used where the trap is of iron, has been recently introduced by J. L. Mott in connection with his side-outlet short-hopper closet. A branch from the soil-pipe has its hub brought to a level with the floor. The end of the trap is formed, as it is usual, with the spigot end of the soil-pipe with a small projecting ring around it. The trap is formed so it can run below the floor, the depth of the hub, into which it is caulked in the usual manner. In either of the cases mentioned, the trap of the closet and the closet itself forms a part of the soil-pipe system, the joints being as perfectly formed as any other joints in the system.

*Safes.*—It is customary and usually necessary to have sheet-lead safes under closets, to catch leakage or splashing that may occur from water or slops thrown into the bowl. These safes are simple shallow pans, formed by the turning of a piece of sheet-lead up about an inch-and-a-quarter high. These pans are placed beneath

the closet, and their size varies, according to circumstances, from a safe which covers the whole of the bath-room floor to one that will only cover a space about two feet square. Safes, to be useful in case of an emergency, must have a waste-pipe. A few years ago (and the same manner of connection may be found in a large number of houses at the present day) it was the custom to connect this waste-pipe with the soil-pipe. In some cases there was not even a siphon-trap between the soil-pipe and the room. In the better class of work it was usual to connect the supply-valve with the trap of this waste by a small quarter-inch pipe, called a "weeping" pipe. This weeping-pipe was intended to carry the water that remained in the short section between the valve and the bowl when the water is cut off, into the trap of the safe-waste. It is positively wrong to connect the safe directly with soil-pipe. The weeping-pipe may be stopped without the knowledge of any one, leaving the water in the trap to evaporate. The trap of the waste would be siphoned by each discharge from the closet. The best practice is to carry a "tell-tale" pipe from the safe into the kitchen or cellar, and leaving it open over a sink that is in constant use, where if anything were wrong, it might be noticed and stopped. In England the custom is to carry this waste-pipe through the wall, where it is allowed to form a drip or overflow. In some instances the outlet is closed by a flap-valve, to be opened by the pressure of water, if the hinge has not in the meantime become rusty. North of the Potomac, in this country, our cold winters would prevent a waste-pipe of this kind from being effective, as a small leakage trickling through would be sure to freeze and stop up the waste-pipe.

Where the floors of bath-rooms are made of impervious materials, such as glazed tile or slate laid in cement, they should be made to drain to a convenient point, and have a waste-pipe similar to the lead safe.

*Slop or Urinal Safes.*—All prominent firms who manufacture plumbers' supplies in this country and England furnish these safes to fit in beneath the seat and over the bowl. These safes are

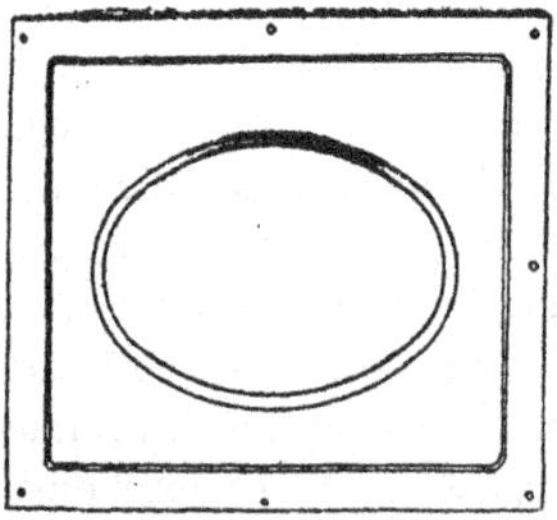

Fig. 220.—Oval Slop Safe.

simply a square, dished covering for the bowl, which has either a circular or oval hole about the size of the hole in the seat, with a surface of an impervious character formed so as to

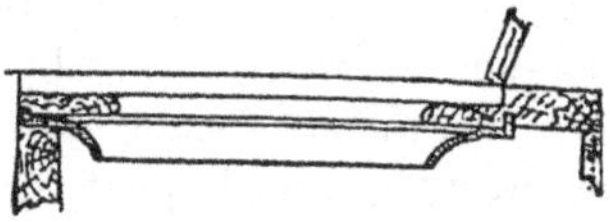

Fig. 221.—Section.

conduct water or urine into the closet bowl. They are made from solid earthenware, glazed, or from iron or copper enamelled. In all cases they are useful and cleanly in fact and in appearance. Thin metal safes are the most convenient for fitting over closets in which the wood-work has already been put in position, but from the liability to chipping, which even the best enamel has, a glazed earthenware safe is to be preferred where a new closet is being fitted up. Where the closet is to be used as the urinal, a slop-safe is very important, as it prevents dripping on the wood-work or into the safe below the closet. In either case the rapid decay of urine would generate dangerous and offensive gases. A poor substitute is sometimes found for a slop-safe, by beating sheet-lead over a wooden piece shaped in a manner similar to the regular slop-safe. Although better than wood, a substitute of this kind is a mere make-shift, and should never be used in place of an enamelled or porcelain one.

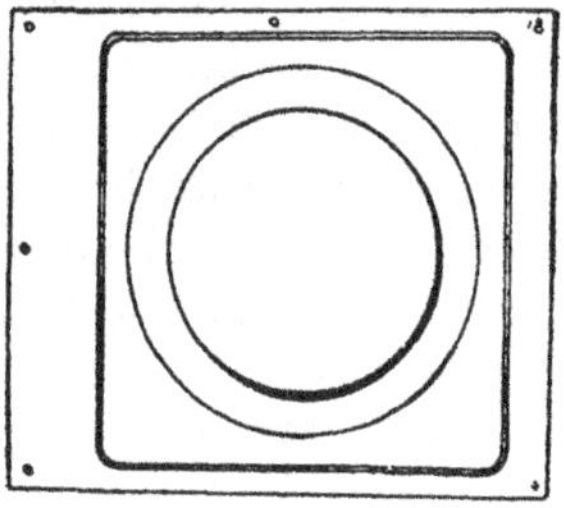

Fig. 222.—Circular Slop Safe.

## WATER-CLOSET SEATS.

All are familiar with common boxed water-closet seats, often fine and expensive pieces of cabinet-work, fair without but foul

within. These boxes secrete dust, dirt, and in many instances leakage and drippings from the closet. Seats of this kind, when made in their best form, have both seat and riser to open upon hinges, stout brass ones being the best: iron rusts easily, and the hinges become useless. When the riser and seat open the housekeeper can easily see that this space is kept clean.

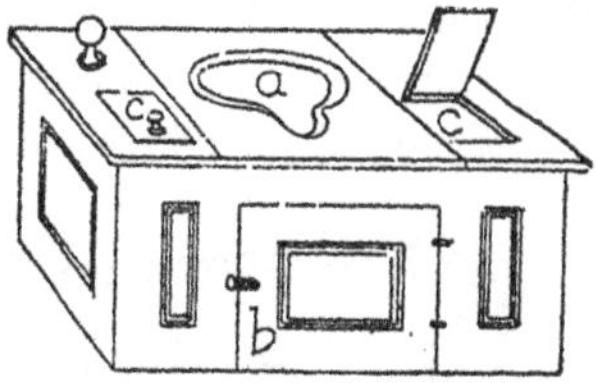

Fig. 223.—French Closet Seat.
*a*, Seat, *b*, Door in riser. *c*, Paper-box.

A convenient shape for the hole in a seat is egg-shaped, eight inches wide and ten inches long, and commencing about three inches and a half from the edge of the seat. This form of hole was patented in England thirty years ago. The manner of such work must be in keeping with the other wood-work in the bath-room. Heavy rubber duck is sometimes used instead of brass for hinges to the seat.

I take an illustration from Liger of a French panelled box water-closet seat. The riser has a panelled, hinged door. The hole in

Fig. 224.—Open Seat.
*a*, Seat. *b*, Legs. *c*, Gas-jet for ventilation. *d*, Ventilating-pipe.

the seat is decidedly pear-shaped, being very much like the ones represented in the closet from Herculaneum, which must have been in use more than eighteen hundred years ago.

I give here two illlustrations of what I consider the best forms of water-closet seat. The seat is supported by legs, which may be as ornamental as desired, and should be of some impervious material. The seat itself consists of a simple hinged leaf or board with a hole in it. A seat of this kind is furnished with the Brighton closet (Fig. 195). When not in use, the seat is turned back against the wall, exposing the white porcelain closet setting on a white or colored glazed tile floor, producing a neat appearance, and it is in fact very cleanly and easily cared for. A seat of this kind may be used with any form of closet except those which have unsightly or complicated parts that require concealment. The illustration shows a ventilating pipe that is intended to create a current of air from and through the closet bowl, carrying off any local offensive odors that may occur.

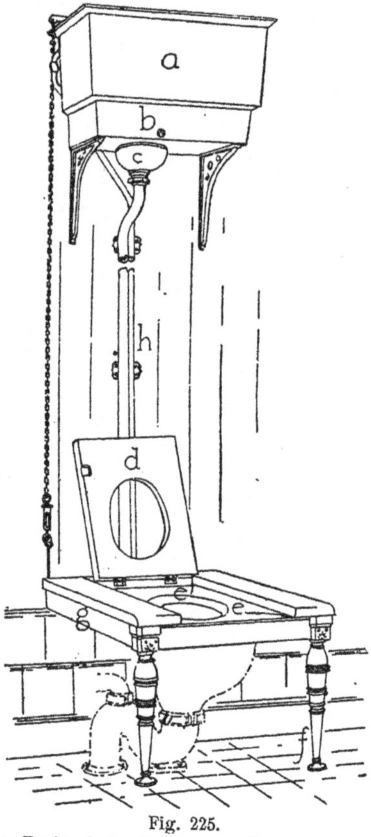


Fig. 225.
*a*, Tank. *b*, Service-box. *c*, Flushing-pipe. *d*, Seat. *e*, Slop-safe. *g*, Seat-frame. *h*, Supply.

In Fig. 225 is shown a seat, the front of which is supported by legs, a slop-safe and tank in position. This arrangement as furnished by J. L. Mott & Co.

has either brass or nickel-plated legs, which are intended to rest upon a tile floor. Similar seats of this kind, having plain iron legs, are furnished by the different manufacturers. The slop-safe is let into the frame of the seat. The seat proper being thrown back against the wall when it is not in use, leaves the enamelled surface of the slop-safe exposed. This closet is flushed by the weight of the person using the closet. The seat presses down on a short lever, which communicates motion to the valve of the tank, by means of a lock chain. The floor and side of the bath-room are intended to be tiled. It is scarcely necessary to say that this makes a clean, healthy and efficient apparátus for the purpose it is intended to fulfil.

For prison use, or in places where the same abuse is likely to occur, it is best to use a jump-up seat like the one illustrated in Figure 226. The seat is oval, formed of wood, and is just large

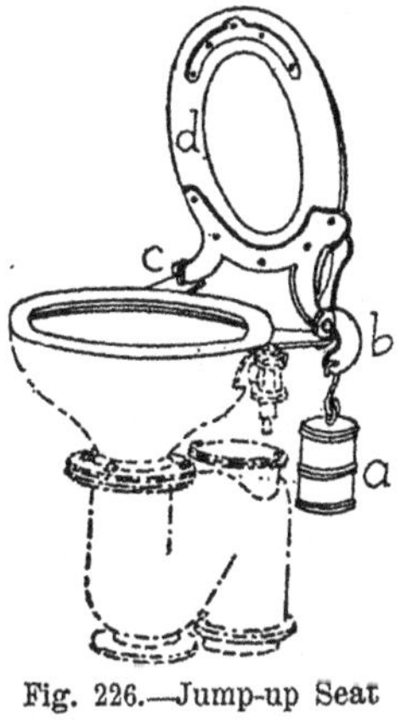


Fig. 226.—Jump-up Seat
*a*, Weight. *b*, Lever.
*c*, Journals. *d*, Seat.

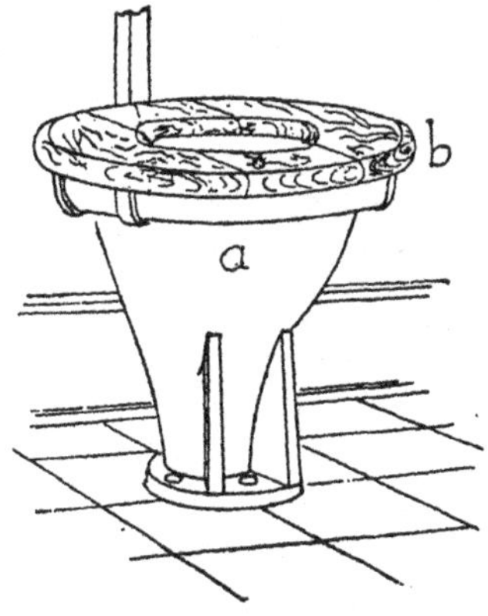


Fig. 227.
*a*, Closet. *b*, Seat. *c*, Clamp.

enough to cover the bowl of the closet. Two iron journals are screwed, by means of an iron plate from which they project, to the wooden seat. One side has a lever and weight attached to the journal. By means of the weight the seat is kept thrown back

against the wall, unless it is held down over the bowl by a greater weight.

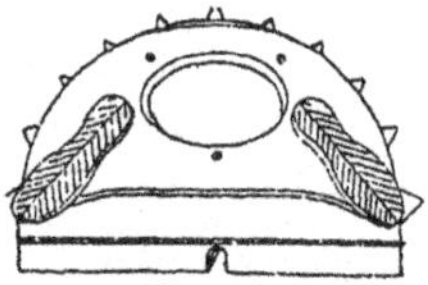

Fig. 228.
Seat for Standing on.

Sometimes the long hopper porcelain closet is set on a tile floor and has a wooden seat which is held in position by iron clamps, that are clamped around the earthenware rim.

Liger describes a closet seat or stand which is intended for use in the crouching or Oriental fashion, instead of as a seat. Places are arranged for the feet, being made rough or ridged, so there will be no danger of slipping.

*Levers.*—It seems advisable to describe the two following simple levers for opening the valve in the flushing-tank by pressure on the seat:

Fig. 229 is a double lever which has its pivots or fulcrums attached to the under side of the seat. Motion is imparted to the first lever by pressure on the seat, while it in turn moves the short arm of the second lever. The long arm of the second lever is connected by a chain with the lever that operates the flushing valve in the tank.

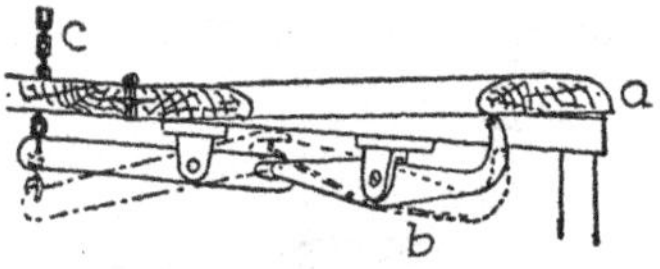


Fig. 229.—Lever for opening Supply-tank.
*a*, Seat. *b*, Lever. *c*, Connecting-chain.

The arrangement used by Henry Huber & Co., of New York, in connection with his "Tidal Wave" closet is simple, and not liable to get out of order. It is a simple lever which moves up or down on two journals, the bearings in which they work being screwed to the floor. The lever is pressed down by rods which are attached to it; these run through guides and come in contact with the under side of the seat. The weight of the tank lever keeps the seat slightly raised

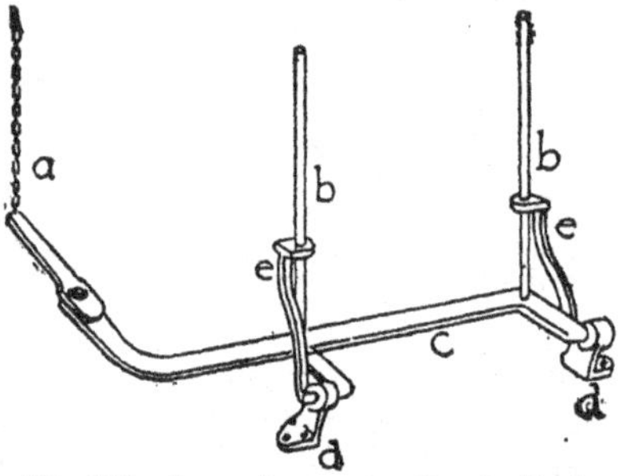


Fig. 230.—Lever for opening Supply-Tank.
*a*, Chain. *b*, Rods. *c*, Lever. *d*, Journal boxes. *e*, Guides.

unless there is sufficient weight on the seat to counteract its effect.

*Bidet Attachment.*—Cragie's bidet attachment is a small branch from the supply that is attached to the inside of the water-closet bowl. When not in use it is turned up under the seat or urinal safe; when in use, it is turned down into the centre of the bowl. From it a small jet or spray of water is thrown up toward the seat. This attachment is important in some diseases.

Quite a number of devices have been invented for the purpose of disinfecting water-closets. These "germicides," as some of them are called, are intended to destroy germs and odors that may be generated in receivers or other concealed compartments that may be connected with badly designed water-closets. Instead of using disinfecting apparatus and a bad or poorly constructed closet, it is far preferable, both on the score of health and comfort, to select a simple, cleanly closet, with its trap properly connected with the soil-pipe and ventilated, and thus have no need of a "germicide."

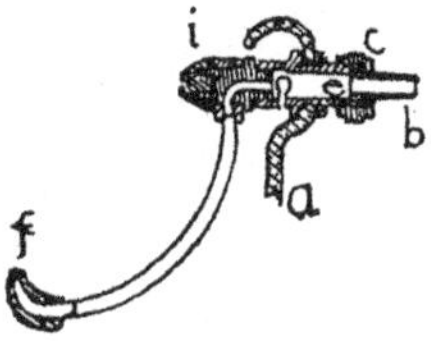


Fig. 231.—Bidet.
*a*. Bowl. *b*, Supply. *c* Nut. *e*, Valve. *f*, Nozzle for jet. *i*, Packing.

# SECTION VIII.

## SUPPLY VALVES AND TANKS.

WATER-CLOSETS are supplied or flushed with water by means of supply valves and tanks, or cisterns.

Supply valves may be automatic in their operation, or opened by the party using the closet.

The simplest form of supply valves are ground and compression cocks, which are opened respectively by a lever and screw. Between these simple valves and the complicated diaphragm valves there are many deviations.

The following valves which I will describe are the ones that have been most extensively used in connection with water-closets.

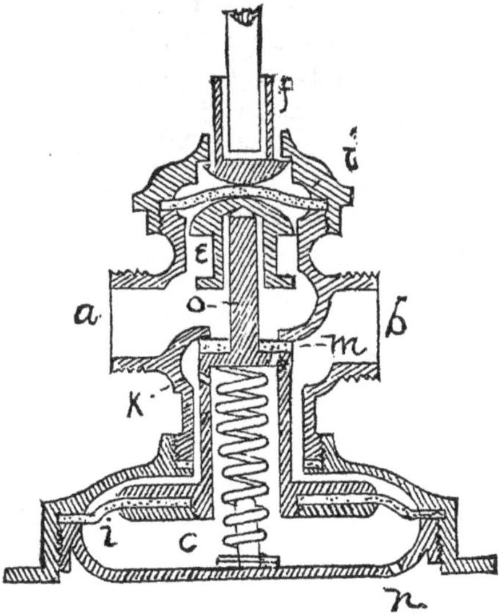


Fig. 232.—Bartholomew's Supply-Valve.
*a*, Inlet. *b*, Outlet. *c* Air chamber. *o*, *f*, Valve rods. *i*. Diaphragm. *k*. Water inlet to air chamber. *m*. Washer. *n*, Leak or drip.

*Bartholomew's Supply-Valve.*—This valve has been extensively used in connection with the pan-closet of the same name (Fig. 120). This supply valve was intended to close itself after a certain amount of water had passed through it into the closet. A projecting disc into which a diaphragm is fitted forms the bottom of

the valve. When the valve is first opened, either by pressure upon the seat or by means of a lever, the chamber above the diaphragm would be immediately filled with water, and it would be impossible for the valve to close until the air chamber beneath the diaphragm fills with water through a small hole made for the purpose.

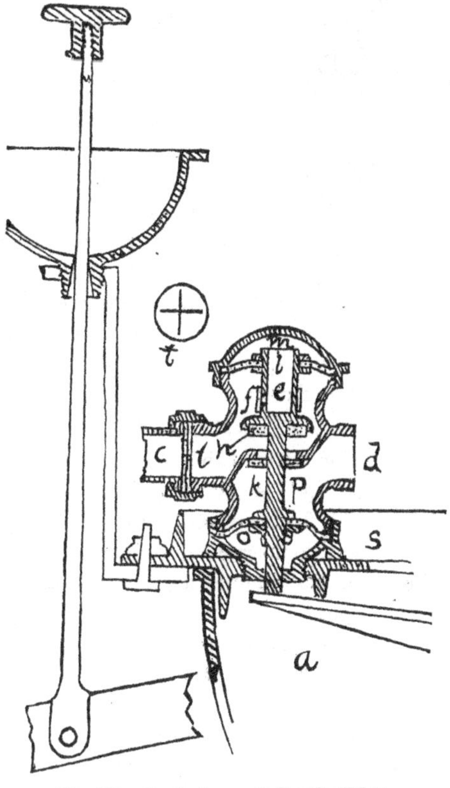


Fig. 233.—Bartholomew's Supply-Valve.
*a*, Water-closet. *d*, Inlet for water. *c*, Outlet. *k* Valve rod. *n*, Valve. *m* Air chamber. *i*, Inlet and *e* Outlet of air. *o* Diaphragms. *t*, Slotted disc. *s*, Drip pan. *k*, Rubber valve. *p*, Washer.

The time the valve occupies in closing is governed by the size of this orifice. A spiral spring forces the valve back against its seat as soon as the pressure above and below the valve has become equal. The water in the bottom compartment wastes through a small hole, and leaves the valve ready again for action. Fig. 233 is a later pattern of Bartholomew's supply-valve. The spring in this one is omitted, and gravity is depended on to close the valve when the pressure in the two compartments has been equalized.

The air chamber is situated in the top of the valve, and the

water passes into it through a hole left for the purpose, and leaks out through a small hole, which is covered by a rubber valve that opens only in an outward direction. To assist the valve by retarding the flow of water, so it will fill the upper chamber, a slotted disc is inserted in the outlet of the valve.

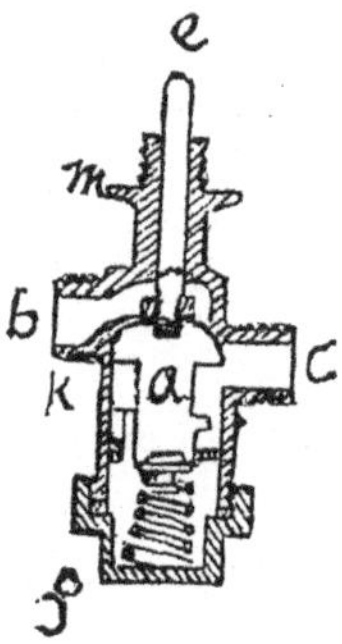


Fig. 234.
*a*, Valve. *b*, Inlet. *c*, Outlet. *e*, Rod. *f*, Spring.

*The Climax Valve.*—The valve used in connection with Demarest's Climax Closet is simple. It consists of a piston-shaped valve, which is held in position, or against its seat, by means of a spiral spring. The valve works between guides. Screwed into the top of the valve is a small rod, which, when pressed down, opens the valve. When the rod is relieved from pressure, the valve is closed by means of the spring in the bottom of the chamber.

*Demarest Closet-Valve.*—This valve is placed within the plunger chamber, and is operated by means of a float situated within the same compartment, when the water has attained a certain fixed height.

The upper part of the valve is perforated, so that the water passing through the same sprinkles, and is supposed to wash the plunger and plunger chamber. It seems hard to keep this valve from leaking in practice. A spring in the bottom of the valve assists in keeping the valve against its seat.

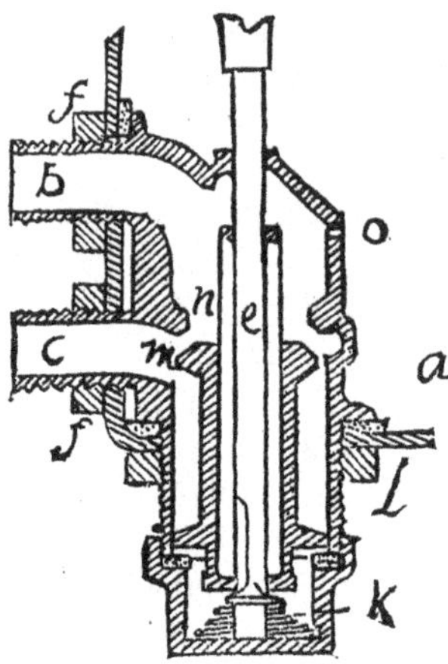


**Fig. 235.—Demarest's Valve.**
*a*. Plunger chamber in closet. *b*, Outlet. *c* Inlet. *e*. Rod connecting with float. *m*. Valve. *n* Valve seat. *f* Nuts to attach to closet. *k* Spring. *o*. Perforation for sprinkling plunger chamber.

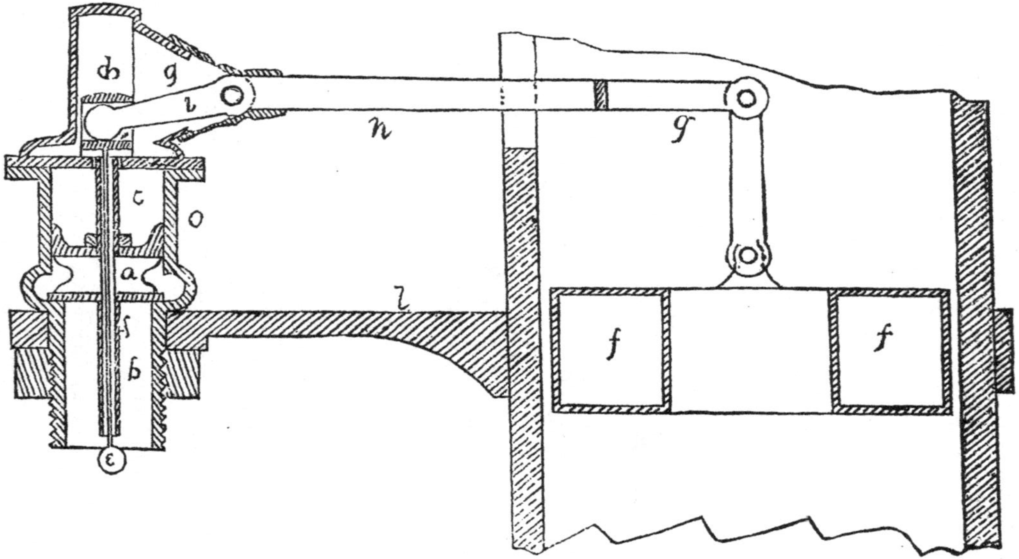


Fig. 236.—Jennings' Hydrostatic Valve.

*a*, Diaphragm. *b*. Water inlet. *c*, Compartment above diaphragm. *d*, Lever compartment. *e*, Ball valve to close inlet to the diaphragm. *f*. Float. *g*, Lever. *l*, Bat connecting valve with closet.

*Jennings' Hydrostatic Valve.*—This valve is also operated by means of a float which encircles the plunger (see Fig. 130). When the water-closet bowl is emptied by raising the plunger, the float drops, the valve is opened, and the closet is flushed. When the float rises it opens an orifice, which lets the water into a diaphragm chamber. The pressure of the water in this chamber closes the valve gradually. The different parts of the hydrostatic valve have to be so nicely adjusted to themselves, and to suit the pressure of the water supply, that they are usually troublesome, and it is hard to keep them from leaking.

*Kelly's Stop and Waste Cock.*—This cock or valve is manufactured in Chicago, and it will be readily understood by reference to the illustration (Fig. 237).

Two plugs or valves are attached to a crank that moves in a

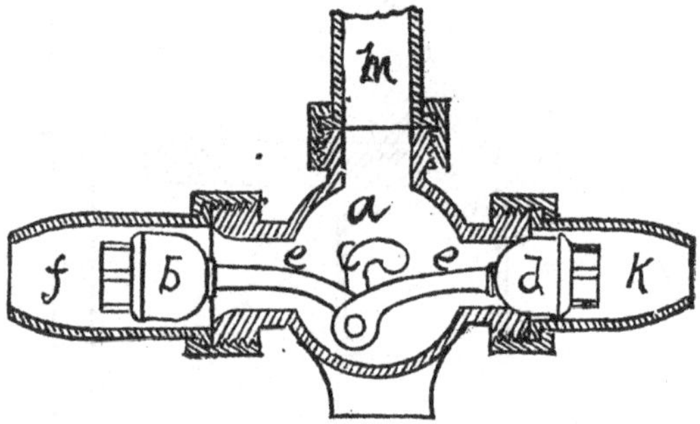


Fig. 237.—Kelly's Valve.
*a*, Inside of valve. *b*, Plug to inlet. *d*, Plug to outlet or waste. *c*, Crank that opens and closes plugs. *e*, Connecting arm. *f*, Inlet for supply. *k*, Waste. *m*, Opening to the flushing outlet.

centre spherical chamber. When the inlet plug is open the waste plug is closed, and the water supply will flow into the closet. Where it is necessary to place a closet in an exposed situation, this supply valve can be used and placed beneath the frost line, when the water in the pipe between the valve and the outlet in the closet would run to waste through the waste valve, and leave no water in the pipe to freeze.

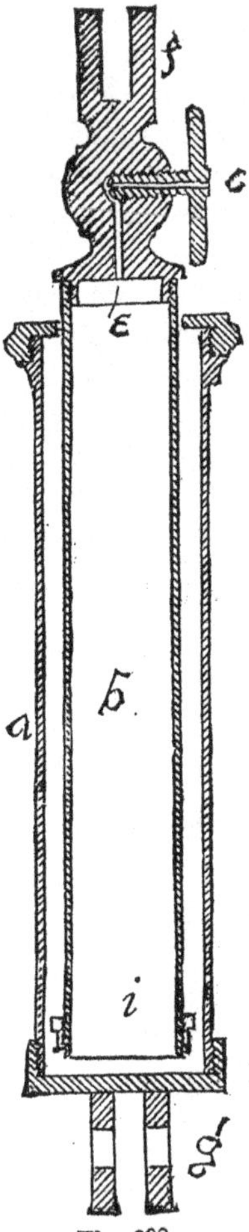


Fig. 238.
Underhay's Regulator.
*a*, Large cylinder. *b*, Small cylinder. *e*, Air hole.
*c*, Thumb-screw with air hole in it. *f*, Point of attachment to valve lever.
*g*, Point of attachment to the floor. *i*, Cupped washer.

*Underhay's Waste Preventer.*—In connection with supply valves I note and illustrate Underhay's Waste Preventer, designed to close supply valves when a certain amount of water has passed through the valve. This regulator consists of two cylindrical chambers. The small one, which is closed at the top, moves up and down freely in the large one that is closed at the bottom.

The small cylinder has on its lower end a cupped leather piston, and on its upper end an air hole; the size of which is regulated by a thumb-screw. The lever that opens and closes the valve is fastened to the regulator by a bolt and slot in the top of the small cylinder. It can be readily raised, as the cupped leather acts as a valve, and lets the air enter the small cylinder freely. When the small cylinder commences to descend by the action of gravity, the air in it would be compressed, and could escape only through the air hole in the top. The leather cup would be pressed tightly against the sides of the large cylinder. The cup dips into a lubricating oil, and in its passage up and down lubricates the side of the large cylinder. The regulator is illustrated in position in Figure 72.

REMARKS.

Water-closet supply valves that are connected directly with the supply pipe

of the building must be condemned, from a sanitary standpoint, because there is a possibility, if not a probability, of the water supply becoming contaminated by suction of either dirty air or water from the bowl of the closet. This may take place when a valve or faucet is opened on a floor below the closet, at the same moment when the water-closet valve is open, the water or air passing back from the closet into the supply pipe and out through the open faucet below.

Valves must also be condemned on the score of economy, as it does not seem possible to prevent them from leaking and keeping up a constant waste of water, and repairs are always needed.

Their principal defect as a water-closet supply is that they do not let the water out at once in a large quantity so as to cleanse the closet thoroughly. Where a large flush can be emptied directly into the bowl it will have the desired cleansing effect, while triple the quantity would be of no avail if supplied in dribblets.

## WATER-CLOSET TANKS.

Tanks have been used at least a hundred years for the purpose of storing water, so as to flush or wash water-closets.

Both the Cummings' and Bramah closets, two of the earliest

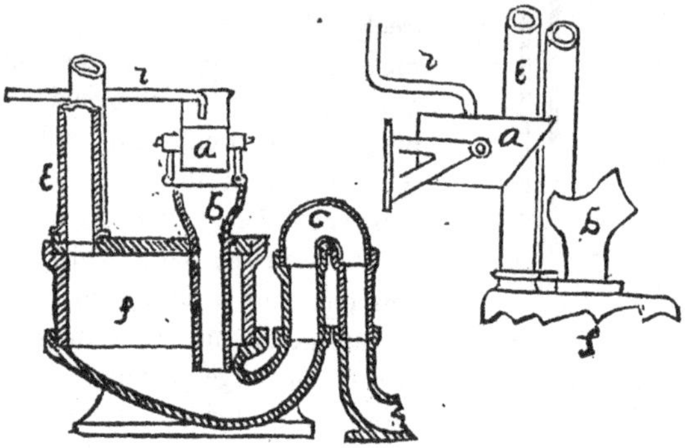

Fig. 239.—Shones' Tank.

*a*, Tumbling tank. *b*, Funnel-shaped inlet. *c*, Syphon. *f*, Reservoir. *i*, Supply pipe. *e*, Waste pipe for utilizing waste water for flushing purposes.

English closets, and patented before the nineteenth century, were flushed by means of tanks.

It has not been many years since tanks were discarded in favor of some of the many patented supply valves.

Tanks for flushing closets are fast coming into general use again as their utility is appreciated. Tanks may be treated under the head of automatic and non-automatic.

*Shones' Syphon Tank.*—This tank is simply a reservoir with an outlet through a syphon, and a funnel-shaped inlet at the top. The reservoir is filled by means of a small tumbling tank. The last time the small tank empties, when the reservoir is nearly full, the syphon will be started, and the whole contents emptied into the closet.

*McFarland's Tumbling Tank* is simply a small receiver or bucket hung on journals. The method of its operation is clearly shown in the cut. By regulating the flow of water from the supply cock the tank can be made to empty at long or short intervals.

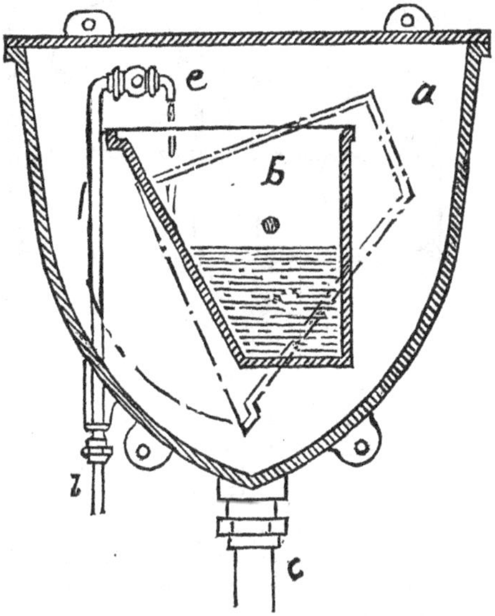

Fig. 240.—McFarland's Tank.

*a*, Receiver or tank. *b*, Tumbling tank. *c*, Outlet to closet. *e*, Supply cock. *i*, Supply pipe

Either of the last-mentioned tanks are simple in their construction, and, if the journals and journal boxes are well made, they will not be liable to get out of order, and will answer well for use in

public places where persons using the closet would be too careless to operate a non-automatic tank.

Among non-automatic tanks I will describe a small number of plug and syphon tanks.

*Tylor & Son's Plug Tank.*—This tank, which is illustrated in Fig. 241, seems a fair example of its class. It has a tank proper and a service box beneath. A plug which is held in position by a weighted lever stops the outgo from the service box. When the lever is pulled down a plug attached to the same lever on the opposite side of the fulcrum closes the communication between the tank and the service box. Unless the plug between the tank and service box fails to take its seat properly, only the water in the service box can be discharged at one time into the closet, although the lever may be held up any length of time. This tank as well as nearly all non-automatic tanks, are supplied with water through a ball cock. A floating copper ball by means of a lever shuts the water off when it has reached a certain height in the tank. This height is generally near the height of the outlet into the overflow. The tank is manufactured with a cover, as is usual in England.

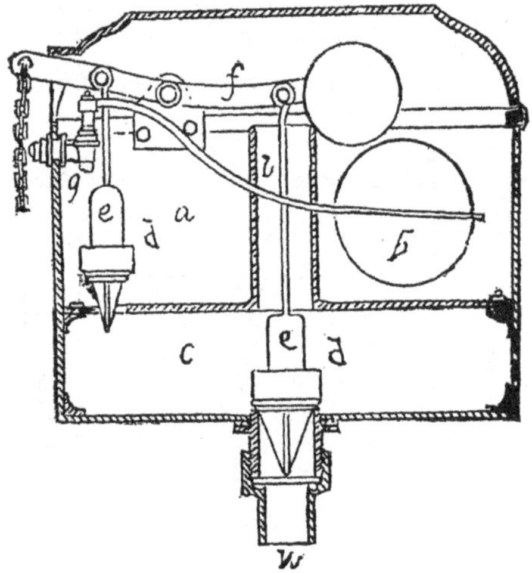

Fig. 241.—Tylor & Son's Tank.

*a*, Tank. *b*. Float for ball-cock. *c*, Service box. *e*, Plugs or valves. *f*, Weighted lever. *i*, Overflow. *g*, Supply valve. *n*, Outlet to closet.

*The Duplex Tank*, manufactured by J. L. Mott & Co., is a good

example of a before and after-wash tank. This tank has three compartments, so arranged that while the service box is being filled the same pull of the lever opens the compartment that gives the

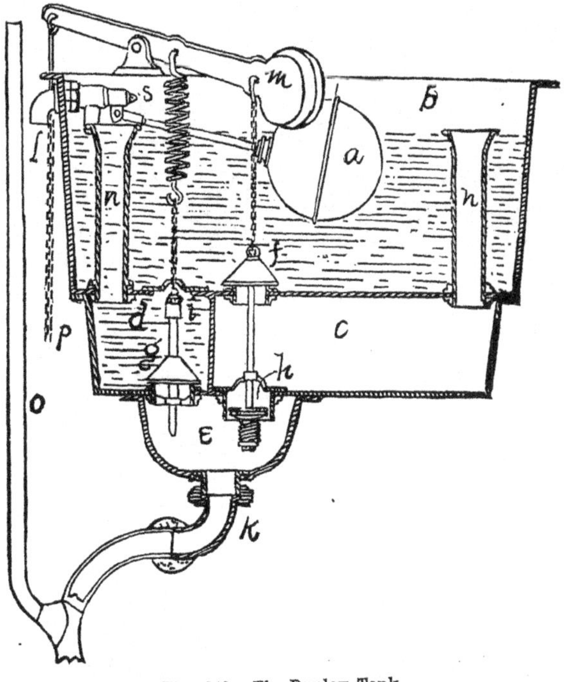


Fig. 242.—The Duplex Tank.

*a*, Float valve. *b*, Tank. *c*, Service box. *d*, Before-wash. *e*, Funnel-shaped outlet. *f*, Valve between tank and service box. *g*, Valve of before-wash. *h*, Valve to close service box while filling. *i*, Valve to close after-wash while emptying. *k*, Outlet to closet. *l*, Lever chain. *m*, Weighted lever. *n*, Overflow. *o*, Air pipe.

before-wash. As soon as the lever is relieved from weight, it takes its normal condition by the act of gravity, and opens the service box which flushes the closet.

The before-wash is of importance, as it prevents fæcal matter from adhering to the bowl. The valves or plugs have all either

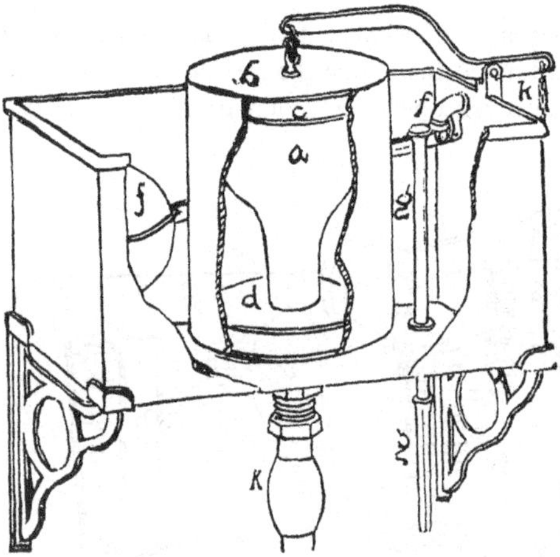


Fig. 243.—Doulton's Syphon Tank.
*a*, Syphon. *b*, Cylinder. *c*, Mouth of Syphon. *d*, Plate nearly the same diameter as cylinder. *f*, Float valve. *g*, Overflow. *h*, Lever. *k*, Outlet.

rubber washers or seats, so that they can easily make a water-tight joint.

An air pipe is shown attached to the outlet pipe below the closet. This prevents air being sucked through the water in the tank, and so prevents noise that would otherwise occur.

*The Doulton Syphon Tank.*—This tank is waste preventing, and its contents are emptied by means of a syphon. An annular syphon is surrounded by a cylinder, air tight on top and connected with the lever, that is operated either by the movement of the seat or by hand. To start the waste preventer the lever must be pulled down; this raises the cylinder and empties enough water into the syphon to start it into action. A two-gallon flush is obtained in this way without the necessity of holding the lever down for a moment.

The supply is governed by the usual float-valve, and there is also an overflow in case the supply valve should leak.

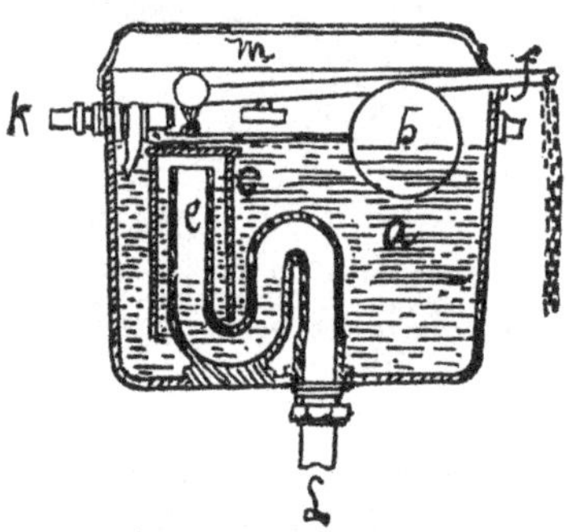


Fig. 244.—Winn's Syphon Tank.
*a*, Tank. *b*, Float. *c*, Cylinder. *e* Syphon. *f*, Lever. *k*, Supply. *l*, Outlet. *m*, Cover.

*Winn's Syphon Tank.*—This tank is manufactured in Birmingham, England. In this tank there is also a cylinder, which, when raised by the lever, empties water into and starts a syphon. When started the syphon would extract the water. The amount of water would be regulated by the size of the tank and the length of the cylinder. The illustration shows the tank covered, a custom almost universal in England, but not usual in this country. It serves a good purpose by preventing dust and other impurities from settling into the tank. Either of the two last-described tanks are simple, and have no valves (except the ball-cock) to get out of order. For this reason they are to be commended.

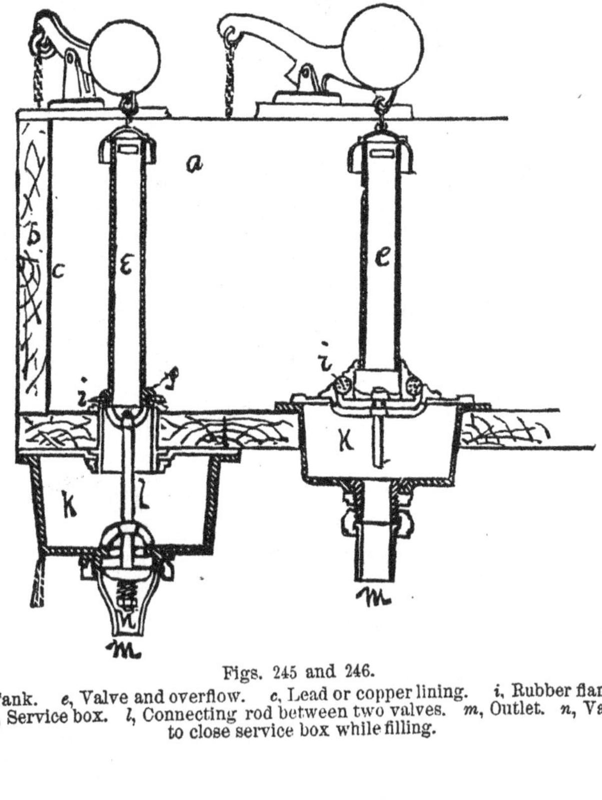


Figs. 245 and 246.
*a*, Tank. *e*, Valve and overflow. *c*, Lead or copper lining. *i*, Rubber flange. *k*, Service box. *l*, Connecting rod between two valves. *m*, Outlet. *n*, Valve to close service box while filling.

Figs. 245 and 246 show a method of using one large tank for a row of closets. The tank can be lined with either copper or sheet lead. The valve or plug is a hollow tube, being open near the top and acting as an overflow. At the bottom it is encircled by a rub-

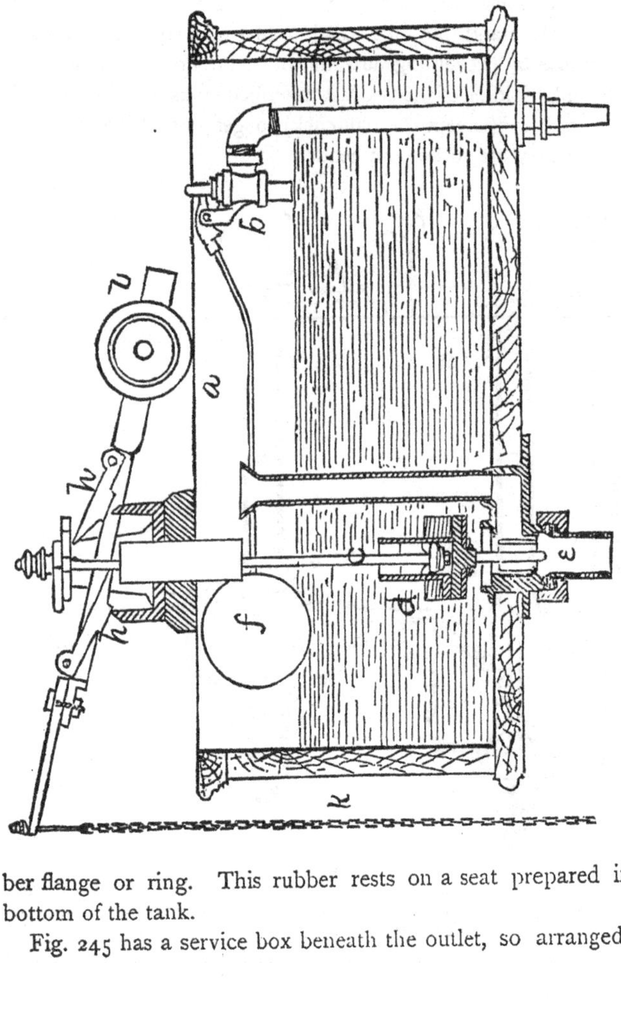


Fig. 247.—Boyle's Tank.
*a*, Tank. *b*, Ball-cock. *c*, Weighted piston-rod. *d*, Valve or plug. *e*, Outlet. *f*, Float. *i*, Weighted lever. *h*, Dogs or short levers. *r*, Chain connecting with hand-pull or seat.

ber flange or ring. This rubber rests on a seat prepared in the bottom of the tank.

Fig. 245 has a service box beneath the outlet, so arranged that

when the plug is raised the outlet to the service box is closed by a second valve, but as soon as the lever is dropped the opening between the tank and service box is closed, the outlet of the service box is opened, and its contents discharged into the closet. Fig. 246 shows a non-waste-preventing plug.

*Boyle's Tank.*—The plug or valve is the distinguishing feature in this tank. A number of plugs may be used in one tank, each connected with a separate closet. By the arrangement of this plug a before and after-wash is obtained, and the cistern is made waste preventing if the mechanism is in proper order.

The lever, which moves on two trunnions in a journal plate, has two dogs or short levers pivoted in a slot of the large lever equidistant from the journals. These dogs are pointed in opposite directions and each rests on independent fulcrums.

In order to obtain the first wash, when the lever is pulled down the dog strikes against a projecting plate on the piston rod and raises the plug, and allows the water to escape and flush the closet. The dog which raises the piston has its fulcrum nearest the end of the lever where the motive power is applied. The valve, which slides up and down on the piston rod, takes its seat by the action of gravity. As soon as the hand-pull is released, the weight on the opposite end of the lever causes the plug to be again raised by the second dog. The position of the fulcrum of this dog causes the plug to be raised higher than it was in the first instance. By the arrangement of these dogs the desired amount of water may be obtained. The plug is a short piece of tubing with a leather or rubber disc on its bottom, so the plug can make a water-tight joint with the seat. In the tube is a piston with a soft leather washer. On the top of the piston rod is a weight, which forces the piston down to the bottom of the

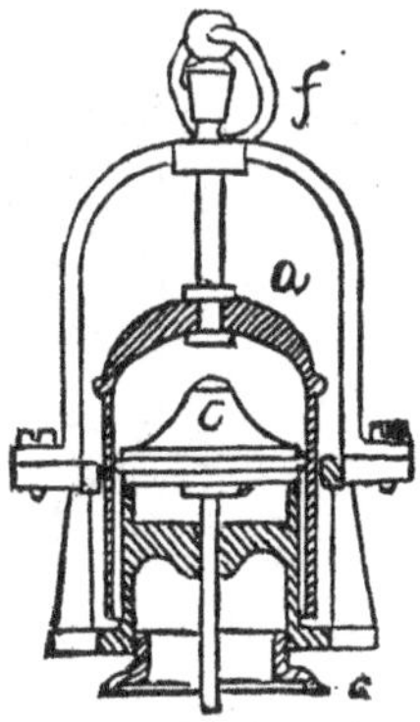

Fig. 248.—Tylor's Valve. *a*, Cup. *c*, Valve. *e*, Bottom of tank. *f*, Attachment for lever.

tube as soon as the long lever is horizontal, the water escaping past the soft leather cup. When the dogs or short levers raise the piston, the plug is lifted at the same time by capillary attraction.

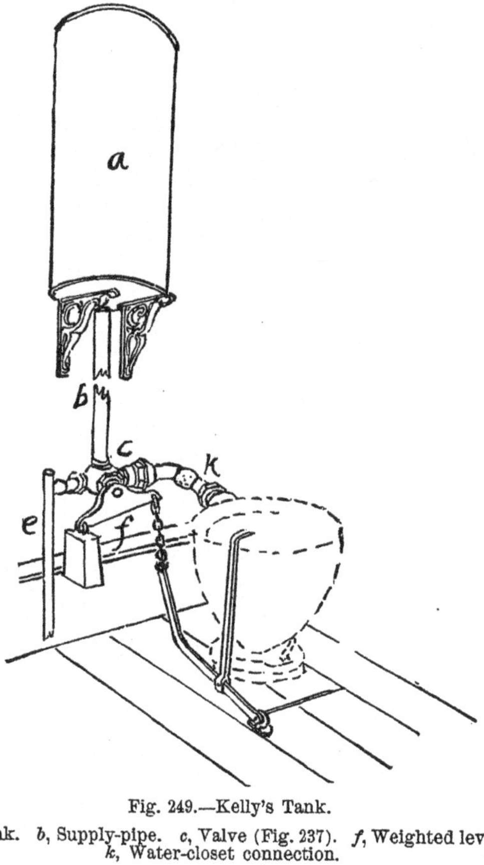


Fig. 249.—Kelly's Tank.
*a*, Tank. *b*, Supply-pipe. *c*, Valve (Fig. 237). *f*, Weighted lever. *k*, Water-closet connection.

*Tylor's Waste-Preventing Plug.*—Fig. 248 represents a waste-preventing plug to be used in connection with a tank, manufactured by Tylor & Son, of London. The valve in this case is within an

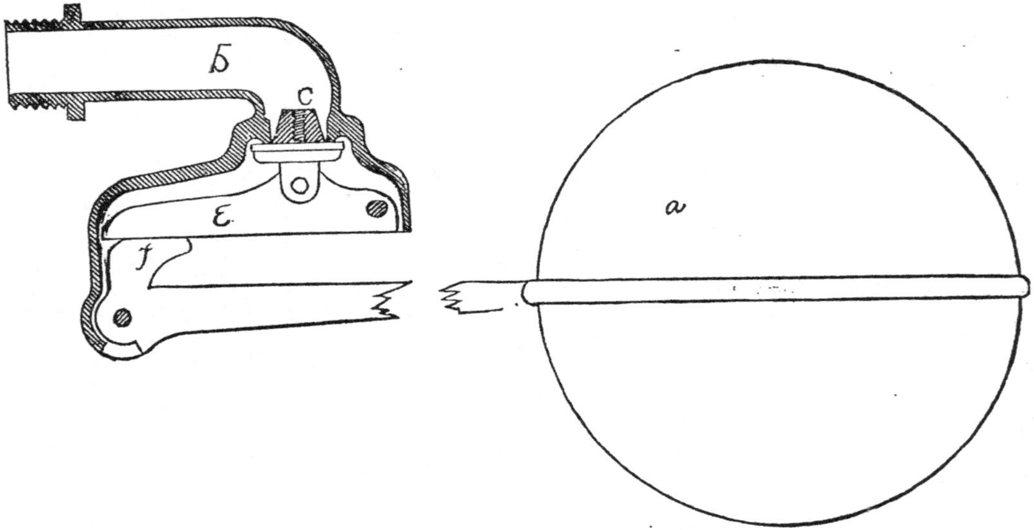


Fig. 250.—Tylor's Ball-Cock

*a*, Copper ball. *b*, Supply. *c*, Valve. *e* and *f*, Levers.

inverted cup. When the cup is raised the plug or valve is raised at the same time. If the cup should be held up for any length of time the plug would descend by the action of gravity, as soon as equilibrium was established between the upper part of the cup and the tank. This valve is generally regulated so the tank will discharge about two gallons of water each time the valve is raised.

*Kelly's Tank.*—In connection with his supply valve, Thomas Kelly, of Chicago, has a tank which is operated by compressed air. This apparatus also professes to be waste preventing. Weight on the water-closet seat by means of a weighted lever opens one side of the valve, when water will rush into the cylindrical and air-tight chamber above the closet. The air would be compressed according to the head of water in the street mains. As soon as the water-closet seat is freed from the imposed weight, the weighted lever would open the outlet side of the valve, when the water under pressure would be driven through the flushing-rim of the closet (see Figs. 237 and 249).

As almost without an exception tanks are supplied with water through ball-cocks, I have illustrated some of the best in the following cuts.

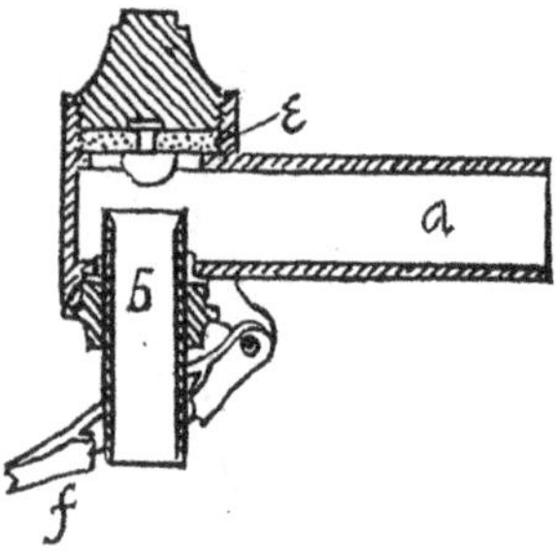

Fig. 251.

The balls are always of spun copper, and, being hollow or filled with air, they float. The ingenuity of inventors has shown itself in the methods adopted in closing the supply pipe with the lever, which is moved by the floating ball. The illustrations scarcely need description as they explain themselves.

Fig. 250 is one of Tylor & Son's, of London. The valve is forced against its seat by a double lever. This is sinple and effective.

Figs. 251 and 252 are taken from Bailey Denton's work. One

is a simple tube, which, when closed, presses against a rubber seating. The other is a valve with a piston and piston rod attached, which aids in keeping the valve against its seat.

Tanks should be connected with the closet by at least an inch-and-a-quarter pipe. The bore of the outlet of the tank, the inside diameter of the pipe, or the inlet to closet should never be diminished in any way. Plumbers, in some cases, diminish the opening by beating the lead pipe together, in bending it, by washers, red lead, and putty used in making the couplings.

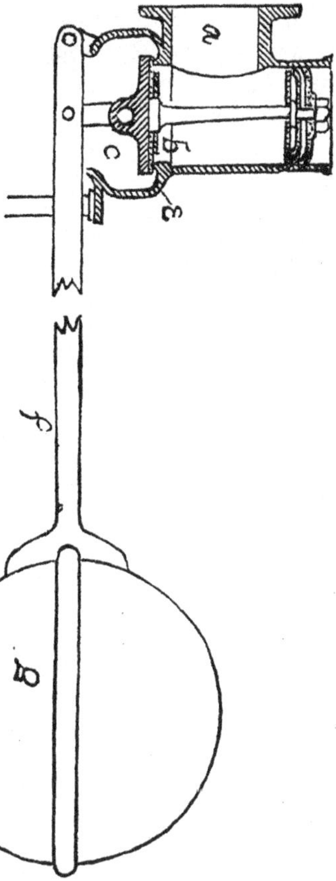


Fig. 252.

Tanks are superior to the valve as a means of flushing closets, because by their means a much larger amount of water can be thrown suddenly into the bowl of the closet, and they disconnect the closet from the main supply-pipe. Neither are they as liable to get out of order as the usual supply valve. All sanitarians agree that tanks are superior to supply-valves as a flushing arrange-

ment for water-closets. The closet should be flushed by at least two gallons of water at each discharge of the tank. Where water is abundant the closet should receive a before-wash of a gallon and an after-wash of three gallons. With a simple closet, properly vented and connected, a simple tank properly connected, the closet all open to inspection, selected as advised in the different parts of this work, there will be no danger from sewer air being generated in or passing through the plumbing fixture which I have treated of in this book.

# INDEX.

Made in the USA
Monee, IL
04 November 2026

67384992R00089